Solid Mechanics and Its Applications

Volume 254

Aims and Scope of the Series

The fundamental questions arising in mechanics are: *Why?*, *How?*, and *How much?* The aim of this series is to provide lucid accounts written by authoritative researchers giving vision and insight in answering these questions on the subject of mechanics as it relates to solids.

The scope of the series covers the entire spectrum of solid mechanics. Thus it includes the foundation of mechanics; variational formulations; computational mechanics; statics, kinematics and dynamics of rigid and elastic bodies; vibrations of solids and structures; dynamical systems and chaos; the theories of elasticity, plasticity and viscoelasticity; composite materials; rods, beams, shells and membranes; structural control and stability; soils, rocks and geomechanics; fracture; tribology; experimental mechanics; biomechanics and machine design.

The median level of presentation is the first year graduate student. Some texts are monographs defining the current state of the field; others are accessible to final year undergraduates; but essentially the emphasis is on readability and clarity.

More information about this series at http://www.springer.com/series/6557

Lester W. Schmerr

Engineering Dynamics 2.0

Fundamentals and Numerical Solutions

Springer

Lester W. Schmerr
Aerospace Engineering
Iowa State University
Ames, IA, USA

ISSN 0925-0042 ISSN 2214-7764 (electronic)
Solid Mechanics and Its Applications
ISBN 978-3-319-98469-8 ISBN 978-3-319-98470-4 (eBook)
https://doi.org/10.1007/978-3-319-98470-4

Library of Congress Control Number: 2018957126

This Springer imprint is published by the registered company Springer Nature Switzerland AG
The registered company address is: Gewerbestrasse 11, 6330 Cham, Switzerland

Preface

One of the striking features of current introductory engineering dynamics texts is that although they are about *dynamics*, nothing moves! By that I mean that calculations for motion parameters and forces are typically done at fixed instances of time. The reason, of course, is that current texts are derived from an era when personal computers (or even calculators) were nonexistent so that fixed time calculations were the only ones feasible to do by hand or slide rule. The content and approach of most engineering dynamics texts has changed little from the 1960s when the use of vectors first became commonplace. Most of the recent attention on the teaching of dynamics has concentrated on using modern learning techniques to effectively deliver the material found in current dynamics courses. As valuable as these improved learning efforts have been, they have had the unintended effect to take the focus away from course content issues. The objective of this book is to provide a new approach to dynamics that significantly changes the content and structure found in most dynamics texts to better align the teaching of this subject to current engineering tools and engineering practice.

This book is written as a second level course in dynamics for engineers. The reason for structuring the book in this way is that there is a tremendous inertia (pun intended) to making changes in entry level engineering texts so that the most effective route is to give students a chance to see a different and more modern approach at a junior/senior or a graduate level where instructors have much more flexibility in choosing content. However, one could also use portions of this book to enhance an existing first course in dynamics. I have called the book *Engineering Dynamics 2.0* (or, more succinctly, *Dynamics 2.0*) rather than "Intermediate Dynamics" or "Advanced Dynamics" since most existing intermediate and advanced dynamics books have the same characteristics as introductory texts in that they rarely focus on obtaining explicit, time-dependent solutions to dynamics problems.

Dynamics 2.0 has three main distinguishing features. First, it emphasizes how to obtain the differential equations of motion for dynamic systems and then how to solve those equations numerically. As part of that approach, understanding constraints and constraint forces is shown to play an integral role. MATLAB® is used to obtain the solutions, primarily through the numerical ordinary differential equation solver ode45. Some higher level dynamics texts have instead used MATLAB® Simulink models to do the numerical calculations. While Simulink is a powerful software tool for performing simulations of complex dynamical systems, that power is not needed for describing the fairly simple problems discussed in most dynamics texts. Also, if students want to use software packages other than MATLAB® to solve the problems discussed in this book, they can replace solvers such as ode45 with similar functions from those packages. Using any "black box" numerical solvers, of course, can lead to numerical inaccuracies if attention is not paid to the propagation and sources of errors. This, however, is really no different than making sure that one keeps sufficient significant figures when doing calculations by hand, a point that was stressed in the past in dynamics courses. In the use of the MATLAB® numerical solvers, I include discussions of the error tolerances one can set with those solvers and show where numerical errors can be important. To understand the behavior of the numerical solutions, animations can also be helpful so that I have demonstrated how to generate animations from the numerical results for a number of the problems considered. Some more recent dynamics books have used MATLAB®'s symbolic algebra capabilities extensively. This book also occasionally takes advantage of symbolic algebra, but it does not play a prominent role because I have wanted to emphasize obtaining numerical solutions, and, as in most dynamics books, the examples are simple enough that one can set up the problems for numerical solution by hand. A second key feature of *Dynamics 2.0* is that it makes extensive use of matrices. Matrices are important for several reasons. First, they are essential both to obtain the equations of motion and to describe the effects of constraints in forms that are directly transferable to numerical computations. They also are very useful in formulating kinematics problems suitable for numerical solutions. Second, matrices are key elements in many of the software tools that engineers will use to solve more complex and practical dynamics problems, such as the multibody codes found in mechanical, aerospace, and biomechanics applications or finite element codes, which have become the basis for solving many structural dynamics problems in industry. A short introduction to matrices is provided in Appendix A that should provide the reader with a sufficient background for the topics covered in this book. Finally, a third feature of *Dynamics 2.0* is that it uses a combination of Newton-Euler and Lagrangian (analytical mechanics) treatments for solving dynamics problems. However, rather than discussing these two treatments separately, as commonly found in many intermediate and advanced dynamics texts, I have used a geometrical approach that ties these two treatments together. Making that connection before introducing concepts such as "virtual displacements" will hopefully take some of the veil off of those aspects of analytical mechanics that students find to be vague and mysterious and provide a solid foundation for higher level dynamics courses and engineering applications.

This book follows the organization of more traditional treatments of dynamics by first considering the motion of a single particle (Chap. 2) and then systems of particles (Chap. 3). Chapter 4 covers kinematics and relative motion using both vectors and matrices. In the discussion of rigid body dynamics problems in Chap. 5

only two-dimensional (planar) motion problems are considered. However, later in Chap. 8 the issues that must be faced in solving three-dimensional rigid body dynamics problems are discussed so that students can transition to that topic in higher level courses. To understand the nonlinear behavior seen in the numerical solution of many dynamical systems, an analysis of the stability of equilibrium states is important. Thus, an entire chapter (Chap. 6) is devoted to static/dynamic stability, where matrices again play an important role. In Chap. 7, the special topic of vibrations is considered, again with a matrix-based approach. Finally, Chap. 9 presents a rather unconventional introduction to the motion of flexible (nonrigid) bodies by examining waves propagating in a thin bar. Solutions are found with the method of finite differences and the finite element method. I have used this last chapter to give the reader a glimpse of the wave equation, which is one of the most important equations of classical mechanics, and I have used that equation as a vehicle to briefly discuss the important numerical methods employed for such partial differential equations.

In *Dynamics 2.0*, the worked-out exercises found in most traditional texts have been replaced by simulations that obtain and discuss the numerical solutions to example problems. I have also not shied away from giving detailed derivations of many important underlying concepts and equations. This was done intentionally so that the book can serve as a foundation for more advanced studies as well as a textbook. The derivations may make some sections seem rather ponderous to students so one can selectively choose how much of this background material to cover in depth.

The book can be used as a text to structure one or two semester courses. Chapters 1–3 and 5 together with the 2-D kinematics of Chap. 4, for example, could serve as a first semester course that treats particles, systems of particles, and 2-D rigid body motions. Chapters 6–9, Appendices D and E, and the 3-D kinematics of Chap. 4 could examine the topics of stability, vibrations, 3-D rigid body motions, and the 1-D dynamics of deformable bodies. Many of the homework problems involve numerical simulations. Some of those simulations are more like student projects than traditional homework problems so one should be aware of that when making assignments. The level of the book lies somewhere between an intermediate and an advanced dynamics text. I have tried to present topics in a direct and simple fashion suitable to an intermediate level, but I have also included details that are traditionally found in advanced texts. In order to keep the book manageable, however, I have been forced to abbreviate or omit many of the discussions of more advanced topics. The book, for example, does not address Hamilton's principle and Hamilton's equations. The discussion given of 3-D rotations, Euler angles, and Euler parameters only scratches the surface of these important topics. The topic of three-dimensional constrained motion of systems of rigid bodies has been left to more advanced treatments of multibody dynamics. I have also not included variable mass problems, orbital motion problems, the use of quasi-coordinates, and more detailed treatments of gyroscopic problems. All of those topics and many others make up the rich tapestry of the field of dynamics that I hope you will study elsewhere.

The MATLAB® m-files for the functions and scripts developed in this book (see Appendix F for code listings) are available on the web at https://doi.org/10.1007/978-3-319-98470-4_2.

For MATLAB and Simulink product information, please contact:

The MathWorks, Inc.
3 Apple Hill Drive
Natick, MA, 01760-2098 USA
Tel: 508-647-7000
Fax: 508-647-7001
E-mail: info@mathworks.com
Web: https://www.mathworks.com
How to buy: https://www.mathworks.com/store
Find your local office: https://www.mathworks.com/company/worldwide

Ames, IA, USA Lester W. Schmerr

Contents

1 Basic Elements of Dynamics 1
1.1 Introduction 1
1.2 Systems of Units 3
1.3 Describing Motion in Different Coordinate Systems 3
1.3.1 Cartesian (Rectangular) Coordinates 4
1.3.2 Cylindrical and Polar Coordinates 5
1.3.3 Spherical Coordinates 7
1.4 Vectors and Matrices 9
1.5 Angular Velocity and the Time Derivative of Unit Vectors 9
1.6 Objective and Organization of the Book 12
1.7 Problems 14

2 Dynamics of a Particle 17
2.1 Governing Equations 17
2.2 The Dynamics of Unconstrained Motion of a Particle 21
2.2.1 Equations of Motion 21
2.2.2 A Projectile Problem 22
2.2.3 Potential Energy 24
2.2.4 Kinetic Energy and Conservative Systems 29
2.2.5 Work–Energy 31
2.2.6 A Projectile Problem with Drag Forces 33
2.3 The Dynamics of Constrained Motion of a Particle 37
2.3.1 Constrained Motion of a Bead on a Wire 37
2.3.2 A Roller-Coaster Problem 53
2.4 Constraints and Equations of Motion: A Matrix Approach 61
2.4.1 Types of Constraints 61
2.4.2 Constraints for Motion in Three Dimensions 67
2.4.3 Augmented Solutions for Ideal Constraint Forces and the Equations of Motion in Cartesian Coordinates 70

2.5 Constraints and Equations of Motion in Polar Coordinates 74
2.5.1 Solutions in Polar Coordinates 74
2.5.2 Unconstrained Motion of a Spring Pendulum 75
2.5.3 Constrained Motion of a Pendulum 85
2.5.4 Constraints and the Motion of the Planets 88
2.6 Generalized Coordinates and the Equations of Motion: A Geometric Approach 94
2.6.1 Embedding of Constraints 94
2.6.2 Augmented Approach with Generalized Coordinates 102
2.7 Lagrange's Equations 105
2.7.1 Generalized Momenta and Ignorable Coordinates 119
2.7.2 Lagrange's Equations: A Summary 121
2.8 Analytical Dynamics and Virtual Work 122
2.9 Other Principles and Virtual Quantities 131
2.10 Nonideal Constraint Forces 136
2.11 Explicit Embedding of Constraints: A General Approach 146
2.12 The Augmented Approach and Constraint Satisfaction 153
2.13 Problems 161
References 169

3 Dynamics of a System of Particles 171
3.1 Internal Forces 171
3.2 Newton-Euler Laws for a System of Particles 176
3.2.1 Motion of the Center of Mass 176
3.2.2 Impulse and Linear Momentum 178
3.2.3 The Moment Equation and Angular Momentum 179
3.2.4 Angular Impulse and Angular Momentum 182
3.2.5 Work and Energy 186
3.2.6 Kinetic Energy and Angular Momentum for a Rigid System of Particles 188
3.2.7 Work-Kinetic Energy for an Elastically Connected System of Particles 191
3.3 Dynamics of a Rigidly Constrained System of Particles (Rigid Body) 194
3.4 Equations of Motion in Generalized Coordinates 200
3.4.1 Motion of a Double Pendulum 204
3.5 A Non-holonomic Constrained System of Particles 221
3.6 Dependent Constraints 227
3.7 Problems 231
References 236

4 Kinematics and Relative Motion 237
4.1 Relative Velocity and Acceleration 237
4.1.1 Relative Motion: Cylindrical and Spherical Coordinates 241

4.2 Relative Motion and the Transport Theorem 243
4.2.1 Relative Velocity and Acceleration: More Explicit Forms 247
4.2.2 Relative Motion for Rigid Bodies 247
4.2.3 The Analysis of Kinematically Driven Systems: I 253
4.2.4 Singular Configurations 258
4.2.5 Numerical Solution of the Position Equations 260
4.2.6 Velocity and Acceleration Constraints 265
4.2.7 The Analysis of Kinematically Driven System: II 268
4.3 Motion on the Rotating Earth 269
4.4 Matrix Kinematics of Rigid Body Planar Motion 279
4.4.1 Positional Analysis 279
4.4.2 Velocity Analysis 282
4.4.3 Acceleration Analysis 285
4.4.4 General Relative Velocity and Acceleration Relations 286
4.4.5 The Transport Theorem Revisited 289
4.5 Matrix-Vector Kinematics of Constraints and Kinematically Driven Systems 291
4.6 Three-Dimension Motion: Finite Rotations and Relative Position 300
4.7 Angular Velocity and Relative Velocity Relations 308
4.8 Angular Velocity and Euler Angles 312
4.9 Acceleration Relations 317
4.10 Euler Parameters 319
4.11 The Commonly Used Euler Angle Sets 325
4.11.1 3-1-3 Set 326
4.11.2 3-2-3 Set 326
4.11.3 3-2-1 Set 327
4.12 Combination Rotations with Body-Fixed and Space-Fixed Frames 327
4.13 Problems 329
References 338

5 Planar Dynamics of Rigid Bodies 339
5.1 Governing Equations for a Rigid Body in Plane Motion 339
5.2 Moment of Inertia 348
5.3 Planar Problems and Constraint Forces 352
5.3.1 Newton-Euler Equations 352
5.3.2 An Augmented Approach 355
5.3.3 Rolling Without Slipping 363
5.4 Kinetic Energy and Work-Energy 369
5.4.1 Kinetic Energy of a Rigid Body in Plane Motion 369
5.4.2 Work-Energy Principle for a Rigid Body in Plane Motion 371

5.5 Angular Momentum and the Moment Equation 375
5.5.1 Motion Relative to a Point that Moves with a Rigid Body 375
5.5.2 Motion Relative to a General Point 377
5.6 Solving Systems of Rigid Bodies in Plane Motion 380
5.6.1 Lagrange's Equations 382
5.7 Problems 386

6 Dynamic and Static Stability 395
6.1 Dynamic Stability 395
6.2 Stability of a Natural, Conservative System Near Equilibrium 397
6.3 Stability of a Nonnatural System Near Equilibrium 402
6.4 Stability Analysis Through Linearization 404
6.5 Static Stability 412
6.6 Bifurcations and Buckling 415
6.7 Limit Load Instability 432
6.8 Snap-Through Instability 439
6.9 Problems 443
References 448

7 Vibrations of Dynamical Systems 449
7.1 An Overview of Linearized Vibrating Systems 449
7.2 Linearized Motion Near Equilibrium 451
7.3 Free Vibrations Without Damping 456
7.4 Forced Vibrations Without Damping 474
7.4.1 Harmonic Driving Forces 476
7.5 Free Vibrations with Damping 477
7.6 Forced Vibrations with Damping 483
7.6.1 Harmonic Driving Forces 483
7.6.2 System Impulse Response 484
7.6.3 Convolution Integrals 486
7.7 Problems 489
Reference 494

8 General Spatial Dynamics of Rigid Bodies 495
8.1 Angular Momentum 495
8.1.1 Angular Momentum About a Body-Fixed Point 495
8.1.2 Angular Momentum About a General Point 499
8.2 Kinetic Energy 500
8.3 Impulse-Momentum and Work-Energy Principles for a Rigid Body 502
8.4 Newton-Euler Equations of Motion 505
8.4.1 Governing Equations: General Case 505
8.4.2 Governing Equations for a Rigid Body: Use of a Body-Fixed Point 507

8.5 Solutions of Euler's Equations for Rotational Motion 512
8.6 Rotational Motion and the Euler Parameter Constraint 525
8.7 Solving Systems of Rigid Bodies 527
8.7.1 Lagrange's Equations 531
8.8 The Rolling Disk 532
8.9 Problems 540
References 548

9 Dynamics of Deformable Bodies 549
9.1 Longitudinal Wave Motion 549
9.1.1 The Method of Finite Differences 553
9.1.2 The Finite Element Method 561
9.2 Problems 567

Appendix A: Matrices 575

Appendix B: Mass Moments and Products of Inertia 589

Appendix C: Numerical Methods 597

Appendix D: Vibrations of One Degree of Freedom Systems 605

Appendix E: Fourier Transforms 621

Appendix F: MATLAB® Functions and Scripts 631

Index 703

Chapter 1
Basic Elements of Dynamics

This chapter discusses some of the basic elements of dynamics, including the Newton-Euler laws, units, description of motion in various coordinate systems, and vector-matrix notation that is used in the book. A short summary is also given of the objectives of this book as well as an outline of the topics covered.

1.1 Introduction

When bodies are in static equilibrium, the forces and moments acting on those bodies must balance, i.e., the resultant vector sum of all the forces, $\mathbf{F}_R$, must be zero and the resultant vector sum, $\mathbf{M}_Q$, of all the moments about any point Q must be zero:

$$\begin{aligned}\mathbf{F}_R &= \sum_{i=1}^{N} \mathbf{F}_i = 0 \\ \mathbf{M}_Q &= \sum_{j=1}^{S} \mathbf{M}_{Qj} = 0\end{aligned} \tag{1.1.1}$$

where $\mathbf{F}_i$ $(i = 1, 2, \ldots, N)$ are the vector forces acting on the body and $\mathbf{M}_{Qj}(j = 1, 2, \ldots, S)$ are the vector moments about Q. If the forces or moments are not balanced, then the body will move. The field of dynamics is concerned with describing those motions and the relationship of the motion to the forces and moments. In dynamics problems the forces and moments can vary with the time, t, and the right-hand sides of the equilibrium equations are no longer zero. Including these motion terms gives the governing equations of dynamics:

L. W. Schmerr, *Engineering Dynamics 2.0*, Solid Mechanics and Its Applications 254,
https://doi.org/10.1007/978-3-319-98470-4_1

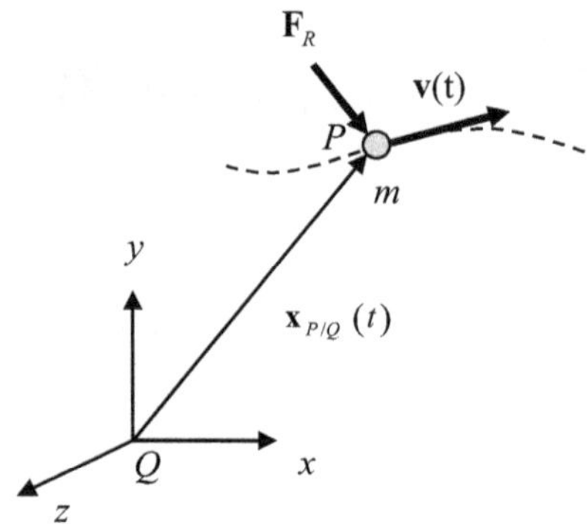

Fig. 1.1 Motion of a particle of mass m under the action of a resultant force, $\mathbf{F}_R(t)$

$$\begin{aligned}\mathbf{F}_R(t) &= \frac{d\mathbf{p}}{dt}\\ \mathbf{M}_Q(t) &= \frac{d\mathbf{L}_Q}{dt}\end{aligned} \tag{1.1.2}$$

where $\mathbf{p}(t)$ is the *linear momentum* and $\mathbf{L}_Q(t)$ is the *angular momentum* about the fixed point Q. In Chap. 2 we will describe these dynamics equations for the motion of a single, small body (particle) of mass m, where the linear momentum $\mathbf{p}(t) = m\mathbf{v}(t)$ in terms of the velocity, $\mathbf{v}(t) = d\mathbf{x}_{P/Q}(t)/dt$ of the particle, and the angular momentum $\mathbf{L}_Q(t) = \mathbf{x}_{P/Q}(t) \times m\mathbf{v}(t)$, i.e., it is the moment of the linear momentum (see Fig. 1.1). [Note that $\mathbf{x}_{P/Q}$ is to be read to mean the position vector $\mathbf{x}$ of point P with respect to point Q.] Similarly, if we use the appropriate expressions for the linear and angular momentum in more complex systems of particles (Chap. 3), rigid bodies (Chaps. 5 and 8), or deformable bodies (Chap. 9) the basic relationships of Eq. (1.1.2) will remain the same (as long as the point Q is a fixed point and the coordinate system axes in which we measure the motion do not rotate—if Q is in motion and/or the coordinate axes rotate then these motions will contribute to changes in the linear momentum and angular momentum as we will see in later chapters). In the single-particle case since the mass, m, is a constant the force equation in Eq. (1.1.2) is $\mathbf{F}_R = m\mathbf{a}$, where $\mathbf{a}(t) = d\mathbf{v}(t)/dt = d^2\mathbf{x}_{P/Q}(t)/dt^2$ is the acceleration of the particle. This is just *Newton's second law* as taught in introductory physics courses. Although Newton's second law takes the center stage when describing the motion of single particles and systems of particles, Newton's law alone is inadequate to describe the complex moving bodies that engineers must deal with such as machinery, automobiles, aircraft, and spacecraft. If the deformation of those complex bodies can be neglected (i.e., they can be assumed to be rigid) Leonhard Euler showed how the combined force and moment equations in Eq. (1.1.2) are sufficient to completely solve for their motions. These *Newton-Euler equations* form the basic foundation for the field of dynamics.

Table 1.1 SI and US customary units

SI (metric) units	US customary units	Conversion
Mass (M): kilogram (kg)	Mass (M): slug	1 slug = 14.5921 kg
Length (L): meter (m)	Length (L): foot (ft)	1 foot = 0.3048 m
Time (T): second (s)	Time (T): second (sec or s)	–
Force (F): newton (N)	Force (F): pound (lb)	1 lb = 4.4488 N

1.2 Systems of Units

In dynamics we need to describe forces and moments and their effects on changes in positions and angles of moving bodies. Table 1.1 shows the units used in both the Système International (SI) and the US customary system.

There is, however, one important difference between the SI and US systems. In the SI system the force is a derived quantity ($F = ML/T^2$) measured in newtons (N), where $1\ \text{N} = 1\ \text{kg m/s}^2$. In the US system the mass is a derived quantity ($M = FT^2/L$) measured in slugs, where $1\ \text{slug} = 1\ \text{lb s}^2/\text{ft}$. Rotational motion in either system is measured either in degrees ($^\circ$) or radians (rad), where 1 radian $= 360/2\pi = 57.2958^\circ$. One force that is commonly encountered in dynamics problems is the force of gravity acting on a body, otherwise known as its weight. The weight, W, of a body on earth at sea level is equal to the mass, m, of the body times the gravitational constant, g, i.e., $W = mg$ where $g = 9.81\ \text{m/s}^2$ in the SI system and $g = 32.2\ \text{ft/s}^2$ in the US system. Thus, a mass of 1 kg weighs 9.81 N and a weight of 1 lb has a mass of 1/32.2 slugs.

1.3 Describing Motion in Different Coordinate Systems

Implicit in our use of the Newton-Euler equations is the assumption that they are valid in a special coordinate system called an *inertial coordinate system*. An inertial system is either fixed or moving at a constant velocity with respect to a fixed system. The choice of what system we consider as "fixed" is dependent on the application. For many engineering problems we can consider the surface of the earth fixed, even though the earth rotates about its own axis (as well as moves around the sun). Although effects of the earth's motion are rather small (see Problem P1.1 where you can explore what "small" means in more precise terms), it is easy to demonstrate the effects of the earth's rotation on the motion of ordinary objects such as a *Foucault pendulum*, which is a swinging mass suspended by a long wire that can be found in many science centers around the world. The mass oscillates in a plane that is continually changing as a result of the effects of earth's rotation and those changes are dependent on the location of the pendulum on the surface of the earth. Many dynamics (and physics) texts derive the equations of motion for a Foucault pendulum as a classical example of describing motion in a non-inertial system. We will examine the Foucault pendulum problem in Chap. 4 when we examine how the Newton-Euler equations change when we measure motion in moving and rotating coordinate systems.

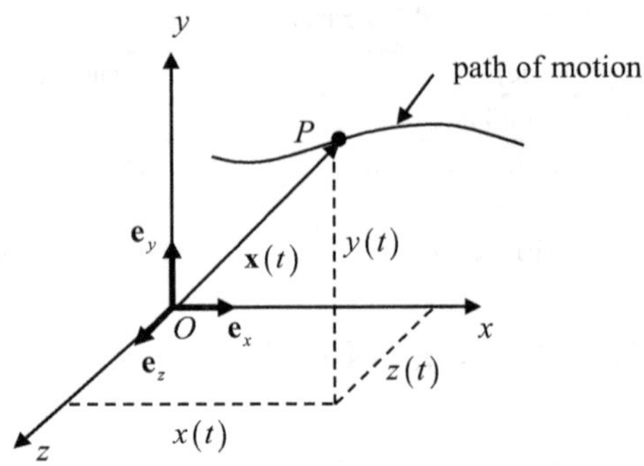

Fig. 1.2 Motion of a point P along a curved path as measured in Cartesian coordinates (x, y, z)

1.3.1 *Cartesian (Rectangular) Coordinates*

In many of our initial discussions of topics in dynamics we will use a Cartesian inertial coordinate system to describe motion. Figure 1.2, for example, describes the motion of a point P moving along a curved path in an (x, y, z) Cartesian coordinate system. The displacement of the point, $\mathbf{x}(t)$, is given by

$$\mathbf{x}(t) = x(t)\mathbf{e}_x + y(t)\mathbf{e}_y + z(t)\mathbf{e}_z \tag{1.3.1}$$

where $(\mathbf{e}_x, \mathbf{e}_y, \mathbf{e}_z)$ are fixed (constant) unit vectors along the coordinate axes and the (x, y, z) Cartesian position vector components are all functions of the time, t. To obtain the velocity, $\mathbf{v}(t)$, of this point we must differentiate this position vector, i.e.,

$$\begin{aligned}\mathbf{v}(t) &= \frac{d\mathbf{x}}{dt} \\ &= \frac{dx}{dt}\mathbf{e}_x + \frac{dy}{dt}\mathbf{e}_y + \frac{dz}{dt}\mathbf{e}_z \\ &= \dot{x}\,\mathbf{e}_x + \dot{y}\,\mathbf{e}_y + \dot{z}\,\mathbf{e}_z \\ &= v_x\mathbf{e}_x + v_y\mathbf{e}_y + v_z\mathbf{e}_z\end{aligned} \tag{1.3.2}$$

In Eq. (1.3.2) we have shown several ways in which we will represent the components of the velocity vector. First, we have shown the explicit time derivatives of each coordinate component (dx/dt, etc.). Second, we have used an abbreviated "overdot" notation to indicate these time derivatives (i.e., $dx/dt = \dot{x}$). Finally, we have simply used a subscript of (x, y, z) to indicate a particular velocity component along a Cartesian axis (v_x = velocity component along the x-axis, etc.). In this book we will, at various times, use these notations interchangeably. To obtain the acceleration, $\mathbf{a}(t)$, of the point P we must differentiate the position vector twice:

$$\begin{aligned}\mathbf{a}(t) &= \frac{d^2\mathbf{x}}{dt^2} \\ &= \frac{d^2x}{dt^2}\mathbf{e}_x + \frac{d^2y}{dt^2}\mathbf{e}_y + \frac{d^2z}{dt^2}\mathbf{e}_z \\ &= \ddot{x}\mathbf{e}_x + \ddot{y}\mathbf{e}_y + \ddot{z}\mathbf{e}_z \\ &= a_x\mathbf{e}_x + a_y\mathbf{e}_y + a_z\mathbf{e}_z\end{aligned} \tag{1.3.3}$$

Again, we have shown several different ways to describe this acceleration where now "two overdots" indicate two derivatives on time ($d^2x/dt^2 = \ddot{x}$, etc.). Since the acceleration is a derivative of the velocity vector, we also have

$$\begin{aligned}\mathbf{a}(t) &= \frac{d\mathbf{v}}{dt} \\ &= \frac{dv_x}{dt}\mathbf{e}_x + \frac{dv_y}{dt}\mathbf{e}_y + \frac{dv_z}{dt}\mathbf{e}_z \\ &= \dot{v}_x\mathbf{e}_x + \dot{v}_y\mathbf{e}_y + \dot{v}_z\mathbf{e}_z\end{aligned} \tag{1.3.4}$$

1.3.2 Cylindrical and Polar Coordinates

One of the nice things about describing motion in Cartesian inertial coordinates is that only the scalar components are functions of time since the unit vectors ($\mathbf{e}_x$, $\mathbf{e}_y$, $\mathbf{e}_z$) are constants. For other choices of coordinates we will lose this feature, making the expressions for the velocity and acceleration more complex-looking. This loss, however, is often more than offset by a simpler ability to describe certain motions in non-Cartesian coordinates. Let us consider here the use of cylindrical coordinates, (r, θ, z), as shown in Fig. 1.3b. This system is defined in terms of the Cartesian coordinate system shown in Fig. 1.3a by

$$r = \sqrt{x^2 + y^2}, \quad \theta = \tan^{-1}\left(\frac{y}{x}\right), \quad z = z \tag{1.3.5}$$

which for $r \neq 0$ can be inverted to give

$$x = r\cos\theta, \quad y = r\sin\theta, \quad z = z \tag{1.3.6}$$

We can write the position vector of the point P shown in Fig. 1.3 as

$$\begin{aligned}\mathbf{x} &= r\cos\theta\mathbf{e}_x + r\sin\theta\mathbf{e}_y + z\mathbf{e}_z \\ &= r\mathbf{e}_r + z\mathbf{e}_z\end{aligned} \tag{1.3.7}$$

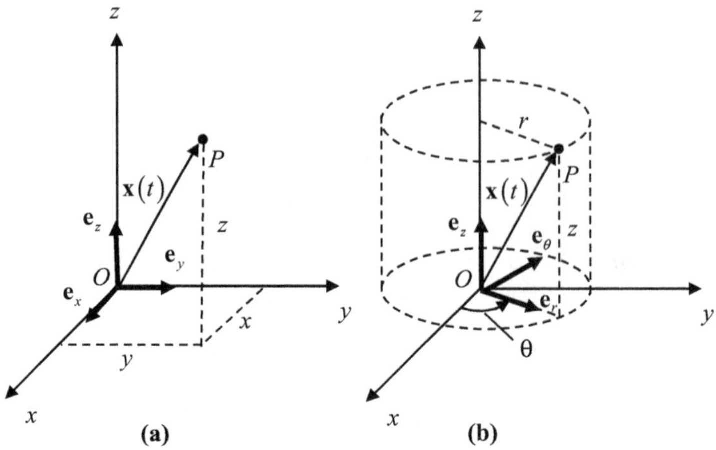

Fig. 1.3 (**a**) A point P as measured in inertial Cartesian (x, y, z) coordinates and (**b**) the same point measured in cylindrical coordinates (r, θ, z)

in terms of two of the three unit vectors $(\mathbf{e}_r, \mathbf{e}_\theta, \mathbf{e}_z)$ along the (r, θ, z) directions, where

$$\begin{aligned} \mathbf{e}_r &= \cos\theta \mathbf{e}_x + \sin\theta \mathbf{e}_y \\ \mathbf{e}_\theta &= -\sin\theta \mathbf{e}_x + \cos\theta \mathbf{e}_y \\ \mathbf{e}_z &= \mathbf{e}_z \end{aligned} \tag{1.3.8}$$

and where from Eq. (1.3.8) we see that $\mathbf{e}_\theta = \mathbf{e}_z \times \mathbf{e}_r$. Unlike the Cartesian unit vectors, however, these cylindrical coordinate unit vectors are not constants because as θ changes the orientations of $\mathbf{e}_r$ and $\mathbf{e}_\theta$ will change. In fact, you can see directly by differentiating the unit vectors in Eq. (1.3.8) that we have

$$\dot{\mathbf{e}}_r = \dot{\theta}\,\mathbf{e}_\theta, \quad \dot{\mathbf{e}}_\theta = -\dot{\theta}\,\mathbf{e}_r, \quad \dot{\mathbf{e}}_z = 0 \tag{1.3.9}$$

If we differentiate the position vector on time, the velocity of point P is then

$$\begin{aligned} \mathbf{v} = \dot{\mathbf{x}} &= \dot{r}\,\mathbf{e}_r + r\dot{\mathbf{e}}_r + \dot{z}\,\mathbf{e}_z \\ &= \dot{r}\,\mathbf{e}_r + r\dot{\theta}\,\mathbf{e}_\theta + \dot{z}\,\mathbf{e}_z \end{aligned} \tag{1.3.10}$$

and differentiating again we get the acceleration

$$\begin{aligned} \mathbf{a} = \ddot{\mathbf{x}} &= \ddot{r}\mathbf{e}_r + \left(r\ddot{\theta} + \dot{r}\,\dot{\theta}\right)\mathbf{e}_\theta + \dot{r}\,\dot{\mathbf{e}}_r + r\dot{\theta}\,\dot{\mathbf{e}}_\theta + \ddot{z}\mathbf{e}_z \\ &= \left(\ddot{r} - r\dot{\theta}^2\right)\mathbf{e}_r + \left(r\ddot{\theta} + 2\dot{r}\,\dot{\theta}\right)\mathbf{e}_\theta + \ddot{z}\mathbf{e}_z \end{aligned} \tag{1.3.11}$$

In addition to terms which, like the acceleration components in Cartesian coordinates, have two time derivatives associated with them, we have the term $-r\dot{\theta}^2$ which is called the centripetal acceleration, and the term $2\dot{r}\dot{\theta}$ which is called the Coriolis acceleration.

Polar coordinates are a special case of cylindrical coordinates where, if we take the motion to occur in the $z = 0$ plane, we have

$$\begin{aligned}\mathbf{x} &= r\mathbf{e}_r \\ \mathbf{v} &= \dot{r}\,\mathbf{e}_r + r\dot{\theta}\,\mathbf{e}_\theta \\ \mathbf{a} &= \left(\ddot{r} - r\dot{\theta}^2\right)\mathbf{e}_r + \left(r\ddot{\theta} + 2\dot{r}\,\dot{\theta}\,\right)\mathbf{e}_\theta\end{aligned} \tag{1.3.12}$$

1.3.3 Spherical Coordinates

In spherical coordinates we define the motion in terms of the variables (R, θ, ϕ) as shown in Fig. 1.4. In this case the position vector of point P is

$$\begin{aligned}\mathbf{x} &= R\sin\theta\cos\phi\mathbf{e}_x + R\sin\theta\sin\phi\mathbf{e}_y + R\cos\theta\mathbf{e}_z \\ &= R\mathbf{e}_R\end{aligned} \tag{1.3.13}$$

so that the unit vector $\mathbf{e}_R$ is given by

$$\mathbf{e}_R = \ \sin\theta\cos\phi\mathbf{e}_x + \ \sin\theta\sin\phi\mathbf{e}_y + \ \cos\theta\mathbf{e}_z \tag{1.3.14}$$

The unit vector, $\mathbf{e}_\phi$, lines in the x–y plane and is in the direction of increasing ϕ (Fig. 1.4b) so that

$$\mathbf{e}_\phi = - \ \sin\phi\mathbf{e}_x + \ \cos\phi\mathbf{e}_y \tag{1.3.15}$$

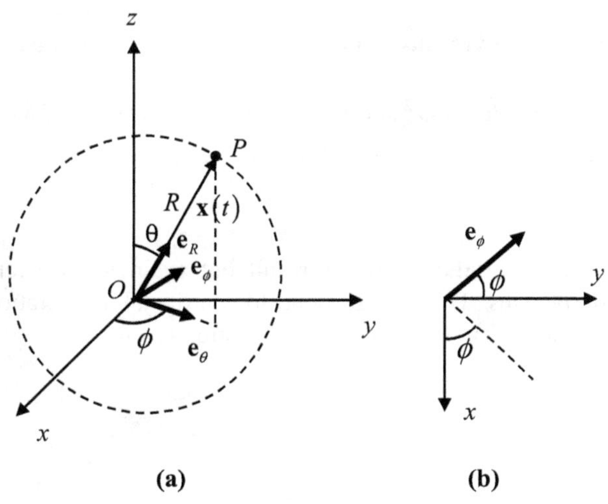

Fig. 1.4 (**a**) Spherical coordinates and the corresponding unit vectors, and (**b**) the unit vector $\mathbf{e}_\phi$ in the x–y plane

and the unit vector, $\mathbf{e}_\theta$, is

$$\mathbf{e}_\theta = \mathbf{e}_\phi \times \mathbf{e}_R = \cos\theta\big(\cos\phi\mathbf{e}_x + \sin\phi\mathbf{e}_y\big) - \sin\theta\mathbf{e}_z \tag{1.3.16}$$

The velocity is

$$\mathbf{v} = \dot{\mathbf{x}} = \dot{R}\,\mathbf{e}_R + R\dot{\mathbf{e}}_R \tag{1.3.17}$$

so that we need to have the time derivative of $\mathbf{e}_R$, which from Eq. (1.3.14) is

$$\begin{aligned}\dot{\mathbf{e}}_R &= \dot{\theta}\,\cos\theta\big(\cos\phi\mathbf{e}_x + \sin\theta\mathbf{e}_y\big) - \dot{\theta}\,\sin\theta\mathbf{e}_z \\ &\quad + \dot{\phi}\,\sin\theta\big(-\sin\phi\mathbf{e}_x + \cos\phi\mathbf{e}_y\big) \\ &= \dot{\theta}\,\mathbf{e}_\theta + \dot{\phi}\,\sin\theta\mathbf{e}_\phi\end{aligned} \tag{1.3.18}$$

and the velocity is

$$\mathbf{v} = \dot{R}\,\mathbf{e}_R + R\dot{\theta}\,\mathbf{e}_\theta + R\dot{\phi}\,\sin\theta\mathbf{e}_\phi \tag{1.3.19}$$

The acceleration is then

$$\begin{aligned}\mathbf{a} = \dot{\mathbf{v}} &= \ddot{R}\mathbf{e}_R + \big(R\ddot{\theta} + \dot{R}\,\dot{\theta}\,\big)\mathbf{e}_\theta + \big(R\ddot{\phi}\sin\theta + \dot{R}\,\dot{\phi}\,\sin\theta + R\dot{\theta}\,\dot{\phi}\,\cos\theta\big)\mathbf{e}_\phi \\ &\quad + \dot{R}\,\dot{\mathbf{e}}_R + R\dot{\theta}\,\dot{\mathbf{e}}_\theta + R\dot{\phi}\,\sin\theta\dot{\mathbf{e}}_\phi\end{aligned} \tag{1.3.20}$$

so we need the time derivatives of ($\mathbf{e}_\theta$, $\mathbf{e}_\phi$) which from Eqs. (1.3.15) and (1.3.16) are

$$\begin{aligned}\dot{\mathbf{e}}_\phi &= -\dot{\phi}\,\big(\sin\theta\mathbf{e}_R + \cos\theta\mathbf{e}_\theta\big) - \dot{\phi}\,\cos\theta\mathbf{e}_\theta \\ \dot{\mathbf{e}}_\theta &= -\dot{\theta}\,\mathbf{e}_R + \dot{\phi}\,\cos\theta\mathbf{e}_\phi\end{aligned} \tag{1.3.21}$$

Using these results in Eq. (1.3.20) the acceleration is, finally,

$$\begin{aligned}\mathbf{a} &= \big(\ddot{R} - R\dot{\phi}^2\sin^2\theta - R\dot{\theta}^2\big)\mathbf{e}_R + \big(R\ddot{\phi}\sin\theta + 2\dot{R}\,\dot{\phi}\,\sin\theta + 2R\dot{\theta}\,\dot{\phi}\,\cos\theta\big)\mathbf{e}_\phi \\ &\quad + \big(R\ddot{\theta} + 2\dot{R}\,\dot{\theta}\, - R\dot{\phi}^2\sin\theta\cos\theta\big)\mathbf{e}_\theta\end{aligned} \tag{1.3.22}$$

This is a rather complex result but again we see terms having squares of first derivatives. These correspond to centripetal acceleration terms. The mixed derivative terms are the Coriolis acceleration terms.

1.4 Vectors and Matrices

Most introductory dynamics books describe variables such as forces and moments, or velocity and acceleration, as vectors. The vector notation seen in the previous sections is typically the one in common use. In the following chapters we will also use matrices and matrix algebra to formulate and solve dynamics problems since some quantities in dynamics are inherently best described by matrices (see the moments of inertia in Appendix B, for example) and matrices are indispensable tools for describing the constrained motion problems found in many engineering applications. Appendix A gives an overview of the properties of matrices that should provide sufficient background for the topics found in this book. As shown in that appendix, vectors can also be treated as special cases of row or column matrices. For example, we can write a three-dimensional vector, **v**, in terms of its components in either of the two matrix forms:

$$\mathbf{v} = [v_1, v_2, v_3], \quad \mathbf{v} = \begin{bmatrix} v_1 \\ v_2 \\ v_3 \end{bmatrix}$$

and these forms are related through the matrix transpose operation as discussed in Appendix A, i.e.,

$$\begin{bmatrix} v_1 \\ v_2 \\ v_2 \end{bmatrix} = [v_1, v_2, v_3]^T, \quad \begin{bmatrix} v_1 \\ v_2 \\ v_2 \end{bmatrix}^T = [v_1, v_2, v_3]$$

To explicitly indicate that a vector quantity is in a column format (without showing the individual components) we will use "squiggly" bracket notation and write a column vector as, for example, $\{v\}$, and its corresponding row vector form as $\{v\}^T$. Matrices, like vectors, will sometimes be written as boldface letters such as **A**, but they can also be written as A or [A], depending on the context. These notations and some basic matrix algebra can be found in Appendix A. MATLAB® is a software package that uses matrices as its fundamental data type so it is especially useful for defining and manipulating matrices and vectors. Appendix A also describes many of the fundamental matrix operations as implemented in MATLAB®. You are strongly encouraged to read Appendix A in order to become familiar with the matrix notation and matrix algebra used throughout the book.

1.5 Angular Velocity and the Time Derivative of Unit Vectors

In both cylindrical and spherical coordinates the time derivatives of the unit vectors were needed to describe the velocity and acceleration in those coordinates. Although we obtained those time derivatives by writing the unit vectors in a fixed Cartesian

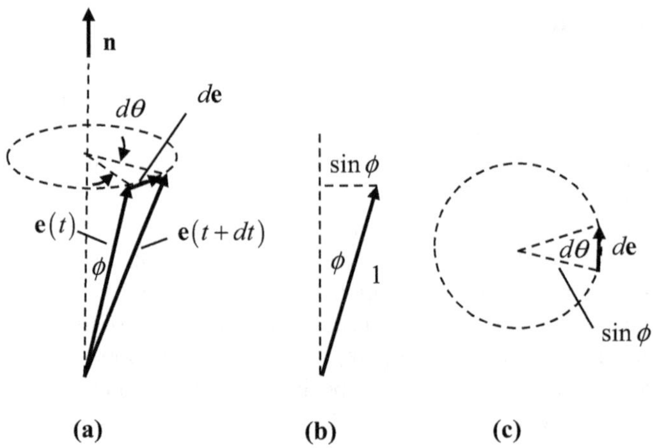

Fig. 1.5 (**a**) A unit vector, **e**, rotating about the **n**-axis; (**b**) a side view of the same geometry; and (**c**) a view of the geometry looking down the **n**-axis

coordinate system and performing the differentiation directly, there is another way to obtain those time derivatives that is more physically motivated. Since the length of unit vectors by definition cannot change, all the changes of unit vectors are produced by the changes of their orientations, i.e., their rotations. Thus, consider a unit vector, **e**, which is rotating about an axis defined by the unit vector, **n**, as shown in Fig. 1.5a. If the small change (in time dt) of the angle of the unit vector in a plane perpendicular to **n** is $d\theta$, as shown, then we can define the angular velocity vector, $\boldsymbol{\omega}$, of the rotating unit vector as $\boldsymbol{\omega} = (d\theta/dt)\mathbf{n}$. The change in **e** during this small rotation is $d\mathbf{e} = \mathbf{e}(t{+}dt) - \mathbf{e}(t)$. From the geometry as shown in Fig. 1.5a–c we see that the magnitude of $d\mathbf{e}$ is given as $|d\mathbf{e}| = \sin\phi d\theta$ so that the magnitude of the time derivative is just $|d\mathbf{e}/dt| = \sin\phi\, d\theta/dt$ and the direction of $d\mathbf{e}/dt$ is perpendicular to both **n** and $\mathbf{e}(t)$ in a direction defined by the right-hand rule when we rotate **n** into **e**. But this is just the definition of the cross product of two vectors **a** and **b**, $\mathbf{a} \times \mathbf{b} = ab \sin\phi\, \mathbf{u}$ where a and b are the magnitudes of the two vectors, ϕ is the acute angle between them, and **u** is a unit vector perpendicular to both **a** and **b** in a direction defined by the right-hand rule when we rotate **a** into **b**. We have, therefore,

$$\frac{d\mathbf{e}}{dt} = \boldsymbol{\omega} \times \mathbf{e} \tag{1.5.1}$$

which is a general relationship that is true for all rotating unit vectors. Equation (1.5.1) is also true for any vector whose magnitude is constant in time since we can multiply Eq. (1.5.1) by a constant u and write for $\mathbf{u} = u\mathbf{e}$

$$\frac{d\mathbf{u}}{dt} = \boldsymbol{\omega} \times \mathbf{u} \tag{1.5.2}$$

which is a particularly useful relationship for rigid bodies where the distance between any two points in the body is fixed.

We can also write Eq. (1.5.1) or (1.5.2) in a matrix form. For example, if we let $\mathbf{e}(t) = e_x(t)\mathbf{e}_x + e_y(t)\mathbf{e}_y + e_z(t)\mathbf{e}_z$ where the unit vectors ($\mathbf{e}_x$, $\mathbf{e}_y$, $\mathbf{e}_z$) are fixed we have

$$\begin{Bmatrix} \dot{e}_x \\ \dot{e}_y \\ \dot{e}_z \end{Bmatrix} = \begin{bmatrix} 0 & -\omega_z & \omega_y \\ \omega_z & 0 & -\omega_x \\ -\omega_y & \omega_x & 0 \end{bmatrix} \begin{Bmatrix} e_x \\ e_y \\ e_z \end{Bmatrix} \tag{1.5.3}$$

which can be written in more compact form as $\{\dot{e}\} = [\Omega]\{e\}$ where the vector cross product has been replaced by a matrix multiplication with a skew matrix $[\Omega]$. In fact, any 3-D vector cross product can be converted into a skew matrix multiplication since if we have $\mathbf{c} = \mathbf{a} \times \mathbf{b}$ then

$$\begin{Bmatrix} c_x \\ c_y \\ c_z \end{Bmatrix} = \begin{bmatrix} 0 & -a_z & a_y \\ a_z & 0 & -a_x \\ -a_y & a_x & 0 \end{bmatrix} \begin{Bmatrix} b_x \\ b_y \\ b_z \end{Bmatrix} \tag{1.5.4}$$

We can now apply Eq. (1.5.1) to our previous coordinate system examples. In cylindrical coordinates the unit vectors ($\mathbf{e}_r$, $\mathbf{e}_\theta$) both rotate around the z-axis with an angular speed, $\dot{\theta}$, so that their angular velocity vector, $\boldsymbol{\omega}$, is $\boldsymbol{\omega} = \dot{\theta}\,\mathbf{e}_z$ and we have

$$\begin{aligned} \dot{\mathbf{e}}_r &= \boldsymbol{\omega} \times \mathbf{e}_r = \dot{\theta}\,\mathbf{e}_z \times \mathbf{e}_r = \dot{\theta}\,\mathbf{e}_\theta \\ \dot{\mathbf{e}}_\theta &= \boldsymbol{\omega} \times \mathbf{e}_\theta = \dot{\theta}\,\mathbf{e}_z \times \mathbf{e}_\theta = -\dot{\theta}\,\mathbf{e}_r \\ \dot{\mathbf{e}}_z &= \boldsymbol{\omega} \times \mathbf{e}_z = \dot{\theta}\,\mathbf{e}_z \times \mathbf{e}_z = 0 \end{aligned} \tag{1.5.5}$$

which agrees with our previous results in Eq. (1.3.9).

The case of spherical coordinates is more complex because we have two angles changing in that coordinate system. The changing angle ϕ in the x–y plane produces an angular velocity of the unit vectors $\boldsymbol{\omega}_1 = \dot{\phi}\,\mathbf{e}_z$ as shown in Fig. 1.6a in a similar fashion to the cylindrical coordinate case. In contrast the changing angle θ occurs in a constant ϕ plane so that it produces an angular velocity that is normal to that plane, $\boldsymbol{\omega}_2 = \dot{\theta}\,\mathbf{e}_\phi$ (Fig. 1.5b). The total angular velocity $\boldsymbol{\omega} = \boldsymbol{\omega}_1 + \boldsymbol{\omega}_2$ is

$$\begin{aligned} \boldsymbol{\omega} &= \dot{\phi}\,\mathbf{e}_z + \dot{\theta}\,\mathbf{e}_\phi \\ &= -\dot{\theta}\sin\phi\mathbf{e}_x + \dot{\theta}\cos\phi\mathbf{e}_y + \dot{\phi}\,\mathbf{e}_z \end{aligned} \tag{1.5.6}$$

Now, consider $\dot{\mathbf{e}}_R = \boldsymbol{\omega} \times \mathbf{e}_R$ which can be written in determinant form as

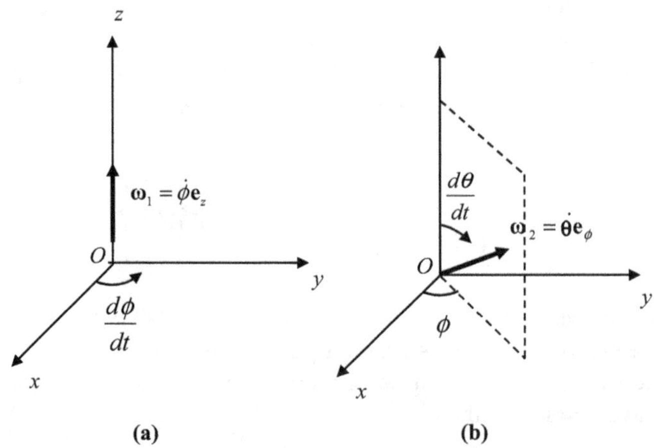

Fig. 1.6 (**a**) The angular velocity, $\boldsymbol{\omega}_1$, due to changes in the angle ϕ in the x–y plane, and (**b**) the angular velocity, $\boldsymbol{\omega}_2$, due to changes in the angle θ in the constant ϕ-plane

$$\begin{aligned}\dot{\mathbf{e}}_R = \boldsymbol{\omega} \times \mathbf{e}_R &= \begin{vmatrix} \mathbf{e}_x & \mathbf{e}_y & \mathbf{e}_z \\ -\dot{\theta}\sin\phi & \dot{\theta}\cos\phi & \dot{\phi} \\ \sin\theta\cos\phi & \sin\theta\sin\phi & \cos\theta \end{vmatrix} \\ &= \mathbf{e}_x\left(\dot{\theta}\cos\phi\cos\theta - \dot{\phi}\sin\theta\sin\phi\right) + \mathbf{e}_y\left(\dot{\theta}\sin\phi\cos\theta + \dot{\phi}\sin\theta\cos\phi\right) \\ &\quad - \mathbf{e}_z\left(\dot{\theta}\sin^2\phi\sin\theta + \dot{\theta}\cos^2\phi\sin\theta\right) \\ &= \dot{\theta}\,\mathbf{e}_\theta + \dot{\phi}\sin\theta\mathbf{e}_\phi \end{aligned} \tag{1.5.7}$$

which agrees with our previous result, Eq. (1.3.18). The other two time derivatives, which we will not show here, can be obtained similarly.

1.6 Objective and Organization of the Book

When the Newton-Euler equations are used to describe the motion of rigid bodies they lead to ordinary differential equations that must be solved in order to obtain the motion of those rigid bodies and the forces and moments acting on the bodies as a function of time. Introductory dynamics books and even many intermediate and advanced texts, however, seldom obtain complete time-dependent solutions except for linear systems where analytical solutions are available. Instead, students are taught to solve a large number of problems for velocities and accelerations or forces and moments at single, specific instants of time. These problems are used to illustrate a variety of fundamental concepts of dynamics such as work-energy or impulse-momentum principles, for example. Although equations of motion are derived, in many cases the discussion stops at that point. This book will not follow that traditional path. Instead, we will obtain the equations of motion governing a dynamical system and then show you how to explicitly solve those equations. In most instances those solutions will be done numerically. Homework problems are designed to reinforce this approach. As a consequence, you will not see "Examples" of motions evaluated at a fixed time worked out in detail, as found in many dynamics books, but instead you will see governing equations obtained and solved numerically as a function of time in "Simulations." We will also weave a variety of advances made in the field of dynamics into our discussions, including some that are quite recent, so that you can learn to apply dynamics in a modern and effective manner. This book is designed not just as an "intermediate" or "advanced" treatment of dynamics that follows an introductory dynamics course but also as a bridge to the areas of engineering that use dynamics and as an introduction to the modern tools used in the engineering practice of dynamics.

The organization of this book is similar in many respects to other dynamics books, but there are some important differences as well. Chapter 2, for example, considers the motion of a particle and discusses classical topics such as potential and kinetic energy and work-energy concepts, as found in most texts. However, Chap. 2 also highlights the important concept of constraints and the numerical methods that can be used to solve for the constrained motion of a particle. Both Newtonian and Lagrangian approaches to dynamics are presented, including the use of virtual displacements and virtual work. A geometrical approach to obtaining the equations of motion of a particle and Lagrange's equations is considered first. This sets the stage for the introduction of virtual quantities. The concept of virtual displacements is also discussed in some detail to help make that concept less mysterious. Chapter 3 considers the motion of systems of particles, including the important special cases of rigidly constrained and elastically connected systems. Examples of both non-holonomic constraints and constraints that are not independent are discussed. Chapter 4 is an introduction to kinematics and relative motion for both two-dimensional and three-dimensional rigid body motion. Both classical vector approaches as well as matrix-vector approaches are covered. Matrix methods are particularly well suited to describing more complex problems where systems of rigid bodies must be considered. The constraint equations for kinematically driven systems are solved numerically using the Newton-Raphson method. Chapter 5 considers the two-dimensional motion of continuous rigid bodies and systems of interconnected bodies. Matrix-vector methods are used to obtain the equations of motion and to solve for the motion and constraint forces. Chapter 6 deals with the dynamic and static stability of systems. Both direct and indirect stability methods are discussed for dynamical systems. Loss of static stability by means of bifurcation, limit load, and snap-through buckling are considered. Chapter 7 considers the small vibrations of multi-degree of freedom systems, an important special topic in dynamics where a matrix-vector approach is shown to lead to elegant and effective solutions. Chapter 8 deals with the three-dimensional motion of rigid bodies, again using a matrix-vector approach. Chapter 9 presents a short introduction to the dynamics of deformable bodies by describing 1-D waves propagating in a bar. It is shown how the method of finite differences and the method of finite elements can be used to numerically solve for the waves present. Those two numerical methods are highlighted since they are frequently applied in real-world solutions of dynamics problems.

There are also a number of appendices in support of the topics covered in the text. Appendix A gives an introduction to matrices and vectors and their manipulation in MATLAB®. Appendix B discusses the mass moments and products of inertia that appear in rigid body problems. The MATLAB® function ode45 and the Newton-Raphson method are two key numerical methods used in this book so Appendix C gives a brief outline of how those methods work. Appendix D summarizes the analytical results one can obtain for the small-amplitude vibrations of one degree of freedom dynamical systems as those simple systems form the basis for many of the results found for the multiple degree of freedom systems considered in Chap. 7. In Appendix E we give a brief introduction to Fourier transforms. Those transforms appear in the vibration problems described in Chap. 7 and Appendix D but they are also useful tools for a much wider range of engineering and scientific problems. Finally, in Appendix F code listings are given of the many MATLAB® functions and scripts used throughout the book.

1.7 Problems

P1.1. A point rotating about a fixed origin follows a circular path of radius R. If the rotational angular velocity, ω, is a constant the point experiences a purely centripetal acceleration of magnitude $R\omega^2$ directed towards the center of rotation. Determine the ratio of this centripetal acceleration to the acceleration of gravity at the earth's surface, g, i.e., compute $R\omega^2/g$ for:

(a) A point on the surface of the earth at the equator. The radius of the earth is 6.38×10^3 km.
(b) A point at the center of the earth in its orbit around the Sun. The radius of the earth's orbit is 149.6×10^6 km.
(c) A point at the center of the Sun in its rotation about the center of the galaxy. The radius of the Sun's orbit about the center of the galaxy is 2.7×10^4 light years and its orbital speed is 225 km/s.

These ratios give some indication of how accurate it is to take these points as "fixed" inertial references.

P1.2. In MATLAB® there is a built-in function named cross that can perform the cross product of two three-dimensional vectors. Write a MATLAB® function, S = vect2skew(v), which turns a three-dimensional vector, v, into a 3×3 skew matrix, S, so that the cross product can be implemented by matrix multiplication. Verify that your function works.

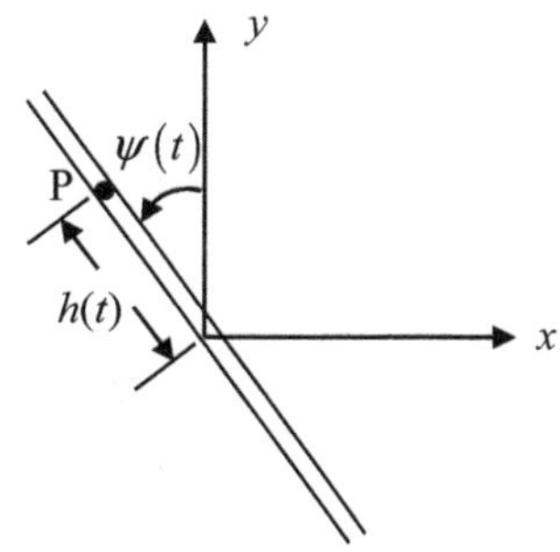

Fig. P1.1 Planar motion of a bead in a rotating tube

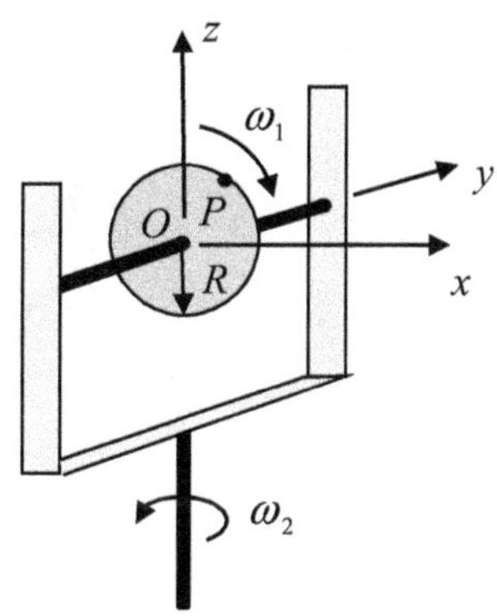

Fig. P1.2 A spinning wheel on a rotating gimbal mount

P1.3. A bead slides in a tube which itself is rotating in an x–y plane (Fig. P1.1). If the tube is rotating about the z-axis with a constant angular velocity $\dot{\psi} = \omega$, and the distance, h, is changing such that $\dot{h} = -v_0(t)$ and $\ddot{h} = -a_0(t)$, where v_0 and a_0 are radial velocity and acceleration terms, determine, using polar coordinates, expressions for the absolute velocity and acceleration of the bead in terms of h, ω, ψ, v_0, and a_0. What are these expressions when the bead passes through the origin of the x–y coordinates ($h = 0$)?

P1.4. A wheel of radius R spins at a constant speed, ω_1, about its own axis which is placed in a gimbal mount which in turn rotates at a constant speed, ω_2, as shown in Fig. P1.2. The axis of the wheel stays in a horizontal plane and the center point, O, of the wheel is stationary. (a) Using spherical coordinates, determine the acceleration of any point, P, on the rim of the wheel at the instant when the wheel is in the position shown, and (b) find the acceleration in the special cases when point P is at the top of the wheel ($\theta = 0$) and in the x–y plane ($\theta = \pi/2$). How would you have to modify these results to obtain an expression for the acceleration of a point P on the rim that is valid for any instant of time?

Chapter 2
Dynamics of a Particle

This chapter contains a detailed treatment of the dynamics of a particle. The main emphasis is on obtaining and solving the equations of motion when the particle is subject to constraints. In most cases the solution will be obtained numerically using MATLAB®. Both Newton-Euler and Lagrangian methods are used to obtain the equations of motion. Constraints are handled by either embedding them into the equations of motion (implicitly or explicitly) or solving an augmented system of equations that also yields an explicit expression for the constraint forces. Both ideal and nonideal constraints are considered. Issues that must be addressed when solving the equations of motion numerically are discussed, including the problem of constraint drift.

2.1 Governing Equations

Most dynamics books begin by analyzing the motion of a single particle, where the particle is considered to be a point mass moving in space. Newton's second law says that the particle's motion is governed by his famous equation, $F = ma$ (force = mass times acceleration), which in vector form is

$$\mathbf{F}_R = \sum_{s=1}^{S} \mathbf{F}_s = m\ddot{\mathbf{x}} \tag{2.1.1}$$

where $\mathbf{F}_R$ is the resultant force of all the forces, $\mathbf{F}_s$, acting on the particle (see Fig. 2.1); m is the mass of the particle; and $\ddot{\mathbf{x}} = d^2\mathbf{x}/dt^2$ is the vector acceleration. In terms of components along Cartesian X–Y–Z-axes this vector equation is

---

Electronic supplementary material: The online version of this chapter (https://doi.org/10.1007/978-3-319-98470-4_2) contains supplementary material, which is available to authorized users.

L. W. Schmerr, *Engineering Dynamics 2.0*, Solid Mechanics and Its Applications 254,
https://doi.org/10.1007/978-3-319-98470-4_2

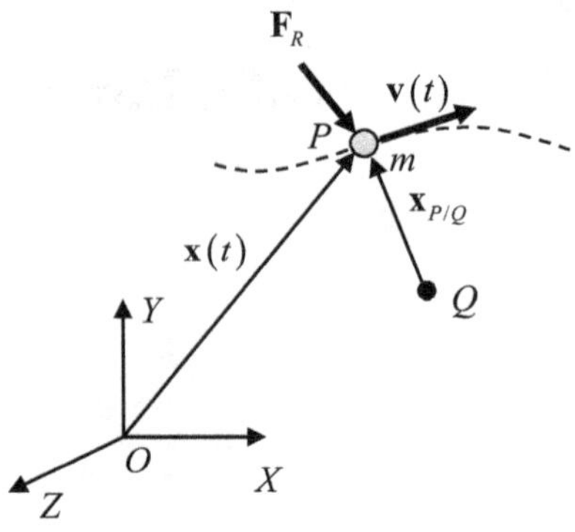

Fig. 2.1 The motion of a particle of mass m in Cartesian coordinates

$$\begin{aligned} F_{Rx} &= m\ddot{X} \\ F_{Ry} &= m\ddot{Y} \\ F_{Rz} &= m\ddot{Z} \end{aligned} \tag{2.1.2}$$

Another way Newton's second law is commonly written is in terms of the *linear momentum*, $\mathbf{p}$, defined as

$$\mathbf{p} = m\dot{\mathbf{x}} = m\mathbf{v}$$

where $\dot{\mathbf{x}} = d\mathbf{x}/dt = \mathbf{v}$ is the particle velocity vector. In this case we have

$$\mathbf{F}_R = \sum_{s=1}^{S} \mathbf{F}_s = \frac{d\mathbf{p}}{dt} \tag{2.1.3}$$

Equation (2.1.3) contains the essence of both Newton's first and second laws which state the following:

Newton's First Law: A body that is not acted upon by a net external force will maintain a state of uniform motion [i.e., if $\mathbf{F}_R = 0$, it follows that $\mathbf{p} =$ constant].

Newton's Second Law: A body subjected to a net external force will undergo a change in momentum whose rate of change with respect to time is equal to the net force.

Note that since the object in motion is idealized here as a mass concentrated at a point P, all the forces acting on the particle are concurrent forces going through that point (Fig. 2.1). Thus, the resultant moment of all the forces about point P, $\mathbf{M}_P$, is automatically zero. Unlike statics, where force and moment equilibrium must both be considered, in dynamics we often do not see the moment equation discussed in conjunction with the force equation (Newton's second law) when considering the motion of point masses yet in some particle problems the moment equation is used in place of Newton's second law. Why is this the case? If we take moments of Newton's second law about another point Q then we have

$$\mathbf{M}_Q = \mathbf{x}_{P/Q} \times \sum_{s=1}^{S} \mathbf{F}_s = \mathbf{x}_{P/Q} \times m\ddot{\mathbf{x}} \tag{2.1.4}$$

so that we have the force and moment equations

$$\begin{aligned} \mathbf{F}_R &= m\ddot{\mathbf{x}} \\ \mathbf{M}_Q &= \mathbf{x}_{P/Q} \times m\ddot{\mathbf{x}} \end{aligned} \tag{2.1.5}$$

which are a total of six equations for a particle moving in three dimensions. But since Eq. (2.1.4) is just the condition

$$\mathbf{x}_{P/Q} \times \left(\sum_{s=1}^{S} \mathbf{F}_s - m\ddot{\mathbf{x}} \right) = 0 \tag{2.1.6}$$

the moment equation is always satisfied if Newton's second law is satisfied so that it is not an independent equation. This is the same situation we find in statics where we can solve a concurrent force system using only force equilibrium and the moment equilibrium equation is not needed. However, we also know from statics that we can still use the moment equation to great advantage when solving concurrent force problems. In fact, we can replace the three force equilibrium equations by three moment equations and solve concurrent force problems simply and directly. Problem P2.1 treats an example of such a statics problem. In dynamics the situation is similar. Although the force and moment equations of Eq. (2.1.5) represent six equations in terms of components, there are only three independent equations out of those six equations that we can use. However, in some cases we can solve particle dynamics problems more efficiently by replacing one or more of the force equations by moment equations.

Just as we can write Newton's second law for the net force in terms of the time rate of change of the linear momentum, we can write the moment equation of Eq. (2.1.5) in terms of the change of the *angular momentum* of the particle about any point Q, $\mathbf{L}_Q$, defined as

$$\mathbf{L}_Q = \mathbf{x}_{P/Q} \times m\dot{\mathbf{x}} = \mathbf{x}_{P/Q} \times m\mathbf{v} \tag{2.1.7}$$

Thus,

$$\begin{aligned} \frac{d\mathbf{L}_Q}{dt} &= \frac{d}{dt}\left(\mathbf{x}_{P/Q} \times m\mathbf{v}\right) = \frac{d\mathbf{x}_{P/Q}}{dt} \times m\mathbf{v} + \mathbf{x}_{P/Q} \times m\frac{d\mathbf{v}}{dt} \\ &= m\mathbf{v} \times \mathbf{v} + \mathbf{x}_{P/Q} \times m\frac{d\mathbf{v}}{dt} \\ &= m\mathbf{x}_{P/Q} \times \ddot{\mathbf{x}} \end{aligned}$$

so that we have

$$\begin{aligned}\mathbf{F}_R &= \frac{d\mathbf{p}}{dt}\\ \mathbf{M}_Q &= \frac{d\mathbf{L}_Q}{dt}\end{aligned} \tag{2.1.8}$$

which is equivalent to the six component equations

$$\begin{aligned}F_{Rx} &= \dot{p}_x = m\ddot{x}\\ F_{Ry} &= \dot{p}_y = m\ddot{y}\\ F_{Rz} &= \dot{p}_z = m\ddot{z}\\ \\ M_{Qx} &= \dot{L}_{Qx}\\ M_{Qy} &= \dot{L}_{Qy}\\ M_{Qz} &= \dot{L}_{Qz}\end{aligned} \tag{2.1.9}$$

Remember, however, that we can only use three independent equations from the six possible equations in Eq. (2.1.9) to obtain the motion of a particle.

We have introduced the force and moment equations of Eq. (2.1.9) in this section because those equations are fundamental to our later discussions of dynamics problems. We will see, for example, that these equations are also valid for a system of interacting particles. Like the single-particle case, a general system of particles can be solved without using the moment equations but the moment equations can also serve as convenient force equation replacements in some cases. However, when we deal with a system of particles that form up a lattice of rigidly constrained particles or when we deal with a rigid body where the mass is distributed continuously, we will see the real value of Eq. (2.1.9). In the rigidly constrained system of N particles, for example, we will see that we do not have to solve N vector force equations but we need to only solve the six equations of Eq. (2.1.9) to obtain the motion of all N particles! In the continuous rigid-body case (where there is actually an infinite number of "particles") again we only need to satisfy the six equations of Eq. (2.1.9) to determine the motion of any point of the rigid body. As mentioned in Chap. 1 these equations were first obtained by Euler, who extended Newton's second law (Eq. (2.1.1)) for particles to Eq. (2.1.9) for moving rigid bodies. Together these *Newton-Euler* force and moment equations have been some of the most important equations of all time. They allow us to predict the motion of everything from molecules to celestial bodies and are an indispensable tool for engineers.

2.2 The Dynamics of Unconstrained Motion of a Particle

2.2.1 Equations of Motion

In dynamics problems engineers are often interested in two results: (1) obtaining an explicit description of the motion, and (2) obtaining the values of all the forces present. As we will see, obtaining the motion for a particle consists first of finding an expression for the acceleration of the particle in terms of *known* forces which, in some cases, may be functions of the particle velocity and position as well as time. This acceleration expression, which represents the *governing equations of motion* for the particle, is in general a second-order vector differential equation of the form

$$\ddot{\mathbf{x}}(t) = \mathbf{f}(\mathbf{x}(t), \dot{\mathbf{x}}(t), t) \tag{2.2.1}$$

where $\mathbf{f}$ is a (vector) function of the position, $\mathbf{x}(t)$, the velocity $\dot{\mathbf{x}}(t) = \mathbf{v}(t)$, and the known forces, $\mathbf{F}$, in the problem. The forces are not shown explicitly in Eq. (2.2.1) since they are either constants or functions of the time, t, $\mathbf{x}(t)$, and $\dot{\mathbf{x}}(t)$. We need to obtain a solution of these equations of motion to find the position as a function of time, $\mathbf{x}(t)$. To obtain such a solution we must also specify initial conditions of the particle at a starting time that is usually taken to be $t = 0$. At that time we have

$$\begin{aligned} \dot{\mathbf{x}}(0) &= \dot{\mathbf{x}}_0 \\ \mathbf{x}(0) &= \mathbf{x}_0 \end{aligned} \tag{2.2.2}$$

in terms of the known constants $(\mathbf{x}_0, \dot{\mathbf{x}}_0)$. In some cases we can find a solution to the governing equations analytically, but in general such solutions must be obtained numerically. In this book, numerical solutions will be obtained with MATLAB® so you should become familiar with that software system.

To obtain the equations of motion with a Newtonian approach we first write down expressions for all the forces, $\mathbf{F}_R$, acting on the particle. If all the forces are known functions of time or known functions of the particle velocity and/or displacement, then from Newton's second law

$$\ddot{\mathbf{x}} = \frac{\mathbf{F}_R(\mathbf{x}, \dot{\mathbf{x}}, t)}{m} \tag{2.2.3}$$

so that Newton's law directly gives us the equations of motion in the form of Eq. (2.2.1). Since Eq. (2.2.1) is a vector equation, in Cartesian (x, y, z) coordinates the equation represents three scalar equations:

$$\begin{aligned} \ddot{x} &= f_x(x, y, z, \dot{x}, \dot{y}, \dot{z}, t) \\ \ddot{y} &= f_y(x, y, z, \dot{x}, \dot{y}, \dot{z}, t) \\ \ddot{z} &= f_z(x, y, z, \dot{x}, \dot{y}, \dot{z}, t) \end{aligned} \tag{2.2.4}$$

and the initial conditions in Cartesian coordinates are

$$
\begin{aligned}
x(0) &= x_0 \\
y(0) &= y_0 \\
z(0) &= z_0 \\
\dot{x}(0) &= \dot{x}_0 = v_{0x} \\
\dot{y}(0) &= \dot{y}_0 = v_{0y} \\
\dot{z}(0) &= \dot{z}_0 = v_{0z}
\end{aligned}
\qquad (2.2.5)
$$

in terms of the known coordinate position constants (x_0, y_0, z_0) and known velocity component constants $(\dot{x}_0, \dot{y}_0, \dot{z}_0) = (v_{0x}, v_{0y}, v_{0z})$. Equations (2.2.4) and (2.2.5) together represent what is called an *initial value problem* for a system of second-order ordinary differential equations. In general these are nonlinear differential equations which must be solved numerically but as we will see in this book there are cases where analytical solutions are possible, particularly in the cases where the system of equations is linear. If we are free to specify the initial conditions arbitrarily and there are no restrictions on the path of the particle then in principle the solution can go through any point in space and we say that this is an *unconstrained* motion problem. An unconstrained motion problem is generally the simplest type of dynamics problem we can consider.

2.2.2 A Projectile Problem

An example of an unconstrained problem that is treated in most physics books and introductory dynamics books (Fig. 2.2a) is where a projectile of mass m is launched

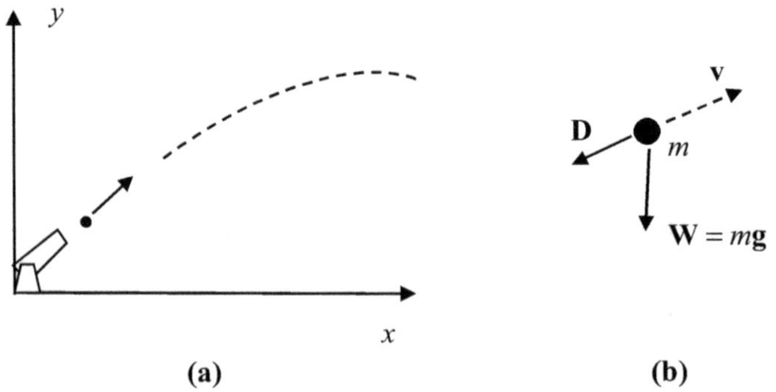

Fig. 2.2 (**a**) A projectile problem where a mass m is launched at a given location with a specified initial velocity, and (**b**) the free body diagram of the projectile when gravity and a drag force are present

with an initial velocity at some specified position and then travels under the action of gravity alone (any drag forces, **D**, as shown in the free body diagram of Fig. 2.2b are for the moment neglected). As shown in Fig. 2.2 we will also assume that all the motion occurs in the *x*–*y* plane so that we can neglect all terms involving the *z*-axis. Thus, Newton's second law generates the equations of motion directly since

$$\begin{aligned} m\frac{d^2x}{dt^2} &= F_x = 0 \\ m\frac{d^2y}{dt^2} &= F_y = -mg \end{aligned} \tag{2.2.6}$$

where $F_y = -mg$ is the force of gravity acting in the negative *y*-direction. These differential equations are particularly easy to solve since after dividing by the mass, *m*, we can just integrate them on time twice and apply the known initial conditions. After the first integration we find

$$\begin{aligned} \frac{dx}{dt} &= v_{0x} \\ \frac{dy}{dt} &= -gt + v_{0y} \end{aligned} \tag{2.2.7}$$

and after the second integration

$$\begin{aligned} x(t) &= v_{0x}t + x_0 \\ y(t) &= -gt^2/2 + v_{0y}t + y_0 \end{aligned} \tag{2.2.8}$$

Since the projectile reaches the top of its flight when $dy/dt = 0$ Eq. (2.2.7) shows that this occurs at the time $t_m = v_{0y}/g$. Placing this time into the equation for *y* in Eq. (2.2.8) gives $y_{\max} = y_0 + v_{0y}^2/2g$. As you likely know, as long as both v_{0x} and v_{0y} are not zero (i.e., we launch the projectile at an angle) the path of the particle is a parabola. This is most easily seen by setting $x_0 = y_0 = 0$ and then computing $y - y_{\max}$. We have

$$\begin{aligned} y - y_{\max} &= -gt^2/2 + v_{0y}t - v_{0y}^2/2g \\ &= -g(t - t_m)^2/2 \\ &= -g[x/v_{0x} - t_m]^2/2 \end{aligned} \tag{2.2.9}$$

which indeed is the equation of a parabola in a standard form.

2.2.3 Potential Energy

You are likely aware from your physics or elementary dynamics course that in this projectile problem the force of gravity is an example where the resultant force can be expressed in terms of a scalar *potential energy* function, $V(\mathbf{x})$, as

$$\mathbf{F}_R = -\boldsymbol{\nabla} V \tag{2.2.10}$$

where $\boldsymbol{\nabla}$ is the vector gradient operator given in Cartesian coordinates by

$$\boldsymbol{\nabla} = \mathbf{e}_x \frac{\partial}{\partial x} + \mathbf{e}_y \frac{\partial}{\partial y} + \mathbf{e}_z \frac{\partial}{\partial z} \tag{2.2.11}$$

with $(\mathbf{e}_x, \mathbf{e}_y, \mathbf{e}_z)$ being unit vectors along the (x, y, z) axes, respectively. Thus, with $V = V(x, y, z)$ we have

$$\begin{aligned} F_{Rx} &= -\frac{\partial V}{\partial x} \\ F_{Ry} &= -\frac{\partial V}{\partial y} \\ F_{Rz} &= -\frac{\partial V}{\partial z} \end{aligned} \tag{2.2.12}$$

The use of a minus sign in Eq. (2.2.10) is arbitrary but it has long been the convention in both the science and engineering community. With a minus sign present we see that the force is pointing in a direction of *decreasing* potential energy.

If the resultant force acting on a particle can be represented in terms of a potential energy function then Newton's second law becomes

$$m\ddot{\mathbf{x}} = -\boldsymbol{\nabla} V \tag{2.2.13}$$

Forces that are expressible in terms of a potential energy include constant forces, spring forces, and gravitational force of attraction between two bodies. In the constant force case if we have a force $\mathbf{F}_R = \mathbf{C} =$ constant acting on a particle, then the potential can be written simply as the dot product of the force with a position vector to that particle, which written in Cartesian coordinates gives

$$V = -\mathbf{C} \cdot \mathbf{x} = -C_x x - C_y y - C_z z \tag{2.2.14a}$$

so that

$$
\begin{aligned}
F_x &= -\frac{\partial V}{\partial x} = C_x \\
F_y &= -\frac{\partial V}{\partial y} = C_y \\
F_z &= -\frac{\partial V}{\partial z} = C_z
\end{aligned}
\tag{2.2.14b}
$$

In the projectile problem just considered, the force of gravity was treated as a constant force $\mathbf{F} = -mg\mathbf{e}_y$, so in that case $V = mgy$.

Note that in all cases we can add or subtract a constant value to the potential energy without affecting the force so we are free to choose that constant to make the potential energy to be zero at a convenient reference location.

For a linear spring, if we assume that the two ends of the spring are located at points $\mathbf{x}_1$ and $\mathbf{x}_2$, where we assume that $\mathbf{x}_1$ is fixed and $\mathbf{x}_2$ is attached to a particle (see Fig. 2.3), we have the potential energy

$$
V = \frac{1}{2}k[|\mathbf{x}_2 - \mathbf{x}_1| - L_0]^2 = \frac{1}{2}k[L - L_0]^2 \tag{2.2.15a}
$$

where k is the spring constant (having dimensions of force/length) and the distance, L, where $L = \sqrt{(x_2 - x_1)^2 + (y_2 - y_1)^2 + (z_2 - z_1)^2}$ is the length of the stretched spring, and L_0 is the unstretched length. In this case the Cartesian force components acting on the particle are

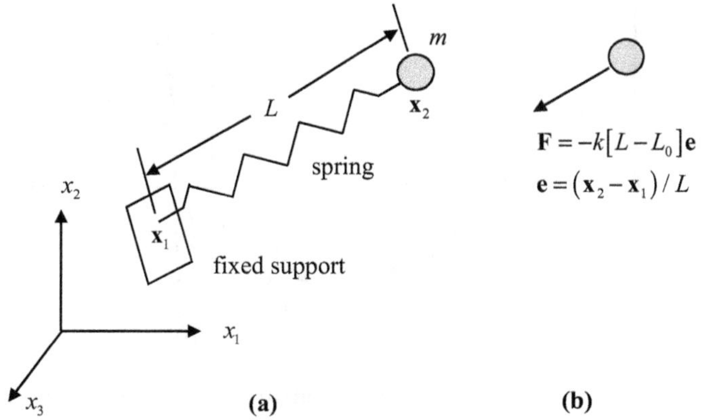

Fig. 2.3 (**a**) A linear spring acting on a particle in three dimensions, and (**b**) the corresponding spring force

$$
\begin{aligned}
F_x &= -\frac{\partial V}{\partial x_2} = -k[L - L_0]\frac{(x_2 - x_1)}{L} \\
F_y &= -\frac{\partial V}{\partial y_2} = -k[L - L_0]\frac{(y_2 - y_1)}{L} \\
F_z &= -\frac{\partial V}{\partial z_2} = -k[L - L_0]\frac{(z_2 - z_1)}{L}
\end{aligned}
\tag{2.2.15b}
$$

which are rather complex expressions when written out explicitly in terms of the Cartesian coordinates. But note that if we had used spherical coordinates (r, θ, ϕ) with origin at $\mathbf{x}_1$ to locate the particle, we would have $r = L$ and the potential energy in those spherical coordinates would just be

$$
V = \frac{1}{2}k[r - L_0]^2 \tag{2.2.16}
$$

We can use Eq. (2.2.16) directly if we know how to take the vector gradient operator in spherical coordinates. You can look in any book on vector calculus to find that operator written as

$$
\nabla = \mathbf{e}_r\frac{\partial}{\partial r} + \mathbf{e}_\theta\frac{1}{r}\frac{\partial}{\partial \theta} + \mathbf{e}_\phi\frac{1}{r\sin\theta}\frac{\partial}{\partial \phi} \tag{2.2.17}
$$

where $(\mathbf{e}_r, \mathbf{e}_\theta, \mathbf{e}_\phi)$ are unit vectors in the (r, θ, ϕ) directions, respectively, for the spherical coordinates shown in Fig. 2.4a. For completeness, we also note that in cylindrical coordinates (r, θ, z) the vector gradient operator is likewise

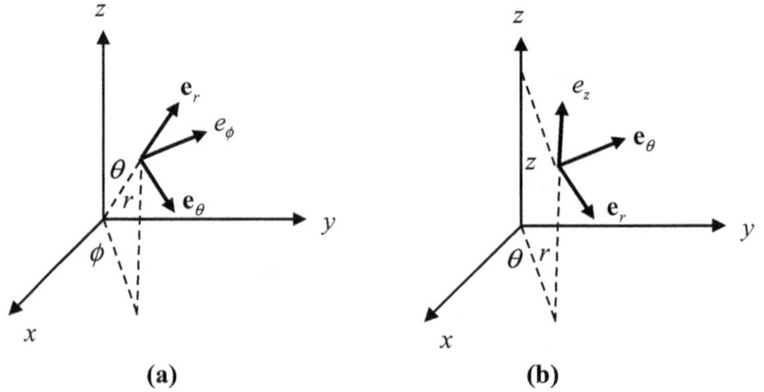

Fig. 2.4 (**a**) Spherical coordinates and unit vectors, and (**b**) cylindrical coordinates and unit vectors

$$\nabla = \mathbf{e}_r \frac{\partial}{\partial r} + \mathbf{e}_\theta \frac{1}{r} \frac{\partial}{\partial \theta} + \mathbf{e}_z \frac{\partial}{\partial z} \tag{2.2.18}$$

for the cylindrical coordinates shown in Fig. 2.4b. For 2-D problems that are described in polar coordinates (r, θ) we can use the gradient

$$\nabla = \mathbf{e}_r \frac{\partial}{\partial r} + \mathbf{e}_\theta \frac{1}{r} \frac{\partial}{\partial \theta} \tag{2.2.19}$$

to represent 2-D forces that can be represented by a potential.

If we use the potential energy of Eq. (2.2.16) to calculate the force components in spherical coordinates we find in component form

$$\begin{aligned} F_r &= -k(r - L_0) \\ F_\theta &= F_\phi = 0 \end{aligned} \tag{2.2.20a}$$

or in vector form

$$\mathbf{F} = -k(r - L_0)\mathbf{e}_r \tag{2.2.20b}$$

which is indeed a simpler form than Eq. (2.2.15b) but completely equivalent to it. Of course, if the spring always lies along one of the Cartesian coordinate axes the potential and force expressions are also in the simpler form of Eqs. (2.2.20a) and (2.2.20b). For example, if the spring is along the x-axis and we let $x = L$ be the stretched length we have

$$\begin{aligned} V &= \frac{1}{2}k(x - L_0)^2 \\ F_x &= -k(x - L_0) \\ F_y &= F_z = 0 \end{aligned} \tag{2.2.21}$$

This is a model of a linear spring, since the force is linearly proportional to the deformation, $x - L_0$, of the spring. Nonlinear springs along the x-axis might have a force-deformation relationship like $F_x = -k(x - L_0)^2$, for example.

As a final example, consider the force of gravitation between two particles of mass m_1 and m_2, respectively (Fig. 2.5). Newton postulated that a particle m_1 experiences a gravitational force of attraction to another mass m_2 given by

$$\mathbf{F} = -\frac{Gm_1m_2}{r^2}\mathbf{e}_r \tag{2.2.22}$$

as measured in a set of spherical coordinates with origin located at m_2 (Fig. 2.5b) where $\mathbf{e}_r$ is a unit vector in the radial r direction and G is the universal gravitational

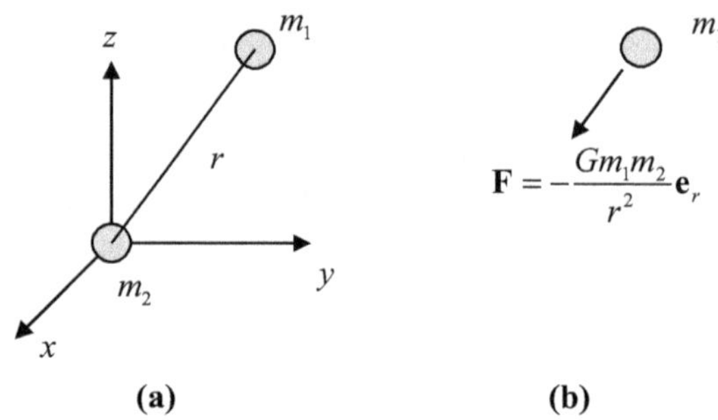

Fig. 2.5 (**a**) Two interacting particles, and (**b**) the gravitational force that mass m_2 exerts on mass m_1

constant, which has been measured to be, approximately, $G = 6.673 \times 10^{-11}$ N m^2/kg^2. It follows that we can define the potential energy of this force as

$$V = -\frac{Gm_1m_2}{r} \tag{2.2.23}$$

Near the surface of the earth, if we let the radius $r = R$, the radius of the earth, and let $m_2 = M$ be the mass of the earth, then

$$\mathbf{F} = -m_1\frac{MG}{R^2}\mathbf{e}_r = -m_1 g\mathbf{e}_r \tag{2.2.24}$$

where g is called the acceleration of gravity at the earth's surface. This acceleration will be a nearly constant value if we have motion over radial distances that do not vary significantly from R, a condition we implicitly assumed when solving our projectile motion problems.

Newton published his universal law of gravitation in 1687 in his Principia, stating

> Every particle in the universe exerts an attractive force on every other particle where the magnitude of the force is proportional to the product of the masses of the two particles and is inversely proportional to the square of the distance between them. The direction of the attractive force lies along the straight line connecting the two particles.

There are mutually attractive forces acting on each of the interacting masses so that just as mass m_2 exerts an attractive force on mass m_1, mass m_1 also exerts an equal and opposite force on mass m_2. Thus, according to Newton's law of gravitation if we consider any two masses, m_i and m_j and write the force that mass m_j exerts on mass m_i as $\mathbf{F}_{ij}$ and, likewise, let the force that mass m_i exerts on mass m_j be $\mathbf{F}_{ji}$, we have (see Fig. 2.6)

$$\mathbf{F}_{ij} = -\mathbf{F}_{ji} = \frac{Gm_im_j}{r_{ij}^2}\left(\frac{\mathbf{r}_{ij}}{r_{ij}}\right) \tag{2.2.25}$$

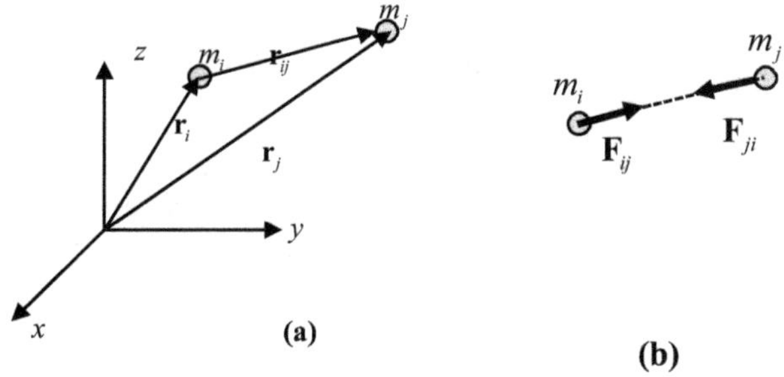

Fig. 2.6 (**a**) Two interacting particle masses and (**b**) the mutual gravitational forces of attraction between those particles

where $\mathbf{r}_{ij} = \mathbf{r}_j - \mathbf{r}_i$ is the position vector directed from mass m_i to mass m_j (Fig. 2.6). Equation (2.2.25), therefore, is a more complete representation of Newton's law than Eq. (2.2.22).

For other forces how do you know if you can associate a potential energy with that force? The answer is: a force can be represented as the gradient of a potential if and only if the curl ($\nabla \times$) of the force is zero, i.e.,

$$\nabla \times \mathbf{F} = 0 \tag{2.2.26}$$

You may have encountered a similar result in vector calculus in a purely mathematical context since for any vector that is represented as the gradient of a scalar function we have the vector identity $\nabla \times \nabla V = 0$, which you can easily verify by carrying out the curl and gradient operations on any scalar, V.

2.2.4 Kinetic Energy and Conservative Systems

Systems governed by Eq. (2.2.13) where the applied force can be represented in terms of a potential energy are called conservative systems since the total energy, E, of the system (which we will define shortly) remains a constant so E is conserved during the motion. To see this, first take the vector dot product of both sides of Eq. (2.2.13) with the velocity vector, $\dot{\mathbf{x}}$, giving

$$m\ddot{\mathbf{x}} \cdot \dot{\mathbf{x}} = -\nabla V \cdot \dot{\mathbf{x}} \tag{2.2.27}$$

This equation can also be written as

$$\frac{m}{2}\frac{d}{dt}\left(\dot{\mathbf{x}} \cdot \dot{\mathbf{x}}\right) = \frac{d(mv^2/2)}{dt} = -\nabla V \cdot \frac{d\mathbf{x}}{dt} \tag{2.2.28}$$

where $v = \sqrt{v_x^2 + v_y^2 + v_z^2}$ is the magnitude of the velocity. Multiplying both sides of Eq. (2.2.28) by dt and integrating from time $t = 0$ to time t we find

$$\begin{aligned}\int_{t=0}^{t} d\left[\frac{m}{2}v^2\right] &= -\int_{\mathbf{x}_0}^{\mathbf{x}(t)} \nabla V \cdot d\mathbf{x} \\ &= -\int_{\mathbf{x}_0}^{\mathbf{x}(t)} \left(\frac{\partial V}{\partial x}dx + \frac{\partial V}{\partial y}dy + \frac{\partial V}{\partial z}dz\right) \\ &= -\int_{\mathbf{x}_0}^{\mathbf{x}(t)} dV\end{aligned} \tag{2.2.29}$$

Since we have perfect differentials on both sides of Eq. (2.2.29) we can perform the integrations. We find, after a rearrangement of terms,

$$\frac{1}{2}mv(t)^2 + V(\mathbf{x}(t)) = \frac{1}{2}mv_0^2 + V(\mathbf{x}_0) \tag{2.2.30}$$

where $v_0^2 = v_{0x}^2 + v_{0y}^2 + v_{0z}^2$. The *kinetic energy*, T, of the particle is defined as

$$T = \frac{1}{2}m\dot{\mathbf{x}} \cdot \dot{\mathbf{x}} = \frac{1}{2}mv^2 \tag{2.2.31}$$

so that Eq. (2.2.30) shows that the *total energy*, E, defined as $E = T+V$, remains a constant during the motion.

For the projectile problem just considered we have

$$\begin{aligned}T &= \frac{m}{2}\left[\left(\frac{dx}{dt}\right)^2 + \left(\frac{dy}{dt}\right)^2\right] \\ V &= mgy\end{aligned} \tag{2.2.32}$$

and using Eqs. (2.2.7) and (2.2.8) we can easily prove that

$$E = T + V = E_0 = \frac{m}{2}\left[v_{0x}^2 + v_{0y}^2\right] + mgy_0 \tag{2.2.33}$$

The conservation of energy equation $T+V =$ constant essentially integrates the equations of motion once since it gives us a relationship for velocities rather than accelerations. This conservation equation, while it does not completely solve the equation of motion, does give us important information about the nature of the motion (see Problem P2.3 for an example). To completely solve dynamics problems analytically is usually only possible in the simplest of cases. Even in the projectile problem if we include drag forces we may find that it is necessary to solve the equations of motion numerically. With modern software tools such as MATLAB®, however, this is not difficult to do. In a following section we will outline all the steps needed in MATLAB® to solve the projectile problem with drag forces included. These steps will be the same when solving much more complex problems so it is important that you study them carefully.

2.2.5 *Work–Energy*

In dynamics problems, the concept of kinetic energy is closely related to work concepts, which you are also probably aware of from an earlier course. For any force, regardless of whether it is obtained from a scalar potential or not, if we dot multiply Newton's second law by the velocity we find

$$m\ddot{\mathbf{x}} \cdot \dot{\mathbf{x}} = \mathbf{F} \cdot \dot{\mathbf{x}} \tag{2.2.34}$$

which, following the same steps as carried out for the potential force case, leads to

$$\int_{t=0}^{t} d\left[\frac{m}{2}v^2\right] = \int_{\mathbf{x}_0}^{\mathbf{x}(t)} \mathbf{F} \cdot d\mathbf{x} \tag{2.2.35}$$

giving

$$W = \int_{\mathbf{x}_0}^{\mathbf{x}(t)} \mathbf{F} \cdot d\mathbf{x} = T(\mathbf{x}(t)) - T(\mathbf{x}_0) = \Delta T \tag{2.2.36}$$

where W is the *work* of the force along the path of the particle and ΔT is the change of the kinetic energy. Some authors prefer to write the work-energy relationship in terms of rate-of-work terms. Since $dW/dt = dT/dt$, where $P = dW/dt$ is the *power* we have

$$W = \int_{t_0}^{t} P dt = \int_{t_0}^{t} \mathbf{F} \cdot \mathbf{v} \, dt = T(t) - T(t_0) = \Delta T \tag{2.2.37}$$

For general forces the work done by the force (and hence the corresponding change in the particle's kinetic energy) will depend on the particular path followed while for potential forces the work done depends only on the starting and ending points of the path since in that case

$$
\begin{aligned}
W &= -\int_{\mathbf{x}_0}^{\mathbf{x}(t)} \nabla V \cdot d\mathbf{x} = -\int_{\mathbf{x}_0}^{\mathbf{x}(t)} \left(\frac{\partial V}{\partial x} dx + \frac{\partial V}{\partial y} dy + \frac{\partial V}{\partial z} dz \right) \\
&= -\int_{\mathbf{x}_0}^{\mathbf{x}(t)} dV = -[V(\mathbf{x}(t)) - V(\mathbf{x}_0)]
\end{aligned} \tag{2.2.38}
$$

We can also use this result to show how Eq. (2.2.26) also arises naturally from work concepts. If we compute the work of a force along any closed path where the starting and ending points coincide, then for potential forces Eq. (2.2.38) shows that the work must be zero, or in other words we have for the closed loop integral around any path

$$
\oint \mathbf{F} \cdot d\mathbf{x} = \oint \nabla V \cdot d\mathbf{x} = 0 \tag{2.2.39}
$$

However, Stokes' theorem states that the closed loop line integral of any vector function, $\mathbf{F}$, is equal to $(\nabla \times \mathbf{F}) \cdot \mathbf{n}dS$ integrated over a surface S surrounded by the closed loop, where $\mathbf{n}$ is a unit normal vector to the surface (see Fig. 2.7), i.e.,

$$
\oint \mathbf{F} \cdot d\mathbf{x} = \int_S (\nabla \times \mathbf{F}) \cdot \mathbf{n}dS \tag{2.2.40}
$$

so that for a force represented by a potential energy, from Eqs. (2.2.39) and (2.2.40) we must also have

$$
\int_S (\nabla \times \mathbf{F}) \cdot \mathbf{n}dS = 0 \tag{2.2.41}
$$

for any surface S, which is only possible if $\nabla \times \mathbf{F} = 0$.

We have introduced work concepts here as a foundation for further discussions in later sections where we will see that these concepts provide an approach to finding dynamic forces, obtaining the equations of motion, and solving dynamics problems that is different from a standard "Newtonian" approach that works directly with Newton's second law. This is probably not surprising as work principles also provide an alternate way to solve statics problems which is different than directly applying the force and moment equilibrium equations (see, for example, discussions of virtual work contained in most statics books).

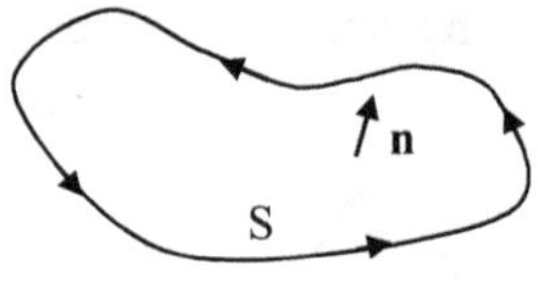

Fig. 2.7 A closed line path enclosing a surface, S, whose unit normal is $\mathbf{n}$

2.2.6 A Projectile Problem with Drag Forces

Consider the projectile problem again but where we include a drag force, **D** (see Fig. 2.2b). We will assume that the magnitude of this drag force is proportional to the square of the magnitude of the particle velocity and the direction of the drag force is opposite to the particle velocity, as shown in Fig. 2.2b. This type of drag force is often used to simulate aerodynamic drag forces. In this case we have

$$\begin{aligned}\mathbf{D} &= -b|\mathbf{v}|\mathbf{v} \\ &= -b\sqrt{\dot{x}^2+\dot{y}^2}\dot{\mathbf{x}}\end{aligned} \tag{2.2.42}$$

where b is a drag constant (it has units of force/velocity^2) and the minus sign is included to ensure that the drag force is always opposite in direction to the particle velocity. In this case the equations of motion for the projectile from Newton's second law become

$$\begin{aligned}m\ddot{x} &= -b\sqrt{\dot{x}^2+\dot{y}^2}\dot{x} \\ m\ddot{y} &= -b\sqrt{\dot{x}^2+\dot{y}^2}\dot{y} - mg\end{aligned} \tag{2.2.43}$$

which we will divide by the mass m to obtain

$$\begin{aligned}\ddot{x} &= -c\sqrt{\dot{x}^2+\dot{y}^2}\dot{x} \\ \ddot{y} &= -c\sqrt{\dot{x}^2+\dot{y}^2}\dot{y} - g\end{aligned} \tag{2.2.44}$$

where the constant $c = b/m$ now has the dimensions of (1/length).

Simulation 2.1: A Projectile with Drag

Note that the equations in Eq. (2.2.44) are in the form of Eq. (2.2.3), which is a system of second-order ordinary differential equations, which we now will solve numerically using MATLAB®. However, the MATLAB® numerical solver we will use assumes that the equations are written instead as a system of first-order equations in the form

$$\dot{\mathbf{x}}(t) = \mathbf{g}(\mathbf{x}(t), t) \tag{2.2.45}$$

To put our equations in this form is easy. We simply take our original two variables (x, y) and add the two velocities $(\dot{x}, \dot{y})$ to define the four variables (in any order we choose) as

$$\begin{aligned} x_1 &= x(t) \\ x_2 &= \dot{x}(t) \\ x_3 &= y(t) \\ x_4 &= \dot{y}(t) \end{aligned} \tag{2.2.46}$$

Then Eq. (2.2.44) can be rewritten as a system of first-order differential equations in these four variables as

$$\begin{aligned} \dot{x}_1 &= x_2 \\ \dot{x}_2 &= -cx_2\sqrt{x_2^2 + x_4^2} \\ \dot{x}_3 &= x_4 \\ \dot{x}_4 &= -cx_4\sqrt{x_2^2 + x_4^2} - g \end{aligned} \tag{2.2.47}$$

which is indeed in the form of Eq. (2.2.45). To solve this system of equations in MATLAB®, we first write a MATLAB® function, projectile_w_drag_eqs, which defines the system of Eq. (2.2.47) and whose calling sequence is

```
[xdot] = projectile_w_drag_eqs (t, x, c, g);
```

The input parameters of this function are the time, t, the vector x containing our four variables, $[x(1), x(2), x(3), x(4)]$, and the constants c and g. This function evaluates the right side of the system of equations in Eq. (2.2.47) and returns the left-hand side as the vector xdot $= [xdot(1), xdot(2), xdot(3), xdot(4)]^T$ which must be written as a *column* vector. The function is given explicitly as

```
function [xdot] = projectile_w_drag_eqs (t, x, c, g)
% [xdot] =projectile_w_drag_eqs (t, x, c, g) implements the equations of
% motion for a
% particle where there is aerodynamic drag force proportional to the
% square
% of the particle velocity. The input parameters are the time, t, a
% vector of system parameters x = [x(1), x(2), x(3), x(4)], and the
% normalized
% drag coefficient c (of dimensions 1/length) and the acceleration of
% gravity, g. The function returns a column vector [ xdot(1), xdot(2),
% xdot(3), xdot(4)]' containing the first derivatives on time of the
% system
% parameters.

% set up a column vector place holder for the derivatives
  xdot = zeros(4,1);
% rename the system variables to make the equations easier to read
x1 = x(1);  % x
x2 = x(2);  % dx/dt
x3 = x(3);  % y
x4 = x(4);  % dy/dt
```

```
% define the system of equations
xdot(1) = x2;
xdot(2) = -c*sqrt(x2.^2 + x4.^2).*x2;
xdot(3) = x4;
xdot(4) = -c*sqrt(x2.^2 + x4.^2).*x4 -g;

end
```

[Note that in this example the time, t, never appears explicitly in the equations of motion so that input parameter is not used. But in other examples where time is present explicitly, we do need the time input to accommodate that behavior. In writing our equations of motion the MATLAB® function we create, like dproj, must always contain t as its first input argument and the first-order system variable, x, is the second argument (a vector), followed by whatever other input parameters may be present].

With the equations of motion for the system so defined we then need to specify three inputs: (1) the time interval tspan = [0, T] (written here in MATLAB® as a row vector) over which we want to calculate the solution, (2) the initial conditions for the position and velocity components in a row vector $x0 = [x(0), \dot{x}(0), y(0), \dot{y}(0)]$ corresponding to our four system variables, and (3) the values for the constants c and g. As an example calculation in MATLAB® let

```
tspan = [0, 7];        % time interval for calculations (sec)
x0 = [0 30 0 30];      % initial conditions (m, m/sec, m, m/sec)
c = 0.006;             % normalized drag coefficient (1/m)
g = 9.81;              % acceleration of gravity (m/sec^2)
```

Using these inputs we then can call the MATLAB® ordinary differential equation solver ode45 with the calling sequence

```
[tout, xout] = ode45( @projectile_w_drag_eqs, tspan, x0, [ ], c, g);
```

The function ode45 solves a system of first-order differential equations and works well for many dynamics problems, so it a good choice. However, solving dynamics problems numerically can sometimes be difficult or time consuming so that there are other solvers available in MATLAB® (and other software packages) that could be used and may be specifically needed in special cases. In this book we will not go through the numerical solution procedures contained in these solvers in detail but in Appendix C we will give a general overview of how solvers such as ode45 work.

The first input argument of the function ode45 is a handle to the projectile_w_drag_eqs function. A function handle is a MATLAB® data type that stores all the information needed about the function that allows one to execute that function. By passing the function handle for projectile_w_drag_eqs to the differential equation solver ode45, that solver can use it to solve our system of equations. The function ode45 solves the system of equations over the time interval tspan with initial conditions x0. The function ode45 also has a vector of optional input

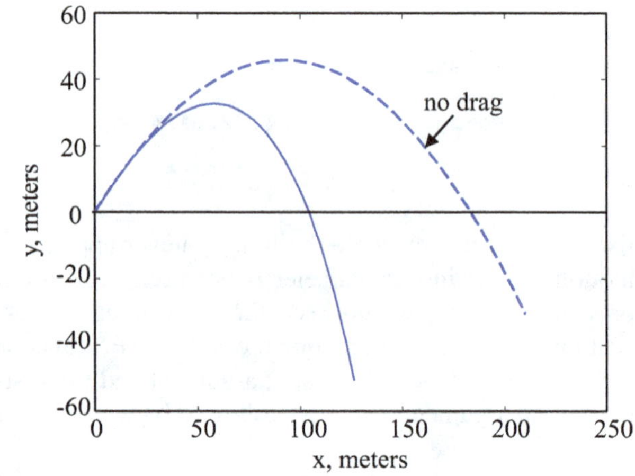

Fig. 2.8 The solution of a projectile problem with initial velocity components of 30 m/s in both the *x*- and *y*-directions at the origin (*x*, *y*) = (0, 0). The case with drag (solid line) and no drag (dashed line)

arguments which are not used here so they are specified by an empty vector, [ ], followed by the names of the input arguments (beyond *t* and *x*) of the function we are using, which in this case are the parameters *c* and *g*. Later, we will replace the empty vector [ ] with options that are used to control the numerical solution. The outputs of ode45 are a column vector of times, tout, at which the solution values were calculated and an array, xout, having the values of $[x_1, x_2, x_3, x_4] = [x, \dot{x}, y, \dot{y}]$ in its four columns at those times. All of these MATLAB® commands have been placed in a MATLAB® script named projectile_w_drag_sim (see Appendix F, where a listing is given of all the functions and scripts used in this book). This script also plots the *y* versus *x* projectile path in this case as well as the case when no drag is present ($c = 0$), as shown in Fig. 2.8. It can be seen that the case with drag (solid line) has considerably less range than the no-drag case (dashed line). The no-drag case follows the expected symmetrical parabolic path which is distorted in the case with drag. For long times, we see that the projectile's *x*-component of the velocity is getting smaller and the particle path is getting more and more vertical. When the vertical drag force is equal and opposite to the force of gravity the vertical acceleration of the projectile is zero and the projectile has reached a constant terminal velocity. From the second equation in Eq. (2.2.44) we see when $\dot{x}$ and $\ddot{y}$ are both zero we find $\dot{y} < 0$ and the magnitude of the terminal velocity is the constant $v_{\text{term}} = |\dot{y}| = \sqrt{g/c}$. In the example just considered we have $v_{\text{term}} = 40.41$ m/s. After executing the script the *y*-velocity is in the vector ydot so that if in the command window we examine the final value of this velocity we find

```
ydot(end)

ans = -31.4088
```

and we have not yet reached the terminal velocity at $t = 7$ s.

2.3 The Dynamics of Constrained Motion of a Particle

In the previous section the 2-D path of motion of the particle was determined by the forces acting on the particle but the particle was free to follow any path that was consistent with those known forces and initial conditions. In most dynamics problems, however, there are conditions (constraints) on the motion path. These constraints must be accounted for when we determine the motion and there are additional forces generated that are present that need to be determined. In this section we examine such constrained motion problems.

2.3.1 *Constrained Motion of a Bead on a Wire*

As our first example of constrained motion, let a particle be acted upon by gravity and a known force, $F(t)$, where the particle is a small bead that is threaded onto a smooth, straight wire that is inclined at an angle θ in the vertical x–y plane, as shown in Fig. 2.9a, and the force F acts along the wire. In this case the bead must move along the straight wire which we can write as the equation of a "surface" in the form

$$f(x, y) = y\cos\theta - x\sin\theta = 0 \tag{2.3.1}$$

[Note: Equation (2.3.1) is really an equation of a line, not a surface because we have restricted the motion to be two-dimensional but we will still refer to Eq. (2.3.1) as a "surface" since it corresponds to the case of the particle moving on a 2-D surface in 3-D in which case the equation $f(x, y, z) = 0$ does indeed describe a surface.]

In this case the initial conditions must satisfy Eq. (2.3.1) and the particle can only pass through points in the (x, y) plane that also satisfy Eq. (2.3.1) so the motion of the particle is *constrained* and Eq. (2.3.1) is the constraint equation. To keep the particle on the wire we must have another force acting in addition to the applied force and the force of gravity. This force we will call a *constraint force*. Newton's second law then takes the form

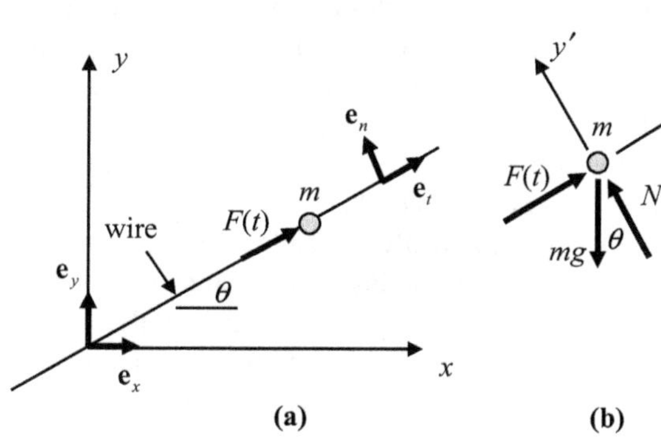

Fig. 2.9 (**a**) A bead sliding on a straight smooth wire in two dimensions, and (**b**) a free body diagram of the bead

$$m\ddot{\mathbf{x}} = \mathbf{F}^e + \mathbf{F}^c \tag{2.3.2}$$

where $\mathbf{F}^e$ is the resultant force from all the external applied forces and $\mathbf{F}^c$ is the resultant of all the constraint forces.

We know from statics that the smooth wire shown in Fig. 2.9a can only exert a force on the particle in a direction that is normal to the wire so the constraint force must be along that normal direction, as shown in the free body diagram for our particle in Fig. 2.9b. Thus, the forces acting are

$$\begin{aligned}\mathbf{F}^e &= F(t)\mathbf{e}_t - mg\mathbf{e}_y \\ \mathbf{F}^c &= N\mathbf{e}_n\end{aligned} \tag{2.3.3}$$

where $(\mathbf{e}_x, \mathbf{e}_y)$ are unit vectors in the x- and y-directions and $(\mathbf{e}_t, \mathbf{e}_n)$ are unit vectors along the wire and normal to the wire (Fig. 2.9). Newton's second law then gives

$$\begin{aligned}m\ddot{\mathbf{x}} &= F(t)\mathbf{e}_t - mg\mathbf{e}_y + N\mathbf{e}_n \\ &= F(t)\left(\cos\theta\mathbf{e}_x + \sin\theta\mathbf{e}_y\right) - mg\mathbf{e}_y + N\left(-\sin\theta\mathbf{e}_x + \cos\theta\mathbf{e}_y\right)\end{aligned} \tag{2.3.4}$$

or in component form

$$\begin{aligned}m\ddot{x} &= F(t)\cos\theta - N\sin\theta \\ m\ddot{y} &= F(t)\sin\theta - mg + N\cos\theta\end{aligned} \tag{2.3.5}$$

Thus, to determine the motion of the bead we must solve the equations of motion, Eq. (2.3.5), jointly with the constraint equation, Eq. (2.3.1), giving three equations

$$\begin{aligned}m\ddot{x} &= F(t)\cos\theta - N\sin\theta \\ m\ddot{y} &= F(t)\sin\theta - mg + N\cos\theta \\ f(x,y) &= y\cos\theta - x\sin\theta = 0\end{aligned} \tag{2.3.6}$$

for the three unknowns $(x(t), y(t), N)$. Can we solve these equations numerically? The answer is yes. Since Newton's second law gives us two second-order differential equations while the constraint equation is an algebraic equation, the system of equations in Eq. (2.3.6) is called a system of differential-algebraic equations (DAEs). More general constrained motion dynamics problems also lead to DAEs. The MATLAB® ode45 solver requires that we have a system of first-order ordinary differential equations (ODEs) so it cannot be used directly to solve Eq. (2.3.6). There are other solvers in MATLAB® that could be used for the DAEs of Eq. (2.3.6) but solving DAEs numerically typically requires more careful analysis of the numerical procedures than does solving ODEs. Since in this book we do not want to focus our attention too heavily on the numerical methods themselves, we will not discuss numerical DAE solvers. For more information on solving DAEs, see [1]. Not using a DAE solver directly means that we will examine ways to express Eq. (2.3.6) that will lead us to a system of ODEs that we can treat with solvers such as ode45. One method that we will use frequently is to differentiate the constraint equation twice on time to write it in terms of the accelerations. In this case we have

$$\ddot{y}\cos\theta - \ddot{x}\sin\theta = 0 \tag{2.3.7}$$

We can then write the equations of motion and this differentiated constraint equation as a system linear equations for the unknowns $(\ddot{x}, \ddot{y}, N)$. The system is

$$\begin{bmatrix} m & 0 & \sin\theta \\ 0 & m & -\cos\theta \\ -\sin\theta & \cos\theta & 0 \end{bmatrix} \begin{Bmatrix} \ddot{x} \\ \ddot{y} \\ N \end{Bmatrix} = \begin{Bmatrix} F(t)\cos\theta \\ F(t)\sin\theta - mg \\ 0 \end{Bmatrix} \tag{2.3.8}$$

which we can write in matrix vector form as $\mathbf{Ku} = \mathbf{d}$ with $\mathbf{u} = [\ddot{x}\ \ddot{y}\ N]^T$. The determinant of this system is just the mass, m, which is never zero so the system always has a solution. It is not difficult to find this solution analytically by hand. In fact, if we multiply the first equation in Eq. (2.3.6) by $-\sin\theta$ and the second equation by $\cos\theta$ and add the resulting two equations together we can eliminate the accelerations and solve for the force N. This force can then be placed back into the first two equations in Eq. (2.3.6) to yield the equations of motion. Another option is to directly solve the system of three equations for three unknowns given in Eq. (2.3.8). That option requires more algebra but we can let MATLAB® do the work for us by using the symbolic math toolbox so we will show that solution. To start, we need to declare the variables m, g, F, and θ in Eq. (2.3.8) as symbolic variables. In MATLAB® this is done through

```
syms m g F theta
```

Next, we need to symbolically form the **K** coefficient matrix and **d** vector in Eq. (2.3.8):

```
K = [m 0 sin(theta);0 m -cos(theta);-sin(theta) cos(theta) 0];
d = [ F*cos(theta); F*sin(theta)-m*g; 0];
```

The formal solution of this linear set of equations is just $\mathbf{u} = \mathbf{K}^{-1}\mathbf{d}$ so we can find the solution symbolically in this fashion and simplify the final expression so that it is more readable:

```
u =inv(K)*d;
simplify(u)
```

```
ans = (F*cos(theta) - (g*m*sin(2*theta))/2)/m
      (sin(theta)*(F - g*m*sin(theta)))/m
               g*m*cos(theta)
```

Using $\sin(2\theta) = 2\sin\theta\cos\theta$ and writing this solution in terms of our original variables we have, with a few minor changes,

$$\begin{aligned}\ddot{x} &= \cos\theta(F(t) - mg\sin\theta)/m \\ \ddot{y} &= \sin\theta(F(t) - mg\sin\theta)/m \\ N &= mg\cos\theta\end{aligned} \tag{2.3.9}$$

Note that we can also solve this problem very directly by looking at Newton's force equations along the wire and perpendicular to the wire, i.e., along the (x', y') coordinates shown in Fig. 2.9b. In this case we find simply

$$\begin{aligned}m\ddot{x}' &= F(t) - mg\sin\theta \\ m\ddot{y}' &= 0 = N - mg\cos\theta\end{aligned} \tag{2.3.10}$$

Obviously the normal force agrees with our previous result. The equations of motion are also in agreement since the transformation from the (x, y) coordinates to the (x', y') coordinates is just

$$\begin{Bmatrix} x' \\ y' \end{Bmatrix} = \begin{bmatrix} \cos\theta & \sin\theta \\ -\sin\theta & \cos\theta \end{bmatrix} \begin{Bmatrix} x \\ y \end{Bmatrix} \tag{2.3.11}$$

which is also true for the acceleration components so that differentiating Eq. (2.3.11) twice on time and using it in Eq. (2.3.9) we do find Eq. (2.3.10).

If we can integrate the $F(t)$ term analytically then we can solve the differential equations in Eq. (2.3.9) analytically and we do not have to perform the solution numerically. However, remember that in performing the solution for (x, y) either analytically or numerically the intial conditions $(x_0, \dot{x}_0, y_0, \dot{y}_0)$ need to satisfy the constraints since the constraints hold for all times t, including the initial time. Thus, in this bead on a straight-wire problem the initial conditions must satisfy

$$\begin{aligned}y_0\cos\theta - x_0\sin\theta &= 0 \\ \dot{y}_0\cos\theta - \dot{x}_0\sin\theta &= 0\end{aligned} \tag{2.3.12}$$

Note that we have examined a two-dimensional motion problem here so at most we could constrain the coordinates of the particle with one constraint of the type given in Eq. (2.3.1). Otherwise, the particle would either be completely constrained or over-constrained and would be in equilibrium. For a particle moving in three dimensions, we could apply either one or two constraints on the coordinates and, as we will see in Chap. 3, for a system of particles we may have many equations of constraint on or between the particles. The procedure we followed in this simple constrained problem gives us a way to solve constrained particle motion problems in general. There are four steps to follow:

1. Apply Newton's second law to obtain the acceleration components in terms of the known applied forces and the unknown constraint forces.
2. Use the constraint equations and differentiate them on time to write those constraints in terms of the accelerations.
3. Solve the combination of Newton's law equations and the differentiated constraint equations for the constraint forces in terms of the position and velocity components and solve for the accelerations in the same fashion. These acceleration expressions will be the equations of motion and the constraint force expression will give us the constraint force values once the equations of motion are solved.
4. Solve the equations of motion to obtain the position and velocity components, using initial conditions that are consistent with the constraints. Use this solution to obtain the constraint forces explicitly.

In our bead-on-the-straight-wire problem the constraint force is a constant so that we can obtain it without having a solution of the equations of motion as the steps listed above would indicate. However, when we solve more complex problems such as where the bead moves on a curved wire (a problem we will consider shortly) we will see that the constraint force will indeed be a function of the coordinates of the bead and its velocity so that the constraint force can be found explicitly only after solving the equations of motion.

Now, let us consider solving this problem numerically. We could turn the equations of motion in Eq. (2.3.9) into a system of four first-order differential equations. But, we can also completely solve the problem by just solving the first equation in Eq. (2.3.9) for $x(t)$ and $\dot{x}(t)$ since the constraint equation and its time derivative then give us the behavior of $y(t)$ and $\dot{y}(t)$ explicitly. This might be expected since this problem is the example of a *single degree of freedom system.* It makes sense that we only need to solve for only one coordinate variable because this is also exactly what we found when we used the (x', y') coordinates and obtained Eq. (2.3.10).

In dynamics problems, if the motion of a completely unconstrained system can be described in terms of n_s *system coordinates* and if we then apply n_h constraints that relate those system coordinates to each other (such as Eq. (2.3.1)), the smallest number of independent coordinates needed to describe the motion of the system is the number of *degrees of freedom*, $n_{\text{dof}} = n_s - n_h$. These n_{dof} coordinates we will call the *independent coordinates* and the remaining coordinates the n_h*-dependent coordinates*. The sum of independent and dependent coordinates equals n_s, the number of system coordinates. We can always partition the system coordinates into dependent and independent coordinates. In our bead problem $n_s = 2$ (the two coordinates x and y), $n_h = 1$ (one constraint on the motion in those coordinates), so $n_{\text{dof}} = 1$. The x-coordinate, for example, can be chosen as an independent coordinate and then the y-coordinate is a dependent coordinate.

If practical, it is nice to solve dynamics problems using only n_{dof} equations of motion for the independent coordinates since then we have as few equations as possible and the constraints are automatically satisfied. However, in complex problems the equations of motion written in terms of the independent coordinates may be much more complex than the larger number of equations of motion we find in terms of the system coordinates (which describe unconstrained motion). But if we use system coordinates we must also explicitly satisfy the constraints between those system coordinates. As we saw with this example of the bead on the wire problem, if we differentiate the constraint equation twice to obtain an expression for the constraint in terms of the accelerations, then we can use those acceleration constraints to find the equations of motion for the system coordinates (which were the coordinates (x, y) for the bead problem). However, because we took two time derivatives of the original constraint equation, satisfying the constraint equation itself and its first time derivative was not explicitly part of the equations of motion for the system coordinates. This is not an issue if we both solve those equations of motion exactly and satisfy the initial conditions exactly, since then the constraint equation and its first derivative (and in fact all derivatives) will also be exactly satisfied. In numerical solutions of the equations of motion for the system coordinates, however, small errors can accumulate over time and cause growing errors in satisfying the constraint equation and its first derivative and we must deal with that problem, which is called the problem of *constraint drift*. We will not discuss explicit solutions to the problem of constraint drift until Sect. 2.12 but another viable approach is to monitor the constraints and their first time derivatives as the numerical solution proceeds to guarantee that constraint errors are indeed small. Of course, this second approach does not really solve the constraint drift problem but simply ensures that it is not present. Thus, in the bead problem, for example, we can choose to either (1) solve the set of two equations of motion for the system coordinates (x, y) and also evaluate the constraint equation and its first derivative to ensure that they are not violated or (2) solve the equation of motion for an independent coordinate, such as x, and use the constraint equation and its derivative to find the dependent coordinate, y, and its derivative, $\dot{y}$. This second choice avoids constraint drift and keeps the constraint equation and its derivative satisfied to within the accuracy of the numerical solution.

Simulation 2.2: Bead on a Straight Wire

We will examine choice (1) and solve numerically for the system variables (x, y) for the problem of the bead sliding on a straight wire. The force F will be taken to be the force

$$F(t) = \begin{cases} 20 - 10t & 0 \leq t \leq 2 \\ 0 & \text{otherwise} \end{cases} \tag{2.3.13}$$

where F is measured in Newtons and t is in seconds. The MATLAB® function bead_on_s_wire_eqs contains the equations of motion written as four first-order differential equations with F given by Eq. (2.3.13). The calling sequence for this function is

```
[xdot] =bead_on_s_wire_eqs(t, x, m, g, ang);
```

where m is the mass, g is the acceleration of gravity, and ang is the angle θ (in rad). Note that in this case we do need the time input variable of this function in order to evaluate F. A script, bead_on_s_wire_sim, sets up default values for the problem as m = 1 kg, g = 9.81 m/s^2, ang = 30*pi/180 ($\theta = 30^\circ$ in rad), tspan = [0, 10] s, and $x0 = [x_0, \dot{x}_0, y_0, \dot{y}_0]$ = [0, 0, 0, 0]. These default values are used unless defined differently by the user in the command window before executing the script.

[Note: You should examine the details of the script bead_on_s_wire_sim in Appendix F to see how the default parameter values are placed in that script using the MATLAB® exist function. By placing the parameters in the script in this fashion they can be changed simply by supplying different values in the command window before executing the script. Scripts with this same structure will be used in many following examples in this book. This approach gives the user great flexibility to rapidly test different conditions. However, this structure also has some risks since if you make a number of changes to the default variables in the command window those changes remain in the current workspace and when you re-execute the script those changes will be present whether or not you wish them to remain. Also, if you have executed other functions or scripts before running this script there may be incorrect values for parameters with the same names in the workspace.

Thus, you should ***clear the workspace*** *(in the MATLAB® command window type clear) before you execute this script (or any other script given in this book) or before you make any changes to the default values in the command window prior to executing the script.*]

With the default parameters in the script bead_on_s_wire_sim we are solving for the motion of the bead for 10 s where the bead starts at the origin with no velocity. The script then calls the MATLAB® solver ode45 and plots the solution for the displacement along the wire, $x'(t) = \cos\theta\, x(t) + \sin\theta\, y(t)$, and the values of the constraint equation and its derivative as a function of time. The results are shown in Fig. 2.10. It can be seen from Fig. 2.10a that the bead is driven up the wire by the force F but after the driving force vanishes the constant gravity force eventually pulls the bead down the wire with ever-increasing displacement (and speed). During this motion both the constraint equation (Fig. 2.10b) and its derivative (Fig. 2.10c) are extremely small so that constraint drift is negligible over this time span.

As a second example of a constrained problem consider the 2-D motion of the bead of mass m on a wire again, where now the only applied force is its own weight and the bead is threaded on a smooth, curved parabolic-shaped wire whose shape is given by $y = ax^2$, where y and x are measured in meters (Fig. 2.11a). The constraint in this case is

$$f(x, y) = y - ax^2 = 0 \tag{2.3.14}$$

The free body diagram of the bead is shown in Fig. 2.11b. Again we have a constraint force, N, which acts normal to the wire, i.e., in the direction defined by the unit vector $\mathbf{e}_n$. From analytical geometry we know that the normal to a "surface"

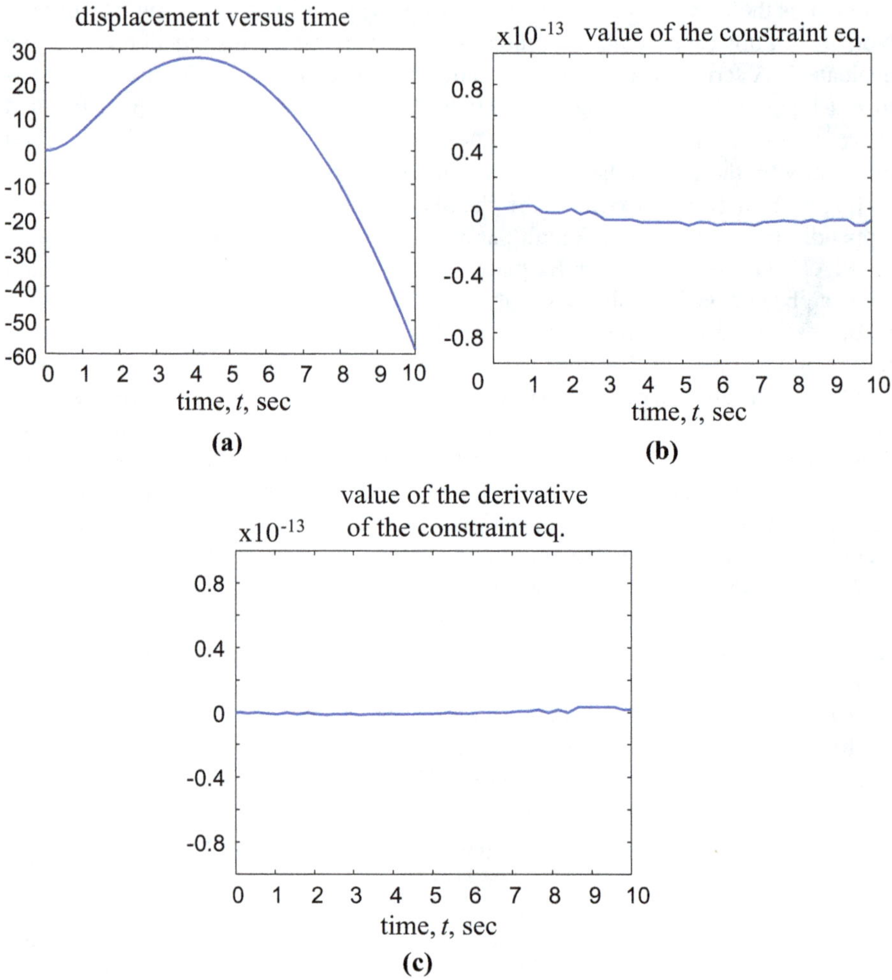

Fig. 2.10 (**a**) The displacement of the bead along the wire versus time, (**b**) the value of the constraint equation during the motion, and (**c**) the value of the derivative on time of the constraint equation

Fig. 2.11 (**a**) Motion of a bead of mass m on a smooth parabolic-shaped wire under the action of gravity, and (**b**) the free body diagram of the forces acting on the bead

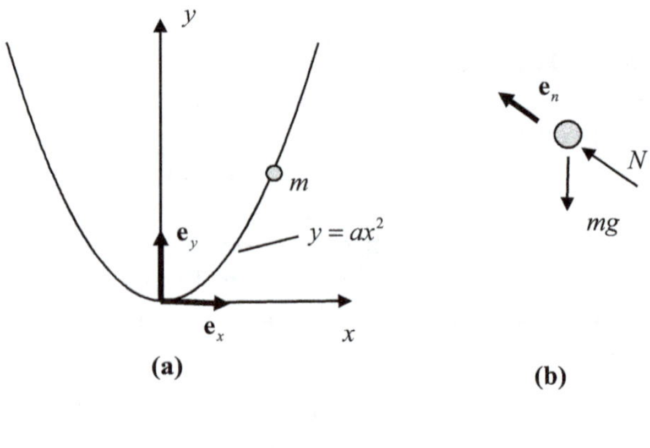

defined by $f(x, y) = 0$ is in the direction of the gradient of the function $f(x, y)$, ∇f, where in this case

$$\nabla f = \frac{\partial f}{\partial x}\mathbf{e}_x + \frac{\partial f}{\partial y}\mathbf{e}_y = -2ax\mathbf{e}_x + \mathbf{e}_y \tag{2.3.15}$$

and, hence, a unit vector normal to the curve is

$$\mathbf{e}_n = \frac{\nabla f}{|\nabla f|} = \frac{-2ax}{\sqrt{1+4a^2x^2}}\mathbf{e}_x + \frac{1}{\sqrt{1+4a^2x^2}}\mathbf{e}_y \tag{2.3.16}$$

The external and constraint forces acting on the bead, therefore, are

$$\begin{aligned}\mathbf{F}^e &= -mg\mathbf{e}_y \\ \mathbf{F}^c &= N\mathbf{e}_n = \frac{-2axN}{\sqrt{1+4a^2x^2}}\mathbf{e}_x + \frac{N}{\sqrt{1+4a^2x^2}}\mathbf{e}_y\end{aligned} \tag{2.3.17}$$

and the x- and y-components of Newton's second law then give

$$\begin{aligned}m\ddot{x} &= \frac{-2ax}{\sqrt{1+4a^2x^2}}N \\ m\ddot{y} &= -mg + \frac{N}{\sqrt{1+4a^2x^2}}\end{aligned} \tag{2.3.18}$$

Taking one derivative and then a second time derivative of the constraint gives

$$\dot{y} - 2ax\dot{x} = 0 \tag{2.3.19a}$$

and

$$\ddot{y} - 2ax\ddot{x} = 2a(\dot{x})^2 \tag{2.3.19b}$$

Placing the acceleration terms of Eq. (2.3.18) into Eq. (2.3.19b) gives

$$-g + \frac{N/m}{\sqrt{1+4a^2x^2}} + \frac{4a^2x^2N/m}{\sqrt{1+4a^2x^2}} = 2a(\dot{x})^2 \tag{2.3.20}$$

which can be solved for N as

$$N = \frac{m(g + 2a\dot{x}^2)}{\sqrt{1+4a^2x^2}} \tag{2.3.21}$$

from which the equations of motion become

$$\begin{aligned}\ddot{x} &= -2ax\frac{\left(g+2a\dot{x}^2\right)}{1+4a^2x^2}\\ \ddot{y} &= -g+\frac{\left(g+2a\dot{x}^2\right)}{1+4a^2x^2}=\frac{2a\left(\dot{x}^2-2ax^2g\right)}{1+4a^2x^2}\end{aligned} \tag{2.3.22}$$

As done before with the bead on the straight wire we could have combined Newton's second law equations, Eq. (2.3.18), with the constraint equation written in terms of the accelerations in Eq. (2.3.19b) to have a linear system of three equations for the three unknowns $\left(\ddot{x},\ddot{y},N\right)$. We can write those three equations succinctly as

$$\begin{bmatrix} m & 0 & -(1/|\nabla f|)\partial f/\partial x\\ 0 & m & -(1/|\nabla f|)\partial f/\partial y\\ \partial f/\partial x & \partial f/\partial y & 0\end{bmatrix}\begin{Bmatrix}\ddot{x}\\ \ddot{y}\\ N\end{Bmatrix}=\begin{Bmatrix}0\\ -mg\\ 2a\dot{x}^2\end{Bmatrix} \tag{2.3.23}$$

and we could again solve this system since the determinant, D, of the matrix of coefficients for this system is never zero where

$$D=\begin{vmatrix} m & 0 & -(1/|\nabla f|)\partial f/\partial x\\ 0 & m & -(1/|\nabla f|)\partial f/\partial y\\ \partial f/\partial x & \partial f/\partial y & 0\end{vmatrix}=m\frac{1}{|\nabla f|}\left(\frac{\partial f}{\partial y}\right)^2+m\frac{1}{|\nabla f|}\left(\frac{\partial f}{\partial x}\right)^2=m|\nabla f| \tag{2.3.24}$$

You can verify that the solution to this system (see Problem P2.5) yields the same equations of motion and normal force expression just obtained so by whichever method we choose we arrive at:

$$\begin{aligned}\ddot{x} &= -2ax\frac{\left(g+2a\dot{x}^2\right)}{1+4a^2x^2}\\ \ddot{y} &= -g+\frac{\left(g+2a\dot{x}^2\right)}{1+4a^2x^2}=\frac{2a\left(\dot{x}^2-2ax^2g\right)}{1+4a^2x^2}\\ N &= \frac{m\left(g+2a\dot{x}^2\right)}{\sqrt{1+4a^2x^2}}\end{aligned} \tag{2.3.25}$$

Here the normal force is not a constant but a function of both the x-position component and x-velocity component of the particle. Thus, we do need to find the solution of the equations of motion before we can evaluate the constraint force. The equations of motion also are more complex than in the straight-wire case and a numerical solution is likely the only option we have. Again, this is a one degree of freedom problem so that the problem can be defined in principle by one independent variable. From Eq. (2.3.25) we see that we could take x as that independent variable and obtain the values of $\left(y,\dot{y}\right)$ from the constraint equation and its derivative and the solutions for $(x, \dot{x})$. Alternatively, we could solve the equations of motion in Eq. (2.3.25) for both x and y and then examine the satisfaction of the constraint equation and its first derivative, as done in the straight-wire case (see Problem P2.6).

Simulation 2.3: Bead on a Parabolic Wire

Now, let us solve the bead on the parabolic wire problem numerically using only the equation for the x-motion. In that case we have only a system of two first-order equations from the first equation of motion in terms of $(x_1, x_2) = (x, \dot{x})$ as

$$\begin{aligned} \dot{x}_1 &= x_2 \\ \dot{x}_2 &= -2ax_1 \frac{\left(g + 2ax_2^2\right)}{1 + 4a^2x_1^2} \end{aligned} \tag{2.3.26}$$

We placed this system of equations in the MATLAB® function bead_on_p_wire_eqs which has the calling sequence

```
[xdot] = bead_on_p_wire_eqs(t, x, a g)
```

The MATLAB® script bead_on_p_wire_sim was then written. The script solves for the motion of the bead for the default parameters a = 0.1 1/m and g = 9.81 m/s^2. All the default input parameters can be specified by the user as different values, if wanted, in the command window before executing the script. At time $t = 0$ we assume that the bead is at the origin and is given an initial velocity of 10 m/s in the plus x-direction so that the default initial conditions are x0 = [0, 10]. The time span for the duration of the numerical solution is given as the default value tspan = [0, 10] s. One difference from the previous bead_on_s_wire_sim script is that we will also specify two error tolerances, rtol and atol, that are used in ode45 to control the numerical accuracy of the results. These error tolerances are placed in the ode45 function as part of an additional input argument to that function called options. To generate this options argument we must use the MATLAB® function odeset. For example, in the script bead_on_p_wire_sim (see Appendix F) we have the statements

```
if ~exist('rtol', 'var'), rtol = 1e-3; end

if ~exist('atol', 'var'), atol = 1e-6; end

options = odeset('RelTol', rtol, 'AbsTol', atol);
```

These generate the options argument with the specified values for rtol and atol unless the user has specified different values before executing the script. There are many other options that can be used by ode45 with the options input. To see these, type

```
help odeset
```

The script calls ode45 as

```
[tout, xout] = ode45(@bead_on_p_wire_eqs, tspan, x0, options, a, g);
```

where now the options input variable is present which will specify that ode45 uses the user-defined values of rtol and atol. Later, we will give more detail on how these tolerances control the numerical solution. The tolerances shown above are the default tolerances that ode45 uses when the options input argument is absent, i.e., when the options specification is the empty vector [ ], as used previously in the straight-wire problem. After setting up default parameters, including these rtol and atol values, and calling ode45 as shown above, the MATLAB® script bead_on_p_wire_sim plots the motion $x(t)$ versus the time, t, as well as the normalized normal force N/mg versus t. These results are shown in Fig. 2.12a, b. Since there is no dissipation modeled in this system the motion of the bead is just a back-and-forth oscillation on the parabolic wire and we see in Fig. 2.12a this oscillatory motion in the x-position. The normal force is also oscillatory, with its maximum values occurring when the bead passes through $x = 0$, which is the equilibrium position of the bead. You might notice that the force curves are not smooth at their peaks where the force is rapidly changing. This suggests that we may need to refine the numerical solution. We can do that here by defining a smaller relative error tolerance, rtol, before executing the script, i.e.,

```
rtol =1e-6;
```

The results are shown in Fig. 2.13, where we see that the force curve is now smoother. Details of exactly how rtol affects the solution will be given in later examples.

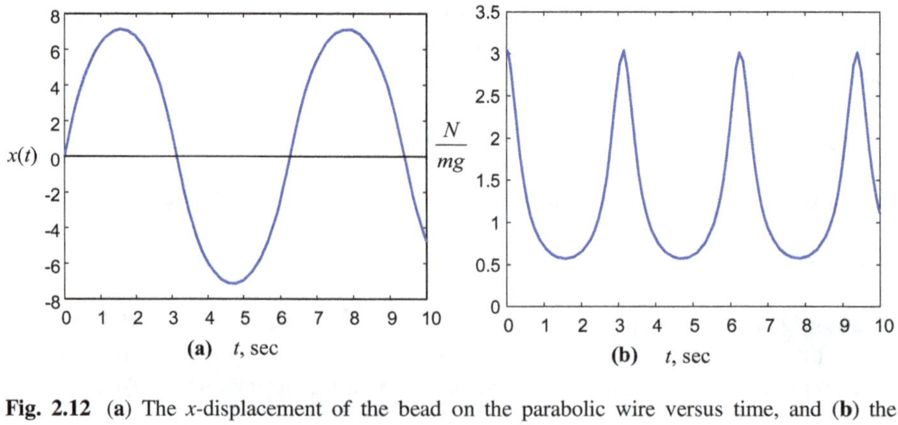

Fig. 2.12 (**a**) The x-displacement of the bead on the parabolic wire versus time, and (**b**) the normalized normal force, N/mg, over the same time interval

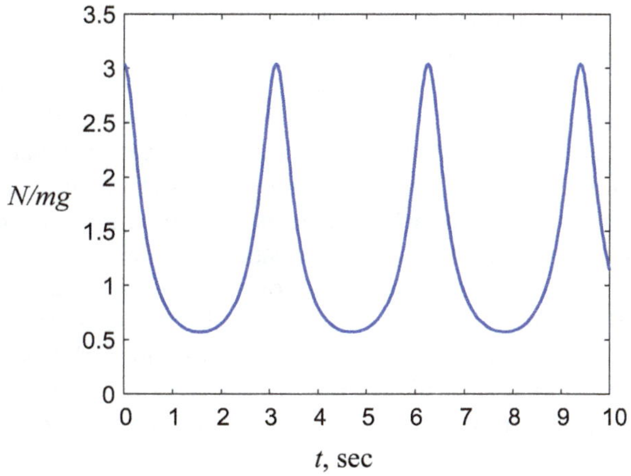

Fig. 2.13 The normalized normal force acting on the bead sliding on a parabolic wire when calculated with a smaller relative error tolerance rtol = 1e-6

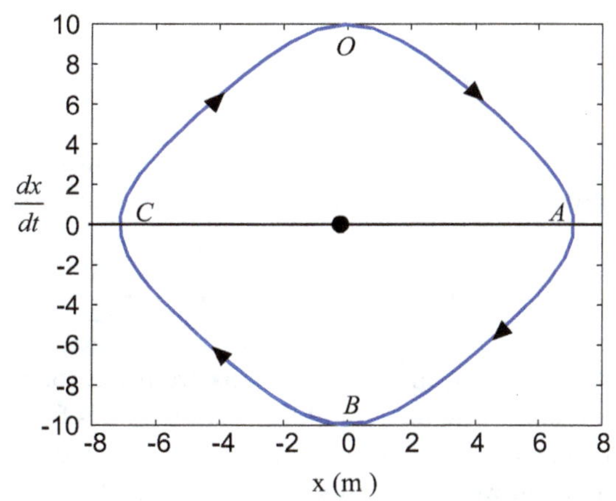

Fig. 2.14 A phase plot of the velocity *dx*/*dt* versus *x* for the motion of the bead on a smooth parabolic wire

It is instructive to examine a plot of the x-velocity, $\dot{x}$, versus the position, x. This is called a *phase plot*. [Note: A phase plot is normally defined as a plot of the *momentum* versus position but since the momentum is proportion to the velocity this distinction will be ignored.] We can obtain a phase plot directly from the numerical solution since both the x-position and x-velocity are in the column vector outputs, xout(:, 1) and xout(:, 2) from ode45. Figure 2.14 shows the phase plot for the bead obtained by running the script bead_on_p_wire and then executing the MATLAB® command

```
plot (xout(:,1), xout(:, 2))
```

The phase plot in this case is a plot of $\dot{x}$ versus x, generating a single closed figure centered about the equilibrium position $x = \dot{x} = 0$ (shown as a black dot in the center of Fig. 2.14). The arrows and other labels seen in Fig. 2.14 have been added manually to this MATLAB® plot. The bead starts at O, moving in the +x-direction until it reaches its maximum x-displacement (and height, y, not shown) on the parabolic wire (point A) with zero velocity. It then continues back to the origin which it reaches with its maximum negative velocity (point B) and then reaches its largest negative x-displacement (and largest height, y, again) at point C. Subsequently, the bead returns to the origin with its original velocity and then repeats the entire motion. Closed figure phase plots of this type are characteristic of repetitive motion problems where there is no damping involved so that the total energy is conserved and the motion repeats indefinitely. Other dynamics problems as physically diverse as the linear motion of a mass attached to a spring and the motion of a planet around the sun thus share similar phase plots to the one shown in Fig. 2.14.

In this problem the normal force does no work since it always acts perpendicular to the motion of the particle. The only other force is the force of gravity, which is derivable from a potential energy so the total energy, E, is conserved in this problem, where

$$\begin{aligned} E &= T + V \\ &= \frac{1}{2}m\dot{x}^2 + \frac{1}{2}m\dot{y}^2 + mgy = \frac{1}{2}m\dot{x}_0^2 + \frac{1}{2}m\dot{y}_0^2 + mgy_0 = \text{constant} \end{aligned} \tag{2.3.27}$$

But since it is a single degree of freedom problem, we can also use the constraint to write the total energy in terms of the single variable, x, and its derivative $\dot{x}$:

$$E = \frac{1}{2}m\left(1 + 4a^2x^2\right)\dot{x}^2 + mgax^2 = \frac{1}{2}m\left(1 + 4a^2x_0^2\right)\dot{x}_0^2 + mgax_0^2 \tag{2.3.28}$$

We can also obtain phase plots from the total energy, which is somewhat more revealing than the plot of Fig. 2.14 which is for just a single set of initial conditions. If we vary the x-position and x-velocity and plot the contours of the total energy function where that function is a constant then we can examine the behavior of the system under a variety of initial conditions. The MATLAB® script sliding_bead_energy evaluates the energy divided by the mass, m, E/m, of Eq. (2.3.28) over a range of $\dot{x}$ values going from -10 to +10 m/s and a range of x values over the range of -8 to +8 m, which are the same limits seen in Fig. 2.14. The MATLAB® function contour is then used to generate the phase plots of the motion for different constant values of E over these ranges. The results are shown in Fig. 2.15. We can see that there are multiple closed curves, all of roughly the same nature as in Fig. 2.14, although some of the curves are cut off at the edges of the plot since they are outside the plot window. Always having closed phase plot curves makes sense since regardless of the initial conditions the bead is always trapped within a finite portion of the parabolic wire and the motion will be repetitive.

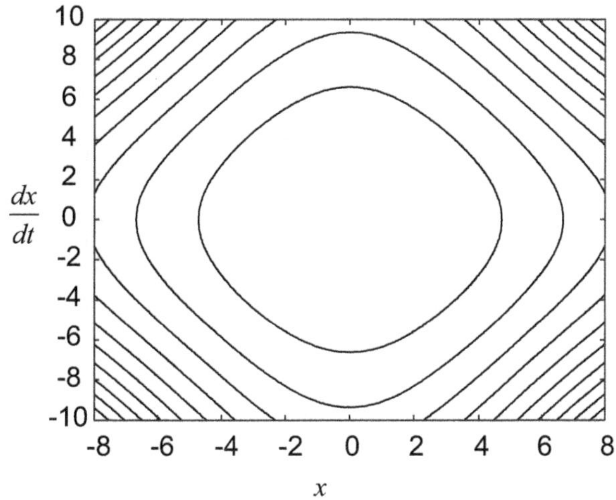

Fig. 2.15 The phase plot curves for the bead on a smooth parabolic wire

We can modify the nature of the motion in this bead on a parabolic wire problem by including a purely viscous-like damping force, $\mathbf{F}_v$, where the magnitude of the damping force is just proportional to the magnitude of the velocity and is always in a direction opposite to the velocity vector. This type of damping force, therefore, is given by

$$\begin{aligned} \mathbf{F}_v &= -c\dot{\mathbf{x}} = -c(\dot{x}\,\mathbf{e}_x + \dot{y}\,\mathbf{e}_y) \\ &= -c\dot{x}\,\mathbf{e}_x - c(2ax)\dot{x}\,\mathbf{e}_y \end{aligned} \tag{2.3.29}$$

where c is a damping constant having the dimensions of mass/seconds. Then the equations of motion of the bead (Eq. (2.3.26)) become

$$\begin{aligned} \ddot{x} &= -2ax\frac{(g+2a\dot{x}^2)}{1+4a^2x^2} - \frac{c}{m}\dot{x} \\ \ddot{y} &= -g + \frac{(g+2a\dot{x}^2)}{1+4a^2x^2} = \frac{2a(\dot{x}^2 - 2ax^2g)}{1+4a^2x^2} - \frac{c}{m}2ax\dot{x} \end{aligned} \tag{2.3.30}$$

Simulation 2.4: Bead on a Wire with Damping

The problem of the bead on the parabolic wire with damping is again a single degree of freedom problem and we can use just the first equation in Eq. (2.3.30) to define the motion. The MATLAB® function bead_on_p_wire_eqs2 differs only from the function bead_on_p_wire_eqs by the addition of an additional input argument, cm = c/m, and by the addition of the $-c\dot{x}/m$ term in the equation of motion. The script bead_on_p_wire_sim2 then uses this new function with the same input parameters as given earlier but with a value cm = 0.5. The x-displacement as a function of time and the corresponding normalized normal force are shown in Fig. 2.16a, b. The motion, while still oscillatory, is seen to be decaying with time,

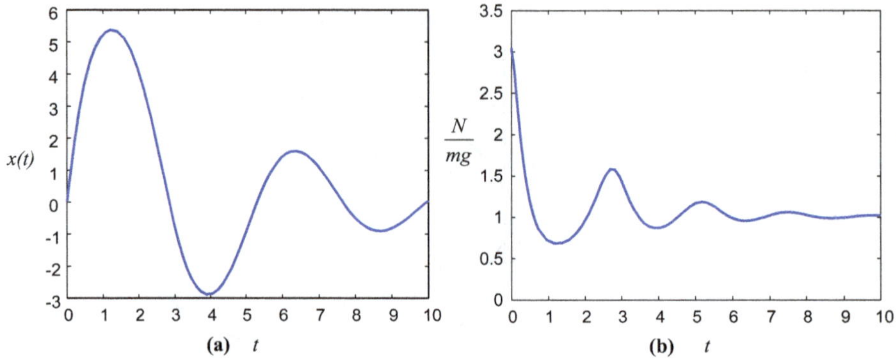

Fig. 2.16 (**a**) The x-displacement of the bead on a smooth parabolic wire versus time with a viscous drag force, and (**b**) the corresponding normalized normal force

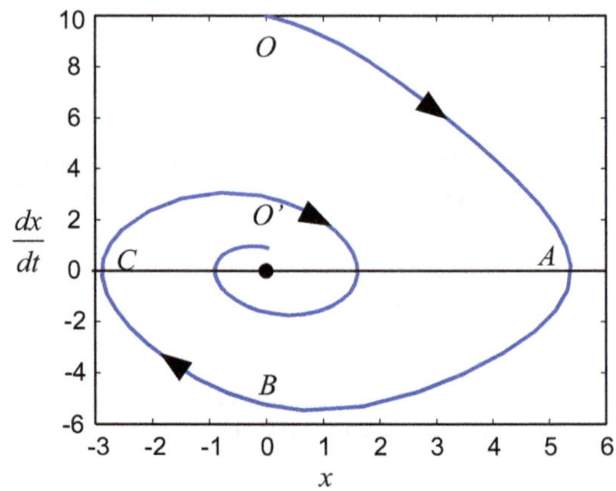

Fig. 2.17 The phase plot of a bead on a smooth parabolic wire when a viscous damping force is present. The points shown correspond to when the bead is at the bottom (origin) starting point (O) moving with a positive velocity (in the +x-direction), when it reaches its maximum height on the right with zero velocity (A), when it returns to the origin, now with a negative velocity (B), then reaches a maximum height again but now to the left (C), and then again returns to the origin with a positive velocity considerably smaller than the starting velocity (O')

as expected, and the normalized normal force is also oscillatory and decaying, approaching the constant value of $N/mg = 1$, which is the value for the normal force when the bead is in equilibrium at the bottom of the parabolic wire. After executing the script the phase plot can again be obtained from the solution by plotting:

```
plot(xout(:,1), xout(:,2))
```

whose output is shown in Fig. 2.17. Now the figure is no longer closed as both the maximum amplitudes and velocities of the bead are decaying and the phase plot path encircles and gradually spirals towards the equilibrium position $(x = \dot{x} = 0)$. Other repetitive motion problems with damping and a stable, decaying dynamic behavior exhibit similar phase plots.

2.3.2 A Roller-Coaster Problem

Another slightly more complex example of a constrained problem is the motion of a cart (particle) on a sinusoidal shaped smooth track of the type shown in Fig. 2.18. We will assume that the particle is launched at the origin with an initial velocity, v_{x0}. The shape of the track shown in Fig. 2.18 is given by

$$y = h(x) = \frac{A}{2}[1 - \cos(\pi x/L)] \tag{2.3.31}$$

and the derivatives of this function are

$$\begin{aligned} \frac{dh}{dx} &= \frac{\pi A}{2L}\sin(\pi x/L) \\ \frac{d^2h}{dx^2} &= \frac{\pi^2 A}{2L^2}\cos(\pi x/L) \end{aligned} \tag{2.3.32}$$

Since the total energy is conserved in this problem we have

$$E = T + V = \frac{1}{2}m(\dot{x}^2 + \dot{y}^2) + mgy = \frac{1}{2}m(\dot{x}_0^2 + \dot{y}_0^2) + mgy_0 \tag{2.3.33}$$

which for the sinusoidal track of Eq. (2.3.31) and initial conditions $x_0 = y_0 = 0$ and $\dot{x}_0 = v_{x0},\ \dot{y}_0 = 0$ gives

$$2E/m = \left[1 + \left(\frac{\pi A}{2L}\sin(\pi x/L)\right)^2\right]\dot{x}^2 + gA[1 - \cos(\pi x/L)] = \dot{x}_0^2 = v_{x0}^2 \tag{2.3.34}$$

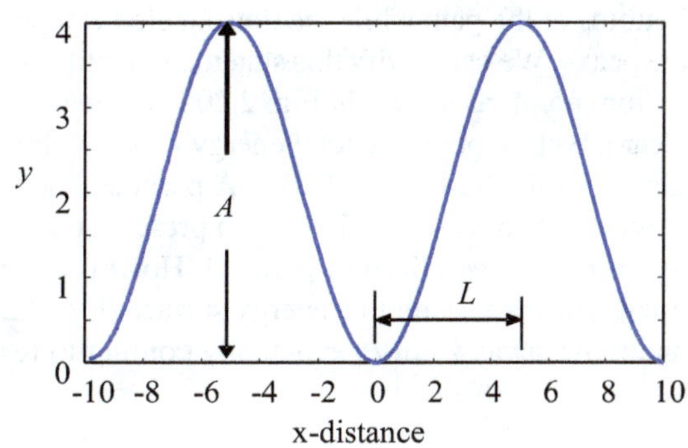

Fig. 2.18 The shape of a sinusoidal roller-coaster track

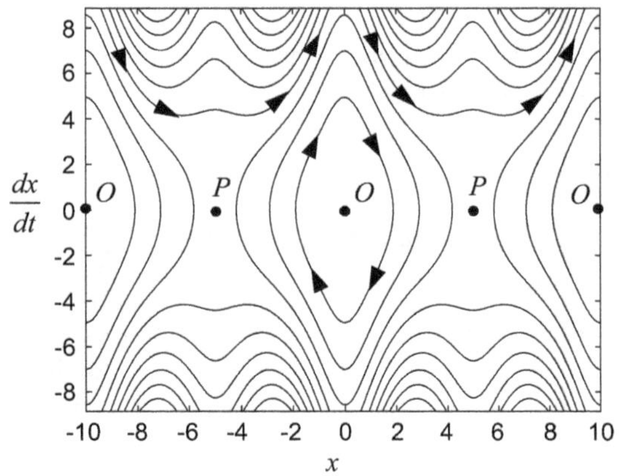

Fig. 2.19 The phase plot for the sinusoidal roller coaster for $A = 4$ m, $L = 5$ m

Now, for the particle to have continuous motion along the roller-coaster track, it must have sufficient initial velocity (to the right or left) to get over the first peaks in the track at $x = \pm L$. If the particle arrives at either of those peaks with zero velocity Eq. (2.3.34) gives the starting velocity, v_m, as $v_m = \pm\sqrt{2gA}$. Thus for $|v_{x0}| < |v_m|$ the particle is trapped in the first trough and can simply oscillate back and forth, as in the parabolic wire case. But if $|v_{x0}| > |v_m|$ then the particle can ride the roller coaster to the right or left forever, assuming that the particle is held to the track by upper and lower rails that can exert either a positive or a negative normal force on the particle. We can see both types of behavior if we use the total energy equation to examine the phase plots of the motion. Figure 2.19 shows the phase plots for $A = 4$ m and $L = 5$ m, with v_x varying over $(-|v_m|, +|v_m|)$ and x varying over two complete cycles of the sinusoidal track. In the center of the figure we see arrows following a closed path about the origin which represents oscillatory motion within the center trough of the roller coaster (between $x = -5$ m and $x = +5$ m). However, we also see arrows that follow an open path where x is continuously increasing as the particle rides the coaster path. The maximum starting velocity for this open path is slightly larger than v_m, a value that occurs outside the plot window. Note that there are entire families of both open and closed curves with each curve depending on the value of the total energy.

All the black dots labeled O in Fig. 2.19 represent stable equilibrium points at the bottoms of the path while the dots labeled P represent unstable equilibrium points at the peaks. We can justify this statement qualitatively if we plot the total energy near a bottom point, as shown in Fig. 2.20a and near a peak point, as shown in Fig. 2.20b. Near a bottom point the total energy forms a "bowl-like" surface with point O at the bottom of the bowl (Fig. 2.20a). A particle that is launched into a small motion near O will, if there is some dissipation present so that $dE/dt < 0$, eventually come to rest at the bottom equilibrium point O. However, a particle set in motion near a peak point will see a saddlelike energy surface (Fig. 2.20b) and tend to slide off the saddle and move away from P, eventually coming to rest, if there is dissipation present, at

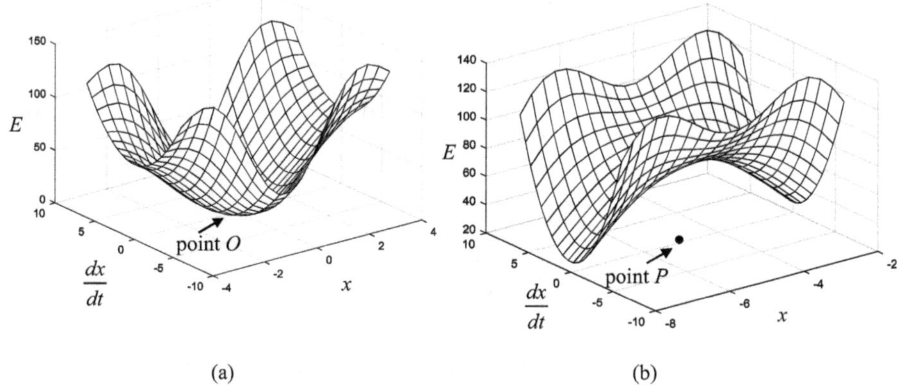

Fig. 2.20 (**a**) The total energy near a bottom point, O, of the roller coaster and (**b**) the total energy near a peak point, P, of the roller coaster

some nearby bottom point on the roller-coaster path. In Chap. 6, when we discuss stability of systems in more detail, we will provide a more rigorous verification of such stability results.

Simulation 2.5: A Roller Coaster

To solve the roller-coaster problem we recall that Newton's second law (Eq. (2.3.17)) for a particle on a 2-D "surface" of any shape defined by the constraint function $f(x, y) = 0$ gives

$$\begin{aligned} \mathbf{F}^e &= -mg\mathbf{e}_y \\ \mathbf{F}^c &= N\mathbf{e}_n \end{aligned} \tag{2.3.35}$$

where (see Eq. (2.3.16))

$$\mathbf{e}_n = \frac{\nabla f}{|\nabla f|} = \frac{1}{|\nabla f|}\left[\frac{\partial f}{\partial x}\mathbf{e}_x + \frac{\partial f}{\partial y}\mathbf{e}_y\right] \tag{2.3.36}$$

so the governing equations for this constrained problem are

$$\begin{aligned} m\ddot{x} &= N\frac{1}{|\nabla f|}\frac{\partial f}{\partial x} \\ m\ddot{y} &= -mg + N\frac{1}{|\nabla f|}\frac{\partial f}{\partial y} \\ f(x, y) &= 0 \end{aligned} \tag{2.3.37}$$

Since the constraint equation must be satisfied for all times, if we differentiate that constraint once on time we have

$$\frac{\partial f}{\partial x}\dot{x} + \frac{\partial f}{\partial y}\dot{y} = 0 \tag{2.3.38}$$

and differentiating once more on time gives

$$\frac{\partial f}{\partial x}\ddot{x} + \frac{\partial f}{\partial y}\ddot{y} + \frac{\partial^2 f}{\partial x^2}(\dot{x})^2 + \frac{\partial^2 f}{\partial y^2}(\dot{y})^2 + 2\frac{\partial^2 f}{\partial x \partial y}\dot{x}\,\dot{y} = 0 \tag{2.3.39}$$

We can simplify writing Eq. (2.3.39) by noting that it is in the form

$$\frac{\partial f}{\partial x}\ddot{x} + \frac{\partial f}{\partial y}\ddot{y} = b(x, y, \dot{x}, \dot{y}) \tag{2.3.40}$$

where

$$b(x, \dot{x}, y, \dot{y}) = -\frac{\partial^2 f}{\partial x^2}(\dot{x})^2 - \frac{\partial^2 f}{\partial y^2}(\dot{y})^2 - 2\frac{\partial^2 f}{\partial x \partial y}\dot{x}\,\dot{y} \tag{2.3.41}$$

is a known function of the position and velocity components. If we combine Eq. (2.3.40) with Newton's second law we have

$$\begin{aligned} m\ddot{x} &= N\frac{1}{|\nabla f|}\frac{\partial f}{\partial x} \\ m\ddot{y} &= -mg + N\frac{1}{|\nabla f|}\frac{\partial f}{\partial y} \\ \frac{\partial f}{\partial x}\ddot{x} + \frac{\partial f}{\partial y}\ddot{y} &= b(x, \dot{x}, y, \dot{y}) \end{aligned} \tag{2.3.42}$$

which are the governing equations for a particle moving in two dimensions on an arbitrary surface. We can solve these three equations for the three unknowns $(\ddot{x}, \ddot{y}, N)$ in the same manner as shown previously for the straight and parbolic wire problems so we will omit the details here and just write down the final results for the normal force and equations of motion as

$$\begin{aligned} N &= \frac{m}{|\nabla f|}\left(b + g\frac{\partial f}{\partial y}\right) \\ \ddot{x} &= \frac{1}{|\nabla f|^2}\frac{\partial f}{\partial x}\left(b + g\frac{\partial f}{\partial y}\right) \\ \ddot{y} &= -g + \frac{1}{|\nabla f|^2}\frac{\partial f}{\partial y}\left(b + g\frac{\partial f}{\partial y}\right) \end{aligned} \tag{2.3.43}$$

Using the explicit form $f(x, y) = y - h(x)$ we find that the constraint force and equations of motion are

$$
\begin{aligned}
N &= \frac{m}{\sqrt{1 + (dh/dx)^2}} \left(g + \frac{d^2 h}{dx^2} \dot{x}^2 \right) \\
\ddot{x} &= \frac{-dh/dx}{1 + (dh/dx)^2} \left(g + \frac{d^2 h}{dx^2} \dot{x}^2 \right) \\
\ddot{y} &= -g + \frac{1}{1 + (dh/dx)^2} \left(g + \frac{d^2 h}{dx^2} \dot{x}^2 \right)
\end{aligned}
\tag{2.3.44}
$$

To obtain the roller-coaster problem we place the values of h and its derivatives from Eqs. (2.3.31) and (2.3.32) into this solution. This leads to rather lengthy expressions so we will not give those explicit forms here. Recall we assumed that the roller coaster is moving on smooth rails that prevent the cart from leaving the track. This assumption is important since if the cart was instead riding on a smooth surface the normal force would always have to be positive, i.e., in the plus $\mathbf{e}_n$ direction. If the normal force ever became zero then the cart could leave the track and we would have an unconstrained problem to solve. The rails prevent that from happening (a good thing!) allowing the normal force to be either positive or negative.

Since this is a one degree of freedom problem we will use the x-coordinate and the x-velocity to define the equations of motion in terms of $(x_1, x_2) = (x, \dot{x})$ which gives us the normal force and equations of motion as

$$
\begin{aligned}
N &= \frac{m}{\sqrt{1 + (dh/dx_1)^2}} \left(g + \frac{d^2 h}{dx_1^2} x_2^2 \right) \\
\dot{x}_1 &= x_2 \\
\dot{x}_2 &= \frac{-dh/dx_1}{1 + (dh/dx_1)^2} \left(g + \frac{d^2 h}{dx_1^2} x_2^2 \right)
\end{aligned}
\tag{2.3.45}
$$

In solving the roller-coaster problem we will show a different way to combine a MATLAB® function which defines these equations of motion with the ode45 function. We will define a function roller_coaster_1D that has the calling sequence

```
[tout, xout, N] = roller_coaster_1D (tspan, x0, A, L,rtol, atol);
```

This function contains both the ode45 numerical integration routine and the function roller_cart that generates the equations of motion for this problem:

```
function [tout, xout]= roller_coaster_1D(tspan,x0,g, A, L,rtol,atol)
% [tout, xout] = roller_coaster_1D(tspan, x0, A,L) solves for the motion of
% a roller-coaster cart on a sinusoidal track whose height, h, is given by
% h= A* (1-cos (pi*x/L)) /2 (the MATLAB® solver is ode45) . tspan is the
% time
```

```
% interval over which we want to obtain the solution and x0 is the vector
% of initial conditions on the x-coordinate, x(0) and x-velocity, vx(0)
% so that x0 = [ x(0) vx(0)] . rtol and atol are input error tolerances.
% The outputs of the function are the times,tout, over which the
% solution is calculated, and the values of x and dx/dt evaluated at
% those times. Both tout and xout are column vectors.
% It is assumed that x is in meters (m) and dx/dt is in m/sec. g is the
% acceleration of gravity.

% place error tolerances into options

 options = odeset('RelTol', rtol, 'AbsTol', atol);

% solve equations of motion. The inputs g, A, L are shared with the
% nested function roller_cart

[tout,xout] = ode45(@roller_cart, tspan, x0,options);

% the nested function to generate the equations of motion

function xdot = roller_cart( t, x)

% generate column vector for outputs
xdot=zeros(2,1);

% rename inputs to make them easier to read
x1 = x(1);
x2 = x(2);

% define various parameters in equations of motion
hp = A*pi*sin(pi*x1/L)/(2*L); % calculate h'(x) = dh/dx
D = 1 + hp.^2;
hpp = A*(pi^2)*cos(pi*x1/L)/(2*L^2); % calculate h"(x)
U = g + hpp.*(x2.^2);

% form equations of motion
xdot(1) =x2;
xdot(2) = -(hp./D).*U;
end

end
```

If you examine the above listing you will see that one difference between this approach and how we constructed solutions previously is that here we define the function that generates the equations of motion, roller_cart, within the function, roller_coaster_1D, i.e., roller_cart is a *nested function*. Functions nested in this manner can share variables without having to pass those variables directly. Thus, the variables g, A, and L are shared by roller_coaster_1D and the nested function roller_cart. Since these variables are not passed as explicit arguments we see that they do not appear as input arguments to roller_cart and when we use ode45 in conjunction with roller_cart these variables do not have to be specified as additional input arguments to that solver as well. This sharing results in very simple calls to both ode45 and the function that defines the equations of motion that ode45 uses so we will frequently use such nested functions in later examples. You can see that we also allow the function roller_coaster_1D to have input arguments, rtol and atol. These input error tolerances, as we have mentioned previously, are used by ode45 to adjust the numerical accuracy of the solution. In the roller-coaster problem we will need to adjust the relative error tolerance, rtol.

A script, roller_coaster_sim, was also written to solve the roller-coaster problem. This script sets up default values for L, A, tspan, x0, rtol, and atol given as L = 5, A = 4, tspan = [0, 10], x0 = [0, 8.5], rtol = 1e-3, and atol = 1e-6, respectively. These default values are used unless a user specifies alternate values in the MATLAB® command window before executing this script, which is something we will need to do for this problem. The values of rtol = 1e-3 and atol = 1e-6 are again the default values of these parameters when the function ode45 is called with the options argument set equal to the empty vector [ ]. After solving the problem the roller_coaster_sim script plots y versus x, the path of the particle; evaluates the normal force divided by the weight, mg; and plots it as a function of time.

If we run the script using the default values, the initial x-velocity of 8.5 m/s is less than the velocity $v_m = 8.86$ m/s needed to get over the first peak. Thus, the motion is just a back-and-forth periodic motion and the normal force is always positive, as shown in the plots of Fig. 2.21a, b. If we change the initial conditions to x0 = [0, 12.0] and run the script again we have the y versus x and normal force curves shown in Fig. 2.22a, b. Now, as expected, the particle can go over the top and will ride the roller coaster forever. We see this behavior in Fig. 2.22a, which appears to have the correct sinusoidal motion along the track. The normal force in Fig. 2.22b now has negative as well as positive values. However, the normal force appears to be growing in time, which is clearly incorrect since we expect that all the motion parameters will

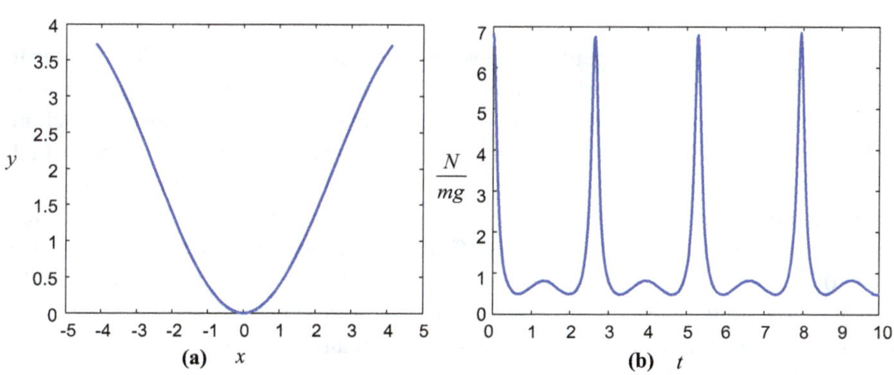

Fig. 2.21 (**a**) The y versus x path of a particle for initial conditions x0 = [0, 8.5] where it is trapped in the first valley of the sinusoidal roller coaster, and (**b**) the corresponding normal force, which is always positive

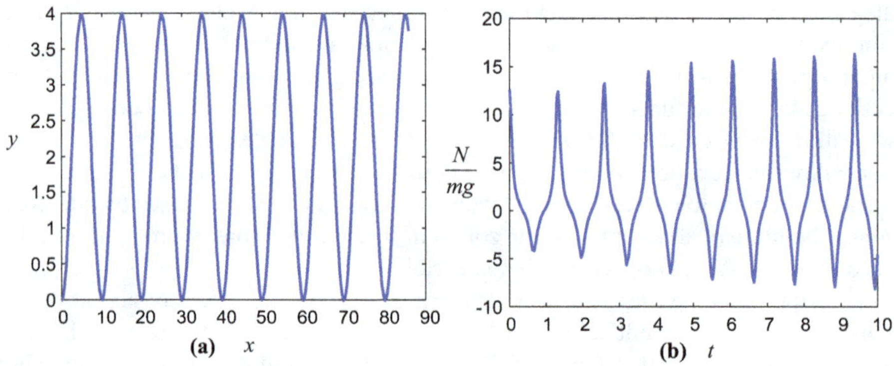

Fig. 2.22 (**a**) The y versus x path of a particle for initial conditions x0 = [0, 12] where it has sufficiently high initial speed to ride the sinusoidal roller coaster and (**b**) the corresponding normal force, which has both positive and negative values and exhibits an unphysical growing behavior due to numerical errors when ode45 is used with its default tolerance

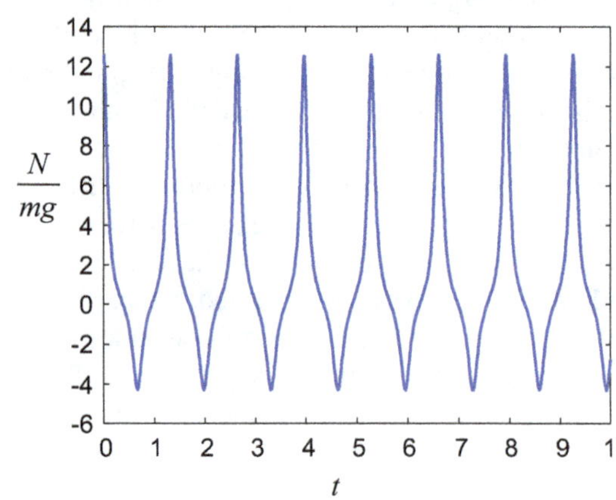

Fig. 2.23 The normal force in the same setup as in Fig. 2.22 when computed with ode45 and a relative tolerance rtol $= 1.0 \times 10^{-5}$

simply repeat as the cart repetitively goes through one cycle of the roller coaster to the next. What is the problem? From Fig. 2.22a we see that while the y-coordinate is always between values of 0 and 4, the x-coordinate is ever increasing, reaching values that are much larger than the maximum y-values. In this case the default relative tolerance of 1×10^{-3} used in ode45 is not sufficient to accurately solve the problem numerically since the relative tolerance is based on the size of the variables, one of which, x, is constantly growing. Thus, if we enter a smaller tolerance, rtol $=$ 1×10^{-5}, in the MATLAB® command window and rerun the script, we see that the normal force now appears to have the proper uniform values, as shown in Fig. 2.23 (the y- versus x-plot does not visually change much and is not shown). This behavior did not show up in Fig. 2.21b because there both y and x varied over a range of similar values. This is a good example to show that when doing numerical solutions we need to be careful that numerical errors do not build up over time and render the solution inaccurate. For conservative problems, one good test is to examine the

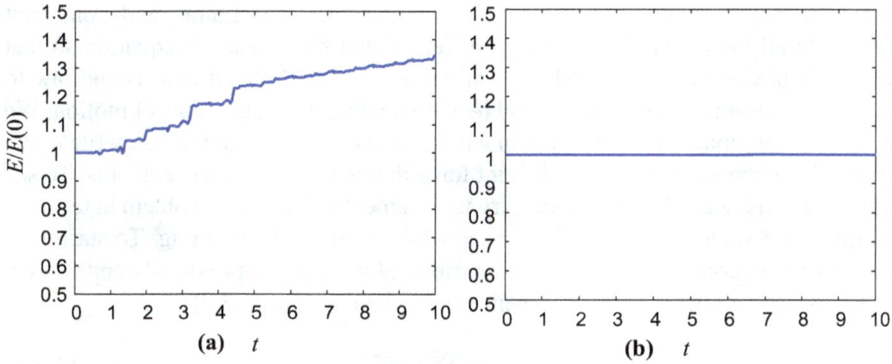

Fig. 2.24 (**a**) $E/E(0)$ for the same setup of Fig. 2.22 plotted versus time, t, for rtol = 1.0×10^{-3} (the default tolerance in ode45), where E is the total energy, and (**b**) the same plot for rtol = 1.0×10^{-5}

numerical values of the total energy. Any significant deviations from its values at the starting time can warn of numerical inaccuracies in the solution. To see what happens in this case we have written a second script, roller_energy, which simply evaluates and plots $E/E(0)$ as a function of time using the workspace values generated by roller_coaster_sim. If the roller_coaster_sim script is run with x0 = [0, 12] initial conditions (and the default rtol value) and then the script roller_energy is run, the total energy as a function of time is shown in Fig. 2.24a, which clearly shows a major error in the solution. However, if the relative tolerance is changed to rtol = 1.0e-5 and the roller_coaster_sim and roller_energy scripts are rerun the error is significantly reduced, as shown in Fig. 2.24b. The reader can verify the error can be reduced even further by choosing rtol = 1.0e-6. The other error tolerance, atol, is not changed from its default value as changes in this error tolerance are needed only when the solution values get quite small, which is not a problem here. We will say more about these two error tolerances later. In problems where energy is not a constant, one can still examine the effects of the tolerances used by ode45 on the numerical solution to help guarantee that the solution is accurate. In fact, it is good practice to help verify the accuracy of *any* numerical solution by running the simulation more than once, if practical, using different error tolerances. If there is no significant change in the values at the end of the simulation (for large times and/or displacements) then one has some confidence that numerical errors have not accumulated and rendered the solution inaccurate.

2.4 Constraints and Equations of Motion: A Matrix Approach

2.4.1 Types of Constraints

In the last section we solved a number of simple constrained motion problems in 2-D. Although we considered very special cases, those cases can be used as stepping stones to better understand the relationship between constraints and constraint forces. In all the cases considered we differentiated the constraint equation so that we could use that differentiated form in conjunction with Newton's second law to find the constraint force explicitly and hence also obtain the equations of motion. We can, however, conduct that process in a more general fashion and in fact write down an explicit expression for the constraint force that is valid in many problems. To see how to do this, we will find it necessary to reframe the dynamics problem in terms of matrices and vectors. Thus, bear with me while we do that reframing. To start, let's go back to the constrained 2-D motion problem for a particle previously considered. Newton's second law for this problem was written as (Eq. (2.3.2))

$$m\ddot{\mathbf{x}} = \mathbf{F}^e + \mathbf{F}^c \tag{2.4.1}$$

This equation can also be written in a matrix-vector form as

$$\mathbf{M}\ddot{\mathbf{x}} = \mathbf{F}^e + \mathbf{F}^c \tag{2.4.2}$$

where the 2×2 mass matrix and acceleration and force vectors are

$$\begin{gathered} \mathbf{M} = \begin{bmatrix} m & 0 \\ 0 & m \end{bmatrix} \\ \ddot{\mathbf{x}} = \begin{Bmatrix} \ddot{x} \\ \ddot{y} \end{Bmatrix}, \quad \mathbf{F}^e = \begin{Bmatrix} F_x^e \\ F_y^e \end{Bmatrix}, \quad \mathbf{F}^c = \begin{Bmatrix} F_x^c \\ F_y^c \end{Bmatrix} \end{gathered} \tag{2.4.3}$$

This law was combined with the twice-differentiated constraint equation $f(x, y) = 0$ given by (see Eq. (2.3.40))

$$\frac{\partial f}{\partial x}\ddot{x} + \frac{\partial f}{\partial y}\ddot{y} = b(x, y, \dot{x}, \dot{y}) \tag{2.4.4}$$

This equation we will also write in matrix-vector form as

$$\mathbf{A}\ddot{\mathbf{x}} = \mathbf{b} \tag{2.4.5}$$

where $\mathbf{A} = \nabla f = [\partial f/\partial x \quad \partial f/\partial y]$ is a 1×2 matrix (a row vector) and $\mathbf{b}$ is a 1×1 matrix (a scalar).

Let us consider a more general case when there are known applied forces $\left(F_x^e, F_y^e\right)$ in the x- and y-directions, respectively. For a particle moving on a smooth wire defined by $f(x, y) = 0$ we have then in place of Eq. (2.3.42)

$$m\ddot{x} = F_x^e + N\frac{1}{|\nabla f|}\frac{\partial f}{\partial x}$$
$$m\ddot{y} = F_y^e + N\frac{1}{|\nabla f|}\frac{\partial f}{\partial y} \tag{2.4.6}$$

which can be written in vector-matrix form as

$$\mathbf{M}\ddot{\mathbf{x}} = \mathbf{F}^e + \mathbf{A}^T\lambda \tag{2.4.7}$$

where $\lambda = N/|\nabla f|$ is a 1×1 matrix (scalar) and $\mathbf{A}^T$ is a 2×1 matrix (column vector). Thus, the constraint force is

$$\mathbf{F}^c = \mathbf{A}^T\lambda \tag{2.4.8}$$

Constraint forces that satisfy the prescription of Eq. (2.4.8), i.e., those that are normal to the "surface" defined by the constraint $f(x, y) = 0$ are called *ideal constraint forces*. Lagrange first used this prescription so it is also called *the Lagrange prescription* and λ is called a *Lagrange multiplier*. Are there other problems where the constraint force is an ideal constraint force so that the form of Eq. (2.4.8) is valid? The answer is definitely yes. In fact, many of the common constraints encountered in the dynamics of rigid bodies such as pins, ball-and-socket joints, and sliding joints produce ideal constraint forces. Here are a few examples of ideal constraint forces involving a single particle and one example of a constraint that is not an ideal constraint:

Example 1 Consider first the case of a moving smooth surface, where the constraint is in the form

$$f(x, y, t) = 0 \tag{2.4.9}$$

An example of such a problem would be the problem shown in Fig. 2.25a, where a block of mass m (treated here as a particle) slides on a moving smooth inclined plane under the action of gravity and a spring force, where the motion of the plane in the x-direction is given as $x_0(t)$, as shown. In this case the constraint equation is

$$y(t) - \tan\alpha(x(t) - x_0(t)) = 0 \tag{2.4.10}$$

which is in the form of Eq. (2.4.9). Taking one and two time derivatives we have

$$\dot{y} - \tan\alpha\dot{x} = -\tan\alpha\dot{x}_0$$
$$\ddot{y} - \tan\alpha\ddot{x} = -\tan\alpha\ddot{x}_0 \tag{2.4.11}$$

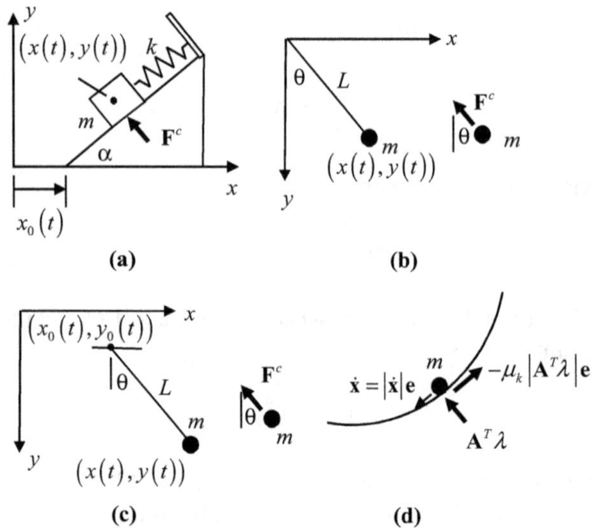

Fig. 2.25 (**a**) A block on a moving smooth inclined surface under the action of gravity, a spring, and the constraint force normal to the smooth surface. (**b**) A pendulum consisting of a mass *m* attached to a massless wire of length *L* to a fixed point and acted upon by gravity and a constraint force along the wire. (**c**) The pendulum problem of (b) but where now the end of the wire is allowed to move. The constraint force is still along the wire. (**d**) A particle moving on a rough surface where there is both a normal constraint force and Coulomb-type friction force

so the second equation in Eq. (2.4.11) is in the form of Eq. (2.4.5) where $\mathbf{A} = [-\tan\,\alpha,\ \ 1]$. The constraint force is still normal to the moving plane, as shown in Fig. 2.25a, so from the geometry we would normally write this force as

$$\mathbf{F}^c = \begin{Bmatrix} -\sin\alpha \\ \cos\alpha \end{Bmatrix} N$$

but we also could write it as

$$\mathbf{F}^c = \begin{Bmatrix} -\tan\alpha \\ 1 \end{Bmatrix} N\cos\alpha = \mathbf{A}^T\lambda$$

where $\lambda = N\cos\,\theta$, so the prescription for this force given in Eq. (2.4.8) is still valid.

Example 2 Now, consider the pendulum problem shown in Fig. 2.25b, where a mass is fixed to the origin by a thin, rigid (inextensible) wire of length *L*. The constraint during the motion of the pendulum is

$$\sqrt{[x(t)]^2 + [y(t)]^2} - L = 0 \tag{2.4.12}$$

and the derivatives are

$$
\begin{aligned}
&\frac{x}{\sqrt{x^2+y^2}}\dot{x}+\frac{y}{\sqrt{x^2+y^2}}\dot{y}=0\\
&\frac{x}{\sqrt{x^2+y^2}}\ddot{x}+\frac{y}{\sqrt{x^2+y^2}}\ddot{y}=-\frac{\dot{x}^2}{\sqrt{x^2+y^2}}-\frac{\dot{y}^2}{\sqrt{x^2+y^2}}+\frac{x^2\dot{x}^2}{(x^2+y^2)^{3/2}}+\frac{y^2\dot{y}^2}{(x^2+y^2)^{3/2}}
\end{aligned}
\tag{2.4.13}
$$

The second equation is again of the form of Eq. (2.4.5) where now (see Fig. 2.25b)

$$
\mathbf{A}=\left[\frac{x}{\sqrt{x^2+y^2}},\frac{y}{\sqrt{x^2+y^2}}\right]=[\sin\theta,\ \cos\theta] \tag{2.4.14}
$$

The tension constraint force in the wire, as shown in Fig. 2.25b, would normally be written as

$$
\mathbf{F}^c=-T\begin{Bmatrix}\sin\theta\\ \cos\theta\end{Bmatrix}=\mathbf{A}^T\lambda
$$

which again satisfies the prescription for this force given in Eq. (2.4.8) where the parameter, λ, in this case is the negative of the tension, T.

Example 3 Next, consider the pendulum problem again as shown in Fig. 2.25c, where now we let one end of the wire move with a specified motion $(x_0(t), y_0(t))$. In this case the constraint is

$$
\sqrt{[x(t)-x_0(t)]^2+[y(t)-y_0(t)]^2}-L=0 \tag{2.4.15}
$$

so differentiating this constraint gives the complex-looking expressions

$$
\begin{aligned}
&\frac{(x-x_0)}{\sqrt{(x-x_0)^2+(y-y_0)^2}}(\dot{x}-\dot{x}_0)+\frac{(y-y_0)}{\sqrt{(x-x_0)^2+(y-y_0)^2}}(\dot{y}-\dot{y}_0)=0\\
&\frac{(x-x_0)}{L}\ddot{x}+\frac{(y-y_0)}{L}\ddot{y}=\frac{(x-x_0)\ddot{x}_0}{L}+\frac{(y-y_0)\ddot{y}_0}{L}\\
&-\frac{(\dot{x}-\dot{x}_0)^2}{L}-\frac{(\dot{y}-\dot{y}_0)^2}{L}+\frac{(x-x_0)^2(\dot{x}-\dot{x}_0)^2}{L^{3/2}}+\frac{(y-y_0)^2(\dot{y}-\dot{y}_0)^2}{L^{3/2}}
\end{aligned}
\tag{2.4.16}
$$

where now

$$\mathbf{A} = \left[\frac{x - x_0}{L}, \frac{y - y_0}{L}\right] = [\sin\theta, \cos\theta] \tag{2.4.17}$$

and the force in the wire obviously once more satisfies the prescription of Eq. (2.4.8). Note that since we are taking derivatives we must be careful to substitute Eq. (2.4.15) into our results only after we are done taking all the derivatives needed, a fact that you may remember (or not) from a calculus course.

Example 4 Do all constraint forces satisfy Eq. (2.4.8)? The answer is no. An example you might be familiar with is where we have a Coulomb friction force present due to sliding on a rough surface (Fig. 2.25d). In this case the friction force is proportional to the magnitude of the normal force, **N**, acting at the surface (and this normal force, as we have seen, does satisfy Eq. (2.4.8)) but the friction force, $\mathbf{F}_f$, acts tangent to the surface and is opposite to the direction in which it is sliding, i.e.,

$$\mathbf{F}_f = -\mu_k |\mathbf{N}| \frac{\dot{\mathbf{x}}}{|\dot{\mathbf{x}}|} \tag{2.4.18}$$

where μ_k is the coefficient of sliding friction. Thus, the total constraint force is in this case

$$\mathbf{F}^c = \mathbf{A}^T\lambda - \mu_k \left|\mathbf{A}^T\lambda\right| \frac{\dot{\mathbf{x}}}{|\dot{\mathbf{x}}|} \tag{2.4.19}$$

and because of the friction force it does not have the form of Eq. (2.4.8). If we have a connection constraint or a contact constraint that is frictionless generally the constraint force will be an ideal constraint force but if there is friction involved the constraint force will not be ideal. In Sect. 2.10 we will discuss an approach that does allow us to consider Coulomb friction and other, more general nonideal constraint forces, but until then we will examine a wide range of problems with ideal constraint forces.

Finally, we note that we could also consider problems where the form of the constraint inherently involves the velocity such as

$$f(x, y, \dot{x}, \dot{y}, t) = 0 \tag{2.4.20}$$

In this case we only need to differentiate the constraint once on time to obtain

$$\frac{\partial f}{\partial \dot{x}}\ddot{x} + \frac{\partial f}{\partial \dot{y}}\ddot{y} = -\frac{\partial f}{\partial x}\dot{x} - \frac{\partial f}{\partial y}\dot{y} - \frac{\partial f}{\partial t} \tag{2.4.21}$$

which again is in the form of Eq. (2.4.5). Such constraints can also be represented by ideal constraint forces, where in this case we have $\mathbf{A} = \left[\partial f/\partial \dot{x} \quad \partial f/\partial \dot{y}\right]$. Note that we can always obtain constraints of the form of Eq. (2.4.20) by simply differentiating a constraint of the form $g(x, y, t) = 0$ once to obtain

$$f(x, y, \dot{x}, \dot{y}, t) = \frac{\partial g}{\partial x}\dot{x} + \frac{\partial g}{\partial y}\dot{y} + \frac{\partial g}{\partial t} = 0 \tag{2.4.22}$$

However, the reverse is not true, namely there are also constraints of the form of Eq. (2.4.20) that cannot be integrated to obtain a constraint of the form $g(x, y, t) = 0$. We will discuss in more depth later the issues surrounding constraints represented by Eq. (2.4.20) and we will give examples of such non-integrable constraints.

2.4.2 Constraints for Motion in Three Dimensions

So far we have only considered problems of a single particle in two-dimensional motion. If we consider a particle moving in three dimensions instead then Newton's second law can again be written in matrix-vector form of Eq. (2.4.2), i.e.,

$$\mathbf{M}\ddot{\mathbf{x}} = \mathbf{F}^e + \mathbf{F}^c \tag{2.4.23}$$

but where now the mass matrix and vectors are

$$\begin{gathered}\mathbf{M} = \begin{bmatrix} m & 0 & 0 \\ 0 & m & 0 \\ 0 & 0 & m \end{bmatrix} \\ \ddot{\mathbf{x}} = \begin{Bmatrix} \ddot{x} \\ \ddot{y} \\ \ddot{z} \end{Bmatrix}, \quad \mathbf{F}^e = \begin{Bmatrix} F_x^e \\ F_y^e \\ F_z^e \end{Bmatrix}, \quad \mathbf{F}^c = \begin{Bmatrix} F_x^c \\ F_y^c \\ F_z^c \end{Bmatrix}\end{gathered} \tag{2.4.24}$$

If we consider a single constraint equation of the types examined in previous 2-D problems, then the main difference is we have to include terms involving the variables $(z, \dot{z}, \ddot{z})$. For a particle moving on a fixed smooth surface in three dimensions, for example, the constraint equation is

$$f(x, y, z) = 0 \tag{2.4.25}$$

and differentiating this equation once and then twice gives

$$\begin{gathered}\frac{\partial f}{\partial x}\dot{x} + \frac{\partial f}{\partial y}\dot{y} + \frac{\partial f}{\partial z}\dot{z} = 0 \\ \frac{\partial f}{\partial x}\ddot{x} + \frac{\partial f}{\partial y}\ddot{y} + \frac{\partial f}{\partial z}\ddot{z} = b(x, y, z, \dot{x}, \dot{y}, \dot{z})\end{gathered} \tag{2.4.26}$$

where

$$b(x,y,z,\dot{x},\dot{y},\dot{z}) = -\frac{\partial^2 f}{\partial x^2}(\dot{x})^2 - \frac{\partial^2 f}{\partial y^2}(\dot{y})^2 - \frac{\partial^2 f}{\partial z^2}(\dot{z})^2 \\ -2\frac{\partial^2 f}{\partial x \partial y}\dot{x}\,\dot{y} - 2\frac{\partial^2 f}{\partial x \partial z}\dot{x}\,\dot{z} - 2\frac{\partial^2 f}{\partial y \partial z}\dot{y}\,\dot{z} \tag{2.4.27}$$

This again is of the form $\mathbf{A}\ddot{\mathbf{x}} = b$ but where now $\mathbf{A} = [\partial f/\partial x, \partial f/\partial y, \partial f/\partial z]$ is a 1×3 matrix (row vector). Thus, a discussion of single-constraint problems of the type discussed for 2-D problems follows along very similar lines so we will not give the details here. However, note that we now could introduce multiple constraints on the particle. For example, we could have two constraints of the form of Eq. (2.4.25), i.e.,

$$f_1(x,y,z) = 0 \\ f_2(x,y,z) = 0 \tag{2.4.28}$$

A physical example of such a case is where a particle being constrained in a smooth V-shaped groove (see Fig. 2.26). In that case, we know that there will be two constraint forces, each acting normal to each side of the groove on the particle. If we differentiate the constraint equations once and then twice on time we obtain

$$\frac{\partial f_1}{\partial x}\dot{x} + \frac{\partial f_1}{\partial y}\dot{y} + \frac{\partial f_1}{\partial z}\dot{z} = 0 \\ \frac{\partial f_2}{\partial x}\dot{x} + \frac{\partial f_2}{\partial y}\dot{y} + \frac{\partial f_2}{\partial z}\dot{z} = 0 \\ \frac{\partial f_1}{\partial x}\ddot{x} + \frac{\partial f_1}{\partial y}\ddot{y} + \frac{\partial f_1}{\partial z}\ddot{z} = b_1(x,y,z,\dot{x},\dot{y},\dot{z}) \\ \frac{\partial f_2}{\partial x}\ddot{x} + \frac{\partial f_2}{\partial y}\ddot{y} + \frac{\partial f_2}{\partial z}\ddot{z} = b_2(x,y,z,\dot{x},\dot{y},\dot{z}) \tag{2.4.29}$$

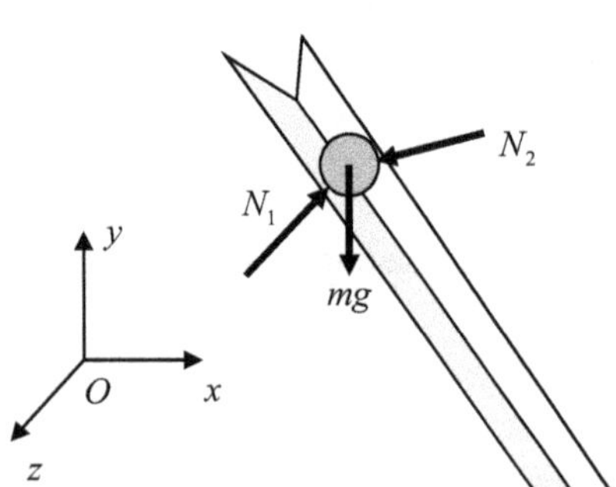

Fig. 2.26 A ball moving in a smooth V-shaped groove, showing the normal forces exerted on the ball from the sides of the groove as well as the weight of the ball

with

$$
\begin{aligned}
b_1(x,y,z,\dot{x},\dot{y},\dot{z}) &= -\frac{\partial^2 f_1}{\partial x^2}(\dot{x})^2 - \frac{\partial^2 f_1}{\partial y^2}(\dot{y})^2 - \frac{\partial^2 f_1}{\partial z^2}(\dot{z})^2 \\
&\quad -2\frac{\partial^2 f_1}{\partial x \partial y}\dot{x}\,\dot{y} - 2\frac{\partial^2 f_1}{\partial x \partial z}\dot{x}\,\dot{z} - 2\frac{\partial^2 f_1}{\partial y \partial z}\dot{y}\,\dot{z} \\
b_2(x,y,z,\dot{x},\dot{y},\dot{z}) &= -\frac{\partial^2 f_2}{\partial x^2}(\dot{x})^2 - \frac{\partial^2 f_2}{\partial y^2}(\dot{y})^2 - \frac{\partial^2 f_2}{\partial z^2}(\dot{z})^2 \\
&\quad -2\frac{\partial^2 f_2}{\partial x \partial y}\dot{x}\,\dot{y} - 2\frac{\partial^2 f_2}{\partial x \partial z}\dot{x}\,\dot{z} - 2\frac{\partial^2 f_2}{\partial y \partial z}\dot{y}\,\dot{z}
\end{aligned}
\tag{2.4.30}
$$

However, the twice-differentiated constraint equation is again in a form very similar to the 2-D case, namely

$$
\mathbf{A}\ddot{\mathbf{x}} = \mathbf{b} \tag{2.4.31}
$$

with the main difference being that $\mathbf{A}$ is now a 2×3 matrix and $\mathbf{b}$ is a 2×1 column vector, where

$$
\begin{aligned}
\mathbf{A} &= \begin{bmatrix} \partial f_1/\partial x & \partial f_1/\partial y & \partial f_1/\partial z \\ \partial f_2/\partial x & \partial f_2/\partial y & \partial f_2/\partial z \end{bmatrix} = \begin{bmatrix} \nabla f_1^T \\ \nabla f_2^T \end{bmatrix} \\
\mathbf{b} &= \begin{Bmatrix} b_1 \\ b_2 \end{Bmatrix}
\end{aligned}
\tag{2.4.32}
$$

The total constraint force is therefore

$$
\mathbf{F}^c = \mathbf{A}^T \boldsymbol{\lambda} \tag{2.4.33}
$$

where $\boldsymbol{\lambda}$ is the 2×1 matrix (column vector)

$$
\boldsymbol{\lambda} = \begin{Bmatrix} \lambda_1 \\ \lambda_2 \end{Bmatrix} \tag{2.4.34}
$$

Multiplying $\mathbf{A}^T$ by $\boldsymbol{\lambda}$ we see that the total constraint force more explicitly is

$$
\begin{aligned}
\mathbf{F}^c &= \begin{Bmatrix} \lambda_1 \partial f_1/\partial x + \lambda_2 \partial f_2/\partial x \\ \lambda_1 \partial f_1/\partial y + \lambda_2 \partial f_2/\partial y \\ \lambda_1 \partial f_1/\partial z + \lambda_2 \partial f_2/\partial z \end{Bmatrix} = \lambda_1 \begin{Bmatrix} \partial f_1/\partial x \\ \partial f_1/\partial y \\ \partial f_1/\partial z \end{Bmatrix} + \lambda_2 \begin{Bmatrix} \partial f_2/\partial x \\ \partial f_2/\partial y \\ \partial f_2/\partial z \end{Bmatrix} \\
&= \lambda_1 \nabla f_1 + \lambda_2 \nabla f_2
\end{aligned}
\tag{2.4.35}
$$

which, as expected, is the sum of the normal forces from each of the smooth surfaces. We can see this more clearly if we let $\lambda_1 = N_1/|\nabla f_1|, \lambda_2 = N_2/|\nabla f_2|$ so that then (N_1, N_2) are just the values of the individual normal forces acting on the particle from each surface.

2.4.3 Augmented Solutions for Ideal Constraint Forces and the Equations of Motion in Cartesian Coordinates

We saw in the last section that for both 2-D and 3-D motions of a particle we are able to write Newton's second law equations and the constraint equations in Cartesian coordinates as

$$\begin{aligned} \mathbf{M}\ddot{\mathbf{x}} &= \mathbf{F}^e + \mathbf{F}^c \\ \mathbf{A}\ddot{\mathbf{x}} &= \mathbf{b} \end{aligned} \tag{2.4.36}$$

where $\mathbf{M}$ is a mass matrix, $(\ddot{\mathbf{x}}, \mathbf{F}^e, \mathbf{F}^c)$ are acceleration and force column vectors, $\mathbf{A}$ is a non-square matrix, and $\mathbf{b}$ is a column vector. Furthermore, when the constraint forces are ideal constraint forces the governing equations are

$$\begin{aligned} \mathbf{M}\ddot{\mathbf{x}} &= \mathbf{F}^e + \mathbf{A}^T\boldsymbol{\lambda} \\ \mathbf{A}\ddot{\mathbf{x}} &= \mathbf{b} \end{aligned} \tag{2.4.37}$$

In later chapters we will see that equations very similar to Eq. (2.4.37) will also be applicable to the motion of multiple particles or the motion of rigid bodies so they represent a much wider range of dynamics problems than the single-particle problems just considered. In the field of dynamics Eq. (2.4.37) is called an *augmentation approach* (or augmented method) to obtaining the equations of motion since we augmented Newton's second law with ideal constraint forces in terms of unknown Lagrange multipliers, $\boldsymbol{\lambda}$, and the differentiated constraint equations. For a very general, high-level discussion of this approach, see Udwadia and Kalaba [2], and for its application to multibody dynamics problems in particular see the texts of Shabana [3, 4]. We will now show how to use this augmentation approach to obtain explicit solutions for the ideal constraint forces and the equations of motion.

Equation (2.4.37) can be solved formally for the unknown $\boldsymbol{\lambda}$ values by first multiplying the first equation in Eq. (2.4.37) by the inverse of the mass matrix. The mass matrix is non-singular so that its inverse always exists. We have

$$\ddot{\mathbf{x}} = \mathbf{M}^{-1}\mathbf{F}^e + \mathbf{M}^{-1}\mathbf{A}^T\boldsymbol{\lambda} \tag{2.4.38}$$

If we place Eq. (2.4.38) into the constraint equations we find

$$\left(\mathbf{AM}^{-1}\mathbf{A}^T\right)\boldsymbol{\lambda} = \mathbf{b} - \mathbf{AM}^{-1}\mathbf{F}^e \tag{2.4.39}$$

Unlike the matrix **A**, the coefficient matrix $\mathbf{C} = \mathbf{AM}^{-1}\mathbf{A}^T$ is a square matrix. For example, for the motion of a particle in two dimensions with one constraint **A** is a 1×2 matrix, $\mathbf{M}^{-1}$ is a 2×2 matrix, and $\mathbf{A}^T$ is a 2×1 matrix. Thus the dimensions of the coefficient **C** matrix, which follows from the product of its components according to the rules of matrix multiplication, are $1 \times 1 = (1 \times 2)(2 \times 2)(2 \times 1)$ which indeed is a square matrix since it is simply a scalar. For the motion of a particle in three dimensions where there are two constraints, **A** is a 2×3 matrix, $\mathbf{M}^{-1}$ is a 3×3 matrix, and $\mathbf{A}^T$ is a 3×2 matrix, yielding a square coefficient **C** matrix with dimensions $2 \times 2 = (2 \times 3)(3 \times 3)(3 \times 2)$. If the constraint equations are all independent (which is always true when we only have a single constraint equation) the coefficient matrix will also have an inverse, so multiplying both sides of Eq. (2.4.39) by $\mathbf{C}^{-1}$ gives

$$\boldsymbol{\lambda} = \left(\mathbf{AM}^{-1}\mathbf{A}^T\right)^{-1}\left(\mathbf{b} - \mathbf{AM}^{-1}\mathbf{F}^e\right) \tag{2.4.40}$$

The ideal constraint force and the equations of motion are then given by

$$\begin{aligned} \mathbf{F}^c &= \mathbf{A}^T\left(\mathbf{AM}^{-1}\mathbf{A}^T\right)^{-1}\left(\mathbf{b} - \mathbf{AM}^{-1}\mathbf{F}^e\right) \\ \ddot{\mathbf{x}} &= \mathbf{M}^{-1}\mathbf{F}^e + \mathbf{M}^{-1}\mathbf{A}^T\left(\mathbf{AM}^{-1}\mathbf{A}^T\right)^{-1}\left(\mathbf{b} - \mathbf{AM}^{-1}\mathbf{F}^e\right) \end{aligned} \tag{2.4.41}$$

What happens when there is more than one constraint and they are not independent (a condition that can happen in more complex dynamics problems such as robotics)? Then the regular inverse of the **C** matrix does not exist. In that case we will see later in Chap. 3 that we can still use Eq. (2.4.41) if the regular inverse of **C** is replaced by a modified matrix operation called a pseudoinverse [2]. For now we will only consider cases where $\mathbf{C}^{-1}$ exists and this is not an issue.

Do not be intimidated by the apparent complexity of Eq. (2.4.41). For a single particle the mass matrix is a diagonal matrix so computing its inverse is trivial (we just take the reciprocal of the diagonal terms). The remainder of the terms all involve only matrix products except for having to take the inverse of the coefficient matrix, **C**. But computing the inverse of the **C** matrix is also not difficult in general. It is true that there are quite a few matrix operations involved in Eq. (2.4.41) so that if we evaluate all the terms analytically by hand there is always a chance for algebraic errors to occur. Thus, it would be nice to have an automated way to perform all the analytical calculations symbolically. We have already seen in this chapter that we can use the MATLAB® symbolic math toolbox for such applications. In Chap. 5 we will use that toolbox to obtain the expressions found in Eq. (2.4.41). Meanwhile, the problems we will consider will be relatively simple so we will do the calculations by hand. We should also point out that while it is nice to evaluate all the terms in Eq. (2.4.41) in symbolic forms, this is not really necessary. In most cases we will be solving the equations of motion numerically anyway, so all we really need for such a numerical solution are the explicit analytical forms for **A**, **M**, **b**, $\mathbf{F}^e$. The corresponding numerical values of all the terms appearing in Eq. (2.4.41) can then be obtained as needed using MATLAB®'s ability to seamlessly deal with matrix operations.

For single-particle problems such as the problem of the bead on the smooth-curved wire considered in Sect. 2.3.1 all of the analytical calculations are extremely simple. Some of the main terms are

$$\begin{aligned}
&\mathbf{M} = \begin{bmatrix} m & 0 \\ 0 & m \end{bmatrix}, \quad \mathbf{M}^{-1} = \begin{bmatrix} 1/m & 0 \\ 0 & 1/m \end{bmatrix}, \quad \mathbf{F}^e = \begin{Bmatrix} 0 \\ -mg \end{Bmatrix} \\
&\mathbf{A} = [\partial f/\partial x, \partial f/\partial y], \quad \mathbf{A}^T = \begin{Bmatrix} \partial f/\partial x \\ \partial f/\partial y \end{Bmatrix}, \quad \mathbf{b} = -\left(\partial^2 f/\partial x^2\right)\dot{x}^2 \\
&\mathbf{C} = \mathbf{A}\mathbf{M}^{-1}\mathbf{A}^T = \left[(\partial f/\partial x)^2 + (\partial f/\partial y)^2\right]/m = |\nabla f|^2/m \\
&\mathbf{C}^{-1} = \left(\mathbf{A}\mathbf{M}^{-1}\mathbf{A}^T\right)^{-1} = m/|\nabla f|^2
\end{aligned} \tag{2.4.42}$$

yielding the constraint force and equations of motion as

$$\begin{aligned}
&\begin{Bmatrix} F_x^c \\ F_y^c \end{Bmatrix} = \begin{Bmatrix} \partial f/\partial x \\ \partial f/\partial y \end{Bmatrix} \left[\frac{m}{|\nabla f|^2}\right] \left[-\left(\partial^2 f/\partial x^2\right)\dot{x}^2 + (\partial f/\partial y)g\right] \\
&\begin{Bmatrix} \ddot{x} \\ \ddot{y} \end{Bmatrix} = \begin{Bmatrix} 0 \\ -g \end{Bmatrix} + \begin{Bmatrix} \partial f/\partial x \\ \partial f/\partial y \end{Bmatrix} \left[\frac{1}{|\nabla f|^2}\right] \left[-\left(\partial^2 f/\partial x^2\right)\dot{x}^2 + (\partial f/\partial y)g\right]
\end{aligned} \tag{2.4.43}$$

which for the specific case of the parabolic wire where $f = y - ax^2$ gives

$$\begin{aligned}
&\begin{Bmatrix} F_x^c \\ F_y^c \end{Bmatrix} = \frac{m(2a\dot{x}^2 + g)}{1 + 4a^2x^2} \begin{Bmatrix} -2ax \\ 1 \end{Bmatrix} \\
&\begin{Bmatrix} \ddot{x} \\ \ddot{y} \end{Bmatrix} = \begin{Bmatrix} 0 \\ -g \end{Bmatrix} + \frac{(2a\dot{x}^2 + g)}{1 + 4a^2x^2} \begin{Bmatrix} -2ax \\ 1 \end{Bmatrix}
\end{aligned} \tag{2.4.44}$$

These equations of motion are identical to those obtained earlier in Eq. (2.3.22). The constraint force is in the direction of a unit vector, $\mathbf{e}_n$, which is normal to the surface and the magnitude of the constraint force is N, where $\mathbf{e}_n$ and N are given by

$$\begin{aligned}
&\mathbf{e}_n = \frac{\mathbf{F}^c}{|\mathbf{F}^c|} = \frac{1}{\sqrt{1 + 4a^2x^2}} \begin{Bmatrix} -2ax \\ 1 \end{Bmatrix} \\
&N = \sqrt{\left(F_x^c\right)^2 + \left(F_y^c\right)^2} = \frac{m(2a\dot{x}^2 + g)}{\sqrt{1 + 4a^2x^2}}
\end{aligned} \tag{2.4.45}$$

The normal force expression is identical to Eq. (2.3.21). See Problem P2.7 which asks you to solve the problem in MATLAB® without having the analytical solutions of Eq. (2.4.42), following a more completely numerical approach, as mentioned previously.

We can reduce the complexity of Eq. (2.4.41) and also gain a more transparent understanding of that equation by rewriting it in different terms. Let $\mathbf{a} = \mathbf{M}^{-1}\mathbf{F}^e$, $\mathbf{K} = \mathbf{A}^T(\mathbf{A}\mathbf{M}^{-1}\mathbf{A}^T)^{-1}$, and $\mathbf{e} = \mathbf{b} - \mathbf{A}\mathbf{a}$. Then

$$\begin{aligned} \mathbf{F}^c &= \mathbf{K}\mathbf{e} \\ \mathbf{M}\ddot{\mathbf{x}} &= \mathbf{M}\mathbf{a} + \mathbf{K}\mathbf{e} \end{aligned} \tag{2.4.46}$$

The $\mathbf{a}$ vector is the acceleration of the particle if it was unconstrained, since if the constraint force was zero ($\mathbf{F}^c = 0$), we would just have $\ddot{\mathbf{x}} = \mathbf{a}$. The "error" vector $\mathbf{e} = \mathbf{b} - \mathbf{A}\mathbf{a}$ is the amount by which the differentiated constraint equations, $\mathbf{A}\ddot{\mathbf{x}} = \mathbf{b}$, are not satisfied by this unconstrained acceleration, $\mathbf{a}$. Equation (2.4.46) then simply says that the constraint force is proportional to this error in satisfying the constraints, where $\mathbf{K}$ is the proportionality matrix.

We cannot use our explicit solution for an ideal constraint force and equations of motion, Eq. (2.4.41), when the prescription for that constraint force, $\mathbf{F}^c = \mathbf{A}^T\boldsymbol{\lambda}$, is not valid. As discussed previously, this happens in problems with friction such as a rough surface characterized by Coulomb friction. But we can still solve such frictional problems by modifying Eq. (2.4.41) to include such nonideal constraints, as shown later in Sect. 2.10. For constrained motion problems involving only ideal constraint forces we can modify the four-step procedure outlined in Sect. 2.3 and state a more explicit five-step procedure, namely:

1. Obtain the resultant of the external applied forces, $\mathbf{F}^e$, and the mass matrix, $\mathbf{M}$.
2. Use the constraint equations and differentiate them on time to write those constraints in terms of the accelerations in the form $\mathbf{A}\ddot{\mathbf{x}} = \mathbf{b}$.
3. Use $\mathbf{F}^e$, $\mathbf{M}$, $\mathbf{A}$, and $\mathbf{b}$ in conjunction with Eq. (2.4.41) to obtain explicit expressions for the resultant of the ideal constraint force, $\mathbf{F}^c$, and the equations of motion.
4. Solve the equations of motion to obtain the position and velocity components, using initial conditions that are consistent with the constraints. Monitor the values of the constraint equations and their first derivatives to ensure that numerical errors do not cause constraint drift.
5. Once we have found the solution for all the system coordinates as a function of time we can use that solution in the ideal constraint force expression obtained in step (3) to obtain the constraint force (as a function of time) explicitly.

Solving for all the system coordinates in this fashion is, as mentioned previously, called an *augmented approach* that combines the Newton-Euler equations with the differentiated constraint equations. Also, as outlined previously, it is possible to use the constraint equations to reduce the number of coordinates to consider. This *embedding approach* involves only n_{dof} second-order equations of motion. If we solve those equations of motion for the n_{dof} independent coordinates we can then use the constraint equations and their first derivatives to find the remaining dependent variables and their first derivatives and constraint drift is not an issue. In Sect. 2.11 the embedding method will be discussed in more detail.

2.5 Constraints and Equations of Motion in Polar Coordinates

2.5.1 *Solutions in Polar Coordinates*

Equation (2.4.41) is a fundamental result for dynamics. It gives us an explicit formulation for the equations of motion and an expression for the ideal constraint forces in many common problems. Remarkably, it is also applicable to a much broader class of problems and conditions than the ones we have so far considered. For example, as presented we have formulated the dynamics problems in Cartesian coordinates. But this is not necessary. Suppose we had expressed Newton's second law for 2-D particle problems in polar coordinates (r, θ) instead. Since the acceleration vector is then (see Chap. 1)

$$\ddot{\mathbf{x}} = \left(\ddot{r} - r\dot{\theta}^2\right)\mathbf{e}_r + \left(r\ddot{\theta} + 2\dot{r}\dot{\theta}\right)\mathbf{e}_\theta \tag{2.5.1}$$

Newton's law gives

$$\begin{aligned} m\ddot{r} &= mr\dot{\theta}^2 + F_r^a + F_r^c \\ mr\ddot{\theta} &= -2m\dot{r}\dot{\theta} + F_\theta^a + F_\theta^c \end{aligned} \tag{2.5.2}$$

where $\left(F_r^a, F_\theta^a\right)$ are components of the known applied force and $\left(F_r^c, F_\theta^c\right)$ are the components of the constraint force. If we define a "mass" matrix, $\mathbf{M}$, and the coordinates, $\mathbf{q}$, as

$$\mathbf{M} = \begin{bmatrix} m & 0 \\ 0 & mr \end{bmatrix}, \quad \mathbf{q} = \begin{Bmatrix} r \\ \theta \end{Bmatrix} \tag{2.5.3}$$

then we can write Newton's second law as

$$\mathbf{M}\ddot{\mathbf{q}} = \tilde{\mathbf{F}}^e + \mathbf{F}^c \tag{2.5.4}$$

where

$$\tilde{\mathbf{F}}^e = \tilde{\mathbf{F}}^e\left(r, \theta, \dot{r}, \dot{\theta}\right) = \mathbf{F}^{in} + \mathbf{F}^e = \begin{Bmatrix} mr\dot{\theta}^2 \\ -2m\dot{r}\dot{\theta} \end{Bmatrix} + \begin{Bmatrix} F_r^e\left(r, \theta, \dot{r}, \dot{\theta}\right) \\ F_\theta^e\left(r, \theta, \dot{r}, \dot{\theta}\right) \end{Bmatrix} \tag{2.5.5}$$

is an explicit function of the coordinates and their derivatives since the applied forces are also known functions of these variables, as indicated in Eq. (2.5.5). Physically, $\tilde{\mathbf{F}}^e$ is the sum of the external forces, $\mathbf{F}^e$, plus the "inertial force term," $\mathbf{F}^{in}$, which represents those parts of the total inertial force, $-\mathbf{M}\ddot{\mathbf{x}}$, that do not involve two time derivatives of the coordinates.

If we have constraints that can be placed, through differentiation on time, in the form $\mathbf{A}\ddot{\mathbf{q}} = \mathbf{b}$ and the constraint force is again prescribed in terms of these constraints as $\mathbf{F}^c = \mathbf{A}^T\boldsymbol{\lambda}$ then constrained motion problems in polar coordinates become

$$\begin{aligned} \mathbf{M}\ddot{\mathbf{q}} &= \tilde{\mathbf{F}}^e + \mathbf{A}^T\boldsymbol{\lambda} \\ \mathbf{A}\ddot{\mathbf{q}} &= \mathbf{b} \end{aligned} \tag{2.5.6}$$

which is identical in form to these same equations written in Cartesian coordinates so that equations of motion and constraint force are in the same form as Eq. (2.4.41), namely

$$\begin{aligned} \mathbf{F}^c &= \mathbf{A}^T\left(\mathbf{A}\mathbf{M}^{-1}\mathbf{A}^T\right)^{-1}\left(\mathbf{b} - \mathbf{A}\mathbf{M}^{-1}\tilde{\mathbf{F}}^e\right) \\ \ddot{\mathbf{q}} &= \mathbf{M}^{-1}\tilde{\mathbf{F}}^e + \mathbf{M}^{-1}\mathbf{A}^T\left(\mathbf{A}\mathbf{M}^{-1}\mathbf{A}^T\right)^{-1}\left(\mathbf{b} - \mathbf{A}\mathbf{M}^{-1}\tilde{\mathbf{F}}^e\right) \end{aligned} \tag{2.5.7}$$

Although we used the case of polar coordinates to justify Eq. (2.5.7), the same arguments can be used for other coordinates such as cylindrical or spherical coordinates. Later we will consider generalized coordinates, $\mathbf{q}$, involving any transformation between coordinates, $\mathbf{x} = \mathbf{x}(\mathbf{q})$, that is well behaved or for a more general transformation $\mathbf{x} = \mathbf{x}(\mathbf{q}, t)$ which can represent motion in a set of moving coordinates.

2.5.2 Unconstrained Motion of a Spring Pendulum

As a first example of a dynamics problem in polar coordinates, consider the motion of a spring pendulum shown in Fig. 2.27a where a mass, m, is attached to a fixed support by a linear spring and is free to move in two dimensions, described by the polar coordinates (r, θ) under the force of gravity. The free body diagram of the mass is shown in Fig. 2.27b. The unstretched length of the spring is L_0. This is an unconstrained problem since the spring can stretch as well as change angle so that the coordinates (r, θ) have no geometric constraining relationship between them. The spring controls the force on the particle for whatever path the particle takes but it does not determine a specific path. From Newton's second law expressed in polar coordinates we have

$$\begin{aligned} m\left(\ddot{r} - r\dot{\theta}^2\right) &= F_r = mg\cos\theta - k(r - L_0) \\ m\left(r\ddot{\theta} + 2\dot{r}\dot{\theta}\right) &= F_\theta = -mg\sin\theta \end{aligned} \tag{2.5.8}$$

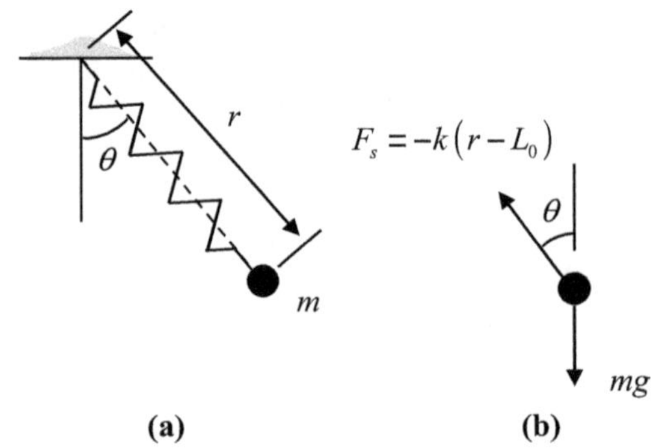

Fig. 2.27 (**a**) A spring-pendulum problem where the mass, *m*, moves in two dimensions under the force of gravity, and (**b**) the free body diagram of the mass

which leads to the equations of motion as

$$\begin{aligned} \ddot{r} &= r\dot{\theta}^2 + g\cos\theta - \frac{k}{m}(r - L_0) \\ \ddot{\theta} &= -2\frac{\dot{r}\dot{\theta}}{r} - g\frac{\sin\theta}{r} \end{aligned} \tag{2.5.9}$$

We can use Eq. (2.5.9) as the basis for a numerical solution but there is some interesting physics associated with this problem that we can highlight better by redefining and normalizing r appropriately, and also normalizing the time, t. First, consider the equilibrium problem when the spring pendulum is simply hanging vertically ($\theta = 0$). In that case the equations of motion become

$$r = L_e = L_0 + mg/k \tag{2.5.10}$$

where L_e is the equilibrium length of the spring under the weight of the mass. We will use this equilibrium length to define a new, nondimensional radius, r_n, as

$$r_n = r/L_e - 1 \tag{2.5.11}$$

so that r_n is physically the elongation of the spring pendulum, as measured from its equilibrium length, L_e, and normalized by that equilibrium length. In terms of r_n the equations of motion become

$$\begin{aligned} \ddot{r}_n &= (1 + r_n)\dot{\theta}^2 - \frac{g}{L_e}(1 - \cos\theta) - \frac{k}{m}r_n \\ \ddot{\theta} &= -2\frac{\dot{r}_n\dot{\theta}}{(1 + r_n)} - \frac{g}{L_e}\frac{\sin\theta}{(1 + r_n)} \end{aligned} \tag{2.5.12}$$

Finally, we will define a new nondimensional time variable, τ, as $\tau = t\sqrt{k/m}$ in terms of which Eq. (2.5.12) becomes

$$\begin{aligned}\frac{d^2 r_n}{d\tau^2} &= (1+r_n)\left(\frac{d\theta}{d\tau}\right)^2 - \frac{mg}{kL_e}(1-\cos\theta) - r_n \\ \frac{d^2\theta}{d\tau^2} &= -\frac{2}{(1+r_n)}\frac{dr_n}{d\tau}\frac{d\theta}{d\tau} - \frac{mg}{kL_e}\frac{\sin\theta}{(1+r_n)}\end{aligned} \tag{2.5.13}$$

Equation (2.5.13) is not in a simpler form than Eq. (2.5.9) but it does show that the spring pendulum behavior in these nondimensional variables is controlled by only a single nondimensional parameter, mg/kL_e. To understand what this constant means, let's consider Eq. (2.5.13) when we are near equilibrium so that both r_n and θ are small. If we expand the trigonometric functions to two terms as $\sin\theta \cong \theta - \theta^3/6$, $\cos\theta \cong 1 - \theta^2/2$ and neglect all terms higher than first order in (r_n, θ) or their derivatives in Eq. (2.5.13) we find

$$\begin{aligned}\frac{d^2 r_n}{d\tau^2} + r_n &= 0 \\ \frac{d^2\theta}{d\tau^2} + \frac{mg}{kL_e}\theta &= 0\end{aligned} \tag{2.5.14}$$

which, in terms of the original time variable, t, become

$$\begin{aligned}\frac{d^2 r_n}{dt^2} + \frac{k}{m} r_n &= 0 \\ \frac{d^2\theta}{dt^2} + \frac{g}{L_e}\theta &= 0\end{aligned} \tag{2.5.15}$$

Equation (2.5.15) represents the linearized equations of motion that are good for small motions about the equilibrium position ($\theta = 0, r_n = 0$) of the spring pendulum. These equations both have analytical solutions representing sinusoidal vibrations in both the r and θ directions given by

$$\begin{aligned}r_n &= A\cos(\omega_r t + \phi_a) \\ \theta &= B\cos(\omega_\theta t + \phi_b)\end{aligned} \tag{2.5.16}$$

where (A, ϕ_a, B, ϕ_b) are all constants that are determined from the initial conditions and $\omega_r = \sqrt{k/m}$ and $\omega_\theta = \sqrt{g/L_e}$ are the natural frequencies of oscillation (in rad/s) in the r- and θ-directions, respectively. You can verify these solutions easily by just plugging them into Eq. (2.5.15). Thus, the single parameter, $mg/kL_e = \omega_\theta^2/\omega_r^2$, is just the square of the ratio of these natural frequencies. Since

$$\frac{mg}{kL_e} = \frac{mg}{(kL_0 + mg)} < 1 \tag{2.5.17}$$

we see that $\omega_\theta < \omega_r$. For small oscillations Eq. (2.5.15) shows that the motions in the r- and θ-directions occur independent of each other, i.e., they are uncoupled. However, as we will see, if the ratio of the natural frequencies is close to a certain value then the nonlinear terms present in our original equations in Eq. (2.5.13) cause an inherent coupling of the r and θ motions, producing what is called a *parametric nonlinear resonance phenomenon* that is missed entirely by the linearized equations, Eq. (2.5.15).

Simulation 2.6: Spring Pendulum

To write Eq. (2.5.13) as a system of first-order ordinary differential equations we will define four variables in the column vector $\mathbf{x}$ as $\mathbf{x} = \left[r_n,\ \dot{r}_n,\ \theta,\ \dot{\theta}\right]^T = \left[x_1,\ x_2, x_3, x_4\right]^T$, where now $\dot{f} \equiv df/d\tau$, i.e., the dot now signifies derivative on the normalized time, τ. Then we have

$$\dot{\mathbf{x}} = \begin{bmatrix} \dot{r}_n \\ (1+r_n)\dot{\theta}^2 - K(1-\cos\theta) - r_n \\ \dot{\theta} \\ \dfrac{-2\dot{r}\dot{\theta}}{(1+r_n)} - K\dfrac{\sin\theta}{(1+r_n)} \end{bmatrix} = \begin{bmatrix} x_2 \\ (1+x_1)x_4^2 - K(1-\cos x_3) - x_1 \\ x_4 \\ \dfrac{-2x_2x_4}{(1+x_1)} - K\dfrac{\sin x_3}{(1+x_1)} \end{bmatrix} \tag{2.5.18}$$

where $K = mg/kL_e$. This system has been placed in the MATLAB® function spring_pend_eqs which together with the MATLAB® solver ode45 is a nested function within the function spring_pendulum, following the same structure as done in the roller-coaster problem. The calling sequence for spring_pendulum is

```
[tout, xout] = spring_pendulum(tspan, x0, K, rtol, atol);
```

where again tspan is the time interval over which we want to compute the solution, x0 is the vector of initial conditions $\left[r_n(0), \dot{r}_n(0), \theta(0), \dot{\theta}(0)\right]$, and K is the nondimensional free parameter of the system. Here the two tolerances are present that are used by the ode45 numerical solver: rtol, a relative tolerance (RelTol), and atol, which is an absolute tolerance (AbsTol). To give you a feel for what these two tolerances represent it is useful to look at their definitions as given in the MATLAB® documentation:

RelTol—This tolerance is a measure of the error relative to the size of each solution component. Roughly, it controls the number of correct digits in all solution components, except those smaller than thresholds AbsTol(i). The default value, 1e-3, corresponds to 0.1% accuracy.

AbsTol—AbsTol(i) is a threshold below which the value of the ith solution component is unimportant. The absolute error tolerances determine the accuracy when the solution approaches zero. The default value is 1e-6.

Thus, it will generally be the relative tolerance that controls the accuracy of the numerical solution except when the solution values are very small, where the absolute tolerance will be important. The ability to control the error tolerances in the spring pendulum is especially important because of some of the unusual behaviors of this seemingly simple problem and the need to distinguish real behavioral changes from numerical errors.

The function spring_pendulum is used as part of the MATLAB® script spring_pendulum_sim to simulate the motion of the spring pendulum and plot the (r_n, θ) values versus time (as well as a normalized energy which we will discuss shortly). The first case considered is for K = 0.1, tspan = [0, 50], and initial conditions x0 = [0.1, 0, 0.05, 0], which are the default values contained in the script unless changed by the user in the command window before executing the script. The default tolerances are also specified in the script as rtol = 1e-5 and atol = 1e-10. The initial condition amplitudes are small enough that the linearized equations of motion, Eq. (2.5.15), might be expected to be valid and Fig. 2.28a, b confirms that this is the case, with nearly sinusoidal oscillations seen for both r_n and θ, although there are some small deviations in the θ versus time plot from the simple constant amplitude sinusoidal solution of Eq. (2.5.16). Those deviations are not, however, numerical errors as the relative tolerance was changed to rtol = 1e-8 and the spring_pendulum_sim rerun without any significant change in the values of both r_n and θ at the final time of the simulation (and no change in the overall curves).

If the workspace is cleared and in the command window the K value is specified as K = 0.25, tspan is specified as tspan = [0, 400], and the initial conditions given as x0 = [0.1, 0, 0.025, 0] (which are still very small) the results of running the simulation script are completely different, as shown in Fig. 2.29a, b. The angular oscillation grows to a much larger amplitude than the original starting value while the extensional vibrations are reduced in amplitude to a very small value. This interchange of energy then reverses with the angular amplitudes diminishing while

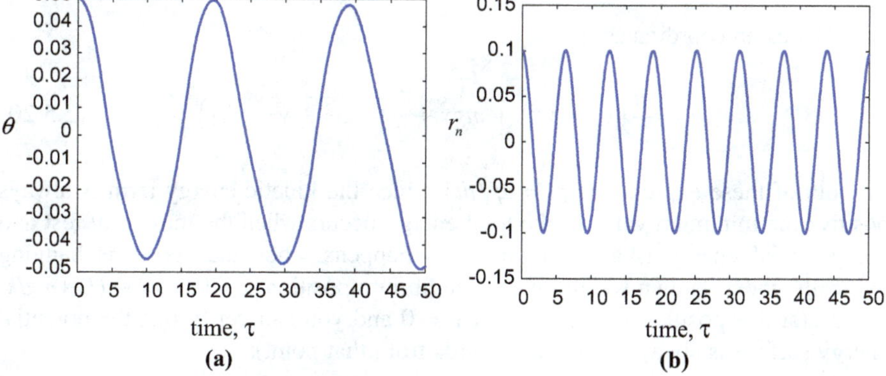

Fig. 2.28 (**a**) The swinging oscillations of the spring pendulum and (**b**) the extensional oscillations of the spring pendulum for $K = 0.1$, $\theta(0) = 0.05$, $r_n(0) = 0.1$, and $\dot{\theta}(0) = \dot{r}_n(0) = 0$

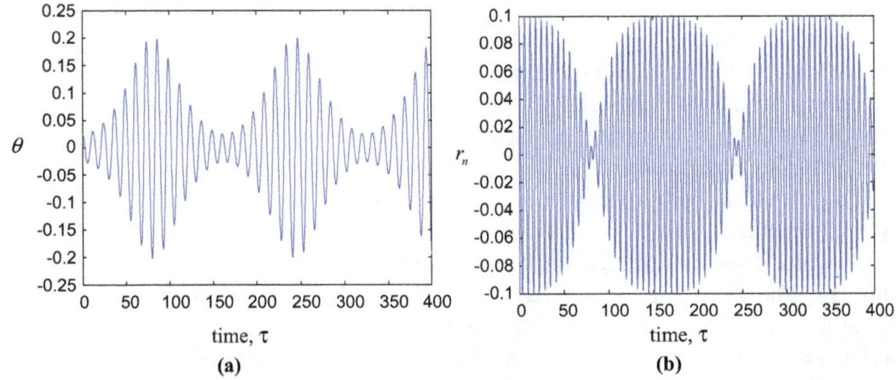

Fig. 2.29 (**a**) The swinging oscillations of the spring pendulum and (**b**) the extensional oscillations of the spring pendulum for $K = 0.25$, $\theta(0) = 0.025$, $r_n(0) = 0.1$, and $\dot{\theta}(0) = \dot{r}_n(0) = 0$

the extensional vibrations grow. This behavior is an example of a phenomenon called *nonlinear parametric resonance.* It only occurs near a range of K values near $K = 0.25$, where $\omega_r = 2\omega_\theta$. The smaller the initial conditions in θ, the longer it takes for the angular amplitude to grow to its maximum value, which is approximately $\theta_{max} \cong 2r_n(0)$ for $K = 0.25$. This is a real phenomenon. If the tolerances are changed to the very small values rtol = atol = 1e-12, there is an insignificant change in the values both r_n and θ at the final time of the simulation (and again no change in the overall curves).

Since the spring pendulum is a conservative system the total energy is a constant. As shown previously, loss in numerical accuracy may also be detected through changes of the total energy. In polar coordinates the total energy, E, is

$$E = T + V = \frac{m}{2}\left(\dot{r}^2 + r^2\dot{\theta}^2\right) - mgr\cos\theta + \frac{k}{2}(r - L_0)^2 \tag{2.5.19}$$

or in Cartesian coordinates

$$E = \frac{m}{2}\left(\dot{x}^2 + \dot{y}^2\right) + mgy + \frac{k}{2}\left(\sqrt{x^2 + y^2} - L_0\right)^2 \tag{2.5.20}$$

[in both of these expressions $\dot{f} = df/dt$]. Since the kinetic energy term is always positive the minimum value of the total energy occurs when the mass is at rest and the potential energy is a minimum. This happens when the spring is hanging vertically under its own weight in the equilibrium position $x = 0$, $y = -(L_0 + mg/k) = -L_e$ (at this position $\partial V/\partial x = \partial V/\partial y = 0$ and you can verify that the potential energy surface is always concave upwards from that point).

Thus,

$$E_{\min} = -mgL_e + \frac{k}{2}(mg/k)^2 = -mgL_e\left(1 - \frac{K}{2}\right) \tag{2.5.21}$$

To facilitate later discussions of the behavior of the spring pendulum we will use a normalized energy, $R = -E/E_{\min}$, written in terms of the initial conditions and the nondimensional parameter $\mu = 1/K$ to evaluate the total energy. This R energy term has values ranging from -1 to infinity while μ ranges from +1 to infinity. The normalized total energy will be written in terms of the nondimensional r_n,θ, τ, and μ variables since the equations of motion, Eq. (2.5.18), are also expressed in those terms. After some algebra, we have

$$R = \frac{1}{2\mu - 1}\left[\mu^2\left(\dot{r}_n^2 + (1 + r_n)^2\dot{\theta}^2\right) - 2\mu(1 + r_n)\cos\theta + \mu^2(r_n + 1/\mu)^2\right] \tag{2.5.22}$$

where now $\dot{f} = df/d\tau$. The script spring_pendulum_sim also plots the value of R versus time so that one can verify that energy conservation is satisfied. In all the examples just discussed, there were insignificant R changes seen in the simulations, confirming those results.

It is useful to see the interchange of energy in action for the spring pendulum so we have written a script spring_pendulum_anim3 that is similar to the spring_pendulum_sim script but where the plotting outputs are replaced with an animation of the motion. To keep the number of times at which the calculations are displayed fixed regardless of the other parameters, the specification of tspan has been changed from a vector consisting only of the starting and ending points to that of a specific set of points. The default value in spring_pendulum_anim3 is tspan = 0:0.1:400. Note that the ode45 function does not calculate the solution with a numerical spacing defined by these specified times. It still uses its own internal error control to determine the number and spacing of the numerical calculations. However, only the interpolated values at the points specified in tspan are returned by the ode45 function. In the script one also has a capability to adjust the speed of the animation. This adjustment can be made with a separate "slowdown" parameter, sldwn, which produces a pause of 0.01*sldwn sec before updating the motion. The default value sldwn = 1 and the 4001 points chosen in tspan should give a reasonably fast animation if the pendulum swings are relatively small. For larger swings one might want to slow the animation down by choosing a larger value of sldwn (see the further discussion on adjusting animation speed in Chap. 3 for the double-pendulum problem). The animation script also contains an additional string parameter named marker. For a value of marker = "on" the path of the mass is traced by a series of dots as the motion proceeds, eventually showing the entire set of positions of the mass at the evaluated times. The default value is marker = "off." The final time frame of an animation using the default values in spring_pendulum_anim3

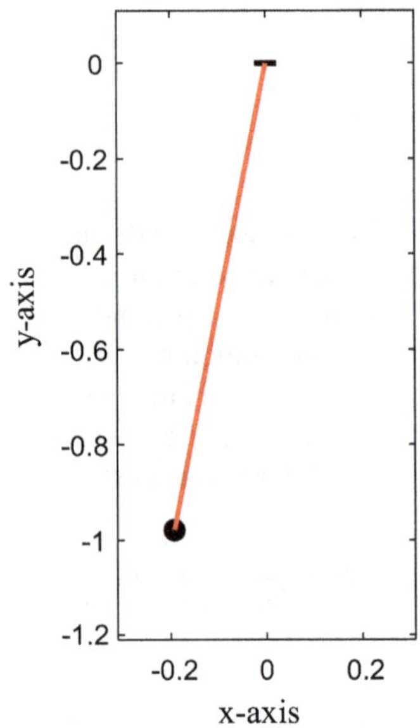

Fig. 2.30 The spring pendulum shown at the final time of a MATLAB® animation that demonstrates nonlinear parametric resonance

is shown in Fig. 2.30. To see the motion of the spring pendulum for the case when parametric resonance is optimally present clear the command window and execute the script spring-pendulum_anim3 with the default values ($K = 0.25$ by default in the script so it is set up to examine the parametric resonance case; you can see the other defaults by looking at the script in Appendix F). You will see the motion of the mass change from primarily up-and-down extensional motion to primarily back-and-forth swinging motion. To also see the path followed by the mass in this case, type marker = 'on' in the command window and re-execute the script.

You can easily experiment with other K values to see the effect on the motion and if parametric resonance is present or not.

Nonlinear parametric resonance is an instability that can cause dangerous motions or even capsizing to occur in ships. Under the right conditions the longitudinal motion of a vessel due to wave action can be parametrically coupled to the roll motion of the ship. In October 1998 the large container vessel APL China encountered a parametric roll resonance on a trip from Taiwan to Seattle. More than 60% of its cargo of containers were lost at sea or damaged. In January 2003 the container vessel Maersk Carolina experienced parametric resonance during a storm in the Atlantic where the rolling increased up to 47° and lasted for minutes. The ship lost 133 containers as well as another 50 that suffered severe water damage, leading to a total cargo loss of almost 4 million dollars. Container ships are not the only ships susceptible to such roll resonance. Destroyers, cruise ships, and roll-on/roll-off passenger (ROPAX) vessels that carry vehicles as well as passengers can also encounter parametric roll resonance.

The spring pendulum is a commonly used example to illustrate parametric resonance. Another less well-known feature of the spring pendulum is that it shows a transition from regular behavior to chaotic behavior and then back to regular behavior as the input parameters are changed. What do we mean by chaotic behavior versus regular motion? The spring pendulum, like all the other dynamical systems we will examine in this book, is deterministic in that the initial conditions and the equations of motion determine the motion of the system for all subsequent times. Regular behavior means that a small change in initial conditions will result in corresponding small changes to the motion at subsequent times. But in chaotic behavior even very small changes in initial conditions eventually result in large differences in the motion. This sensitivity to initial conditions renders the prediction of the future behavior of chaotic systems impossible.

Chaos is exhibited by many types of systems—dynamical systems, chemical systems, biological systems, and electrical systems to name just a few. In the case of the spring pendulum its regular or chaotic behavior is controlled by two nondimensional parameters: the constant μ (or, equivalently, K) and the normalized total energy, R, which we have seen is a function of μ and the initial conditions. Studies of the spring pendulum have shown that for extreme values of R and μ (for either values near the smallest possible parameter values or large values of the parameters) the motion is regular but for intermediate values the motion can become chaotic. A simplified picture of this behavior is shown in Fig. 2.31, where the intermediate values when chaotic behavior is observed are shown to lie roughly in an elliptical shaped shaded region in the μ, R plane (there is no suggestion here that the shape of the chaotic region is a simple shape or that chaos is observed at all points within the ellipse—the ellipse is only shown to give you a rough idea of where in the (R, μ) plane numerical experiments have revealed chaotic behavior). The parametric resonance case just considered where K = 0.25 ($\mu = 4$); x0 = [0.1, 0, 0.025, 0] gives $R = -0.9768$ (point A in Fig. 2.31) which lies outside the elliptical region so we might expect the motion will be regular and not sensitive to the initial conditions. To test this assumption we set the tolerances rtol = atol = 1e-12 to keep the numerical errors very small. *These same very small tolerances were then kept for the series of parametric studies we will now describe.* First, we ran spring_pendulum_anim3 again for this parametric resonance case, now using these very small tolerances. At the final time of the calculation the end values were $r_n = -0.0016$, $\theta = -0.1942$ (in the MATLAB® command window type rn(end) and th(end) after running the script). If the initial conditions are changed to x0 = [0.100001, 0, 0.0250, 0], and the animation is rerun, the final values are unchanged (to the four displayed figures). Now, consider what happens when the initial conditions are changed to x0 = [0.8, 0, 0.025, 0], with all other parameters unchanged. The R value becomes 0.4635, which is well within the chaotic region (point B in Fig. 2.31). The animation is rather interesting to watch as it shows a highly variable path for the mass, part of which is shown in Fig. 2.32. At the end of this simulation one finds the final values $r_n = 0.4713, \theta = -0.9113$. However, if the initial conditions are changed to x0 = [0.800001, 0, 0.025, 0], with all other parameters unchanged the final values are $r_n = 0.1114$, $\theta = -0.2623$, showing a strong sensitivity to these very small

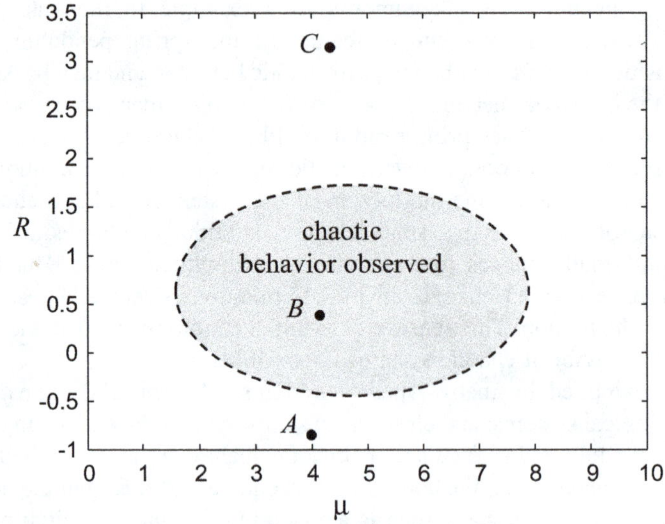

Fig. 2.31 Plot showing the approximate range of (R, μ) values (shaded region) where the motion of the spring pendulum is observed to be chaotic and points A, B, and C where the motion changes from regular to chaotic and back to regular

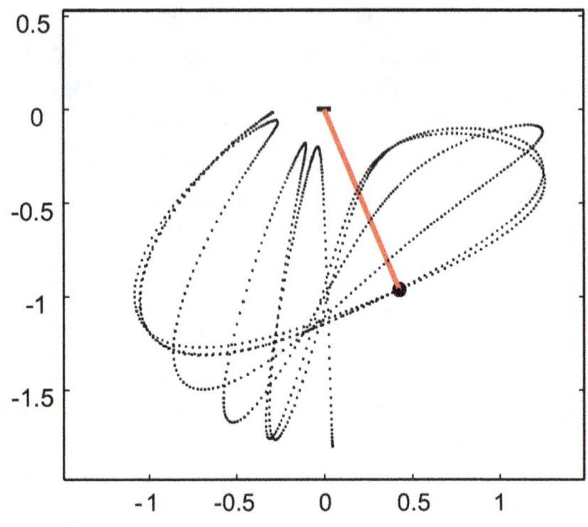

Fig. 2.32 Part of the path of the spring pendulum for $R = 0.4635$, $\mu = 4$, which is in the chaotic region of the (R, μ) plane

changes in the initial conditions. Now, change the initial conditions to x0 = [0.8, 0, 0.025, 0.6], and set sldwn = 5 to see a smoother motion, with all the other parameters unchanged. This gives $R = 3.13$ which is again outside the shaded region in Fig. 2.31 (point C). The energy is large enough that the spring pendulum exerts complete 360-degree rotations around the support. At the end of the motion we find $r_n = 0.7957$ and $\theta = 276.5937$ (note that the animation in this case runs for some time so you might want to terminate it by typing ctrl-c). Changing just the initial conditions to x0 = [0.800001, 0, 0.025, 0.6] and executing the animation again leave these final values essentially unchanged, demonstrating the change from regular motion to chaos and then back to regular motion.

Chaos is a complex topic. We do not intend to deal with chaos in this book other than to illustrate it in some simple examples and point the reader to a number of sources that cover chaotic behavior in more depth (see [5, 6], for example). Engineers often analyze dynamical systems by considering very small motions where the equations of motion are linearized and where one can often find exact analytical solutions. Since chaos is inherently a nonlinear phenomenon, such linearization eliminates the possibility of chaos and all of the interesting features and important consequences of its existence. Analytic solutions for nonlinear problems are generally nonexistent so studies of chaos also require detailed numerical calculations.

2.5.3 Constrained Motion of a Pendulum

Our first example of constrained motion in polar coordinates will be a classic problem that is covered in all dynamics texts—the motion of a simple pendulum of fixed length L under the action of gravity (Fig. 2.33). This is, of course, a special case of the spring pendulum where the spring constant is infinitely large so that r is a constant. In polar coordinates, the one constraint for this problem is

$$f(r,\theta) = r - L = 0 \tag{2.5.23}$$

which, differentiated once and twice, gives

$$\begin{aligned} \dot{r} &= 0 \\ \ddot{r} &= 0 \end{aligned} \tag{2.5.24}$$

We can write this last equation as $\mathbf{A}\ddot{\mathbf{q}} = \mathbf{b}$, where

$$\mathbf{A} = [1 \quad 0], \quad \ddot{\mathbf{q}} = \begin{Bmatrix} \ddot{r} \\ \ddot{\theta} \end{Bmatrix}, \quad \mathbf{b} = 0 \tag{2.5.25}$$

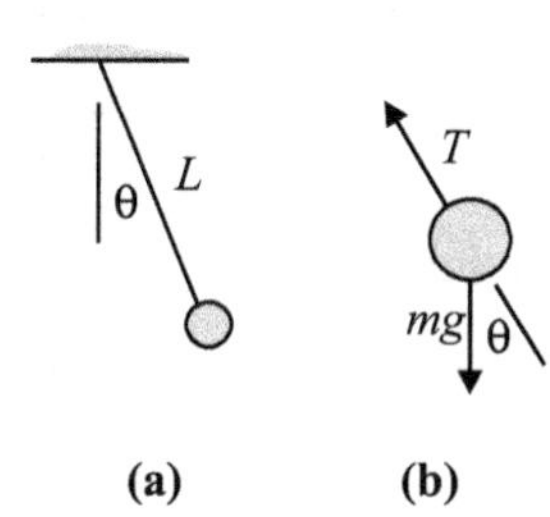

Fig. 2.33 (**a**) A pendulum where a mass m moves in two dimensions under the action of gravity and (**b**) the free body diagram of that mass

In this case, since $r = L, \dot{r} = 0$ we find the intermediate results

$$\mathbf{M} = \begin{bmatrix} m & 0 \\ 0 & mL \end{bmatrix}, \quad \tilde{\mathbf{F}}^e = \begin{Bmatrix} mL\dot{\theta}^2 + mg\cos\theta \\ -mg\sin\theta \end{Bmatrix}$$
$$\mathbf{AM}^{-1}\tilde{\mathbf{F}}^e = L\dot{\theta}^2 + g\cos\theta, \quad \left(\mathbf{AM}^{-1}\mathbf{A}^T\right)^{-1} = m \tag{2.5.26}$$

Using the augmented approach with Eq. (2.5.7) leads finally to

$$\mathbf{F}^c = -\begin{Bmatrix} 1 \\ 0 \end{Bmatrix}\left(mL\dot{\theta}^2 + mg\cos\theta\right)$$
$$\begin{Bmatrix} \ddot{r} \\ \ddot{\theta} \end{Bmatrix} = \begin{Bmatrix} L\dot{\theta}^2 + g\cos\theta \\ -g\sin\theta/L \end{Bmatrix} - \begin{Bmatrix} 1 \\ 0 \end{Bmatrix}\left(L\dot{\theta}^2 + g\cos\theta\right) \tag{2.5.27}$$

Here the first equation of motion is $\ddot{r} = 0$, with initial conditions $r(0) = L, \dot{r}(0) = 0$ which just has as its solution the original constraint equation, $r = L$, leaving only the equation of motion in terms of θ and the radial constraint force, $F_r^c = -T$, given, respectively, as

$$\ddot{\theta} + \frac{g}{L}\sin\theta = 0 \tag{2.5.28}$$

and

$$T = m\left(L\dot{\theta}^2 + g\cos\theta\right) \tag{2.5.29}$$

This problem can be more easily solved directly with Newton's second law, as is done in many texts. In fact, if we start with the force equations from Newton's second law we have

$$m\ddot{r} = mr\dot{\theta}^2 + mg\cos\theta - T$$
$$mr\ddot{\theta} = -2m\dot{r}\dot{\theta} - mg\sin\theta \tag{2.5.30}$$

We can apply the constraint (and its derivatives) $r = L, \dot{r} = 0, \ddot{r} = 0$ directly to these force equations to recover Eqs. (2.5.28) and (2.5.29). This is also a problem where some authors choose to replace the θ component of Newton's law with the moment equation about the fixed support point, O:

$$\mathbf{M}_0 = \mathbf{r} \times m\ddot{\mathbf{x}} \tag{2.5.31}$$

giving

$$\begin{aligned} L\mathbf{e}_r \times [-T\mathbf{e}_r + mg\cos\theta\mathbf{e}_r - mg\sin\theta\mathbf{e}_\theta] \\ = L\mathbf{e}_r \times \left[(m\ddot{r} - mL\dot{\theta}^2)\mathbf{e}_r + (mL\ddot{\theta})\mathbf{e}_\theta\right] \end{aligned} \tag{2.5.32}$$

or

$$-mgL\sin\theta\mathbf{e}_z = mL^2\ddot{\theta}\mathbf{e}_z \tag{2.5.33}$$

which is just the force equation of motion in the θ direction in Eq. (2.5.30) multiplied by L.

This pendulum problem shows that in dynamics problems there are often multiple ways to arrive at a solution. Thus, the expressions for the constraint force and equations of motion in Eq. (2.5.7) may not always be the easiest to implement in very simple problems like the pendulum. However, the fact that Eq. (2.5.7) gives us both the constraint force and the equations of motion is important for engineering purposes. Alternate methods, like the use of the moment equation, Eq. (2.5.31), for example, may effectively give us the equations of motion but they do not give us the values of the constraint forces present. The real value of using Eq. (2.5.7) over other methods, however, lies in its generality. It can be applied in a consistent manner for many different problems whereas other methods often involve specific steps that are unique to the problem being solved.

The radial constraint force does no work since it always acts perpendicular to the motion and the force of gravity is a potential force so the total energy is conserved for the simple pendulum. In polar coordinates the velocity is given as

$$\mathbf{v} = \dot{r}\,\mathbf{e}_r + r\dot{\theta}\,\mathbf{e}_\theta \tag{2.5.34}$$

so the kinetic energy is

$$T = \frac{m}{2}\left(\dot{r}^2 + r^2\dot{\theta}^2\right) \tag{2.5.35}$$

and for the pendulum the total energy is

$$E = \frac{mL^2\dot{\theta}^2}{2} - mgL\cos\theta \tag{2.5.36}$$

A contour plot of E/m for $L = 0.4$ m and θ ranging from $(-2\pi, 2\pi)$ rad and $\dot{\theta}$ ranging from $(-10, +10)$ rad/s gives the phase plot shown in Fig. 2.34. Although the shapes of the curves are somewhat different from the roller-coaster phase plot (Fig. 2.19) the overall behavior is very similar. From Eq. (2.5.36) we see if the particle starts at the bottom ($\theta = 0$) with an initial angular velocity, $\dot{\theta}_0$, it will just reach the top ($\theta = \pi$) with zero angular velocity if we have $mL^2\dot{\theta}_0^2/2 - mgL = mgL$, which gives $\dot{\theta}_0 = \sqrt{4g/L}$. Thus, if $\dot{\theta}_0 < \sqrt{4g/L}$ the mass m will never go "over the top" ($|\theta_{max}| < \pi$) and the mass will just oscillate back and forth, forming a closed

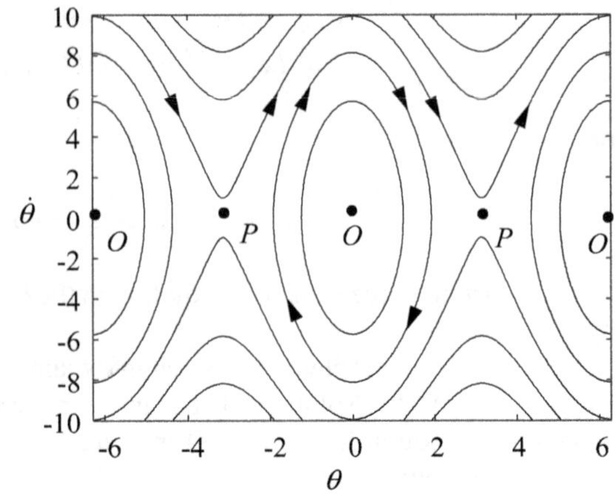

Fig. 2.34 Phase plot for the pendulum problem

curve in the phase plot about the equilibrium position $\theta = 0$. However, if $\dot{\theta}_0 > \sqrt{4g/L}$ the mass will execute full rotations around the pivot point with θ ever increasing as shown in Fig. 2.33. For the parameters chosen in Fig. 2.33 we have $\sqrt{4g/L} = 9.9$ rad/s, approximately. The points labeled O in Fig. 2.34 are the stable equilibrium points if the particle is resting at the bottom of a swing and the points labeled P in Fig. 2.34 are the unstable equilibrium points when the mass is resting at a top point of its swing. These stability assertions, which are rather obvious in this simple example, will be discussed in more detail in Chap. 6.

2.5.4 Constraints and the Motion of the Planets

The second constrained motion example we will consider in polar coordinates is a historically famous dynamics problem involving the motion of the planets around the sun. Kepler, in the 1600s, developed his three "laws" about planetary motion based on his analysis of the observational data of Galileo and Tycho Brahe. These laws are as follows:

Law 1: Each planet revolves in an elliptical orbit about the sun, with the sun located at one focus of the ellipse.
Law 2: Equal areas are swept per unit time by the line joining the sun to the planet.
Law 3: The squares of the periods of the planets are proportional to the cubes of the semimajor axes of the ellipses.

Newton showed that using his law of gravitation, Eq. (2.2.2), produced a solution for the motion of a planet around the sun that satisfied all three of Kepler's laws. Many dynamics texts give the details of this problem as it is one of Newton's major accomplishments in science. Here, instead of repeating what you can find in many other treatments, we are going to use our knowledge of how constraints and

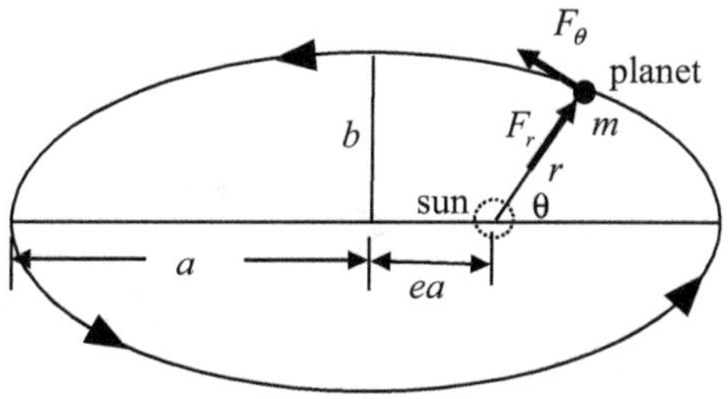

Fig. 2.35 Motion of a planet in an elliptical path around the sun as defined in polar coordinates with the origin at the center of the sun that is also located at one of the foci of the ellipse

constraint forces are related to consider the problem in reverse. In particular, we will consider Kepler's first and second laws as constraints on the motion of a planet and ask the following question: What type of constraint force is needed to satisfy those two laws? We will take the motion of the planet to lie in a fixed plane where the motion is described in polar coordinates (r, θ) with the origin placed at the center of the sun (Fig. 2.35). The planet will be treated as a particle of mass, m, that is acted upon only by the constraint force components (F_r, F_θ). The equation of an ellipse, written in polar coordinates with the origin of the coordinates taken at one of the foci of the ellipse (as required by Kepler's first law), is just

$$r = \frac{a(1-e^2)}{1+e\cos\theta} \tag{2.5.37}$$

where a is the semimajor axis of the ellipse and e is called the eccentricity of the ellipse. The other semiminor axis of the ellipse, b, can be shown to be given as $b = a\sqrt{1-e^2}$. We can write Eq. (2.5.37) as a constraint equation of the form $f_1(r, \theta) = 0$, where

$$f_1(r,\theta) = r(1+e\cos\theta) - a\left(1-e^2\right) = 0 \tag{2.5.38}$$

Differentiating this equation once on time, we find

$$(1+e\cos\theta)\dot{r} - er\sin\theta\dot{\theta} = 0 \tag{2.5.39}$$

Another differentiation gives

$$(1+e\cos\theta)\ddot{r} - er\sin\theta\ddot{\theta} = 2e\sin\theta\dot{r}\dot{\theta} + er\cos\theta\dot{\theta}^2 \tag{2.5.40}$$

Now, let us turn to Kepler's second law. As the planet moves in its orbit, in a small time dt it traces out a small amount of "sectorial area," dA, as shown in Fig. 2.36. This sectorial area is easily shown to be $dA = r^2 d\theta/2$ so that we can define the sectorial area velocity as $dA/dt = (r^2/2)d\theta/dt$. Kepler's second law, therefore, can be phrased mathematically as a constraint of the form $f_2\left(r,\dot{r},\theta,\dot{\theta}\right) = 0$ where

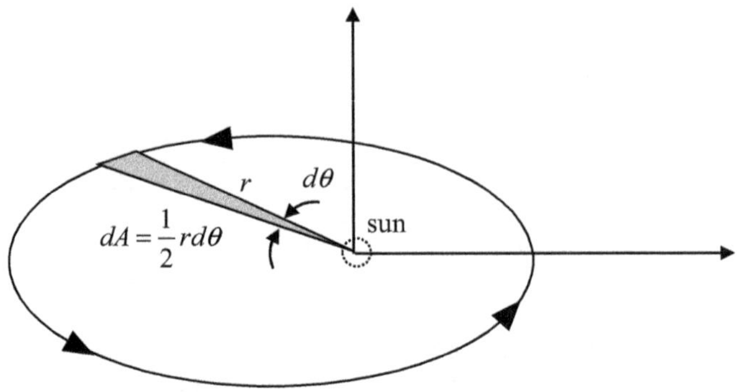

Fig. 2.36 The area, *dA*, swept out by a planet in a small time, *dt*, as it orbits the sun

$$f_2(r,\dot{r},\theta,\dot{\theta}) = \frac{r^2\dot{\theta}}{2} - C = 0 \tag{2.5.41}$$

where C is a constant. If we differentiate this equation once on time we find

$$r^2\ddot{\theta} = -2r\dot{r}\dot{\theta} \tag{2.5.42}$$

Instead of using Eqs. (2.5.42) and (2.5.40) directly, if we divide both sides in Eq. (2.5.42) by r and place that relationship into Eq. (2.5.40) we find

$$(1 + e\cos\theta)\ddot{r} = er\cos\theta\dot{\theta}^2 \tag{2.5.43}$$

Then by using Eqs. (2.5.42) and (2.5.43) as our two constraints in the form $\mathbf{A}\ddot{\mathbf{q}} = \mathbf{b}$, we find

$$\ddot{\mathbf{q}} = \begin{Bmatrix} \ddot{r} \\ \ddot{\theta} \end{Bmatrix}, \quad \mathbf{A} = \begin{bmatrix} (1 + e\cos\theta) & 0 \\ 0 & r^2 \end{bmatrix}$$
$$\mathbf{b} = \begin{Bmatrix} er\cos\theta\dot{\theta}^2 \\ -4r\dot{r}\dot{\theta} \end{Bmatrix} \tag{2.5.44}$$

The use of Eq. (2.5.43) instead of Eq. (2.5.40) makes the **A** matrix diagonal so, as we will see, all the calculations become very simple. Since we have assumed that there are no applied forces here, $\tilde{\mathbf{F}}^e$ only includes the inertial forces (see Eq. (2.5.5)) so

$$\tilde{\mathbf{F}}^e = \begin{Bmatrix} mr\dot{\theta}^2 \\ -2m\dot{r}\dot{\theta} \end{Bmatrix} \tag{2.5.45}$$

Now, we can compute the constraint force compatible with these two constraints as

$$\mathbf{F}^c = \mathbf{A}^T\left(\mathbf{A}\mathbf{M}^{-1}\mathbf{A}^T\right)^{-1}\left(\mathbf{b} - \mathbf{A}\mathbf{M}^{-1}\tilde{\mathbf{F}}^e\right) \tag{2.5.46}$$

Some of the intermediate results, which you can easily verify, are

$$\begin{aligned}
&\mathbf{A} = \mathbf{A}^T = \begin{bmatrix} (1+e\cos\theta) & 0 \\ 0 & r \end{bmatrix}, \quad \mathbf{M}^{-1}\mathbf{A}^T = \begin{bmatrix} (1+e\cos\theta)/m & 0 \\ 0 & 1/m \end{bmatrix} \\
&\mathbf{A}\mathbf{M}^{-1}\mathbf{A}^T = \begin{bmatrix} (1+e\cos\theta)^2/m & 0 \\ 0 & r/m \end{bmatrix}, \quad \left[\mathbf{A}\mathbf{M}^{-1}\mathbf{A}^T\right]^{-1} = \begin{bmatrix} m/(1+e\cos\theta)^2 & 0 \\ 0 & m/r \end{bmatrix} \\
&\mathbf{A}^T\left[\mathbf{A}\mathbf{M}^{-1}\mathbf{A}^T\right]^{-1} = \begin{bmatrix} m/(1+e\cos\theta) & 0 \\ 0 & m \end{bmatrix} \\
&\mathbf{b} - \mathbf{A}\mathbf{M}^{-1}\tilde{\mathbf{F}}^e = \begin{Bmatrix} er\cos\theta\dot{\theta}^2 \\ -2\dot{r}\dot{\theta} \end{Bmatrix} - \begin{bmatrix} (1+e\cos\theta) & 0 \\ 0 & r \end{bmatrix}\begin{Bmatrix} r\dot{\theta}^2 \\ -2\dot{r}\dot{\theta}/r \end{Bmatrix} = \begin{Bmatrix} -r\dot{\theta}^2 \\ 0 \end{Bmatrix}
\end{aligned} \tag{2.5.47}$$

so that the constraint force is just

$$\mathbf{F}^c = \begin{bmatrix} m/(1+e\cos\theta) & 0 \\ 0 & m \end{bmatrix}\begin{Bmatrix} -r\dot{\theta}^2 \\ 0 \end{Bmatrix} = \begin{Bmatrix} -mr\dot{\theta}^2/(1+e\cos\theta) \\ 0 \end{Bmatrix} \tag{2.5.48}$$

But from the constraint equation, Eq. (2.5.41), and the equation of the ellipse, Eq. (2.5.37), we can rewrite the constraint force as

$$\mathbf{F}^c = \begin{Bmatrix} \dfrac{-4mC^2}{a(1-e^2)r^2} \\ 0 \end{Bmatrix} \tag{2.5.49}$$

Thus, we have shown that the constraint force that generates Kepler's first two laws must be a purely radial attractive force whose strength varies like $1/r^2$. This is nothing more than Newton's law of gravitational attraction! Newton wrote his law in terms of the mass of the sun, M, and the universal gravitational constant, G, but as part of the solution for the motion one also finds $4C^2/a(1-e^2) = GM$. Of course, Newton came up with the universal law of gravitational attraction and then showed that this force in conjunction with his second law, $\mathbf{F} = m\ddot{\mathbf{x}}$, was compatible with all three of Kepler's laws, a true testament to his insight and genius.

This planetary motion problem is also an example where we can replace one of the force equations in Newton's second law with a moment equation. In this case Newton's second law gives

$$\begin{aligned} m\ddot{r} - mr\dot{\theta}^2 &= F_r = GMm/r^2 \\ mr\ddot{\theta} + 2m\dot{r}\,\dot{\theta} &= 0 \end{aligned} \tag{2.5.50}$$

Note that the second equation in Eq. (2.5.50), the force equation in the θ-direction, is also just a statement of the constraint (2.5.41) since it can also be written as

$$2m\frac{d}{dt}\left(\frac{1}{2}r^2\dot{\theta}\right) = 0 \tag{2.5.51}$$

However, if we examine the moment equation about the origin we also recover the same equation since

$$\begin{aligned} \mathbf{M}_0 = \mathbf{x} \times \mathbf{F} &= \frac{d}{dt}(\mathbf{x} \times m\mathbf{v}) \\ r\mathbf{e}_r \times F_r\mathbf{e}_r &= \frac{d}{dt}\left[r\mathbf{e}_r \times m\left(\dot{r}\,\mathbf{e}_r + r\dot{\theta}\,\mathbf{e}_\theta\right)\right] \\ 0 &= \frac{d}{dt}\left(mr^2\dot{\theta}\,\mathbf{e}_z\right) \end{aligned} \tag{2.5.52}$$

where $\mathbf{e}_z$ is a unit vector in the z-direction. But Eq. (2.5.52) says that the angular momentum of the particle about O, $\mathbf{L}_0 = mr^2\dot{\theta}\,\mathbf{e}_z$, must be a constant, which written in terms of the sectorial area velocity gives $\mathbf{L}_0 = 2m\dot{A}\,\mathbf{e}_z = \text{constant}$. Thus, the constraint where we force the sectorial area velocity to be a constant can also be interpreted as a statement of the conservation of angular momentum of the particle during its orbital motion.

You might have noticed that one of the constraints, the sectorial area velocity constraint, was a constraint involving velocities as well as coordinates as a special case of the form

$$f\left(r, \theta, \dot{r}, \dot{\theta}, t\right) = 0 \tag{2.5.53}$$

(in this case $f = r^2\dot{\theta}/2 - C$). We discussed such constraints briefly before in terms of Cartesian coordinates. This sectorial area velocity constraint cannot be integrated directly to obtain a constraint between the coordinates themselves and time in the form

$$g(r, \theta, t) = 0 \tag{2.5.54}$$

Non-integrable constraints involving both the coordinates and their time derivatives (velocities) in the form seen in Eq. (2.5.53) are called *non-holonomic constraints* while constraints which either directly or indirectly imply a relationship between the coordinates themselves in the form seen in Eq. (2.5.54) are called *holonomic* constraints. However, recall we mentioned that not all velocity constraints in the form of Eq. (2.5.53) are non-holonomic since when we differentiate a holonomic constraint such as Eq. (2.5.54) we get

$$f(r,\theta,\dot{r},\dot{\theta},t) = \frac{\partial g}{\partial r}\dot{r} + \frac{\partial g}{\partial \theta}\dot{\theta} + \frac{\partial g}{\partial t} = 0 \tag{2.5.55}$$

which is in the form of Eq. (2.5.53) but it can be integrated back to the form seen in Eq. (2.5.54) so we would label it a holonomic constraint.

How do we determine if a constraint involving velocities is integrable (holonomic) or not integrable (non-holonomic)? One way for a constraint of the form of Eq. (2.5.53) to be integrable is if it is expressible in terms of a perfect differential. However, this is not necessary and the more general necessary and sufficient conditions for integrability are called the Frobenius conditions. We will not state such general conditions here since when we write our constraints in the form $\mathbf{A}\ddot{\mathbf{x}} = \mathbf{b}$, regardless of whether the underlying constraint is holonomic or non-holonomic, we can still solve for the ideal constraint force as our planetary motion example showed. Thus, we can treat holonomic and non-holonomic constraints in the same fashion (if you are interested in the details of the requirements for identifying non-holonomic constraints, see [7], for example).

There is however another reason to want to know if a constraint is holonomic or non-holonomic. Since a holonomic constraint places a condition on the coordinates themselves, each holonomic constraint reduces the number of degrees of freedom of a problem by one (if all the constraints are independent, as we have been assuming; we will discuss the issue of nonindependent constraints later), allowing us to solve a dynamics problem with fewer independent coordinates. However, a non-holonomic constraint between the coordinates and their derivatives does not affect the degrees of freedom. To see this, go back to our example of this section. Requiring the sectorial area function velocity to be a constant does not constrain that particle to follow a specific type of path where r and θ are related; it simply forces the angular momentum of the particle along the path the particle takes, as defined by the solution of the equations of motion and initial conditions, to be conserved. If we specify a holonomic constraint such as specifying the orbit to be an ellipse, we do reduce the degrees of freedom by one as the particle now is constrained to follow a path where r and θ are related to one another. This is why it is possible to place two constraints on a particle moving in two dimensions and still have a dynamics problem where the particle can move, as long as at least one of the constraints is non-holonomic. Specifying the sectorial area velocity to be a constant and the path of the particle to be an ellipse is an example of just such a situation.

2.6 Generalized Coordinates and the Equations of Motion: A Geometric Approach

2.6.1 Embedding of Constraints

The previous section used polar coordinates as an example of describing dynamics problems in coordinates other than Cartesian coordinates. In this section we want to discuss the use of other "generalized" coordinates in more depth. Recall that we defined system coordinates to be the coordinates needed to completely describe unconstrained motion of the system we are considering. When using generalized coordinates we will label such coordinates as *system generalized coordinates*. For a single particle moving in three dimensions, for example, we need three system coordinates while the same particle moving in two dimensions requires only two system coordinates. Later, when we talk about a collection of N particles moving in two or three dimensions, we will have $2N$ or $3N$ system coordinates, respectively, to completely describe their unconstrained dynamics. Likewise, for N rigid bodies moving in two or three dimensions we will show that we will need $3N$ or $6N$ system coordinates, respectively. Of course, in most problems there will be constraints present so that not all of the system coordinates will be independent. Recall that if we let n_s be the number of system coordinates and n_h be the number of independent holonomic constraints present, then the number of degrees of freedom of the system, n_{dof}, is given by $n_{\text{dof}} = n_s - n_h$, where the number of degrees of freedom represents the smallest number of independent coordinates needed to describe the constrained motion. Ideally we would like to express our equations of motion in terms of only the independent coordinates so that we have to solve only n_{dof} equations. In simple problems such as the case of the bead moving on the straight wire considered in Sect. 2.3.1 we saw that we could choose as the independent coordinate an x'-coordinate along the wire, thus eliminating the constraint and yielding the necessary equation of motion (see Eq. (2.3.10)). Here we will use a geometric approach to describe how constraints can be automatically eliminated in the same fashion by the choice of generalized coordinates in more complex problems. To be specific, let us consider a particle moving in three dimensions under the action of one or more holonomic constraints. In Cartesian coordinates we can of course write Newton's second law as

$$m\ddot{\mathbf{x}} = \mathbf{F}^e + \mathbf{F}^c \tag{2.6.1}$$

where $\ddot{\mathbf{x}}(x_1, x_2, x_3)$ is the acceleration, which is a function of the three system coordinates (x_1, x_2, x_3), and $(\mathbf{F}^e, \mathbf{F}^c)$ are the sum of all the external forces and constraint forces, respectively. The constraint forces are assumed to be generated by holonomic constraints of the form

$$f_m(\mathbf{x}, t) = 0 \tag{2.6.2}$$

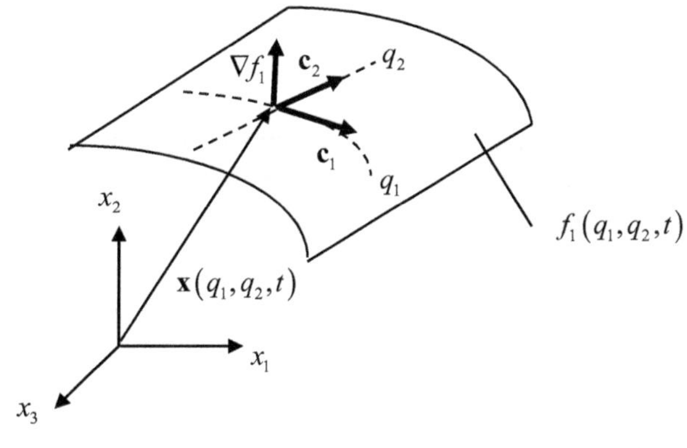

Fig. 2.37 A moving constraint surface at a fixed instant of time

($m = 1$ or 2) and for ideal constraint forces we assumed that those forces could be written as

$$\mathbf{F}_m^c = \lambda_m \nabla f_m \tag{2.6.3}$$

Consider for the moment one constraint ($m = 1$). We saw previously that we can interpret both the constraint equation and the constraint force geometrically. The constraint $f_1 = 0$ is the equation of a moving surface in three dimensions and the force $\mathbf{F}_1^c$ is proportional to the gradient of that surface so it is parallel to the normal to the surface (at a fixed instance of time) on which the particle must move (see Fig. 2.37). Since there is one constraint there are two degrees of freedom for the particle so that the motion of the particle can be described by two independent generalized coordinates ($q_1(t)$,$q_2(t)$) and the time, t. Any point on the surface can be described as

$$\mathbf{x} = \mathbf{x}(q_1, q_2, t) \tag{2.6.4}$$

The family of curves $q_1 =$ constant and $q_2 =$ constant define the surface parametrically and the two vectors

$$\mathbf{c}_1 = \frac{\partial \mathbf{x}}{\partial q_1}, \quad \mathbf{c}_2 = \frac{\partial \mathbf{x}}{\partial q_2} \tag{2.6.5}$$

are tangent to the coordinate curves and normal to ∇f_1 (Fig. 2.37). To give a more concrete example, suppose a particle is constrained to move on the time-dependent ellipsoidal surface (see Fig. 2.38) represented by the constraint

$$f_1(\mathbf{x}, t) = \frac{x_1^2}{a_1^2(t)} + \frac{x_2^2}{a_2^2(t)} + \frac{x_3^2}{a_3^2(t)} = 1 \tag{2.6.6}$$

Then the gradient of f_1, ∇f_1, is given by

$$\nabla f_1 = \frac{2x_1}{a_1(t)^2}\mathbf{e}_1 + \frac{2x_2}{a_2(t)^2}\mathbf{e}_2 + \frac{2x_3}{a_3(t)^2}\mathbf{e}_3 \tag{2.6.7}$$

If we write the Cartesian components of the position vector, $\mathbf{x}$, in terms of two generalized coordinates (q_1, q_2) through

$$\begin{aligned} x_1 &= a_1(t)\cos q_1 \sin q_2 \\ x_2 &= a_2(t)\sin q_1 \sin q_2 \\ x_3 &= a_3(t)\cos q_2 \end{aligned} \tag{2.6.8}$$

the constraint equation, Eq. (2.6.6), will be automatically satisfied so that the position vector will lie on the time-dependent ellipsoid. The vectors $(\mathbf{c}_1, \mathbf{c}_2)$ are then given by

$$\begin{aligned} \mathbf{c}_1 &= \frac{\partial \mathbf{x}}{\partial q_1} = -a_1(t)\sin q_1 \sin q_2 \mathbf{e}_1 + a_2(t)\cos q_1 \sin q_2 \mathbf{e}_2 \\ \mathbf{c}_2 &= \frac{\partial \mathbf{x}}{\partial q_2} = a_1(t)\cos q_1 \cos q_2 \mathbf{e}_1 + a_2(t)\sin q_1 \cos q_2 \mathbf{e}_2 - a_3(t)\sin q_2 \mathbf{e}_3 \end{aligned} \tag{2.6.9}$$

It is easy to verify that these vectors are indeed orthogonal to the gradient vector in this example, as illustrated in Fig. 2.38b. Note that $(\mathbf{c}_1, \mathbf{c}_2)$ are not unit vectors and they are not orthogonal to each other in general.

The $(\mathbf{c}_1, \mathbf{c}_2)$ vectors play a key role in describing the n_{dof} independent equations of motion (here $n_{\mathrm{dof}} = 2$) since if we take the dot product of these vectors with the equations of motion we have

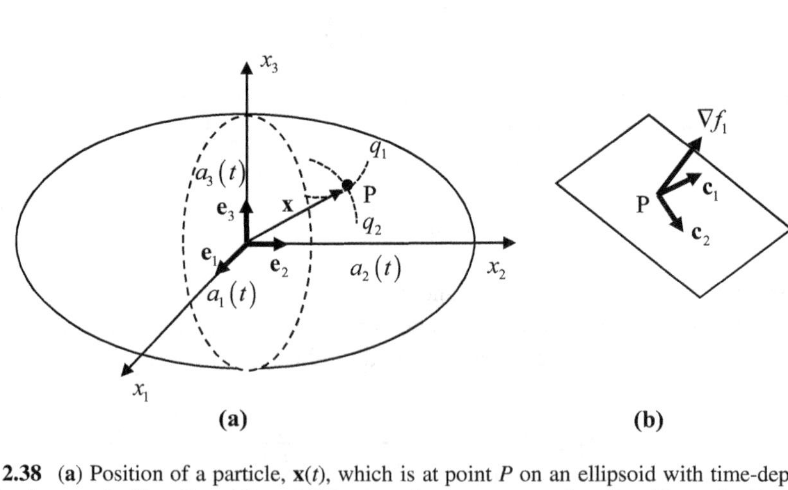

Fig. 2.38 (**a**) Position of a particle, $\mathbf{x}(t)$, which is at point P on an ellipsoid with time-dependent semimajor axes $(a_1(t), a_2(t), a_3(t))$, and (**b**) a plane that is tangent to the ellipsoid at point P, showing the vectors $(\mathbf{c}_1, \mathbf{c}_2)$ that lie in this tangent plane and the gradient vector, ∇f_1, which is normal to the tangent plane

$$
\begin{aligned}
m\ddot{\mathbf{x}} \cdot \mathbf{c}_n &= \mathbf{F}^e \cdot \mathbf{c}_n + \mathbf{F}^c \cdot \mathbf{c}_n \\
&= \mathbf{F}^e \cdot \mathbf{c}_n + \lambda_1 \nabla f_1 \cdot \mathbf{c}_n \quad (n = 1, 2) \\
&= \mathbf{F}^e \cdot \mathbf{c}_n
\end{aligned} \tag{2.6.10}
$$

where we have used the fact that the gradients of the constraint are orthogonal to both $\mathbf{c}_n$ tangent vectors. To obtain the two equations of motion in terms of the (q_1, q_2) coordinates we can differentiate Eq. (2.6.4) twice to get the acceleration:

$$
\begin{aligned}
\mathbf{x} &= \mathbf{x}(q_1, q_2, t) \\
\mathbf{v} = \dot{\mathbf{x}} &= \sum_{m=1}^{2} \frac{\partial \mathbf{x}}{\partial q_m} \dot{q}_m + \frac{\partial \mathbf{x}}{\partial t} = \sum_{m=1}^{2} \mathbf{c}_m \dot{q}_m + \frac{\partial \mathbf{x}}{\partial t} \\
\mathbf{a} = \ddot{\mathbf{x}} &= \sum_{m=1}^{2} \mathbf{c}_m \ddot{q}_m + \sum_{m=1}^{2} \sum_{n=1}^{2} \frac{\partial \mathbf{c}_m}{\partial q_n} \dot{q}_m \dot{q}_n + 2 \sum_{m=1}^{2} \frac{\partial \mathbf{c}_m}{\partial t} \dot{q}_m + \frac{\partial^2 \mathbf{x}}{\partial t^2}
\end{aligned} \tag{2.6.11}
$$

so that

$$
\begin{aligned}
m\ddot{\mathbf{x}} \cdot \mathbf{c}_k = {} & m \sum_{m=1}^{2} [\mathbf{c}_k \cdot \mathbf{c}_m] \ddot{q}_m + m \sum_{m=1}^{2} \sum_{n=1}^{2} \left[\mathbf{c}_k \cdot \frac{\partial \mathbf{c}_m}{\partial q_n} \right] \dot{q}_m \dot{q}_n \\
& + 2m \sum_{m=1}^{2} \left[\mathbf{c}_k \cdot \frac{\partial \mathbf{c}_m}{\partial t} \right] \dot{q}_m + m\mathbf{c}_k \cdot \frac{\partial^2 \mathbf{x}}{\partial t^2} = \sum_{m=1}^{2} M_{km} \ddot{q}_m \\
& + \sum_{m=1}^{2} \sum_{n=1}^{2} K_{kmn} \dot{q}_m \dot{q}_n + \sum_{m=1}^{2} J_{km} \dot{q}_m + B_k
\end{aligned} \tag{2.6.12}
$$

where

$$
\begin{aligned}
M_{km} &= m\mathbf{c}_k \cdot \mathbf{c}_m, \quad K_{kmn} = m\mathbf{c}_k \cdot \frac{\partial \mathbf{c}_m}{\partial q_n} \\
J_{km} &= 2m\mathbf{c}_k \cdot \frac{\partial \mathbf{c}_m}{\partial t}, \quad B_k = m\mathbf{c}_k \cdot \frac{\partial^2 \mathbf{x}}{\partial t^2}
\end{aligned} \tag{2.6.13}
$$

The equations of motion are then

$$
\sum_{m=1}^{2} M_{nm} \ddot{q}_m + \sum_{m=1}^{2} \sum_{k=1}^{2} K_{nmk} \dot{q}_m \dot{q}_k + \sum_{m=1}^{2} J_{nm} \dot{q}_m + B_n = \mathbf{F}^e \cdot \mathbf{c}_n \quad (n = 1, 2) \tag{2.6.14}
$$

Since we can write $\mathbf{x}$ in terms of its Cartesian components, x_m, as

$$\mathbf{x} = \sum_{m=1}^{3} x_m(q_1, q_2, t)\mathbf{e}_m \tag{2.6.15}$$

we have

$$\mathbf{c}_n = \frac{\partial \mathbf{x}}{\partial q_n} = \sum_{m=1}^{3} \frac{\partial x_m}{\partial q_n}\mathbf{e}_m \tag{2.6.16}$$

and in terms of the Cartesian components of the external force the *generalized external force*, $Q_n^e = \mathbf{F}^e \cdot \mathbf{c}_n$, is given by

$$Q_n^e = \mathbf{F}^e \cdot \mathbf{c}_n = \sum_{p=1}^{3}\sum_{m=1}^{3} F_p^e \mathbf{e}_p \cdot \frac{\partial x_m}{\partial q_n}\mathbf{e}_m = \sum_{p=1}^{3} F_p^e \frac{\partial x_p}{\partial q_n} \tag{2.6.17}$$

Calling this a generalized "force" is somewhat misleading since the $\mathbf{c}_n$ vectors are not dimensionless and when the generalized coordinate is an angular coordinate the generalized force has the dimensions of a moment. When the external force can be written in terms of a potential $F_p^e = -\partial V/\partial x_p$ a similar relation holds for the generalized force since

$$Q_n^e = \mathbf{F}^e \cdot \mathbf{c}_n = \sum_{p=1}^{3} F_p^e \frac{\partial x_p}{\partial q_n} = -\sum_{p=1}^{3} \frac{\partial V}{\partial x_p}\frac{\partial x_p}{\partial q_n} = -\frac{\partial V}{\partial q_n} \tag{2.6.18}$$

Equation (2.6.14) has acceleration terms other than those involving the second derivatives on time of the generalized coordinates. These are analogous to the inertial force terms we saw in polar coordinates. Here, the generalized inertial forces, Q_n^{in}, are given by

$$Q_n^{in} = -\sum_{m=1}^{2}\sum_{k=1}^{2} K_{nmk}\dot{q}_m\dot{q}_k - \sum_{m=1}^{2} J_{nm}\dot{q}_m - B_n \tag{2.6.19}$$

Although we derived all of these results explicitly for a 3-D problem with one constraint, they are also applicable to a 2-D problem with one constraint by making all the terms involving (x_1, x_2, x_3) reduce to expressions of (x_1, x_2) only.

In the example of the particle moving on the time-dependent ellipsoidal surface, we did not have to use the constraint equation explicitly since the generalized coordinates were chosen so that the constraint was automatically satisfied. In that case we could say that the constraints were *implicitly embedded* into the equations of motion. However, we can also define a set of n_{dof} generalized coordinates as simply a subset of the system generalized coordinates and use the constraint equations and/or their derivatives to express the equations of motion only in terms of those coordinates. This is called *explicit embedding* of the constraints. For example, consider the use of Eq. (2.6.14) for the very simple problem of the bead on the smooth parabolic wire we considered previously. In that two-dimensional problem there was only one degree of freedom and one generalized coordinate needed to describe the motion so Eq. (2.6.14) reduces to the single equation

$$M_{11}\ddot{q}_1 + K_{111}\dot{q}_1\dot{q}_1 + J_{11}\dot{q}_1 + B_1 = -\frac{\partial V}{\partial q_1} \tag{2.6.20}$$

Here, if we take x as our one independent generalized coordinate q_1

$$\begin{aligned} f_1 &= y - ax^2 = 0 \\ \nabla f_1 &= -2ax\mathbf{e}_x + \mathbf{e}_y \\ \mathbf{x}(x) &= x\mathbf{e}_x + y(x)\mathbf{e}_y = x\mathbf{e}_x + ax^2\mathbf{e}_y \\ \mathbf{c}_1 &= \frac{\partial \mathbf{x}(x)}{\partial x} = \mathbf{e}_x + 2ax\mathbf{e}_y \end{aligned} \tag{2.6.21}$$

from which you can verify directly that $\nabla f_1 \cdot \mathbf{c}_1 = 0$ and we have

$$\begin{aligned} M_{11} &= m\mathbf{c}_1 \cdot \mathbf{c}_1 = m(1 + 4a^2x^2) \\ K_{111} &= m\mathbf{c}_1 \cdot \frac{\partial \mathbf{c}_1}{\partial x} = 4a^2xm \\ J_{11} &= 2m\mathbf{c}_1 \cdot \frac{\partial \mathbf{c}_1}{\partial t} = 0 \\ B_1 &= m\mathbf{c}_1 \cdot \frac{\partial^2 \mathbf{x}}{\partial t^2} = 0 \\ V(x) &= mgy(x) = mg\,ax^2 \end{aligned} \tag{2.6.22}$$

which yields a result identical to what we obtained earlier with Newton's second law (see Eq. (2.3.25)), namely

$$(1 + 4a^2x^2)\ddot{x} + 4a^2x\dot{x}^2 + 2gax = 0 \tag{2.6.23}$$

Notice that we had to use the constraint equation to obtain both $\mathbf{x}(x)$ and $V(x)$ in terms of x only since both of these quantities are inherently functions of the system generalized coordinates (x, y). This process is where we explicitly embedded the constraints to yield a set of n_{dof} equations of motion in terms of only n_{dof} independent coordinates. This process was easy and could be done by inspection in this simple problem. In Sect. 2.11 we will examine the processs of explicit embedding from a more general standpoint.

In the explicit embedding approach we can reduce the complexity of these calculations somewhat by noting that in most problems it is relatively easy to write the velocity of the particle in terms of the independent generalized coordinates and their derivatives. In the case of a particle moving in three dimensions with one constraint $\mathbf{v} = \mathbf{v}(q_1, q_2, \dot{q}_1, \dot{q}_2, t)$. Then we have the acceleration given by

$$\mathbf{a} = \frac{d\mathbf{v}}{dt} = \sum_{m=1}^{2} \frac{\partial \mathbf{v}}{\partial \dot{q}_m} \ddot{q}_m + \sum_{m=1}^{2} \frac{\partial \mathbf{v}}{\partial q_m} \dot{q}_m + \frac{\partial \mathbf{v}}{\partial t} \tag{2.6.24}$$

But the vector $\mathbf{c}_n$ can also be calculated directly from the velocity rather than the position vector $\mathbf{x}$ because recall

$$\mathbf{v} = \dot{\mathbf{x}} = \sum_{m=1}^{2} \frac{\partial \mathbf{x}}{\partial q_m} \dot{q}_m + \frac{\partial \mathbf{x}}{\partial t} = \sum_{m=1}^{2} \mathbf{c}_m \dot{q}_m + \frac{\partial \mathbf{x}}{\partial t} \tag{2.6.25}$$

so that $\mathbf{c}_n = \partial \mathbf{v} / \partial \dot{q}_n$. Thus

$$m\mathbf{a} \cdot \mathbf{c}_n = m \sum_{m=1}^{2} [\mathbf{c}_n \cdot \mathbf{c}_m]\ \ddot{q}_m + m \sum_{m=1}^{2} \left[\mathbf{c}_n \cdot \frac{\partial \mathbf{v}}{\partial q_m}\right] \dot{q}_m + m \left[\mathbf{c}_n \cdot \frac{\partial \mathbf{v}}{\partial t}\right] \tag{2.6.26}$$

which can be written as

$$m\mathbf{a} \cdot \mathbf{c}_n = \sum_{m=1}^{2} M_{nm}\ \ddot{q}_m + \sum_{m=1}^{2} D_{nm} \dot{q}_m + E_n \tag{2.6.27}$$

where

$$\begin{aligned} M_{nm} &= m\mathbf{c}_n \cdot \mathbf{c}_m \\ D_{nm} &= m\mathbf{c}_n \cdot \frac{\partial \mathbf{v}}{\partial q_m} \\ E_n &= m\mathbf{c}_n \cdot \frac{\partial \mathbf{v}}{\partial t} \end{aligned} \tag{2.6.28}$$

giving the equations of motion in the alternate form

$$\sum_{m=1}^{2} M_{nm} \ddot{q}_m + \sum_{m=1}^{2} D_{nm} \dot{q}_m + E_n = Q_n^e \tag{2.6.29a}$$

which unlike Eq. (2.6.14) can also be written in standard matrix-vector form as

$$[M]\{\ddot{q}\} + [D]\{\dot{q}\} + \{E\} = \{Q^e\} \tag{2.6.29b}$$

The generalized inertia force can be expressed in component form as

$$Q_n^{in} = -\sum_{m=1}^{2} D_{nm}\dot{q}_m - E_n \qquad (2.6.30a)$$

or in matrix-vector form as

$$\{Q^{in}\} = -[D]\{\dot{q}\} - \{E\} \qquad (2.6.30b)$$

If we go back to our bead sliding on the smooth parabolic wire problem with the independent coordinate $q_1 = x$ again we have the one equation of motion

$$M_{11}\ddot{q}_1 + D_{11}\dot{q}_1 + E_1 = -\frac{\partial V}{\partial q_1} \qquad (2.6.31)$$

with

$$\begin{aligned}
&\mathbf{v} = \dot{\mathbf{x}} = \dot{x}\,\mathbf{e}_x + \dot{y}\left(x,\dot{x}\right)\mathbf{e}_y = \dot{x}\,\mathbf{e}_x + 2ax\dot{x}\,\mathbf{e}_y \\
&\frac{\partial \mathbf{v}}{\partial x} = 2a\dot{x}\,\mathbf{e}_y \\
&\mathbf{c}_1 = \frac{\partial \mathbf{v}}{\partial \dot{x}} = \mathbf{e}_x + 2ax\mathbf{e}_y \\
&M_{11} = m\mathbf{c}_1 \cdot \mathbf{c}_1 = m(1 + 4a^2x^2) \\
&D_{11} = m\mathbf{c}_1 \cdot \frac{\partial \mathbf{v}}{\partial x} = 4ma^2x\dot{x} \\
&E_1 = m\mathbf{c}_1 \cdot \frac{\partial \mathbf{v}}{\partial t} = 0 \\
&Q_1^e = -\frac{\partial V}{\partial x} = -2mgax
\end{aligned} \qquad (2.6.32)$$

which again yields the equation of motion, Eq. (2.6.23). Notice that in this case we had to use the derivative of the constraint equation to get the value of $\dot{y}\left(x,\dot{x}\right)$. Once we have solved the equation of motion for $x(t)$ we will also have to use the constraint equation itself to get $y(t)$. Thus, as we will see in the general discussion of explicit embedding in Sect. 2.11 both the constraint equation and its derivative will normally be needed in the explicit embedding process.

Now, consider a 3-D problem where there are two holonomic constraints defined by the functions (f_1, f_2) (see Eq. (2.6.2)). There is then only one degree of freedom so that we can write the position vector of the particle in terms of one independent generalized coordinate, q_1:

$$\mathbf{x} = \mathbf{x}(q_1, t) \qquad (2.6.33)$$

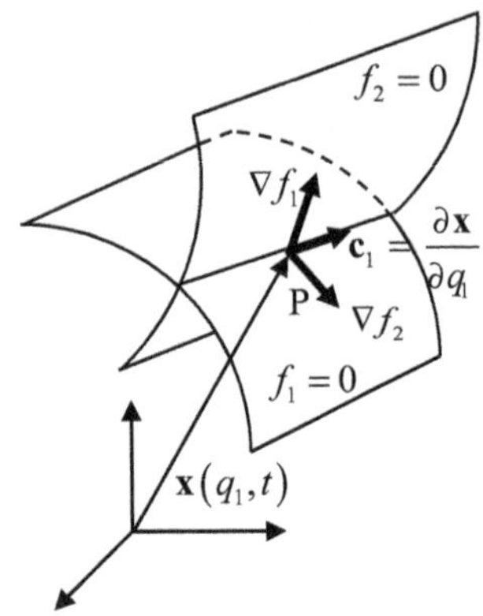

Fig. 2.39 Two constraint surfaces at a fixed time and the vector $\mathbf{c}_1$ along the path of intersection of those surfaces

The vector $\mathbf{c}_1 = \partial \mathbf{x}/\partial q_1$ is along a line defining the intersection of the two surfaces (Fig. 2.39) at a fixed time and is perpendicular to both the gradient vectors (and therefore is perpendicular to the two ideal constraint forces as well). Thus all of our previous results for defining the equations of motion hold true when multiple holonomic constraints are present.

Since our discussion here was for a very simple problem, we will illustrate explicit embedding of holonomic constraints in a more general context later in Sect. 2.11. When we embed the constraints we eliminate the constraint forces from the equations of motion. This is nice from the standpoint of generating a more compact set of equations of motion but we also lose a direct evaluation of the constraint forces in the process. Since engineers are often interested in the constraint forces as well as a description of the motion, when using an explicit embedding approach one must find the constraint forces separately. Embedding in the manner in which we have just described can only be done with holonomic constraints since if we have, say, n_{nh} non-holonomic constraints we can use those non-holonomic constraints on the velocities to reduce the number of independent velocities (and accelerations) by n_{nh} but we cannot do the same for the generalized coordinates themselves since we cannot integrate those non-holonomic constraints to find n_{nh} comparable constraints on the generalized coordinates. We can, however, still embed non-holonomic constraints and eliminate the constraint forces. The details of that type of embedding, however, are left to Sect. 2.11.

2.6.2 Augmented Approach with Generalized Coordinates

In an embedding approach with n_s system generalized coordinates and n_h holonomic constraints we generate only $n_{\text{dof}} = n_s - n_h$ equations of motion in terms of a set of n_{dof} independent coordinates and in the process we eliminate the ideal constraint forces. In Sect. 2.4.3 we saw that for a system described by n_s system coordinates where there were n_h holonomic constraints and a similar number of ideal constraint forces present, instead of embedding those constraints we could use the constraints to solve explicitly for those n_h constraint forces explicitly and so obtain n_s equations of motion for the n_s system coordinates. In Cartesian coordinates, for example, we found the n_s equations of motion as

$$\ddot{\mathbf{x}} = \mathbf{M}^{-1}\mathbf{F}^e + \mathbf{M}^{-1}\mathbf{A}^T\left(\mathbf{A}\mathbf{M}^{-1}\mathbf{A}^T\right)^{-1}\left(\mathbf{b} - \mathbf{A}\mathbf{M}^{-1}\mathbf{F}^e\right) \tag{2.6.34}$$

and the n_h constraint forces were

$$\mathbf{F}^c = \mathbf{A}^T\left(\mathbf{A}\mathbf{M}^{-1}\mathbf{A}^T\right)^{-1}\left(\mathbf{b} - \mathbf{A}\mathbf{M}^{-1}\mathbf{F}^e\right) \tag{2.6.35}$$

Obtaining the equations of motion via Eq. (2.6.34) is, as mentioned previously, called an *augmented approach* since Eqs. (2.6.34) and (2.6.35) give the result of augmenting the n_s equations of motion with the n_h twice-differentiated constraint equations and then solving that augmented set of (n_s+n_h) equations to obtain n_s equations of motion as well as the n_h ideal constraint forces. This augmented approach can be performed in the same manner for non-holonomic constraints where the non-holonomic constraints only need to be differentiated once so that the constraint equations written in terms of accelerations can be solved in conjunction with the differential equations of Newton's second law. The augmented method obviously leads to a larger number of equations of motion than with an embedding approach.

We extended this augmented approach to polar coordinates in Sect. 2.5.1. However, our geometric approach in this section for generalized coordinates differs from that obtained previously for polar coordinates. To see how these differences arise, recall that in this section we have written the equations of motion by taking the dot product of the vectors $\mathbf{c}_k = \partial\mathbf{r}/\partial q_k = \partial\mathbf{v}/\partial\dot{q}_k$ with Newton's second law, i.e.,

$$\left(\mathbf{M}\ddot{\mathbf{x}} - \mathbf{F}^e - \mathbf{F}^c\right)\cdot\mathbf{c}_k = 0 \tag{2.6.36}$$

If the generalized coordinates chosen are independent coordinates that satisfy the constraints we saw that the constraint force in Eq. (2.6.36) was eliminated. However, if in the coordinate transformation $\mathbf{x} = \mathbf{x}(\mathbf{q}, t)$ we simply change from a set of system Cartesian coordinates to a set of system generalized coordinates, we can still apply Eq. (2.6.36) but now the constraint forces will remain in the equations of motion and we find

$$\mathbf{M}\ddot{\mathbf{q}} = \bar{\mathbf{Q}}^e + \mathbf{Q}^c \tag{2.6.37}$$

where (using the velocity vector to define these terms—see Eq. (2.6.30b))

$$\begin{aligned}[\mathbf{M}]_{kj} &= [m\mathbf{c}_k \cdot \mathbf{c}_j] \\ \bar{\mathbf{Q}}^e &= \mathbf{Q}^{in} + \mathbf{Q}^e = -\mathbf{D}\dot{\mathbf{q}} - \mathbf{E} + \mathbf{Q}^e\end{aligned} \tag{2.6.38}$$

Here, $\mathbf{M}$ is the generalized mass matrix in these generalized coordinates, $\mathbf{Q}^{in} = -\mathbf{D}\dot{\mathbf{q}} - \mathbf{E}$ is the generalized inertial force, and $Q_k^e = \mathbf{F}^e \cdot \mathbf{c}_k, Q_k^c = \mathbf{F}^c \cdot \mathbf{c}_k$ are components of the generalized external force, $\mathbf{Q}^e$, and constraint force, $\mathbf{Q}^c$, respectively. Thus, the total effective external force, $\bar{\mathbf{Q}}^e$, is the sum of the inertial and external forces.

When working in polar coordinates in Sect. 2.5.1 we let $\mathbf{x} = \mathbf{x}(\mathbf{q})$ where $\mathbf{q} = (r, \theta)$. But instead of taking the dot products as seen in Eq. (2.6.36) in Sect. 2.5.1 we took the dot product of Newton's second law with unit vectors along the r and θ coordinate directions, i.e.,

$$\left(\mathbf{M}\ddot{\mathbf{x}} - \mathbf{F}^e - \mathbf{F}^c\right) \cdot \mathbf{u}_k = 0 \tag{2.6.39}$$

where $\mathbf{u}_1 = \mathbf{e}_r$, $\mathbf{u}_2 = \mathbf{e}_\theta$ are unit vectors in the r- and the θ-directions, respectively. In contrast, $\mathbf{c}_1 = \partial\mathbf{x}/\partial r = \mathbf{e}_r, \mathbf{c}_2 = \partial\mathbf{x}/\partial\theta = r\mathbf{e}_\theta$ so there will be some differences in the generalized mass and inertia matrices and in some of the components of all the generalized forces depending on whether we choose to generate the equations of motion with $\mathbf{c}_k$ or $\mathbf{u}_k$. We will summarize those differences here:

Equations of motion using $(\mathbf{e}_r, \mathbf{e}_\theta)$:

$$\begin{bmatrix} m & 0 \\ 0 & mr \end{bmatrix}\begin{Bmatrix} \ddot{r} \\ \ddot{\theta} \end{Bmatrix} + \begin{bmatrix} 0 & mr\dot{\theta} \\ m\dot{\theta} & m\dot{r} \end{bmatrix}\begin{Bmatrix} \dot{r} \\ \dot{\theta} \end{Bmatrix} = \begin{Bmatrix} F_r^e \\ F_\theta^e \end{Bmatrix} + \begin{Bmatrix} F_r^c \\ F_\theta^c \end{Bmatrix} \tag{2.6.40a}$$

Equations of motion using $(\mathbf{c}_1, \mathbf{c}_2)$:

$$\begin{aligned}\begin{bmatrix} m & 0 \\ 0 & mr^2 \end{bmatrix}\begin{Bmatrix} \ddot{r} \\ \ddot{\theta} \end{Bmatrix} + \begin{bmatrix} 0 & mr\dot{\theta} \\ mr\dot{\theta} & mr\dot{r} \end{bmatrix}\begin{Bmatrix} \dot{r} \\ \dot{\theta} \end{Bmatrix} &= \begin{Bmatrix} Q_r^e \\ Q_\theta^e \end{Bmatrix} + \begin{Bmatrix} Q_r^c \\ Q_\theta^c \end{Bmatrix} \\ &= \begin{Bmatrix} F_r^e \\ rF_\theta^e \end{Bmatrix} + \begin{Bmatrix} F_r^c \\ rF_\theta^c \end{Bmatrix}\end{aligned} \tag{2.6.40b}$$

Ultimately, however, these reduce to the same equations. Note that in polar coordinates (using $(\mathbf{e}_r, \mathbf{e}_\theta)$) the inertial, external, and constraint forces appearing in the equations of motion are simply those forces in the (r, θ) directions while when using $(\mathbf{c}_1, \mathbf{c}_2)$, the equation of motion in the θ-direction is actually a moment equation and the generalized "force" in that direction is likewise a moment. In matrix-vector form we can acknowledge the differences in the two cases by writing them in the two distinct forms:

$$\begin{aligned}\tilde{\mathbf{M}}\ddot{\mathbf{q}} &= \mathbf{F}^{in} + \mathbf{F}^e + \mathbf{F}^c = \tilde{\mathbf{F}}^e + \mathbf{F}^c \\ \mathbf{M}\ddot{\mathbf{q}} &= \mathbf{Q}^{in} + \mathbf{Q}^e + \mathbf{Q}^c = \bar{\mathbf{Q}}^e + \mathbf{Q}^c\end{aligned} \tag{2.6.41}$$

In either case, for constrained motion the constraint equations are written in the same form:

$$\mathbf{A}\ddot{\mathbf{q}} = \mathbf{b} \tag{2.6.42}$$

so that when we solve explicitly for the constraint force and equations of motion, as done previously, we would have

$$\begin{aligned} \mathbf{F}^c &= \mathbf{A}^T\left(\mathbf{A}\tilde{\mathbf{M}}^{-1}\mathbf{A}^T\right)^{-1}\left(\mathbf{b} - \mathbf{A}\tilde{\mathbf{M}}^{-1}\tilde{\mathbf{F}}^e\right) \\ \ddot{\mathbf{q}} &= \tilde{\mathbf{M}}^{-1}\tilde{\mathbf{F}}^e + \tilde{\mathbf{M}}^{-1}\mathbf{A}^T\left(\mathbf{A}\tilde{\mathbf{M}}^{-1}\mathbf{A}^T\right)^{-1}\left(\mathbf{b} - \mathbf{A}\tilde{\mathbf{M}}^{-1}\tilde{\mathbf{F}}^e\right) \end{aligned} \tag{2.6.43}$$

when using the polar coordinate ($\mathbf{e}_r$, $\mathbf{e}_\theta$) vectors and

$$\begin{aligned} \mathbf{Q}^c &= \mathbf{A}^T\left(\mathbf{A}\mathbf{M}^{-1}\mathbf{A}^T\right)^{-1}\left(\mathbf{b} - \mathbf{A}\mathbf{M}^{-1}\bar{\mathbf{Q}}^e\right) \\ \ddot{\mathbf{q}} &= \mathbf{M}^{-1}\bar{\mathbf{Q}}^e + \mathbf{M}^{-1}\mathbf{A}^T\left(\mathbf{A}\mathbf{M}^{-1}\mathbf{A}^T\right)^{-1}\left(\mathbf{b} - \mathbf{A}\mathbf{M}^{-1}\bar{\mathbf{Q}}^e\right) \end{aligned} \tag{2.6.44}$$

using the $(\mathbf{c}_1, \mathbf{c}_2)$ vectors. Using the $(\mathbf{c}_1, \mathbf{c}_2)$ vectors when forming the equations of motion in generalized coordinates is more general, allowing us to use any types of generalized coordinates we wish. *Thus, Eq. (2.6.44) is the augmented approach in its most general form.* As a consequence, from now on we will only employ the ($\mathbf{c}_1$, $\mathbf{c}_2$) vectors. It is important to recall, as mentioned previously, that the $\mathbf{c}_k$ vectors are not necessarily unit vectors nor are they orthogonal in general.

2.7 Lagrange's Equations

We have seen how we can generate the equations of motion for a particle moving in 3-D with one or two holonomic constraints directly from components of Newton's laws in the form

$$m\ddot{\mathbf{x}} \cdot \mathbf{c}_n = Q_n^e \tag{2.7.1}$$

We wrote either the position vector or the velocity vector of a particle in terms of the independent generalized coordinates and differentiated those vectors to get the acceleration. There is, however, another way to compute the left side of Eq. (2.7.1) that is the basis of a different approach to dynamics called *analytical mechanics*. First, we rewrite the left side of Eq. (2.7.1) as

$$m\ddot{\mathbf{x}} \cdot \mathbf{c}_n = m\frac{d\mathbf{v}}{dt} \cdot \frac{\partial \mathbf{v}}{\partial \dot{q}_n} = m\frac{d}{dt}\left(\mathbf{v} \cdot \frac{\partial \mathbf{v}}{\partial \dot{q}_n}\right) - m\mathbf{v} \cdot \frac{d}{dt}\left(\frac{\partial \mathbf{v}}{\partial \dot{q}_n}\right) \tag{2.7.2}$$

But if we use the definition of the kinetic energy as $T(\mathbf{q}, \dot{\mathbf{q}}, t) = m\mathbf{v} \cdot \mathbf{v}/2$ and use the relationships

$$\frac{\partial \mathbf{v}}{\partial \dot{q}_n} = \frac{\partial \mathbf{x}}{\partial q_n} \tag{2.7.3}$$

we have

$$\begin{aligned} m\ddot{\mathbf{x}} \cdot \mathbf{c}_n &= \frac{d}{dt}\left(\frac{\partial T}{\partial \dot{q}_n}\right) - m\mathbf{v} \cdot \frac{d}{dt}\left(\frac{\partial \mathbf{x}}{\partial q_n}\right) \\ &= \frac{d}{dt}\left(\frac{\partial T}{\partial \dot{q}_n}\right) - m\mathbf{v} \cdot \frac{\partial}{\partial q_n}\left(\frac{d\mathbf{x}}{dt}\right) \\ &= \frac{d}{dt}\left(\frac{\partial T}{\partial \dot{q}_n}\right) - m\mathbf{v} \cdot \frac{\partial \mathbf{v}}{\partial q_n} \\ &= \frac{d}{dt}\left(\frac{\partial T}{\partial \dot{q}_n}\right) - \frac{\partial T}{\partial q_n} \end{aligned} \tag{2.7.4}$$

Thus, from Eq. (2.7.1) we have the equations of motion as

$$\frac{d}{dt}\left(\frac{\partial T}{\partial \dot{q}_n}\right) - \frac{\partial T}{\partial q_n} = Q_n^e \tag{2.7.5}$$

This form of the equations of motion is called *Lagrange's equations*. It shows the importance that the kinetic energy has in terms of generating the equations of motion. It is valid for both unconstrained motion of a particle in 3-D problems and constrained motion with holonomic constraints since in the unconstrained case there is no constraint force and we have three independent generalized coordinates (three degrees of freedom) and the kinetic energy is a function of all three of those coordinates and their derivatives, i.e., $T = T(q_1, q_2, q_3, \dot{q}_1, \dot{q}_2, \dot{q}_3, t)$. In the constrained case with two degrees of freedom if we can use the constraint equation and its derivative to explicitly embed the constraints in the kinetic energy, we can write the kinetic energy as $T = \tilde{T}(q_1, q_2, \dot{q}_1, \dot{q}_2, t)$ in terms of two independent generalized coordinates and their derivatives. In the one degree of freedom case if we can explicitly embed two constraint equations and their derivatives we have $T = \tilde{T}(q_1, \dot{q}_1, t)$ instead. In either constrained case ideal constraint forces do not appear in Eq. (2.7.5) because of their orthogonality to the vectors $\mathbf{c}_n$. Note that since the generalized forces may be a function of all the unconstrained system coordinates and their derivatives, in the presence of constraints those constraints must be embedded in the generalized force expressions as well so that those expressions are only functions of the independent coordinates present.

If the external forces are derivable from a potential energy and that potential energy, $\tilde{V}$, is written in terms of the independent coordinates then we can write Eq. (2.7.5) as

$$\frac{d}{dt}\left(\frac{\partial \tilde{L}}{\partial \dot{q}_n}\right) - \frac{\partial \tilde{L}}{\partial q_n} = 0 \tag{2.7.6}$$

where the Lagrangian, $\tilde{L}$, is defined as $\tilde{L} = \tilde{T} - \tilde{V}$.

This is a rather remarkable result since it shows that we can generate the equations of motion of the particle directly from our knowledge of the single scalar, $\tilde{L}$. Although it comes directly from Newton's second law, Eq. (2.7.6) does not involve forces, only scalar energy quantities like the kinetic energy and potential energy. The use of Lagrange's equations is called an *analytical mechanics* approach since just as *analytical geometry* uses calculus to deal with geometry problems, an analytical mechanics approach allows one to use calculus as the basic tool for treating dynamics problems. When we consider more complex systems of interacting particles and rigid bodies, we will see another difference between a Newton-Euler approach to dynamics and an analytical mechanics formulation. Applying Newton-Euler methods typically involves breaking a system into parts, drawing free body diagrams, and using a vector-matrix application of the Newton-Euler equations to those parts. In contrast, an analytical mechanics approach uses the kinetic and potential energy of the entire system and generates the equations of motion directly. Physicists find Lagrange's equations particularly attractive as a means of transitioning from classical mechanics to other areas such as quantum mechanics where energy concepts rather than force and moment concepts are more fundamental. Engineers also find Lagrange's equation useful in treating the fundamental concepts of dynamics and in developing applications, but because an engineering focus is often on evaluating the constraint forces and moments acting on a system in addition to obtaining and solving the equations of motion engineers often take a more eclectic view on the role and importance of Lagrange's equations, where the constraint forces, either by choice of coordinates or by explicit embedding, are usually absent.

To see a simple application of Lagrange's equations consider the bead sliding on the parabolic wire again. In that case our one equation of motion in terms of the one independent generalized coordinate, x, is

$$\frac{d}{dt}\left(\frac{\partial \tilde{L}}{\partial \dot{x}}\right) - \frac{\partial \tilde{L}}{\partial x} = 0 \tag{2.7.7}$$

where, with embedding of the constraint, we have

$$\begin{aligned}
\tilde{\mathbf{v}} &= \dot{x}\,\mathbf{e}_x + 2ax\dot{x}\,\mathbf{e}_y \\
\tilde{T} &= \frac{1}{2}m\mathbf{v}\cdot\mathbf{v} = \frac{m}{2}(1+4a^2x^2)\dot{x}^2 \\
\tilde{V} &= mgax^2 \\
\tilde{L} &= \tilde{T} - \tilde{V} = \frac{m}{2}(1+4a^2x^2)\dot{x}^2 - mgax^2
\end{aligned} \tag{2.7.8}$$

These give

$$\begin{aligned}
\frac{\partial \tilde{L}}{\partial x} &= 4ma^2x\dot{x}^2 - 2mgax \\
\frac{\partial \tilde{L}}{\partial \dot{x}} &= m(1+4a^2x^2)\dot{x} \\
\frac{d}{dt}\left(\frac{\partial \tilde{L}}{\partial \dot{x}}\right) &= m(1+4a^2x^2)\ddot{x} + 8ma^2x\dot{x}^2
\end{aligned} \tag{2.7.9}$$

so that Eq. (2.7.7) again gives the equation of motion, Eq. (2.6.23).

Since we derived Lagrange's equations from Newton's second law, one might think that they are simply different forms of the same law for the forces. This, however, is not quite true as you might expect from our discussion in the previous subsection. Previously we considered the case of unconstrained two-dimensional motion of a particle in terms of polar coordinates, i.e., $(q_1, q_2) = (r, \theta)$. Then

$$\begin{aligned}
\mathbf{v} &= \dot{r}\,\mathbf{e}_r + r\dot{\theta}\,\mathbf{e}_\theta \\
T &= \frac{m}{2}\mathbf{v}\cdot\mathbf{v} = \frac{m}{2}(\dot{r}^2 + r^2\dot{\theta}^2) \\
\mathbf{F}^e &= F_r\mathbf{e}_r + F_\theta\mathbf{e}_\theta \\
\mathbf{c}_1 &= \frac{\partial \mathbf{v}}{\partial \dot{r}} = \mathbf{e}_r, \quad \mathbf{c}_2 = \frac{\partial \mathbf{v}}{\partial \dot{\theta}} = r\mathbf{e}_\theta \\
Q_1^e &= \mathbf{F}^e\cdot\mathbf{c}_1 = F_r, \quad Q_2^e = \mathbf{F}^e\cdot\mathbf{c}_2 = rF_\theta
\end{aligned} \tag{2.7.10}$$

so that from Eq. (2.7.5) we find (see Eq. (2.6.29b))

$$\begin{aligned}
m(\ddot{r} - r\dot{\theta}^2) &= F_r \\
m(r^2\ddot{\theta} + 2r\dot{r}\dot{\theta}) &= rF_\theta
\end{aligned} \tag{2.7.11}$$

The first of these equations is just the component of Newton's second law in the r-direction, but the second equation is actually a moment equation, which in this case is also r times Newton's second law force equation in the θ-direction. Our discussion at the beginning of this chapter showed that the interchangeability of force and moment equations is just a consequence of the Newton-Euler equations. This interchange occurs automatically in a Lagrangian approach when dealing with generalized coordinates that are angular variables rather than positional variables.

Since the kinetic energy plays such a key role in Lagrange's equations we want to examine closer the structure of the terms in the kinetic energy expression. For the case of a particle moving in 3-D described by p independent generalized coordinates (p could in this case could be one or two depending on the number of constraints present) we have

$$\begin{aligned} T &= \frac{1}{2} m \sum_{i=1}^{3} \dot{x}_i^2 \\ x_i &= x_i(q_1, \ldots, q_p, t) \\ \dot{x}_i &= \sum_{j=1}^{p} \frac{\partial x_i}{\partial q_j} \dot{q}_j + \frac{\partial x_i}{\partial t} \end{aligned} \tag{2.7.12}$$

(for economy of notation we will stop placing a tilde over these functions when constraints are present and those constraints are embedded to generate functions of only the independent coordinates). Substituting the $\dot{x}_i$ in terms of $(q_j, \dot{q}_j)$ gives

$$T = T_2 + T_1 + T_0 \tag{2.7.13}$$

where

$$T_2 = \frac{1}{2} \sum_{i=1}^{p} \sum_{j=1}^{p} M_{ij}(\mathbf{q}, t) \dot{q}_i \dot{q}_j, \quad T_1 = \sum_{i=1}^{p} \mu_i(\mathbf{q}, t) \dot{q}_i, \quad T_0 = w_0(\mathbf{q}, t) \tag{2.7.14}$$

with

$$\begin{aligned} M_{ij} = M_{ji} &= m \sum_{k=1}^{3} \frac{\partial x_k}{\partial q_i} \frac{\partial x_k}{\partial q_j} = m \left[\frac{\partial x}{\partial q} \right]^T \left[\frac{\partial x}{\partial q} \right] \\ \mu_i &= m \sum_{k=1}^{3} \frac{\partial x_k}{\partial q_i} \frac{\partial x_k}{\partial t} = m \left[\frac{\partial x}{\partial q} \right]^T \left\{ \frac{\partial x}{\partial t} \right\} \\ w_0 &= \frac{m}{2} \sum_{k=1}^{3} \frac{\partial x_k}{\partial t} \frac{\partial x_k}{\partial t} = \frac{m}{2} \left\{ \frac{\partial x}{\partial t} \right\}^T \left\{ \frac{\partial x}{\partial t} \right\} \end{aligned} \tag{2.7.15}$$

The separation of the kinetic energy into the three parts identified in Eq. (2.7.13) and Eq. (2.7.14) identifies three separate types of dynamics effects. The first term, T_2, is quadratic in the generalized velocities and is typically present in all problems since it is analogous to the kinetic energy expression in Cartesian coordinates (see the kinetic energy expression in Eq. (2.7.12)) but where the mass m is replaced by the matrix $M_{ij}(\mathbf{q}, t)$. Systems where $T_1 = T_0 = 0$ are called *natural systems* and all other systems are called *nonnatural systems*. As we will see, nonnatural systems are often where one part of the motion is treated as known. This happens, for example, when

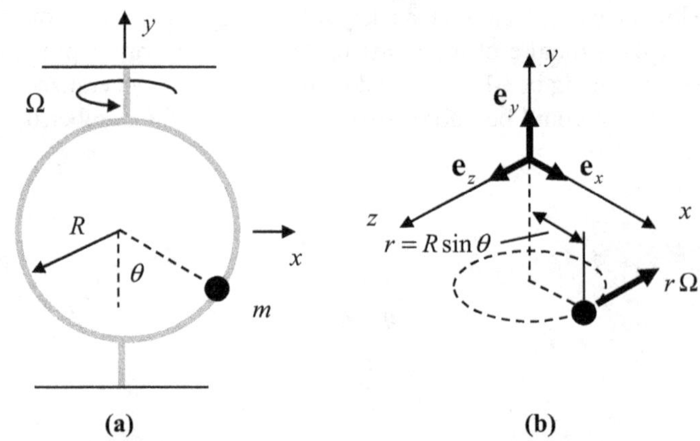

Fig. 2.40 (**a**) A particle sliding on a smooth hoop which is rotating at a constant angular velocity, Ω, about the *y*-axis, and (**b**) the velocity of the particle due to the rotation of the hoop for a set of (*x*, *y*, *z*) coordinates that rotate with the hoop

we are describing the system in terms of moving coordinates where the motion of the coordinates is specified. One example of a nonnatural system is where the coordinates are fixed to the surface of the rotating earth. The T_1 component of the kinetic energy in that case is related to so-called *Coriolis effects* and the T_0 component to *centrifugal effects*. A simple example where one can see the centrifugal effects in particular is the problem shown in Fig. 2.40a where a bead slides on a smooth circular ring of radius R that is rotating with a constant angular velocity, Ω. If we choose a set of moving coordinates (x, y, z) that rotate with the ring so that the ring is always in the $x - y$ plane (see Fig. 2.40b), and use polar coordinates (R, θ) in that plane to describe the motion, this is a one degree of freedom problem that can be described by the single independent generalized coordinate, θ. The position and velocity of the particle are

$$\begin{aligned} \mathbf{x} &= R\sin\theta\ \mathbf{e}_x(t) - R\cos\theta\ \mathbf{e}_y \\ \mathbf{v} = \dot{\mathbf{x}}\ &= R\cos\theta\dot{\theta}\ \mathbf{e}_x + R\sin\theta\,\dot{\mathbf{e}}_x + R\sin\theta\dot{\theta}\ \mathbf{e}_y \end{aligned} \tag{2.7.16}$$

where since the unit vectors $(\mathbf{e}_x, \mathbf{e}_y, \mathbf{e}_z)$ are rotating with the angular velocity, $\boldsymbol{\omega} = \Omega\mathbf{e}_y$, of the ring their time derivatives are (see Sect. 1.5)

$$\begin{aligned} \dot{\mathbf{e}}_x &= \boldsymbol{\omega}\times\mathbf{e}_x = \Omega\,\mathbf{e}_y\times\mathbf{e}_x = -\Omega\mathbf{e}_z \\ \dot{\mathbf{e}}_y &= \boldsymbol{\omega}\times\mathbf{e}_y = \Omega\,\mathbf{e}_y\times\mathbf{e}_y = 0 \\ \dot{\mathbf{e}}_z &= \boldsymbol{\omega}\times\mathbf{e}_z = \Omega\,\mathbf{e}_y\times\mathbf{e}_z = \Omega\mathbf{e}_x \end{aligned} \tag{2.7.17}$$

so that the velocity and kinetic energy are

$$\begin{aligned}\mathbf{v} &= R\cos\theta\dot{\theta}\,\mathbf{e}_x + R\sin\theta\dot{\theta}\,\mathbf{e}_y - R\Omega\sin\theta\,\mathbf{e}_z \\ T &= \frac{1}{2}m\mathbf{v}\cdot\mathbf{v} = \frac{1}{2}m\left[R^2\dot{\theta}^2 + R^2\Omega^2\sin^2\theta\right]\end{aligned} \tag{2.7.18}$$

and

$$T_2 = \frac{1}{2}mR^2\dot{\theta}^2,\;\; T_1 = 0,\;\; T_0 = \frac{1}{2}m(R\sin\theta)^2\Omega^2 \tag{2.7.19}$$

from which we can recognize the T_0 term as due to the velocity of the ring at the particle position caused by the ring rotation (see Fig. 2.40b) and the T_2 term as due to the velocity of the particle as seen in the rotating x- and z-axes. Even when there is no motion of the bead relative to the ring so that $\dot{\theta} = \ddot{\theta} = 0$ if we differentiate the velocity in Eq. (2.7.18) on time under those conditions we find that the acceleration is

$$\mathbf{a} = \dot{\mathbf{v}} = -R\sin\theta\;\Omega^2\mathbf{e}_x = -r\Omega^2\mathbf{e}_x \tag{2.7.20}$$

which we recognize as an acceleration term due to the mass rotating at a radius $r = R\sin\theta$ with an angular velocity, Ω (Fig. 2.40b) so indeed T_0 comes from those centrifugal effects.

For systems like this one where $T_1 = 0$ and the applied forces are derivable from a potential energy, the Lagrangian becomes

$$L = T - V = T_2 + T_0 - V = T_2 - U \tag{2.7.21}$$

where $U = V - T_0$ is called the *modified potential energy* or the *dynamic potential.* When we discuss equilibrium and stability conditions in Chap. 6 we will see the fundamental role that U plays in stability analyses.

We can get the equations of motion by using Newton's second law in the form

$$m\ddot{\mathbf{x}}\cdot\mathbf{c}_n = Q_n^e \tag{2.7.22}$$

and computing the acceleration directly, as shown in the last section, or we can use Lagrange's equations

$$\frac{d}{dt}\left(\frac{\partial T}{\partial\dot{q}_n}\right) - \frac{\partial T}{\partial q_n} = Q_n^e \tag{2.7.23}$$

and compute the equations of motion from the kinetic energy. To see an application of both of these approaches together consider the bead sliding on the rotating ring example we just discussed briefly. This is a very simple problem since it is a one degree of freedom system where we will take as our single independent generalized coordinates $q_1 = \theta$. First, determine the equations of motion using Eq. (2.6.14) where we need the position vector and its derivatives:

$$\begin{aligned}
\mathbf{x} &= R\sin\theta\;\; \mathbf{e}_x(t) - R\cos\theta\;\; \mathbf{e}_y \\
\mathbf{c} &= \frac{\partial \mathbf{x}}{\partial \theta} = R\cos\theta \mathbf{e}_x(t) + R\sin\theta \mathbf{e}_y \\
\frac{\partial \mathbf{c}}{\partial \theta} &= -R\sin\theta \mathbf{e}_x(t) + R\cos\theta \mathbf{e}_y \\
\frac{\partial \mathbf{c}}{\partial t} &= -R\Omega\cos\theta \mathbf{e}_z(t) \\
\frac{\partial \mathbf{x}}{\partial t} &= -R\Omega\sin\theta \mathbf{e}_z(t) \\
\frac{\partial^2 \mathbf{x}}{\partial t^2} &= -R\Omega^2\sin\theta \mathbf{e}_x(t)
\end{aligned} \tag{2.7.24}$$

which gives

$$\begin{aligned}
M_{11} &= m\mathbf{c}\cdot\mathbf{c} = mR^2 \\
K_{111} &= m\mathbf{c}\cdot\frac{\partial \mathbf{c}}{\partial \theta} = 0 \\
J_{11} &= 2m\mathbf{c}\cdot\frac{\partial \mathbf{c}}{\partial t} = 0 \\
B_1 &= m\mathbf{c}\cdot\frac{\partial^2 \mathbf{x}}{\partial t^2} = -mR^2\Omega^2\sin\theta\cos\theta
\end{aligned} \tag{2.7.25}$$

The only applied force is that of gravity whose potential energy is $V = -\; mgR\,\cos\, q_1 = -\; mgR\,\cos\,\theta$, giving the generalized force $Q_1^e = -\partial V/\partial\theta = -mgR\sin\theta$. Thus the equation of motion is

$$M_{11}\ddot{\theta} + B_1 = -\frac{\partial V}{\partial \theta} \tag{2.7.26}$$

or explicitly

$$mR^2\ddot{\theta} - mR^2\Omega^2\sin\theta\cos\theta + mgR\sin\theta = 0 \tag{2.7.27}$$

If, however, we work with the velocity vector and use the form of Eqs. (2.6.29a) and (2.6.29b) we only need the terms

$$\begin{aligned}
\mathbf{v} &= R\cos\theta\dot{\theta}\;\mathbf{e}_x(t) + R\sin\theta\dot{\theta}\;\mathbf{e}_y - R\Omega\sin\theta\,\mathbf{e}_z(t) \\
\mathbf{c} &= \frac{\partial \mathbf{v}}{\partial \dot{\theta}} = R\cos\theta \mathbf{e}_x(t) + R\sin\theta \mathbf{e}_y \\
\frac{\partial \mathbf{v}}{\partial \theta} &= -R\sin\theta\dot{\theta}\;\mathbf{e}_x(t) + R\cos\theta\dot{\theta}\;\mathbf{e}_y - R\Omega\cos\theta \mathbf{e}_z(t) \\
\frac{\partial \mathbf{v}}{\partial t} &= -R\Omega^2\sin\theta \mathbf{e}_x(t) - R\Omega\cos\theta\dot{\theta}\;\mathbf{e}_z(t)
\end{aligned} \tag{2.7.28}$$

so that

$$
\begin{aligned}
M_{11} &= mR^2 \\
D_{11} &= m\mathbf{c} \cdot \frac{\partial \mathbf{v}}{\partial \theta} = 0 \\
E_1 &= m\mathbf{c} \cdot \frac{\partial \mathbf{v}}{\partial t} = -mR^2\Omega^2 \sin\theta \cos\theta
\end{aligned} \tag{2.7.29}
$$

and the equation of motion is

$$
M_{11}\ddot{\theta} + E_1 = -\frac{\partial V}{\partial \theta} \tag{2.7.30}
$$

which again yields Eq. (2.7.27). Now, let us use Lagrange's equations in terms of the kinetic energy, Eq. (2.7.23), to get the equations of motion in this case. From Eq. (2.7.5) and the potential energy expression we have

$$
\begin{aligned}
T &= \frac{1}{2}m\left[R^2\dot{\theta}^2 + R^2\Omega^2 \sin^2\theta\right] \\
Q^e &= -\frac{\partial V}{\partial \theta} = -mgR\sin\theta \\
\frac{\partial T}{\partial \dot{\theta}} &= mR^2\dot{\theta} \\
\frac{d}{dt}\left(\frac{\partial T}{\partial \dot{\theta}}\right) &= mR^2\ddot{\theta} \\
\frac{\partial T}{\partial \theta} &= mR^2 \sin\theta \cos\theta
\end{aligned} \tag{2.7.31}
$$

Lagrange's equation in this problem is

$$
\frac{d}{dt}\left(\frac{\partial T}{\partial \dot{\theta}}\right) - \frac{\partial T}{\partial \theta} = -\frac{\partial V}{\partial \theta} \tag{2.7.32}
$$

which once again gives Eq. (2.7.27). Note that the negative of the potential appears in the Lagrangian and there is a negative gradient of the Lagrangian in Lagrange's equations so that if one uses Lagrange's equations in the form of Eq. (2.7.6) it is easy to make sign errors in the calculations. Thus, it is often better to use Lagrange's equations written in terms of the kinetic energy alone (see Eq. (2.7.5)), as done here, so that the negative sign on the gradient of the potential appears only once in the expression for the generalized force.

As we have seen before there are multiple ways to get the equations of motion. For the simple problems considered in most dynamics books there often is little difference in the method we choose. For more complex problems found in engineering practice there may be important differences in the approach used so that it is important to understand a variety of methods.

In this rotating hoop example, the total energy is not conserved even though the only external force acting (gravity) is derivable from a potential energy. We can see this since the total energy, E, is given by

$$E = T + V = \frac{1}{2} mR^2 \dot{\theta}^2 + \frac{1}{2} m (R \sin\theta)^2 \Omega^2 - mgR\cos\theta \tag{2.7.33}$$

so that its derivative in time is a nonzero value given by

$$\begin{aligned} \frac{dE}{dt} &= mR^2 \dot{\theta}\ddot{\theta} + mR^2 \Omega^2 \sin\theta\cos\theta\dot{\theta} + mgR\sin\theta\dot{\theta} \\ &= 2mR^2\Omega^2 \sin\theta\cos\theta\dot{\theta} \end{aligned} \tag{2.7.34}$$

where use has been made of the equation of motion, Eq. (2.7.27). This result makes sense from a physical standpoint since in order to produce a constant rotational velocity, Ω, of the hoop, there must be some torque driving the shaft of the hoop, and this torque will do work, changing the total energy of the system.

Just because the total energy is not conserved, however, does not mean that one cannot find a different integral of the equations of motion that is a constant. Consider, for example, a system with p degrees of freedom where there are only forces derivable from potentials and the Lagrangian, L, is not an explicit function of time. In that case we can write $L = L(\mathbf{q}, \dot{\mathbf{q}})$ in terms of p independent generalized coordinates and their derivatives. In the one degree of freedom rotating hoop case we see that this form is indeed true as we have

$$L = T - V = \frac{1}{2} m \left[R^2 \dot{\theta}^2 + R^2 \Omega^2 \sin^2\theta \right] + mgR\cos\theta \tag{2.7.35}$$

so that $L = L(\theta, \dot{\theta})$. But it follows that in the more general case

$$\frac{dL}{dt} = \sum_{k=1}^{p} \frac{\partial L}{\partial q_k} \dot{q}_k + \sum_{k=1}^{p} \frac{\partial L}{\partial \dot{q}_k} \frac{d}{dt} (\dot{q}_k) \tag{2.7.36}$$

and from Lagrange's equation, we have

$$\frac{\partial L}{\partial q_k} = \frac{d}{dt} \left(\frac{\partial L}{\partial \dot{q}_k} \right) \quad (k = 1, \ldots, p) \tag{2.7.37}$$

so that placing Eq. (2.7.37) into Eq. (2.7.36) we find

$$\frac{dL}{dt} = \sum_{k=1}^{p} \frac{d}{dt} \left(\frac{\partial L}{\partial \dot{q}_k} \right) \dot{q}_k + \sum_{k=1}^{p} \frac{\partial L}{\partial \dot{q}_k} \frac{d}{dt} (\dot{q}_k) = \sum_{k=1}^{p} \frac{d}{dt} \left(\frac{\partial L}{\partial \dot{q}_k} \dot{q}_k \right) \tag{2.7.38}$$

which can be rewritten as

$$\frac{d}{dt}\left[\sum_{k=1}^{p}\frac{\partial L}{\partial \dot{q}_k}\dot{q}_k - L\right] = 0 \tag{2.7.39}$$

showing that

$$h = \sum_{k=1}^{p}\frac{\partial L}{\partial \dot{q}_k}\dot{q}_k - L = \text{constant} \tag{2.7.40}$$

The quantity h is an integral of the equations of motion called the *Jacobi integral.* To put this integral in a more convenient form note that $T = T_2+T_1+T_0$ (see Eqs. (2.7.13) and (2.7.14)) and

$$\begin{aligned}\sum_{k=1}^{p}\frac{\partial L}{\partial \dot{q}_k}\dot{q}_k &= \sum_{k=1}^{p}\frac{\partial T}{\partial \dot{q}_k}\dot{q}_k = \sum_{i=1}^{p}\sum_{j=1}^{p}M_{ij}(\mathbf{q},t)\dot{q}_i\dot{q}_j + \sum_{i=1}^{p}\mu_i(\mathbf{q},t)\dot{q}_i \\ &= 2T_2 + T_1\end{aligned} \tag{2.7.41}$$

so that

$$\begin{aligned}h &= (2T_2 + T_1) - (T_2 + T_1 + T_0 - V) = T_2 - T_0 + V \\ &= T_2 + U\end{aligned} \tag{2.7.42}$$

where $U = V - T_0$ is the dynamic potential mentioned previously. Notice that the T_1 terms in Eq. (2.7.42) cancel out so that it is not necessary that T_1 be identically zero (as it is for the rotating hoop problem) for h to be conserved. In the rotating hoop case we see that the conservation of h gives

$$\begin{aligned}h &= \frac{1}{2}mR^2\dot{\theta}^2 - \frac{1}{2}m(R\sin\theta)^2\Omega^2 - mgR\cos\theta \\ &= \frac{1}{2}mR^2\dot{\theta}_0^2 - \frac{1}{2}m(R\sin\theta_0)^2\Omega^2 - mgR\cos\theta_0\end{aligned} \tag{2.7.43}$$

For a natural system $T_1 = T_0 = 0$ so that $T_2 = T$ and then the Jacobi integral simply becomes

$$h = E = T + V \tag{2.7.44}$$

and we have conservation of the total energy integral. For the rotating hoop example this obviously occurs if $\Omega = 0$.

Simulation 2.7: Rotating Hoop

Now, let's obtain a numerical solution for the rotating hoop problem. From Eq. (2.7.27) the equation of motion can be written as

$$\ddot{\theta} + \left(\frac{g}{R} - \Omega^2 \cos\theta\right) \sin\theta = 0 \tag{2.7.45}$$

The dynamics of this problem has some interesting behavior that is controlled by the static equilibrium solutions. If we set $\ddot{\theta} = 0$ we see those equilibrium solutions must satisfy

$$\left(\frac{g}{R} - \Omega^2 \cos\theta\right) \sin\theta = 0 \tag{2.7.46}$$

which has multiple possible solutions:

$$\begin{aligned} \theta_1 &= \pm 2n\pi \\ \theta_2 &= \pm(2n+1)\pi \\ \theta_3 &= \pm\cos^{-1}\left(\frac{g}{R\Omega^2}\right) \pm 2n\pi \quad \text{for } \frac{g}{R\Omega^2} \leq 1 \end{aligned} \tag{2.7.47}$$

for $n = 0, 1, 2, \ldots$ but it is sufficient to examine these solutions for $n = 0$ as the behavior for other values of n is identical. The solutions $\theta_1 = 0$, $\theta_2 = \pm\pi$ are when the bead is at the bottom or top of the hoop, respectively, while the solution $\theta_3 = \pm\cos^{-1}(g/R\Omega^2)$ is an intermediate equilibrium position that is dependent on the rotational speed. In Chap. 6, when we discuss stability conditions, we will see that the equilibrium position $\theta_2 = \pm\pi$ is always unstable while the stability of the other two positions depends on the angular velocity, Ω. Specifically, we will find that the $\theta_1 = 0$ position is stable (and there is no real θ_3 equilibrium position) as long as $\Omega^2 < g/R$ while for $\Omega^2 > g/R$ the $\theta_1 = 0$ position is unstable and the θ_3 positions are stable.

These properties, as we will see, are reflected in the dynamical behavior of the system. The system of first-order differential equations for this system is for $(x_1, x_2) = (\theta, \dot{\theta})$:

$$\begin{Bmatrix} \dot{x}_1 \\ \dot{x}_2 \end{Bmatrix} = \begin{Bmatrix} x_2 \\ \sin x_1 \left(\Omega^2 \cos x_1 - g/R\right) \end{Bmatrix} \tag{2.7.48}$$

This system has been placed in the MATLAB® function rotating_hoop_eqs which together with the MATLAB® solver ode45 is a nested function within the function rotating_hoop. The calling sequence for the rotating_hoop function is

```
[tout, xout] = rotating_hoop(tspan, x0, omega, R, rtol, atol);
```

This function is evaluated in the rotating_hoop_sim script which sets default values for the input parameters given as R = 0.5 m, omega = 2 rad/s, tspan = [0, 10], x0 = [0, 0.1], rtol = 1e-6, and atol = 1e-6 and then calls the rotating_hoop function. The script then plots the angular position, θ, of the bead versus time and the phase plot of θ versus $\dot{\theta}$. The default parameters satisfy $\Omega^2 < g/R$ so the $\theta = 0$ equilibrium position is stable and we expect to see harmonic oscillations about that position. This is indeed the case as seen in Figs. 2.41 and 2.42. If we change the angular velocity of the hoop to omega = 7.0 rad/s, where $\Omega^2 > g/R$, and then run the script the behavior of the system is shown in Figs. 2.43 and 2.44. Although the motion is still harmonic, Fig. 2.43 shows that the nature of these oscillations has changed significantly. From the phase plot of Fig. 2.44 it can be easily seen what is happening. The bead would

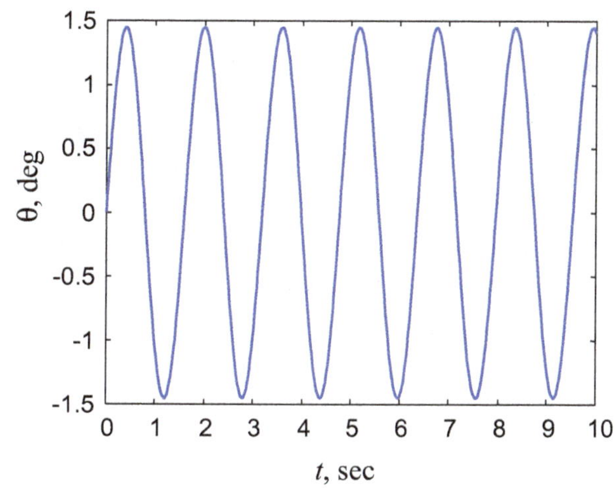

Fig. 2.41 Angular motion of the bead on a rotating hoop for $\Omega^2 < g/R$

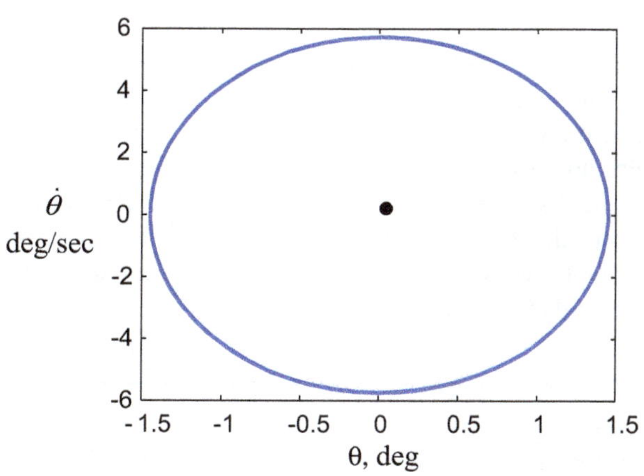

Fig. 2.42 Phase plot for the bead on a rotating hoop for $\Omega^2 < g/R$

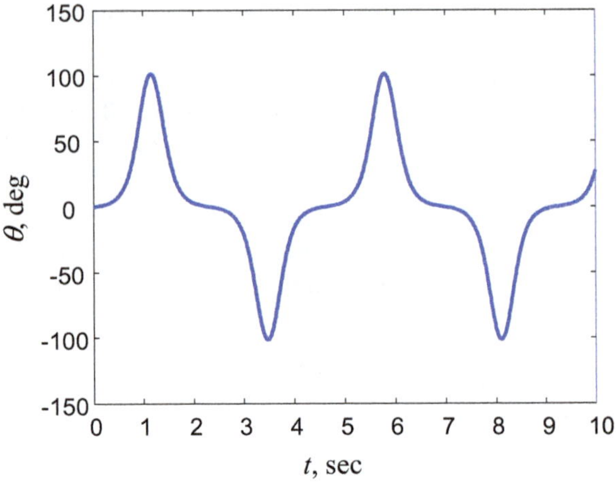

Fig. 2.43 Angular motion of the bead on a rotating hoop for $\Omega^2 > g/R$

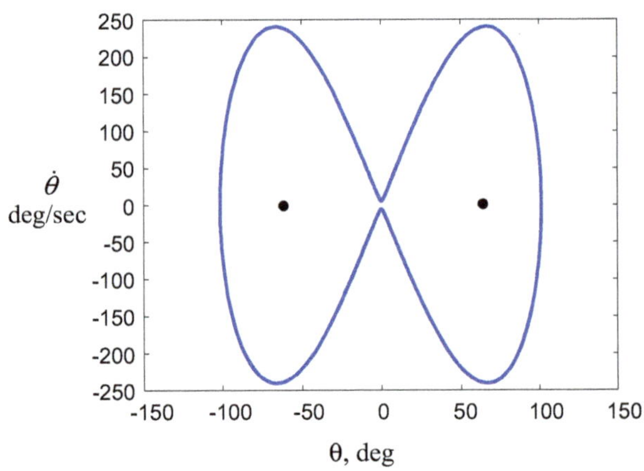

Fig. 2.44 Phase plot of the bead on a rotating hoop for $\Omega^2 > g/R$

now like to oscillate about the equilibrium positions of $\theta_3 = \pm \cos^{-1}(g/R\Omega^2)$ which in this case are $\theta_3 = \pm 66.4^{\circ}$, approximately. Thus, the motion alternately moves about one of these equilibrium positions, passes through the unstable $\theta = 0$ equilibrium position, and then moves about the other equilibrium position. If, however, we had set the initial conditions to x0 = [1.16, 0.5] and keep omega = 7.0, which starts the bead out near the 66.4° equilibrium position with a relatively small velocity, we see from Figs. 2.45 and 2.46 that the bead is now trapped and oscillates about that equilibrium position only.

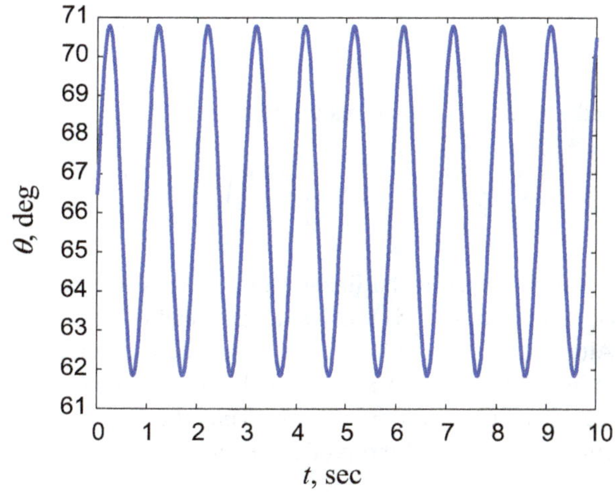

Fig. 2.45 Angular motion of the bead on the rotating hoop for $\Omega^2 > g/R$ where the bead is initially started near the $\theta = 66.4°$ equilibrium position with a small velocity

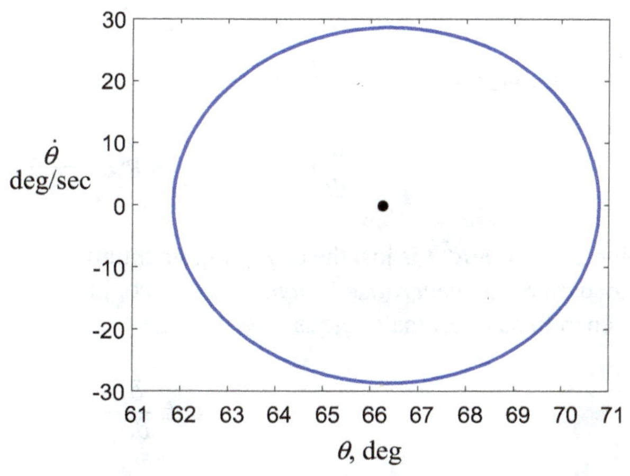

Fig. 2.46 Phase plot of the bead on a rotating hoop for $\Omega^2 > g/R$ where the bead is initially started near the $\theta = 66.4°$ equilibrium position with a small velocity

2.7.1 Generalized Momenta and Ignorable Coordinates

Lagrange's equations for a problem where the forces are derivable from a potential can be written in terms of an independent set of generalized coordinates $(q_1, q_2, \ldots, q_n)$ as

$$\frac{d}{dt}\left(\frac{\partial L}{\partial \dot{q}_k}\right) - \frac{\partial L}{\partial q_k} = 0 \quad (k = 1, 2, \ldots, n) \tag{2.7.49}$$

The $\partial L/\partial \dot{q}_k$ terms are called the generalized momenta of the system. This naming is consistent with the linear and angular momenta you are likely familiar with from various particle problems. For example, consider a single particle of mass m freely in the x-direction, i.e., where it is not subjected to any forces. Then we have

$$L = T - V = T = \frac{1}{2} m\dot{x}^2 \tag{2.7.50}$$

and Lagrange's equations are just

$$\frac{d}{dt}\left(\frac{\partial L}{\partial \dot{x}}\right) = \frac{d}{dt}(m\dot{x}) = \dot{p}_x = 0 \tag{2.7.51}$$

where $p_x = m\dot{x}$ is the linear momentum of the particle and we see that Lagrange's equation, Eq. (2.7.51), is just a statement of conservation of this linear momentum. Alternatively, consider a particle attached to string and moving in a circular path of radius R in a horizontal plane with no other force except the ideal constraint force of the string acting on the particle. Then the Lagrangian can be written in terms of the angular velocity of the particle, $\dot{\theta}$, as

$$L = \frac{1}{2} mR^2\dot{\theta}^2 \tag{2.7.52}$$

and Lagrange's equations are

$$\frac{d}{dt}\left(\frac{\partial L}{\partial \dot{\theta}}\right) = \frac{d}{dt}(mR^2\dot{\theta}) = \dot{p}_\theta = 0 \tag{2.7.53}$$

where $p_\theta = mR^2\dot{\theta}$ is just the angular momentum of the particle. Thus, it makes sense to define the *generalized momenta*, p_k, conjugated to (i.e., associated with) the generalized coordinate, q_k, as

$$p_k \triangleq \frac{\partial L}{\partial \dot{q}_k} \tag{2.7.54}$$

When a generalized coordinate, q_k, is absent from the Lagrangian then that q_k coordinate is said to be an *ignorable coordinate*. In that case we have

$$\dot{p}_k = \frac{\partial L}{\partial q_k} = 0 \tag{2.7.55}$$

so that the generalized momentum, p_k, is a constant. In more complex problems we can use information of this type to deduce properties of the motion without having to numerically solve the equations of motion since for a missing coordinate, q_k, the condition $p_k =$ constant is an integral of the equations in the same fashion that $E = T+V =$ constant is an integral for problems where all the forces are derivable from a potential.

2.7.2 Lagrange's Equations: A Summary

Lagrange's equations, written in terms of the kinetic energy, are (see Eq. (2.7.5))

$$\frac{d}{dt}\left(\frac{\partial T}{\partial \dot{q}_n}\right) - \frac{\partial T}{\partial q_n} = Q_n^e \tag{2.7.56}$$

Note that the generalized coordinates here were assumed to be independent so that one way in which that can happen, we have seen, is if all constraints are implicitly satisfied by the choice of coordinates. For example, in the pendulum problem there is one independent coordinate which can be taken as the angle θ and the constraint that the length of the string, L, is fixed can be satisfied by simply setting $r = L$. However, another way to obtain these equations in terms of a set of independent coordinates, as we have seen, is by explicitly embedding the constraints and/or their derivatives into the kinetic energy and generalized force (or equivalently the potential energy) expressions so that again we have all the terms in Eq. (2.7.56) written in terms of a set of independent coordinates. We have seen this explicit imbedding in a simple context in the bead-on-a-wire problems.

We can also write the position of a particle as $\mathbf{x} = \mathbf{x}(q_1, q_2, \ldots, q_N)$ in terms of N generalized coordinates that are not independent and form up N Lagrange's equations from Newton's second law by dotting it with the $\mathbf{c}_n = \partial \mathbf{x}/\partial q_n = \partial \mathbf{v}/\partial \dot{q}_n$ vectors. However, in this case the constraint forces, $\mathbf{F}^c$, will not be eliminated and Lagrange's equations will become

$$\frac{d}{dt}\left(\frac{\partial T}{\partial \dot{q}_n}\right) - \frac{\partial T}{\partial q_n} = Q_n^e + Q_n^c \quad (n = 1, 2, \ldots, N) \tag{2.7.57}$$

where $Q_n^c = \mathbf{F}^c \cdot \mathbf{c}_n$. If, as done previously, we write the constraint equations in terms of the generalized accelerations and assume that the constraint forces are ideal constraints, then in terms of the constrained generalized coordinates we have

$$\begin{aligned} &\frac{d}{dt}\left(\frac{\partial T}{\partial \dot{q}_n}\right) - \frac{\partial T}{\partial q_n} = Q_n^e + \sum_{m=1}^{M} A_{mn}\lambda_m \quad (n = 1, 2, \ldots, N) \\ &\sum_{j=1}^{N} A_{mj}\ddot{q}_j = b_m \quad (m = 1, 2, \ldots, M) \end{aligned} \tag{2.7.58}$$

where we have assumed that there are N nonindependent coordinates and M constraints. We can solve these equations with an augmented approach once we write the kinetic energy terms explicitly in terms of the accelerations. For example, if we compute the $\mathbf{c}_n$ vectors and the acceleration from the velocity, as we have shown previously, then the kinetic energy terms in Lagrange's equations yields (see Eq. (2.6.29a))

$$\begin{aligned}&\sum_{k=1}^{N} M_{nk}\ddot{q}_k + \sum_{k=1}^{N} D_{nk}\dot{q}_k + E_n = Q_n^e + \sum_{m=1}^{M} A_{mn}\lambda_m \quad (n = 1, 2, \ldots, N)\\&\sum_{k=1}^{N} A_{mk}\ddot{q}_k = b_m \quad (m = 1, 2, \ldots, M)\end{aligned} \tag{2.7.59}$$

or, equivalently,

$$\begin{aligned}&\sum_{k=1}^{N} M_{nk}\ddot{q}_k = \bar{Q}_n^e + \sum_{m=1}^{M} A_{mn}\lambda_m \quad (n = 1, 2, \ldots, N)\\&\sum_{k=1}^{N} A_{mk}\ddot{q}_k = b_m \quad (m = 1, 2, \ldots, M)\end{aligned} \tag{2.7.60}$$

where

$$\bar{Q}_n^e = Q_n^e - \sum_{m=1}^{N} D_{nm}\dot{q}_m - E_n \tag{2.7.61}$$

is the sum of the generalized external force components and the generalized inertial force components. Equation (2.7.60) is in a form we have seen before and leads directly to the explicit equations for the constraint force and the equation of motion in the augmented approach.

Thus, Lagrange's equations can be used with either an embedding approach or an augmented approach. The augmented approach we have outlined is a general method but the embedding approach has been done so far on a case-by-case basis. In Sect. 2.11 the embedding method will be considered in more general terms.

In closing, it is important to note that we can only embed holonomic constraints and their derivatives in the kinetic energy expression and then use that kinetic energy expression in Lagrange's equations. If one attempts to embed non-holonomic constraints in the same fashion, one will obtain incorrect equations of motion (see Problem P2.19). This should not be surprising since non-holonomic constraints do not imply similar constraints on the coordinates themselves.

2.8 Analytical Dynamics and Virtual Work

In all of our discussions to date we have generated the equations of motion for systems with ideal constraint forces by using geometrical arguments, including the derivation of Lagrange's equations. However, analytical dynamics has traditionally been developed using "virtual" quantities such as virtual displacements, which you may be familiar with from statics. We have not followed that traditional path because virtual changes appear at first glance to be rather vague and mysterious concepts and students encountering them for the first time are often not comfortable with them. Since we have already discussed many of the concepts in analytical dynamics from a geometrical standpoint, in this section we will tie that foundation to an approach that uses virtual displacements and other virtual quantities and try to remove some of the ambiguity surrounding their use.

Consider first a particle moving in 3-D. If we have holonomic constraints acting on that particle we can write them in vector form as

$$\mathbf{f}(\mathbf{x}, t) = 0 \tag{2.8.1}$$

If we differentiate those constraints once we find

$$\mathbf{A}\dot{\mathbf{x}} = \mathbf{a}_t \tag{2.8.2}$$

where $\mathbf{A} = \mathbf{A}(\mathbf{x}, t) = [\nabla\mathbf{f}]$ is a matrix of gradients of the constraints (called the *Jacobian matrix*) and $\mathbf{a}_t = \mathbf{a}_t(\mathbf{x}, t) = \partial\mathbf{f}/\partial t$ is a vector of time derivatives of those constraints. For example, in Cartesian coordinates if there are two constraints we have seen that

$$\begin{aligned} \mathbf{A}(\mathbf{x}, t) &= \begin{bmatrix} \partial f_1/\partial x & \partial f_1/\partial y & \partial f_1/\partial z \\ \partial f_2/\partial x & \partial f_2/\partial y & \partial f_2/\partial z \end{bmatrix} \\ \mathbf{a}_t(\mathbf{x}, t) &= -\begin{Bmatrix} \partial f_1/\partial t \\ \partial f_2/\partial t \end{Bmatrix} \end{aligned} \tag{2.8.3}$$

Note that this $\mathbf{A}$ matrix is the same matrix appearing in the twice-differentiated constraint equation $\mathbf{A}\ddot{\mathbf{x}} = \mathbf{b}$. In the presence of constraints Newton's second law gives

$$m\ddot{\mathbf{x}} = \mathbf{F}^e + \mathbf{F}^c \tag{2.8.4}$$

where $\mathbf{F}^e$ is the net applied external force and $\mathbf{F}^c$ is the net constraint force. If we assume that the constraint forces are ideal constraint forces we have

$$m\ddot{\mathbf{x}} = \mathbf{F}^e + \mathbf{A}^T\boldsymbol{\lambda} \tag{2.8.5}$$

We can write the constraint equation, Eq. (2.8.2), in differential form as

$$\mathbf{A}d\mathbf{x} - \mathbf{a}_t dt = 0 \tag{2.8.6}$$

There are infinitely many differentials, $d\mathbf{x}(t)$, that satisfy these constraint equations, including the differential for the actual motion of the particle. If we let

$(d\mathbf{x}', d\mathbf{x}'')$ be two possible differential displacements that satisfy the constraints at the same position and time then

$$\begin{aligned} \mathbf{A}(\mathbf{x}, t)d\mathbf{x}' - \mathbf{a}_t(\mathbf{x}, t)dt = 0 \\ \mathbf{A}(\mathbf{x}, t)d\mathbf{x}'' - \mathbf{a}_t(\mathbf{x}, t)dt = 0 \end{aligned} \tag{2.8.7}$$

We will call differential displacements such as $(d\mathbf{x}', d\mathbf{x}'')$ *allowed differential displacements*. By taking the difference of the two equations in Eq. (2.8.7) we see that the difference of the two allowed differential displacements, $\delta\mathbf{x}(t) = d\mathbf{x}'(t) - d\mathbf{x}''(t)$, satisfies

$$\mathbf{A}(\mathbf{x}, t)\delta\mathbf{x}(t) = 0 \tag{2.8.8}$$

The quantity $\delta\mathbf{x}(t)$ is what most authors call a *virtual displacement.* Using the term "virtual" has led to much confusion since it implies that somehow this quantity is not real. However, as our above discussion shows, a virtual displacement can be defined simply as the difference between any two allowed differential displacements the particle can have that satisfy the constraints, i.e., $\delta\mathbf{x} \triangleq d\mathbf{x}' - d\mathbf{x}''$, and there is nothing that is not real in that definition. Part of the confusion comes from authors making vague statements like "a virtual displacement takes place without any passage of time" which make no sense since as we have seen the virtual displacements are in fact themselves functions of time, just as are the allowed differential displacements. Also, many times virtual displacements are defined simply as having to satisfy the geometrical constraints but without giving any explicit meaning to what that satisfaction looks like. As we will see the great value of the virtual displacements is that they satisfy the constraints in the form of Eq. (2.8.8), unlike allowed differential displacements which satisfy Eq. (2.8.6). Operationally, we can distinguish virtual displacements, $\delta\mathbf{x}$, from ordinary differential displacements, $d\mathbf{x}$, by noting that if, say, we have any function, $g(y_1(t), y_2(t), t)$ (which could be a vector function), then the difference between the differential of g and the virtual change of g is embodied in the relations

$$\begin{aligned} dg &= \sum_{k=1}^{2} \frac{\partial g}{\partial y_k} dy_k + \frac{\partial g}{\partial t} dt \\ \delta g &= \sum_{k=1}^{2} \frac{\partial g}{\partial y_k} \delta y_k \end{aligned} \tag{2.8.9}$$

where we see that there is no corresponding $(\partial g/\partial t)\delta t$ term in the expression for δg. Thus, one could obtain δg formally by including the $(\partial g/\partial t)\delta t$ term and then setting $\delta t = 0$. However, note that although using the formalism of setting $\delta t = 0$ does lead to the correct expression for the virtual change of g it can also lead to confusion, as noted above, as to the meaning of that virtual change. It is in the elimination of the term that involves changes of g with respect to its explicit dependency on time that distinguishes a "virtual" change of the function from an ordinary differential change. Thus, if g does not have any explicit dependency on time we can always take a virtual change to be an allowed differential change of the function.

The value of introducing the virtual displacement is that if we dot Newton's second law with a virtual displacement, we have

$$\delta\mathbf{x}^T\left(m\ddot{\mathbf{x}} - \mathbf{F}^e\right) = \delta\mathbf{x}^T\mathbf{A}^T\boldsymbol{\lambda} \tag{2.8.10}$$

But taking the transpose of Eq. (2.8.8) gives $\delta\mathbf{x}^T\mathbf{A}^T = 0$ so that we have in matrix-vector notation

$$\delta\mathbf{x}^T\left(m\ddot{\mathbf{x}} - \mathbf{F}^e\right) \tag{2.8.11a}$$

or, equivalently, in terms of the vector dot product

$$\left(m\ddot{\mathbf{x}} - \mathbf{F}^e\right)\cdot\delta\mathbf{x} = 0 \tag{2.8.11b}$$

which is called the *D'Alembert-Lagrange principle.* To show the connection between this principle and our previous results suppose that we write the position vector of the particle in terms of p independent generalized coordinates, q_m, for a particle moving in 3-D. Then we have

$$\delta\mathbf{x} = \sum_{k=1}^{p}\frac{\partial\mathbf{x}}{\partial q_k}\delta q_k = \sum_{k=1}^{p}\mathbf{c}_k\delta q_k \tag{2.8.12}$$

so that we have from the D'Alembert-Lagrange principle

$$\sum_{k=1}^{p}\left(m\ddot{\mathbf{x}}\cdot\mathbf{c}_k - \mathbf{F}^e\cdot\mathbf{c}_k\right)\delta q_k = 0 \tag{2.8.13}$$

But if the p generalized coordinates are unconstrained then we are free to choose the δq_k independently so that we must have

$$m\ddot{\mathbf{x}}\cdot\mathbf{c}_k = \mathbf{F}^e\cdot\mathbf{c}_k \quad (k = 1,\ldots,p) \tag{2.8.14}$$

which is identical to the results we obtained previously (see Eq. (2.6.10)) and directly leads to Lagrange's equation because of Eq. (2.7.4).

If you follow the above steps in obtaining Eq. (2.8.14) you might notice that if we had simply defined an ideal constraint force to be one where

$$\mathbf{F}^c\cdot\delta\mathbf{x} = 0 \tag{2.8.15}$$

we obtain the D'Alembert-Lagrange principle. Then if we use the expression for the virtual displacement given by Eq. (2.8.12), and the independence of the generalized coordinates, we would arrive at Eq. (2.8.14), or equivalently Lagrange's equations. This is in fact the way in which Lagrange originally obtained his equations. Equation (2.8.15) is simply a statement that the *virtual work*, δW, defined as $\delta W = \mathbf{F}^c \cdot \delta\mathbf{x}$, of an ideal constraint force when taken through the virtual displacement, $\delta\mathbf{x}$, is zero. Note that from Eq. (2.8.15) we have

$$\delta W = \mathbf{F}^c \cdot \delta\mathbf{x} = \mathbf{F}^c \cdot \sum_{k=1}^{p} \frac{\partial \mathbf{x}}{\partial q_k} \delta q_k = \sum_{k=1}^{p} Q_k^c \delta q_k = 0 \tag{2.8.16}$$

which is the statement of zero virtual work in terms of the generalized ideal constraint forces.

To see in more explicit terms why the virtual work of an ideal constraint force is zero, consider the motion of a particle on an inclined surface of a smooth wedge whose motion along the horizontal direction (Fig. 2.47) is specified. Let $\mathbf{u}(t)$ be the given displacement of the wedge and $\mathbf{s}(t)$ be the displacement along the inclined surface as shown in Fig. 2.47. Then the actual velocity, $\mathbf{v}(t)$, of the particle is just $\mathbf{v}(t) = \dot{\mathbf{u}}(t) + \dot{\mathbf{s}}(t)$ (velocity of the particle = velocity of the wedge + velocity of the particle with respect to the wedge) and the actual differential displacement of the particle in a time interval dt is $d\mathbf{r} = \mathbf{v}(t)dt$ (see Fig. 2.48a). In this problem, the constraint force, $\mathbf{F}^c$, of the smooth surface acting on the particle is normal to the inclined surface (Fig. 2.48a). Obviously $d\mathbf{r}$ is an allowed differential displacement since it is the displacement in the actual motion. Now, consider another allowable velocity of the particle $\mathbf{v}'(t) = \dot{\mathbf{u}}(t) + \dot{\mathbf{s}}'(t)$. This velocity must have the same velocity of the wedge as in the actual motion since the motion of the wedge is assumed to be given, but it can have a different velocity along the inclined surface, relative to the wedge, $\dot{\mathbf{s}}'(t)$, without violating the constraint of having to move on that surface. Thus, this allowable velocity produces an allowable differential displacement $d\mathbf{r}'(t) = \mathbf{v}'(t)dt$ in the time interval dt (Fig. 2.48b). We can take as a virtual displacement, $\delta\mathbf{r}$, the difference of these two allowable displacements, i.e., $\delta\mathbf{r} = d\mathbf{r}'(t) - d\mathbf{r}(t)$ where $\delta\mathbf{r}$ must be along the inclined plane since $d\mathbf{r}'(t) - d\mathbf{r}(t) = \dot{\mathbf{s}}'(t)dt - \dot{\mathbf{s}}(t)dt$ and both $(\dot{\mathbf{s}}'(t), \dot{\mathbf{s}}(t))$ are along the inclined plane (Fig. 2.48c). Thus, while $\mathbf{F}^c \cdot d\mathbf{r} \neq 0$ (the work of the constraint force acting through the actual differential displacement is not zero) we do have $\mathbf{F}^c \cdot \delta\mathbf{r} = 0$, i.e., the virtual work of the constraint force is indeed zero.

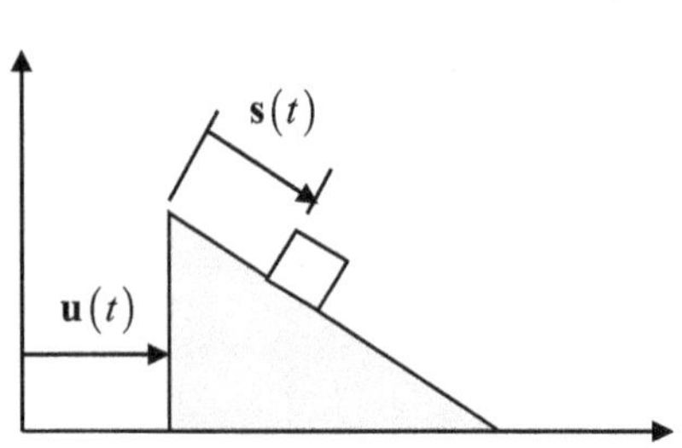

Fig. 2.47 A particle sliding on a smooth inclined plane which itself is in motion

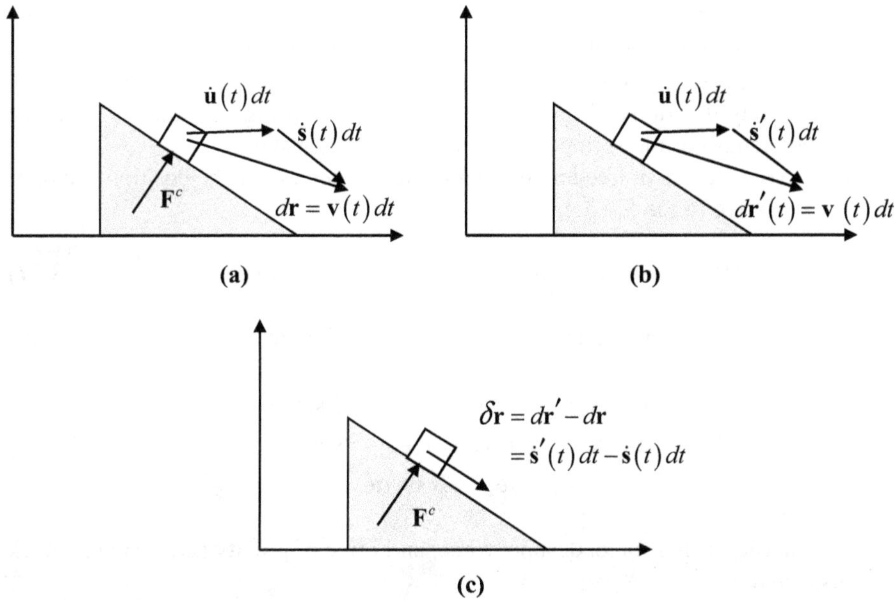

Fig. 2.48 (**a**) The velocity and differential displacement of the particle in the actual motion, and (**b**) a different allowed velocity and differential displacement of particle, producing (**c**) a virtual displacement that is the difference of the allowed motions of (a) and (b), showing that the ideal constraint force is perpendicular to the virtual displacement and so does no virtual work

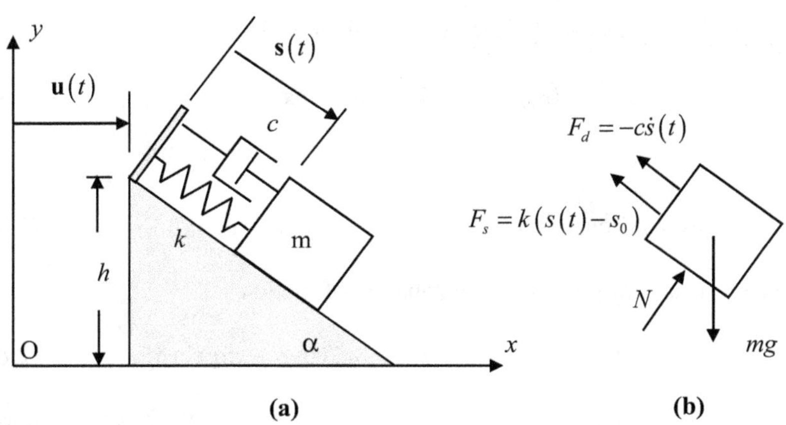

Fig. 2.49 (**a**) A block moving on a smooth wedge which has a specified displacement, **u**(t), where the block is attached to a linear spring and a viscous dashpot. Gravity acts in the negative y-direction. (**b**) The free body diagram of the block

We will use the problem of the moving wedge to solve the problem shown in Fig. 2.49a using the D'Alembert-Lagrange principle, where a block of mass m slides on a smooth wedge which has a specified displacement, **u**(t). The block is attached to a linear spring and a dashpot that provides a viscous damping force that is proportional to the velocity of the block relative to the moving wedge. These forces as well as the normal constraint force, N, acting on the block and its own weight are shown in the free body diagram in Fig. 2.49b, where k is the spring constant and s_0 is the unstretched length of the spring and c is the dashpot constant.

This is a one degree of freedom problem where we can write the position vector of the block relative to the fixed origin O as

$$\mathbf{x}(s(t),t) = u(t)\mathbf{e}_x + h\mathbf{e}_y + s(t)\cos\alpha\mathbf{e}_x - s(t)\sin\alpha\mathbf{e}_y \tag{2.8.17}$$

which gives on differentiation the velocity, acceleration, and virtual displacement:

$$\begin{aligned}\mathbf{v} = \dot{\mathbf{x}} &= (\dot{u} + \dot{s}\cos\alpha)\mathbf{e}_x - \dot{s}\sin\alpha\mathbf{e}_y \\ \mathbf{a} = \ddot{\mathbf{x}} &= (\ddot{u} + \ddot{s}\cos\alpha)\mathbf{e}_x - \ddot{s}\sin\alpha\mathbf{e}_y \\ \delta\mathbf{x} &= \delta s\cos\alpha\mathbf{e}_x - \delta s\sin\alpha\mathbf{e}_y\end{aligned} \tag{2.8.18}$$

and the external spring force, dashpot force, and force of gravity acting on the block are, respectively,

$$\begin{aligned}\mathbf{F}_s &= -k(s - s_0)\cos\alpha\mathbf{e}_x + k(s - s_0)\sin\alpha\mathbf{e}_y \\ \mathbf{F}_d &= -c\dot{s}\cos\alpha\mathbf{e}_x + c\dot{s}\sin\alpha\mathbf{e}_y \\ \mathbf{F}_g &= -mg\mathbf{e}_y\end{aligned} \tag{2.8.19}$$

From the D'Alembert-Lagrange principle then we have

$$(m\mathbf{a} - \mathbf{F}_s - \mathbf{F}_d - \mathbf{F}_g) \cdot \delta\mathbf{x} = 0 \tag{2.8.20}$$

which gives

$$[m\ddot{s} + m\ddot{u}\cos\alpha + k(s - s_0) + c\dot{s} - mg\sin\alpha]\delta s = 0 \tag{2.8.21}$$

and since δs is arbitrary we have the equation of motion

$$m\ddot{s} + c\dot{s} + k(s - s_0) = -m\ddot{u}\cos\alpha + mg\sin\alpha \tag{2.8.22}$$

Note that when there is no motion (equilibrium) of either the block or the wedge the equation of motion reduces to

$$s_e = s_0 + mg\sin\alpha/k \tag{2.8.23}$$

which gives the equilibrium displacement of the spring, $s_e = s_0 + mg\sin/\alpha k$. Thus, if instead of s we use the displacement of the block from this equilibrium position, $d = s - s_e$ our equation of motion in terms of d becomes simply

$$m\ddot{d} + c\dot{d} + kd = -m\ddot{u}\cos\alpha \tag{2.8.24}$$

Simulation 2.8: Forced Vibration of a Mass Attached to a Spring and Dashpot
Equation (2.8.24) is the well-studied equation for the damped vibration of a particle connected to a spring and dashpot and subjected to a driving force term $F(t) = -m\ddot{u}(t)\cos\alpha$ (which in this case is due to the acceleration of the wedge). Many dynamics books discuss the solutions of this equation in detail since it appears in many vibration problems when the motion is assumed to have a small amplitude (see Chap. 7). At such small amplitudes the original nonlinear equations of motion often become linear, allowing one to solve the equations of motion analytically. See Appendix D for a description of such analytical solutions. Here, we will obtain a numerical solution as done previously with the other problems considered. The specific driving force we will use is

$$F(t) = \begin{cases} A & 0 < t < t_0 \\ 0 & \text{otherwise} \end{cases} \tag{2.8.25}$$

which is the "box" function shown in Fig. 2.50. Functions like this one that have discontinuous changes in behavior are easy to simulate in MATLAB® with the use of logical vectors. For example, if we define in MATLAB® the function f as

```
f = A*(t >= 0 & t <= t0);
```

the term in parentheses is a logical expression which evaluates to one whenever the expression is true and zero otherwise. Thus if t is a vector of different time values, we will just get a set of f values from this term that are ones for $0 \leq t \leq t_0$ and zeros otherwise. Multiplying this logical vector by A then gives values that generate a sampled version of the box function $F(t)$.

This equation of motion has a term with an explicit function of time, $F(t)$, as part of its definition, but as we have seen with the problem of a bead moving on the straight wire such a function can be easily incorporated into the equations of motion. In this case we have generated a MATLAB® function moving_wedge which has the calling sequence

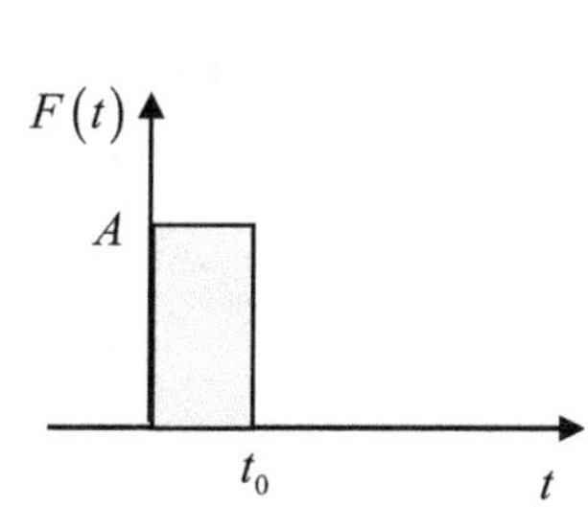

Fig. 2.50 A simple function that simulates the force term $F(t) = m\ddot{u}(t)\cos\alpha$ for the problem of the block of Fig. 2.49

```
[xout, tout] = moving_wedge(tspan, x0, km, cm, Am, t0,rtol,atol);
```

where, as in our previous MATLAB® examples, tspan is the range over which we want to simulate the motion of the block; x0 is a vector of initial conditions; km = *k/m* is the spring constant, *k*, normalized by the mass, *m*, with the dimensions of 1/s^2; cm = *c/m* is likewise the normalized dashpot coefficient, with dimensions of 1/s; Am = *A/m* is the amplitude of *F*(*t*)/*m* with dimensions of acceleration in m/s^2; t0 is the duration of the driving force (Eq. (2.8.24)) in s; and (rtol, atol) are the relative and absolute error tolerances used by ode45 as we have described previously. The function moving_wedge contains a nested function wedge_func that simulates the equations of motion as a system of first-order differential equations, and uses that nested function and ode45 to generate the outputs xout and tout, where the matrix xout contains the simulated displacement, *d*, and velocity, $\dot{d}$, of the block in its two columns and tout is a column vector of times at which the motion was evaluated. The MATLAB® script moving_wedge_sim has a set of default values for all the input parameters that are used unless you enter different values in the command window before executing the script. These values are used to evaluate the function moving_wedge and the resulting (xout, tout) values are then used by the script to generate a plot of *d* versus the time, *t*, and a phase plot of $\dot{d}$ versus *d*. Figure 2.51a plots the displacement of the block versus time, showing how the behavior changes depending on whether the driving force is present or not. The phase plot of Fig. 2.51b can explain that changing behavior. During the first 3 s of the motion the force is constant. If we look at the equilibrium solution for a constant driving force, we see that the equilibrium position is $d_e = A/k$. The parameters chosen for Fig. 2.51 are the default values given in the moving_wedge_sim script, k/m = 10, Am = 10, so the equilibrium position is $d_e = 1$ as long as Am = 10. The phase plot begins with a slowing decaying oscillation about that equilibrium position (point *P* in Fig. 2.51b), but when the amplitude of the driving force drops to zero the

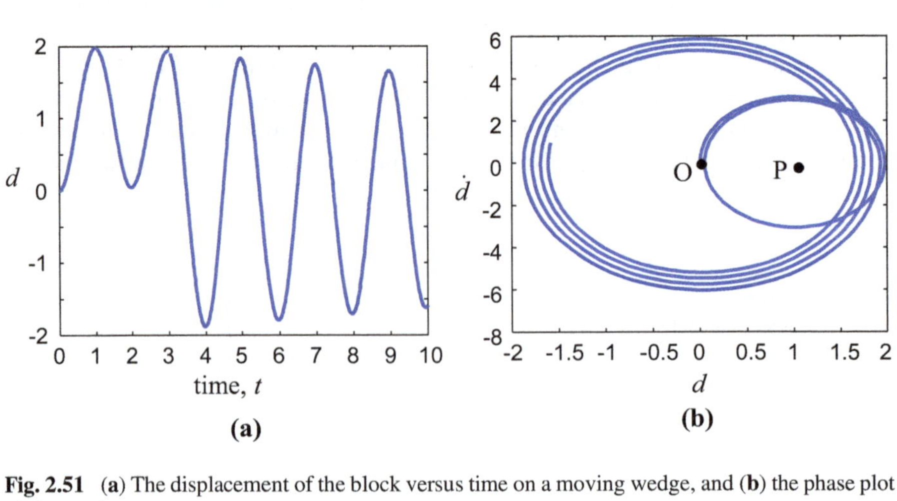

Fig. 2.51 (**a**) The displacement of the block versus time on a moving wedge, and (**b**) the phase plot of the motion

equilibrium position changes to $d_e = 0$ so the phase plot transforms to that of a slowly decaying oscillation about that new equilibrium position (point O). This behavior is also evident in Fig. 2.51a.

2.9 Other Principles and Virtual Quantities

The last section showed that if we have an ideal constraint force given in terms of the constraints as $\mathbf{F}^c = \mathbf{A}^T\boldsymbol{\lambda}$ and if we can find any vector, $\mathbf{u}$, such that it satisfies the constraints in the form

$$\mathbf{A}\mathbf{u} = 0 \tag{2.9.1}$$

then the virtual "work" of the constraint force taken through this vector is zero since

$$\mathbf{u}^T\mathbf{F}^c = \mathbf{u}^T\mathbf{A}^T\boldsymbol{\lambda} = (\mathbf{A}\mathbf{u})^T\boldsymbol{\lambda} = 0 \tag{2.9.2}$$

If the vector $\mathbf{u}$ is a displacement, then Eq. (2.9.2) indeed has the dimensions of work but we will still refer to Eq. (2.9.2) as a statement of virtual "work" for other choices of $\mathbf{u}$. The virtual displacement, $\delta\mathbf{u}$, defined in that section as the difference between two possible differential displacements taken at the same time, t, obviously qualifies as such a vector, $\mathbf{u}$. However, we will now examine other vectors $\mathbf{u}$ that can be used as well. For example, consider non-holonomic constraints of the form

$$\mathbf{f}(\mathbf{x}, \dot{\mathbf{x}}, t) = \mathbf{A}(\mathbf{x}, t)\dot{\mathbf{x}} - \mathbf{a}_t(\mathbf{x}, t) = 0 \tag{2.9.3}$$

If we have two possible velocities $(\dot{\mathbf{x}}', \dot{\mathbf{x}}'')$ that satisfy these constraints at the same position $\mathbf{x}$, and time, t, then

$$\begin{aligned} \mathbf{A}(\mathbf{x}, t)\dot{\mathbf{x}}' - \mathbf{a}_t(\mathbf{x}, t) &= 0 \\ \mathbf{A}(\mathbf{x}, t)\dot{\mathbf{x}}'' - \mathbf{a}_t(\mathbf{x}, t) &= 0 \end{aligned} \tag{2.9.4}$$

Then taking the difference of these equations and defining a virtual velocity as $\delta_J\dot{\mathbf{x}} \triangleq (\dot{\mathbf{x}}'' - \dot{\mathbf{x}}')$, we have

$$\begin{aligned} \mathbf{A}\delta_J\dot{\mathbf{x}} &= 0 \\ \delta_J\dot{\mathbf{x}}^T\mathbf{F}^c &= 0 \end{aligned} \tag{2.9.5}$$

We will call this velocity difference a *Jourdain virtual velocity*. Note that the difference of the two possible velocities need not be small so that the Jourdain virtual velocity can be of arbitrary size. Note also that holonomic constraints can be differentiated and placed into the same form as Eq. (2.9.3), so that for both holonomic and non-holonomic constraints of this form we can use the Jourdain velocity in place of the previously defined virtual displacement. Since the virtual "work" done by the constraint forces is zero, if we dot Newton's second law with the Jourdain virtual velocity we obtain the *Jourdain variational equation*

$$\left(m\ddot{\mathbf{x}} - \mathbf{F}^e\right) \cdot \delta_J \dot{\mathbf{x}} = 0 \quad (\delta_J \mathbf{x} = 0, \ \ \delta_J t = 0) \tag{2.9.6}$$

in contrast to the D'Alembert-Lagrange equation, which we rewrite here as

$$\left(m\ddot{\mathbf{x}} - \mathbf{F}^e\right) \cdot \delta \mathbf{x} = 0 \quad (\delta t = 0) \tag{2.9.7}$$

The Jourdain equation is very useful for considering non-holonomic problems in the form of Eq. (2.9.3), which is the most commonly encountered form for non-holonomic constraints, and can be used to embed those non-holonomic constraints in the equations of motion. We will not pursue that topic further in this book but the reader can see some such examples of this type of use of the Jourdain equation in [8]. However, recall that if we differentiate holonomic constraints twice on time, or differentiate non-holonomic constraints of the form given by Eq. (2.9.3) once on time, we arrive at a common form that we have used frequently in past sections, i.e.,

$$\mathbf{A}\left(\mathbf{x}, \dot{\mathbf{x}}, t\right)\ddot{\mathbf{x}} - \mathbf{b}\left(\mathbf{x}, \dot{\mathbf{x}}, t\right) = 0 \tag{2.9.8}$$

As mentioned previously, Eq. (2.9.8) is also obtained if we differentiate once on time more general non-holonomic constraints of the form

$$\mathbf{f}\left(\mathbf{x}, \dot{\mathbf{x}}, t\right) = 0 \tag{2.9.9}$$

Now consider two possible accelerations $\left(\ddot{\mathbf{x}}'', \ddot{\mathbf{x}}'\right)$ that satisfy Eq. (2.9.8) at the same position $\mathbf{x}$, time t, and velocity $\dot{\mathbf{x}}$. Then

$$\begin{aligned} \mathbf{A}\left(\mathbf{x}, \dot{\mathbf{x}}, t\right)\ddot{\mathbf{x}}'' - \mathbf{b}\left(\mathbf{x}, \dot{\mathbf{x}}, t\right) = 0 \\ \mathbf{A}\left(\mathbf{x}, \dot{\mathbf{x}}, t\right)\ddot{\mathbf{x}}' - \mathbf{b}\left(\mathbf{x}, \dot{\mathbf{x}}, t\right) = 0 \end{aligned} \tag{2.9.10}$$

Then taking the difference and defining a new virtual acceleration as $\delta_G \ddot{\mathbf{x}} \triangleq \left(\ddot{\mathbf{x}}'' - \ddot{\mathbf{x}}'\right)$, we have

$$\begin{aligned} \mathbf{A}\delta_G \ddot{\mathbf{x}} = 0 \\ \delta_G \ddot{\mathbf{x}}^T \mathbf{F}^c = 0 \end{aligned} \tag{2.9.11}$$

We will call this acceleration a Gaussian virtual acceleration. Like the Jourdain virtual velocity this virtual acceleration need not be a differentially small quantity since it is based on the finite difference between two possible accelerations. Again, since the virtual "work" done by this virtual acceleration is zero when we dot Newton's second law with this acceleration we obtain *Gibbs-Gauss variational equation*:

$$(m\ddot{\mathbf{x}} - \mathbf{F}^e) \cdot \delta_G \ddot{\mathbf{x}} = 0 \quad (\delta_G \mathbf{x} = 0, \delta_G \dot{\mathbf{x}} = 0, \delta_G t = 0) \tag{2.9.12}$$

This Gibbs-Gauss equation can handle a wide range of holonomic and non-holonomic constraints but its real value is that it can be used to formulate a fundamental law of dynamics called *Gauss's law of least constraint*. To see this, first multiply Newton's second law by the inverse of the mass matrix, $\mathbf{M}^{-1}$, which gives

$$\ddot{\mathbf{x}} = \mathbf{M}^{-1}\mathbf{F}^e + \mathbf{M}^{-1}\mathbf{F}^c \tag{2.9.13}$$

and place this expression for $\ddot{\mathbf{x}}$ into the constraint equation, Eq. (2.9.8), and we have

$$\mathbf{A}\mathbf{M}^{-1}\mathbf{F}^c = \mathbf{b} - \mathbf{A}\mathbf{M}^{-1}\mathbf{F}^e \tag{2.9.14}$$

Even though Eq. (2.9.14) has multiple solutions for the constraint force, $\mathbf{F}^c$, if we place some conditions on the solution of Eq. (2.9.14) we can actually solve it for a unique constraint force associated with the constrained motion. This is possible because Gauss's law of least constraint provides just such a condition. Gauss's law that says that for any possible acceleration, $\ddot{\mathbf{x}}'$, that satisfies the constraints at a fixed position, velocity, and time, if we form up a scalar function of the acceleration, $G(\ddot{\mathbf{x}}')$, where G is the *Gaussian function*

$$G(\ddot{\mathbf{x}}') = \frac{1}{2}(\ddot{\mathbf{x}}' - \mathbf{M}^{-1}\mathbf{F}^e)^T \mathbf{M}(\ddot{\mathbf{x}}' - \mathbf{M}^{-1}\mathbf{F}^e) \tag{2.9.15}$$

then G is a minimum for the actual acceleration, $\ddot{\mathbf{x}}$. To show this, let $\ddot{\mathbf{x}} + \Delta\ddot{\mathbf{x}}$ be a possible acceleration, i.e., one that satisfies Eq. (2.9.8), where $\ddot{\mathbf{x}}$ is the actual acceleration. Then $\Delta\ddot{\mathbf{x}}$ must also satisfy the constraints. Now, consider $G(\ddot{\mathbf{x}} + \Delta\ddot{\mathbf{x}})$. We have

$$\begin{aligned} G(\ddot{\mathbf{x}} + \Delta\ddot{\mathbf{x}}) &= \frac{1}{2}(\ddot{\mathbf{x}} + \Delta\ddot{\mathbf{x}} - \mathbf{M}^{-1}\mathbf{F}^e)^T \mathbf{M}(\ddot{\mathbf{x}} + \Delta\ddot{\mathbf{x}} - \mathbf{M}^{-1}\mathbf{F}^e) \\ &= \frac{1}{2}(\ddot{\mathbf{x}} - \mathbf{M}^{-1}\mathbf{F}^e)^T \mathbf{M}(\ddot{\mathbf{x}} - \mathbf{M}^{-1}\mathbf{F}^e) + \Delta\ddot{\mathbf{x}}^T \mathbf{M}(\ddot{\mathbf{x}} - \mathbf{M}^{-1}\mathbf{F}^e) + \frac{1}{2}\Delta\ddot{\mathbf{x}}^T \mathbf{M}\Delta\ddot{\mathbf{x}} \\ &= G(\ddot{\mathbf{x}}) + \Delta\ddot{\mathbf{x}}^T \mathbf{F}^c + \frac{1}{2}\Delta\ddot{\mathbf{x}}^T \mathbf{M}\Delta\ddot{\mathbf{x}} \end{aligned} \tag{2.9.16}$$

But the virtual "work" term $\Delta\ddot{\mathbf{x}}^T \mathbf{F}^c$ is zero since $\Delta\ddot{\mathbf{x}}$ is a possible acceleration so that it can serve as the virtual acceleration in Eq. (2.9.12). This shows that for any $\Delta\ddot{\mathbf{x}} \neq 0$

$$G(\ddot{\mathbf{x}} + \Delta\ddot{\mathbf{x}}) - G(\ddot{\mathbf{x}}) = \frac{1}{2}\Delta\ddot{\mathbf{x}}^T \mathbf{M}\Delta\ddot{\mathbf{x}} > 0 \tag{2.9.17}$$

which is true since the mass matrix $\mathbf{M}$ is a positive definite matrix. Thus G always has its smallest value for the actual motion.

Gauss's law is called the law of *least constraint* since we can write it in terms of the constraint force (and the mass matrix, which acts as a weighting factor):

$$G = \left(\mathbf{M}^{-1}\mathbf{F}^c\right)^T\mathbf{M}\left(\mathbf{M}^{-1}\mathbf{F}^c\right) = \left(\mathbf{F}^c\right)^T\mathbf{M}^{-1}\left(\mathbf{F}^c\right) \tag{2.9.18}$$

so that when we minimize G we are minimizing the constraint force in this weighted fashion.

Now, let

$$\mathbf{y} = \mathbf{M}^{-1/2}\mathbf{F}^c \tag{2.9.19}$$

Normally, we cannot take square roots or inverse square roots of matrices, but because $\mathbf{M}$ is a positive definite matrix this operation has meaning. For example, if $\mathbf{M}$ is just a diagonal matrix of positive masses such as

$$\mathbf{M} = \begin{bmatrix} m & 0 & 0 \\ 0 & m & 0 \\ 0 & 0 & m \end{bmatrix} \tag{2.9.20}$$

then

$$\mathbf{M}^{1/2} = \begin{bmatrix} \sqrt{m} & 0 & 0 \\ 0 & \sqrt{m} & 0 \\ 0 & 0 & \sqrt{m} \end{bmatrix}, \quad \mathbf{M}^{-1/2} = \begin{bmatrix} 1/\sqrt{m} & 0 & 0 \\ 0 & 1/\sqrt{m} & 0 \\ 0 & 0 & 1/\sqrt{m} \end{bmatrix} \tag{2.9.21}$$

Positive definite matrices that are not diagonal can be put into these same forms and their square roots or inverse square roots defined similarly. In terms of the vector $\mathbf{y}$ the Gaussian function G is

$$G = \mathbf{y}^T\mathbf{y} = |\mathbf{y}|^2 \tag{2.9.22}$$

and the constraint equations (see Eq. (2.9.13)) become

$$\mathbf{B}\mathbf{y} = \mathbf{c} \tag{2.9.23}$$

where $\mathbf{B} = \mathbf{A}\mathbf{M}^{-1/2}, \mathbf{c} = \mathbf{b} - \mathbf{A}\mathbf{M}^{-1}\mathbf{F}^e$. Thus, Gauss's law of least constraint written in terms of $\mathbf{y}$ states that for all $\mathbf{y}$ that satisfy the constraint equation, Eq. (2.9.23), we want to find a $\mathbf{y}$ where the norm of the vector $\mathbf{y}$ is a minimum. In this form Gauss's law is a well-known mathematical problem that has a solution for $\mathbf{y}$. If the constraint equations are linearly independent the unique solution for $\mathbf{y}$ that satisfies Eq. (2.9.23) and minimizes the norm of the vector $\mathbf{y}$ can be shown to be given by [2]

$$\mathbf{y} = \mathbf{B}^T\left(\mathbf{B}\mathbf{B}^T\right)^{-1}\mathbf{c} \tag{2.9.24}$$

or, more explicitly,

$$\begin{aligned}\mathbf{M}^{-1/2}\mathbf{F}^c &= \left(\mathbf{A}\mathbf{M}^{-1/2}\right)^T\left[\mathbf{A}\mathbf{M}^{-1/2}\left(\mathbf{A}\mathbf{M}^{-1/2}\right)^T\right]^{-1}\left(\mathbf{b} - \mathbf{A}\mathbf{M}^{-1}\mathbf{F}^e\right) \\ &= \mathbf{M}^{-1/2}\mathbf{A}^T\left[\mathbf{A}\mathbf{M}^{-1}\mathbf{A}^T\right]^{-1}\left(\mathbf{b} - \mathbf{A}\mathbf{M}^{-1}\mathbf{F}^e\right)\end{aligned} \tag{2.9.25}$$

which can be multiplied by $\mathbf{M}^{1/2}$ on both sides to give the constraint force as

$$\mathbf{F}^c = \mathbf{A}^T\left[\mathbf{A}\mathbf{M}^{-1}\mathbf{A}^T\right]^{-1}\left(\mathbf{b} - \mathbf{A}\mathbf{M}^{-1}\mathbf{F}^e\right) \tag{2.9.26}$$

which is just the solution for an ideal constraint force that we have been using previously. This value for the constraint force also gives the previously obtained equations of motion explicitly as

$$\ddot{\mathbf{x}} = \mathbf{M}^{-1}\mathbf{F}^e + \mathbf{M}^{-1}\mathbf{A}^T\left[\mathbf{A}\mathbf{M}^{-1}\mathbf{A}^T\right]^{-1}\left(\mathbf{b} - \mathbf{A}\mathbf{M}^{-1}\mathbf{F}^e\right) \tag{2.9.27}$$

Satisfying Gauss's law of least constraint requires that all constraint forces that come from independent constraints of the form $\mathbf{A}\ddot{\mathbf{x}} = \mathbf{b}$ must be ideal constraint forces of the form $\mathbf{F}^c = \mathbf{A}^T\boldsymbol{\lambda}$, as we have assumed previously. Furthermore, the constraint forces are explicitly given by Eq. (2.9.26) and the equations of constrained motion must satisfy Eq. (2.9.27). Even Newton's second law for unconstrained motion follows from this fundamental law. One way to see this is to simply set $\mathbf{F}^c = 0$ in Eq. (2.9.27) which then reduces to Newton's second law for the unconstrained acceleration. Alternatively, we could examine the variation of G due to virtual changes of the acceleration. When G is a minimum we must have

$$\begin{aligned}\delta_G G &= \delta_G\left[\frac{1}{2}\left(\ddot{\mathbf{x}} - \mathbf{M}^{-1}\mathbf{F}^e\right)^T\mathbf{M}\left(\ddot{\mathbf{x}} - \mathbf{M}^{-1}\mathbf{F}^e\right)\right] \\ &= \frac{1}{2}\delta_G\ddot{\mathbf{x}}^T\left(\mathbf{M}\ddot{\mathbf{x}} - \mathbf{F}^e\right) + \frac{1}{2}\left(\mathbf{M}\ddot{\mathbf{x}} - \mathbf{F}^e\right)^T\delta_G\ddot{\mathbf{x}} \\ &= \delta_G\ddot{\mathbf{x}}^T\left(\mathbf{M}\ddot{\mathbf{x}} - \mathbf{F}^e\right) = 0\end{aligned} \tag{2.9.28}$$

for the actual motion (where we have used the fact that the mass matrix is symmetric, i.e., $\mathbf{M} = \mathbf{M}^T$, and the general property of the transpose operation $(\mathbf{AB})^T = \mathbf{B}^T\mathbf{A}^T$). But if the motion is unconstrained $\delta_G\ddot{\mathbf{x}}$ is arbitrary so that we must also have

$$\mathbf{M}\ddot{\mathbf{x}} - \mathbf{F}^e = 0 \tag{2.9.29}$$

which is just Newton's second law. It is remarkable that Gauss's law of least constraint provides the foundation for much of the field of dynamics that we have discussed in previous sections.

In Sect. 2.7 we showed that in terms of generalized coordinates we could write components of Newton's second law in the form

$$\mathbf{M}\ddot{\mathbf{q}} = \bar{\mathbf{Q}}^e + \mathbf{Q}^c = \mathbf{Q}^{in} + \mathbf{Q}^e + \mathbf{Q}^c \tag{2.9.30}$$

We can write Gauss's law of least constraint in generalized coordinates as

$$\begin{aligned} G &= \frac{1}{2}(\ddot{\mathbf{q}} - \mathbf{M}^{-1}\bar{\mathbf{Q}}^e)^T\mathbf{M}(\ddot{\mathbf{q}} - \mathbf{M}^{-1}\bar{\mathbf{Q}}^e) \\ &= \frac{1}{2}(\mathbf{Q}^c)^T\mathbf{M}^{-1}(\mathbf{Q}^c) \end{aligned} \tag{2.9.31}$$

which is in the same form as Eq. (2.9.15) so all of our previous results obtained for Cartesian coordinates are true for generalized coordinates by making the replacements $\mathbf{x} \rightarrow \mathbf{q}$, $\mathbf{F}^e \rightarrow \bar{\mathbf{Q}}^e$, and $\mathbf{F}^c \rightarrow \mathbf{Q}^c$ and noting that the mass matrix in Cartesian coordinates becomes the generalized mass matrix for generalized coordinates.

2.10 Nonideal Constraint Forces

So far we have only considered ideal constraint forces when formulating the equations of motion. However, we can handle other constraint forces without much difficulty if we follow the approach outlined recently by Udwadia [9]. Specifically, we will show that by generalizing Gauss's law of least constraint appropriately we can obtain explicit expressions for both the ideal and nonideal constraint forces and generate the equations of motion.

First, consider Newton's second law for a particle in Cartesian coordinates when constraint forces are present, written in matrix-vector form:

$$\mathbf{M}\ddot{\mathbf{x}} - \mathbf{F}^e - \mathbf{F}^c - \mathbf{F}^{c;ni} = 0 \tag{2.10.1}$$

where $\mathbf{M} = m\mathbf{I}$ is the mass matrix, $\mathbf{F}^e$ is the known net external force, and $\mathbf{F}^c$ is the total ideal constraint force present, and we now also allow a total nonideal constraint force, $\mathbf{F}^{c;ni}$. Let the virtual "work" done by the ideal and nonideal constraint forces be specified explicitly as

$$\begin{aligned} \delta_G\ddot{\mathbf{x}}^T\mathbf{F}^c &= 0 \\ \delta_G\ddot{\mathbf{x}}^T\mathbf{F}^{c;ni} &= \delta_G\ddot{\mathbf{x}}^T\mathbf{K} \end{aligned} \tag{2.10.2}$$

where $\mathbf{K}$ is a known vector. In cases such as Coulomb friction, $\mathbf{K}$ could be a function of the ideal constraint forces present. Note that Eq. (2.10.10) does not require that $\mathbf{F}^{c;ni} = \mathbf{K}$ since in principle there could be parts of the nonideal constraint force that are also workless. If we dot Newton's second law with the Gaussian virtual acceleration and use Eq. (2.10.2) we find

$$\delta_G \ddot{\mathbf{x}}^T \left(\mathbf{M}\ddot{\mathbf{x}} - \mathbf{F}^e - \mathbf{K} \right) = 0 \tag{2.10.3}$$

Recall that for the case of all ideal constraints, in place of Eq. (2.10.3) we obtained

$$\delta_G \ddot{\mathbf{x}}^T \left(\mathbf{M}\ddot{\mathbf{x}} - \mathbf{F}^e \right) = 0 \tag{2.10.4}$$

and we showed that for any possible acceleration, $\ddot{\mathbf{x}}'$, that satisfies the constraints the Gaussian function, $G(\ddot{\mathbf{x}}')$, was a minimum, where

$$G(\ddot{\mathbf{x}}') = \frac{1}{2} \left(\ddot{\mathbf{x}}' - \mathbf{M}^{-1}\mathbf{F}^e \right)^T \mathbf{M} \left(\ddot{\mathbf{x}}' - \mathbf{M}^{-1}\mathbf{F}^e \right) \tag{2.10.5}$$

This suggests that when nonideal constraint forces are present that satisfy Eq. (2.10.2) a modified Gaussian function, $\tilde{G}(\ddot{\mathbf{x}}')$, is a minimum, among all possible accelerations, at the actual acceleration, $\ddot{\mathbf{x}}$, where

$$\tilde{G}(\ddot{\mathbf{x}}') = \frac{1}{2} \left(\ddot{\mathbf{x}}' - \mathbf{M}^{-1}\mathbf{F}^e - \mathbf{M}^{-1}\mathbf{K} \right)^T \mathbf{M} \left(\ddot{\mathbf{x}}' - \mathbf{M}^{-1}\mathbf{F}^e - \mathbf{M}^{-1}\mathbf{K} \right) \tag{2.10.6}$$

To show this extended version of Gauss's law, as in the ideal constraint force case let $\ddot{\mathbf{x}} + \Delta\ddot{\mathbf{x}}$ be a possible acceleration, where $\ddot{\mathbf{x}}$ is the actual acceleration. Then $\Delta\ddot{\mathbf{x}}$ must also satisfy the constraints so that

$$\Delta\ddot{\mathbf{x}}^T \mathbf{F}^c = 0$$

Now, consider $G(\ddot{\mathbf{x}} + \Delta\ddot{\mathbf{x}})$. To keep the complexity of the expressions to a minimum, let $\mathbf{M}^{-1}\mathbf{F}^e = \mathbf{a}$ and $\mathbf{M}^{-1}\mathbf{K} = \mathbf{k}$. Then

$$\tilde{G}(\ddot{\mathbf{x}} + \Delta\ddot{\mathbf{x}}) = \frac{1}{2} \left(\ddot{\mathbf{x}} + \Delta\ddot{\mathbf{x}} - \mathbf{a} - \mathbf{k} \right)^T \mathbf{M} \left(\ddot{\mathbf{x}} + \Delta\ddot{\mathbf{x}} - \mathbf{a} - \mathbf{k} \right) \tag{2.10.7a}$$

or, equivalently,

$$\begin{aligned}\tilde{G}(\ddot{\mathbf{x}} + \Delta\ddot{\mathbf{x}}) &= \frac{1}{2}(\ddot{\mathbf{x}} - \mathbf{a} - \mathbf{k})^T\mathbf{M}(\ddot{\mathbf{x}} + \Delta\ddot{\mathbf{x}} - \mathbf{a} - \mathbf{k}) \\ &\quad + \frac{1}{2}\Delta\ddot{\mathbf{x}}^T\mathbf{M}(\ddot{\mathbf{x}} + \Delta\ddot{\mathbf{x}} - \mathbf{a} - \mathbf{k}) \\ &= \frac{1}{2}(\ddot{\mathbf{x}} - \mathbf{a} - \mathbf{k})^T\mathbf{M}(\ddot{\mathbf{x}} - \mathbf{a} - \mathbf{k}) + \frac{1}{2}(\ddot{\mathbf{x}} - \mathbf{a} - \mathbf{k})^T\mathbf{M}\Delta\ddot{\mathbf{x}} \\ &\quad + \frac{1}{2}\Delta\ddot{\mathbf{x}}^T\mathbf{M}(\ddot{\mathbf{x}} + \Delta\ddot{\mathbf{x}} - \mathbf{a} - \mathbf{k}) \\ &= \tilde{G}(\ddot{\mathbf{x}}) + \frac{1}{2}\Delta\ddot{\mathbf{x}}^T\mathbf{M}\Delta\ddot{\mathbf{x}} + \Delta\ddot{\mathbf{x}}^T\mathbf{M}(\ddot{\mathbf{x}} - \mathbf{a} - \mathbf{k})\end{aligned} \tag{2.10.7b}$$

But

$$\Delta\ddot{\mathbf{x}}^T\mathbf{M}(\ddot{\mathbf{x}} - \mathbf{a} - \mathbf{k}) = \Delta\ddot{\mathbf{x}}^T(\mathbf{M}\ddot{\mathbf{x}} - \mathbf{F}^e - \mathbf{K}) = 0 \tag{2.10.8}$$

since $\Delta\ddot{\mathbf{x}}$ can serve as a virtual acceleration in Eq. (2.10.3) and we find

$$\tilde{G}(\ddot{\mathbf{x}} + \Delta\ddot{\mathbf{x}}) - \tilde{G}(\ddot{\mathbf{x}}) = \frac{1}{2}\Delta\ddot{\mathbf{x}}^T\mathbf{M}\Delta\ddot{\mathbf{x}} > 0 \tag{2.10.9}$$

which is true since the mass matrix $\mathbf{M}$ is a positive definite matrix. Thus $\tilde{G}$, like G, always has its smallest value for the actual motion. If we let

$$\mathbf{y} = \mathbf{M}^{1/2}(\ddot{\mathbf{x}} - \mathbf{M}^{-1}\mathbf{F}^e - \mathbf{M}^{-1}\mathbf{K}) \tag{2.10.10}$$

then $\tilde{G} = \mathbf{y}^T\mathbf{y}$, as in the ideal constraint force only case, and the constraint equation $\mathbf{A}\ddot{\mathbf{x}} = \mathbf{b}$ becomes

$$\mathbf{B}\mathbf{y} = \mathbf{c} \tag{2.10.11}$$

where now

$$\begin{aligned}\mathbf{B} &= \mathbf{A}\mathbf{M}^{-1/2} \\ \mathbf{c} &= \mathbf{b} - \mathbf{A}\mathbf{M}^{-1}\mathbf{F}^e - \mathbf{A}\mathbf{M}^{-1}\mathbf{K}\end{aligned} \tag{2.10.12}$$

As in the ideal constraint case if the constraint equations are linearly independent the unique solution for $\mathbf{y}$ that satisfies Eq. (2.10.11) and minimizes the norm of the vector $\mathbf{y}$ can be shown as

$$\mathbf{y} = \mathbf{B}^T(\mathbf{B}\mathbf{B}^T)^{-1}\mathbf{c} \tag{2.10.13}$$

Placing the expressions for **B** and **c** into Eq. (2.10.13), after some algebraic steps very similar to the ideal constraint case, we can write the equations of motion as

$$\begin{aligned}\ddot{\mathbf{x}} &= \mathbf{M}^{-1}\mathbf{F}^e + \mathbf{M}^{-1}\mathbf{A}^T\left(\mathbf{A}\mathbf{M}^{-1}\mathbf{A}^T\right)^{-1}\left(\mathbf{b} - \mathbf{A}\mathbf{M}^{-1}\mathbf{F}^e\right) \\ &\quad + \mathbf{M}^{-1}\left[\mathbf{I} - \mathbf{A}^T\left(\mathbf{A}\mathbf{M}^{-1}\mathbf{A}^T\right)^{-1}\mathbf{A}\mathbf{M}^{-1}\right]\mathbf{K}\end{aligned} \tag{2.10.14}$$

where **I** is the identity matrix. There is now an ideal constraint force present in Eq. (2.10.14) in exactly the same form as before as well as a nonideal constraint force:

$$\begin{aligned}\mathbf{F}^c &= \mathbf{A}^T\left(\mathbf{A}\mathbf{M}^{-1}\mathbf{A}^T\right)^{-1}\left(\mathbf{b} - \mathbf{A}\mathbf{M}^{-1}\mathbf{F}^e\right) \\ \mathbf{F}^{c;ni} &= \left[\mathbf{I} - \mathbf{A}^T\left(\mathbf{A}\mathbf{M}^{-1}\mathbf{A}^T\right)\mathbf{A}\mathbf{M}^{-1}\right]\mathbf{K}\end{aligned} \tag{2.10.15}$$

These constraints do satisfy $\delta_G\ddot{\mathbf{x}}^T\mathbf{F}^c = 0$ and $\delta_G\ddot{\mathbf{x}}^T\mathbf{F}^{c;ni} = \delta_G\ddot{\mathbf{x}}^T\mathbf{K}$. We see that the part of the nonideal constraint force given by $\mathbf{A}^T(\mathbf{A}\mathbf{M}^{-1}\mathbf{A}^T)^{-1}\mathbf{A}\mathbf{M}^{-1}\mathbf{K}$ is a workless part since $\delta_G\ddot{\mathbf{x}}^T\mathbf{A}^T = \left(\mathbf{A}\delta_G\ddot{\mathbf{x}}\right)^T = 0$. If this workless part is zero then the nonideal constraint force is just $\mathbf{F}^{c;ni} = \mathbf{K}$. This can occur, for example, in the important special case of Coulomb friction when a particle of mass m moves on a rough surface. In this case, if we specify **K** as the vector

$$\mathbf{K} = -\mu|\mathbf{F}^c|\frac{\dot{\mathbf{x}}}{|\dot{\mathbf{x}}|} \tag{2.10.16}$$

where $\mathbf{F}^c$ is the ideal normal constraint force exerted by a smooth surface on the particle and μ is the coefficient of sliding friction then

$$\mathbf{M}^{-1}\mathbf{K} = -\frac{\mu}{m}|\mathbf{F}^c|\frac{\dot{\mathbf{x}}}{|\dot{\mathbf{x}}|} \tag{2.10.17}$$

which is an acceleration that is tangential to the surface (and always opposite to the direction of motion). However, recall that **A** is also normal to the surface so that $\mathbf{A}\mathbf{M}^{-1}\mathbf{K} = 0$ and from Eqs. (2.10.15) and (2.10.16) it follows that

$$\mathbf{F}^{c,ni} = -\mu|\mathbf{F}^c|\frac{\dot{\mathbf{x}}}{|\dot{\mathbf{x}}|} \tag{2.10.18}$$

which is just the ordinary Coulomb frictional force. In cases like these we see that we can simply add the **K** vector into the equation of motion, using the same ideal constraint force generated by the constraint equations in the case of ideal constraints only in the expression for **K** (assuming that **K** depends on the ideal constraints).

We have obtained all our results so far in terms of Cartesian components but as with the ideal constraint force case very similar forms can be found in terms of n_s independent generalized coordinates where the Gibbs-Gauss virtual acceleration and generalized force are

$$\begin{aligned}\delta_G \ddot{\mathbf{x}} &= \sum_{k=1}^{n_s} \partial \ddot{\mathbf{x}} / \partial \ddot{q}_k \delta_G \ddot{q}_k = \sum_{k=1}^{n_s} \mathbf{c}_k \delta_G \ddot{q}_k \\ Q_k &= \mathbf{F} \cdot \mathbf{c}_k\end{aligned} \tag{2.10.19}$$

from which the virtual "work" of the nonideal generalized constraint force can be expressed in terms of a known vector, $\bar{\mathbf{K}}$, whose components are $\bar{K}_k$, as

$$\delta_G \ddot{\mathbf{x}}^T \mathbf{F}^{c;ni} = \sum_{k=1}^{n_s} Q_k^{c;ni} \delta_G \ddot{q}_k = \sum_{k=1}^{n_s} \bar{K}_k \delta_G \ddot{q}_k \tag{2.10.20}$$

and

$$\bar{K}_k = \mathbf{K} \cdot \mathbf{c}_k \tag{2.10.21}$$

All of the steps then follow in a similar fashion as the Cartesian component case, as discussed in the last section, so we only show the final results for the equations of motion in generalized coordinates and generalized constraint forces as

$$\begin{aligned}\ddot{\mathbf{q}} &= \mathbf{M}^{-1}\bar{\mathbf{Q}}^e - \mathbf{M}^{-1}\mathbf{A}^T\left(\mathbf{A}\mathbf{M}^{-1}\mathbf{A}^T\right)^{-1}\left(\mathbf{b} - \mathbf{A}\mathbf{M}^{-1}\bar{\mathbf{Q}}^e\right) \\ &\quad + \mathbf{M}^{-1}\left[\mathbf{I} - \mathbf{A}^T\left(\mathbf{A}\mathbf{M}^{-1}\mathbf{A}^T\right)^{-1}\mathbf{A}\mathbf{M}^{-1}\right]\overline{\mathbf{K}} \\ \mathbf{Q}^c &= \mathbf{A}^T\left(\mathbf{A}\mathbf{M}^{-1}\mathbf{A}^T\right)^{-1}\left(\mathbf{b} - \mathbf{A}\mathbf{M}^{-1}\bar{\mathbf{Q}}^e\right) \\ \mathbf{Q}^{c;ni} &= \left[\mathbf{I} - \mathbf{A}^T\left(\mathbf{A}\mathbf{M}^{-1}\mathbf{A}^T\right)^{-1}\mathbf{A}\mathbf{M}^{-1}\right]\overline{\mathbf{K}}\end{aligned} \tag{2.10.22}$$

where we have lumped the known generalized forces and inertial forces into the $\bar{\mathbf{Q}}^e$ force term, as done before.

As an example, let's consider the roller-coaster problem of Sect. 2.3.2 but now assume that the track is rough with a coefficient of sliding Coulomb friction, μ. The roller coaster was a special case of the general problem of motion on a 2-D smooth surface of the form $f(x, y) = y - h(x)$, where $h(x) = \frac{A}{2}[1 - \cos(\pi x/\mathrm{L})]$. Some of the relevant terms appearing in Eq. (2.10.14) are now (see Eq. (2.3.44))

$$
\begin{aligned}
\mathbf{A} &= [\partial f/\partial x, \partial f/\partial y] = [-dh/dx, 1] \\
\mathbf{M} &= \begin{bmatrix} m & 0 \\ 0 & m \end{bmatrix} \\
\mathbf{K} &= -\frac{\mu|\mathbf{F}^c|}{\sqrt{\dot{x}^2+\dot{y}^2}}\begin{bmatrix} \dot{x} \\ \dot{y} \end{bmatrix} = -\frac{\mu|\mathbf{F}^c|\dot{x}}{|\dot{x}|\sqrt{1+(dh/dx)^2}}\begin{bmatrix} 1 \\ dh/dx \end{bmatrix} \\
N &= \frac{m}{\sqrt{1+(dh/dx)^2}}\left(g+\frac{d^2h}{dx^2}\dot{x}^2\right) \\
\mathbf{F}^c &= N\mathbf{e}_n = \frac{N}{\sqrt{1+(dh/dx)^2}}\begin{bmatrix} -dh/dx \\ 1 \end{bmatrix}
\end{aligned}
\tag{2.10.23}
$$

It is easy to see that $\mathbf{A}\mathbf{M}^{-1}\mathbf{K} = 0$ and so indeed $\mathbf{F}^{c;ni} = \mathbf{K}$. Thus, the original equations of motion for the case of no friction simply are modified by the addition of the x- and y-components of $\mathbf{K}$, giving

$$
\ddot{x} = \frac{-dh/dx}{1+(dh/dx)^2}\left(g+\frac{d^2h}{dx^2}\dot{x}^2\right) - \frac{\mu}{1+(dh/dx)^2}\left|\left(g+\frac{d^2h}{dx^2}\dot{x}^2\right)\right|\frac{\dot{x}}{|\dot{x}|} \tag{2.10.24}
$$

$$
\ddot{y} = -g + \frac{1}{1+(dh/dx)^2}\left(g+\frac{d^2h}{dx^2}\dot{x}^2\right) - \frac{\mu(dh/dx)}{1+(dh/dx)^2}\left|\left(g+\frac{d^2h}{dx^2}\dot{x}^2\right)\right|\frac{\dot{x}}{|\dot{x}|} \tag{2.10.25}
$$

Coulomb frictional forces, like the drag forces we considered earlier, slow the particle down and eventually bring it to rest. However, from statics we know that once at rest the particle will remain at rest unless the tangential component of the net external force acting along the surface exceeds a threshold. For Coulomb friction, if the particle comes to rest it will remain at rest if

$$
|\mathbf{F}^e \cdot \mathbf{e}_t| \leq \mu|\mathbf{F}^c| \tag{2.10.26}
$$

where $\mathbf{e}_t$ is a unit vector tangent to the rough surface. In our example

$$
\begin{aligned}
\mathbf{e}_t &= \frac{1}{\sqrt{1+(dh/dx)^2}}\begin{bmatrix} 1 \\ dh/dx \end{bmatrix} \\
\mathbf{F}^e &= \begin{bmatrix} 0 \\ mg \end{bmatrix}
\end{aligned}
\tag{2.10.27}
$$

which yields

$$\left|\frac{dh}{dx}\right| \leq \mu \tag{2.10.28}$$

You may recognize Eq. (2.10.28) from statics where it is normally shown that a mass sitting on a rough inclined surface under the external force of its own weight will remain in equilibrium as long as the slope of the surface (which here is dh/dx) is less than the static coefficient of friction. [Note that we have let the dynamic coefficient of sliding friction appearing in Eqs. (2.10.24) and (2.10.25) to be the same as the static coefficient in Eqs. (2.10.26) and (2.10.28). In general the dynamic and static values of the friction coefficient might be slightly different. We can easily take any difference into account, if needed.] For the roller-coaster problem Eq. (2.10.28) is

$$\left|\frac{\pi A}{2L}\sin\left(\frac{\pi x}{L}\right)\right| \leq \mu \tag{2.10.29}$$

which determines the "sticking regions" on the roller coaster. For $\mu \geq \pi A/2L$ the mass can stick anywhere along the roller-coaster track.

Simulation 2.9: A Roller Coaster with Coulomb Friction
Dynamics problems involving slipstick behavior are more complex since we need to know when to stop the motion (and in some cases to determine if the motion can start up again). In the MATLAB® implementation of the roller-coaster problem with friction we have again written a function, roller_coaster_f2, and a script, roller_coaster_f2_sim. The script sets up default input parameters for the problem, calls roller_coaster_f2 to solve the problem numerically with ode45, and then plots results. The function roller_coaster_f2 contains a nested function, roller_cartf, which implements the x-equation of motion in Eq. (2.10.24), written as a system of first-order differential equations. There are several changes needed for this frictional case, however, in the way in which we use the numerical solver ode45. First, we need to know when the cart "sticks" and the motion ceases. In the use of the ode45 solver, this sticking is called an event. In this case the numerical integration should cease when the event sticking occurs. To define an event like this and place it into ode45 so it can recognize when to stop we must define an event function and put it into the options, similar to what was done with the error tolerances, rtol, and atol. We again use the odeset function to place an event into the options, which in this case looks like

```
options = odeset( 'Events', @stick, 'RelTol', rtol, 'AbsTol', atol);
```

The event function stick must describe the event and the actions which have to be taken by ode45 when the event occurs. This function in our case is also nested in the function roller_coaster_f2 in addition to the function roller_cartf. The general form for defining the event function stick is

```
function [value, isterminal, direction] = stick(t, x)
value = {expression};
isterminal = { 0 or 1};
direction = { -1 or 0 or 1};
end
```

The inputs (t, x) are the values of the time and first-order differential equation unknowns, which in our case are the vector components x(1) = x, x(2) = dx/dt. Value is a mathematical expression describing the event, which in our case is sticking. This event is defined to occur when value is equal to zero. The parameter isterminal is set to 1 if the integration is to terminate when the event occurs or set to 0 if the integration is allowed to continue. In our case motion stops so we want to set isterminal = 1. The parameter direction can be 0, 1, or − 1. In the case direction = 0, which is the default, whenever a zero of value is detected it is labeled as an event, while in the case of direction = −1, only those zeros of value are labeled as an event if the event function value is decreasing, while if direction = 1, only those zeros of value where that event function is decreasing. In our case value is chosen to be the expression:

$$\text{value} = 1 - \text{real}\left(\sqrt{\dot{x}^2 + \dot{y}^2} < \text{vtol} \;\; \& \;\; |dh/dx| < \mu\right)$$

(written here as text pseudo-code rather than a MATLAB® expression; see the function stick for the actual MATLAB® version). Within the parentheses in the above expression is a logical statement which says that the magnitude of the cart velocity must be less than some small velocity tolerance, vtol, and the slope of the track must be less than the coefficient of friction. The ideal sticking criterion is that the velocity must be zero but in a numerical solution this condition may never be satisfied exactly so we must replace it with a small user-defined tolerance. Thus, vtol is another parameter that must be defined in the script roller-coaster_f2_sim and passed to the function roller_coaster_f2, which then shares the parameter with the nested function stick. As long as the sticking criterion is not satisfied the logical expression is false and has a logical value of 0. The MATLAB® function real converts this logical expression to a real value of 0, and so value = 1 − 0 = 1. When the sticking criterion is met the logical expression is true and has a logical value of 1 which is converted to a real value of 1, giving value = 1 – 1 = 0. Thus, for the roller-coaster problem when sticking occurs value changes from 1 to 0, which is an event. With isterminal = 1, the event function tells the numerical integrator to stop, and direction = −1 in our case since value always changes from 1 to 0 (i.e., value is decreasing) when the event sticking occurs, although we could also use the default direction = 0 since value always changes from 1 to 0.

When an event function is present, there are some additional output parameters available when calling ode45. These outputs are shown here:

```
[tout, xout, te, xe, ie] = ode45(@roller_cartf, tspan, x0, options);
```

where again tout is a column vector containing the times at which the numerical solution is evaluated and xout is an array containing the numerical solutions for x and dx/dt, again placed into columns. The new output parameters due to the presence of the event function are te, xe, and ie. The value of te is the time within tspan at which sticking occurs and the integration stops, and xe is the vector xe = $[x, dx/dt]$ at time te. The output ie is an indicator of which event (if there is more than one event) that the numerical solver detected. In our case there is only one event and ie = 1 always so this output is not used. As you might suspect, if needed we can handle multiple events of different types using this same approach, where there can be multiple value functions, and different event conditions. If you are interested in more general event cases, you can look at the MATLAB® documentation or see various examples shown on the Web.

In the roller-coaster example the default values placed in the roller_coaster_f2_sim script are A = 1 m; L = 5 m; mu = 0.1 for the coefficient of friction; tspan = [0, 30] s; x0 = [0, 8] in m and m/s, respectively; rtol = 1e-5; atol = 1e-6; and vtol = 1e-3. This is not a very steep roller coaster and a mild friction coefficient where we can see some interesting behavior. The outputs of the script are (1) a plot of y versus x during the motion; (2) a plot of the normal force divided by the weight, mg, versus time; and (3) a plot of dx/dt versus x (a phase plot). These plots are shown in Figs. 2.52, 2.53 and 2.54. After executing the script the output te = 10.57 s, which is the time at which sticking occurred, and xe = [20.37 m, 9.97e-4 m/s], which is the corresponding x-location and x-velocity component when sticking occurred. Figure 2.52 shows that the cart has passed over two maxima in the

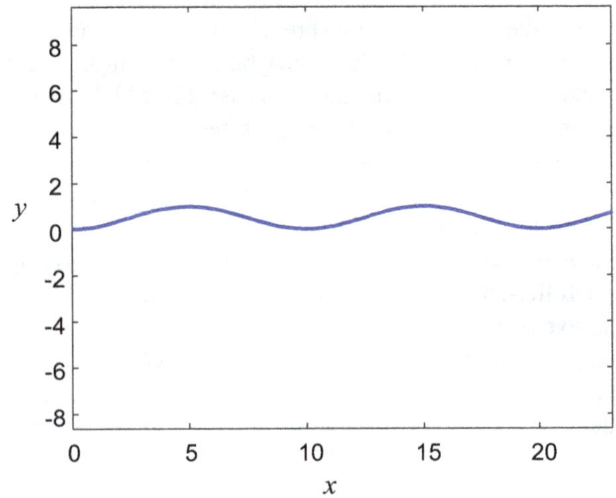

Fig. 2.52 A plot of the y versus x motion of the roller-cart problem with friction

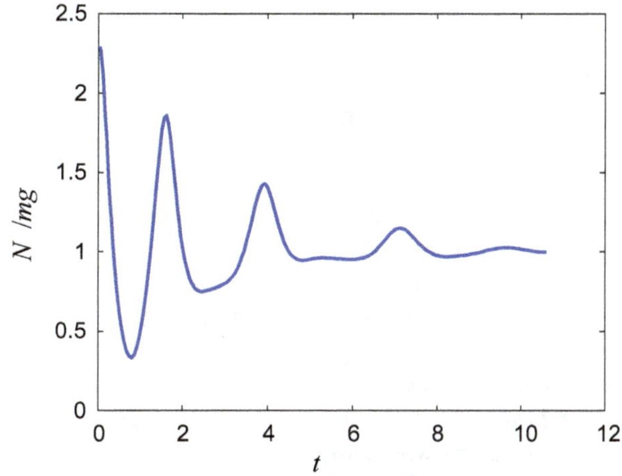

Fig. 2.53 The normal force divided by the weight, *mg*, as a function of time for the roller-cart problem with friction

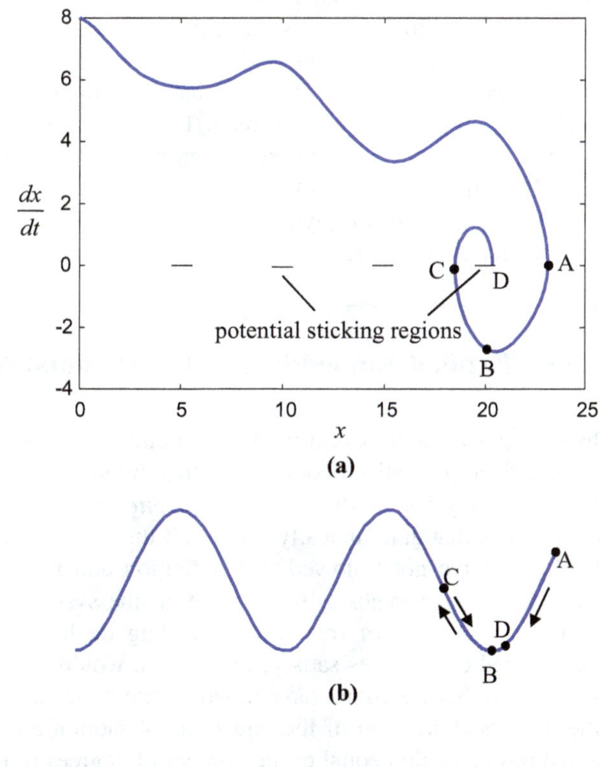

Fig. 2.54 (**a**) A phase plot of the motion of the cart on the frictional track, also showing the sticking regions along the track, and (**b**) the details of the motion along the track at the points labeled in (a). The vertical scale of the track in (b) is exaggerated to better show the motion

track before coming to rest and sticking. The maximum x-value in Fig. 2.52 is $x = 23.16$ m, which is larger than the x-value at sticking, xe $= 20.37$ m, so the cart must have slid partly up the track and then slipped back before coming to rest (although this behavior cannot be seen in Fig. 2.52 because of the nature of the plot). We can see details of this behavior in the phase plot. The normal force divided by the mass versus time plot in Fig. 2.53 shows that the effects of friction are to slow the cart down and also to reduce the intensity of the normal force oscillations. The ending time when the cart sticks is te = 10.57 s, which is much less than the total 30-s interval over which we were going to conduct the integration, showing the existence of the termination due to sticking. The phase plot in Fig. 2.54a shows us another view of the motion that is valuable. We see that the motion of the cart slows on the average as it rides over the first two peaks in the track and then is trapped in the final track valley. Point A occurs at the largest value of x, which is when the cart has stopped and it is about to start sliding back into the final valley, as shown in Fig. 2.54b. The cart reaches the bottom of the valley at point B and then continues to point C where it stops again and then reverses its motion again, passes through the bottom of the valley, and eventually comes to rest at a distance $x = 20.37$ m (point D) which is slightly to the right of the bottom at $x = 20$ m. There are potential sticking regions (regions where $|dh/dx| \leq \mu$) centered around all the maxima and minima of the track. Figure 2.54a shows the sticking regions for the first two maxima and subsequent minima as small horizontal lines in that figure. These equilibrium regions replace the equilibrium points that exist in the case without friction. The motion of the cart intersects the sticking region centered at $x = 20$ m and stops in that region at D. You should be aware of the fact that depending on the parameters chosen these sticking regions might be very small. Thus, during the numerical solution it may be possible to miss the sticking regions entirely due to numerical errors and time step size. One may need to experiment with different choices of the fineness of the numerical time steps through the setting of the error tolerances or through changes to the velocity tolerance, vtol, to obtain the proper event behavior.

2.11 Explicit Embedding of Constraints: A General Approach

In Sect. 2.6.1 we discussed the implicit embedding of holonomic constraints using independent generalized coordinates that automatically satisfied the constraints. In fact, many physics and dynamics texts *define* generalized coordinates as independent coordinates that automatically satisfy all the holonomic constraints present. In this book, we have not followed that definition and we have allowed the number of generalized coordinates to be as many as the system coordinates or as few as the number of degrees of freedom, depending on how many of the constraints the generalized coordinates satisfy. Ideally, we would like (1) all the constraints to be satisfied so that we do not have to worry about the constraints being violated during the numerical solution of the equations of motion and (2) to have the number of equations of motion equal to the number of degrees of freedom, since then we only have to solve the smallest number of equations of motion possible. Both of these conditions can be satisfied even if we use generalized coordinates that do not satisfy all the constraints by using an explicit embedding approach that is an extension of some of the simple examples we have already seen in Sects. 2.6.1 and 2.7.

To demonstrate how explicit embedding can be done in general let's begin by assuming that all the constraints are holonomic so that they can be expressed as

$$\mathbf{f}(\mathbf{q}, t) = 0 \tag{2.11.1}$$

where $\mathbf{q}$ is a $n_s \times 1$ column vector of generalized coordinates, where n_s is the number of system coordinates and $\mathbf{f}$ is a column vector of $n_h \times 1$ holonomic constraints. If we differentiate these constraints on time we have in matrix-vector form

$$\mathbf{A}\dot{\mathbf{q}} + \mathbf{f}_t = 0 \tag{2.11.2a}$$

or in equivalent component form

$$\sum_{j=1}^{n_s} \frac{\partial f_i}{\partial q_j} \dot{q}_j = -\frac{\partial f_i}{\partial t} \quad (i = 1, 2, \ldots, n_h) \tag{2.11.2b}$$

where $\mathbf{A} \equiv \mathbf{f_q}$ is an $n_h \times n_s$ matrix of derivatives on the components of $\mathbf{q}$ called the constraint Jacobian matrix given by

$$\mathbf{A} \equiv \mathbf{f_q} = \begin{bmatrix} \partial f_1/\partial q_1 & \partial f_1/\partial q_2 & \ldots & \partial f_1/\partial q_{n_s} \\ \partial f_2/\partial q_1 & \partial f_2/\partial q_2 & \ldots & \partial f_2/\partial q_{n_s} \\ & & & \\ \partial f_{n_h}/\partial q_1 & \partial f_{n_n}/\partial q_2 & \ldots & \partial f_{n_h}/\partial q_{n_s} \end{bmatrix} \tag{2.11.3}$$

and $\mathbf{f}_t \equiv \partial \mathbf{f}/\partial t$ is the $n_h \times 1$ column vector of the time derivatives of the components of $\mathbf{f}$. As we have seen before if we take another time derivative of Eq. (2.11.3) we arrive at constraint equations expressed in terms of the accelerations, given in matrix-vector form by

$$\mathbf{A}\ddot{\mathbf{q}} = \mathbf{b} \tag{2.11.4a}$$

equivalent to the component form

$$\sum_{j=1}^{n_s} \frac{\partial f_i}{\partial q_j} \ddot{q}_j = b_i \quad (i = 1, 2, \ldots, n_h) \tag{2.11.4b}$$

where the $n_h \times 1$ vector $\mathbf{b}$ is in matrix-vector form

$$\mathbf{b} = -\left[(\mathbf{A}\dot{\mathbf{q}})_{\mathbf{q}}\dot{\mathbf{q}} + 2\mathbf{A}_t\dot{\mathbf{q}} + \mathbf{f}_{tt}\right] \tag{2.11.5a}$$

or in equivalent component form

$$b_i = -\left[\sum_{j=1}^{n_s}\sum_{k=1}^{n_s}\frac{\partial^2 f_i}{\partial q_j \partial q_k}\dot{q}_j\dot{q}_k + 2\sum_{j=1}^{n_s}\frac{\partial^2 f_i}{\partial q_j \partial t}\dot{q}_j + \frac{\partial^2 f_i}{\partial t^2}\right] \quad (i = 1, 2, \ldots, n_h) \tag{2.11.5b}$$

Here $(\mathbf{A}\dot{\mathbf{q}})_{\mathbf{q}}$ represents a matrix of partial derivatives of the vector $\mathbf{A}\dot{\mathbf{q}}$ on the components of $\mathbf{q}$ (holding $\dot{\mathbf{q}}$ and t fixed), following the same notation seen in Eq. (2.11.3), and t subscripts again represent partial derivatives on time.

To discuss embedding we will again use the Gibbs-Gauss variational principle. Let us begin by considering a system where there are n_s independent generalized system coordinates and no constraints present. From our discussion in the last section we have in Cartesian coordinates

$$\delta_G \ddot{\mathbf{x}}^T\left(m\ddot{\mathbf{x}} - \mathbf{F}^e\right) = 0 \tag{2.11.6}$$

or, in terms of the generalized coordinates, using Eq. (2.10.19),

$$\sum_{k=1}^{n_s}\left[m\ddot{\mathbf{x}}\cdot\mathbf{c}_k - \mathbf{F}^e\cdot\mathbf{c}_k\right]\delta_G\ddot{q}_k = 0 \tag{2.11.7a}$$

which can be written as

$$\sum_{k=1}^{n_s}\left[m\ddot{\mathbf{x}}\cdot\mathbf{c}_k - Q_k^e\right]\delta_G\ddot{q}_k = 0 \tag{2.11.7b}$$

by using the definition of the generalized external force, $\mathbf{Q}^e$. But, recall from Eqs. (2.6.27) and (2.6.30a) (written in terms of our current notation)

$$\begin{aligned} m\ddot{\mathbf{x}}\cdot\mathbf{c}_k &= m\sum_{m=1}^{n_s}[\mathbf{c}_k\cdot\mathbf{c}_m]\ \ddot{q}_m + m\sum_{m=1}^{n_s}\left[\mathbf{c}_k\cdot\frac{\partial\mathbf{v}}{\partial q_m}\right]\dot{q}_m + m\left[\mathbf{c}_k\cdot\frac{\partial\mathbf{v}}{\partial t}\right] \\ &= \sum_{m=1}^{n_s}M_{km}\ \ddot{q}_m + \sum_{k=1}^{n_s}D_{nm}\dot{q}_m + E_k \\ &= \sum_{m=1}^{n_s}M_{km}\ \ddot{q}_m - Q_k^{in} \end{aligned} \tag{2.11.8}$$

where $\mathbf{Q}^{in}$ is an inertial force term. Thus, the variational principle in terms of generalized coordinates and generalized forces becomes

$$\sum_{k=1}^{n_s}\left[\sum_{k=1}^{n_s}M_{km}\ddot{q}_m - \bar{Q}_k^e\right]\delta_G\ddot{q}_k = 0 \tag{2.11.9}$$

or, equivalently, in matrix-vector form

$$\delta_G \ddot{\mathbf{q}}^T \left(\mathbf{M}\ddot{\mathbf{q}} - \bar{\mathbf{Q}}^e \right) = 0 \tag{2.11.10}$$

where $\bar{\mathbf{Q}}^e = \mathbf{Q}^e + \mathbf{Q}^{in}$ is the generalized external force due to both applied forces and inertial forces. Now, consider the presence of n_h holonomic constraints as given by Eqs. (2.11.4a) and (2.11.4b). These constraints must satisfy

$$\mathbf{A}\delta_G \ddot{\mathbf{q}} = 0 \tag{2.11.11}$$

so that if we dot Eq. (2.11.11) with a vector of n_h Lagrangian multipliers, $\boldsymbol{\lambda}$, and combine it with Eq. (2.11.10) we have

$$\delta_G \ddot{\mathbf{q}}^T \left(\mathbf{M}\ddot{\mathbf{q}} - \bar{\mathbf{Q}}^e - \mathbf{A}^T \boldsymbol{\lambda} \right) = 0 \tag{2.11.12}$$

In the augmented approach we use the arbitrariness of the Lagrange multipliers to generate n_s equations of motion consisting of the expression in parentheses in Eq. (2.11.12) and then solve the combination of these equations of motion and the constraint equations, Eqs. (2.11.4a) and (2.11.4b), for $\boldsymbol{\lambda}$ (which yields the constraint force) and n_s equations of motion. Here, in an embedding approach we will instead eliminate the $\boldsymbol{\lambda}$ (and hence the constraint forces) and generate a set of $n_{\text{dof}} = n_s - n_h$ equations of motion. To do this we first separate the original n_s generalized coordinates into n_h dependent coordinates and $n_{\text{dof}} = (n_s - n_h)$ independent coordinates, i.e., we write $\mathbf{q}$ in the partitioned form $\mathbf{q} = \left[\mathbf{q}_{dp}^T \ \mathbf{q}_{in}^T \right]^T = \left[q_{dp;1}, q_{dp;2}, \ldots, q_{dp;n_h}, q_{in;1}, q_{in;2}, \ldots, q_{in;n_{\text{dof}}} \right]^T$. Then we can also write Eq. (2.11.11) in partitioned form as

$$\mathbf{A}_{dp}\delta_G \ddot{\mathbf{q}}_{dp} + \mathbf{A}_{in}\delta_G \ddot{\mathbf{q}}_{in} = 0 \tag{2.11.13}$$

where $\mathbf{A}_{in}$ is a $n_h \times n_{\text{dof}}$ matrix and we assume that the choice of the dependent variables is made so that the $n_h \times n_h$ matrix $\mathbf{A}_{dp}$ is non-singular and its inverse exists. We will discuss in more detail later how this partitioning can be done, but for now we will assume that we have Eq. (2.11.13). Multiplying both sides of Eq. (2.11.13) by the inverse of $\mathbf{A}_{dp}$ we can solve for the dependent virtual accelerations in terms of the independent accelerations:

$$\delta_G \ddot{\mathbf{q}}_{dp} = -\mathbf{A}_{dp}^{-1} \mathbf{A}_{in} \delta_G \ddot{\mathbf{q}}_{in} \tag{2.11.14}$$

and the total vector of system virtual accelerations can then be expressed in terms of the independent virtual accelerations only as

$$\delta_G \ddot{\mathbf{q}} = \begin{bmatrix} \delta_G \ddot{\mathbf{q}}_{dp} \\ \delta_G \ddot{\mathbf{q}}_{in} \end{bmatrix} = \begin{bmatrix} -\mathbf{A}_{dp}^{-1} \mathbf{A}_{in} \\ \mathbf{I} \end{bmatrix} \delta_G \ddot{\mathbf{q}}_{in} \tag{2.11.15}$$

where $\mathbf{I}$ is the $n_{\text{dof}} \times n_{\text{dof}}$ identity matrix. This can be written as

$$\delta_G \ddot{\mathbf{q}} = \mathbf{B}_{in} \delta_G \ddot{\mathbf{q}}_{in} \tag{2.11.16}$$

where the $n_s \times n_{\text{dof}}$ $\mathbf{B}_{in}$ matrix is

$$\mathbf{B}_{in} = \begin{bmatrix} -\mathbf{A}_{dp}^{-1}\mathbf{A}_{in} \\ \mathbf{I} \end{bmatrix} \tag{2.11.17}$$

Now, if we place Eq. (2.11.16) into Eq. (2.11.12) we obtain

$$\delta_G \ddot{\mathbf{q}}_{in}^T \mathbf{B}_{in}^T \left(\mathbf{M}\ddot{\mathbf{q}} - \bar{\mathbf{Q}}^e - \mathbf{A}^T \boldsymbol{\lambda} \right) = 0 \tag{2.11.18}$$

However,

$$\mathbf{B}_{in}^T \mathbf{A}^T = (\mathbf{A}\mathbf{B}_{in})^T = \left(\begin{bmatrix} \mathbf{A}_{dp} & \mathbf{A}_{in} \end{bmatrix} \begin{bmatrix} -\mathbf{A}_{dp}^{-1}\mathbf{A}_{in} \\ \mathbf{I} \end{bmatrix} \right)^T = 0 \tag{2.11.19}$$

so that the Lagrange multiplier terms (and the corresponding constraint forces) are eliminated from Eq. (2.11.12) and because of the independence of the $\delta_G \ddot{\mathbf{q}}_{in}$ we obtain the n_{dof} equations of motion

$$\mathbf{B}_{in}^T \mathbf{M}\ddot{\mathbf{q}} - \mathbf{B}_{in}^T \bar{\mathbf{Q}}^e = 0 \tag{2.11.20}$$

If we partition the first and second derivatives of the constraints, Eqs. (2.11.2a), (2.11.2b), (2.11.4a) and (2.11.4b) into dependent and independent terms, we have

$$\begin{aligned} \mathbf{A}_{in}\ddot{\mathbf{q}}_{in} + \mathbf{A}_{dp}\ddot{\mathbf{q}}_{dp} &= \mathbf{b} \\ \mathbf{A}_{in}\dot{\mathbf{q}}_{in} + \mathbf{A}_{dp}\ddot{\mathbf{q}}_{dp} &= -\mathbf{f}_t \end{aligned} \tag{2.11.21}$$

which we can also solve for the dependent terms:

$$\begin{aligned} \ddot{\mathbf{q}}_{dp} &= \mathbf{A}_{dp}^{-1}\mathbf{b} - \mathbf{A}_{dp}^{-1}\mathbf{A}_{in}\ddot{\mathbf{q}}_{in} \\ \dot{\mathbf{q}}_{dp} &= -\mathbf{A}_{dp}^{-1}\mathbf{f}_t - \mathbf{A}_{dp}^{-1}\mathbf{A}_{in}\dot{\mathbf{q}}_{in} \end{aligned} \tag{2.11.22}$$

so that

$$\begin{aligned} \ddot{\mathbf{q}} &= \begin{bmatrix} -\mathbf{A}_{dp}^{-1}\mathbf{A}_{in} \\ \mathbf{I} \end{bmatrix} \ddot{\mathbf{q}}_{in} + \begin{bmatrix} \mathbf{A}_{dp}^{-1}\mathbf{b} \\ \mathbf{0} \end{bmatrix} = \mathbf{B}_{in}\ddot{\mathbf{q}}_{in} + \mathbf{d} \\ \dot{\mathbf{q}} &= \begin{bmatrix} -\mathbf{A}_{dp}^{-1}\mathbf{A}_{in} \\ \mathbf{I} \end{bmatrix} \dot{\mathbf{q}}_{in} + \begin{bmatrix} -\mathbf{A}_{dp}^{-1}\mathbf{f}_t \\ \mathbf{0} \end{bmatrix} = \mathbf{B}_{in}\dot{\mathbf{q}}_{in} + \mathbf{e} \end{aligned} \tag{2.11.23}$$

Placing the generalized acceleration expression of Eq. (2.11.23) into Eq. (2.11.20) we find

$$\mathbf{B}_{in}^T\mathbf{M}\mathbf{B}_{in}\ddot{\mathbf{q}}_{in} - \mathbf{B}_{in}^T\bar{\mathbf{Q}}^e + \mathbf{B}_{in}^T\mathbf{M}\mathbf{d} = 0 \tag{2.11.24}$$

which is of the form

$$\widehat{\mathbf{M}}_{in}\ddot{\mathbf{q}}_{in} = \widehat{\mathbf{Q}}_{in} \tag{2.11.25}$$

where

$$\begin{aligned} \widehat{\mathbf{M}}_{in} &= \mathbf{B}_{in}^T\mathbf{M}\mathbf{B}_{in} \\ \widehat{\mathbf{Q}}_{in} &= \mathbf{B}_{in}^T\bar{\mathbf{Q}}_e - \mathbf{B}_{in}^T\mathbf{M}\mathbf{d} \end{aligned} \tag{2.11.26}$$

Equation (2.11.25) is $n_{\text{dof}} = (n_s - n_h)$ equations for the n_{dof} accelerations, $\ddot{\mathbf{q}}_{in}$, but it may also include all the generalized velocities and coordinates since in general we may have $\widehat{\mathbf{M}}_{in} = \widehat{\mathbf{M}}_{in}(\mathbf{q}, t)$ and $\widehat{\mathbf{Q}}_{in} = \widehat{\mathbf{Q}}_{in}(\mathbf{q}, \dot{\mathbf{q}}, t)$. To write Eq. (2.11.25) in terms of independent coordinates and their derivatives only we must also write both $\dot{\mathbf{q}}$ and $\mathbf{q}$ in terms of independent quantities only. The velocity relation of Eq. (2.11.22) can be used to write $\dot{\mathbf{q}}$ in terms of $\dot{\mathbf{q}}_{in}$ but to eliminate the dependent coordinates we need to solve the constraint equations of Eq. (2.11.1) for those dependent coordinates. Symbolically we can write these solutions as $\mathbf{q}_{dp} = \mathbf{g}(\mathbf{q}_{in}, t)$ but in practice the constraint equations are often nonlinear so this solution must be done numerically using methods such as the *Newton-Raphson method* (see the discussion in Appendix C). If the constraint equations are linearly independent the "mass" matrix, $\widehat{\mathbf{M}}_{in}$, is non-singular so that we can obtain the second-order equations of motion for the independent accelerations, finally, as

$$\ddot{\mathbf{q}}_{in} = \widehat{\mathbf{M}}_{in}^{-1}\widehat{\mathbf{Q}}_{in} \tag{2.11.27}$$

You might have noticed that in obtaining Eq. (2.11.25) we did not at any point use the fact that the constraints were holonomic since we obtained that equation through the Gibbs-Gauss variational principle and only worked with the constraint equations written in terms of accelerations in a form that can be obtained from either holonomic or non-holonomic constraints. However, suppose that the constraints of Eqs. (2.11.4a) and (2.11.4b) came from differentiating n_{nh} non-holonomic constraints. Then we could still partition the accelerations as shown and obtain Eq. (2.11.25), which would be $n_s - n_{nh}$ equations of motion, but the number of degrees of freedom would still be n_s since the non-holonomic constraints do not reduce the number of independent coordinates. Thus, we can eliminate the constraint forces for these non-holonomic constraints from the equations of motion but we still need n_{nh} additional equations to solve for the n_s independent generalized coordinates. If the non-holonomic equations are of the form of Eqs. (2.11.2a) and (2.11.2b), for example, then we can solve the $n_s - n_{nh}$ embedded equations of Eq. (2.11.25) in conjunction with the n_{nh} non-holonomic constraints for the velocity written in the form shown in Eq. (2.11.23) as

$$\begin{aligned} \widehat{\mathbf{M}}_{in}\ddot{\mathbf{q}}_{in} &= \widehat{\mathbf{Q}}_{in} \\ \dot{\mathbf{q}} &= \mathbf{B}_{in}\dot{\mathbf{q}}_{in} + \mathbf{e} \end{aligned} \tag{2.11.28}$$

which are a total of n_s differential equations of motion for the n_s generalized coordinates, $\mathbf{q}$. These can be placed in the form of a system of first-order differential equations and solved as shown previously. This is the same number of differential equations as found in an augmented approach where we solve explicitly for the Lagrange multipliers rather than eliminating them but the major difference between the embedded equations of Eq. (2.11.27) and the corresponding augmented system which are n_s equations of the form

$$\ddot{\mathbf{q}} = \mathbf{M}^{-1}\bar{\mathbf{Q}}^e + \mathbf{M}^{-1}\mathbf{A}^T\left(\mathbf{A}\mathbf{M}^{-1}\mathbf{A}^T\right)^{-1}\left(\mathbf{b} - \mathbf{A}\mathbf{M}^{-1}\bar{\mathbf{Q}}^e\right) \tag{2.11.29}$$

is that the n_{nh} non-holonomic constraints are being explicitly satisfied in Eq. (2.11.28) while Eq. (2.11.29) only explicitly satisfies any constraints (holonomic or non-holonomic) written in terms of the accelerations. As mentioned previously, satisfying constraints written in terms of accelerations would not be a problem if the equations of motion were being solved exactly, but with a numerical solution errors in satisfying the constraint equations written in terms of the generalized velocities or generalized coordinates (for holonomic constraints) can accumulate over time with the augmented approach and must be dealt with, as we will discuss in the next section. In embedding holonomic or non-holonomic constraints this is not an issue since the constraints are being satisfied or are part of the final equations to be solved.

Partitioning the generalized coordinates into dependent and independent coordinates is obviously a crucial part of the embedding approach. There are several methods available to do the partitioning, including the method of Gaussian elimination, the pseudo upper triangular decomposition method, the zero-eigenvalue method, and others. You may be familiar with the Gaussian elimination method in solving linear systems of equations. It can also be used for partitioning and it can also serve as a tool for detecting if the constraint equations are linearly independent, as we have been assuming throughout our discussions so far. We will not discuss any of these methods further here. A complete discussion is covered in greater detail in more advanced texts on multibody dynamics [3, 10, 11].

Explicit embedding, as we have shown, is a rather involved process in general but it does lead to as few equations of motion as possible and in some cases may be an effective method. Many dynamics texts only discuss very simple problems where embedding is accomplished "by inspection" through a judicious choice of the generalized coordinates, so that one may be led to believe that it is the preferred method. The two-dimensional pendulum problem is a good example where by using polar coordinates (r, θ) one can directly embed the constraint $r = l$ (where l is the length of the pendulum) and obtain the equation of motion in terms of θ only by a variety of methods. In contrast, if we use Cartesian coordinates to describe the motion of the pendulum embedding is more problematic. In this case the constraint equation becomes

$$x^2 + y^2 - l^2 = 0 \tag{2.11.30}$$

and the differentiated constraint equation in terms of the accelerations becomes

$$2x\ddot{x} + 2y\ddot{y} = -2(\dot{x}^2 + \dot{y}^2) \tag{2.11.31}$$

Here if we choose x as the independent coordinate then the dependent and dependent matrices are just scalars, i.e., $(\mathbf{A}_{dp}, \mathbf{A}_{in}) = (y, x)$ so that their inverses are just $\left(\mathbf{A}_{dp}^{-1}, \mathbf{A}_{in}^{-1}\right) = (1/y, 1/x)$. Thus, if $y = 0$ during the motion the inverse of $\mathbf{A}_{dp}$ will be infinite and our embedding solution will fail. Switching to y as the independent variable will lead to the same issue when $x = 0$. We could, of course, make one of these choices to describe the motion as long as the inverse of the dependent matrix is not singular and then switch to the other choice to continue the motion as long as it is well behaved but this will lead to a more complex solution procedure that can be avoided by simply using a more appropriate choice of generalized coordinates such as (r, θ). In specific problems it may be impossible to guarantee that an initial choice of partitioning of the coordinates that is non-singular will generate an embedded solution that remains well behaved during the entire motion so that one must monitor $\mathbf{A}_{dp}$ to ensure that it is non-singular and, if it does become singular, identify the source of the problem so that the solution process can be modified and the determination of the motion can be continued. Because of these issues with embedding as a general approach (and others, including the issue of computational efficiency), in this book we emphasize the use of the augmentation method instead.

2.12 The Augmented Approach and Constraint Satisfaction

The augmented approach is a direct approach for obtaining both the equations of motion and the constraint forces. The two major limitations of the formulation of this approach as we have presented it so far are that (1) the inverse, $(\mathbf{A}\mathbf{M}^{-1}\mathbf{A}^T)^{-1}$, must exist and (2) the constraint equations and their first derivatives may "drift" during the numerical calculations, leading to violation of those constraints. In the next chapter when we examine systems of particles we will discuss how to implement the augmented approach when this inverse does not exist. Here we will only examine the constraint violation issue.

Consider a system with holonomic constraints of the form $\mathbf{f}(\mathbf{q},t) = 0$. These constraints must be satisfied at all times so that in particular we must have

$$\begin{aligned} &\mathbf{f}(\mathbf{q},t) = 0 \\ &\dot{\mathbf{f}}\left(\mathbf{q},\dot{\mathbf{q}},t\right) = \mathbf{A}\dot{\mathbf{q}} - \mathbf{b}_q = 0 \\ &\ddot{\mathbf{f}}\left(\mathbf{q},\dot{\mathbf{q}},\ddot{\mathbf{q}},t\right) = \mathbf{A}\ddot{\mathbf{q}} - \mathbf{b}_v = 0 \end{aligned} \tag{2.12.1}$$

where, in terms of our previous notation,

$$\begin{aligned} &\mathbf{b}_q = -\mathbf{f}_t \\ &\mathbf{b}_v = \mathbf{b} = -\left[\left(\mathbf{A}\dot{\mathbf{q}}\right)_{\mathbf{q}}\dot{\mathbf{q}} + 2\mathbf{A}_t\dot{\mathbf{q}} + \mathbf{f}_{tt}\right] \end{aligned} \tag{2.12.2}$$

However, in formulating the equations of motion in the augmented approach we only used the twice-differentiated form in Eq. (2.12.1) so that in a numerical solution errors can accumulate over time in the satisfaction of the constraints themselves and their first derivatives. There are a number of approaches to dealing with this issue that have been proposed. To date, there is not one method that is the preferred choice and there is active research on the topic. See [10], for example, for a summary of a number of possible approaches. We will describe a recently developed method that fits well with the level and approach of this book [12]. During the numerical solution of the equations of motion between time t and $t+\Delta t$, where Δt is the step size, satisfaction of the constraints and their first derivatives requires that to first order

$$\begin{aligned} &\mathbf{f}(\mathbf{q},t+\Delta t) = \mathbf{f}(\mathbf{q},t) + \left(\mathbf{A}\dot{\mathbf{q}} - \mathbf{b}_q\right)\Delta t + O(\Delta t^2) \\ &\dot{\mathbf{f}}(\mathbf{q},\mathbf{q},t+\Delta t) = \dot{\mathbf{f}}\left(\mathbf{q},\dot{\mathbf{q}},t\right) + \left(\mathbf{A}\ddot{\mathbf{q}} - \mathbf{b}_v\right)\Delta t + O(\Delta t^2) \end{aligned} \tag{2.12.3}$$

so that if we are going to maintain the constraints and their first derivatives at time $t+\Delta t$ to first order we must have

$$\mathbf{A}\dot{\mathbf{q}} = \mathbf{b}_q - \mathbf{f}/\Delta t \tag{2.12.4a}$$

$$\mathbf{A}\ddot{\mathbf{q}} = \mathbf{b}_v - \dot{\mathbf{f}}/\Delta t \tag{2.12.4b}$$

Consider first the equation in Eq. (2.12.4b) involving accelerations. This equation is a modification of the ideal constraint equation in the absence of numerical errors, with a correction term, $-\dot{\mathbf{f}}/\Delta t$, that accounts for errors in satisfaction of the first derivative of the original constraint equation over the time step Δt. If this corrected constraint equation generates an ideal constraint force $\mathbf{A}^T\boldsymbol{\lambda}$ then we arrive at the augmented system of equations:

$$\begin{aligned}\ddot{\mathbf{q}} &= \mathbf{M}^{-1}\bar{\mathbf{Q}}^e + \mathbf{M}^{-1}\mathbf{A}^T\boldsymbol{\lambda} \\ \mathbf{A}\ddot{\mathbf{q}} &= \mathbf{b}_v - \dot{\mathbf{f}}/\Delta t\end{aligned} \tag{2.12.5}$$

which can be solved as before for the Lagrange multiplier, constraint force, and equations of motion, as demonstrated previously:

$$\begin{aligned}\boldsymbol{\lambda} &= \left(\mathbf{A}\mathbf{M}^{-1}\mathbf{A}^T\right)^{-1}\left(\mathbf{b}_v - \mathbf{A}\mathbf{M}^{-1}\bar{\mathbf{Q}}^e - \dot{\mathbf{f}}/\Delta t\right) \\ \mathbf{Q}^c &= \mathbf{A}^T\left(\mathbf{A}\mathbf{M}^{-1}\mathbf{A}^T\right)^{-1}\left(\mathbf{b}_v - \mathbf{A}\mathbf{M}^{-1}\bar{\mathbf{Q}}^e - \dot{\mathbf{f}}/\Delta t\right) \\ \ddot{\mathbf{q}} &= \mathbf{M}^{-1}\bar{\mathbf{Q}}^e + \mathbf{M}^{-1}\mathbf{A}^T\left(\mathbf{A}\mathbf{M}^{-1}\mathbf{A}^T\right)^{-1}\left(\mathbf{b}_v - \mathbf{A}\mathbf{M}^{-1}\bar{\mathbf{Q}}^e - \dot{\mathbf{f}}/\Delta t\right)\end{aligned} \tag{2.12.6}$$

Thus, we have taken into account, to first order, any loss of satisfaction of the first derivative of the constraint equation $\dot{\mathbf{f}} = 0$. To include the constraint equation itself in the solution process we write the equations of motion in first-order form as

$$\begin{aligned}\dot{\mathbf{q}} &= \mathbf{v} \\ \dot{\mathbf{v}} &= \mathbf{M}^{-1}\bar{\mathbf{Q}}^e + \mathbf{M}^{-1}\mathbf{A}^T\left(\mathbf{A}\mathbf{M}^{-1}\mathbf{A}^T\right)^{-1}\left(\mathbf{b}_v - \mathbf{A}\mathbf{M}^{-1}\bar{\mathbf{Q}}^e - \dot{\mathbf{f}}/\Delta t\right)\end{aligned} \tag{2.12.7}$$

We can consider Eq. (2.12.4a) as a constraint on the solution for the generalized velocities, $\dot{\mathbf{q}}$, in Eq. (2.12.7) that accounts for errors in satisfaction of the constraint equation. To include this constraint we can multiply this constraint by a Lagrange multiplier vector, $\boldsymbol{\mu}$, and place it in the equations of motion for $\dot{\mathbf{q}}$ (in the same form as done with $\boldsymbol{\lambda}$ and the equations for $\ddot{\mathbf{q}}$), to obtain

$$\begin{aligned}\dot{\mathbf{q}} &= \mathbf{v} + \mathbf{M}^{-1}\mathbf{A}^T\boldsymbol{\mu} \\ \mathbf{A}\dot{\mathbf{q}} &= \mathbf{b}_q - \mathbf{f}/\Delta t\end{aligned} \tag{2.12.8}$$

By substituting the expression for $\dot{\mathbf{q}}$ into the constraint equation in Eq. (2.12.8) and taking the inverse of $\mathbf{A}\mathbf{M}^{-1}\mathbf{A}^T$ we can solve for $\boldsymbol{\mu}$ as

$$\boldsymbol{\mu} = \left(\mathbf{A}\mathbf{M}^{-1}\mathbf{A}^T\right)^{-1}\left(\mathbf{b}_q - \mathbf{A}\mathbf{v} - \mathbf{f}/\Delta t\right) \tag{2.12.9}$$

so that Eq. (2.12.7) becomes

$$\begin{aligned}\dot{\mathbf{q}} &= \mathbf{v} + \mathbf{M}^{-1}\mathbf{A}^T\left(\mathbf{A}\mathbf{M}^{-1}\mathbf{A}^T\right)^{-1}\left(\mathbf{b}_q - \mathbf{A}\mathbf{v} - \mathbf{f}/\Delta t\right) \\ \dot{\mathbf{v}} &= \mathbf{M}^{-1}\bar{\mathbf{Q}}^e + \mathbf{M}^{-1}\mathbf{A}^T\left(\mathbf{A}\mathbf{M}^{-1}\mathbf{A}^T\right)^{-1}\left(\mathbf{b}_v - \mathbf{A}\mathbf{M}^{-1}\bar{\mathbf{Q}}^e - \dot{\mathbf{f}}/\Delta t\right)\end{aligned} \tag{2.12.10}$$

This can be placed into a somewhat less complex looking and more symmetrical form by defining the unconstrained acceleration, $\mathbf{a}$, as $\mathbf{a} = \mathbf{M}^{-1}\bar{\mathbf{Q}}^e$ to give

$$\begin{aligned}\dot{\mathbf{q}} &= \mathbf{v} + \mathbf{M}^{-1}\mathbf{A}^T\left(\mathbf{A}\mathbf{M}^{-1}\mathbf{A}^T\right)^{-1}\left(\mathbf{b}_q - \mathbf{A}\mathbf{v} - \mathbf{f}/\Delta t\right)\\ \dot{\mathbf{v}} &= \mathbf{a} + \mathbf{M}^{-1}\mathbf{A}^T\left(\mathbf{A}\mathbf{M}^{-1}\mathbf{A}^T\right)^{-1}\left(\mathbf{b}_v - \mathbf{A}\mathbf{a} - \dot{\mathbf{f}}/\Delta t\right)\end{aligned} \tag{2.12.11}$$

The additional constraint terms can be viewed as adding small generalized constraint forces and generalized impulses (a generalized impulse is a time integral of a generalized force) to account for errors in satisfying the constraints and their first derivatives. This interpretation follows from noting that if we write Newton's second law as

$$\mathbf{M}\dot{\mathbf{v}} = \mathbf{Q}$$

where $\dot{\mathbf{v}} = \ddot{\mathbf{q}}$ then integrating on time (assuming that $\mathbf{M}$ is not a function of time) gives

$$\mathbf{M}\dot{\mathbf{q}} = \int \mathbf{Q}dt = \mathbf{I}_Q$$

where $\mathbf{I}_Q$ is the generalized impulse. Thus we have

$$\begin{aligned}\dot{\mathbf{q}} &= \mathbf{M}^{-1}\mathbf{I}_Q\\ \dot{\mathbf{v}} &= \mathbf{M}^{-1}\mathbf{Q}\end{aligned}$$

and the terms added to ensure constraint satisfaction have these forms. However, these are not real forces or impulses as they are present only to account for the imperfect conditions present when we solve the equations of motion numerically that cause the constraints to be violated. Unfortunately, one cannot use Eq. (2.12.11) directly with solvers such as ode45 since that solver has a variable time step size that is adjusted during the integration to meet the error tolerances and the Δt used in the solver is not available. One could write a special solver to include these Δt with a variable step size or one can use a solver that has a fixed time step size. A MATLAB® function, ode4, is available that implements a fourth-order Runge-Kutta method with a fixed step size. This function, like ode45, solves the system of first-order differential equations $\dot{\mathbf{x}} = \mathbf{f}(\mathbf{x}, \dot{\mathbf{x}}, t)$ and is called in a very similar fashion:

```
[xout] = ode4(@ode_function, tspan, x0, p1, p2,..., pn)
```

where ode_function(t, x, p1, p2, ..., pn) is a function written for the first-order system in the same form required by ode45, tspan = [t1, t2,..., tn] is a vector of given times (spaced equally apart by the time step, Δt) at which the solution is to be obtained, x0 is a vector containing the initial conditions, and p1, p2, ..., pn are any additional input parameters needed, if any, when calling the function ode_function. Note that tspan here specifies all the times at which the calculations are to be performed by ode4 whereas in using ode45 the actual times at which the solution is found are determined by the ode45 algorithm itself and ode45 then interpolates those results to obtain the values specified in tspan when it contains a vector of given times. The output matrix, xout, contains the solution for the vector x at the times given in tspan, with the components of x written in the columns of xout, which is identical to the form used by ode45. Unlike ode45, there are no options like error tolerances or events available when using ode4. Solvers with a variable time step are generally more versatile than fixed step size solvers since the use of a variable step size that adjusts to the rate at which the system solution is changing is generally computationally more efficient. However, fixed step size solvers can also be effective in many problems.

Simulation 2.10: Motion of a Pendulum with Constraint Satisfaction

To illustrate the method of constraint satisfaction described, consider a simple pendulum (Fig. 2.55) that is modeled using the (x, y) coordinates. In this case the governing equations are Newton's second law, written in matrix-vector form:

$$\mathbf{M}\ddot{\mathbf{x}} = \mathbf{F}^e + \mathbf{F}^c \tag{2.12.12}$$

or

$$\begin{bmatrix} m & 0 \\ 0 & m \end{bmatrix} \begin{Bmatrix} \ddot{x} \\ \ddot{y} \end{Bmatrix} = \begin{Bmatrix} 0 \\ mg \end{Bmatrix} + \begin{Bmatrix} F_x^c \\ F_y^c \end{Bmatrix} \tag{2.12.13}$$

where $\mathbf{M}$ is the mass matrix and $(\mathbf{F}^e, \mathbf{F}^c)$ are the external and constraint forces, respectively. The constraint equation is simply

$$f(x, y) = x^2 + y^2 - L^2 = 0 \tag{2.12.14a}$$

and its first and second derivatives on time are

$$\dot{f} = 2x\dot{x} + 2y\dot{y} = 0 \tag{2.12.14b}$$

and

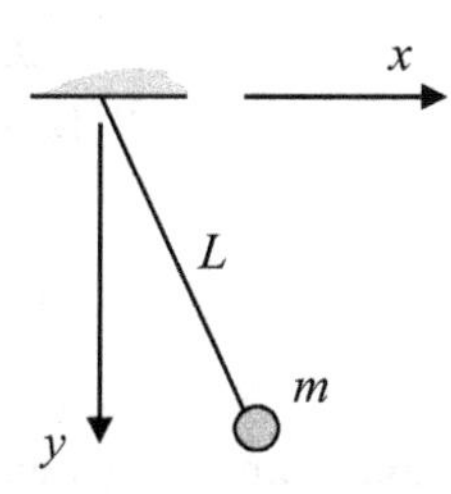

Fig. 2.55 The simple pendulum described in Cartesian coordinates

$$\ddot{f} = 2x\ddot{x} + 2y\ddot{y} + 2\left(\dot{x}^2 + \dot{y}^2\right) = 0 \tag{2.12.14c}$$

which leads to the following results for $\mathbf{A}$, $\mathbf{b}_q$, and $\mathbf{b}_v$:

$$\begin{aligned} \mathbf{A} &= [2x \quad 2y] \\ \mathbf{b}_q &= 0 \\ \mathbf{b}_v &= -2\left(\dot{x}^2 + \dot{y}^2\right) \end{aligned} \tag{2.12.15}$$

In the first-order system of Eq. (2.12.11) we have the system coordinates $\mathbf{x} = (\mathbf{q}, \mathbf{v}) = \left(x, y, \dot{x}, \dot{y}\right)$ which we will write as $\mathbf{x} = (x_1, x_2, x_3, x_4)$. To evaluate the system the terms we need are very simple to evaluate:

$$\begin{aligned} \mathbf{M}^{-1}\mathbf{A}^T &= \begin{Bmatrix} 2x/m \\ 2y/m \end{Bmatrix}, \quad \left(\mathbf{A}\mathbf{M}^{-1}\mathbf{A}^T\right)^{-1} = m/4L^2 \\ \mathbf{a} = \mathbf{F}^e/m &= \begin{Bmatrix} 0 \\ g \end{Bmatrix} \end{aligned} \tag{2.12.16}$$

yielding

$$\begin{aligned} \dot{x}_1 &= x_3 + \frac{x_1}{2L^2}\left[-2(x_1x_3 + x_2x_4) - \left(x_1^2 + x_2^2 - L^2\right)/\Delta t\right] \\ \dot{x}_2 &= x_4 + \frac{x_2}{2L^2}\left[-2(x_1x_3 + x_2x_4) - \left(x_1^2 + x_2^2 - L^2\right)/\Delta t\right] \\ \dot{x}_3 &= \frac{x_1}{2L^2}\left[-2\left(x_3^2 + x_4^2\right) - 2x_2g - 2(x_1x_3 + x_2x_4)/\Delta t\right] \\ \dot{x}_4 &= g + \frac{x_2}{2L^2}\left[-2\left(x_3^2 + x_4^2\right) - 2x_2g - 2(x_1x_3 + x_2x_4)/\Delta t\right] \end{aligned} \tag{2.12.17}$$

This gives us the first-order system of equations we can use with ode4. However, we will also want to examine the equations of motion without constraint satisfaction corrections. We can do this by simply omitting the additional constraint terms via a parameter, e, where $e = 0$ for no constraint corrections and $e = 1$ with corrections:

$$\begin{aligned} \dot{x}_1 &= x_3 + \frac{ex_1}{2L^2}\left[-2(x_1x_3 + x_2x_4) - \left(x_1^2 + x_2^2 - L^2\right)/\Delta t\right] \\ \dot{x}_2 &= x_4 + \frac{ex_2}{2L^2}\left[-2(x_1x_3 + x_2x_4) - \left(x_1^2 + x_2^2 - L^2\right)/\Delta t\right] \\ \dot{x}_3 &= \frac{x_1}{2L^2}\left[-2\left(x_3^2 + x_4^2\right) - 2x_2g - 2e(x_1x_3 + x_2x_4)/\Delta t\right] \\ \dot{x}_4 &= g + \frac{x_2}{2L^2}\left[-2\left(x_3^2 + x_4^2\right) - 2x_2g - 2e(x_1x_3 + x_2x_4)/\Delta t\right] \end{aligned} \tag{2.12.18}$$

This system of equations has been placed in the MATLAB® function pendulum_cs_eqs which has the calling sequence

```
[xdot] = pendulum_cs_eqs(t, x, g, L, dt, e);
```

where g is the acceleration of gravity, L is the length of the pendulum, dt is the time step size, and e is the parameter controlling if constraint satisfaction is present or not. The script pendulum_cs_sim gives default values of these parameters as g = 9.81 m/s^2, L = 1 m, tspan = 0:0.001:900 s, x0 = [L, 0, 0, 0], and e = 0 (no constraint satisfaction). The default dt value is thus dt = 0.001 s. As usual, these defaults can be changed by the user in the command window before executing the script. With this choice of parameters we are releasing the pendulum at rest in a horizontal position and then solving for its motion over a very long time span of about 15 min, which corresponds to 900,000 time steps. The script uses ode4 in conjunction with pendulum_cs_eqs to solve for the pendulum motion. The value of the constraint and its first derivative are then plotted along with the total energy divided by the mass m. These plots are shown in Fig. 2.56. Figure 2.56a, b shows

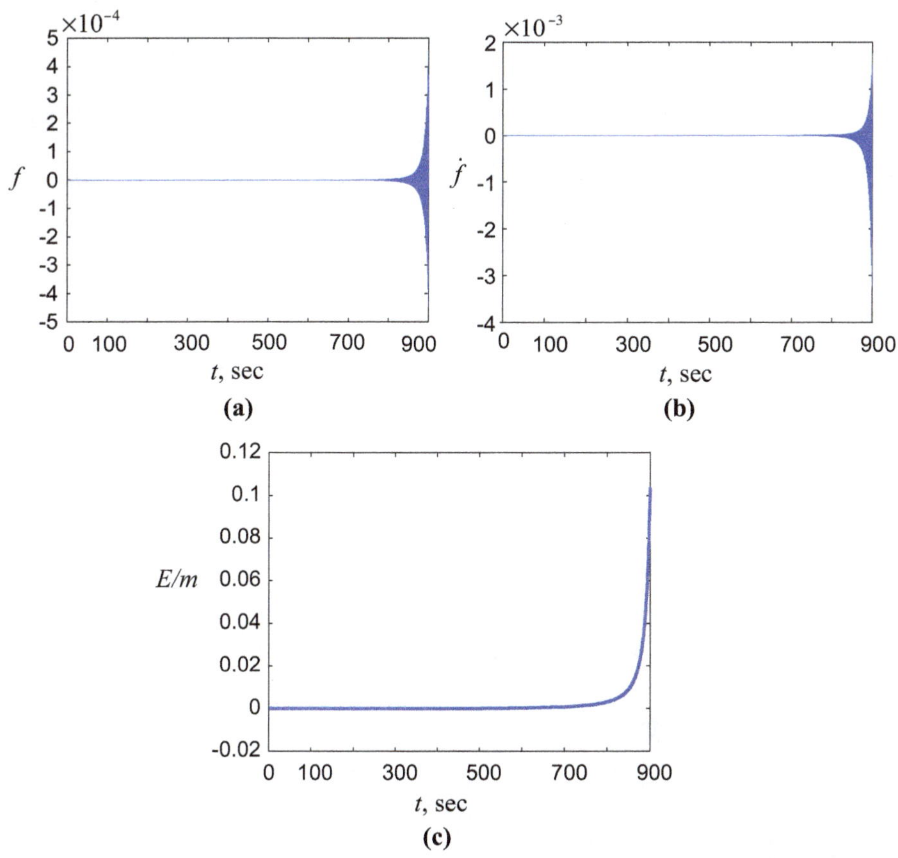

Fig. 2.56 Solution for the pendulum motion without satisfaction of constraints. (**a**) The value of the constraint function versus time, and (**b**) the corresponding derivative of the constraint function versus time. (**c**) The total energy divided by the mass as a function of time

that the constraint function and its derivative remain small until about t = 750 s when the oscillations of their values start to grow exponentially (the oscillations are very closely spaced and so appear as a solid-filled figure). In fact if the simulation is continued after t = 900 s the numerical solution quickly fails. This can also be seen in the behavior of the total energy in Fig. 2.56c which remains nearly a constant (zero) until about t = 750 s when it also begins to grow exponentially. If the value of e is changed to e = 1 so that constraint satisfaction is present, the results are shown in Fig. 2.57 (the plotting limits on the vertical axis were changed in all the plots shown). As Fig. 2.57a, b shows, the constraint and its derivative are both well satisfied over the full time span. Typically the values seen in Fig. 2.57a, b are on the order of 10^{-6}. Note that the vertical scales in Fig. 2.57a, b are smaller than those in Fig. 2.56a, b and are not the scales seen when executing the script. The total energy, as seen in Fig. 2.57c, now remains nearly constant. If you examine the total energy on the

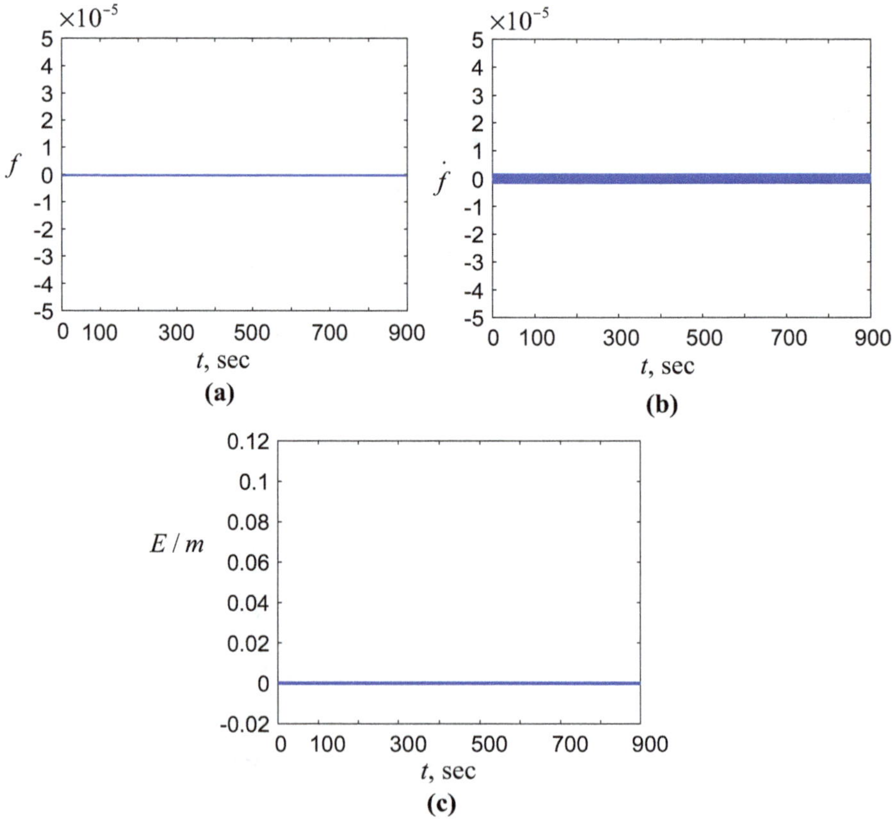

Fig. 2.57 Solution for the pendulum motion with first-order satisfaction of the constraint and its first derivative. (**a**) The value of the constraint function versus time, and (**b**) the corresponding derivative of the constraint function versus time. (**c**) The total energy divided by the mass as a function of time

default vertical scale you will see that while it has a roughly linearly increasing growth it remains very small (on the order of 10^{-5}) over this long time interval. Since the constraint errors were small and the energy nearly a constant when we used constraint satisfaction, how accurate were the solutions we obtained for such long time spans with ode4 and a fixed time step of 0.001 s? To answer that question we solved the same problem with ode45 using very small relative and absolute error tolerances of both 1e-12 and compared the results to that obtained with ode4 (with constraint correction) at $t = 700$ s, which was before any significant constraint errors were seen. The ode4 solution with the 0.001 s step size gave $x = -0.9036$ m and $y = 0.4283$ m while ode45 gave $x = -0.9058$ m and $y = 0.4237$ m, corresponding to differences of less than 0.3% and 1.1%, respectively. Thus, if we treat the ode45 solution as the "trusted solution" we can say that the ode4 solutions were reasonably accurate for this time step size even for these very long times.

This method for eliminating constraint drift is effective and relatively simple but other methods may be preferred depending on requirements such as the need for speed or accuracy in a particular application. However, this is no different than other choices one must make in solving dynamics problems numerically such as the choice of an ODE solver, for example. In this book we have tried to illustrate some of the fundamental issues involved in numerical solutions but we will also be forced to omit some of the important details and alternatives that can only be filled in with higher level and more specialized texts and courses.

2.13 Problems

P2.1. A weight is supported by three cables, as shown in Fig. P2.1. The tensions in the cables can be obtained by solving the force equilibrium equations

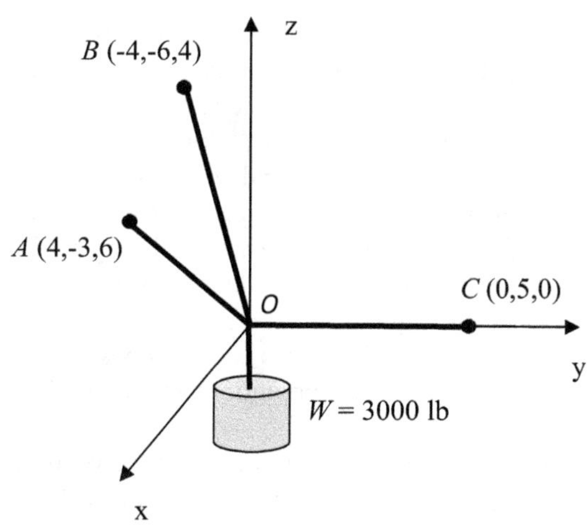

Fig. P2.1 A weight supported by three cables whose attachment points are (*A*, *B*, *C*)

$$\sum \mathbf{F} = 0$$
$$T_A\mathbf{e}_A + T_B\mathbf{e}_B + T_C\mathbf{e}_C - W\mathbf{e}_z = 0 \tag{P2.1}$$

where (T_A, T_B, T_C) are the tensions along the wires (OA, OB, OC), respectively, and $(\mathbf{e}_A, \mathbf{e}_B, \mathbf{e}_C)$ are unit vectors along those same wires. The weight $W = 3000$ lb and the unit vector $\mathbf{e}_z$ are along the z-axis. Equation (P2.1) represents three simultaneous equations for the three unknown tensions. However, we can solve for the tensions directly without solving simultaneous equations by replacing the three force equations by three moment equilibrium equations. For example, let $\mathbf{r} = \mathbf{e}_B \times \mathbf{e}_C$, $\mathbf{s} = \mathbf{e}_A \times \mathbf{e}_C$, and $\mathbf{t} = \mathbf{e}_A \times \mathbf{e}_B$. Take the cross product of the forces in Eq. (P2.1) with $\mathbf{e}_C$ and then dot that result with $\mathbf{e}_b$. Show that we obtain the tension T_A explicitly as

$$T_A = \frac{\mathbf{r} \cdot W\mathbf{e}_z}{\mathbf{r} \cdot \mathbf{e}_A}$$

Similarly show that we can find the other tensions as

$$T_B = \frac{\mathbf{s} \cdot W\mathbf{e}_z}{\mathbf{s} \cdot \mathbf{e}_B}, \quad T_C = \frac{\mathbf{t} \cdot W\mathbf{e}_z}{\mathbf{t} \cdot \mathbf{e}_C}$$

Solve these equations for the tensions and show that the solution agrees with the solution of the simultaneous equations of Eq. (P2.1).

P2.2. The equations of motion for the projectile without drag problem were given as two second-order differential equations in Eq. (2.2.6). Although we can obtain an analytical solution to these equations, write these equations of motion as a system of four first-order differential equations for the four variables $(x, \dot{x}, y, \dot{y})$ and solve that first-order system using the MATLAB® ode45 solver. Choosing a particular set of initial conditions, verify your numerical solution agrees with the analytical solution given by Eq. (2.2.8).

P2.3. A bead of mass $m = 1$ kg slides on a smooth parabolic wire given by the equation $y = ax^2$ and is connected to a linear spring whose other end is attached at the fixed point O as shown in Fig. P2.2. The spring constant $k = 10$ N/m. The unstretched length of the spring is $L_0 = d = 2m$ and the constant $a = 0.5$ 1/m. The bead starts at the bottom of the wire with a velocity to the right $v_x = \dot{x} = 5$ m/s. Since this is a conservative system the total energy, T+V, of the system is conserved. Write expressions for $T(x)$ and $V(x)$ and use the conservation of energy to determine the maximum height reached by the bead.

P2.4. Set up and solve numerically the bead sliding on a smooth, straight wire as considered in Simulation 2.2 but treat it as a one degree of freedom problem where we define a first-order system of equations for $(x, \dot{x})$ and solve that

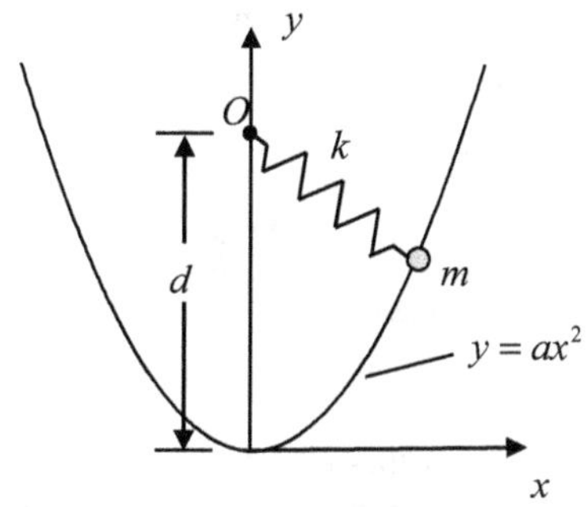

Fig. P2.2 A bead of mass m sliding on a smooth, parabolic wire and connected to a linear spring

system with ode45, using the same input parameters as used to generate Fig. 2.10a. Verify you obtain the same solution shown in that figure. In this case we do not have to worry about the constraint or its derivative being satisfied since that constraint and its derivative are used to generate the $(y, \dot{y})$ values for the motion.

P2.5. Solve the system of equations in Eq. (2.3.23) for the bead sliding on the parabolic wire problem either by hand or with the use of the symbolic math toolbox. Verify you obtain the same solution given in Eq. (2.3.25).

P2.6. Set up and solve numerically the bead sliding on the smooth parabolic wire problem in Simulation 2.3 as a system of four first-order equations for $(x, \dot{x}, y, \dot{y})$ using the same input parameters used in the script bead_on_p_wire_sim. Examine how well the constraint and its derivative are satisfied, using the default tolerances for rtol, atol. Examine the effect that a smaller rtol value has on the satisfaction of the constraint and its derivative.

P2.7. Set up and solve numerically the bead sliding on the smooth parabolic wire problem of Simulation 2.3 by using the augmentation approach of Sect. 2.4.3 where we obtain explicit expressions for the mass matrix, $\mathbf{M}$; the constraint matrix, $\mathbf{A}$; the vector, $\mathbf{b}$; and the external force, $\mathbf{F}^e$, and then use those directly to form up the equations of motion, as seen in Eq. (2.4.41), where all the needed terms are obtained through numerical matrix-vector operations in MATLAB®.

P2.8. Set up and solve numerically the bead sliding on a parabolic wire where the bead is connected to a linear spring, as considered in Problem P2.3. Use a system of two first-order equations for $(x, \dot{x})$ and the same initial conditions of Problem P2.3 to obtain the x-displacement of the bead and the normalized normal force versus time. Show that the total energy of the system is conserved during the motion. Verify that the maximum height of the bead is the same as predicted by the conservation of energy.

P2.9. Modify Problem P2.6 by including a viscous-like damping force of the type given by Eq. (2.3.29). Again, numerically solve this problem for the x- and y-displacements of the bead versus time and for the normalized normal force. Take $c/m = 0.5$ (1/s) and compare your results to Simulation 2.4.

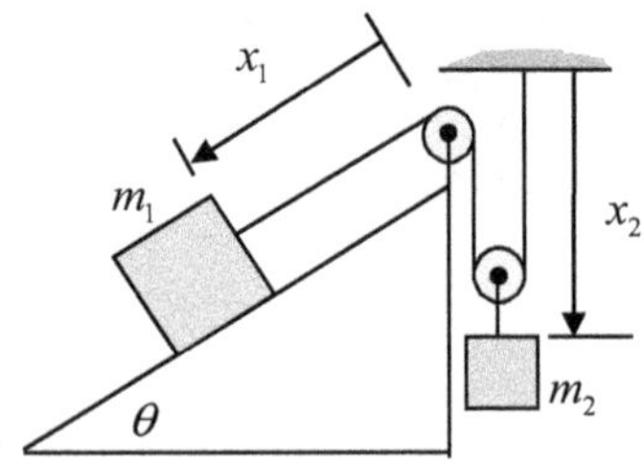

Fig. P2.3 A block of mass m_1 sliding on a fixed, smooth inclined plane and connected by an inextensible string to another block of mass m_2. The pulleys are weightless and frictionless

P2.10. In this chapter we have only considered ideal constraint forces acting on a single particle, but we will see that such ideal constraint forces appear when considering forces acting between multiple particles as well. We will examine such cases in detail in the following chapter. As a simple introduction to such cases, consider the two blocks shown in Fig. P2.3. (a) Determine the equations of motion for each of the two interconnected blocks by drawing free body diagrams of the two blocks and then applying Newton's second law for each block along the x_1- and x_2-directions, respectively. What constraint does the inextensible string impose on (1) the positions, (2) the velocities, and (3) the accelerations of the two blocks? Using the relationship between the accelerations of the blocks, solve for the tension, T, in the string. (b) Now use an augmented approach for the system of two particles, where from Eq. (2.4.41) we have

$$\begin{aligned}\mathbf{F}^c &= \mathbf{A}^T\left(\mathbf{A}\mathbf{M}^{-1}\mathbf{A}^T\right)^{-1}\left(\mathbf{b}-\mathbf{A}\mathbf{M}^{-1}\mathbf{F}^e\right)\\ \ddot{\mathbf{x}} &= \mathbf{M}^{-1}\mathbf{F}^e+\mathbf{M}^{-1}\mathbf{A}^T\left(\mathbf{A}\mathbf{M}^{-1}\mathbf{A}^T\right)^{-1}\left(\mathbf{b}-\mathbf{A}\mathbf{M}^{-1}\mathbf{F}^e\right)\end{aligned} \tag{P2.2}$$

In this case $\mathbf{x} = [x_1\ x_2]^T$, $\mathbf{M} = \begin{bmatrix} m_1 & 0 \\ 0 & m_2 \end{bmatrix}$, and the force of constraint is $\mathbf{F}^c = \mathbf{A}^T\lambda$, where λ is a Lagrange multiplier. Determine the corresponding $\mathbf{F}^e$, $\mathbf{A}$,$\mathbf{b}$ for this problem, noting that there is only one constraint between the two particles, and solve for the accelerations and constraint force from Eq. (P2.2), showing that your results agree with part (a). What is the physical meaning of the Lagrange multiplier, λ, in this problem?

P2.11. Examine the spring-pendulum problem of Sect. 2.5.2 for small, moderate, and large values of mu, demonstrating the transition from regular to chaotic and back to regular behavior.

P2.12. Consider the problem shown in Fig. P2.4 where a bead slides on a smooth parabolic wire that is rotating at a constant angular velocity, Ω, about the y-axis. Determine the equations of motion (a) using explicit embedding of the constraint and the $\mathbf{c}_m$ vectors to obtain Eqs. (2.6.29a) and (2.6.29b), and (b) using explicit embedding of the constraint in Lagrange's equations (note that you can do the embedding by inspection so that you do not have to

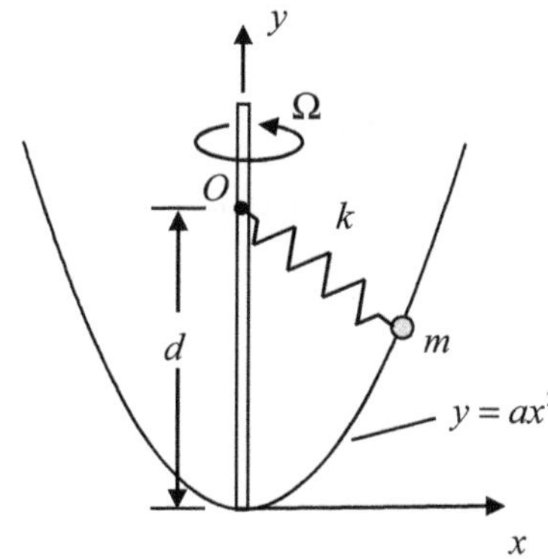

Fig. P2.4 A bead of mass m slides on a smooth parabolic wire and is attached to a spring. The wire and the shaft that the spring is attached to lie in a vertical plane and rotate about the y-axis with a constant angular velocity of Ω rad/s

follow the general embedding approach outlined in Sect. 2.11). Obtain the Jacobi integral for this problem. What are the equilibrium positions for the bead in this problem? Numerically solve the equations of motion using the same parameters given in Problem P2.3 and for different angular velocities. At what value of the angular velocity does the $x = y = 0$ equilibrium position become unstable?

P2.13. Obtain phase plots for the problem of a particle sliding on a rotating hoop (considered in Sect. 2.7) by using the Jacobi integral, h, and the MATLAB® contour function for two cases: $\Omega = 2$ rad/s and $\Omega = 7$ rad/s. Use the same limits for $(\theta, \dot{\theta})$ used in Fig. 2.44 and the same parameters as used in Simulation 2.7. Identify the stable and unstable equilibrium points on the phase plots.

P2.14. Consider the spherical pendulum shown in Fig. P2.5 in spherical coordinates where the θ angle is taken with respect to the negative z-axis as shown. Show that the kinetic energy and the potential energy are given for the pendulum as

$$\begin{aligned} T &= \frac{m}{2}\left(R^2\dot{\theta}^2 + R^2\dot{\phi}^2\sin^2\theta\right) \\ V &= -mgR\cos\theta \end{aligned} \tag{P2.3}$$

Also show that since the angle ϕ is missing from the Lagrangian, $L = T - V$, that coordinate is an ignorable coordinate and the generalized momentum, p_ϕ, given by $p_\phi = mR^2\dot{\phi}\,\sin^2\theta$, is a constant. Show that the equations for the spherical pendulum therefore can be written as

$$\begin{aligned} \dot{\phi} &= p_\phi / mR^2\sin^2\theta \\ \ddot{\theta} &= \dot{\phi}^2\sin\theta\cos\theta - g\sin\theta/R \end{aligned} \tag{P2.4}$$

where the constant p_ϕ is determined from the initial conditions. It would appear that there is a singularity in these equations at $\theta = 0,\pi$ because of the

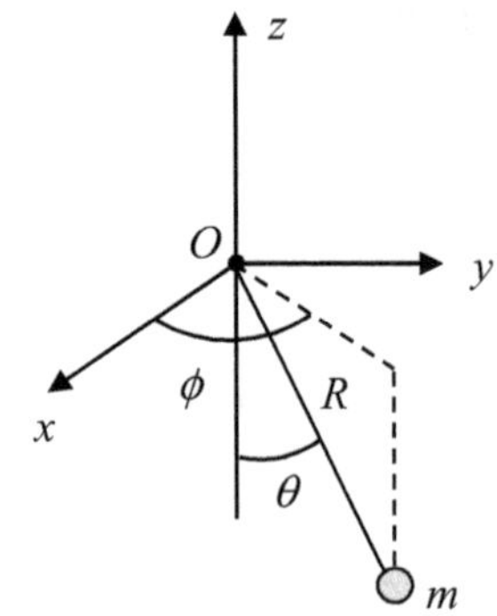

Fig. P2.5 A spherical pendulum consisting of a mass m connected to a fixed point O by a massless, inextensible string. Note that the angle θ is measured from the negative z-axis

$1/\sin^2\theta$ term in these equations. However the total energy $E = T + V$ can be written as

$$E = \frac{m}{2}R^2\dot{\theta}^2 + V_{\text{eff}}(\theta) \tag{P2.5}$$

where

$$V_{\text{eff}}(\theta) = \frac{p_\phi^2}{2mR^2\sin^2\theta} - mgR\cos\theta \tag{P2.6}$$

is called the effective potential. Although the effective potential goes to infinity at $\theta = 0,\pi$, the total energy is a constant during the motion and from Eq. (P2.5) we see that $E \geq V_{\text{eff}}$. Thus, as shown in Fig. P2.6, there are always maxima and minima values for θ during the motion and the singularity is never encountered. Turn Eq. (P2.4) into a system of first-order equations and solve for the motion of the pendulum numerically under different initial conditions with $R = 2$ m, $m = 0.5$ kg, and $g = 9.81$ m/s^2. Include an animation of the motion. Show that for an initial condition $\dot{\phi}_0 = 0$ we just obtain the motion of a simple pendulum in a vertical plane and for $\dot{\phi}_0 = \sqrt{g/R\cos\theta_0}$ the mass executes a purely conical motion about the z-axis at a constant angle θ_0.

P2.15. Consider the bead-on-a-parabolic wire problem considered in Simulation 2.3 where there is Coulomb friction between the bead and the wire. Let a = 0.05 1/m and let the coefficient of friction be $\mu = 0.3$. Obtain the equation of motion and solve for the motion of the bead numerically, using different initial starting velocities for the bead at the origin. Determine the sticking region for the bead and verify that the bead stops within that region by examining a phase plot of the motion.

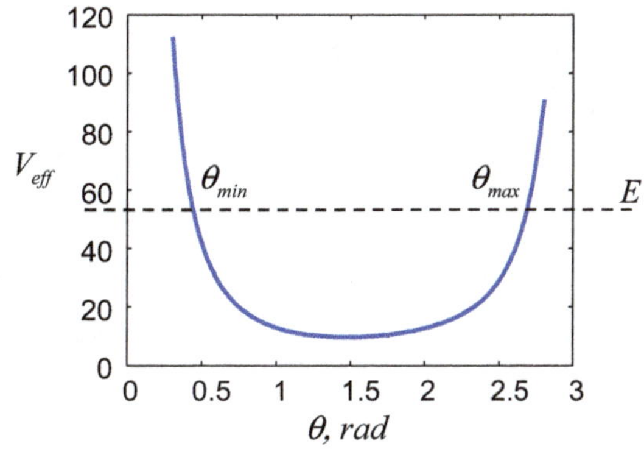

Fig. P2.6 A plot of the effective potential for the spherical pendulum, showing that there are maximum and minimum values for the angle θ at a constant total energy, E

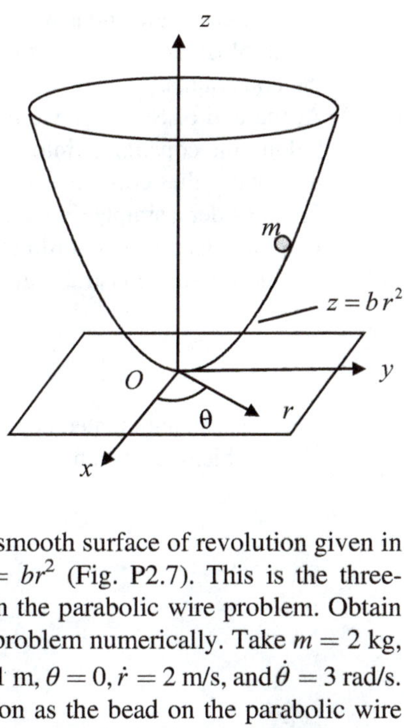

Fig. P2.7 A particle of mass m moving on a smooth paraboloid surface

P2.16. Consider the motion of a particle on a smooth surface of revolution given in cylindrical coordinates (r, θ, z) by $z = br^2$ (Fig. P2.7). This is the three-dimensional counterpart of the bead on the parabolic wire problem. Obtain the equations of motion and solve the problem numerically. Take $m = 2$ kg, $b = 0.1$ 1/m with initial conditions $r = 1$ m, $\theta = 0$, $\dot{r} = 2$ m/s, and $\dot{\theta} = 3$ rad/s. Verify that you obtain the same solution as the bead on the parabolic wire when the initial motion of the particle has no velocity in the θ-direction.

P2.17. Examine the bead on a parabolic wire problem considered in Problem P2.6 for very long simulation times where the constraint and its first derivative may not be well satisfied. Use ode4 and a time interval $\Delta t = 0.001$ s. Implement the approach of Sect. 2.12 to control the constraint drift.

P2.18. A block of mass m is attached to a linear spring with spring constant k, and slides on a smooth inclined plane as shown in Fig. P2.8. Obtain the equations of motion in terms of the x- and y-coordinates using an augmented approach and solve the equations numerically for a time interval of 20 s. Let $m = 3$ kg,

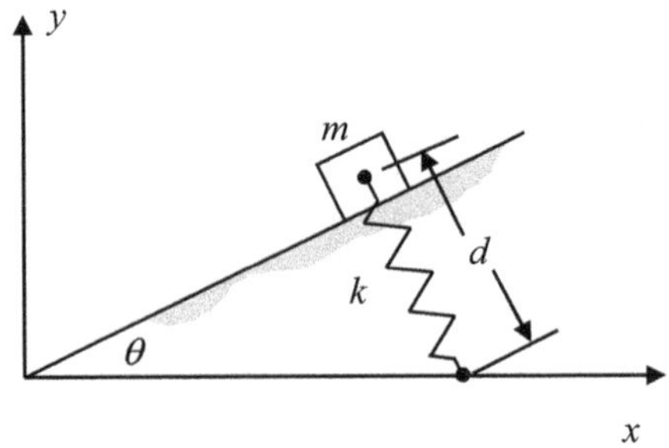

Fig. P2.8 A block sliding on a smooth inclined surface and attached to a linear spring

$k = 10$ N/m, and $\theta = 30°$. The spring is unstretched when it is perpendicular to the plane and its unstretched length $d = 2$ m in that position. The block starts at the origin and is released with no velocity. Show that the total energy of the block is conserved during the motion and examine the satisfaction of the constraints.

P2.19. At the end of Sect. 2.7 we mentioned that one cannot directly embed non-holonomic constraints into the expression for the kinetic energy and expect to obtain the correct equations of motion from Lagrange's equation. (a) Consider a simple (but admittedly artificial) example of a particle moving in (x, y, z) Cartesian coordinates with no external forces but where there is a non-holonomic constraint given by

$$\dot{y} - z\dot{x} = 0 \tag{P2.7}$$

First, use an augmented approach and show that the equations of motion for this problem are given as

$$\begin{aligned}\ddot{x} &= -\frac{z\dot{x}\,\dot{z}}{1+z^2}\\ \ddot{y} &= \frac{\dot{x}\,\dot{z}}{1+z^2}\\ \ddot{z} &= 0\end{aligned} \tag{P2.8}$$

Second, embed the constraint equation into the kinetic energy expression to write it as a function of $(z, \dot{x}, \dot{z})$ only. With this kinetic energy show that Lagrange's equations produce incorrect equations of motion. (b) In contrast, show that the same problem with the holonomic constraint

$$\dot{y} - c\dot{z} = 0 \tag{P2.9}$$

(where c is a constant) produces the same equations of motion from either the augmented approach or the embedding approach.

P2.20. The bead sliding on a straight, smooth wire considered in Sect. 2.3.1 was initially modeled as the set of differential algebraic equations (DAEs) given in Eq. (2.3.6). Explore the solution of these equations using a DAE solver, and compare with the solution to the problem considered in Simulation 2.2. There are some MATLAB® DAE solvers available as well as others that you can find online.

References

1. P. Kunkel, V. Mehrmann, *Differential-Algebraic Equations* (European Mathematical Society, Zürich, 2006)
2. F.E. Udwadia, R.E. Kalaba, *Analytical Dynamics* (Cambridge University Press, Cambridge, 1996)
3. A.A. Shabana, *Computational Dynamics*, 3rd edn. (Wiley, Chichester, 2010)
4. A.A. Shabana, *Dynamics of Multibody Systems*, 4th edn. (Cambridge University Press, New York, 2013)
5. S.H. Strogatz, *Nonlinear Dynamics and Chaos*, 2nd edn. (Westview Press, Boulder, CO, 2014)
6. K.T. Alligood, T.D. Sauer, J.A. Yorke, *Chaos—An Introduction to Dynamical Systems* (Springer, New York, 2000)
7. J.G. Papastavridis, *Analytical Mechanics, A Comprehensive Treatise on the Dynamics of Constrained Systems* (New York, Oxford, 2001)
8. L.S. Wang, Y.H. Pao, Jourdain's variational equation and Appell's equation of motion for non-holonomic dynamical systems. Am. J. Phys. **71**, 72–82 (2003)
9. F.E. Udwadia, Fundamental principles of Lagrangian dynamics: mechanical systems with non-ideal, holonomic, and nonholonomic constraints. J. Math. Anal. Appl. **251**, 341–355 (2000)
10. O.A. Bauchau, *Flexible Multibody Dynamics* (Springer, New York, 2011)
11. F. Amirouche, *Fundamentals of Multibody Dynamics* (Birkhäuser, Boston, MA, 2006)
12. D.J. Braun, M. Goldfarb, Eliminating constraint drift in the numerical simulation of constrained dynamical systems. Comput. Methods Appl. Mech. Eng. **198**, 3151–3160 (2009)

Chapter 3
Dynamics of a System of Particles

In Chap. 2 we used the motion of a single particle to discuss a wide range of topics including force versus moment equations, potential energy, work-energy concepts, constraint forces, generalized coordinates, Lagrange's equations, and others. In this chapter we consider those topics and others for the case where a system of multiple, interacting particles is moving under the action of a set of forces.

3.1 Internal Forces

A system of particles is shown in Fig. 3.1a. As in the single-particle case we can have a total applied force, $\mathbf{F}_i^a$, acting on the ith particle, where the applied force may be a combination of known external forces and externally applied constraint forces, i.e., $\mathbf{F}_i^a = \mathbf{F}_i^e + \mathbf{F}_i^c$. But a system of particles may also have internal forces which are forces of interaction between particles, such as the force $\mathbf{F}_{ij}$ that particle j exerts on particle i, and the force $\mathbf{F}_{ji}$ that particle i exerts on particle j (Fig. 3.1a). Newton formulated a third law to deal with these internal forces (see Fig. 3.1b, where $\mathbf{x}_{ij} = \mathbf{x}_j - \mathbf{x}_i$):

Newton's Third Law: The mutual forces of interaction between any two particles are equal in magnitude and opposite in direction (i.e., $\mathbf{F}_{ji} = -\mathbf{F}_{ij}$) and they act along the line between those two particles (i.e., $\mathbf{F}_{ij} = F_{ij}\mathbf{x}_{ij}/|\mathbf{x}_{ij}|$, $\mathbf{F}_{ji} = F_{ji}\mathbf{x}_{ji}/|\mathbf{x}_{ji}| = -F_{ji}\mathbf{x}_{ij}/|\mathbf{x}_{ij}|$) so that both forces act along $\mathbf{x}_{ij}$ (Fig. 3.1b).

This law is often given more succinctly in an alternate version as follows: For every action (of one body on another) there is an equal and opposite reaction.

Our original statement of the third law is sometimes called the "strong" version, since it restricts both the forces to be equal and opposite and to lie along the line separating the two particles. The alternate version given above is called the "weak" version since it simply requires the forces to be equal and opposite. Although there

L. W. Schmerr, *Engineering Dynamics 2.0*, Solid Mechanics and Its Applications 254,
https://doi.org/10.1007/978-3-319-98470-4_3

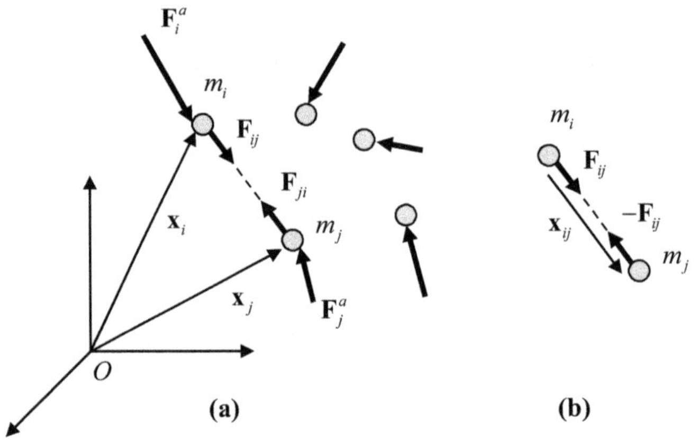

Fig. 3.1 (**a**) A system of particles acted upon by external and internal forces, and (**b**) a pair of particles showing the equal and opposite internal forces

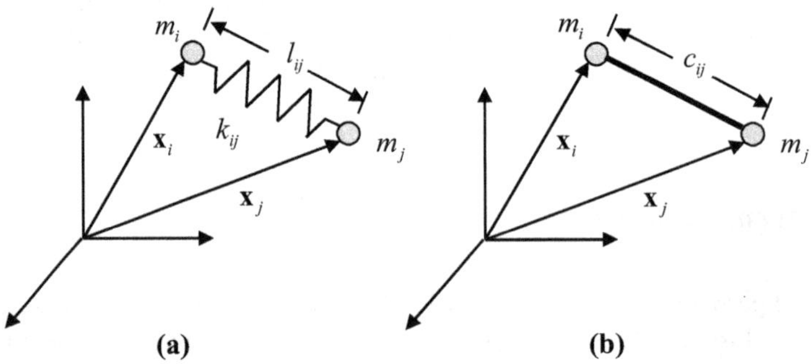

Fig. 3.2 (**a**) A pair of particles connected by a linear spring, and (**b**) the particles connected by a rigid rod (of negligible mass)

are electromagnetic force interactions that violate the strong version of this "law," all the examples of internal forces seen in this book will satisfy the strong version.

For example, consider two particles m_i and m_j that are connected to a linear spring with spring constant k_{ij} and unstretched length l_{ij}^0 (Fig. 3.2a). From Chap. 2 we know that the spring force is derivable from a potential energy, V_{ij}, where

$$V_{ij} = \frac{1}{2}k_{ij}\left[\left|\mathbf{x}_j - \mathbf{x}_i\right| - l_{ij}^0\right]^2 \tag{3.1.1}$$

where we will let $\mathbf{x}_j = (x_{j;1}, x_{j;2}, x_{j;3})$ and $\mathbf{x}_i = (x_{i;1}, x_{i;2}, x_{i;3})$. Then the kth component $(k = 1, 2, 3)$ of the force that mass j exerts on mass i through the spring, $F_{ij;k}$, is

$$F_{ij;k} = -\frac{\partial V_{ij}}{\partial x_{i;k}} = k_{ij}\left[\left|\mathbf{x}_j - \mathbf{x}_i\right| - l_{ij}^0\right]\frac{x_{j;k} - x_{i;k}}{\left|\mathbf{x}_j - \mathbf{x}_i\right|} \tag{3.1.2a}$$

and the kth component of the force that mass i exerts on mass j through the spring, $F_{ji;k}$, is

$$F_{ji;k} = -\frac{\partial V_{ij}}{\partial x_{j;k}} = k_{ij}\left[\left|\mathbf{x}_j - \mathbf{x}_i\right| - \ell_{ij}^0\right]\frac{x_{i;k} - x_{j;k}}{\left|\mathbf{x}_j - \mathbf{x}_i\right|} \tag{3.1.2b}$$

and obviously $F_{ji;k} = -F_{ij;k}$ and we have from Eqs. (3.1.2a) and (3.1.2b)

$$\begin{aligned}\mathbf{F}_{ij} &= k_{ij}\left[l_{ij} - \ell_{ij}^0\right]\frac{\mathbf{x}_j - \mathbf{x}_i}{l_{ij}} = k_{ij}\left[l_{ij} - \ell_{ij}^0\right]\frac{\mathbf{x}_{ij}}{l_{ij}} \\ \mathbf{F}_{ji} &= k_{ij}\left[l_{ij} - \ell_{ij}^0\right]\frac{\mathbf{x}_i - \mathbf{x}_j}{l_{ij}} = k_{ij}\left[l_{ij} - \ell_{ij}^0\right]\frac{\mathbf{x}_{ji}}{l_{ij}}\end{aligned} \tag{3.1.3}$$

(see Fig. 3.1b) where $l_{ij} = |\mathbf{x}_j - \mathbf{x}_i| = |\mathbf{x}_{ij}|$ is the stretched length of the spring. Thus, both internal forces are also directed along the line between the two masses as specified by Newton's third law. It is interesting to note that any set of internal forces between a pair of masses i and j that are derivable from a potential that only depends on the distance between those particles, i.e.,

$$V_{ij} = V_{ij}\left(\left|\mathbf{x}_i - \mathbf{x}_j\right|\right) \tag{3.1.4}$$

will, upon differentiation, produce forces that automatically satisfy Newton's third law. The mutual forces of gravitation attraction between two masses that we discussed in Chap. 2 is an example of forces which come from a potential of the form given in Eq. (3.1.4) and so they also satisfy Newton's third law.

Internal forces may or may not be ideal constraint forces. Internal spring forces (Fig. 3.2a), for example, are not forces of constraint since the spring does not make the attached particles follow a specified path but rather the spring controls the force between particles as a function of whatever path the particles take.

One of the most important cases of an internal force that is an ideal constraint force is shown in Fig. 3.2b where the two particles are separated by a thin, massless rigid rod so that the distance between the particles remains fixed, i.e., we have the constraint

$$\left|\mathbf{x}_j - \mathbf{x}_i\right| - c_{ij} = \sqrt{\left(x_{j;1} - x_{i;1}\right)^2 + \left(x_{j;2} - x_{i;2}\right)^2 + \left(x_{j;3} - x_{i;3}\right)^2} - c_{ij} = 0 \tag{3.1.5}$$

where $c_{ij} = c_{ji}$ is a constant. This is a holonomic constraint of the form

$$f_{ij}\left(\left|\mathbf{x}_j - \mathbf{x}_i\right|\right) = 0 \tag{3.1.6}$$

which, as we saw in the last chapter, will generate an ideal constraint forces acting on particles i and j with components given by

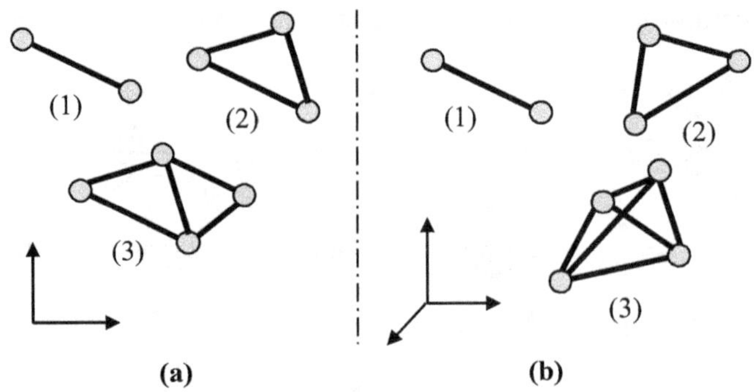

Fig. 3.3 (**a**) Particles moving in two dimensions connected by weightless rigid rods. Each particle has two degrees of freedom and each rod acts as a single constraint. Thus the number of degrees of freedom, n_{dof}, are (1) $n_{\text{dof}} = 2(2) - 1 = 3$, (2) $n_{\text{dof}} = 3(2) - 3 = 3$, and (3) $n_{\text{dof}} = 4(2) - 5 = 3$. (**b**) Particles moving in 3-D connected by weightless rigid rods. Each particle has three degrees of freedom and each rod acts as a single constraint. Thus, the number of degrees of freedom are (1) $n_{\text{dof}} = 2(3) - 1 = 5$, (2) $n_{\text{dof}} = 3(3) - 3 = 6$, and (3) $n_{\text{dof}} = 4(3) - 6 = 6$

$$
\begin{aligned}
F_{ij;k} &= \lambda_{ij}\frac{\partial f_{ij}}{\partial x_{i;k}} = \frac{\lambda_{ij}\left(x_{i;k} - x_{j;k}\right)}{\left|\mathbf{x}_j - \mathbf{x}_i\right|} \\
F_{ji;k} &= \lambda_{ij}\frac{\partial f_{ij}}{\partial x_{j;k}} = \frac{\lambda_{ij}\left(x_{j;k} - x_{i;k}\right)}{\left|\mathbf{x}_j - \mathbf{x}_i\right|} = -F_{ji;k}
\end{aligned}
\tag{3.1.7}
$$

Like the spring forces, these constraint forces satisfy Newton's third law.

If two masses connected by a rigid rod (see Fig. 3.3a, case (1)) are moving only in two dimensions, then this system has three degrees of freedom. In fact, any number of particles moving in 2-D with rigid connections between all particles also have only three degrees of freedom as shown in Fig. 3.3a for cases of up to four particles. For particles moving in three dimensions (Fig. 3.3b), as long as there are three or more rigidly connected particles there will always be only six degrees of freedom. [The case of two rigidly connected particles moving in three dimensions is different because this system has no mass distributed off the axis connecting those particles so the system has no angular momentum about that axis. Thus, the system is always in moment equilibrium about that axis. This eliminates any possibility of having rotational motion about that axis for the system and reduces the number of degrees of freedom by one.] A lattice of rigidly connected particles is a simple example of a "rigid system" or "rigid body." In Chap. 5, where we will consider rigid bodies with continuously distributed masses, it will also be true that those rigid bodies moving in two or three dimensions will have at most only three degrees of freedom and six degrees of freedom, respectively.

The limited number of degrees of freedom for even a large number of rigidly constrained particles means that the motion of a rigid system can be characterized in terms of only a few geometrical variables. Consider the rigid system of Fig. 3.4a, for

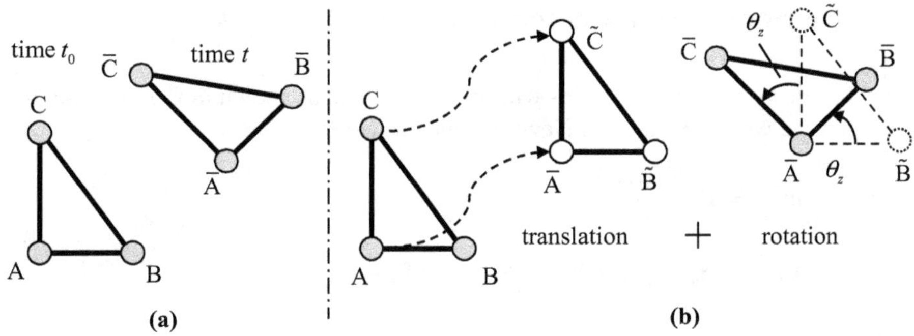

Fig. 3.4 (**a**) The 2-D motion of a rigidly connected system of particles from time t_0 to a later time t, and (**b**) the decomposition of that motion into (1) a translation of the system, where every particle follows the same path as a reference particle, such as A, for example, and (2) a rotation about the z-axis (in this case shown as a rotation θ_z about Ā) where every line in the rigid system experiences the same rotation as shown for lines $\bar{A}\tilde{B} \rightarrow \bar{A}\bar{B}$ and $\bar{A}\tilde{C} \rightarrow \bar{A}\bar{C}$

example, which is a rigid system of three particles moving in two dimensions and has three degrees of freedom. Figure 3.4a shows the configuration of the system at two times (t_0, t). We can take as those degrees of freedom: (1) the motion of any one of the particles, such as the one at point A, described by its coordinates $(x_A(t), y_A(t))$ relative to a fixed origin, O, and (2) a rotation about that point, $\theta_z(t)$. Physically, this means that any motion of a rigid system of particles can be decomposed into a translation and a rotation, as shown in Fig. 3.4b. Suppose, for example we let all the particles have the same motion as point A. Then the system moves as shown in Fig. 3.4b to a configuration where point A is in its correct final position and the lines connecting the particles are all unchanged in their lengths and orientations. This is a translational motion. We can, however, after the translation rotate all the lines between particles in the system about point A through an angle θ_z, as shown in Fig. 3.4b, to also bring points B and C into their correct final positions. This is possible since in a rigidly constrained system of particles the angles between lines connecting the particles must remain fixed (in Fig. 3.4, for example, the angles inside the triangle ABC are all fixed) so that the rotation experienced by any one line is experienced by all the lines. A rotation about a given point also does not change the length of the lines connecting that point to all the other points so that the rigid constraints of the system remain intact. In three dimensions a similar decomposition of the motion of a rigid system into a translation and rotation can be made, but the motion of the reference point A used to define the translation has three components (x_A, y_A, z_A) and we can rotate the lines in the body about all three axes $(\theta_x, \theta_y, \theta_z)$ to produce the six degrees of freedom needed to describe the motion of the system.

3.2 Newton-Euler Laws for a System of Particles

If we have a system of N particles with forces on them as shown in Fig. 3.5 then for each particle we have from the Newton-Euler equations

$$\begin{aligned} m_i \ddot{\mathbf{x}}_i &= \mathbf{F}_i^a + \sum_{j \neq i} \mathbf{F}_{ij} \quad (i = 1, \ldots, N) \\ \mathbf{x}_{Bi} \times m_i \ddot{\mathbf{x}}_i &= \mathbf{x}_{Bi} \times \mathbf{F}_i^a + \sum_{j \neq i} \mathbf{x}_{Bi} \times \mathbf{F}_{ij} \quad (i = 1, \ldots, N) \end{aligned} \tag{3.2.1}$$

where $\mathbf{F}_i^a$ is the net external and constraint force acting on mass m_i and $\mathbf{x}_{Bi}$ is the position vector of m_i with respect to point B, which we will now let be either a stationary or a moving point. We can use at most N of these equations but, as in the single-particle case, we could obtain those N equations from combinations of the force and moment equations if that makes the generation of the equations of motion easier. In the sum over the internal forces the case $j = i$ must be excluded since $\mathbf{F}_{ii} = 0$, i.e., a mass does not exert a force on itself. In general if there are N particles moving in 3-D and n_h independent holonomic internal and/or external constraints, the number of degrees of freedom will be $n_{\text{dof}} = 3N - n_h$. This is the smallest number of equations of motion we will have to solve. In two-dimensional motion we would have $n_{\text{dof}} = 2N - n_h$ instead.

3.2.1 *Motion of the Center of Mass*

Rather than treating a system of particles separately for each particle it is useful to also consider the Newton-Euler laws for the entire system of particles. Let's examine the force equations first. If we sum up Newton's second law equations for the entire system we have

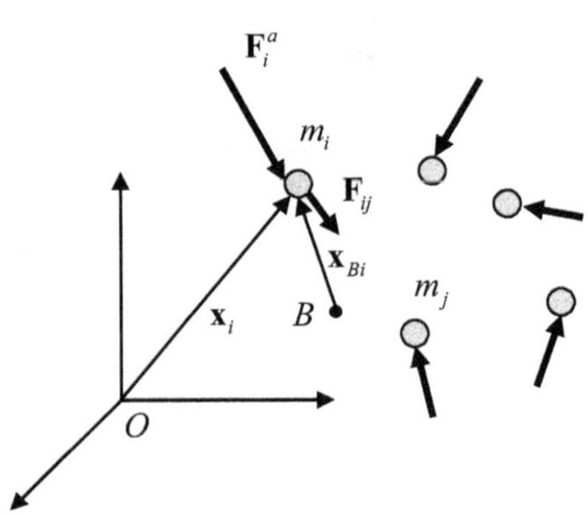

Fig. 3.5 A system of particles moving under external and internal forces

$$\sum_{i=1}^{N} m_i \ddot{\mathbf{x}}_i = \sum_{i=1}^{N} \mathbf{F}_i^a + \sum_i \sum_{j \neq i} \mathbf{F}_{ij} \tag{3.2.2}$$

The sums over all the internal forces need to be made more explicit. These sums can be written as

$$\sum_i \sum_{j \neq i} \mathbf{F}_{ij} = \sum_{i=1}^{N} \sum_{j=i+1}^{N} \mathbf{F}_{ij} + \sum_{i=1}^{N} \sum_{j=i+1}^{N} \mathbf{F}_{ji} \tag{3.2.3}$$

You can verify this form by looking at a simple case such as $N = 3$. If you carry out all the sums you will see you obtain all the values of the 3×3 internal force matrix $\mathbf{F}_{ij}$ except the diagonal terms, $\mathbf{F}_{ii}$, which are zero anyway. However, from Newton's third law $\mathbf{F}_{ji} = -\mathbf{F}_{ij}$ so that

$$\sum_i \sum_{j \neq i} \mathbf{F}_{ij} = \sum_{i=1}^{N} \sum_{j=i+1}^{N} \mathbf{F}_{ij} + \sum_{i=1}^{N} \sum_{j=i+1}^{N} (-\mathbf{F}_{ij}) = 0 \tag{3.2.4}$$

and Eq. (3.2.2) becomes

$$\sum_{i=1}^{N} m_i \ddot{\mathbf{x}}_i = \mathbf{F}^{\text{sys}} \tag{3.2.5}$$

where $\mathbf{F}^{\text{sys}}$ is the total sum of all the external applied and constraint forces acting on the entire system. On the left side of Eq. (3.2.5) we see a weighted sum of the accelerations of the particles. Consider the same weighted sum of the position vectors of the particles, and define a point, $\mathbf{x}_C$ called the *center of mass* of the system of particles, as

$$\begin{aligned} m\mathbf{x}_C &= \sum_{i=1}^{N} m_i \mathbf{x}_i \\ m &= \sum_{i=1}^{N} m_i \end{aligned} \tag{3.2.6}$$

where m is the total mass of the system. It then follows that the velocity and acceleration of this center of mass point are given by

$$\begin{aligned} \mathbf{v}_C = \dot{\mathbf{x}}_C &= \frac{1}{m} \sum_{i=1}^{N} m_i \dot{\mathbf{x}}_i \\ \mathbf{a}_C = \ddot{\mathbf{x}}_C &= \frac{1}{m} \sum_{i=1}^{N} m_i \ddot{\mathbf{x}}_i \end{aligned} \tag{3.2.7}$$

Equation (3.2.5) becomes

$$m\mathbf{a}_C = \mathbf{F}^{\text{sys}} \tag{3.2.8}$$

so that the entire system of particles acts like a single particle of mass m at the center of mass, where the motion of this single particle is determined by the total external forces acting on the system. If we define the linear momentum of this single particle, $\mathbf{p}_C$, as

$$\mathbf{p}_C = m\mathbf{v}_C = \sum_{i=1}^{N} m_i \dot{\mathbf{x}}_i = \sum_{i=1}^{N} \mathbf{p}_i \tag{3.2.9}$$

then Eq. (3.2.8) can also be written as

$$\frac{d\mathbf{p}_C}{dt} = \mathbf{F}^{\text{sys}} \tag{3.2.10}$$

3.2.2 *Impulse and Linear Momentum*

Equation (3.2.10) can be integrated on time to get the impulse-momentum relation:

$$\Delta\mathbf{p}_C = \mathbf{p}_C(t_2) - \mathbf{p}_C(t_1) = \int_{t_1}^{t_2} \mathbf{F}^{\text{sys}}(t)dt \tag{3.2.11}$$

where $\Delta\mathbf{p}_C$ is the change of the linear momentum of the system and the integral of the force is called the *impulse* (or *linear impulse*) of that force. Thus, we see if the net external force acting on a system of particles is zero (or if the impulse of those external forces is zero) the linear momentum of the system remains unchanged. This is a statement of the principle of *conservation of linear momentum* for a system of particles. In general this principle applies only to the behavior of the entire system, but in some cases it can also give us information about the individual particles. For example, suppose we have a mass, m_1, at rest hanging vertically from a string under its own weight that is struck by a projectile of mass, m_2, that is moving horizontally with a velocity, $\mathbf{v}_{\text{in}}$, as shown in Fig. 3.6a. If the projectile imbeds itself into the hanging mass, we can determine the common velocity, $\mathbf{v}_{\text{out}}$, immediately after the collision (Fig. 3.6b) since there is no external horizontal force acting on the system of two particles during the collision, only the internal forces acting between them (the tension in the string and the weight of the hanging mass both act in the vertical direction so do not enter into the x-momentum calculations). Thus, the net external impulse on the system in the x-direction is zero and the x-components of the momenta satisfy the conservation law:

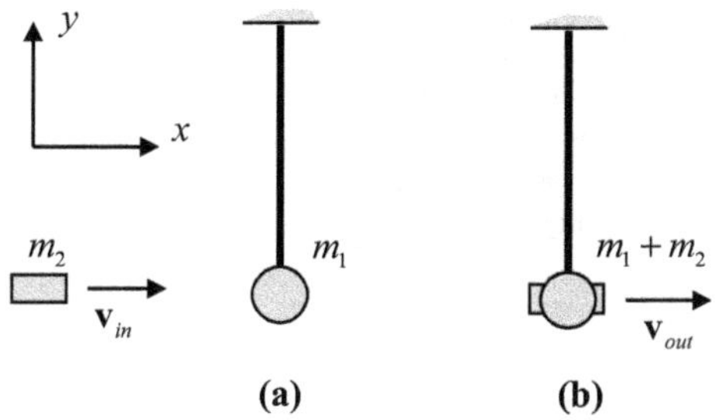

Fig. 3.6 (**a**) A projectile striking a stationary mass with velocity, $\mathbf{v}_{\text{in}}$, and (**b**) immediately after the collision where both masses travel together with velocity, $\mathbf{v}_{\text{out}}$

$$m_2 v_{\text{in}} = (m_1 + m_2) v_{\text{out}} \tag{3.2.12}$$

giving $v_{\text{out}} = m_2 v_{\text{in}}/(m_1 + m_2)$.

3.2.3 The Moment Equation and Angular Momentum

Now, consider the moment equation for each particle in Eq. (3.2.1) and sum over all the particles in the entire system:

$$\sum_{i=1}^{N} \mathbf{x}_{Bi} \times m_i \ddot{\mathbf{x}}_i = \sum_{i=1}^{N} \mathbf{x}_{Bi} \times \mathbf{F}_i^a + \sum_{i=1}^{N} \sum_{j \neq i} \mathbf{x}_{Bi} \times \mathbf{F}_{ij} \tag{3.2.13}$$

We will let B be a general moving point and let C be the center of mass of the system (Fig. 3.7). Keeping the motion of B general will make our analysis a bit more involved but it will allow us to make some important choices for B later. First, consider the sum over the moments of the internal forces. As with the force equation we can write this term more explicitly and we obtain

$$\begin{aligned}
\sum_{i=1}^{N} \sum_{j \neq i} \mathbf{x}_{Bi} \times \mathbf{F}_{ij} &= \sum_{i=1}^{N} \sum_{j=i+1}^{N} \mathbf{x}_{Bi} \times \mathbf{F}_{ij} + \sum_{i=1}^{N} \sum_{j=i+1}^{N} \mathbf{x}_{Bj} \times \mathbf{F}_{ji} \\
&= \sum_{i=1}^{N} \sum_{j=i+1}^{N} \left(\mathbf{x}_{Bi} - \mathbf{x}_{Bj}\right) \times \mathbf{F}_{ij} \\
&= \sum_{i=1}^{N} \sum_{j=i+1}^{N} \mathbf{x}_{ji} \times \mathbf{F}_{ij} = 0
\end{aligned}$$

where from Newton's third law we have used the fact that $\mathbf{F}_{ji} = -\mathbf{F}_{ij}$ and that $\mathbf{x}_{ji} = \mathbf{x}_{Bi} - \mathbf{x}_{Bj}$ (Fig. 3.7) is parallel to $\mathbf{F}_{ij}$. Thus, the internal moment contributions to Eq. (3.2.13) cancel and we find

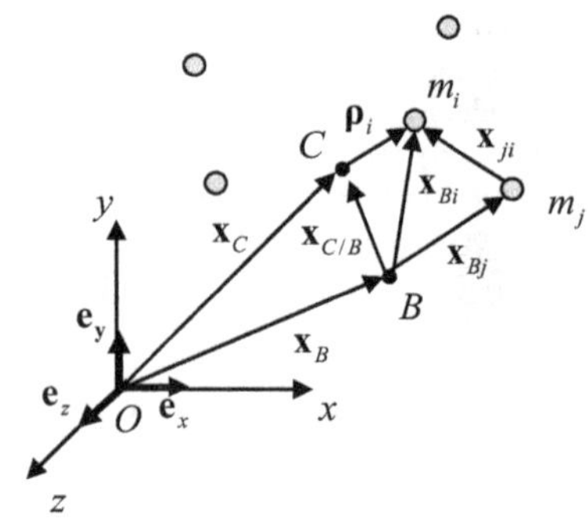

Fig. 3.7 Particles in a moving system as measured from a fixed origin, O; a moving point, B; and the center of mass, C, which is also moving

$$\sum_{i=1}^{N} \mathbf{x}_{Bi} \times m_i \ddot{\mathbf{x}}_i = \mathbf{M}_B^{\text{sys}} \tag{3.2.14}$$

where $\mathbf{M}_B^{\text{sys}}$ is the moment about point B of all the externally applied forces, including forces of constraint, acting on the system. We can write this moment in terms of the acceleration of the center of mass of the system in the form

$$\begin{aligned} \mathbf{M}_B^{\text{sys}} &= \sum_{i=1}^{N} \mathbf{x}_{Bi} \times m_i \ddot{\mathbf{x}}_i = \sum_{i=1}^{N} \left(\mathbf{x}_{C/B} + \boldsymbol{\rho}_i\right) \times m_i \ddot{\mathbf{x}}_i \\ &= m\mathbf{x}_{C/B} \times \mathbf{a}_C + \sum_{i=1}^{N} \boldsymbol{\rho}_i \times m_i \ddot{\mathbf{x}}_i \end{aligned} \tag{3.2.15}$$

where $\mathbf{x}_{C/B}$ is the position vector of C with respect to point B (see Fig. 3.7) and we obtained the first term in Eq. (3.2.15) by writing it in terms of the acceleration of the center of mass, $\mathbf{a}_C$ as defined in Eq. (3.2.7). For the second term in Eq. (3.2.15) since from the geometry $\mathbf{x}_i = \mathbf{x}_C + \boldsymbol{\rho}_i$ (Fig. 3.7), differentiation on time of this expression yields

$$\begin{aligned} \mathbf{x}_i &= \mathbf{x}_C + \boldsymbol{\rho}_i \\ \dot{\mathbf{x}}_i &= \mathbf{v}_C + \dot{\boldsymbol{\rho}}_i \\ \ddot{\mathbf{x}}_i &= \mathbf{a}_C + \ddot{\boldsymbol{\rho}}_i \end{aligned} \tag{3.2.16}$$

and, using the expression for the acceleration, Eq. (3.2.15) becomes

$$\begin{aligned} \mathbf{M}_B^{\text{sys}} &= m\mathbf{x}_{C/B} \times \mathbf{a}_C + \sum_{i=1}^{N} \boldsymbol{\rho}_i \times m_i \left(\mathbf{a}_C + \ddot{\boldsymbol{\rho}}_i\right) \\ &= m\mathbf{x}_{C/B} \times \mathbf{a}_C + \sum_{i=1}^{N} m_i \boldsymbol{\rho}_i \times \mathbf{a}_C + \sum_{i=1}^{N} \boldsymbol{\rho}_i \times \ m_i \ddot{\boldsymbol{\rho}}_i \end{aligned} \tag{3.2.17}$$

But because $\boldsymbol{\rho}_i$ is measured with respect to the center of mass, the weighted sum of this position vector and its derivatives are zero, i.e.,

$$\begin{aligned}\sum_{i=1}^{N} m_i \boldsymbol{\rho}_i &= 0 \\ \sum_{i=1}^{N} m_i \dot{\boldsymbol{\rho}}_i &= 0 \\ \sum_{i=1}^{N} m_i \ddot{\boldsymbol{\rho}}_i &= 0\end{aligned} \tag{3.2.18}$$

and Eq. (3.2.17) becomes

$$\mathbf{M}_B^{\text{sys}} = m\mathbf{x}_{C/B} \times \mathbf{a}_C + \sum_{i=1}^{N} \boldsymbol{\rho}_i \times \; m_i \ddot{\boldsymbol{\rho}}_i \tag{3.2.19}$$

Now, consider the angular momentum of the system of particles about point B, $\mathbf{L}_B$, where

$$\mathbf{L}_B = \sum_{i=1}^{N} \mathbf{x}_{Bi} \times m_i \dot{\mathbf{x}}_i \tag{3.2.20}$$

From the geometry of Fig. 3.7 and Eq. (3.2.18) we obtain

$$\begin{aligned}\mathbf{L}_B &= \sum_{i=1}^{N} \left(\mathbf{x}_{C/B} + \boldsymbol{\rho}_i\right) \times m_i \left(\mathbf{v}_C + \dot{\boldsymbol{\rho}}_i\right) \\ &= m\mathbf{x}_{C/B} \times \mathbf{v}_C + \mathbf{x}_{C/B} \times \sum_{i=1}^{N} m_i \dot{\boldsymbol{\rho}}_i + \sum_{i=1}^{N} m_i \boldsymbol{\rho}_i \times \mathbf{v}_C + \sum_{i=1}^{N} \boldsymbol{\rho}_i \times m_i \dot{\boldsymbol{\rho}}_i \\ &= m\mathbf{x}_{C/B} \times \mathbf{v}_C + \sum_{i=1}^{N} \boldsymbol{\rho}_i \times m_i \dot{\boldsymbol{\rho}}_i\end{aligned} \tag{3.2.21}$$

so that the derivative on time of the angular momentum of the system is

$$\dot{\mathbf{L}}_B = m\mathbf{v}_{C/B} \times \mathbf{v}_C + m\mathbf{x}_{C/B} \times \mathbf{a}_C + \sum_{i=1}^{N} \boldsymbol{\rho}_i \times m_i \ddot{\boldsymbol{\rho}}_i \tag{3.2.22}$$

where we have used the property of the cross product to set $\dot{\boldsymbol{\rho}}_i \times m_i \dot{\boldsymbol{\rho}}_i = 0$.

If we compare Eq. (3.2.19) for the moment to the derivative of the angular momentum, Eq. (3.2.22), we find

$$\begin{aligned}\mathbf{M}_B^{\text{sys}} &= \dot{\mathbf{L}}_B - m\mathbf{v}_{C/B} \times \mathbf{v}_C \\ &= \dot{\mathbf{L}}_B - m\mathbf{v}_C \times \mathbf{v}_B\end{aligned} \tag{3.2.23}$$

which is the general form of the Newton-Euler equation for the applied moments acting on a general system of particles for moments taken about a general moving point B. Two special cases of particular importance are as follows:

1. When point B is a zero velocity point, say, O. Then $\mathbf{v}_B = \mathbf{v}_0 = 0$ and we have

$$\mathbf{M}_0^{\text{sys}} = \dot{\mathbf{L}}_0 \tag{3.2.24}$$

2. When point B is the center of mass, C. Then $\mathbf{v}_C \times \mathbf{v}_B = 0$, giving

$$\mathbf{M}_C^{\text{sys}} = \dot{\mathbf{L}}_C \tag{3.2.25}$$

3.2.4 Angular Impulse and Angular Momentum

If a point P is either a fixed point or the center of mass the moment equation of motion for the entire system is the same and can be integrated on time to give an angular impulse-angular momentum relationship:

$$\Delta \mathbf{L}_P = \mathbf{L}_P(t_2) - \mathbf{L}_P(t_1) = \int_{t_1}^{t_2} \mathbf{M}_P^{\text{sys}}(t)dt \tag{3.2.26}$$

where the integral of the moment is called the *angular impulse*. If the net moment on the system of particles is zero or if the angular impulse is zero, then $\Delta\,\mathbf{L}_P = 0$ so that the angular momentum with respect to point P is a constant, which is a statement of the principle of *conservation of angular momentum.*

The linear impulse-linear momentum and angular impulse-angular momentum relations for the entire system can be used jointly to find the changes in motion in a system of rigidly constrained particles due to known external impulses since the number of degrees of freedom is equal to the number of integrated Newton-Euler equations for the system. For example, consider the two-dimensional problem shown in Fig. 3.8a where a pair of particles are rigidly connected to each other and can move on a smooth x–y plane. If a force $\mathbf{F}(t)$ acts in the y-direction for a very short time interval, Δt, on mass m_1 and the system is initially at rest, find the velocities of the masses immediately after the impulse $\mathbf{R} = \int_0^{\Delta t} \mathbf{F}(t)dt$ is applied. Here it is convenient to use the center of mass coordinates (x_C, y_C) and the angle θ as

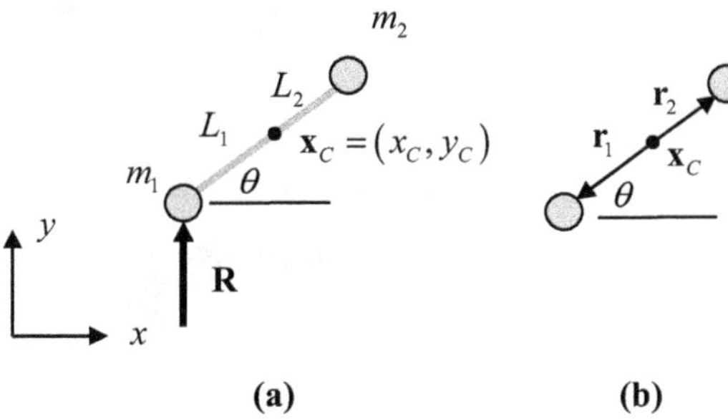

Fig. 3.8 (**a**) A pair of particles of masses (m_1, m_2) that are held together by a thin, weightless rigid rod and where mass m_1 is subjected to an impulse, $\mathbf{R}(t)$, acting in the y-direction. (**b**) The position vectors of the two masses with respect to the center of mass

our three independent coordinates. It is easy to show that if the total length between particles is L, then the distances (L_1, L_2) to the center of mass (Fig. 3.8) are given by

$$L_1 = \frac{m_2 L}{m_1 + m_2}, \quad L_2 = \frac{m_1 L}{m_1 + m_2} \tag{3.2.27}$$

The coordinates of the two masses are then

$$\begin{aligned} x_1 &= x_C - L_1 \cos\theta, \quad y_1 = y_C - L_1 \sin\theta \\ x_2 &= x_C + L_2 \cos\theta, \quad y_2 = y_C + L_2 \sin\theta \end{aligned} \tag{3.2.28}$$

and the velocity components of the masses are

$$\begin{aligned} \mathbf{v}_1 &= \left(\dot{x}_C + L_1 \sin\theta\dot{\theta}\right)\mathbf{e}_x + \left(\dot{y}_C - L_1 \cos\theta\dot{\theta}\right)\mathbf{e}_y \\ \mathbf{v}_2 &= \left(\dot{x}_C - L_2 \sin\theta\dot{\theta}\right)\mathbf{e}_x + \left(\dot{y}_C + L_2 \cos\theta\dot{\theta}\right)\mathbf{e}_y \end{aligned} \tag{3.2.29}$$

From the linear impulse-linear momentum relation we have

$$\Delta(m\mathbf{v}_C) = \mathbf{R} \tag{3.2.30}$$

so that immediately after the impulsive force is applied we have

$$\begin{aligned} &\dot{x}_C = 0 \\ &(m_1 + m_2)\dot{y}_C = R \end{aligned} \tag{3.2.31}$$

The angular impulse-angular momentum relationship with respect to the center of mass is

$$\Delta \mathbf{L}_C = \mathbf{r}_1 \times m_1 \mathbf{v}_1 + \mathbf{r}_2 \times m_2 \mathbf{v}_2 = \mathbf{r}_1 \times \mathbf{R} \tag{3.2.32}$$

where (see Fig. 3.8b)

$$\begin{aligned} \mathbf{r}_1 &= -L_1 \cos\theta \mathbf{e}_x - L_1 \sin\theta \mathbf{e}_y \\ \mathbf{r}_2 &= L_2 \cos\theta \mathbf{e}_x + L_2 \sin\theta \mathbf{e}_y \end{aligned} \tag{3.2.33}$$

so that computing the cross products we have

$$\begin{aligned} \Delta \mathbf{L}_C &= \begin{vmatrix} \mathbf{e}_x & \mathbf{e}_y & \mathbf{e}_z \\ -L_1 \cos\theta & -L_1 \sin\theta & 0 \\ m_1 L_1 \sin\theta\dot{\theta} & m_1\left(\dot{y}_C - L_1 \cos\theta\dot{\theta}\right) & 0 \end{vmatrix} \\ &+ \begin{vmatrix} \mathbf{e}_x & \mathbf{e}_y & e_z \\ L_2 \cos\theta & L_2 \sin\theta & 0 \\ -m_2 L_2 \sin\theta\dot{\theta} & m_2\left(\dot{y}_C + L_2 \cos\theta\dot{\theta}\right) & 0 \end{vmatrix} \\ &= \mathbf{e}_z\left[m_1 L_1^2 \dot{\theta} - m_1 L_1 \cos\theta \dot{y}_C\right] + \mathbf{e}_z\left[m_2 L_2^2 \dot{\theta} + m_2 L_2 \cos\theta \dot{y}_C\right] \\ &= \mathbf{e}_z\left(m_1 L_1^2 + m_2 L_2^2\right)\dot{\theta} + \mathbf{e}_z \cos\theta \dot{y}_C\left(-\frac{m_1 m_2 L}{m_1 + m_2} + \frac{m_1 m_2 L}{m_1 + m_2}\right) \\ &= \mathbf{e}_z\left(m_1 L_1^2 + m_2 L_2^2\right)\dot{\theta} \end{aligned} \tag{3.2.34}$$

and

$$\begin{aligned} \mathbf{r}_1 \times \mathbf{R} = &\begin{vmatrix} \mathbf{e}_x & \mathbf{e}_y & \mathbf{e}_z \\ -L_1 \cos\theta & -L_1 \sin\theta & 0 \\ 0 & R & 0 \end{vmatrix} \\ &= -R L_1 \cos\theta \mathbf{e}_z \end{aligned} \tag{3.2.35}$$

which gives

$$\dot{\theta} = \frac{-R L_1 \cos\theta}{m_1 L_1^2 + m_2 L_2^2} \tag{3.2.36}$$

and the velocities of the particles are finally

$$\begin{aligned} \mathbf{v}_1 &= R\left(\frac{-L_1^2 \sin\theta \cos\theta}{m_1 L_1^2 + m_2 L_2^2}\right)\mathbf{e}_x + R\left(\frac{1}{m_1 + m_2} + \frac{L_1^2 \cos^2\theta}{m_1 L_1^2 + m_2 L_2^2}\right)\mathbf{e}_y \\ \mathbf{v}_2 &= R\left(\frac{L_1 L_2 \sin\theta \cos\theta}{m_1 L_1^2 + m_2 L_2^2}\right)\mathbf{e}_x + R\left(\frac{1}{m_1 + m_2} - \frac{L_1 L_2 \cos^2\theta}{m_1 L_1^2 + m_2 L_2^2}\right)\mathbf{e}_y \end{aligned} \tag{3.2.37}$$

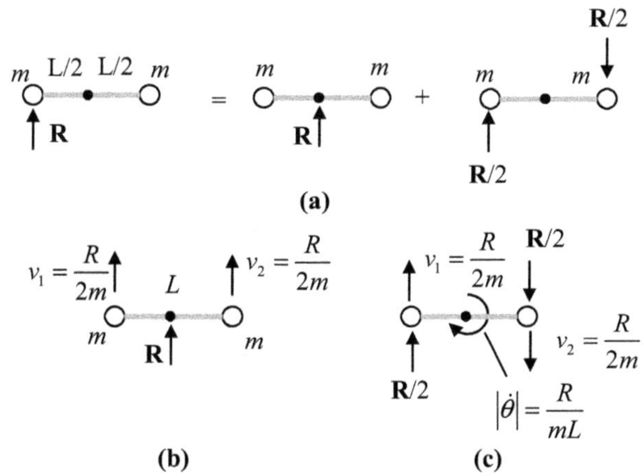

Fig. 3.9 (**a**) An impulsive force applied to a pair of rigidly connected masses is equivalent to applying the impulse at the center of mass and an angular impulsive moment about that center. (**b**) The velocities of the masses when the impulse is applied at the center of mass, resulting in no angular motion, and (**c**) the velocities of the masses resulting from the angular motion generated when the angular impulsive moment is applied

Equation (3.2.37) is a rather complicated looking expression so consider a very simple case where $m_1 = m_2 = m$, $L_1 = L_2 = L/2$, and $\theta = 0$ (Fig. 3.9a). In that case Eq. (3.2.37) reduces to

$$\begin{aligned} \mathbf{v}_1 &= \frac{R}{m}\mathbf{e}_y \\ \mathbf{v}_2 &= 0 \\ \dot{\theta} &= \frac{-R}{mL} \end{aligned} \tag{3.2.38}$$

We can interpret this result as follows. As Fig. 3.9a shows applying the impulse to the left end of the rigidly connected masses is equivalent to applying the impulse at the center of mass plus applying an impulsive moment about the center of mass. [By "equivalent" we mean that the system on the left side in Fig. 3.9a has the same force and moment (and hence the same impulse and angular impulse) as the sum of the two systems on the right side of Fig. 3.9a. This is the same definition of equivalence as used in statics to equate two different systems of forces and couples.] If the impulse was applied to the center of mass, there would be no angular impulse about the center of mass so we would have $\dot{\theta} = 0$ and $\dot{x}_C = 0$, $\dot{y}_C = R/2m$. From Eq. (3.2.29) we would find $\mathbf{v}_1 = \mathbf{v}_2 = (R/2m)\mathbf{e}_y$ (see Fig. 3.9b). If instead the angular impulsive moment shown in Fig. 3.9c is applied, we would have $\dot{x}_C = \dot{y}_C = 0$ and $\dot{\theta} = -R/mL$. This angular velocity, when placed into Eq. (3.2.29), produces velocities of the masses given by $\mathbf{v}_1 = (R/2m)\mathbf{e}_y$ and $\mathbf{v}_2 = -(R/2m)\mathbf{e}_y$ (Fig. 3.9c). Adding both effects together we obtain the velocities of the masses given in Eq. (3.2.38).

3.2.5 Work and Energy

The work dW_i done on the ith particle in a system of N particles when that particle is given a differential displacement, $d\mathbf{x}_i$, is

$$dW_i = \mathbf{F}_i^a \cdot d\mathbf{x}_i + \sum_{j \neq i} \mathbf{F}_{ij} \cdot d\mathbf{x}_i \tag{3.2.39}$$

where $\mathbf{F}_i^a$ is the sum of all the known external applied forces and constraint forces and $\mathbf{F}_{ij}$ is the internal force that the jth particle exerts on the ith particle. From the results of the last chapter we can write this work in terms of the differential change of the kinetic energy through the *work-energy theorem*:

$$dW_i = dT_i \tag{3.2.40}$$

where T_i is the kinetic energy of the ith particle given by

$$T_i = \frac{1}{2} m_i \mathbf{v}_i \cdot \mathbf{v}_i \tag{3.2.41}$$

in terms of the velocity, $\mathbf{v}_i = \dot{\mathbf{x}}_i$, of the ith particle. If we sum over all the particles in the system then the total kinetic energy, T, is given by

$$T = \sum_{i=1}^{N} T_i \tag{3.2.42}$$

so that if we integrate the work-energy relationship for the entire system between two configurations (1) and (2) where the particle positions are $\mathbf{x}_i^{(1)}$ and $\mathbf{x}_i^{(2)}$, respectively, we have

$$\begin{aligned} \Delta T = T^{(2)} - T^{(1)} &= \sum_{i=1}^{N} \int_{\mathbf{x}_i^{(1)}}^{\mathbf{x}_i^{(2)}} \mathbf{F}_i^a \cdot d\mathbf{x}_i + \sum_{i=1}^{N} \sum_{j=i+1}^{N} \int_{\mathbf{x}_i^{(1)}}^{\mathbf{x}_i^{(2)}} \mathbf{F}_{ij} \cdot d\mathbf{x}_i + \sum_{i=1}^{N} \sum_{j=i+1}^{N} \int_{\mathbf{x}_i^{(1)}}^{\mathbf{x}_i^{(2)}} \mathbf{F}_{ji} \cdot d\mathbf{x}_j \\ &= \sum_{i=1}^{N} \int_{\mathbf{x}_i^{(1)}}^{\mathbf{x}_i^{(2)}} \mathbf{F}_i^a \cdot d\mathbf{x}_i + \sum_{i=1}^{N} \sum_{j=i+1}^{N} \int_{\mathbf{x}_i^{(1)}}^{\mathbf{x}_i^{(2)}} \mathbf{F}_{ij} \cdot \left(d\mathbf{x}_i - d\mathbf{x}_j\right) \end{aligned} \tag{3.2.43}$$

where we have used Newton's third law. Just as the external constraint forces can do work, any internal constraint forces may also do work. One important exception is the case of internal rigid constraints between the particles. Such ideal holonomic constraints satisfy Eq. (3.1.6), so letting $l_{ij} = |\mathbf{x}_j - \mathbf{x}_i|$ and $u_{ij;k} = x_{i;k} - x_{j;k}$, the ideal constraint forces satisfy

$$F_{ij;k} = \lambda_{ij}\frac{\partial f_{ij}}{\partial x_{i;k}} = \lambda_{ij}\frac{\partial f_{ij}}{\partial l_{ij}}\frac{\partial l_{ij}}{\partial u_{ij;k}}\frac{\partial u_{ij;k}}{\partial x_{i;k}} = \lambda_{ij}\frac{\partial f_{ij}}{\partial l_{ij}}\frac{\partial l_{ij}}{\partial u_{ij;k}} \tag{3.2.44}$$

so that Eq. (3.2.43) becomes

$$\begin{aligned}\Delta T &= \sum_{i=1}^{N}\int_{\mathbf{x}_i^{(1)}}^{\mathbf{x}_i^{(2)}} \mathbf{F}_i^a \cdot d\mathbf{x}_i + \sum_{i=1}^{N}\sum_{j=i+1}^{N}\int_{\mathbf{x}_i^{(1)}}^{\mathbf{x}_i^{(2)}}\sum_{k=1}^{3}\lambda_{ij}\frac{\partial f_{ij}}{\partial l_{ij}}\frac{\partial l_{ij}}{\partial u_{ij;k}}du_{ij;k} \\ &= \sum_{i=1}^{N}\int_{\mathbf{x}_i^{(1)}}^{\mathbf{x}_i^{(2)}} \mathbf{F}_i^a \cdot d\mathbf{x}_i + \sum_{i=1}^{N}\sum_{j=i+1}^{N}\int_{\mathbf{x}_i^{(1)}}^{\mathbf{x}_i^{(2)}}\sum_{k=1}^{3}\lambda_{ij}df_{ij}(l_{ij})\end{aligned} \tag{3.2.45}$$

where $du_{ij;k} = (dx_{i;k} - dx_{j;k})$. But the total differential of all the constraints between the particles must satisfy $df_{ij} = 0$ so for any system of particles rigidly connected together, as shown in Fig. 3.10a, the internal forces will not only cancel for the entire system but will also not do any work, and the work-kinetic energy relationship for the rigid body of Fig. 3.10a will always be

$$\Delta T = T^{(2)} - T^{(1)} = \sum_{i=1}^{N}\int_{\mathbf{x}_i^{(1)}}^{\mathbf{x}_i^{(2)}} \mathbf{F}_i^a \cdot d\mathbf{x}_i \tag{3.2.46a}$$

If all of the applied forces are derivable from potentials, then $\mathbf{F}_i^a = -\nabla_{\mathbf{x}_i} V_i$, and Eq. (3.2.46a) is instead given as

$$\Delta T = -\sum_{i=1}^{N}\Delta V_i = -\sum_{i=1}^{N}\left[V_i\left(\mathbf{x}_i^{(2)}\right) - V_i\left(\mathbf{x}_i^{(1)}\right)\right] = -\Delta V \tag{3.2.46b}$$

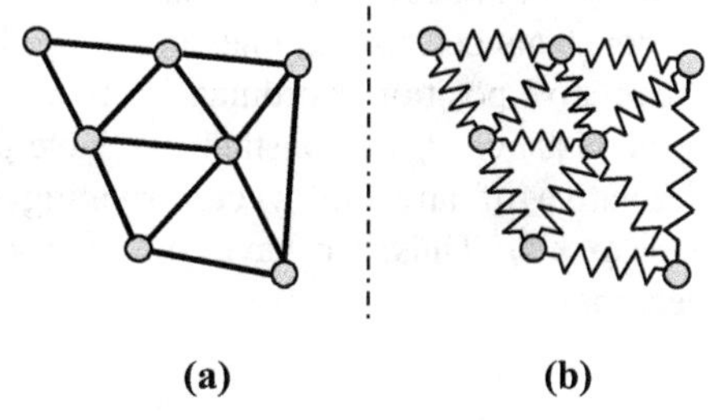

Fig. 3.10 (**a**) A system of particles connected by massless, rigid links to form a rigid body, and (**b**) a system of particles connected by springs to form a deformable body

where V is the total potential energy of all the external forces acting on the system. This leads to the conservation of total energy of the entire system, i.e.,

$$T^{(2)} + V^{(2)} = T^{(1)} + V^{(1)} \tag{3.2.47}$$

However, note that Eq. (3.2.47) is only true for systems of particles where the internal forces do no work, like a rigidly constrained system.

For a general system of particles the total kinetic energy can be written in terms of the center of mass velocity and the velocity with respect to the center of mass as

$$\mathbf{v}_i = \mathbf{v}_C + \dot{\boldsymbol{\rho}}_i \tag{3.2.48}$$

(see Fig. 3.7) so that

$$\begin{aligned} T &= \frac{1}{2}\sum_{i=1}^{N} m_i(\mathbf{v}_C + \dot{\boldsymbol{\rho}}_i) \cdot (\mathbf{v}_C + \dot{\boldsymbol{\rho}}_i) \\ &= \frac{1}{2}m\mathbf{v}_C \cdot \mathbf{v}_C + \mathbf{v}_C \cdot \sum_{i=1}^{N} m_i\dot{\boldsymbol{\rho}}_i + \frac{1}{2}\sum_{i=1}^{N} m_i\dot{\boldsymbol{\rho}}_i \cdot \dot{\boldsymbol{\rho}}_i \\ &= \frac{1}{2}m\mathbf{v}_C \cdot \mathbf{v}_C + \frac{1}{2}\sum_{i=1}^{N} m_i\dot{\boldsymbol{\rho}}_i \cdot \dot{\boldsymbol{\rho}}_i \end{aligned} \tag{3.2.49}$$

where m is the total mass of the system and we have used the second equation in Eq. (3.2.18) to arrive at this result. The first term in Eq. (3.2.49) is the kinetic energy of a mass m moving with the center of mass and the second term is the kinetic energy due to motion of the particles relative to the center of mass.

3.2.6 Kinetic Energy and Angular Momentum for a Rigid System of Particles

For a rigid system of particles (rigid body) we can get a more explicit form for the relative motion part of the total kinetic energy. In the position vectors $\boldsymbol{\rho}_i = l_i\mathbf{e}_i$ from the center of mass to the individual particles, let the unit vectors $\mathbf{e}_i$ rotate with the rigid system angular velocity $\boldsymbol{\omega}$. Then the distances l_i in those position vectors will be constants because all lines in the rigid system have the same angular motion and the relative position coordinates between any two points in the system are fixed. But if the distances l_i are constants we have $\dot{\boldsymbol{\rho}}_i = l_i\dot{\mathbf{e}}_i$. In Chap. 1 we found that the time derivative of any unit vector rotating with an angular velocity $\boldsymbol{\omega}$ is given by $\dot{\mathbf{e}}_i = \boldsymbol{\omega} \times \mathbf{e}_i$. Thus, we have $\dot{\boldsymbol{\rho}}_i = l_i\boldsymbol{\omega} \times \mathbf{e}_i = \boldsymbol{\omega} \times \boldsymbol{\rho}_i$ and the total kinetic energy becomes

$$
\begin{aligned}
T &= \frac{1}{2}m\mathbf{v}_C \cdot \mathbf{v}_C + \frac{1}{2}\sum_{i=1}^{N} m_i(\boldsymbol{\omega} \times \boldsymbol{\rho}_i) \cdot (\boldsymbol{\omega} \times \boldsymbol{\rho}_i) \\
&= T_C + T_{\text{rot}}
\end{aligned}
\tag{3.2.50}
$$

Physically, Eq. (3.2.50) says that the total kinetic energy of a rigid system is equal to the translational kinetic energy, T_C, of the center of mass of the system plus the rotational kinetic energy, T_{rot}, about the center of mass. The rotational energy can be written in terms of components by letting $\boldsymbol{\omega} = \omega_x\mathbf{e}_x + \omega_y\mathbf{e}_y + \omega_z\mathbf{e}_z$, $\boldsymbol{\rho}_i = x_{Ci}\mathbf{e}_x + y_{Ci}\mathbf{e}_y + z_{Ci}\mathbf{e}_z$ (where the notation means that these position coordinates are measured from the center of mass C to particle i; we could also write such position coordinates as $x_{i/C}$, etc.) and expanding the cross and dot products. The process yields an end result which is best placed in a vector-matrix form. By expanding the two cross products in Eq. (3.2.50) and collecting terms, we obtain

$$
T_{\text{rot}} = \frac{1}{2}\{\omega\}^T[I_C]\{\omega\}
\tag{3.2.51}
$$

with $\{\omega\} = [\omega_x \ \omega_y \ \omega_z]^T$ and $[I_C]$ is the mass moment of inertia matrix:

$$
[I_C] = \begin{bmatrix}
\sum_i m_i\left(y_{Ci}^2 + z_{Ci}^2\right) & -\sum_i m_i x_{Ci} y_{Ci} & -\sum_i m_i x_{Ci} z_{Ci} \\
-\sum_i m_i x_{Ci} y_{Ci} & \sum_i m_i\left(x_{Ci}^2 + z_{Ci}^2\right) & -\sum_i m_i y_{Ci} z_{Ci} \\
-\sum_i m_i x_{Ci} z_{Ci} & -\sum_i m_i y_{Ci} z_{Ci} & \sum_i m_i\left(x_{Ci}^2 + y_{Ci}^2\right)
\end{bmatrix}
\tag{3.2.52}
$$

In the case of two-dimensional motion of the rigid system in the x–y plane the kinetic energy expression is much simpler as there is only translational motion in that plane and rotational motion is only about the z-axis so that

$$
\begin{aligned}
T_C &= \frac{1}{2}m\left(v_{Cx}^2 + v_{Cy}^2\right) \\
T_{\text{rot}} &= \frac{1}{2}I_C\omega_z^2, \quad I_C = \sum_{i=1}^{N} m_i\left(x_{Ci}^2 + y_{Ci}^2\right)
\end{aligned}
\tag{3.2.53}
$$

The mass moment of inertia matrix also appears in the expression for the angular momentum about the center of mass since from Eq. (3.2.21)

$$
\mathbf{L}_C = \sum_{i=1}^{N} \boldsymbol{\rho}_i \times m_i \dot{\boldsymbol{\rho}}_i = \sum_{i=1}^{N} m_i \boldsymbol{\rho}_i \times (\boldsymbol{\omega} \times \boldsymbol{\rho}_i)
\tag{3.2.54}
$$

You can either expand the two cross products or use the vector identity $\boldsymbol{\rho}_i \times (\boldsymbol{\omega} \times \boldsymbol{\rho}_i) = (\boldsymbol{\rho}_i \cdot \boldsymbol{\rho}_i)\boldsymbol{\omega} - (\boldsymbol{\rho}_i \cdot \boldsymbol{\omega})\boldsymbol{\rho}_i$ and expand the dot products (which is much easier) to obtain in matrix form

$$\{L_C\} = [I_C]\{\omega\} \tag{3.2.55}$$

where $\{L_C\} = [L_{Cx}\ L_{Cy}\ L_{Cz}]^T$. We can obtain a similar relationship for the angular momentum of the system about a fixed point, O, since from Eq. (3.2.21)

$$\begin{aligned}
\mathbf{L}_O &= \sum_{i=1}^{N} m_i(\mathbf{x}_{C/O} \times \mathbf{v}_C) + \mathbf{x}_{C/O} \times \sum_{i=1}^{N} m_i(\boldsymbol{\omega} \times \boldsymbol{\rho}_i) + \sum_{i=1}^{N} m_i\boldsymbol{\rho}_i \times \mathbf{v}_C + \sum_{i=1}^{N} m_i\boldsymbol{\rho}_i \times (\boldsymbol{\omega} \times \boldsymbol{\rho}_i) \\
&= \sum_{i=1}^{N} m_i(\mathbf{x}_{C/O} + \boldsymbol{\rho}_i) \times \mathbf{v}_C + \sum_{i=1}^{N} m_i(\mathbf{x}_{C/O} + \boldsymbol{\rho}_i) \times (\boldsymbol{\omega} \times \boldsymbol{\rho}_i) \\
&= \sum_{i=1}^{N} m_i(\mathbf{x}_{C/O} + \boldsymbol{\rho}_i) \times (\mathbf{v}_C + \boldsymbol{\omega} \times \boldsymbol{\rho}_i)
\end{aligned} \tag{3.2.56}$$

But because the center of mass is rotating about O, $\mathbf{x}_{C/O} = l\mathbf{e}_{C/O}$ and $\mathbf{v}_C = \dot{\mathbf{x}}_{C/O} = l\,\dot{\mathbf{e}}_{C/O} = l\boldsymbol{\omega} \times \mathbf{e}_{C/O}$ we have $\mathbf{v}_C = \boldsymbol{\omega} \times \mathbf{x}_{C/O}$ and

$$\begin{aligned}
\mathbf{L}_O &= \sum_{i=1}^{N} m_i(\mathbf{x}_{C/O} + \boldsymbol{\rho}_i) \times \left[\boldsymbol{\omega} \times (\mathbf{x}_{C/O} + \boldsymbol{\rho}_i)\right] \\
&= \sum_{i=1}^{N} m_i\mathbf{x}_i \times (\boldsymbol{\omega} \times \mathbf{x}_i)
\end{aligned} \tag{3.2.57}$$

which is identical to Eq. (3.2.54) except that the components of the position vectors $\mathbf{x}_i = x_{Oi}\mathbf{e}_x + y_{Oi}\mathbf{e}_y + z_{Oi}\mathbf{e}_z$ of the particles are measured from the fixed point O to each mass while the components of the position vectors $\boldsymbol{\rho}_i = x_{Ci}\mathbf{e}_x + y_{Ci}\mathbf{e}_y + z_{Ci}\mathbf{e}_z$ are measured with respect to the center of mass. Thus, $[I_O]$ has exactly the same form as $[I_C]$ (see Eq. (3.2.52)) except that (x_{Ci}, y_{Ci}, z_{Ci}) in Eq. (3.2.52) are replaced by (x_{Oi}, y_{Oi}, z_{Oi}) and we have

$$\{L_O\} = [I_O]\{\omega\} \tag{3.2.58}$$

In the case of planar motion where there is only a rotation about the z-axis this matrix relation reduces to the scalar relation

$$\begin{aligned}
L_{Oz} &= I_O\omega_z \\
I_O &= \sum_{i=1}^{N} m_i(x_{Oi}^2 + y_{Oi}^2)
\end{aligned} \tag{3.2.59}$$

3.2.7 Work-Kinetic Energy for an Elastically Connected System of Particles

Another important case for which we want to examine the internal forces and work-energy relations is where all the particles are connected by linear springs, as shown in Fig. 3.10b. Just as the lattice of rigidly connected particles represents a rigid body composed of discrete masses, this lattice of springs represents an elastic body, i.e., one that can deform. If, as before, we let V_{ij} be the potential energy of the spring connecting the ith and jth particle then

$$V_{ij} = \frac{k_{ij}}{2}\left[l_{ij} - l_{ij}^0\right]^2 \tag{3.2.60}$$

where k_{ij} is the spring constant and $\left(l_{ij}, l_{ij}^0\right)$ are the current and the unstretched spring lengths and $l_{ij} = |\mathbf{x}_j - \mathbf{x}_i|$. [Note: Here we will assume that all particles can be connected to all the other particles by springs. In Fig. 3.10b only nearest-neighbor connections are shown. However, the case shown can be obtained from this general case by simply setting appropriate spring constants to zero.] Then

$$\begin{aligned} F_{ij;k} &= -\frac{\partial V_{ij}}{\partial x_{i;k}} = -\frac{\partial V_{ij}}{\partial l_{ij}}\frac{\partial l_{ij}}{\partial u_{ij;k}}\frac{\partial u_{ij;k}}{\partial x_{i;k}} \\ &= -\frac{\partial V_{ij}}{\partial l_{ij}}\frac{\partial l_{ij}}{\partial u_{ij;k}} \end{aligned} \tag{3.2.61}$$

where $u_{ij;k} = x_{i;k} - x_{j;k}$. The work done on the entire system is

$$\begin{aligned} \sum_{i=1}^{N} dW_i &= \sum_{i=1}^{N} \mathbf{F}_i^a \cdot d\mathbf{x}_i + \sum_{i=1}^{N}\sum_{j\neq i}^{N} \mathbf{F}_{ij} \cdot d\mathbf{x}_i \\ &= \sum_{i=1}^{N} \mathbf{F}_i^a \cdot d\mathbf{x}_i + \sum_{i=1}^{N}\sum_{j=i+1}^{N} \mathbf{F}_{ij} \cdot d\mathbf{x}_i + \sum_{i=1}^{N}\sum_{j=i+1}^{N} \mathbf{F}_{ji} \cdot d\mathbf{x}_j \\ &= \sum_{i=1}^{N} \mathbf{F}_i^a \cdot d\mathbf{x}_i + \sum_{i=1}^{N}\sum_{j=i+1}^{N} \mathbf{F}_{ij} \cdot \left(d\mathbf{x}_i - d\mathbf{x}_j\right) \end{aligned} \tag{3.2.62}$$

where we have again used Newton's third law $\mathbf{F}_{ji} = -\mathbf{F}_{ij}$. However, assuming motion in 3-D we have

$$\begin{aligned}\mathbf{F}_{ij} \cdot \left(d\mathbf{x}_i - d\mathbf{x}_j\right) &= \sum_{k=1}^{3} F_{ij;k} d\left(x_{i;k} - x_{j;k}\right) \\ &= \sum_{k=1}^{3} F_{ij;k} d\left(u_{ij;k}\right)\end{aligned} \tag{3.2.63}$$

so using Eq. (3.2.63) and Eq. (3.2.61) in Eq. (3.2.62) we find

$$\begin{aligned}\sum_{i=1}^{N} dW_i &= \sum_{i=1}^{N} \mathbf{F}_i^a \cdot d\mathbf{x}_i - \sum_{i=1}^{N}\sum_{j=i+1}^{N} \left(\sum_{k=1}^{3} \frac{\partial V_{ij}}{\partial l_{ij}} \frac{\partial l_{ij}}{\partial u_{ij,k}} du_{ij;k} \right) \\ &= \sum_{i=1}^{N} \mathbf{F}_i^a \cdot d\mathbf{x}_i - \sum_{i=1}^{N}\sum_{j=i+1}^{N} dV_{ij} \\ &= \sum_{i=1}^{N} \mathbf{F}_i^a \cdot d\mathbf{x}_i - dV^{\text{int}}\end{aligned} \tag{3.2.64}$$

where we have let dV^{int} be the change in the *internal potential energy* (also often called the *strain energy*) of all the springs to distinguish it from any potential energy, V, coming from the external forces. Writing the differential of the work in terms of the differential of the kinetic energy (Eq. (3.2.40)) gives

$$\sum_{i=1}^{N} dT_i = \sum_{i=1}^{N} \mathbf{F}_i^a \cdot d\mathbf{x}_i - dV^{\text{int}} \tag{3.2.65}$$

and integrating this work-energy relationship between two states, as before, we obtain finally

$$\Delta T = \sum_{i=1}^{N} \int_{\mathbf{r}_i^{(1)}}^{\mathbf{r}_i^{(2)}} \mathbf{F}_i^a \cdot d\mathbf{x}_i - \Delta V^{\text{int}} \tag{3.2.66}$$

where, apart from an arbitrary constant,

$$V^{\text{int}} = \frac{1}{2} \sum_{i=1}^{N} \sum_{j=i+1}^{N} k_{ij} \left[l_{ij} - l_{ij}^0 \right]^2 \tag{3.2.67}$$

Equation (3.2.66) is the work-kinetic energy relationship for a deformable body. The internal potential energy for a lattice of springs is given by Eq. (3.2.67). This internal potential energy is absent for a rigid body since work done by the external forces cannot be stored internally, so that all the external work must go into changing the kinetic energy of the rigid body. If the external forces are also derivable from a potential, and V^e is the total potential energy of those forces, then Eq. (3.2.66) becomes

$$\Delta T = -\Delta V^e - \Delta V^{\text{int}} \tag{3.2.68}$$

or, equivalently,

$$\Delta\left(T + V^e + V^{\text{int}}\right) = 0 \tag{3.2.69}$$

which is a statement of the conservation of energy for a deformable body.

As in the case of a single particle we can also write these work-energy relations in terms of power. From Eq. (3.2.65) we have

$$\begin{aligned} dT &= \sum_{i=1}^{N} dT_i = \sum_{i=1}^{N} \mathbf{F}_i^a \cdot d\mathbf{x}_i - dV^{\text{int}} \\ &= dW^e - dV^{\text{int}} \end{aligned} \tag{3.2.70}$$

where dT is the differential of the total kinetic energy and dW^e is a differential work done by the external forces. Dividing by the differential time dt we have

$$\begin{aligned} \frac{dT}{dt} &= \sum_{i=1}^{N} \mathbf{F}_i^a \cdot \mathbf{v}_i - \frac{dV^{\text{int}}}{dt} \\ &= P^e - \frac{dV^{\text{int}}}{dt} \end{aligned} \tag{3.2.71}$$

in terms of the power $P^e = dW^e/dt$. Integrating over a time interval (t_0, t) gives

$$\begin{aligned} &T(t) - T(t_0) = \Delta T = \int_{t_0}^{t} P^e\,dt - \left(V^{\text{int}}(t) - V^{\text{int}}(t_0)\right) \\ &\text{or} \\ &\Delta T = W^e - \Delta V^{\text{int}} \end{aligned} \tag{3.2.72}$$

When the applied external forces are all derivable from potential energy functions $W^e = -\Delta V^e$ and we again obtain Eq. (3.2.69).

3.3 Dynamics of a Rigidly Constrained System of Particles (Rigid Body)

For unconstrained motion of a system of N particles there are a total of only N independent equations we can use from the N force equations and the N moment equations. But as we showed previously for a rigidly constrained system (rigid body) there are always only three degrees of freedom for two-dimensional motion of the body and six degrees of freedom for three-dimensional motion. We also showed that the general motion of a rigid body could always be obtained as the superposition of the translational motion of a reference point plus a rotation about that point. It can be shown (see [1], for example) that the center of mass always moves as if it were part of the rigid body even if it is not a point on the body itself [for example, the center of mass of a hollow disk (washer) is at the center of the hole in the disk]. Thus, we can always take the center of mass point C as the reference point to describe the translational motion of a rigid body. This translational motion of the center of mass is directly related to the total external forces acting on the system, Eq. (3.2.8). The angular momentum of a rigid body is also described by the angular velocity vector so that the moment equation for the entire system of particles, Eq. (3.2.25), which relates the net moment acting on that system to the derivative on time of the total angular momentum of the system, is an equation for the rotational motion. Thus, a rigid body of particles is a special collection of particles where the location of the center of mass of the system and the rotation about that center of mass can be used as independent generalized coordinates to completely define the motion of the entire system and we can use all of the Newton-Euler equations to obtain those translational and rotational parts, i.e., we have the governing equations

$$\begin{aligned} \mathbf{F}^{\text{sys}} &= m\ddot{\mathbf{x}}_C \\ \mathbf{M}_C^{\text{sys}} &= \dot{\mathbf{L}}_C \end{aligned} \tag{3.3.1}$$

which are six equations for the six degrees of freedom in the case of three-dimensional motion. For two-dimensional motion (in the x–y plane), the Newton-Euler equations (in component form) are

$$\begin{aligned} m\frac{d^2x_C}{dt^2} &= F_x^{\text{sys}} \\ m\frac{d^2y_C}{dt^2} &= F_y^{\text{sys}} \\ \frac{d}{dt}(I_C\omega_z) = I_C\frac{d^2\theta_z}{dt^2} &= M_{Cz}^{\text{sys}} \end{aligned} \tag{3.3.2}$$

for the three degrees of freedom of the rigid body. Consider the 2-D case. If, under specified initial conditions $(x_{C0}, y_{C0}, \theta_{z0})$ at some time t_0, we solve Eq. (3.3.2) for the position of the center of mass $\mathbf{x}_C(t) = (x_C(t; x_{C0}), y_C(t; y_{C0}))$ and rotation $\theta_z(t; \theta_{z0})$

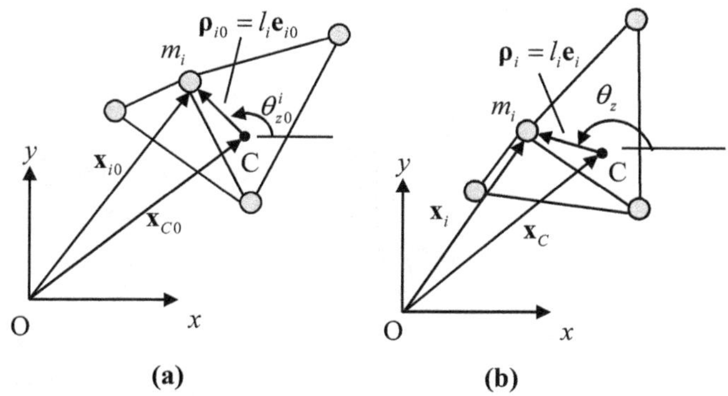

Fig. 3.11 Geometry for a system of rigidly constrained particles moving in two dimensions at (**a**) some initial time, t_0 and (**b**) some arbitrary time, t, during the motion

about C at later times t we have in fact completely solved for the motion of all the particles in the system. We can show this explicitly by examining in detail the two-dimensional motion of the system in Fig. 3.11, which shows a particular particle in the rigid system at some initial and current time. The position of each particle at time t is $\mathbf{x}_i(t) = \mathbf{x}_C(t)+\boldsymbol{\rho}_i(t)$ or $\mathbf{x}_i(t) = \mathbf{x}_C(t)+l_i\mathbf{e}_i(t)$. The unit vectors $\mathbf{e}_i$ all have the same angular motion in the rigid system but they start at time t_0 with different orientations θ^i_{z0}, i.e., $\mathbf{e}_i(t) = \mathbf{e}_i\left(\theta_z; \theta^i_{z0}\right)$ (Fig. 3.11). Thus, solving the Newton-Euler equations for $(\mathbf{x}_C, \theta_z)$ and using the known starting values $\left(l_i, \theta^i_{z0}\right)$for each particle we can find the position vector of any particle. A similar result, which we will not show here, follows for 3-D problems as well.

In Chap. 5 we will consider the motion of rigid bodies with continuously distributed mass. This is a generalization of the rigidly constrained system of particles considered here and has very similar behavior in terms of both kinematics and dynamics. Thus, in this chapter we limit our treatment of rigidly constrained particles. However, to set the stage for the discussions in Chap. 5 we will give an example in this section of the two-dimensional motion of a system of rigidly constrained particles. Specifically, consider the case of two masses, each of mass m, connected by a thin, massless rigid rod of length l_1, where the center of the rod (which is also the center of mass, C, of the two particles) is attached by a pin to another massless rigid rod of length l_2 that rotates about a fixed point O as shown in Fig. 3.12a. A rotational damper acts at the connection point C, producing a resisting moment that is proportional to the total angular velocity, $\dot{\theta} + \dot{\phi}$, between the two rods as shown in the free body diagram of Fig. 3.12b, where the proportionality constant is c_1. A similar rotational damper exists at O producing a resisting moment that is proportional angular velocity, $\dot{\phi}$, with proportionality constant, c_2 (see Fig. 3.12c). This is a two degree of freedom problem since the center of mass position of the two particles is a function only of the angle ϕ and the angular motion of the rigidly connected pair of particles can be described by the angle θ. This is also

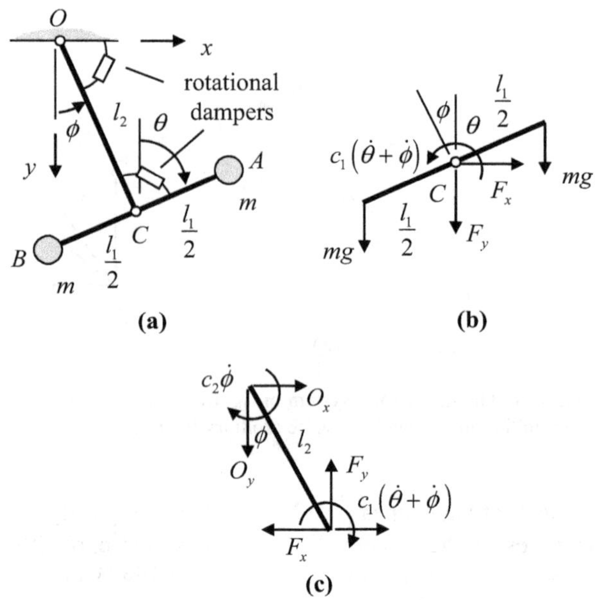

Fig. 3.12 (**a**) An equal pair of masses at *A* and *B* connected to a rigid rod that is pinned to another rod *OC* that rotates about a fixed point *O*. (**b**) The free body diagram of the two masses, and (**c**) the free body of the connecting rod *OC*

a problem of two rigid bodies since the dumbbell rigid body *AB* is connected to the rigid rod *OC*. The Newton-Euler equations for *AB* are given by

$$\begin{aligned} 2m\frac{d\mathbf{v}_C}{dt} &= \mathbf{F}^{AB} \\ \frac{d\mathbf{L}_C}{dt} &= \mathbf{M}_C^{AB} \end{aligned} \tag{3.3.3}$$

where $(\mathbf{F}^{AB}, \mathbf{M}_C^{AB})$ are the net force and moment acting on *AB*. Because the mass and angular moment of inertia of rod *OC* are neglected, this rod is in "dynamic equilibrium," i.e., the Newton-Euler equations are just

$$\begin{aligned} \mathbf{F}^{OC} &= 0 \\ \mathbf{M}_O^{OC} &= 0 \end{aligned} \tag{3.3.4}$$

where $(\mathbf{F}^{OC}, \mathbf{M}_O^{OC})$ are the net force and moment acting on *OC*, and where in this case we have taken the moments about the fixed point *O*. In terms of components for this two-dimensional motion problem then the Newton-Euler equations for *AB* are (see the free body diagram in Fig. 3.12b)

$$
\begin{aligned}
&2m\frac{d^2x_C}{dt^2} = F_x \\
&2m\frac{d^2y_C}{dt^2} = F_y + 2mg \\
&\frac{d}{dt}(I_C\omega_z) = I_C\frac{d^2\theta_z}{dt^2} = -c_1\left(\dot{\theta} + \dot{\phi}\right)
\end{aligned}
\tag{3.3.5}
$$

where the moment of inertia, I_C, is (see Eq. (3.2.53))

$$
I_C = m\left(\frac{l_1}{2}\right)^2 + m\left(\frac{l_1}{2}\right)^2 = \frac{ml_1^2}{2}
\tag{3.3.6}
$$

For rod OC (Fig. 3.12c) we have from Eq. (3.3.4)

$$
\begin{aligned}
&O_x - F_x = 0 \\
&O_y - F_y = 0 \\
&F_x l_2 \cos\phi - F_y l_2 \sin\phi + c_2\dot{\phi} + c_1\left(\dot{\theta} + \dot{\phi}\right) = 0
\end{aligned}
\tag{3.3.7}
$$

Now, consider the motion of the center of mass of AB in Fig. 3.11a in terms of the angle ϕ. Defining that motion with respect to the fixed origin O we have

$$
\begin{aligned}
&\mathbf{x}_C = l_2 \sin\phi\mathbf{e}_x + l_2\cos\phi\mathbf{e}_y \\
&\mathbf{v}_C = \dot{\mathbf{x}}_C = l_2\cos\phi\dot{\phi}\,\mathbf{e}_x - l_2\sin\phi\dot{\phi}\,\mathbf{e}_y \\
&\mathbf{a}_C = \ddot{\mathbf{x}}_C = l_2\left(-\sin\phi\dot{\phi}^2 + \cos\phi\ddot{\phi}\right)\mathbf{e}_x - l_2\left(\cos\phi\dot{\phi}^2 + \sin\phi\ddot{\phi}\right)\mathbf{e}_y
\end{aligned}
\tag{3.3.8}
$$

where $(\mathbf{e}_x, \mathbf{e}_y)$ are unit vectors in the x- and y-directions, respectively. Placing Eq. (3.3.8) into Eq. (3.3.5) gives

$$
\begin{aligned}
&2ml_2\left(-\sin\phi\dot{\phi}^2 + \cos\phi\ddot{\phi}\right) = F_x \\
&-2ml_2\left(\cos\phi\dot{\phi}^2 + \sin\phi\ddot{\phi}\right) = F_y + 2mg \\
&m\frac{l_1^2}{2}\ddot{\theta} = -c_1\left(\dot{\theta} + \dot{\phi}\right)
\end{aligned}
\tag{3.3.9}
$$

To eliminate the unknown forces from these equations we multiply the first equation in Eq. (3.3.9) by $l_2 \cos\phi$ and the second equation by $-l_2 \sin\phi$ and add the two resulting equations, giving

$$
2ml_2^2\ddot{\phi} = F_x l_2\cos\phi - F_y l_2 \sin\phi - 2mg\sin\phi
\tag{3.3.10}
$$

But using the moment equation in Eq. (3.3.7) we obtain the equations of motion

$$\begin{aligned} 2ml_2^2\ddot{\phi} &= -c_2\dot{\phi} - c_1(\dot{\theta} + \dot{\phi}) - 2mgl_2\sin\phi \\ \frac{ml_1^2}{2}\ddot{\theta} &= -c_1(\dot{\theta} + \dot{\phi}) \end{aligned} \tag{3.3.11}$$

which can also be rewritten as

$$\ddot{\phi} + \frac{c_2}{2ml_2^2}\dot{\phi} + \frac{c_1}{2ml_2^2}(\dot{\theta} + \dot{\phi}) + \frac{g}{l_2}\sin\phi = 0 \quad \ddot{\theta} + \frac{2c_1}{ml_1^2}(\dot{\theta} + \dot{\phi}) = 0 \tag{3.3.12}$$

These equations involve four constant coefficients. The constant g/l_2 has the dimensions of $1/\text{s}^2$ and is the square of the natural frequency, ω, of a simple pendulum of length l_2, i.e., $\omega = \sqrt{g/l_2}$. Thus, as we have done before, it is useful to define a nondimensional time $\tau = \omega t$ and write our equations in terms of τ to obtain

$$\begin{aligned} &\frac{d^2\phi}{d\tau^2} + \bar{c}_2\frac{d\phi}{d\tau} + \bar{c}_1\left(\frac{d\theta}{d\tau} + \frac{d\phi}{d\tau}\right) + \sin\phi = 0 \\ &\frac{d^2\theta}{d\tau^2} + \frac{4l_2^2}{l_1^2}\bar{c}_1\left(\frac{d\theta}{d\tau} + \frac{d\phi}{d\tau}\right) = 0 \end{aligned} \tag{3.3.13}$$

where $\bar{c}_2 = c_2/(2ml_2^2\omega)$ and $\bar{c}_1 = c_1/(2ml_2^2\omega)$ are nondimensional damping coefficients.

Simulation 3.1: Multiple-Particle Rigid-Body Pendulum

A MATLAB® function MP_pendulum has been written to implement this multiple-particle pendulum problem. The calling sequence for this function is

```
[tout, xout] = MP_pendulum(x0, tspan, c1, c2, lr, rtol, atol);
```

where x0 = $[\theta(0), \dot{\theta}(0), \phi(0), \dot{\phi}(0)]$ is a vector of initial conditions (in rad and rad/s), tspan is the time interval over which we want to compute the solution, c1 = $\bar{c}_1$, c2 = $\bar{c}_2$, lr = l_2/l_1, and (rtol, atol) are the relative and absolute tolerances used in the solution. This function contains a nested function MP_eqs which implements the equations of motion of Eq. (3.3.13) as a system of first-order differential equations. The function MP_pendulum uses MP_eqs in conjunction with ode45 to solve the problem numerically and outputs the times at which the motion is calculated in the column vector tout and the matrix xout = $[\theta, \dot{\theta}, \phi, \dot{\phi}]$ for the motion variables calculated at those times, organized in separate columns of the matrix. The function MP_pendulum is called by the MATLAB® script MP_pendulum_sim which has default values of the parameters given by x0 = [30, 10, 20, 0] (measured in deg and deg/s), tspan = [0, 50], c1 = 0.2,

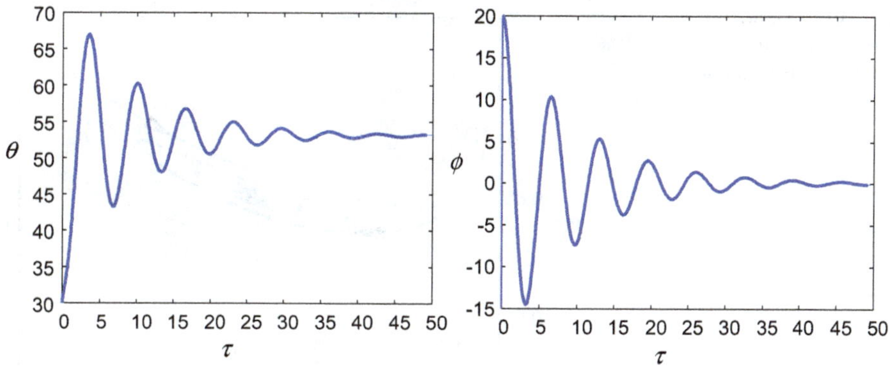

Fig. 3.13 The angles θ and ϕ (measured in degrees) versus time for the multiparticle pendulum

c2 = 0.2, lr = 2.0, rtol = 1e-6, and atol = 1e-6 which are used unless the user changes these values in the command window before executing the script. The script also plots the θ versus time and ϕ versus time values for the pendulum. Figure 3.13 shows these outputs for the default values of the script. Both angles exhibit decaying oscillations, eventually reaching equilibrium values. If we look for equilibrium values from the equations of motion, Eq. (3.3.13), by setting all the acceleration and velocity terms to zero, we find $\sin\phi = 0$ which has equilibrium values of 0 and π. Only the zero value represents a stable equilibrium position which is also the ϕ value that the system approaches to asymptotically in Fig. 3.13 as time increases. But what about the final equilibrium value of θ? If we integrate the second equation in Eq. (3.3.13) on time, we find

$$\frac{d\theta}{d\tau} + \frac{4l_2^2}{l_1^2}\bar{c}_1(\theta + \phi) = \left.\frac{d\theta}{d\tau}\right|_{\tau=0} + \frac{4l_2^2}{l_1^2}\bar{c}_1(\theta_0 + \phi_0) \tag{3.3.14}$$

so at when the motion ceases we have $\phi = 0$ and we find an equilibrium angle, θ_{eq}, given by

$$\theta_{eq} = (\theta_0 + \phi_0) + \frac{l_1^2}{4l_2^2\bar{c}_1}\left.\frac{d\theta}{d\tau}\right|_{\tau=0} \tag{3.3.15}$$

For the parameters used in simulating Fig. 3.13 we have $\theta_{eq} = 53.125^{\circ}$ which is close to the final value seen in Fig. 3.13. Note that this is not a problem where the total energy is a constant so we cannot use energy conservation to check that our numerical solution is accurate, but we can use Eq. (3.3.14) in the same manner since, like the total energy for conservative problems, it is an exact integral of the equations of motion.

Once we find the solution of the equations of motion for (θ, ϕ) we have the motion of the translation of the center of mass $\mathbf{x}_C = \mathbf{x}_C(\phi)$ and the rotation $\theta_z = \theta$

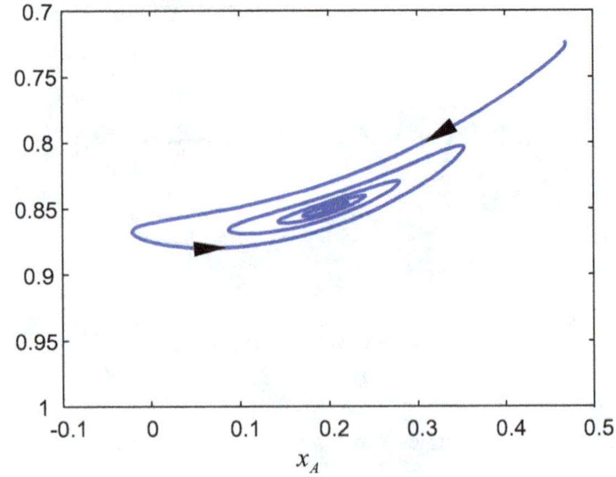

Fig. 3.14 Motion of particle *A* for the multiparticle pendulum

about the center of mass for the rigid body *AB*. As we have seen, from this information we can determine the motion of any particle in the rigid system. For the particle at *A* in Fig. 3.12, for example, we have its (x_A, y_A) coordinates given by

$$\begin{aligned} x_A &= l_2 \sin\phi + (l_1/2)\sin\theta \\ y_A &= l_2 \cos\phi - (l_1/2)\cos\theta \end{aligned} \tag{3.3.16}$$

which are plotted in Fig. 3.14 for motion defined by the default parameters in MP_pendulum_sim and with $l_2 = 1.0$ and $l_1 = 0.5$ (the arrows in Fig. 3.14 give the direction of motion) showing the 2-D oscillatory and decaying motion for this point.

3.4 Equations of Motion in Generalized Coordinates

In the last section we solved the multiparticle pendulum problem with a classical "Newtonian" approach that directly used the Newton-Euler equations for a rigidly constrained system and used those equations to eliminate the unknown forces (F_x, F_y) and obtain the equations of motion. Those forces are actually forces of constraint between the two rigid bodies. As seen in the last chapter for the motion of a single particle there are a number of other ways to eliminate or evaluate those constraint forces. In Chap. 5, when we discuss the motion of rigid bodies in depth we will examine those alternate approaches. In this section we will examine the problem of obtaining the equations of motion for a more general constrained system of particles.

To begin, we will follow the approach of Sect. 2.8, where we used n independent generalized coordinates and the D'Alembert-Lagrange principle to embed the constraints and generate $n = n_{\text{dof}}$ equations of motion. If a constrained system of N particles is subjected to holonomic constraints such that the system can be described in terms of n independent generalized coordinates $(q_1, q_2, \ldots, q_n)$ then we can write the positions of the particles in terms of those generalized coordinates as $\mathbf{x}_i = \mathbf{x}_i(q_1, \ldots, q_n, t)$. The virtual displacements of those particles are then given by

$$\delta \mathbf{x}_i = \sum_{k=1}^{n} \frac{\partial \mathbf{x}_i}{\partial q_k} \delta q_k \tag{3.4.1}$$

The D'Alembert-Lagrange principle for the system of particles is

$$\sum_{i=1}^{N} \Big(m_i \ddot{\mathbf{x}}_i - \mathbf{F}_i^e - \sum_{j \neq i}^{N} \mathbf{F}_{ij}^{\text{int}} \Big) \cdot \delta \mathbf{x}_i = 0 \tag{3.4.2}$$

If the virtual displacements satisfy all the holonomic constraints that are present then the associated ideal constraint forces, both external and internal, are absent from Eq. (3.4.2) since, as discussed in the last chapter, they do no virtual work. Most books do not distinguish between the remaining external and internal forces because they are both "external" to an individual particle, but here we will keep that distinction and also assume that the internal forces are derivable from potentials. Then following the same steps as when dealing with the real work done by the internal forces, we have

$$\sum_{i=1}^{N} \sum_{j \neq i}^{N} \mathbf{F}_{ij}^{\text{int}} \cdot \delta \mathbf{x}_i = -\delta V^{\text{int}} = -\sum_{k=1}^{n} \frac{\partial V^{\text{int}}}{\partial q_k} \delta q_k \tag{3.4.3}$$

where V^{int} is the total internal energy of the system. Then, using Eq. (3.4.1) the D'Alembert-Lagrange principle becomes

$$\sum_{k=1}^{n} \left[\sum_{i=1}^{N} m_i \ddot{\mathbf{x}}_i \cdot \frac{\partial \mathbf{x}_i}{\partial q_k} - Q_k^e + \frac{\partial V^{\text{int}}}{\partial q_k} \right] \delta q_k = 0 \tag{3.4.4}$$

where the generalized force components, Q_k^e, are

$$Q_k^e = \sum_{i=1}^{N} \mathbf{F}_i^e \cdot \frac{\partial \mathbf{x}_i}{\partial q_k} \tag{3.4.5}$$

Since the q_k generalized coordinates are assumed to be independent, from Eq. (3.4.5) we find

$$\sum_{i=1}^{N} m_i \ddot{\mathbf{x}}_i \cdot \frac{\partial \mathbf{x}_i}{\partial q_k} - Q_k^e + \frac{\partial V^{\text{int}}}{\partial q_k} = 0 \tag{3.4.6}$$

Following the same steps as followed for the single particle we can write the position vector and its derivatives as

$$\begin{aligned}
\mathbf{x}_i &= \mathbf{x}_i(q_1,\ldots,q_n,t) \\
\mathbf{v}_i &= \dot{\mathbf{x}}_i = \sum_{m=1}^{n}\frac{\partial \mathbf{x}_i}{\partial q_m}\dot{q}_m + \frac{\partial \mathbf{x}_i}{\partial t} = \sum_{m=1}^{n}\mathbf{c}_{i;m}\dot{q}_m + \frac{\partial \mathbf{x}_i}{\partial t} \\
\mathbf{a}_i &= \ddot{\mathbf{x}}_i = \sum_{m=1}^{n}\mathbf{c}_{i;m}\ddot{q}_m + \sum_{m=1}^{n}\sum_{n=1}^{n}\frac{\partial \mathbf{c}_{i;m}}{\partial q_n}\dot{q}_m\dot{q}_n + 2\sum_{m=1}^{n}\frac{\partial \mathbf{c}_{i;m}}{\partial t}\dot{q}_m + \frac{\partial^2 \mathbf{x}_i}{\partial t^2}
\end{aligned} \tag{3.4.7}$$

where

$$\mathbf{c}_{i;m} = \partial \mathbf{x}_i/\partial q_m \tag{3.4.8}$$

so that

$$\begin{aligned}
\sum_{i=1}^{N} m_i\ddot{\mathbf{x}}_i\cdot\mathbf{c}_{i;k} &= \sum_{m=1}^{n}\sum_{i=1}^{N} m_i\left[\mathbf{c}_{i;k}\cdot\mathbf{c}_{i;m}\right]\ddot{q}_m + \sum_{m=1}^{n}\sum_{p=1}^{n}\sum_{i=1}^{N} m_i\left[\mathbf{c}_{i;k}\cdot\frac{\partial \mathbf{c}_{i;m}}{\partial q_p}\right]\dot{q}_m\dot{q}_p \\
&\quad + 2\sum_{m=1}^{n}\sum_{i=1}^{N} m_i\left[\mathbf{c}_{i;k}\cdot\frac{\partial \mathbf{c}_{i;m}}{\partial t}\right]\dot{q}_m + \sum_{i=1}^{N} m_i\mathbf{c}_{i;k}\cdot\frac{\partial^2 \mathbf{x}_i}{\partial t^2} \\
&= \sum_{m=1}^{n} M_{km}\ddot{q}_m + \sum_{m=1}^{n}\sum_{p=1}^{n} K_{kmp}\dot{q}_m\dot{q}_p + \sum_{m=1}^{n} J_{km}\dot{q}_m + B_k
\end{aligned} \tag{3.4.9}$$

where we have used the same notation as the single-particle case for these coefficients since they are just sums of the single-particle coefficients over the entire system:

$$\begin{aligned}
M_{km} &= \sum_{i=1}^{N} m_i\mathbf{c}_{i;k}\cdot\mathbf{c}_{i;m}, \quad K_{kmp} = \sum_{i=1}^{N} m_i\mathbf{c}_{i;k}\cdot\frac{\partial \mathbf{c}_{i;m}}{\partial q_p} \\
J_{km} &= \sum_{i=1}^{N} 2m_i\mathbf{c}_{i;k}\cdot\frac{\partial \mathbf{c}_{i;m}}{\partial t}, \quad B_k = \sum_{i=1}^{N} m_i\mathbf{c}_{i;k}\cdot\frac{\partial^2 \mathbf{x}_i}{\partial t^2}
\end{aligned} \tag{3.4.10}$$

and the equations of motion are

$$\sum_{m=1}^{n} M_{km}\ddot{q}_m + \sum_{m=1}^{n}\sum_{p=1}^{n} K_{kmp}\dot{q}_m\dot{q}_p + \sum_{m=1}^{n} J_{km}\dot{q}_m + B_k = Q_k^e - \frac{\partial V^{\text{int}}}{\partial q_k} \tag{3.4.11}$$

If, instead we used the velocity of the particles, $\mathbf{v}_i = \mathbf{v}_i(q_1,\ldots,q_n,\dot{q}_1,\ldots,\dot{q}_n,t)$ to calculate the equations of motion, as in the single-particle case we find

$$\begin{aligned}\mathbf{a}_i=\frac{d\mathbf{v}_i}{dt}&=\sum_{m=1}^{n}\frac{\partial\mathbf{v}_i}{\partial\dot{q}_m}\ddot{q}_m+\sum_{m=1}^{n}\frac{\partial\mathbf{v}_i}{\partial q_m}\dot{q}_m+\frac{\partial\mathbf{v}_i}{\partial t}\\&=\sum_{m=1}^{n}\mathbf{c}_{i;m}\ddot{q}_m+\sum_{m=1}^{n}\frac{\partial\mathbf{v}_i}{\partial q_m}\dot{q}_m+\frac{\partial\mathbf{v}_i}{\partial t}\end{aligned}\tag{3.4.12}$$

since $\partial\mathbf{v}_i/\partial\dot{q}_m=\partial\mathbf{x}_i/\partial q_m=\mathbf{c}_{i;m}$, leading to

$$\begin{aligned}\sum_{i=1}^{N}m_i\mathbf{a}_i\cdot\mathbf{c}_{i;k}&=\sum_{m=1}^{n}\sum_{i=1}^{N}m_i\left[\mathbf{c}_{i;k}\cdot\mathbf{c}_{i;m}\right]\ddot{q}_m+\sum_{m=1}^{n}\sum_{i=1}^{N}m_i\left[\mathbf{c}_{i;k}\cdot\frac{\partial\mathbf{v}_i}{\partial q_m}\right]\dot{q}_m+\sum_{i=1}^{N}m_i\left[\mathbf{c}_{i;k}\cdot\frac{\partial\mathbf{v}_i}{\partial t}\right]\\&=\sum_{m=1}^{n}M_{km}\,\ddot{q}_m+\sum_{m=1}^{n}D_{km}\dot{q}_m+E_k\end{aligned}\tag{3.4.13}$$

where

$$\begin{aligned}M_{km}&=\sum_{i=1}^{N}m_i\,\mathbf{c}_{i;k}\cdot\mathbf{c}_{i;m}\\D_{km}&=\sum_{i=1}^{N}m_i\,\mathbf{c}_{i;k}\cdot\frac{\partial\mathbf{v}_i}{\partial q_m}\\E_k&=\sum_{i=1}^{N}m_i\,\mathbf{c}_{i;\mathrm{k}}\cdot\frac{\partial\mathbf{v}_i}{\partial t}\end{aligned}\tag{3.4.14}$$

and giving the equations of motion

$$\sum_{m=1}^{n}M_{km}\ddot{q}_m+\sum_{m=1}^{n}D_{km}\dot{q}_m+E_k=Q_k^e-\frac{\partial V^{\text{int}}}{\partial q_k}\tag{3.4.15}$$

For a single particle we also showed (see Eq. (2.7.4)) that

$$m_i\ddot{\mathbf{x}}_i\cdot\mathbf{c}_{i;k}=\frac{d}{dt}\left(\frac{\partial T_i}{\partial\dot{q}_k}\right)-\frac{\partial T_i}{\partial q_k}\tag{3.4.16}$$

where $T_i=m_i\mathbf{v}_i\cdot\mathbf{v}_i/2$ is the kinetic energy of the ith particle so that

$$\sum_{i=1}^{N}m_i\ddot{\mathbf{x}}_i\cdot\mathbf{c}_{i;k}=\frac{d}{dt}\left(\frac{\partial T}{\partial\dot{q}_k}\right)-\frac{\partial T}{\partial q_k}\tag{3.4.17}$$

where $T = \sum_{i=1}^{N} T_i$ is the total kinetic energy of the system. This leads us then to Lagrange's equation for the system of particles, namely,

$$\frac{d}{dt}\left(\frac{\partial T}{\partial \dot{q}_k}\right) - \frac{\partial T}{\partial q_k} = Q_k^e - \frac{\partial V^{\text{int}}}{\partial q_k} \tag{3.4.18}$$

As in the single-particle case we can write the kinetic energy as a combination of three parts, i.e.,

$$T = T_2 + T_1 + T_0 \tag{3.4.19}$$

where for the system of particles since

$$T = \frac{1}{2}\sum_{i=1}^{N} m_i \mathbf{v}_i \cdot \mathbf{v}_i = \frac{1}{2}\sum_{i=1}^{N} m_i \left(\sum_{k=1}^{n} \frac{\partial \mathbf{x}_i}{\partial q_k}\dot{q}_k + \frac{\partial \mathbf{x}_i}{\partial t}\right) \cdot \left(\sum_{m=1}^{n} \frac{\partial \mathbf{x}_i}{\partial q_m}\dot{q}_m + \frac{\partial \mathbf{x}_i}{\partial t}\right) \tag{3.4.20}$$

we can let

$$T_2 = \frac{1}{2}\sum_{k=1}^{n}\sum_{m=1}^{n} M_{km}\dot{q}_k\dot{q}_m, \quad T_1 = \sum_{m=1}^{n} \mu_m \dot{q}_m, \quad T_0 = \frac{1}{2}\sum_{i=1}^{N} m_i \frac{\partial \mathbf{x}_i}{\partial t} \cdot \frac{\partial \mathbf{x}_i}{\partial t} \tag{3.4.21}$$

with

$$\begin{aligned} M_{km} = M_{mk} &= \sum_{i=1}^{N} m_i \frac{\partial \mathbf{x}_i}{\partial q_m} \cdot \frac{\partial \mathbf{x}_i}{\partial q_k} \\ \mu_m &= \sum_{i=1}^{N} m_i \frac{\partial \mathbf{x}_i}{\partial q_m} \cdot \frac{\partial \mathbf{x}_i}{\partial t} \end{aligned} \tag{3.4.22}$$

3.4.1 *Motion of a Double Pendulum*

One of the most frequently used examples of multiple-particle systems in dynamics books is the double pendulum shown in Fig. 3.15. Despite that popularity, most treatments do little more than deriving the equations of motion. We will try to

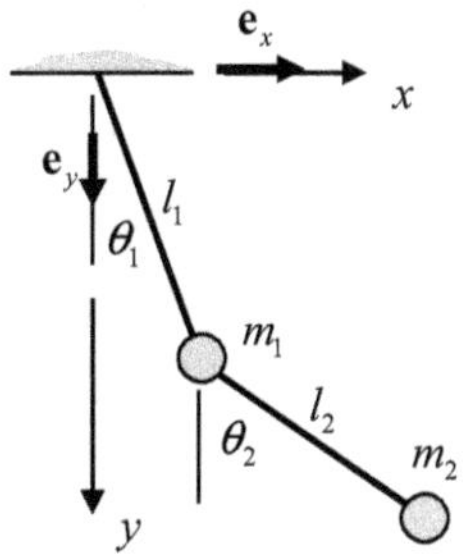

Fig. 3.15 Geometry of the double pendulum

provide a more detailed discussion here. First, let's obtain the equations of motion using several of the methods described previously. This is a two degree of freedom problem since each mass has two degrees of freedom and there are two holonomic constraints l_1,l_2 = constants. The motion of the masses can be completely described in terms of the two independent generalized coordinates $(q_1, q_2) = (\theta_1, \theta_2)$. If we write the position vectors of the two masses in terms of these angles we have

$$\begin{aligned} \mathbf{x}_1 &= l_1 \sin\theta_1 \mathbf{e}_x + l_1 \cos\theta_1 \mathbf{e}_y \\ \mathbf{x}_2 &= (l_1 \sin\theta_1 + l_2 \sin\theta_2)\mathbf{e}_x + (l_1 \cos\theta_1 + l_2 \cos\theta_2)\mathbf{e}_y \end{aligned} \tag{3.4.23}$$

so the velocities are

$$\begin{aligned} \mathbf{v}_1 &= l_1 \cos\theta_1 \dot{\theta}_1 \mathbf{e}_x - l_1 \sin\theta_1 \dot{\theta}_1 \mathbf{e}_y \\ \mathbf{v}_2 &= \left(l_1 \cos\theta_1 \dot{\theta}_1 + l_2 \cos\theta_2 \dot{\theta}_2\right)\mathbf{e}_x - \left(l_1 \sin\theta_1 \dot{\theta}_1 + l_2 \sin\theta_2 \dot{\theta}_2\right)\mathbf{e}_y \end{aligned} \tag{3.4.24}$$

Using the velocity and its derivatives (Eq. (3.4.14)) to form the equations of motion, we need

$$\begin{aligned} \mathbf{c}_{1;1} &= \frac{\partial \mathbf{v}_1}{\partial \dot{\theta}_1} = l_1 \cos\theta_1 \mathbf{e}_x - l_1 \sin\theta_1 \mathbf{e}_y \\ \mathbf{c}_{1;2} &= \frac{\partial \mathbf{v}_1}{\partial \dot{\theta}_2} = 0 \\ \mathbf{c}_{2;1} &= \frac{\partial \mathbf{v}_2}{\partial \dot{\theta}_1} = l_1 \cos\theta_1 \mathbf{e}_x - l_1 \sin\theta_1 \mathbf{e}_y \\ \mathbf{c}_{2;2} &= \frac{\partial \mathbf{v}_2}{\partial \dot{\theta}_2} = l_2 \cos\theta_2 \mathbf{e}_x - l_2 \sin\theta_2 \mathbf{e}_y \end{aligned} \tag{3.4.25}$$

and

$$\begin{aligned}\frac{\partial \mathbf{v}_1}{\partial \theta_1} &= -l_1 \sin\theta_1 \dot{\theta}_1 \mathbf{e}_x - l_1 \cos\theta_1 \dot{\theta}_1 \mathbf{e}_y \\ \frac{\partial \mathbf{v}_1}{\partial \theta_2} &= 0 \\ \frac{\partial \mathbf{v}_2}{\partial \theta_1} &= -l_1 \sin\theta_1 \dot{\theta}_1 \mathbf{e}_x - l_1 \cos\theta_1 \dot{\theta}_1 \mathbf{e}_y \\ \frac{\partial \mathbf{v}_2}{\partial \theta_2} &= -l_2 \sin\theta_2 \dot{\theta}_2 \mathbf{e}_x - l_2 \cos\theta_2 \dot{\theta}_2 \mathbf{e}_y\end{aligned} \tag{3.4.26}$$

and, finally, $\partial \mathbf{v}_1/\partial t = \partial \mathbf{v}_2/\partial t = 0$. This gives the matrices

$$\begin{aligned}M_{km} &= \sum_{i=1}^{N} m_i \mathbf{c}_{i;k} \cdot \mathbf{c}_{i;m} \rightarrow [M] = \begin{bmatrix} (m_1+m_2)l_1^2 & m_2 l_1 l_2 \cos(\theta_1-\theta_2) \\ m_2 l_1 l_2 \cos(\theta_1-\theta_2) & m_2 l_2^2 \end{bmatrix} \\ D_{km} &= \sum_{i=1}^{N} m_i \mathbf{c}_{i;k} \cdot \frac{\partial \mathbf{v}_i}{\partial q_m} \rightarrow [D] = \begin{bmatrix} 0 & m_2 l_1 l_2 \sin(\theta_1-\theta_2)\dot{\theta}_2 \\ -m_2 l_1 l_2 \sin(\theta_1-\theta_2)\dot{\theta}_1 & 0 \end{bmatrix} \\ E_k &= \sum_{i=1}^{N} m_i \mathbf{c}_{i;k} \cdot \frac{\partial \mathbf{v}_i}{\partial t} \rightarrow \{E\} = \begin{Bmatrix} 0 \\ 0 \end{Bmatrix}\end{aligned} \tag{3.4.27}$$

The generalized external forces are due to the gravitational forces $\mathbf{F}_1 = m_1 g \mathbf{e}_y$, and $\mathbf{F}_2 = m_2 g \mathbf{e}_y$ shown in Fig. 3.15, and $V^{\text{int}} = 0$ giving

$$\begin{aligned}Q_1 &= \sum_{i=1}^{2} \mathbf{F}_i \cdot \mathbf{c}_{i;1} = -(m_1+m_2) g l_1 \sin\theta_1 \\ Q_2 &= \sum_{i=1}^{2} \mathbf{F}_i \cdot \mathbf{c}_{i;2} = -m_2 g l_2 \sin\theta_2\end{aligned} \tag{3.4.28}$$

Collecting all these results the equations of motion (see Eq. (3.4.15)) are

$$\begin{aligned}&(m_1+m_2)l_1^2\ddot{\theta}_1 + m_2 l_1 l_2 \cos(\theta_1-\theta_2)\ddot{\theta}_2 + m_2 l_1 l_2 \sin(\theta_1-\theta_2)\dot{\theta}_2^2 + (m_1+m_2) \\ &\quad \times g l_1 \sin\theta_1 = 0 \\ &m_2 l_1 l_2 \cos(\theta_1-\theta_2)\ddot{\theta}_1 + m_2 l_2^2 \ddot{\theta}_2 - m_2 l_1 l_2 \sin(\theta_1-\theta_2)\dot{\theta}_1^2 + m_2 g l_2 \sin\theta_2 = 0\end{aligned} \tag{3.4.29}$$

We can divide the first equation by l_1 and the second equation by l_2 to obtain the form of these equations found in most texts:

$$
\begin{aligned}
&(m_1+m_2)l_1\ddot{\theta}_1+m_2l_2\cos(\theta_1-\theta_2)\ddot{\theta}_2+m_2l_2\sin(\theta_1-\theta_2)\dot{\theta}_2^2\\
&\quad+(m_1+m_2)g\sin\theta_1=0\\
&m_2l_1\cos(\theta_1-\theta_2)\ddot{\theta}_1+m_2l_2\ddot{\theta}_2-m_2l_1\sin(\theta_1-\theta_2)\dot{\theta}_1^2+m_2g\sin\theta_2=0
\end{aligned}
\tag{3.4.30}
$$

If we want to derive these equations instead using Lagrange's equations we need to compute the total kinetic energy:

$$
T=\frac{1}{2}(m_1\mathbf{v}_1\cdot\mathbf{v}_1+m_2\mathbf{v}_2\cdot\mathbf{v}_2)
\tag{3.4.31}
$$

But we already have the velocities in Eq. (3.4.24) so we find

$$
T=\frac{m_1}{2}l_1^2\dot{\theta}_1^2+\frac{m_2}{2}\left(l_1^2\dot{\theta}_1^2+l_2^2\dot{\theta}_2^2+2l_1l_2\cos(\theta_1-\theta_2)\dot{\theta}_1\dot{\theta}_2\right)
\tag{3.4.32}
$$

from which we obtain

$$
\begin{aligned}
\frac{\partial T}{\partial\dot{\theta}_1}&=m_1l_1^2\dot{\theta}_1+m_2\left(l_1^2\dot{\theta}_1+l_1l_2\cos(\theta_1-\theta_2)\dot{\theta}_2\right)\\
\frac{\partial T}{\partial\theta_1}&=-m_2l_1l_2\sin(\theta_1-\theta_2)\dot{\theta}_1\dot{\theta}_2\\
\frac{\partial T}{\partial\dot{\theta}_2}&=m_2l_2^2\dot{\theta}_2+m_2l_1l_2\cos(\theta_1-\theta_2)\dot{\theta}_1\\
\frac{\partial T}{\partial\theta_2}&=m_2l_1l_2\sin(\theta_1-\theta_2)\dot{\theta}_1\dot{\theta}_2
\end{aligned}
\tag{3.4.33}
$$

so that $\frac{d}{dt}\left(\frac{\partial T}{\partial\dot{\theta}_1}\right)-\frac{\partial T}{\partial\theta_1}-Q_1=0$ becomes

$$
\begin{aligned}
&(m_1+m_2)l_1^2\ddot{\theta}_1+m_2(l_1l_2\cos(\theta_1-\theta_2))\ddot{\theta}_2+m_2l_1l_2\sin(\theta_1-\theta_2)\dot{\theta}_2^2\\
&\quad+(m_1+m_2)gl_1\sin\theta_1=0
\end{aligned}
\tag{3.4.34}
$$

which is the same as the first equation in Eq. (3.4.29). Also, $\frac{d}{dt}\left(\frac{\partial T}{\partial\dot{\theta}_2}\right)-\frac{\partial T}{\partial\theta_2}-Q_2=0$ becomes

$$
m_2l_2^2\ddot{\theta}_2+m_1l_1l_2\cos(\theta_1-\theta_2)\ddot{\theta}_1-m_2l_1l_2\sin(\theta_1-\theta_2)\dot{\theta}_1^2+m_2gl_2\sin\theta_2=0
\tag{3.4.35}
$$

which is the same as the second equation in Eq. (3.4.29).

The equations of motion in Eq. (3.4.29) are in the matrix form

$$[M]\{\ddot{q}\} + [D]\{\dot{q}\} - \{Q^e\} = 0 \tag{3.4.36}$$

but to put these equations into the standard form we have used before for a system of first-order differential equations, we need to first remove the generalized mass matrix, **M**. We can do this by multiplying Eq. (3.4.36) by the inverse of the mass matrix to obtain

$$\{\ddot{q}\} = [M]^{-1}(\{Q^e\} - [D]\{\dot{q}\}) \tag{3.4.37}$$

which is not difficult to do but the resulting expressions are rather complex so to make the forms more manageable we will leave the inverse matrix written in terms of the components of $[M]$, i.e.,

$$[M]^{-1} = \frac{1}{\Delta}\begin{bmatrix} M_{22} & -M_{12} \\ -M_{21} & M_{11} \end{bmatrix} \qquad \Delta = M_{11}M_{22} - M_{12}M_{21} \tag{3.4.38}$$

where (see Eq. (3.4.27))

$$\begin{aligned} M_{11} &= (m_1 + m_2)l_1^2, \quad M_{12} = m_2 l_1 l_2 \cos(\theta_1 - \theta_2) \\ M_{21} &= m_2 l_1 l_2 \cos(\theta_1 - \theta_2), \quad M_{22} = m_2 l_2^2 \end{aligned} \tag{3.4.39}$$

Then, since

$$[D]\{\dot{q}\} = m_2 l_1 l_2 \sin(\theta_1 - \theta_2)\begin{Bmatrix} \dot{\theta}_2^2 \\ -\dot{\theta}_1^2 \end{Bmatrix} \tag{3.4.40}$$

and

$$\{Q^e\} = \begin{Bmatrix} -(m_1 + m_2)g l_1 \sin\theta_1 \\ -m_2 g l_2 \sin\theta_2 \end{Bmatrix} \tag{3.4.41}$$

we have

$$[M]^{-1}(\{Q^e\} - [D]\{\dot{q}\}) = [M]^{-1}\begin{Bmatrix} -(m_1 + m_2)g l_1 \sin\theta_1 - m_2 l_1 l_2 \sin(\theta_1 - \theta_2)\dot{\theta}_2^2 \\ -m_2 g l_2 \sin\theta_2 + m_2 l_1 l_2 \sin(\theta_1 - \theta_2)\dot{\theta}_1^2 \end{Bmatrix} \tag{3.4.42}$$

which we will write as

$$[M]^{-1}(\{Q^e\} - [D]\{\dot{q}\}) = [M]^{-1}\begin{Bmatrix} t_1 \\ t_2 \end{Bmatrix} \tag{3.4.43}$$

with

$$\begin{Bmatrix} t_1 \\ t_2 \end{Bmatrix} = \begin{Bmatrix} -(m_1 + m_2)gl_1 \sin\theta_1 - m_2 l_1 l_2 \sin(\theta_1 - \theta_2)\dot{\theta}_2^2 \\ -m_2 g l_2 \sin\theta_2 + m_2 l_1 l_2 \sin(\theta_1 - \theta_2)\dot{\theta}_1^2 \end{Bmatrix} \tag{3.4.44}$$

Thus, we find

$$\begin{Bmatrix} \ddot{\theta}_1 \\ \ddot{\theta}_2 \end{Bmatrix} = \frac{1}{\Delta}\begin{bmatrix} M_{22} & -M_{12} \\ -M_{21} & M_{11} \end{bmatrix}\begin{Bmatrix} t_1 \\ t_2 \end{Bmatrix} \tag{3.4.45}$$

which gives

$$\begin{aligned} \ddot{\theta}_1 &= \frac{M_{22}t_1 - M_{12}t_2}{\Delta} \\ \ddot{\theta}_2 &= \frac{M_{11}t_2 - M_{12}t_1}{\Delta} \end{aligned} \tag{3.4.46}$$

and by letting $[x_1, x_2, x_3, x_4] = \left[\theta_1, \dot{\theta}_1, \theta_2, \dot{\theta}_2\right]$ the first-order equations of motion are

$$\begin{aligned} \dot{x}_1 &= x_2 \\ \dot{x}_2 &= \frac{M_{22}t_1 - M_{12}t_2}{\Delta} \\ \dot{x}_3 &= x_4 \\ \dot{x}_4 &= \frac{M_{11}t_2 - M_{21}t_1}{\Delta} \end{aligned} \tag{3.4.47}$$

where

$$\begin{aligned} t_1 &= \left(-m_2 l_1 l_2 \sin(x_1 - x_3)x_4^2 - (m_1 + m_2)gl_1 \sin x_1\right) \\ t_2 &= \left(m_2 l_1 l_2 \sin(x_1 - x_3)x_2^2 - m_2 g l_2 \sin x_3\right) \\ \Delta &= M_{11}M_{22} - M_{12}^2 \\ &= (m_1 + m_2)m_2 l_1^2 l_2^2 - m_2^2 l_1^2 l_2^2 \cos^2(x_1 - x_3) \end{aligned} \tag{3.4.48}$$

and

$$\begin{aligned} &M_{11} = (m_1 + m_2)l_1^2, \quad M_{12} = m_2 l_1 l_2 \cos(x_1 - x_3) \\ &M_{21} = M_{12}, \quad M_{22} = m_2 l_2^2 \end{aligned} \tag{3.4.49}$$

Simulation 3.2: The Double Pendulum

A MATLAB® function double_pend has been written to implement the equations of motion for the double pendulum. The calling sequence for this function is

```
[tout, xout] =double_pend(x0, tspan, m1, m2, l1 , l2, g, rtol, atol);
```

where x0 $= [\theta_1(0),\ \dot{\theta}_1(0),\ \theta_2(0),\ \dot{\theta}_2(0)]$ is the vector of initial conditions (in rad and rad/s)), tspan is the time interval over which the numerical calculations take place, (m1, m2) and (l1, l2) are the masses and lengths of the connected pendulums, g is the acceleration of gravity, and rtol and atol are the relative and absolute error tolerances used by ode45 to solve the system of equations. The function double_pend has a nested function dp_eqs that implements the first-order system of Eq. (3.4.47) and the numerical solver ode45 that uses dp_eqs to obtain the numerical solution. The MATLAB® script double_pendulum_sim contains default values of all the input parameters needed by double_pend which are executed unless changed by the user in the MATLAB® command window before executing the script. The script plots the angles (θ_1, θ_2) versus time as well as the y_2 versus x_2 motion of mass m_2. The script also shows a plot of the total energy, normalized by its starting value, versus the time, t, since this is a conservative system and the total energy, E, can be used as a check on the accuracy of the numerical solution, where

$$\begin{aligned} E = & \frac{m_1}{2} l_1^2 \dot{\theta}_1^2 + \frac{m_2}{2} \left(l_1^2 \dot{\theta}_1^2 + l_2^2 \dot{\theta}_2^2 + 2 l_1 l_2 \cos(\theta_1 - \theta_2) \dot{\theta}_1 \dot{\theta}_2 \right) \\ & - (m_1 + m_2) g \cos\theta_1 - m_2 g \cos\theta_2 \end{aligned} \tag{3.4.50}$$

The double pendulum can have chaotic behavior so that the default error tolerances in the script are set to the very small values rtol = 1e-10, atol = 1e-10.

For different starting angles the double pendulum can exhibit a variety of behaviors. Figure 3.16 shows the angles (in radians) versus time for the default values in the double_pendulum_sim script which are l1 = l2 = 0.1 m, m1 = m2 = 0.1 kg, $\theta_1(0) = 50^\circ$, $\theta_2(0) = 25^\circ$ (which must be converted to radians), $\dot{\theta}_1(0) = \dot{\theta}_2(0) = 0$, and tspan = [0, 4] s. The angular motion is somewhat periodic as is the (x_2, y_2) motion of the mass m_2 as shown in Fig. 3.17. The double pendulum can exhibit a rich variety of behaviors so we have also written a MATLAB® script double_pendulum_anim2 that replaces the plotting outputs of the double_pendulum_sim script with an animation of the motion so that you can visualize the entire motion. In addition to the ordinary inputs this animation script, like the animation script described in Chap. 2, has an additional input, sldwn, that produces a slow down of the animation, when needed, by placing a pause of 0.01*sldwn seconds in the animation between draws and a marker parameter input for following the path of mass m_2, as found in the animation script for the spring pendulum. There is no built-in parameter for speeding up the animation, but the easiest way to obtain a speedup of any of these animations is to simply reduce the

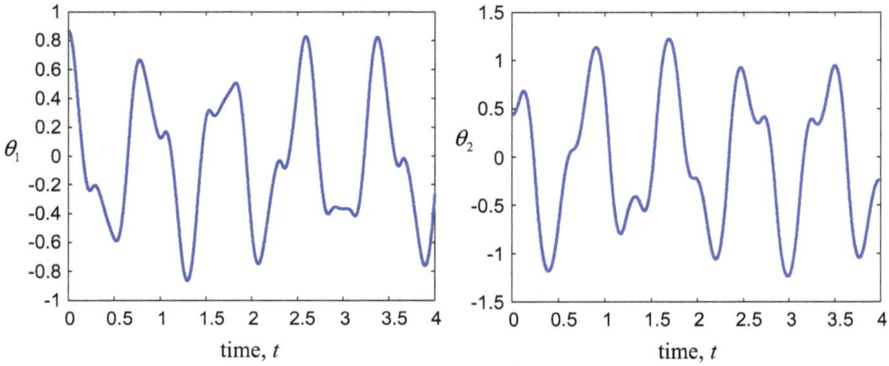

Fig. 3.16 The angles θ_1 and θ_2 versus time for the double pendulum, for l1 = l2 = 0.1 m, m1 = m2 = 0.1 kg, with starting angles $\theta_1 = 50^\circ$, $\theta_2 = 25^\circ$, and $\dot{\theta}_1 = \dot{\theta}_2 = 0$

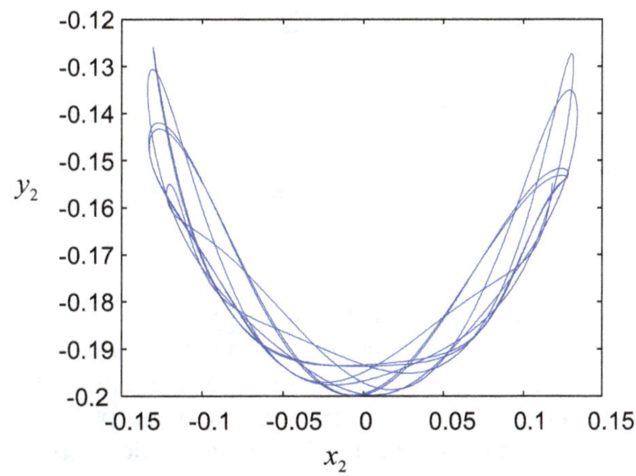

Fig. 3.17 Motion of the mass m_2 for the same parameters as given in Fig. 3.16

number of points being calculated by specifying the number of points to be returned in tspan, since typically such slowness in the animation is caused by small time intervals in the numerical solution, leading to a large number of points (specifying the points in tspan, however, does not change the number of underlying evaluation points which is controlled by ode45, as mentioned previously). Such a speedup is needed with double_pendulum_anim2 since the default error tolerances have been set to the same very small values used in double_pendulum_sim, which indeed causes very small time intervals. You can see such a speedup by first running double_pendulum_anim2 with the default tspan specification of tspan = [0, 4] and then clearing the workspace and changing the tspan specification to tspan = linspace (0, 4, 1000) which reduces the number of points being plotted by about a factor of 8 and results in a much faster animation. One must be careful, however, not to reduce the number of points too much or you will lose the smoothness of the animation. In most cases with some experimentation you can adjust the number of points in tspan and the value of sldwn to reach a reasonably fast but smooth animation.

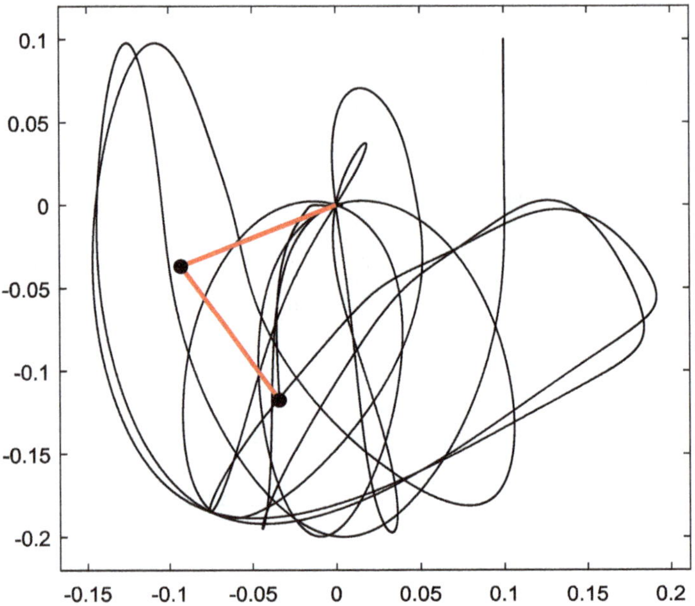

Fig. 3.18 An animation of the motion of the double pendulum, showing the path followed by mass m_2 and the final configuration of the pendulum with tspan = linspace(0, 4.5, 1000) and x0 = [pi/2, 0, pi, 0]

Figure 3.18 shows the final configuration of the double pendulum and the path followed by the marker by clearing the workspace, specifying initial conditions x0 = [pi/2, 0, pi, 0], and setting tspan = linspace(0, 4.5, 1000) to speed up the animation. You can see the very complex path followed in this example. It is instructive to experiment with this animation script to see the different types of motion possible. For example, if you use the same inputs as in Fig. 3.18 except with slightly different initial conditions x0 = [pi/2 + 0.01, 0, pi, 0] the path followed (see Fig. 3.19) is initially the same as seen in Fig. 3.18 but differences grow with time, ending in a final configuration that is not close to the one seen in Fig. 3.18. This sensitivity of the motion to initial conditions is the trademark of chaotic behavior.

Both at very small and very large energy levels the double pendulum shows regular periodic behavior. The low-energy case corresponds to very-small-angle vibrations about $\theta_1 = \theta_2 = 0$ while the high-energy case is when the masses are executing complete revolutions with ever-increasing or -decreasing angles. Let's consider the low-energy case. At small angles, we can neglect all the nonlinear terms in both Eq. (3.4.34) and (3.4.35) and write the approximate linear equations of motion as

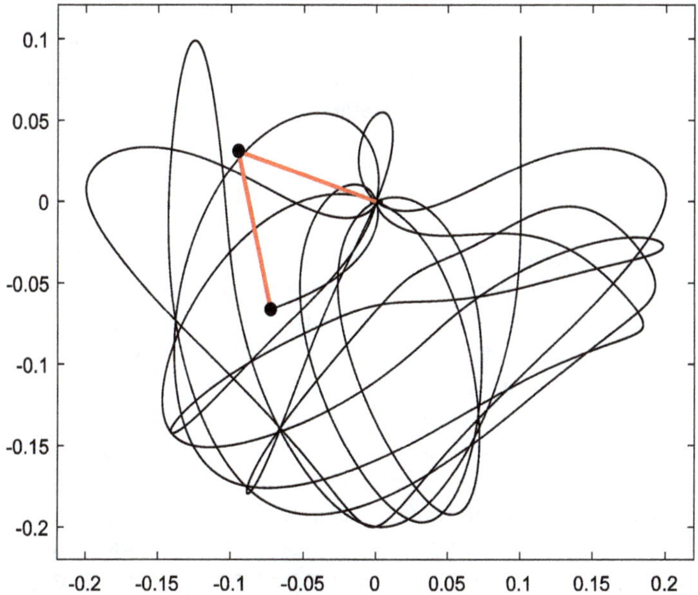

Fig. 3.19 An animation of the motion of the double pendulum, showing the path followed by mass m_2 and the final configuration of the pendulum with slightly different initial conditions of x0 = [pi/2 + 0.01, 0, pi, 0] from those used in Fig. 3.18

$$\begin{aligned}\ddot{\theta}_1 + \frac{\mu g}{l_1}\left(\theta_1 - \frac{\mu - 1}{\mu}\theta_2\right) &= 0\\ \ddot{\theta}_2 + \frac{\mu g}{l_2}(\theta_2 - \theta_1) &= 0\end{aligned} \tag{3.4.51}$$

where

$$\mu = 1 + \frac{m_2}{m_1} \tag{3.4.52}$$

Even at small amplitudes we see that the motions of the two masses are intimately coupled. We will study such coupled vibratory motions in more detail in Chap. 7, but here we can learn some important aspects of such coupled systems by examining the solutions of Eq. (3.4.51). To keep things simple algebraically, we will let $l_1 = l_2 = l$ and $m_1 = m_2 = m$. Then our equations of motion become

$$\begin{aligned}\ddot{\theta}_1 + \frac{2g}{l}\left(\theta_1 - \frac{1}{2}\theta_2\right) &= 0\\ \ddot{\theta}_2 + \frac{2g}{l}(\theta_2 - \theta_1) &= 0\end{aligned} \tag{3.4.53}$$

The quantity $\omega_0^2 = g/l$ is physically just the square of the natural frequency, ω_0 (in rad/s) for small oscillations of a simple pendulum of length l. To see this, first note that the simple pendulum problem is contained in Eq. (3.4.51) by simply setting $m_2 = 0$ ($\mu = 1$) which gives the equation of motion for θ_1 as

$$\ddot{\theta}_1 + \frac{g}{l}\theta_1 = 0 \tag{3.4.54}$$

This is the same equation of motion we found for the simple pendulum in the last chapter (see Eq. (2.5.28)) under the small-angle approximation where $\sin\theta_1 \cong \theta_1$. To obtain a solution of Eq. (3.4.54) that represents harmonic oscillations we will let θ_1 be represented in the form of a complex function, $\theta_1 = a_1 \exp(i\omega t)$, where ω is a circular frequency and a_1 is also complex, i.e., we can represent it in terms of a magnitude and phase as $a_1 = A_1 \exp(i\phi_1)$ so that $\theta_1 = A_1 \exp[i(\omega t+\phi_1)]$. Since θ_1 must be a real angle we can take the real part of this complex expression to find $\theta_1 = A_1 \cos(\omega t+\phi_1)$. If instead we take the imaginary part we find $\theta_1 = A_1 \sin(\omega t+\phi_1)$. Any of these forms can be used to represent real solutions of Eq. (3.4.54) where the constants (A_1, ϕ_1) can be found from the initial conditions. Thus, the complex function form is an alternate (and convenient) way to express these real oscillatory solutions. Placing $\theta_1 = a_1 \exp(i\omega t)$ into Eq. (3.4.54) gives

$$\left(\frac{g}{l} - \omega^2\right)a_1\exp(i\omega t) = 0$$

which indeed is satisfied by a frequency $\omega = \omega_0 = \sqrt{g/l}$ for the pendulum. The equations for the double pendulum, Eq. (3.4.53), are more complicated but we can again seek complex solutions given by $\theta_k = a_k \exp(i\omega t)$ $(k = 1, 2)$. The two equations in this case can be put in matrix form (dividing out the common $\exp(i\omega t)$ term) as

$$\begin{bmatrix} 2g/l - \omega^2 & -g/l \\ -2g/l & 2g/l - \omega^2 \end{bmatrix} \begin{Bmatrix} a_1 \\ a_2 \end{Bmatrix} = \begin{Bmatrix} 0 \\ 0 \end{Bmatrix} \tag{3.4.55a}$$

or, equivalently,

$$\begin{bmatrix} 2\omega_0^2 - \omega^2 & -\omega_0^2 \\ -2\omega_0^2 & 2\omega_0^2 - \omega^2 \end{bmatrix} \begin{Bmatrix} a_1 \\ a_2 \end{Bmatrix} = \begin{Bmatrix} 0 \\ 0 \end{Bmatrix} \tag{3.4.55b}$$

This is a homogeneous system of linear equations which only has the trivial solution of $a_1 = a_2 = 0$ unless the determinant of the matrix of coefficients vanishes, giving (in terms of ω_0)

$$\omega^4 - 4\omega_0^2\omega^2 + 2\omega_0^4 = 0 \tag{3.4.56}$$

which is just a quadratic equation for ω^2. Solving this quadratic equation for its two roots we find

$$\omega^2 = 2\omega_0^2 \pm \sqrt{2}\,\omega_0^2 \tag{3.4.57}$$

so taking the positive square root of this result gives, finally, two natural frequencies at which the system can oscillate:

$$\begin{aligned}\omega_1 &= \sqrt{2-\sqrt{2}}\ \omega_0 \cong 0.765\omega_0 \\ \omega_2 &= \sqrt{2+\sqrt{2}}\ \omega_0 \cong 1.848\omega_0\end{aligned} \tag{3.4.58}$$

Because the determinant was set equal to zero the two equations in Eq. (3.4.55b) are not independent and we cannot solve for the amplitude constants (a_1, a_2) but we can use one of those equations to find a relationship between those constants. For the lowest natural frequency, ω_1, for example, we obtain $a_2 = \sqrt{2}\,a_1$ so the two masses oscillate in phase as shown in Fig. 3.20a. At the higher natural frequency, ω_2, we have instead $a_2 = -\sqrt{2}\,a_1$, so the masses oscillate completely out of phase as shown in Fig. 3.20b. These amplitudes define the two *normal modes of oscillation* of the double pendulum at these two natural frequencies. We can write these amplitudes as components of a column vector. For ω_1, for example, if we normalize the amplitudes to produce a unit vector, $\mathbf{e}^{(1)}$, we have

$$\mathbf{e}^{(1)} = \frac{1}{\sqrt{a_1^2 + a_2^2}}\begin{Bmatrix} a_1 \\ a_2 \end{Bmatrix} = \begin{Bmatrix} \sqrt{1/3} \\ \sqrt{2/3} \end{Bmatrix} \tag{3.4.59a}$$

whereas for ω_2 we find a corresponding unit vector, $\mathbf{e}^{(2)}$, given by

$$\mathbf{e}^{(2)} = \frac{1}{\sqrt{a_1^2 + a_2^2}}\begin{Bmatrix} a_1 \\ a_2 \end{Bmatrix} = \begin{Bmatrix} \sqrt{1/3} \\ -\sqrt{2/3} \end{Bmatrix} \tag{3.4.59b}$$

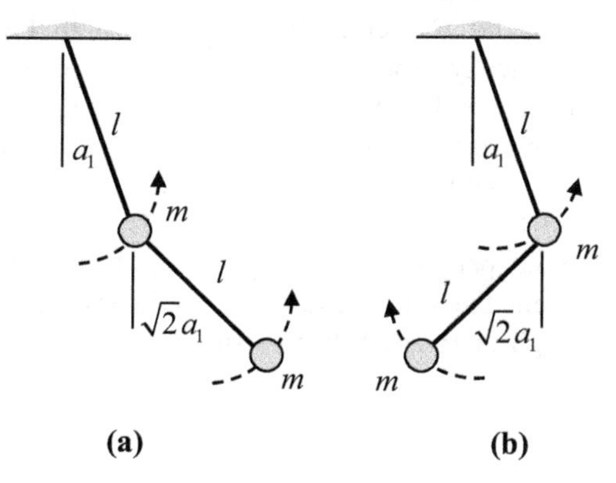

Fig. 3.20 (**a**) The motion of the double pendulum oscillating in the first mode at the natural frequency ω_1, and (**b**) the motion in the second mode at the natural frequency, ω_2

Since our small-amplitude equations, Eq. (3.4.53), are linear and homogeneous we can take as a general solution of those equations the sum of these two normal mode solutions with unknown (complex) constant amplitudes, i.e.,

$$\boldsymbol{\theta} = \begin{Bmatrix} \theta_1 \\ \theta_2 \end{Bmatrix} = A_1 \exp(i\omega_1 t + i\delta_1)\mathbf{e}^{(1)} + A_2 \exp(i\omega_2 t + i\delta_2)\mathbf{e}^{(2)} \tag{3.4.60}$$

or, if we take the real part,

$$\boldsymbol{\theta} = A_1 \cos(\omega_1 t + \delta_1)\mathbf{e}^{(1)} + A_2 \cos(\omega_2 t + \delta_2)\mathbf{e}^{(2)} \tag{3.4.61}$$

where the real constants $(A_1, \delta_1, A_2, \delta_2)$ are obtained from the given initial conditions of the motion $[\theta_1(0), \dot{\theta}_1(0), \theta_2(0), \dot{\theta}_2(0)]$. Specifically, we have from Eq. (3.4.61)

$$\begin{aligned} \theta_1(0) &= \frac{A_1}{\sqrt{3}} \cos\delta_1 + \frac{A_2}{\sqrt{3}} \cos\delta_2 \\ \theta_2(0) &= \frac{\sqrt{2}A_1}{\sqrt{3}} \cos\delta_1 - \frac{\sqrt{2}A_2}{\sqrt{3}} \cos\delta_2 \end{aligned} \tag{3.4.62a}$$

and by differentiating Eq. (3.4.61) we find

$$\begin{aligned} \dot{\theta}_1(0) &= -\frac{\omega_1 A_1}{\sqrt{3}} \sin\delta_1 - \frac{\omega_2 A_2}{\sqrt{3}} \sin\delta_2 \\ \dot{\theta}_2(0) &= -\frac{\sqrt{2}\omega_1 A_1}{\sqrt{3}} \sin\delta_1 + \frac{\sqrt{2}\omega_2 A_2}{\sqrt{3}} \sin\delta_2 \end{aligned} \tag{3.4.62b}$$

Simulation 3.3: Linear Oscillations of the Double Pendulum

A MATLAB® script linear_dp has been written that implements the linear oscillatory solutions of Eq. (3.4.61) for the case $m_1 = m_2 = 0.1$ kg and $l_1 = l_2 = 0.2$ m, with $A_1 = 0.1$ rad, $A_2 = 0.07$ rad, and $\delta_1 = 0$,$\delta_2 = \pi/2$ rad. The solutions for (θ_1, θ_2) are calculated in the script over the time interval tspan = [0, 10] s and then plotted. The results for θ_2 versus time are shown in Fig. 3.21a. The same result calculated numerically for the full nonlinear equations from the MATLAB® script double_pendulum_sim is shown in Fig. 3.21b. This result was obtained by first running the script linear_dp, saving the figure, and then immediately running the script double_pendulum_sim so that the input parameters used by linear_dp are also used by double_pendulum_sim. The vector of initial conditions, x0, needed by double_pendulum_sim is generated by linear_dp. For the small-amplitude motions present the linear equations obviously agree well with the full equations. Even though there are only two simple sinusoidal modal solutions that contribute to the

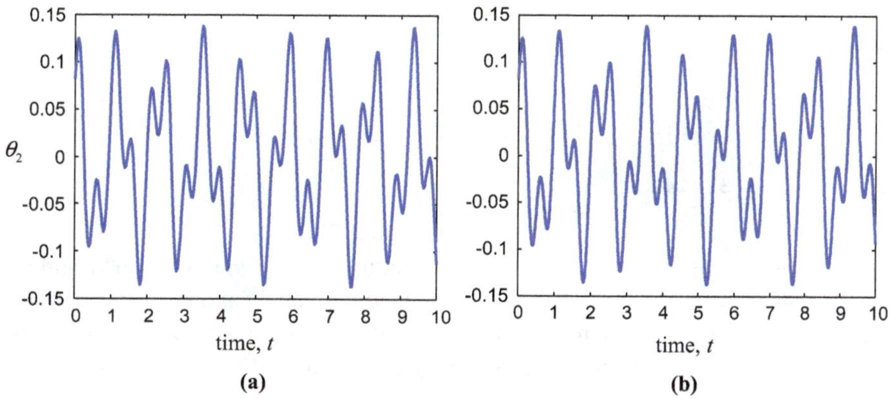

Fig. 3.21 (**a**) The angle θ_2 versus time for the double pendulum from the linearized solution in terms of the normal modes, and (**b**) the numerical solution of the double pendulum as found from the MATLAB® script double_pendulum_sim

complete solution, the complete solution behavior is certainly not simple, even at these small amplitudes.

When we discuss vibration problems in more detail in Chap. 7, we will see that the determination of the natural frequencies and normal modes is the solution to what is called an eigenvalue problem. In the current example, if we let $\lambda = \omega^2/\omega_0^2$ we can rewrite Eq. (3.4.55b) as

$$\begin{bmatrix} 2-\lambda & -1 \\ -2 & 2-\lambda \end{bmatrix} \begin{Bmatrix} a_1 \\ a_2 \end{Bmatrix} = \begin{Bmatrix} 0 \\ 0 \end{Bmatrix} \tag{3.4.63}$$

or as the matrix equation

$$\begin{bmatrix} 2 & -1 \\ -2 & 2 \end{bmatrix} \begin{Bmatrix} a_1 \\ a_2 \end{Bmatrix} = \lambda \begin{Bmatrix} a_1 \\ a_2 \end{Bmatrix} \tag{3.4.64}$$

which is in the standard form of an eigenvalue problem where the eigenvalues are the solutions for λ (i.e., the squares of the normalized natural frequencies) and the eigenvectors are corresponding modal vectors of Eqs. (3.4.59a) and (3.4.59b). There is a MATLAB® function eig that can directly solve such eigenvalue problems. One merely needs to define the matrix of coefficients in Eq. (3.4.64) and give that matrix as an argument to the eig function to get both the eigenvalues and the eigenvectors. For the example considered in the linear_dp script, the natural frequencies are approximately $\omega_1 = 5.36$ rad/s and $\omega_2 = 12.94$ rad/s and $\omega_0 = 7.00$ rad/s. If in the MATLAB® command window we place the 2 × 2 matrix of coefficients in Eq. (3.4.64) into a matrix A and call the eig function we find

```
A = [ 2 -1; -2 2];
[eigvectors, eigvals] = eig(A)

eigvectors =   0.5774   0.5774
              -0.8165   0.8165


eigvals =     3.4142      0
                0       0.5858
```

The first column in the matrix eigvectors is just the normal mode vector $\mathbf{e}^{(2)}$ and the second column is the modal vector $\mathbf{e}^{(1)}$ as defined previously and the corresponding eigenvalues in the eigvals matrix are the values of $\lambda_2 = \omega_2^2/\omega_0^2 = 3.4142$ and $\lambda_1 = \omega_1^2/\omega_0^2 = 0.5858$, which you can verify are the correct squares of the normalized natural frequencies.

We also want to use the double-pendulum problem to show a simpler way to set up and solve this problem, using an approach that is very useful for more complex multiple degree of freedom problems. Our solution with ode45 used a system of equations where the mass matrix was first eliminated so that we could put the first-order system of equations in the standard form of $\dot{\mathbf{x}} = \mathbf{f}(x, \dot{x}, t)$. This entailed doing some algebra where there is always a chance of introducing errors. However, ode45 can also solve a first-order system of ordinary differential equations in the form $\mathbf{M}(\mathbf{x}, t)\dot{\mathbf{x}} = \mathbf{f}(\mathbf{x}, \dot{\mathbf{x}}, t)$ where $\mathbf{M}$ is a "mass" matrix. [Note: This is not the same "mass" matrix found in the equations of motion written as second-order differential equations.] Our original second-order system of equations, Eqs. (3.4.34) and (3.4.35), can be easily placed in this form by simply collecting terms appropriately, so that no additional algebra is needed. From those two equations, and recalling that $[x_1, x_2, x_3, x_4] = [\theta_1, \dot{\theta}_1, \theta_2, \dot{\theta}_2]$, the mass matrix of the first-order system is

$$\mathbf{M} = \begin{bmatrix} 1 & 0 & 0 & 0 \\ 0 & (m_1 + m_2)l_1 & 0 & m_2 l_2 \cos(x_1 - x_3) \\ 0 & 0 & 1 & 0 \\ 0 & m_1 l_1 \cos(x_1 - x_3) & 0 & m_2 l_2 \end{bmatrix} \tag{3.4.65}$$

and the $\mathbf{f}$ vector is

$$\mathbf{f} = \begin{bmatrix} x_2 \\ -m_2 l_2 x_4^2 \sin(x_1 - x_3) - (m_1 + m_2) g \sin x_1 \\ x_4 \\ m_2 l_1 x_2^2 \sin(x_1 - x_3) - m_2 g \sin x_3 \end{bmatrix} \tag{3.4.66}$$

where a common factor of l_1 in Eq. (3.4.34) and a common factor of l_2 in Eq. (3.4.35) have been eliminated before generating these $\mathbf{M}$ and $\mathbf{f}$ expressions.

Simulation 3.4: The Double Pendulum with a Mass Matrix
The function double_pend and script double_pendulum_sim previously used have been modified slightly to use Eqs. (3.4.65) and (3.4.66). First, the double_pend function has been renamed double_pend_m. The inputs and outputs of these functions are the same so that a new script double_pendulum_m_sim simply calls double_pend_m rather than double_pend. The function double_pend_m is changed in three ways. First, in order for ode45 to recognize that a mass matrix is present, it must be included in the options given to ode45. This can be done with the odeset function before calling ode45:

```
options = odeset('RelTol', rtol, 'AbsTol', atol, 'Mass', @MM);
```

where MM is the name of a function which must return in this case the 4×4 matrix of Eq. (3.4.65). Thus, the second change we must make is to include a nested function MM. This function has input arguments (t, x) and in this case is given explicitly as

```
function  M =MM(t, x)
r1 = [1 0 0 0];
r2=[ 0 (m1+m2)*l1 0 m2*l2*cos(x(1)-x(3))];
r3 = [ 0 0 1 0];
r4 = [0 m1*l1*cos(x(1)-x(3)) 0 m2*l2];
M=[r1;r2;r3;r4];
end
```

The other arguments appearing in this nested function such as m1 do not have to be passed explicitly since they are shared with the function double_pend_m, where they are defined. Third, and finally, we must include another nested function within double_pend_m, called dp_m_eqs, which returns the **f** vector of Eq. (3.4.66) and is

```
function [f] = dp_m_eqs(t, x)

f = zeros(4,1);
x1 =x(1); %theta1
x2 = x(2);%theta1dot
x3 = x(3); %theta2
x4 =x(4); %theta2dot

f(1)= x2;
f(2) = -(m2*l2*x4^2)*sin(x1-x3)-(m1+m2)*g*sin(x1);
f(3) = x4;
f(4) =( m2*l1*x2^2)*sin(x1-x3) -m2*g*sin(x3);
end
```

where again all the other parameters besides the (t, x) input parameters are shared with the function double_pend_m.

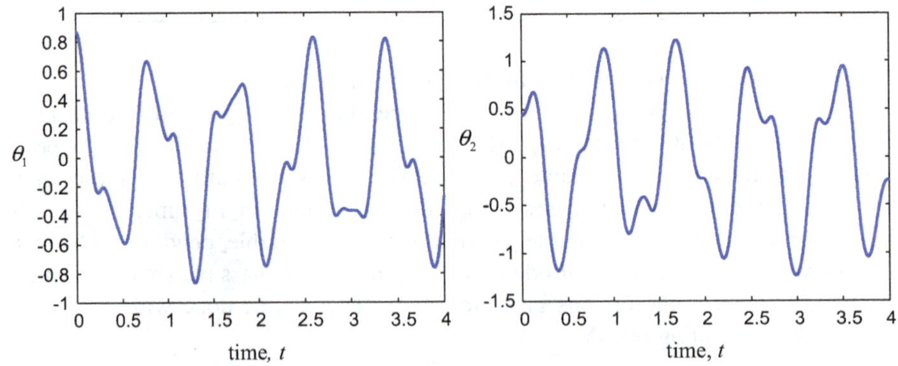

Fig. 3.22 Use of a mass matrix for solving the double-pendulum problem. The angles θ_1 and θ_2 versus time for the double pendulum, for l1 = l2 = 0.1 m and m1 = m2 = 0.1 kg, with starting angles $\theta_1 = 50^\circ$, $\theta_2 = 25^\circ$, and $\dot{\theta}_1 = \dot{\theta}_2 = 0$. These are the same parameters used to generate Fig. 3.16

Running the double_pendulum_m_sim script with the same default parameters as used in double_pendulum_sim script yields identical results. For example, Fig. 3.22 shows the angles θ_1 and θ_2 versus time, which can be compared to the same angles found previously when the mass matrix was absent, as shown in Fig. 3.16.

Before leaving the double-pendulum problem we want to introduce a MATLAB® function, animate_lines, that produces an animation of a set of lines, defined by the motion of their endpoints. One calls this function as

```
animate_lines (x_array, y_array, sldwn, marker)
```

The x-coordinates of the endpoints of the lines to be animated are held in the N columns of the matrix, x_array(:, n) (n = 1, N), where N is the total number of endpoints to be animated. Note that we must include any fixed points, such as the base point of the double pendulum, in this array. Similarly, y_array(:, n) holds the y-coordinates of the endpoints. The input argument sldwn is a "slow down" factor as described previously. There is not a speedup factor included in this animation function, but, as mentioned before, this can be accomplished by specifying fewer points in tspan. The input argument marker is a string ("on" or "off") that determines if the path of the endpoint of the last line is to be plotted or not plotted as a series of dots, as done previously in animation plots. The animate_lines function is a simple function for doing 2-D animation of any number of lines, so it can be used in many 2-D simulations without having to write animation coding for each new problem, as we have been doing up to this point.

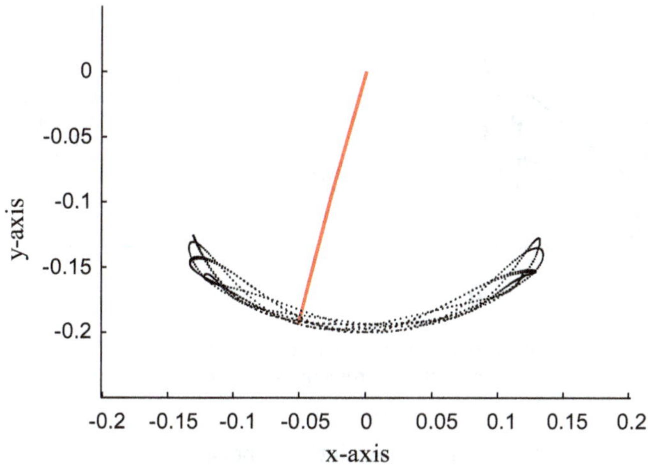

Fig. 3.23 Final frame of the animation of the double-pendulum motion in Simulation 3.5

Simulation 3.5: Animation Function for the Double Pendulum
To illustrate the use of animate_lines, we have written a MATLAB® script, double_pendulum_m_anim, that is very similar to double_pendulum_m_sim. It includes default values for sldwn (sdwn = 1.0) and marker (marker = "off") and, unlike double_pendulum_m_sim, the time interval, span, is specified as tspan = linspace(0, 4, 1000) to ensure that the animation is not slowed down by an excessive number of plotting points. The script performs the numerical solution as done previously and then sets up the x_array and y_array matrices and calls the animate_lines function. Clearing the workspace, setting marker = "on," and running this script, the final frame of the animation is shown in Fig. 3.23. The dotted path of the endpoint of the double pendulum is the same as shown in Fig. 3.17 but plotted on a different scale.

The double pendulum is a very simple system but as we have seen it can be used as the vehicle to describe a number of issues including chaotic motion, effects of linearization, vibrations of a coupled two degree of freedom system, and use of a mass matrix in the numerical calculations.

3.5 A Non-holonomic Constrained System of Particles

Almost all of the constraints considered so far have been holonomic constraints, as these are typically the types of constraints found in single particle and systems of particle problems. Problems involving the rolling of rigid bodies are one class of

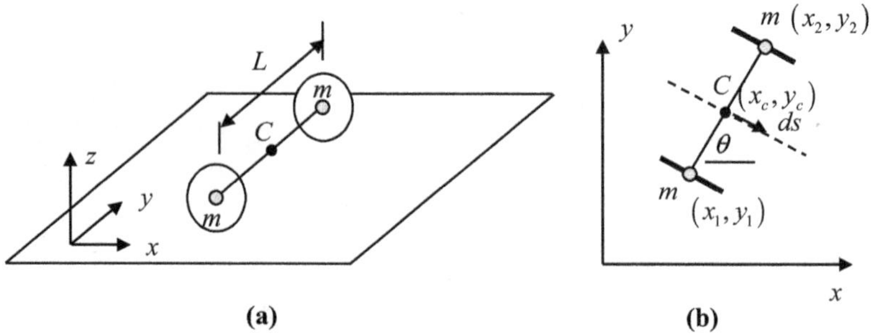

Fig. 3.24 (**a**) A pair of wheels connected by a rigid axle of length L and rolling without slipping on the x–y plane, and (**b**) the geometry of this problem as seen in the x–y plane

problems where non-holonomic constraints do appear. We can examine that type of non-holonomic constraint by considering the problem shown in Fig. 3.24a, where two wheels are connected by a rigid axle of length L and the wheels roll without slipping in the x–y plane. We will assume that the mass of the system is concentrated at the ends of the axle in two particles of mass m each so that we now have a rigidly constrained system of particles problem. This is a two-dimensional problem since all the motion occurs parallel to the x–y plane. Thus, there are two degrees of freedom for each particle but the constraint of the rigid axle connecting the particles means that there are only three degrees of freedom for the entire system. We could take the two coordinates that describe the motion of the center of mass, at point C, and the angular orientation of the axle, θ, as shown in Fig. 3.24b, as the three independent generalized coordinates for this problem. The fact that the wheels roll without slipping, however, means that there is another constraint on the motion, namely each particle must move only in a direction perpendicular to the axle. The center of mass also must have this same motion. Figure 3.24b shows a small allowable displacement, ds, of the center of mass (ds could also be shown in the opposite sense, as we will see the final result is unaffected by our choice of ds). In terms of the displacements of the center of mass we must then have

$$\begin{aligned} dx_c &= \sin\theta\, ds \\ dy_c &= -\cos\theta\, ds \end{aligned} \tag{3.5.1}$$

But we can eliminate the ds and divide by the time differential dt to write Eq. (3.5.1) instead as a single-velocity constraint in the form

$$\cos\theta\dot{x}_c + \sin\theta\dot{y}_c = 0 \tag{3.5.2}$$

which is a non-holonomic constraint. To justify the assertion that this constraint is non-holonomic note that if it was holonomic we could integrate the constraint and arrive at a positional constraint of the form $f(x_c, y_c, \theta) = 0$ but this is clearly not possible since the system could start out at some values $(x_{c0}, y_{c0}, \theta_0)$ and roll around the plane, coming back to the original point (x_{c0}, y_{c0}) but at any orientation, θ_1, so we cannot hope to find an equivalent holonomic constraint that relates all three of these variables. Since Eq. (3.5.2) is a non-holonomic constraint, there are still only three degrees of freedom which we will take as (x_c, y_c, θ). To keep the problem simple so that we can analyze it easily, we will assume that there are no external forces acting on the system but that at time $t = 0$ we have initial conditions $x_{c0} = y_{c0} = 0$, $\dot{x}_{c0} = 0, \dot{y}_{c0} = v_0$, and $\theta_0 = 0, \dot{\theta}_0 = \omega$, i.e., the center of mass starts at the origin moving with a velocity, v_0, in the y-direction and with the axle oriented along the x-axis and spinning with an angular velocity, ω. To obtain the equations of motion, let's use Lagrange's equations since it is easy to obtain the kinetic energy, T, which is the sum of the translational and rotational parts:

$$\begin{aligned} T &= T_{\text{trans}} + T_{\text{rot}} \\ &= \frac{1}{2}(2m)\left(\dot{x}_c^2 + \dot{y}_c^2\right) + \frac{1}{2}I_c\dot{\theta}^2 \end{aligned} \tag{3.5.3}$$

and $I_c = m(L/2)^2 + m(L/2)^2 = mL^2/2$ so that

$$T = m\left(\dot{x}_c^2 + \dot{y}_c^2\right) + \frac{1}{4}mL^2\dot{\theta}^2 \tag{3.5.4}$$

There is no potential energy since the masses move only parallel to the x–y plane. The non-holonomic constraint, Eq. (3.5.2), can be differentiated once to obtain

$$\cos\theta\ddot{x}_c + \sin\theta\ddot{y}_c = \dot{\theta}\left(\sin\theta\dot{x}_c - \cos\theta\dot{y}_c\right) \tag{3.5.5}$$

which is of the standard form we have used before, $\mathbf{A}\ddot{\mathbf{q}} = \mathbf{b}$, where here $\mathbf{q} = (x_c, y_c, \theta)$ and

$$\begin{aligned} \mathbf{A} &= [\cos\theta, \sin\theta, 0] \\ \mathbf{b} &= \dot{\theta}\left(\sin\theta\dot{x}_c - \cos\theta\dot{y}_c\right) \end{aligned} \tag{3.5.6}$$

and where $\mathbf{b}$ in this case is just a scalar. This problem is a specific example of where we have a combination of Lagrange's equations for a set of independent generalized coordinates, q_k, and one or more non-holonomic constraints. If these non-holonomic constraints generate ideal constraint forces of the form

$$Q_k^c = \sum_j A_{jk}\lambda_j \tag{3.5.7}$$

where λ_j are Lagrange multipliers, then we can use an augmented approach and write the solution as a combination of Lagrange's equations and the non-holonomic constraint conditions as

$$\begin{aligned} &\frac{d}{dt}\left(\frac{\partial T}{\partial \dot{q}_k}\right)-\frac{\partial T}{\partial q_k}=Q_k^e+Q_k^c=Q_k^e+\sum_i A_{ik}\lambda_i \\ &\sum_j A_{ij}\ddot{q}_j=b_i \end{aligned} \tag{3.5.8}$$

In our problem the external forces do not have any components in the x–y plane and the external forces do not generate any moments about the z-axis so that the external generalized forces are all zero and there is only one constraint so that we have

$$\begin{aligned} &\frac{d}{dt}\left(\frac{\partial T}{\partial \dot{x}_c}\right)-\frac{\partial T}{\partial x_c}=2m\ddot{x}_c= \cos\theta\lambda \\ &\frac{d}{dt}\left(\frac{\partial T}{\partial \dot{y}_c}\right)-\frac{\partial T}{\partial y_c}=2m\ddot{y}_c= \sin\theta\lambda \\ &\frac{d}{dt}\left(\frac{\partial T}{\partial \dot{\theta}}\right)-\frac{\partial T}{\partial \theta}=\frac{1}{2}mL^2\ddot{\theta}=0 \end{aligned} \tag{3.5.9}$$

and Eq. (3.5.5), which we write again here as

$$\cos\theta\ddot{x}_c+\sin\theta\ddot{y}_c=\dot{\theta}\left(\sin\theta\dot{x}_c-\cos\theta\dot{y}_c\right) \tag{3.5.10}$$

We can use the general solution we wrote earlier for augmented problems of the type given by Eq. (3.5.9) and (3.5.10) (see Eq. (2.6.44), for example) to find the equations of motion and the constraint force (or, equivalently, the Lagrange multiplier, λ) but here we can easily solve for λ directly by just multiplying the constraint equation, Eq. (3.5.10), by $2m$ and placing the first two equations in Eq. (3.5.9) into that constraint equation to obtain the Lagrange multiplier as

$$\lambda=2m\dot{\theta}\left(\sin\theta\dot{x}_c-\cos\theta\dot{y}_c\right) \tag{3.5.11}$$

One can substitute this result back into Eq. (3.5.9) to obtain the second-order equations of motion as

$$\begin{aligned} &\ddot{x}_c=\cos\theta\dot{\theta}\left(\sin\theta\dot{x}_c-\cos\theta\dot{y}_c\right) \\ &\ddot{y}_c=\sin\theta\dot{\theta}\left(\sin\theta\dot{x}_c-\cos\theta\dot{y}_c\right) \\ &\ddot{\theta}=0 \end{aligned} \tag{3.5.12}$$

which we then could rewrite as a system of first-order equations and solve them numerically. However, we can also solve this problem analytically by noting that the Lagrange multiplier in this case is actually a constant. This is not immediately obvious but if we compute the time derivative of this multiplier we have

$$\begin{aligned}\frac{d\lambda}{dt} &= 2m\ddot{\theta}\left(\sin\theta\dot{x}_c - \cos\theta\dot{y}_c\right) + 2m\dot{\theta}\left(\sin\theta\ddot{x}_c - \cos\theta\ddot{y}_c\right)\\ &\quad +2m\dot{\theta}\left(\cos\theta\dot{x}_c + \sin\theta\dot{y}_c\right)\\ &= 0\end{aligned}$$

which follows by using the equations of motion, Eq. (3.5.12), and the constraint equation, Eq. (3.5.2). Thus, from Eq. (3.5.11) and the initial conditions we have $\lambda = -2m\omega v_0$ and we can integrate the third of the equations in Eq. (3.5.12), using the initial conditions, to obtain $\theta = \omega t$, which gives from Eq. (3.5.9)

$$\begin{aligned}\ddot{x}_c &= -\omega v_0 \cos(\omega t)\\ \ddot{y}_c &= -\omega v_0 \sin(\omega t)\end{aligned} \tag{3.5.13}$$

Using the initial conditions, Eq. (3.5.13) can easily be integrated twice to find

$$\begin{aligned}\dot{x}_c &= -v_0 \sin(\omega t)\\ \dot{y}_c &= v_0 \cos(\omega t)\end{aligned} \tag{3.5.14}$$

and

$$\begin{aligned}x_c &= \frac{v_0}{\omega}\cos(\omega t) - \frac{v_0}{\omega}\\ y_c &= \frac{v_0}{\omega}\sin(\omega t)\end{aligned} \tag{3.5.15}$$

It follows from Eq. (3.5.15) that

$$\left(x_c + \frac{v_0}{\omega}\right)^2 + y_c^2 = \frac{v_0^2}{\omega^2} \tag{3.5.16}$$

so that the center of mass follows a circular path of radius v_0/ω where the center of the circle is at $(-v_0/\omega, 0)$ (Fig. 3.25a). One mass follows a circular path of radius $r_1 = v_0/\omega - L/2$ and the other mass follows a circle of radius $r_2 = v_0/\omega + L/2$ so these masses will experience inward centripetal accelerations of $r_1\omega^2$ and $r_2\omega^2$, respectively. Thus, there must be constraint forces $F_1 = -mr_1\omega^2$ and $F_2 = -mr_2\omega^2$ that act on the particles (actually, the forces act on the contact points between the wheels and the x–y plane) to keep them on their circular paths (Fig. 3.25b). The total constraint force is just $\mathbf{Q}^c = \mathbf{A}^T\lambda = -2mv_0\omega[\cos\theta, \sin\theta, 0]^T$ which we can also write in terms of the forces (F_1, F_2) as $\mathbf{Q}^c = F_1[\cos\theta, \sin\theta, 0]^T + F_2[\cos\theta, \sin\theta, 0]^T$, so we see that the total force $F_1 + F_2 = -mr_1\omega^2 - mr_2\omega^2 = \lambda$. Another way to obtain these results is to use the definition of the generalized constraint force (Eq. (3.4.6)) in terms of the applied constraint forces

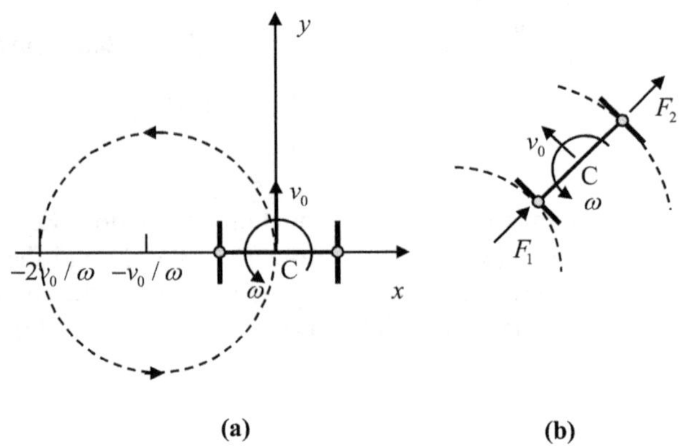

Fig. 3.25 (**a**) The circular path followed by a system of two masses on wheels that roll without slipping and (**b**) the sidewise constraint forces acting on the particles

$$Q_k^c = \sum_{i=1}^{N} \mathbf{F}_i^c \cdot \frac{\partial \mathbf{x}_i}{\partial q_k} \tag{3.5.17}$$

where (see Fig. 3.25b)

$$\begin{aligned}
\mathbf{F}_1^c &= F_1[\cos\theta,\ \sin\theta]^T \\
\mathbf{F}_2^c &= F_2[\cos\theta,\ \sin\theta]^T \\
\mathbf{x}_1(x_c, y_c, \theta) &= [(x_c - L\cos\theta/2), (y_c - L\sin\theta/2)]^T \\
\mathbf{x}_2(x_c, y_c, \theta) &= [(x_c + L\cos\theta/2), (y_c + L\sin\theta/2)]^T
\end{aligned} \tag{3.5.18}$$

so that with $(q_1, q_2, q_3) = (x_c, y_c, \theta)$ we have

$$\begin{aligned}
Q_1^c &= \sum_{i=1}^{2} \mathbf{F}_i^c \cdot \frac{\partial \mathbf{x}_i}{\partial q_1} = (F_1 + F_2)\cos\theta \\
Q_2^c &= \sum_{i=1}^{2} \mathbf{F}_i^c \cdot \frac{\partial \mathbf{x}_i}{\partial q_2} = (F_1 + F_2)\sin\theta \\
Q_3^c &= \sum_{i=1}^{2} \mathbf{F}_i^c \cdot \frac{\partial \mathbf{x}_i}{\partial q_3} = F_1[\cos\theta,\ \sin\theta][L\sin\theta/2, -L\cos\theta/2]^T \\
&\quad + F_2[\cos\theta,\ \sin\theta][-L\sin\theta/2, +L\cos\theta/2]^T \\
&= 0
\end{aligned} \tag{3.5.19}$$

and since $\mathbf{Q}^c = \mathbf{A}^T\lambda = -2mv_0\omega[\cos\theta, \sin\theta, 0]^T$ we again arrive at $F_1 + F_2 = \lambda$.

Note that this is a conservative problem where the potential energy is zero so that the total kinetic energy is conserved. But the rotational part of the total kinetic energy is also a constant so that the translational kinetic energy of the center of mass itself is conserved and we have

$$\dot{x}_c^2 + \dot{y}_c^2 = v_0^2 \tag{3.5.20}$$

which can also be seen to be true directly from Eq. (3.5.14).

3.6 Dependent Constraints

The rolling axle problem of the previous section can also be used to give an example of where the constraints might not all be independent. For example, suppose we take the generalized coordinates to be the individual coordinates of the two particles, $\mathbf{X} = (x_1, y_1, x_2, y_2)$. The distance between the mass must be fixed, but also both masses, like the center of mass, must move perpendicular to the axle so that we have the three constraints

$$\begin{aligned} (x_2 - x_1)^2 + (y_2 - y_1)^2 &= L^2 \\ (x_2 - x_1)\dot{x}_1/L + (y_2 - y_1)\dot{y}_1/L &= 0 \\ (x_2 - x_1)\dot{x}_2/L + (y_2 - y_1)\dot{y}_2/L &= 0 \end{aligned} \tag{3.6.1}$$

since $\cos\theta = (x_2 - x_1)/L$ and $\sin\theta = (y_2 - y_1)/L$. The first constraint is a holonomic constraint but the other two constraints are non-holonomic so that the system indeed has three degrees of freedom. If we differentiate these constraints to write them in terms of the accelerations we have

$$\begin{aligned} (x_2 - x_1)(\ddot{x}_2 - \ddot{x}_1) + (y_2 - y_1)(\ddot{y}_2 - \ddot{y}_1) &= -(\dot{x}_2 - \dot{x}_1)^2 - (\dot{y}_2 - \dot{y}_1)^2 \\ (x_2 - x_1)\ddot{x}_1 + (y_2 - y_1)\ddot{y}_1 &= -(\dot{x}_2 - \dot{x}_1)\dot{x}_1 - (\dot{y}_2 - \dot{y}_1)\dot{y}_1 \\ (x_2 - x_1)\ddot{x}_2 + (y_2 - y_1)\ddot{y}_2 &= -(\dot{x}_2 - \dot{x}_1)\dot{x}_2 - (\dot{y}_2 - \dot{y}_1)\dot{y}_2 \end{aligned} \tag{3.6.2}$$

Equation (3.6.2) can be placed in the vector-matrix form $\mathbf{A}\ddot{\mathbf{X}} = \mathbf{b}$, where

$$\mathbf{A} = \begin{bmatrix} -(x_2 - x_1) & -(y_2 - y_1) & (x_2 - x_1) & (y_2 - y_1) \\ (x_2 - x_1) & (y_2 - y_1) & 0 & 0 \\ 0 & 0 & (x_2 - x_1) & (y_2 - y_1) \end{bmatrix} \tag{3.6.3}$$

and

$$\mathbf{b} = \begin{Bmatrix} -(\dot{x}_2 - \dot{x}_1)^2 - (\dot{y}_2 - \dot{y}_1)^2 \\ -(\dot{x}_2 - \dot{x}_1)\dot{x}_1 - (\dot{y}_2 - \dot{y}_1)\dot{y}_1 \\ -(\dot{x}_2 - \dot{x}_1)\dot{x}_2 - (\dot{y}_2 - \dot{y}_1)\dot{y}_2 \end{Bmatrix} \tag{3.6.4}$$

It is clear that the constraints of Eq. (3.6.2) are not independent since the first equation is equal to the difference of the third equation and the second equation. Physically this makes sense since if the velocities of the masses are both always perpendicular to the axle, then the distance between the masses along the axle also cannot change, which is just our first constraint. Here it is easy to recognize the reason for this dependency, but in more complex systems it may not be this obvious. We could eliminate one of the redundant constraints or, as shown in the last section, we can avoid it entirely by simply changing to a new set of generalized coordinates. However it may be simpler to try to directly solve the problem with a dependent set of constraints. Can we do that? The answer is yes. To see what is needed, note that in the Cartesian coordinates we are using here the equations of motion and constraint equations for the system of particles can be written as

$$\begin{aligned} \mathbf{M}\ddot{\mathbf{X}} &= \mathbf{F}^e + \mathbf{F}^c \\ \mathbf{A}\ddot{\mathbf{X}} &= \mathbf{b} \end{aligned} \tag{3.6.5}$$

In Chap. 2 we showed that a solution to these equations was equivalent to Gauss's law of least constraint. That law can be stated as follows: find a vector, $\mathbf{y}$, which has a minimum norm, i.e., where $G = \mathbf{y}^T\mathbf{y} = |\mathbf{y}|^2$ is a minimum, subject to the constraint equations written as

$$\mathbf{B}\mathbf{y} = \mathbf{c} \tag{3.6.6}$$

where in terms of the original dynamics variables

$$\begin{aligned} \mathbf{y} &= \mathbf{M}^{-1/2}\mathbf{F}^c \\ \mathbf{B} &= \mathbf{A}\mathbf{M}^{-1/2} \\ \mathbf{c} &= \mathbf{b} - \mathbf{A}\mathbf{M}^{-1}\mathbf{F}^e \end{aligned} \tag{3.6.7}$$

If the constraints are all independent there is unique solution for $\mathbf{y}$ given by

$$\mathbf{y} = \mathbf{B}^T\left(\mathbf{B}\mathbf{B}^T\right)^{-1}\mathbf{c} \tag{3.6.8}$$

which, in terms of the original variables of Eq. (3.6.5), leads to the solutions

$$\begin{aligned} \ddot{\mathbf{X}} &= \mathbf{M}^{-1}\mathbf{F}^e + \mathbf{M}^{-1}\mathbf{A}^T\left[\mathbf{A}\mathbf{M}^{-1}\mathbf{A}^T\right]^{-1}\left(\mathbf{b} - \mathbf{A}\mathbf{M}^{-1}\mathbf{F}^e\right) \\ \mathbf{F}^c &= \mathbf{A}^T\left[\mathbf{A}\mathbf{M}^{-1}\mathbf{A}^T\right]^{-1}\left(\mathbf{b} - \mathbf{A}\mathbf{M}^{-1}\mathbf{F}^e\right) \end{aligned} \tag{3.6.9}$$

for the equations of motion and the ideal constraint force that we have already used. These solutions fail if the constraints are not independent because then the ordinary inverse of the $\mathbf{A}\mathbf{M}^{-1}\mathbf{A}^T$ matrix does not exist. However, even in the case of dependent constraints there is a unique solution for $\mathbf{y}$ which has a minimum norm and satisfies the constraint equations of Eq. (3.6.6) given by

$$\mathbf{y} = \mathbf{B}^T\left(\mathbf{B}\mathbf{B}^T\right)^{\dagger}\mathbf{c} \tag{3.6.10}$$

where $(\ )^{\dagger}$ denotes taking the Moore-Penrose pseudoinverse. This pseudoinverse acts as a generalization of the ordinary inverse. Just as in MATLAB® we can numerically compute the ordinary inverse, A^{-1}, of a matrix A as inv(A), the Moore-Penrose pseudoinverse of A, $A^{\dagger}$, can be calculated in MATLAB® as pinv (A). The book by Udwadia and Kalaba [2] gives an extensive treatment of this and other generalizations of the ordinary inverse that can be applied to the problems of constrained motion in dynamics. The equations of motion and constraint force, therefore, become

$$\begin{aligned}\ddot{\mathbf{X}} &= \mathbf{M}^{-1}\mathbf{F}^e + \mathbf{M}^{-1}\mathbf{A}^T\left[\mathbf{A}\mathbf{M}^{-1}\mathbf{A}^T\right]^{\dagger}\left(\mathbf{b} - \mathbf{A}\mathbf{M}^{-1}\mathbf{F}^e\right)\\ \mathbf{F}^c &= \mathbf{A}^T\left[\mathbf{A}\mathbf{M}^{-1}\mathbf{A}^T\right]^{\dagger}\left(\mathbf{b} - \mathbf{A}\mathbf{M}^{-1}\mathbf{F}^e\right)\end{aligned} \tag{3.6.11}$$

We can also solve problems in this same way when we have dependent constraints involving generalized coordinates, $\mathbf{q}$. In this case the equations of motion and constraints become

$$\begin{aligned}\mathbf{M}\ddot{\mathbf{q}} &= \bar{\mathbf{Q}}^e + \mathbf{Q}^c\\ \mathbf{A}\ddot{\mathbf{q}} &= \mathbf{b}\end{aligned} \tag{3.6.12}$$

so the equations of motion and generalized constraint force are

$$\begin{aligned}\ddot{\mathbf{q}} &= \mathbf{M}^{-1}\bar{\mathbf{Q}}^e + \mathbf{M}^{-1}\mathbf{A}^T\left[\mathbf{A}\mathbf{M}^{-1}\mathbf{A}^T\right]^{\dagger}\left(\mathbf{b} - \mathbf{A}\mathbf{M}^{-1}\bar{\mathbf{Q}}^e\right)\\ \mathbf{Q}^c &= \mathbf{A}^T\left[\mathbf{A}\mathbf{M}^{-1}\mathbf{A}^T\right]^{\dagger}\left(\mathbf{b} - \mathbf{A}\mathbf{M}^{-1}\bar{\mathbf{Q}}^e\right)\end{aligned} \tag{3.6.13}$$

where the main difference with the Cartesian case, Eq. (3.6.11), is that the generalized external force, $\bar{\mathbf{Q}}^e$, includes both the external applied forces and the inertial forces present in terms of the generalized coordinates and that the mass matrix, $\mathbf{M}$, in Eq. (3.6.13) is generally a function of the generalized coordinates as well as the masses present.

Simulation 3.6: The Rolling Axle

Let's apply these pseudoinverse solutions to the rolling axle problem considered in the last section, using the Cartesian coordinates $\mathbf{X} = (x_1, y_1, x_2, y_2)$ and the dependent constraints, Eq. (3.6.2). Since we are working directly with the Cartesian components we can write down Newton's second law and the constraint equations for the two particles as

$$\begin{aligned}\mathbf{M}\ddot{\mathbf{X}} &= \mathbf{F}^c = \mathbf{A}^T\boldsymbol{\lambda}\\ \mathbf{A}\ddot{\mathbf{X}} &= \mathbf{b}\end{aligned} \tag{3.6.14}$$

since there are no external applied forces with components in the x–y plane. In this case the mass matrix is just the diagonal matrix $\mathbf{M} = m\mathbf{I}$ so $\mathbf{M}^{-1} = (1/m)\mathbf{I}$ and the equations of motion are just

$$\ddot{\mathbf{X}} = \mathbf{A}^T\left[\mathbf{A}\mathbf{A}^T\right]^{\dagger}\mathbf{b} \tag{3.6.15}$$

where we have used the fact that $\left(\frac{1}{m}\mathbf{B}\right)^{\dagger} = m\mathbf{B}^{\dagger}$ to cancel out the mass terms (which we know are absent from the problem—see the exact solution obtained in the last section). If we let $\mathbf{x} = \left[\mathbf{X}^T, \dot{\mathbf{X}}^T\right]^T$ then we have the first-order system

$$\dot{\mathbf{x}} = \left\{\begin{matrix} \dot{\mathbf{X}} \\ \mathbf{A}^{\mathrm{T}}\left[\mathbf{A}\mathbf{A}^{\mathrm{T}}\right]^{\dagger}\mathbf{b} \end{matrix}\right\} \tag{3.6.16}$$

To solve these equations with initial conditions consistent with those used in the last section ($\mathbf{v}_C = v_0\mathbf{e}_y$, $\boldsymbol{\omega} = \omega\mathbf{e}_z$) we must take

$$\mathbf{x}(0) = [-L/2 \;\; 0 \;\; L/2 \;\; 0 \;\; 0 \;\; (v_0 - \omega L/2) \;\; 0 \;\; (v_0 + \omega L/2)]^T \tag{3.6.17}$$

since the particles are located at $(-L/2, 0)$ and $(L/2, 0)$ at $t = 0$ and if we decompose the motion at this time into a translational and rotational part (Fig. 3.26a) each mass (and in fact any point on the axle) will have the same vertical velocity, v_0, in the translational motion, while for the rotational motion the velocity of the center of mass will be zero while the mass on the left side will move down with a velocity $\omega L/2$ while the mass on the right side will move up with a velocity $\omega L/2$ (Fig. 3.26b) so that the total velocity components of the mass on the left side will be $(0, v_0 - \omega L/2)$ while the mass on the right side will have total velocity components $(0, v_0 + \omega L/2)$.

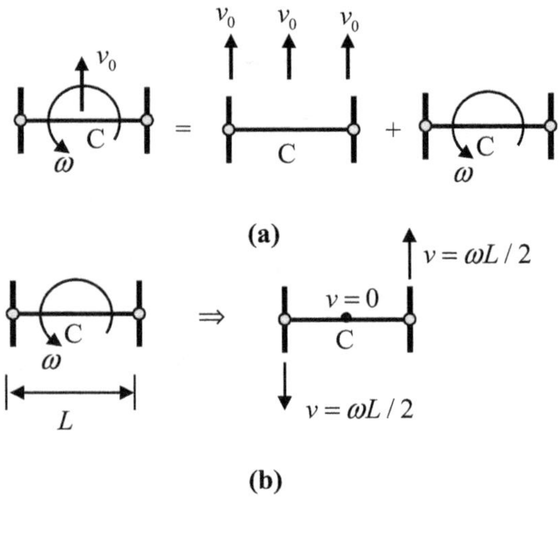

Fig. 3.26 (**a**) The initial motion of the axle decomposed into a translation motion and a rotation about the center of mass, and (**b**) velocities generated by the rotational motion

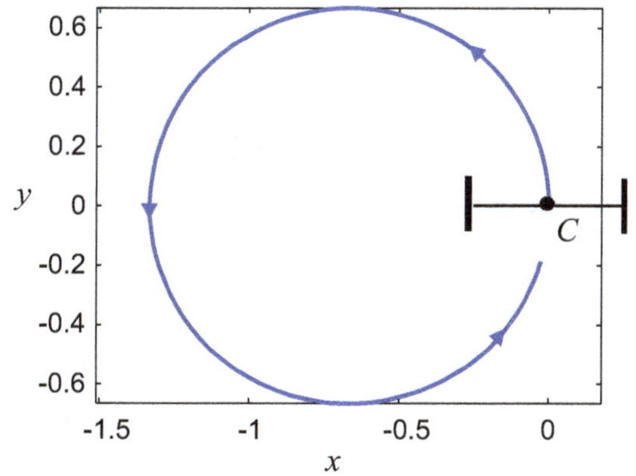

Fig. 3.27 The numerical solution for the motion of the center of mass of a rotating axle, using a set of dependent constraints. The axle is shown in its original position at $t = 0$

A MATLAB® script rolling_axle_sim sets up default parameters of $L = 0.5$ m, $v_0 = 2$ m/s, $\omega = 3$ rad/s, rtol = 1e-5, atol = 1e-6, and tspan = [0, 2] s. The initial conditions are those given in Eq. (3.6.17). The time span was chosen so that the axle makes almost one complete revolution. The script calls the function rolling_axle which contains a nested function axle2 that implements the first-order system of Eq. (3.6.16) and the ode45 solver. This function returns the solution and the script then uses that solution to plot the x–y motion of the center of mass. Figure 3.27 shows the results. The circular motion predicted by the analytical solution is obviously captured quite well by the numerical solution.

3.7 Problems

P3.1. A block of wood of mass M is suspended by two wires. A bullet of mass m is fired from a gun with a velocity, v, as shown in Fig. P3.1a. The bullet is embedded in the block, which rises to a height, h, as shown in Fig. P3.1b. Determine an expression for v in terms of m, M, and h. This arrangement is called a ballistic pendulum.

P3.2. A mass m_1 traveling downwards with a velocity v_1 strikes and sticks in a bucket of mass m_2 which is attached to weightless arm that is allowed to pivot around its center at point O as seen in Fig. P3.2. The other end of the arm is attached to a linear spring of spring constant k. The arm is initially horizontal when in equilibrium. Determine an expression for the maximum distance, d, moved down by the bucket.

P3.3. Two masses, m_1 and m_2, are attached to a weightless rigid rod which is free to rotate about its center, point O, as shown in Fig. P3.3. Each mass is also connected to a liner spring with spring constant, k. Assume $m_1 = m_2 = m$ and that the springs are both unstretched when the arm is horizontal (i.e., the unstretched length is d). (a) Using the angle, θ, that the arm makes with respect to the positive x-axis as the generalized coordinate for this one degree

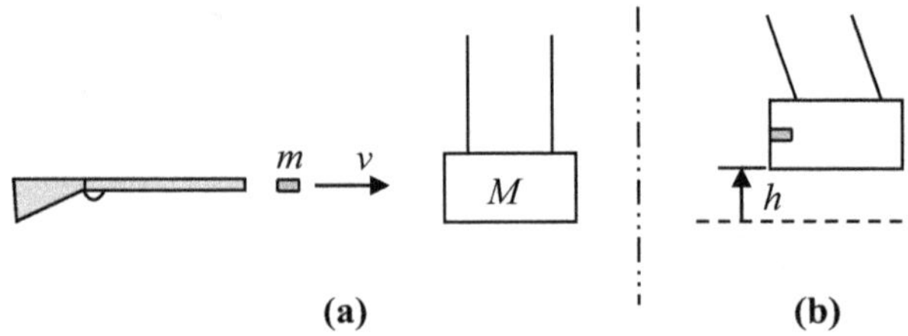

Fig. P3.1 A ballistic pendulum where (**a**) a bullet of mass m is fired at a velocity v into a block of wood of mass M, and (**b**) where the combined bullet/block rises to a maximum height, h

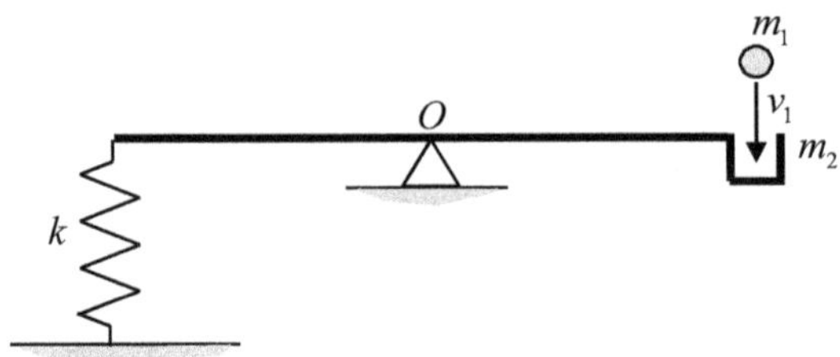

Fig. P3.2 A mass, m_1, strikes and sticks in a bucket of mass, m_2, which then rotates around the center of a weightless arm whose other end is attached to a linear spring with spring constant, k

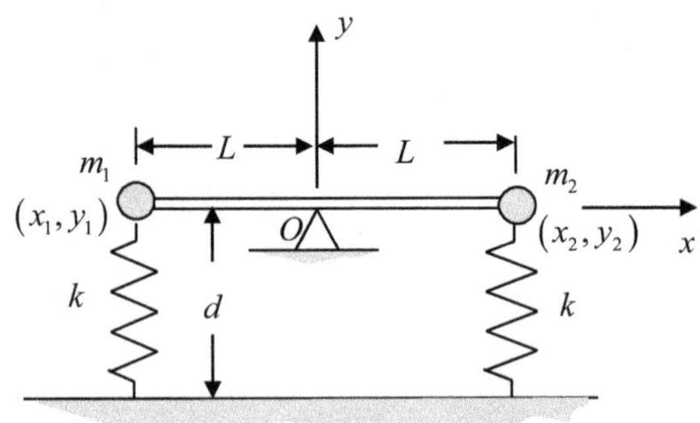

Fig. P3.3 Two masses that are attached to an arm of negligible weight and to two linear springs with spring constant k

of freedom system, determine the equation of motion. Do not assume that angle θ is small so that the masses move in both the x- and y-directions. Solve this equation of motion numerically for 20 s taking $m = 2$ kg, $k = 50$ N/m, $L = 0.5$ m, and $d = 1$ m with initial conditions $\theta(0) = 45^\circ$ and $\dot{\theta}(0) = 0$. (b) What is the approximate linear equation of motion for the system when the angle θ is small? What is the natural frequency of small oscillations for the system? (c) Using the Cartesian coordinates (x_1, y_1) for mass m_1 and (x_2, y_2) for m_2 as shown in Fig. P3.3, what are the three constraints on the motion of this system? Using an augmented approach, determine the equations of motion and the constraint forces acting on the masses and solve for their motions using the same parameters and initial conditions of part (a). How are the constraint forces acting on the masses related to the forces acting at O on the pivot point? (Hint: Draw free body diagrams of the masses and the rod.)

P3.4. A particle of mass m_1 rides in a smooth vertical track and is connected by a rigid, weightless rod of length L to a particle of mass m_2 that rides in a smooth horizontal track (Fig. P3.4). (a) Using the angle θ as a generalized coordinate,

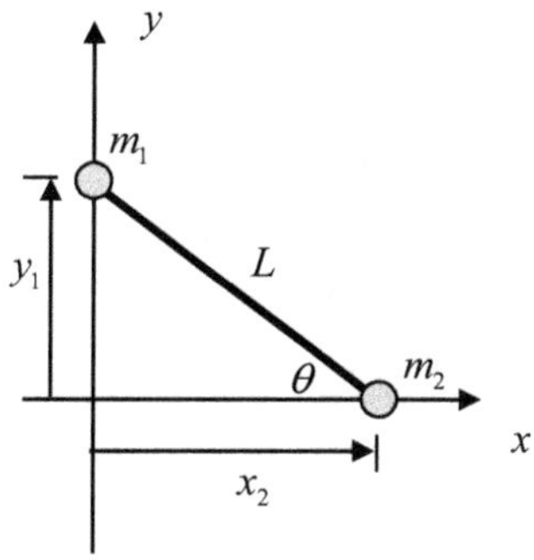

Fig. P3.4 Two masses, separated by a weightless rigid rod of length L, ride in smooth vertical and horizontal tracks

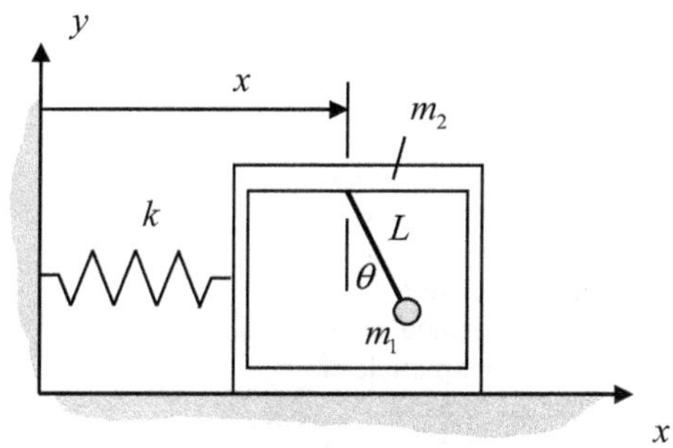

Fig. P3.5 A pendulum consisting of a mass m_1 and a weightless wire of length L is attached to a box of mass m_2 that slides on a smooth horizontal surface and is attached to a spring with spring constant k

obtain the equation of motion for the system and solve the equation of motion numerically for 10 s with the initial conditions $\theta = \pi/2$, $\dot{\theta} = -5$ rad/s, $m_1 = 2m_2 = 0.5$ kg, and $L = 2$ m. (b) Using the (y_1, x_2) coordinates as shown in Fig. P3.4 determine the equations of motion for this system and solve the equations of motion numerically using the augmented method for the same initial conditions and parameters used in part (a). Also, determine the tension in the rod as a function of time.

P3.5. A mass m_1 and a weightless, inextensible wire of length L form a simple pendulum whose base is attached to a box of mass m_2 (Fig. P3.5). The box is free to slide on a smooth horizontal surface and is attached to a linear spring with a spring constant, k. The unstretched length of the spring is x_u. (a) Determine the equations of motion in terms of the generalized coordinates (x, θ). Solve these equations of motion numerically for 20 s for $x_u = 0.5$ m, $L = 0.25$ m, $m_1 = 3m_2 = 3$ kg, and $k = 20$ N/m, and initial conditions $x(0) = 1$ m, $\dot{x}(0) = 0$, and $\theta(0) = \dot{\theta}(0) = 0$. (b) Using the x-coordinate of the box and the (x_1, y_1) coordinates of the mass m_1 as generalized coordinates, determine the equations of motion of the systems and solve those equations numerically using an augmented approach for the same parameters and initial conditions used in part (a). Also, determine the tension in the wire as a function of time.

P3.6. Two particles of mass m_1 and m_2 move on a smooth horizontal surface. The two particles are connected together by a weightless, inextensible rod of

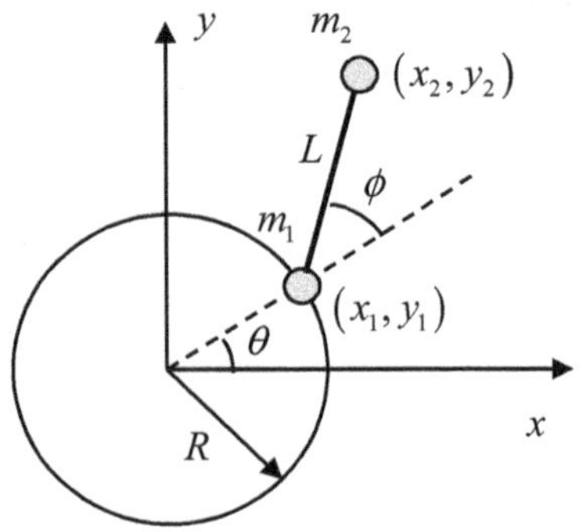

Fig. P3.6 Two masses moving on a smooth horizontal plane connected by a weightless, inextensible rod of length L. Mass m_1 rides in a smooth, circular track of radius R

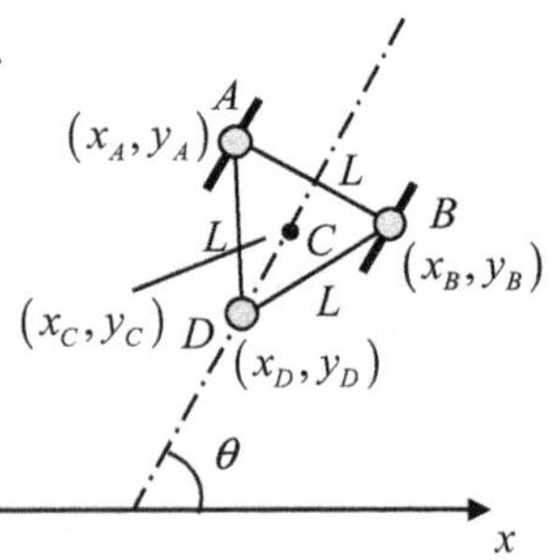

Fig. P3.7 Three particles, each of mass m, that are connected by rigid, weightless rods of lengths L. The particles at A and B are connected to wheels that roll without slipping in a direction perpendicular to the rod AB. The particle at D is attached to a caster

length L and mass m_1 is constrained to move along a smooth, circular track of radius R (Fig. P3.6). (a) Using the angles (θ, ϕ) as generalized coordinates, determine the equations of motion for this system. Solve these equations of motion for 10 s for $m_1 = 2$ kg, $m_2 = 1$ kg, $R = 0.5$ m, and $L = 1.5$ m, and initial conditions $\phi(0) = \dot{\phi}(0) = 0$, $\theta(0) = 0$, and $\dot{\theta}(0) = 3$ rad/s. (b) Using the coordinates (x_1, y_1) of mass m_1 and coordinates (x_2, y_2) of mass m_2, determine the equations of motion for the system and solve those equations for this system numerically using the augmented approach for the same parameters and initial conditions as used in part (a). Determine the force that the circular track exerts on mass m_1.

P3.7. Three particles, each of mass m, move in a horizontal plane as shown in Fig. P3.7. The masses are connected by rigid rods, each of length, L, whose weights can be neglected and that form an equilateral triangle. The masses at A and B are attached to wheels that roll without slipping and that permit these masses to only move in a direction perpendicular to the axis of the rod AB. The third mass, at D, is attached to a caster wheel that permits the mass to move in any direction consistent with the other constraints. (a) Using the coordinates, (x_C, y_C) of the center of mass C and the angle, θ, between the x-axis and the plane of the wheels at A and B, determine the equations of motion of this system. There are three holonomic constraints in this problem so that there are three degrees of freedom for this problem and we can take (x_C, y_C, θ) as independent generalized coordinates. However, there are also two non-holonomic constraints that relate the generalized velocities $(\dot{x}_C, \dot{y}_C, \dot{\theta})$. Show that those two non-holonomic constraints are not independent. Solve this system of equations numerically for 15 s with the initial conditions $x_C = y_C = \theta = 0$ and $\dot{x}_C = 0$, $\dot{y}_C = 2$ rad/s at time $t = 0$, taking $m = 3$ kg and $L = 0.5$ m. What is the initial angular velocity, $\dot{\theta}$, consistent with the non-holonomic constraints? (b) Use the coordinates of the three masses (x_A, y_A), (x_B, y_B), and (x_D, y_D) as the generalized coordinates, subject to three holonomic constraints and two non-independent, non-holonomic constraints. Determine the equations of motion and solve this system numerically using the augmented approach using the same initial conditions and parameters of part (a).

P3.8. In Chap. 2 we saw that with a judicious choice of generalized coordinates for a single particle we can implicitly imbed the constraints of a problem so that the constraint forces are not explicitly a part of the problem. The same is true for a system of particles. Consider, for example, three masses connected by inextensible strings to a two-pulley system as shown in Fig. P3.8. (a) Determine the constraint equation satisfied by the (x_1, x_2, x_3) coordinates. (b) Express the three x_n coordinates in terms of the two q_m coordinates and show that the equation of constraint of part (a) is satisfied by the q_m so that they are independent generalized coordinates for this problem.

P3.9. Use the animate_lines function to animate the motion of the Simulation 3.1 multiple-particle pendulum problem, which we can treat as the motion of the three connected lines OC, CA, and AB (see Fig. 3.12).

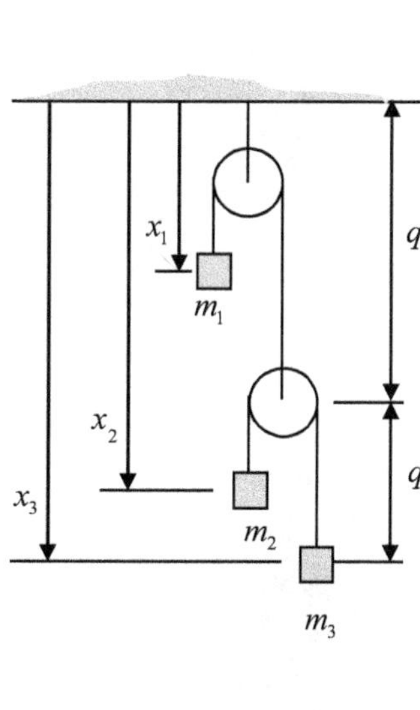

Fig. P3.8 A two-pulley system defined in terms of the coordinates of the three masses (x_1, x_2, x_3) and the independent generalized coordinates (q_1, q_2)

References

1. O.M. O'Reilly, *Intermediate Dynamics for Engineers* (Cambridge University Press, New York, 2008)
2. F.E. Udwadia, R.E. Kalaba, *Analytical Dynamics* (Cambridge University Press, Cambridge, 1996)

Chapter 4
Kinematics and Relative Motion

This chapter describes the kinematics of point masses and rigid bodies when non-inertial coordinate systems (frames) are used to describe their motion. We obtain relative velocity and acceleration expressions for moving frames and then apply those expressions to find the velocities and accelerations of constrained systems of rigid bodies at specific instances of time, similar to what is done in many elementary dynamics texts. However, we also show that it is possible to determine the kinematics of a constrained system of rigid bodies more completely as a function of time. In some cases this can be done analytically but in general the solution must be done numerically since the positional constraints are normally nonlinear. To solve the positional constraints we use the Newton-Raphson method. Kinematics is treated in this chapter both by the traditional vector approach and by an equivalent matrix-vector method that is more readily suited to dealing with complex systems. The matrix-vector approach for planar problems is covered in Sects. 4.4 and 4.5 while more general three-dimensional problems are treated in Sect. 4.6 and those that follow. Three-dimensional rotations are described in terms of both Euler angles and Euler parameters as these are the most commonly used generalized rotational coordinates. Some classical examples of the dynamics effects seen in rotating coordinate systems, such as the Foucault pendulum, are also given.

4.1 Relative Velocity and Acceleration

The Newton-Euler equations involve time derivative terms such as the acceleration and rate of change of the angular momentum. As discussed in Chap. 1 these time derivatives must be computed in an inertial coordinate system whose origin is "fixed" and whose axes do not rotate. If, however, we make measurements of the time derivatives in a coordinate system where either the origin translates and rotates then those measurements relative to such a moving coordinate system are only a part of the total time derivatives as seen in an inertial system and we must include

L. W. Schmerr, *Engineering Dynamics 2.0*, Solid Mechanics and Its Applications 254,
https://doi.org/10.1007/978-3-319-98470-4_4

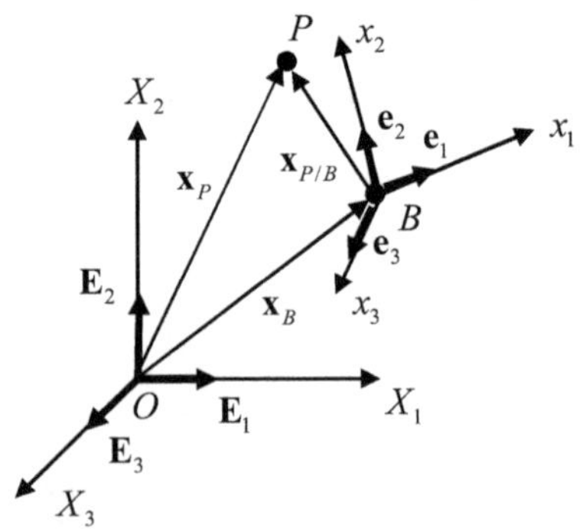

Fig. 4.1 Position vector, $\mathbf{x}_P$, of a point P relative to a fixed origin, O, of an inertial coordinate system and its relationship to the relative position vector of P with respect to a point B, $\mathbf{x}_{P/B}$, where the position vector of B is $\mathbf{x}_B$. The origin of the (x_1, x_2, x_3) coordinate system moves with point B

additional terms that account for the translation and rotation of the coordinate system itself. For example, consider the position vector, $\mathbf{x}_P = X_1\mathbf{E}_1 + X_2\mathbf{E}_2 + X_3\mathbf{E}_3$, of a point, P, as measured in a Cartesian inertial coordinate system whose origin, O, is fixed and where the vectors $(\mathbf{E}_1, \mathbf{E}_2, \mathbf{E}_3)$ along the (X_1, X_2, X_3) axes are constant (i.e., nonrotating) unit vectors (see Fig. 4.1). Then computing the velocity and acceleration of the point P is very simple since

$$\begin{aligned} \mathbf{v}_P &= \dot{\mathbf{x}}_P = \dot{X}_1\mathbf{E}_1 + \dot{X}_2\mathbf{E}_2 + \dot{X}_3\mathbf{E}_3 \\ \mathbf{a}_P &= \ddot{\mathbf{x}}_P = \ddot{X}_1\mathbf{E}_1 + \ddot{X}_2\mathbf{E}_2 + \ddot{X}_3\mathbf{E}_3 \end{aligned} \tag{4.1.1}$$

which only involve the time derivatives of the inertial coordinates. However, if we have a coordinate system (x_1, x_2, x_3) whose origin B moves, then we can let the position vector of the point B be $\mathbf{x}_B$ and the position vector of point P relative to this moving origin B be $\mathbf{x}_{P/B}$, as shown in Fig. 4.1. From the geometry we have directly

$$\mathbf{x}_P = \mathbf{x}_B + \mathbf{x}_{P/B} \tag{4.1.2}$$

and taking time derivatives we have

$$\begin{aligned} \mathbf{v}_P &= \mathbf{v}_B + \mathbf{v}_{P/B} \\ \mathbf{a}_P &= \mathbf{a}_B + \mathbf{a}_{P/B} \end{aligned} \tag{4.1.3}$$

These relations simply state that the velocity of P, $\mathbf{v}_P$, equals the velocity of B, $\mathbf{v}_B$, plus the velocity of P with respect to B, $\mathbf{v}_{P/B}$, and the acceleration of P, $\mathbf{a}_P$, equals the acceleration of B, $\mathbf{a}_B$, plus the acceleration of P with respect to B, $\mathbf{a}_{P/B}$. We can, of course, measure all of the terms appearing in Eqs. (4.1.2) and (4.1.3) in the inertial (X_1, X_2, X_3) coordinate system in which case only the components of these terms will vary with time. However, since $\mathbf{v}_{P/B}$ and $\mathbf{a}_{P/B}$ are the velocity and acceleration of P relative to point B, it is advantageous to measure these relative motion terms with respect to a coordinate system (x_1, x_2, x_3) whose origin moves with B and whose unit vectors are $(\mathbf{e}_1, \mathbf{e}_2, \mathbf{e}_3)$ as shown in Fig. 4.1. In terms of this moving coordinate

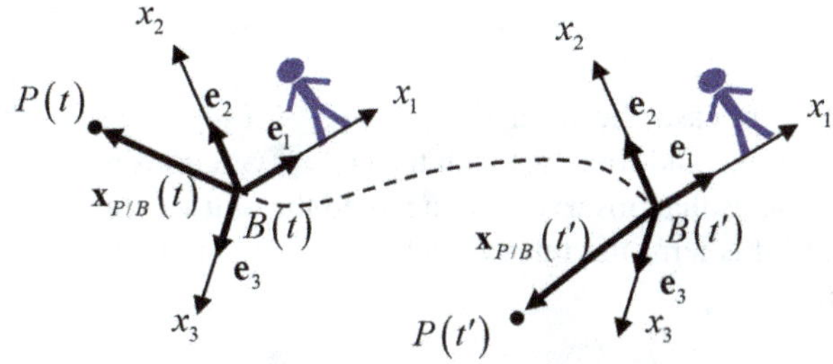

Fig. 4.2 The motion of a point P (shown at times t and t') as seen in a coordinate system that translates with point B but does not rotate. The dashed line is the path of motion of point B

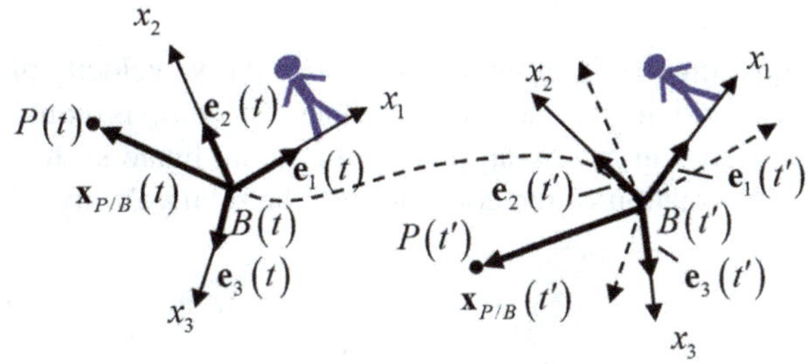

Fig. 4.3 The motion of a point P (shown at times t and t') as seen in a coordinate system that translates with point B and rotates

system we could write the relative position vector as $\mathbf{x}_{P/B} = x_1\mathbf{e}_1 + x_2\mathbf{e}_2 + x_3\mathbf{e}_3$. Now, consider the case when the unit vectors are constants, i.e., the coordinate system translates with the motion of point B but does not rotate. In this case we simply have

$$\begin{aligned} \mathbf{v}_{P/B} &= \dot{x}_1\mathbf{e}_1 + \dot{x}_2\mathbf{e}_2 + \dot{x}_3\mathbf{e}_3 \\ \mathbf{a}_{P/B} &= \ddot{x}_1\mathbf{e}_1 + \ddot{x}_2\mathbf{e}_2 + \ddot{x}_3\mathbf{e}_3 \end{aligned} \tag{4.1.4}$$

An observer who translates with the coordinate system (see Fig. 4.2) would measure the position vector of point P relative to B as $\mathbf{x}_{P/B} = x_1\mathbf{e}_1 + x_2\mathbf{e}_2 + x_3\mathbf{e}_3$ and would see the motion of point P relative to B simply as changes of his/her measurements of the (x_1, x_2, x_3) coordinates. These changes would give the total relative velocity or acceleration of P with respect to B since in this case the $(\mathbf{e}_1, \mathbf{e}_2, \mathbf{e}_3)$ unit vectors, like the $(\mathbf{E}_1, \mathbf{E}_2, \mathbf{E}_3)$ unit vectors, do not change with time.

Now, consider the case when the moving coordinate system translates with point B and rotates, as shown in Fig. 4.3. In that case if we again write the relative position vector as $\mathbf{x}_{P/B} = x_1\mathbf{e}_1 + x_2\mathbf{e}_2 + x_3\mathbf{e}_3$ we have for the relative velocity

$$\frac{d\mathbf{x}_{P/B}}{dt} = \mathbf{v}_{P/B} = \dot{x}_1\mathbf{e}_1 + \dot{x}_2\mathbf{e}_2 + \dot{x}_3\mathbf{e}_3 + x_1\dot{\mathbf{e}}_1 + x_2\dot{\mathbf{e}}_2 + x_3\dot{\mathbf{e}}_3 \tag{4.1.5}$$

But from Chap. 1 we saw that if these unit vectors $(\mathbf{e}_1, \mathbf{e}_2, \mathbf{e}_3)$ rotate with respect to the inertial system with an angular velocity vector, $\boldsymbol{\omega}$, we have $\dot{\mathbf{e}}_k = \boldsymbol{\omega} \times \mathbf{e}_k \ (k = 1, 2, 3)$ and so

$$\mathbf{v}_{P/B} = \dot{x}_1\mathbf{e}_1 + \dot{x}_2\mathbf{e}_2 + \dot{x}_3\mathbf{e}_3 + \boldsymbol{\omega} \times \mathbf{x}_{P/B} \tag{4.1.6}$$

In this case the term $\dot{x}_1\mathbf{e}_1 + \dot{x}_2\mathbf{e}_2 + \dot{x}_3\mathbf{e}_3$ is again the velocity that an observer moving (and rotating) with the (x_1, x_2, x_3) coordinates would measure (Fig. 4.3) since as far as that observer is concerned the unit vectors $(\mathbf{e}_1, \mathbf{e}_2, \mathbf{e}_3)$ do not change. We will call this term the relative velocity of P with respect to B, $\mathbf{v}_{P/B}^{\mathrm{rel}}$, and Eq. (4.1.6) then becomes

$$\mathbf{v}_{P/B} = \mathbf{v}_{P/B}^{\mathrm{rel}} + \boldsymbol{\omega} \times \mathbf{x}_{P/B} \tag{4.1.7}$$

Equation (4.1.7) shows that the total relative velocity of P with respect to B, which is measured in an inertial coordinate system, is equal to the relative velocity as measured in a moving and rotating coordinate system plus a term, $\boldsymbol{\omega} \times \mathbf{x}_{P/B}$, due to the rotation of the coordinates. The total velocity of point P is then given as

$$\mathbf{v}_P = \mathbf{v}_B + \mathbf{v}_{P/B}^{\mathrm{rel}} + \boldsymbol{\omega} \times \mathbf{x}_{P/B} \tag{4.1.8}$$

To obtain the acceleration of point P we have to differentiate Eq. (4.1.8) on time, to obtain

$$\frac{d\mathbf{v}_P}{dt} = \frac{d\mathbf{v}_B}{dt} + \frac{d}{dt}\left(\mathbf{v}_{P/B}^{\mathrm{rel}}\right) + \frac{d\boldsymbol{\omega}}{dt} \times \mathbf{x}_{P/B} + \boldsymbol{\omega} \times \frac{d\mathbf{x}_{P/B}}{dt} \tag{4.1.9}$$

Since $\mathbf{v}_{P/B}^{\mathrm{rel}} = \dot{x}_1\mathbf{e}_1 + \dot{x}_2\mathbf{e}_2 + \dot{x}_3\mathbf{e}_3$ it follows that

$$\frac{d\mathbf{v}_{P/B}^{\mathrm{rel}}}{dt} = \mathbf{a}_{P/B}^{\mathrm{rel}} + \boldsymbol{\omega} \times \mathbf{v}_{P/B}^{\mathrm{rel}} \tag{4.1.10}$$

where $\mathbf{a}_{P/B}^{\mathrm{rel}} = \ddot{x}_1\mathbf{e}_1 + \ddot{x}_2\mathbf{e}_2 + \ddot{x}_3\mathbf{e}_3$ is the relative acceleration seen by an observer moving and rotating with the (x_1, x_2, x_3) coordinates and from the changes of the unit vectors we have

$$\begin{aligned} \dot{x}_1\dot{\mathbf{e}}_1 + \dot{x}_2\dot{\mathbf{e}}_2 + \dot{x}_3\dot{\mathbf{e}}_3 &= \dot{x}_1\boldsymbol{\omega} \times \mathbf{e}_1 + \dot{x}_2\boldsymbol{\omega} \times \mathbf{e}_2 + \dot{x}_3\boldsymbol{\omega} \times \mathbf{e}_3 \\ &= \boldsymbol{\omega} \times \left(\dot{x}_1\mathbf{e}_1 + \dot{x}_2\mathbf{e}_2 + \dot{x}_3\mathbf{e}_3\right) \\ &= \boldsymbol{\omega} \times \mathbf{v}_{P/B}^{\mathrm{rel}} \end{aligned} \tag{4.1.11}$$

which gives us the second term in Eq. (4.1.10). Also, $\dot{\boldsymbol{\omega}} = \boldsymbol{\alpha}$, where $\boldsymbol{\alpha}$ is the angular acceleration of the rotating coordinates and finally we have $\dot{\mathbf{x}}_{P/B} = \mathbf{v}_{P/B} = \mathbf{v}_{P/B}^{\mathrm{rel}} + \boldsymbol{\omega} \times \mathbf{x}_{P/B}$ so collecting all these results we can write Eq. (4.1.9) as

$$\mathbf{a}_P = \mathbf{a}_B + \boldsymbol{\alpha} \times \mathbf{x}_{P/B} + \boldsymbol{\omega} \times \left(\boldsymbol{\omega} \times \mathbf{x}_{P/B}\right) + \mathbf{a}_{P/B}^{\mathrm{rel}} + 2\boldsymbol{\omega} \times \mathbf{v}_{P/B}^{\mathrm{rel}} \tag{4.1.12}$$

where $\mathbf{a}_P$ and $\mathbf{a}_B$ are the accelerations of P and B, respectively, as seen in an inertial coordinate system. The first three terms on the right-hand side of Eq. (4.1.12) are due to the motion of the coordinate system. The first term is due to the acceleration of the origin of the coordinates, and the second and third terms are due to the angular acceleration and angular velocity of the coordinates, where the third term is the *centripetal acceleration*. The last two terms in Eq. (4.1.12) are due to the motion seen relative to the moving coordinates, where the last term is the *Coriolis acceleration*. We have seen these acceleration terms before in Chap. 1 when we considered cylindrical and spherical coordinates. In fact, the results we obtained in Chap. 1 for those coordinate systems are merely a special case of Eqs. (4.1.8) and (4.1.12), as we will show in the next section.

4.1.1 Relative Motion: Cylindrical and Spherical Coordinates

Consider first the case of cylindrical coordinates. The origin, O, of this coordinate system is fixed ($\mathbf{v}_O = \mathbf{a}_O = 0$) but the unit vectors ($\mathbf{e}_r$, $\mathbf{e}_\theta$, $\mathbf{e}_z$) rotate about the z-axis with an angular velocity $\boldsymbol{\omega} = \dot{\theta}\,\mathbf{e}_z$. Thus, since in the rotating coordinate system where the position vector to a point P is $\mathbf{x}_{P/O} = r\mathbf{e}_r + z\mathbf{e}_z$ we have

$$\begin{aligned}\mathbf{v}_P &= \mathbf{v}_O + \mathbf{v}_{P/O}^{\text{rel}} + \boldsymbol{\omega} \times \mathbf{x}_{P/O} \\ &= 0 + \left(\dot{r}\,\mathbf{e}_r + \dot{z}\,\mathbf{e}_z\right) + \dot{\theta}\,\mathbf{e}_z \times \left(r\mathbf{e}_r + z\mathbf{e}_z\right) \\ &= \dot{r}\,\mathbf{e}_r + r\dot{\theta}\,\mathbf{e}_\theta + \dot{z}\,\mathbf{e}_z\end{aligned} \tag{4.1.13}$$

which is just Eq. (1.3.10). Similarly, to calculate the acceleration we have $\boldsymbol{\alpha} = \ddot{\theta}\mathbf{e}_z$, $\mathbf{v}_{P/O}^{\text{rel}} = \dot{r}\,\mathbf{e}_r + \dot{z}\,\mathbf{e}_z$, $\mathbf{a}_{P/O}^{\text{rel}} = \ddot{r}\mathbf{e}_r + \ddot{z}\mathbf{e}_z$ so that

$$\begin{aligned}\mathbf{a}_P &= \mathbf{a}_O + \boldsymbol{\alpha} \times \mathbf{x}_{P/O} + \boldsymbol{\omega} \times \left(\boldsymbol{\omega} \times \mathbf{x}_{P/O}\right) + \mathbf{a}_{P/O}^{\text{rel}} + 2\boldsymbol{\omega} \times \mathbf{v}_{P/O}^{\text{rel}} \\ &= 0 + \ddot{\theta}\mathbf{e}_z \times \left(r\mathbf{e}_r + z\mathbf{e}_z\right) + \dot{\theta}\,\mathbf{e}_z \times \left[\dot{\theta}\,\mathbf{e}_z \times \left(r\mathbf{e}_r + z\mathbf{e}_z\right)\right] \\ &\quad + \left(\ddot{r}\mathbf{e}_r + \ddot{z}\mathbf{e}_z\right) + 2\dot{\theta}\,\mathbf{e}_z \times \left(\dot{r}\,\mathbf{e}_r + \dot{z}\,\mathbf{e}_z\right)\end{aligned} \tag{4.1.14}$$

which yields when the cross products are calculated

$$\mathbf{a}_P = \left(\ddot{r} - r\dot{\theta}^2\right)\mathbf{e}_r + \left(r\ddot{\theta} + 2\dot{r}\dot{\theta}\right)\mathbf{e}_\theta + \ddot{z}\mathbf{e}_z \tag{4.1.15}$$

in agreement with Eq. (1.3.11).

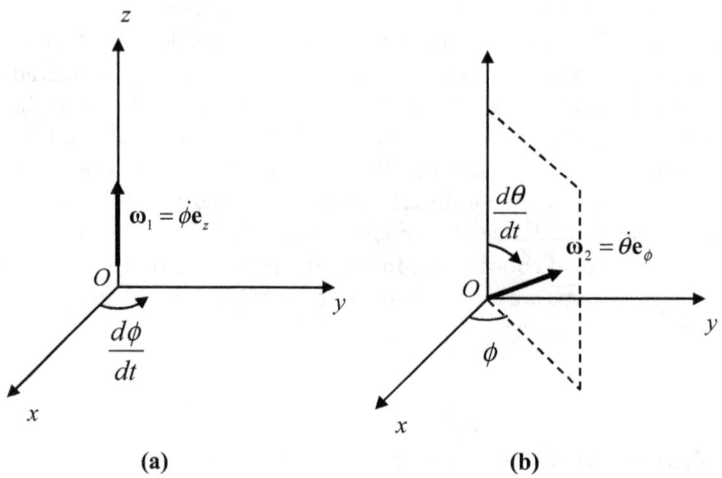

Fig. 4.4 In spherical coordinates, the total angular velocity of the coordinates is composed of (**a**) an angular velocity, $\dot{\phi}$, about the z-axis, and (**b**) an angular velocity, $\dot{\theta}$, in the $\mathbf{e}_\phi$ direction (normal to the $\phi - z$ plane)

Spherical coordinates are more complicated since the total angular velocity of the $(\mathbf{e}_R, \mathbf{e}_\theta, \mathbf{e}_\phi)$ unit vectors (see Fig. 4.4) is

$$\begin{aligned}\boldsymbol{\omega} &= \dot{\phi}\,\mathbf{e}_z + \dot{\theta}\,\mathbf{e}_\phi \\ &= \dot{\phi}\,(\cos\theta\mathbf{e}_R - \sin\theta\mathbf{e}_\theta) + \dot{\theta}\,\mathbf{e}_\phi \\ &= \dot{\phi}\,\cos\theta\mathbf{e}_R + \dot{\theta}\,\mathbf{e}_\phi - \dot{\phi}\,\sin\theta\mathbf{e}_\theta\end{aligned} \tag{4.1.16}$$

In spherical coordinates $\mathbf{x}_{P/O} = R\mathbf{e}_R$. To calculate the velocity we have

$$\begin{aligned}\mathbf{v}_P &= \mathbf{v}_O + \mathbf{v}_{P/O}^{\text{rel}} + \boldsymbol{\omega} \times \mathbf{x}_{P/O} \\ &= 0 + \dot{R}\,\mathbf{e}_R + \left[\dot{\phi}\,\cos\theta\mathbf{e}_R + \dot{\theta}\,\mathbf{e}_\phi - \dot{\phi}\,\sin\theta\mathbf{e}_\theta\right] \times R\mathbf{e}_R \\ &= \dot{R}\,\mathbf{e}_R + R\dot{\theta}\,\mathbf{e}_\theta + R\dot{\phi}\,\sin\theta\mathbf{e}_\phi\end{aligned} \tag{4.1.17}$$

which is the same as Eq. (1.3.19). For the acceleration we need $\mathbf{v}_{P/O}^{\text{rel}} = \dot{R}\,\mathbf{e}_R$, $\mathbf{a}_{P/O}^{\text{rel}} = \ddot{R}\mathbf{e}_R$. Also,

$$\begin{aligned}\boldsymbol{\alpha} &= \ddot{\phi}\mathbf{e}_z + \ddot{\theta}\mathbf{e}_\phi + \dot{\theta}\,\boldsymbol{\omega} \times \mathbf{e}_\phi \\ &= \ddot{\phi}\mathbf{e}_z + \ddot{\theta}\mathbf{e}_\phi + \dot{\theta}\left(-\dot{\phi}\,\cos\theta\mathbf{e}_\theta - \dot{\phi}\,\sin\theta\mathbf{e}_R\right) \\ &= \left(\ddot{\phi}\cos\theta - \dot{\theta}\,\dot{\phi}\,\sin\theta\right)\mathbf{e}_R + \ddot{\theta}\mathbf{e}_\phi + \left(-\ddot{\phi}\sin\theta - \dot{\theta}\,\dot{\phi}\,\cos\theta\right)\mathbf{e}_\theta\end{aligned} \tag{4.1.18}$$

so that

$$\begin{aligned}\mathbf{a}_P &= \mathbf{a}_O + \boldsymbol{\alpha} \times \mathbf{x}_{P/O} + \boldsymbol{\omega} \times \left(\boldsymbol{\omega} \times \mathbf{x}_{P/O}\right) + \mathbf{a}^{\text{rel}}_{P/O} + 2\boldsymbol{\omega} \times \mathbf{v}^{\text{rel}}_{P/O} \\ &= 0 + \left[\ddot{\theta}\mathbf{e}_\phi + \left(-\ddot{\phi}\sin\theta - \dot{\theta}\dot{\phi}\cos\theta\right)\mathbf{e}_\theta\right] \times R\mathbf{e}_R \\ &\quad + \left(\dot{\phi}\cos\theta\mathbf{e}_R + \dot{\theta}\,\mathbf{e}_\phi - \dot{\phi}\sin\theta\mathbf{e}_\theta\right) \times \left(R\dot{\theta}\,\mathbf{e}_\theta + R\dot{\phi}\sin\theta\mathbf{e}_\phi\right) \\ &\quad + \ddot{R}\mathbf{e}_R + 2\left(\dot{\phi}\cos\theta\mathbf{e}_R + \dot{\theta}\,\mathbf{e}_\phi - \dot{\phi}\sin\theta\mathbf{e}_\theta\right) \times \dot{R}\,\mathbf{e}_R\end{aligned} \tag{4.1.19}$$

Carrying out all the cross products and grouping terms we have

$$\begin{aligned}\mathbf{a}_P &= R\ddot{\theta}\mathbf{e}_\theta + \left(R\ddot{\phi}\sin\theta + R\dot{\theta}\dot{\phi}\cos\theta\right)\mathbf{e}_\phi \\ &\quad + R\dot{\theta}\dot{\phi}\cos\theta\mathbf{e}_\phi - R\dot{\phi}^2\sin\theta\cos\theta\mathbf{e}_\theta - R\dot{\theta}^2\mathbf{e}_R - R\dot{\phi}^2\sin^2\theta\mathbf{e}_R \\ &\quad + \ddot{R}\mathbf{e}_R + 2\dot{R}\dot{\theta}\,\mathbf{e}_\theta + 2\dot{R}\dot{\phi}\sin\theta\mathbf{e}_\phi \\ &= \left(\ddot{R} - R\dot{\theta}^2 - R\dot{\phi}^2\sin^2\theta\right)\mathbf{e}_R \\ &\quad + \left(R\ddot{\theta} + 2\dot{R}\dot{\theta} - R\dot{\phi}^2\sin\theta\cos\theta\right)\mathbf{e}_\theta \\ &\quad + \left(R\ddot{\phi}\sin\theta + 2R\dot{\theta}\dot{\phi}\cos\theta + 2\dot{R}\dot{\phi}\sin\theta\right)\mathbf{e}_\phi\end{aligned} \tag{4.1.20}$$

which is the same as Eq. (1.3.22).

4.2 Relative Motion and the Transport Theorem

The results shown in the previous section for the velocity and acceleration as computed in fixed (inertial) and moving coordinate systems are based on a relation that is valid for vectors in general when we take derivatives of those vectors. Any vector, $\mathbf{p}$, for example, can be measured in a fixed (inertial) $I = (X, Y, Z)$ system as

$$\mathbf{p} = p_X\mathbf{E}_x + p_Y\mathbf{E}_Y + p_Z\mathbf{E}_Z \tag{4.2.1}$$

and the time derivative of this vector, as measured in the inertial system, I, $(d\mathbf{p}/dt)_I$, would just be

$$\begin{aligned}\left(\frac{d\mathbf{p}}{dt}\right)_I &= \frac{d}{dt}(p_X\mathbf{E}_X + p_Y\mathbf{E}_Y + p_Z\mathbf{E}_Z) \\ &= \frac{dp_X}{dt}\mathbf{E}_X + \frac{dp_Y}{dt}\mathbf{E}_Y + \frac{dp_Z}{dt}\mathbf{E}_Z\end{aligned} \tag{4.2.2}$$

However, if we write this vector in a moving coordinate system $R = (x, y, z)$ that rotates with an angular velocity, $\boldsymbol{\omega}$, then we have $\mathbf{p} = p_x\mathbf{e}_x + p_y\mathbf{e}_y + p_z\mathbf{e}_z$ and

$$\begin{aligned}\left(\frac{d\mathbf{p}}{dt}\right)_I &= \frac{d}{dt}(p_x\mathbf{e}_x + p_y\mathbf{e}_y + p_z\mathbf{e}_z) \\ &= \left(\frac{dp_x}{dt}\mathbf{e}_x + \frac{dp_y}{dt}\mathbf{e}_y + \frac{dp_z}{dt}\mathbf{e}_z\right) + \left(p_x\frac{d\mathbf{e}_x}{dt} + p_y\frac{d\mathbf{e}_y}{dt} + p_z\frac{d\mathbf{e}_z}{dt}\right) \\ &= \left(\frac{d\mathbf{p}}{dt}\right)_R + (p_x\boldsymbol{\omega}\times\mathbf{e}_x + p_y\boldsymbol{\omega}\times\mathbf{e}_y + p_z\boldsymbol{\omega}\times\mathbf{e}_z)\end{aligned} \tag{4.2.3}$$

where $(d\mathbf{p}/dt)_R$ is the derivative of $\mathbf{p}$ as seen in the rotating coordinates, R ($\mathbf{p}$ here can be any vector so $(d\mathbf{p}/dt)_R$ is the generalization of the velocity and acceleration terms labeled as "relative" earlier and denoted by a "rel" superscript). Rearranging the second term on the right side of Eq. (4.2.3) we can write that relationship as

$$\left(\frac{d\mathbf{p}}{dt}\right)_I = \left(\frac{d\mathbf{p}}{dt}\right)_R + \boldsymbol{\omega}\times\mathbf{p} \tag{4.2.4}$$

which is called the *transport theorem*. In many cases authors will assume that a time derivative term taken where the $(\ )_I$ subscript is missing will implicitly mean that such a derivative is with respect to an inertial system and write the transport theorem simply as

$$\frac{d\mathbf{p}}{dt} = \left(\frac{d\mathbf{p}}{dt}\right)_R + \boldsymbol{\omega}\times\mathbf{p} \tag{4.2.5}$$

However, in some complex relative motion problems it is advisable to keep the more explicit form of Eq. (4.2.4). There is another change in the notation we will use to write the transport theorem that is important. The angular velocity, $\boldsymbol{\omega}$, is the angular velocity of the rotating R coordinates relative to the inertial I coordinates. To make this more explicit we will write $\boldsymbol{\omega} = {}_I\boldsymbol{\omega}_R$ so that Eq. (4.2.4) then becomes

$$\left(\frac{d\mathbf{p}}{dt}\right)_I = \left(\frac{d\mathbf{p}}{dt}\right)_R + {}_I\boldsymbol{\omega}_R\times\mathbf{p} \tag{4.2.6}$$

This notation is a bit more cumbersome but useful since the transport equation in the form of Eq. (4.2.6) is also valid for any two coordinate systems (R_1, R_2), i.e.,

$$\left(\frac{d\mathbf{p}}{dt}\right)_{R_1} = \left(\frac{d\mathbf{p}}{dt}\right)_{R_2} + {}_{R_1}\boldsymbol{\omega}_{R_2}\times\mathbf{p} \tag{4.2.7}$$

where the time derivatives are those seen in coordinates (R_1, R_2), and ${}_{R_1}\boldsymbol{\omega}_{R_2}$ is the angular velocity of coordinates R_2 relative to the R_1 coordinates. These coordinate systems are also sometimes called *frames*. Thus, Eq. (4.2.7) relates the time derivative taken in the R_1 frame to the time derivative taken in the R_2 frame and ${}_{R_1}\boldsymbol{\omega}_{R_2}$ is the angular velocity of the R_2 frame as seen in the R_1 frame. The proof of Eq. (4.2.7) follows the same steps used to derive Eq. (4.2.6) so we will not show them here.

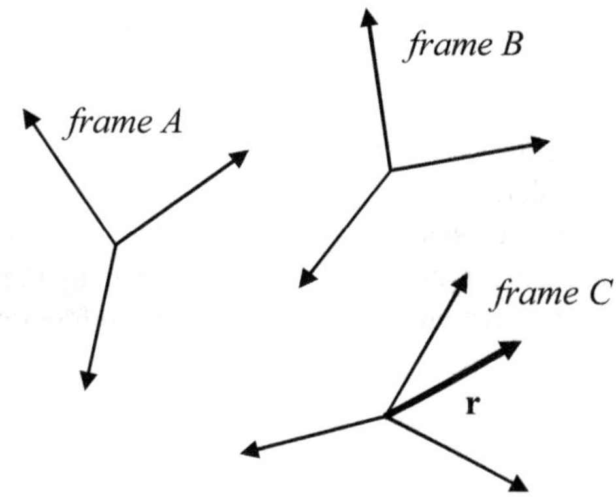

Fig. 4.5 Three coordinate frames (A, B, C) and a vector, $\mathbf{r}$, that is fixed in both magnitude and direction relative to the C frame

Equation (4.2.7) is convenient for solving complex problems when we want to use more than one set of rotating coordinates to describe the motion. A key calculation is obtaining the relative angular velocity term appearing in Eq. (4.2.7). This calculation can often be made simpler by using the addition theorem for angular velocities. We can obtain this theorem by simply applying Eq. (4.2.7). To show this, consider the case where we have three coordinate systems (frames) labeled A, B, and C in Fig. 4.5 and a vector, $\mathbf{r}$, which is fixed in both magnitude and direction relative to the C frame. Then

$$\begin{aligned} \left(\frac{d\mathbf{r}}{dt}\right)_B &= \left(\frac{d\mathbf{r}}{dt}\right)_C + {}_B\boldsymbol{\omega}_C \times \mathbf{r} = {}_B\boldsymbol{\omega}_C \times \mathbf{r} \\ \left(\frac{d\mathbf{r}}{dt}\right)_A &= \left(\frac{d\mathbf{r}}{dt}\right)_B + {}_A\boldsymbol{\omega}_B \times \mathbf{r} \end{aligned} \tag{4.2.8}$$

where we have used the fact that $\mathbf{r}$ is fixed in C so that its time derivative is zero in that frame. Combining the results in Eq. (4.2.8) and relating the motion in frame A to the motion in frame C we have

$$\begin{aligned} \left(\frac{d\mathbf{r}}{dt}\right)_A &= \left({}_A\boldsymbol{\omega}_B + {}_B\boldsymbol{\omega}_C\right) \times \mathbf{r} \\ \left(\frac{d\mathbf{r}}{dt}\right)_A &= \left(\frac{d\mathbf{r}}{dt}\right)_C + {}_A\boldsymbol{\omega}_C \times \mathbf{r} = {}_A\boldsymbol{\omega}_C \times \mathbf{r} \end{aligned} \tag{4.2.9}$$

Equating these two time derivatives gives

$$\left({}_A\boldsymbol{\omega}_C - {}_A\boldsymbol{\omega}_B - {}_B\boldsymbol{\omega}_C\right) \times \mathbf{r} = 0 \tag{4.2.10}$$

which, since $\mathbf{r}$ is an arbitrary fixed vector, yields the *addition theorem* for angular velocities in this case as

$$ {}_A\boldsymbol{\omega}_C = {}_A\boldsymbol{\omega}_B + {}_B\boldsymbol{\omega}_C \tag{4.2.11} $$

This result is easily generalized to very complex cases. For example, when there are N multiple coordinate systems $(R_1, R_2, \ldots, R_N)$ and we want to calculate, say, the time derivative of a vector in an inertial coordinate system I in terms of the derivative in the R_N system using the transport theorem, then we need to find the relative angular velocity, ${}_I\boldsymbol{\omega}_{R_N}$, that appears in that theorem. We can obtain this velocity from the relative angular velocity addition theorem as

$$ {}_I\boldsymbol{\omega}_{R_N} = {}_I\boldsymbol{\omega}_{R_1} + {}_{R_1}\boldsymbol{\omega}_{R_2} + \cdots + {}_{R_{N-1}}\boldsymbol{\omega}_{R_N} \tag{4.2.12} $$

With the right choice of the coordinate frames computing and then summing the angular velocity terms on the right side of Eq. (4.2.12) are usually easier than trying to calculate ${}_I\boldsymbol{\omega}_{R_N}$ itself directly.

We can differentiate the relative angular velocity addition theorem of Eq. (4.2.12) and use the transport theorem to obtain a relation between the angular accelerations. If, for example, there is the inertial frame I and two other frames A and B we have

$$ \begin{aligned} {}_I\boldsymbol{\alpha}_B &\equiv \left(\frac{d_I\boldsymbol{\omega}_B}{dt}\right)_I \\ &= \left(\frac{d_I\boldsymbol{\omega}_A}{dt}\right)_I + \left(\frac{d_A\boldsymbol{\omega}_B}{dt}\right)_I \\ &= \left(\frac{d_I\boldsymbol{\omega}_A}{dt}\right)_I + \left(\frac{d_A\boldsymbol{\omega}_B}{dt}\right)_A + {}_I\boldsymbol{\omega}_A \times {}_A\boldsymbol{\omega}_B \\ &= {}_I\boldsymbol{\alpha}_A + {}_A\boldsymbol{\alpha}_B + {}_I\boldsymbol{\omega}_A \times {}_A\boldsymbol{\omega}_B \end{aligned} \tag{4.2.13} $$

We can also use the transport theorem to relate the angular velocities ${}_A\boldsymbol{\omega}_B$ and ${}_B\boldsymbol{\omega}_A$ themselves since for any vector $\mathbf{p}$

$$ \begin{aligned} \left(\frac{d\mathbf{p}}{dt}\right)_A &= \left(\frac{d\mathbf{p}}{dt}\right)_B + {}_A\boldsymbol{\omega}_B \times \mathbf{p} \\ \left(\frac{d\mathbf{p}}{dt}\right)_B &= \left(\frac{d\mathbf{p}}{dt}\right)_A + {}_B\boldsymbol{\omega}_A \times \mathbf{p} \end{aligned} \tag{4.2.14} $$

so that combining these two relations we find

$$ ({}_B\boldsymbol{\omega}_A + {}_A\boldsymbol{\omega}_B) \times \mathbf{p} = 0 \tag{4.2.15} $$

which must be true for any vector $\mathbf{p}$ so that we have

$$ {}_B\boldsymbol{\omega}_A = -{}_A\boldsymbol{\omega}_B \tag{4.2.16} $$

which makes physical sense since, for example, if you are standing on a merry-go-round and rotating with it you will see the stationary bystanders as if they are rotating in the opposite sense of the merry-go-round.

4.2.1 Relative Velocity and Acceleration: More Explicit Forms

We can express all of the results of Sect. 4.4.1 in terms of the notation used in Eq. (4.2.6). For example, the relative velocity relationship of Eq. (4.1.7) can also be written as

$$\left(\dot{\mathbf{x}}_{P/B}\right)_I = \left(\dot{\mathbf{x}}_{P/B}\right)_R + {}_I\boldsymbol{\omega}_R \times \mathbf{x}_{P/B} \tag{4.2.17}$$

so that Eq. (4.1.8) can be written as

$$\left(\dot{\mathbf{x}}_{P/O}\right)_I = \left(\dot{\mathbf{x}}_{B/O}\right)_I + \left(\dot{\mathbf{x}}_{P/B}\right)_R + {}_I\boldsymbol{\omega}_R \times \mathbf{x}_{P/B} \tag{4.2.18}$$

where we have set $\left(\dot{\mathbf{x}}_P\right)_I = \left(\dot{\mathbf{x}}_{P/O}\right)_I$ and $\left(\dot{\mathbf{x}}_B\right)_I = \left(\dot{\mathbf{x}}_{B/O}\right)_I$ to make it clear that these terms are the velocities calculated with derivatives of position vectors measured relative to O. The acceleration relationship, Eq. (4.1.12), can likewise be expressed as

$$\begin{aligned}\left(\ddot{\mathbf{x}}_{P/O}\right)_I = \left(\ddot{\mathbf{x}}_{B/O}\right)_I &+ {}_I\boldsymbol{\alpha}_R \times \mathbf{x}_{P/B} + {}_I\boldsymbol{\omega}_R \times \left({}_I\boldsymbol{\omega}_R \times \mathbf{x}_{P/B}\right) + \left(\ddot{\mathbf{x}}_{P/B}\right)_R \\ &+ 2{}_I\boldsymbol{\omega}_R \times \left(\dot{\mathbf{x}}_{P/B}\right)_R\end{aligned} \tag{4.2.19}$$

Note that if the angular velocity of the R frame, ${}_I\boldsymbol{\omega}_R$, is expressed in terms of unit vectors of another frame, a, that is rotating with an angular velocity, ${}_I\boldsymbol{\omega}_a$, we would have

$${}_I\boldsymbol{\alpha}_R \equiv \left(\frac{d_I\boldsymbol{\omega}_R}{dt}\right)_I = \left(\frac{d_I\boldsymbol{\omega}_R}{dt}\right)_a + {}_I\boldsymbol{\omega}_a \times {}_I\boldsymbol{\omega}_R \tag{4.2.20}$$

We will see the use of this result in Chap. 8 when we examine the motion of a rigid body in a frame other than one attached to the body itself.

4.2.2 Relative Motion for Rigid Bodies

An important special case of the relative velocity and acceleration expressions that is applicable to rigid bodies is where we take a moving coordinate system to have the motion of a particular point, say C, of the rigid body and we let the coordinate system also rotate with the rigid body. Then for any other point A in the rigid body there is no velocity or acceleration of point A relative to C as measured in a coordinate system moving with the rigid body since the distance must remain constant between any two points in a rigid body. Thus, Eqs. (4.2.18) and (4.2.19) become, for a rigid body,

$$\left(\mathbf{v}_{A/O}\right)_I = \left(\mathbf{v}_{C/O}\right)_I + \boldsymbol{\omega} \times \mathbf{x}_{A/C} \tag{4.2.21a}$$

and

$$\left(\mathbf{a}_{A/O}\right)_I = \left(\mathbf{a}_{C/O}\right)_I + \boldsymbol{\alpha} \times \mathbf{x}_{A/C} + \boldsymbol{\omega} \times \left(\boldsymbol{\omega} \times \mathbf{x}_{A/C}\right) \tag{4.2.21b}$$

where $\boldsymbol{\omega}$ and $\boldsymbol{\alpha}$ are the angular velocity and angular acceleration of the rigid body as measured in the inertial frame I. These relations show that if we can find the motion of a point C of the rigid body and the rotational motion of the rigid body we then use these relations to find the motion of any other point A in the rigid body. From Chap. 3 we saw that the Newton-Euler equations relate the acceleration of the center of mass point, C, of a rigidly constrained system of particles to the net external force acting on the system and also relate the net external moment to the angular acceleration of the system. Thus, if we solve the equations of motion for the rigid system we can, with the use of Eqs. (4.2.21a) and (4.2.21b) determine the velocity and acceleration of all particles in the rigid system. This is also true for rigid bodies composed of a continuously distributed mass.

In the special case where we consider the motion of a two-dimensional rigid body in planar motion, these relative velocity and acceleration expressions can be made more explicit. Consider, for example, a rigid body that resides in the X–Y plane is translating in that plane and rotating about the Z-axis in an inertial X, Y, Z coordinate system with origin at O (Fig. 4.6). In this case the angular velocity of the body with respect to the inertial system is $\boldsymbol{\omega} = \omega \mathbf{E}_Z$ and the angular acceleration $\boldsymbol{\alpha} = \alpha \mathbf{E}_Z$,

Fig. 4.6 Planar motion of a rigid body in the X–Y plane

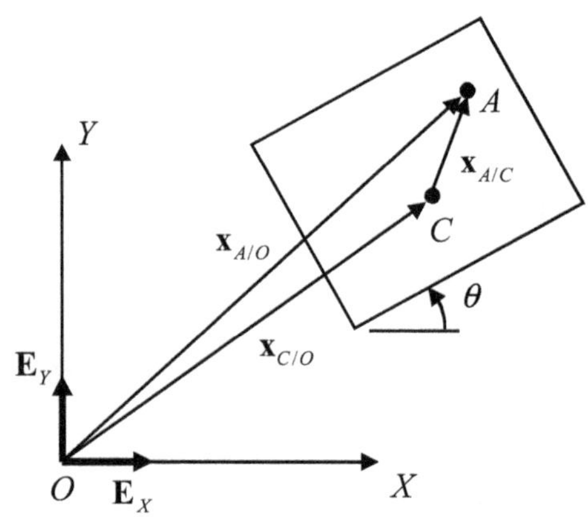

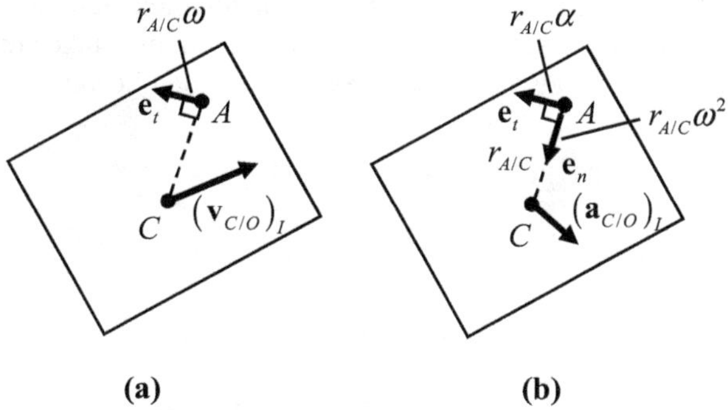

Fig. 4.7 For the rigid body of Fig. 4.6 in planar motion: (**a**) the velocity of point C, $(\mathbf{v}_{C/O})_I$, and the relative velocity of point A with respect to point C, $(\mathbf{v}_{A/C})_I = r_{A/C}\omega\mathbf{e}_t$, and (**b**) the acceleration of point C, $(\mathbf{a}_{C/O})_I$, and the relative acceleration of point A with respect to point C, $(\mathbf{a}_{A/C})_I = r_{A/C}\alpha\mathbf{e}_t + r_{A/C}\omega^2\mathbf{e}_n$

where $\omega = \dot{\theta}$, $\alpha = \ddot{\theta}$, in terms of the angle θ that the body makes with respect to the X-axis (Fig. 4.6). The relative velocity term, $\boldsymbol{\omega} \times \mathbf{x}_{A/C}$, is given by

$$\boldsymbol{\omega} \times \mathbf{x}_{A/C} = \omega\mathbf{E}_Z \times r_{A/C}\mathbf{e}_{A/C} = r_{A/C}\omega\,\mathbf{e}_t \tag{4.2.22}$$

where $r_{A/C} = |\mathbf{x}_{A/C}|$ is the distance from C to A, $\mathbf{e}_{A/C}$ is a unit vector directed along the line from C to A, and $\mathbf{e}_t = \mathbf{E}_Z \times \mathbf{e}_{A/C}$ is a unit vector in the X–Y plane perpendicular to the line from C to A (Fig. 4.7a). Since the distance from C to A is fixed, the relative motion of C with respect to A at any given instant is a circular motion of radius $r_{A/C}$ about point A and $\mathbf{e}_t$ is a unit vector tangent to that circular path. Thus, the relative velocity expression is

$$(\mathbf{v}_{A/O})_I = (\mathbf{v}_{C/O})_I + r_{A/C}\omega\mathbf{e}_t \tag{4.2.23a}$$

If we implicitly assume that the velocities of A and C are measured with respect to an inertial system we can write Eq. (4.2.23a) in the simpler form

$$\mathbf{v}_A = \mathbf{v}_C + r_{A/C}\omega\,\mathbf{e}_t \tag{4.2.23b}$$

The relative acceleration expression, Eq. (4.2.21b), also simplifies for planar motion. The relative acceleration term $\boldsymbol{\alpha} \times \mathbf{x}_{A/C}$ reduces like the relative velocity term to (Fig. 4.7b)

$$\boldsymbol{\alpha} \times \mathbf{x}_{A/C} = \alpha\mathbf{E}_Z \times r_{A/C}\mathbf{e}_{A/C} = r_{A/C}\alpha\,\mathbf{e}_t \tag{4.2.24}$$

while the double cross product term is

$$\boldsymbol{\omega} \times (\boldsymbol{\omega} \times \mathbf{x}_{A/C}) = \omega\mathbf{E}_Z \times r_{A/C}\omega\mathbf{e}_t = r_{A/C}\omega^2\,\mathbf{e}_n \tag{4.2.25}$$

where $\mathbf{e}_n = \mathbf{E}_Z \times \mathbf{e}_t$ is a unit vector in the X–Y plane directed along the line from A to C (Fig. 4.7b), i.e., it is normal to the circular path that point A takes relative to point C at that instant. The relative acceleration expression then becomes

$$\left(\mathbf{a}_{A/O}\right)_I = \left(\mathbf{a}_{C/O}\right)_I + r_{A/C}\alpha\,\mathbf{e}_t + r_{A/C}\omega^2\,\mathbf{e}_n \tag{4.2.26a}$$

or in the simpler form

$$\mathbf{a}_A = \mathbf{a}_C + r_{A/C}\alpha\mathbf{e}_t + r_{A/C}\omega^2\,\mathbf{e}_n \tag{4.2.26b}$$

These reduced and explicit terms for the relative velocity and acceleration terms for the planar motion of a rigid body allow one to examine the planar motion kinematics of interconnected systems of rigid bodies in a very simple manner using graphical or vector-based approaches. Many introductory dynamics books apply these equations to find the velocities and accelerations or angular velocities and accelerations of rigid bodies, usually at fixed instances of time, so that all the calculations can be done by hand. For example, consider the slider-crank mechanism shown in Fig. 4.8 where a slider (piston) P is connected to two rods. Rod AB is the crankshaft and rod BC the connecting rod. We will label bar AB as body (1), the rod BC as body (2), and the sliding piston P as body (3). The labels of these bodies are shown in Fig. 4.8. In this example we specify that the crankshaft is driven with a constant clockwise angular velocity of 6 rad/s, as shown in Fig. 4.8, and we wish to determine the velocity and acceleration of the piston at the time when the mechanism is in the position shown in Fig. 4.8. We can solve problems of this type by successively applying the relative velocity and acceleration expressions just developed for planar motion. Consider first the velocity analysis, where the velocities and relative velocities needed are shown in Fig. 4.9a. For the crankshaft (body 1) we have for the relative velocity relationship between points A and B

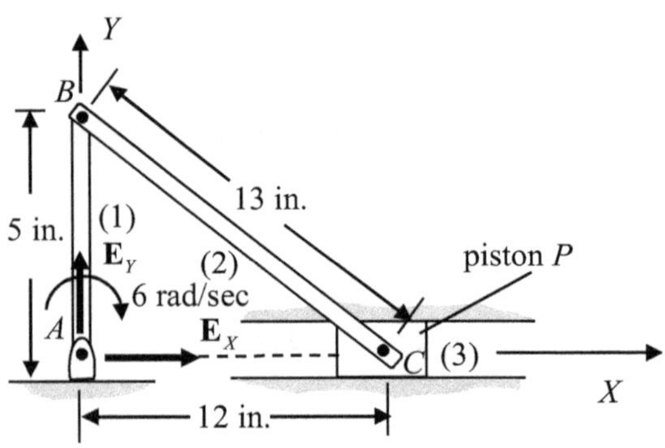

Fig. 4.8 A slider-crank mechanism

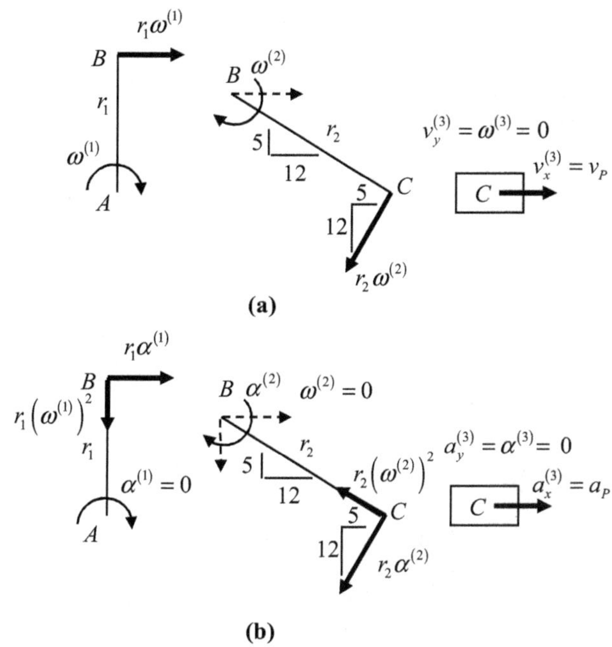

Fig. 4.9 The three bodies of the slider-crank mechanism seen in Fig. 4.8, showing (**a**) the linear and angular velocities needed to determine the velocity of the piston, *P*, and (**b**) the linear and angular accelerations (and angular velocities) needed to determine the acceleration of the piston

$$
\begin{aligned}
\mathbf{v}_B^{(1)} &= \mathbf{v}_A^{(1)} + \boldsymbol{\omega}^{(1)} \times \mathbf{x}_{B/A}^{(1)} \\
&= \mathbf{v}_A^{(1)} + r_1\omega^{(1)}\mathbf{e}_t^{(1)} \\
&= 0 + (5)(6)\mathbf{E}_X \\
&= 30\mathbf{E}_X \,\text{in./s}
\end{aligned}
$$

where we have used the fact that the crankshaft is pinned to the fixed ground at *A* and so has no velocity, and we let $\mathbf{e}_t^{(1)} = \mathbf{E}_X$. [Note that if we had chosen to calculate the cross product expression for the relative velocity term directly we would have to set $\boldsymbol{\omega}^{(1)} = -6\mathbf{E}_Z$ and $\mathbf{x}_{B/A} = 5\mathbf{E}_Y$.] For the connecting rod, relating points *B* and *C* we have

$$
\begin{aligned}
\mathbf{v}_C^{(2)} &= \mathbf{v}_B^{(2)} + \boldsymbol{\omega}^{(2)} \times \mathbf{x}_{C/B}^{(2)} \\
&= \mathbf{v}_B^{(1)} + r_2\omega^{(2)}\mathbf{e}_t^{(2)} \\
&= 30\mathbf{E}_X + 13\omega^{(2)}\left[-\frac{5}{13}\mathbf{E}_X - \frac{12}{13}\mathbf{E}_Y\right] \\
&= \left(30 - 5\omega^{(2)}\right)\mathbf{E}_X - 12\omega^{(2)}\mathbf{E}_Y \;\text{ in./s}
\end{aligned}
$$

where we have assumed that the unknown angular velocity of the rod is $\omega^{(2)}$ in the direction shown in Fig. 4.9a and have used the fact that the rods are pinned to each other at point B and so have the same velocity. [To calculate the relative velocity term with the cross product we would use $\boldsymbol{\omega}^{(2)} = -\omega^{(2)}\mathbf{E}_Z$ and $\mathbf{x}^{(2)}_{C/B} = 12\mathbf{E}_X - 5\mathbf{E}_Y$.] But the velocity of the piston is along the X-direction only and is the same as that of point C so we have

$$\mathbf{v}^{(2)}_C = v_P\mathbf{E}_X = \left(30 - 5\omega^{(2)}\right)\mathbf{E}_X - 12\omega^{(2)}\mathbf{E}_Y$$

which gives $\omega^{(2)} = 0$, $v_P = 30$ in./s, yielding the answer for the velocity of the piston as $\mathbf{v}_P = 30\mathbf{E}_X$ in./s. At this instant of time the angular velocity of the connecting rod is zero, but at earlier or later times where the positions have changed this will not be true so that the angular acceleration of the connecting rod is not necessarily zero, as we will see. To obtain the acceleration of the piston we proceed in a similar fashion, using the relative acceleration expressions (see Fig. 4.9b). For the crankshaft, since its angular velocity is a constant, $\alpha^{(1)} = 0$ and we have

$$\begin{aligned}\mathbf{a}^{(1)}_B &= \mathbf{a}^{(1)}_A + \boldsymbol{\alpha}^{(1)} \times \mathbf{x}^{(1)}_{B/A} + \boldsymbol{\omega}^{(1)} \times \left(\boldsymbol{\omega}^{(1)} \times \mathbf{x}^{(1)}_{B/A}\right)\\ &= \mathbf{a}^{(1)}_A + r_1\alpha^{(1)}\mathbf{e}^{(1)}_t + r_1\left(\omega^{(1)}\right)^2\mathbf{e}^{(1)}_n\\ &= 0 + (5)(0)\mathbf{E}_X + (5)(36)(-\mathbf{E}_Y)\\ &= -180\mathbf{E}_Y \ \text{in./s}\end{aligned}$$

For the connecting rod (see Fig. 4.9b)

$$\begin{aligned}\mathbf{a}^{(2)}_C &= \mathbf{a}^{(2)}_B + \boldsymbol{\alpha}^{(2)} \times \mathbf{x}^{(2)}_{C/B} + \boldsymbol{\omega}^{(2)} \times \left(\boldsymbol{\omega}^{(2)} \times \mathbf{x}^{(2)}_{C/B}\right)\\ &= \mathbf{a}^{(1)}_B + r_2\alpha^{(2)}\mathbf{e}^{(2)}_t + r_2\left(\omega^{(1)}\right)^2\mathbf{e}^{(2)}_n\\ &= -180\mathbf{E}_Y + 13\alpha^{(2)}\left[-\frac{5}{13}\mathbf{E}_X - \frac{12}{13}\mathbf{E}_Y\right] + (13)(0)\left[-\frac{12}{13}\mathbf{E}_X + \frac{5}{13}\mathbf{E}_Y\right]\\ &= -5\alpha^{(2)}\mathbf{E}_X - \left(180 + 12\alpha^{(2)}\right)\mathbf{E}_Y\end{aligned}$$

where we have used the fact that the acceleration is the same for bodies 1 and 2 at point B since they are connected by a pin at that point. Since the acceleration of the piston is the same as the acceleration of point C and is along the X-direction we find

$$\mathbf{a}^{(2)}_C = a_P\mathbf{E}_X = -5\alpha^{(2)}\mathbf{E}_X - \left(180 + 12\alpha^{(2)}\right)\mathbf{E}_Y$$

which yields $\alpha^{(2)} = -15 \ \text{rad/s}^2$ and the answer for the acceleration of the piston as $\mathbf{a}_P = 75\mathbf{E}_X \ \text{in./s}^2$.

There are several important features of this kinematics problem that we should mention. First, we see that we need to solve for the unknown linear and/or angular velocities from the relative velocity relations before we can perform the acceleration analysis. These relative velocity and relative acceleration relations are linear, so that it is easy to obtain their solution. Note, however, that we did not have to perform a positional analysis since the positions and angles of the bodies were given at the particular time at which the calculations were performed. Positional relations, as we will see, are generally nonlinear and so may require a numerical solution. Second, this problem was an example of a one degree of freedom kinematically driven mechanism, which means that if we specify the motion, say, of the single generalized coordinate, $\theta_1(t)$, of the crankshaft, for example (which also means we know the angular velocity, $\dot{\theta}_1$, and angular acceleration, $\ddot{\theta}_1$, of the crankshaft) then we can solve for the motion of any other parts in the mechanism such as the piston. We did not state that fact up front but simply assumed that we could find the motion of the piston from the given information. Clearly, this fixed time type of a kinematic analysis gives us a very limited view of the actual dynamical motion of the mechanism so that it is important to determine the motion of kinematically driven systems at any time, which we will now do.

4.2.3 The Analysis of Kinematically Driven Systems: I

Consider the slider-crank problem again, but now let us examine the mechanism when it is in a general position and let the lengths of the crankshaft and connecting rod be (l_1, l_2), respectively (Fig. 4.10). The geometry of this mechanism at any time can be completely described if we know the values of the angles $(\theta_1(t), \theta_2(t))$ and the displacement of the piston, $X_P(t)$, at any time t. These generalized coordinates for the mechanism, however, are not independent, so we must examine the constraints present. Note that regardless of these angles and displacement the vector from A to B plus the vector from B to C plus the vector from C to A must add up to zero, i.e., **AB**+**BC**+**CA** = 0. This is called the *loop closure equation*. Explicitly, in terms of geometry of Fig. 4.10 we have

$$(l_1 \cos\theta_1 \mathbf{E}_X + l_1 \sin\theta_1 \mathbf{E}_Y) + (l_2 \cos\theta_2 \mathbf{E}_X - l_2 \sin\theta_2 \mathbf{E}_Y) - X_P \mathbf{E}_X = 0 \quad (4.2.27)$$

or, in terms of the X, Y components,

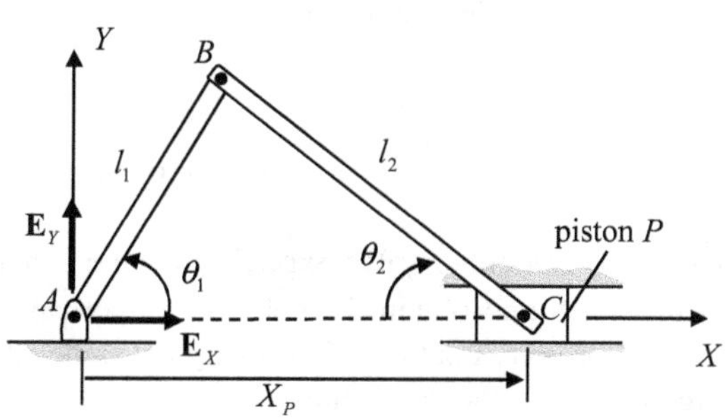

Fig. 4.10 The geometry for a slider-crank mechanism in a general position

$$\begin{aligned} &l_1 \cos\theta_1 + l_2 \cos\theta_2 - X_P = 0 \\ &l_1 \sin\theta_1 - l_2 \sin\theta_2 = 0 \end{aligned} \tag{4.2.28}$$

which are two constraint equations for the three generalized coordinates $(\theta_1, \theta_2, X_P)$. This clearly shows that the slider-crank is indeed a one degree of freedom system. If we specify, say, the crankshaft motion, $\theta_1(t)$, then we see that in principle Eq. (4.2.28) can be solved for the other two coordinates in terms of θ_1. However, as mentioned previously, solving these positional relations is not as easy as using the velocity and acceleration relations since the positional relations are generally nonlinear. In this simple case this is not a particular problem since we can rewrite these equations as

$$\begin{aligned} X_P &= l_1 \cos\theta_1 \pm l_2\sqrt{1 - \sin^2\theta_2} \\ \sin\theta_2 &= l_1 \sin\theta_1 / l_2 \end{aligned} \tag{4.2.29}$$

Placing the second equation in Eq. (4.2.29) into the first equation and solving the second equation directly give

$$\begin{aligned} X_P &= l_1 \cos\theta_1 \pm \sqrt{l_2^2 - l_1^2 \sin^2\theta_1} \\ \theta_2 &= \sin^{-1}\left(\frac{l_1 \sin\theta_1}{l_2}\right) \end{aligned} \tag{4.2.30}$$

where the plus or minus sign in the first equation in Eq. (4.2.30) depends on whether the piston is started out to the right or left of point B. Note that we must have $|l_1 \sin\theta_1/l_2| \leq 1$ in order to have a real solution for θ_2. Obviously, we can always guarantee this is the case if $l_1/l_2 < 1$. While we do have an analytical solution available for this problem, more complex problems will typically require a numerical solution. In Sect. 4.2.5 we will examine how to obtain numerical solutions to such systems of nonlinear algebraic equations.

Now, let us consider the velocity and acceleration analysis for this problem where we obtain expressions that will allow us to examine the motion of the slider-crank mechanism for all times t. One way to do this analysis is to differentiate the positional equations of Eq. (4.2.28) on time. Taking one derivative of these equations, for example, yields the set of linear equations

$$\begin{bmatrix} 1 & l_2 \sin\theta_2 \\ 0 & l_2 \cos\theta_2 \end{bmatrix} \begin{Bmatrix} \dot{X}_P \\ \dot{\theta}_2 \end{Bmatrix} = \begin{Bmatrix} -l_1\dot{\theta}_1 \sin\theta_1 \\ l_1\dot{\theta}_1 \cos\theta_1 \end{Bmatrix} \tag{4.2.31}$$

which we can either solve explicitly for the generalized velocities $\left(\dot{X}_P, \dot{\theta}_2\right)$ or leave in this form and obtain the solution numerically. A second approach is to use the relative velocity relations. Following the same steps as used previously we have (see Fig. 4.11a) for the crankshaft

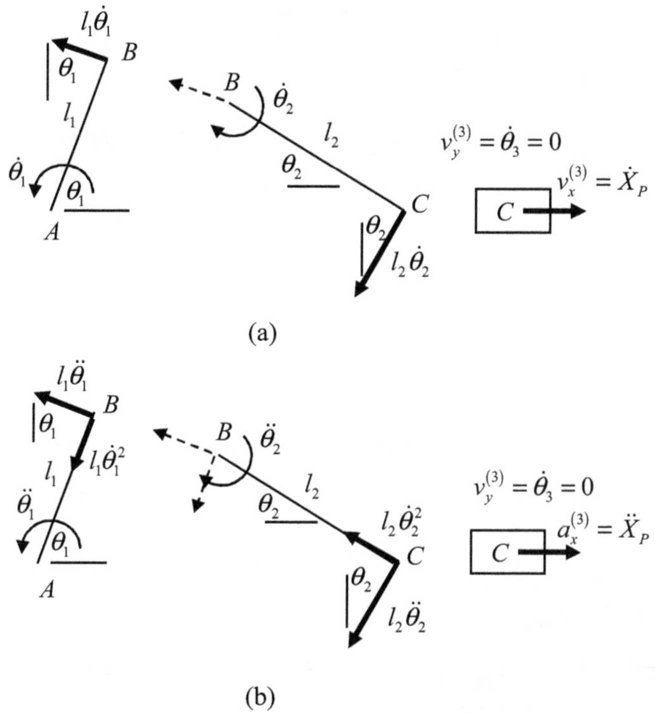

Fig. 4.11 (**a**) The three bodies of the slider-crank mechanism shown in Fig. 4.10, showing (a) the linear and angular velocities needed to determine the velocities of the system, and (**b**) the linear and angular accelerations (and angular velocities) needed to determine the accelerations of the system

$$\begin{aligned}\mathbf{v}_B^{(1)} &= \mathbf{v}_A^{(1)} + l_1\dot{\theta}_1\mathbf{e}_t^{(1)} \\ &= 0 + l_1\dot{\theta}_1(-\sin\theta_1\mathbf{E}_X + \cos\theta_1\mathbf{E}_Y)\end{aligned} \tag{4.2.32}$$

and for the connecting rod and piston

$$\begin{aligned}\mathbf{v}_C^{(2)} &= \mathbf{v}_B^{(2)} + l_2\dot{\theta}_2\mathbf{e}_t^{(2)} \\ &= l_1\dot{\theta}_1(-\sin\theta_1\mathbf{E}_X + \cos\theta_1\mathbf{E}_Y) + l_2\dot{\theta}_2(-\sin\theta_2\mathbf{E}_X - \cos\theta_2\mathbf{E}_Y) \\ &= \dot{X}_P\mathbf{E}_X\end{aligned} \tag{4.2.33}$$

which yields the two equations

$$\begin{aligned}\dot{X}_P + l_2\sin\theta_2\dot{\theta}_2 &= -l_1\dot{\theta}_1\sin\theta_1 \\ l_2\cos\theta_2\dot{\theta}_2 &= l_1\dot{\theta}_1\cos\theta_1\end{aligned} \tag{4.2.34}$$

which are identical to the matrix system of equations in Eq. (4.2.31). For the acceleration analysis we can differentiate these velocity equations once again on time to yield the linear equations

$$\begin{bmatrix} 1 & l_2 \sin\theta_2 \\ 0 & l_2 \cos\theta_2 \end{bmatrix} \begin{Bmatrix} \ddot{X}_P \\ \ddot{\theta}_2 \end{Bmatrix} = \begin{Bmatrix} -l_2\dot{\theta}_2^2 \cos\theta_2 - l_1\ddot{\theta}_1 \sin\theta_1 - l_1\dot{\theta}_1^2 \cos\theta_1 \\ l_2\dot{\theta}_2^2 \sin\theta_2 + l_1\ddot{\theta}_1 \cos\theta_1 - l_1\dot{\theta}_1^2 \sin\theta_1 \end{Bmatrix} \tag{4.2.35}$$

which can be solved for $\left(\ddot{X}_P, \ddot{\theta}_2\right)$. Alternatively, we can use the relative acceleration expressions. For body 1 we have

$$\begin{aligned} \mathbf{a}_B^{(1)} &= \mathbf{a}_A^{(1)} + l_1\ddot{\theta}_1\mathbf{e}_t^{(1)} + l_1\dot{\theta}_1^2\mathbf{e}_n^{(1)} \\ &= 0 + l_1\ddot{\theta}_1(-\sin\theta_1\mathbf{E}_X + \cos\theta_1\mathbf{E}_Y) + l_1\dot{\theta}_1^2(-\cos\theta_1\mathbf{E}_X - \sin\theta_1\mathbf{E}_Y) \\ &= \left(-l_1\ddot{\theta}_1 \sin\theta_1 - l_1\dot{\theta}_1^2 \cos\theta_1\right)\mathbf{E}_X + \left(l_1\ddot{\theta}_1 \cos\theta_1 - l_1\dot{\theta}_1^2 \sin\theta_1\right)\mathbf{E}_Y \end{aligned} \tag{4.2.36}$$

and for body 2 and body 3

$$\begin{aligned} \mathbf{a}_C^{(2)} &= \mathbf{a}_B^{(2)} + l_2\ddot{\theta}_2\mathbf{e}_t^{(2)} + l_2\dot{\theta}_2^2\mathbf{e}_n^{(2)} \\ &= \mathbf{a}_B^{(1)} + l_2\ddot{\theta}_2(-\sin\theta_2\mathbf{E}_X - \cos\theta_2\mathbf{E}_Y) + l_2\dot{\theta}_2^2(-\cos\theta_2\mathbf{E}_X + \sin\theta_2\mathbf{E}_Y) \\ &= \left(-l_2\ddot{\theta}_2 \sin\theta_2 - l_2\dot{\theta}_2^2 \cos\theta_2 - l_1\ddot{\theta}_1 \sin\theta_1 - l_1\dot{\theta}_1^2 \cos\theta_1\right)\mathbf{E}_X \\ &\quad + \left(-l_2\ddot{\theta}_2 \cos\theta_2 + l_2\dot{\theta}_2^2 \sin\theta_2 + l_1\ddot{\theta}_1 \cos\theta_1 - l_1\dot{\theta}_1^2 \sin\theta_1\right)\mathbf{E}_Y \\ &= \ddot{X}_P\mathbf{E}_X \end{aligned} \tag{4.2.37}$$

which yield the two equations

$$\begin{aligned} \ddot{X}_p + l_2 \sin\theta_2\ddot{\theta}_2 &= -l_2\dot{\theta}_2^2 \cos\theta_2 - l_1\ddot{\theta}_1 \sin\theta_1 - l_1\dot{\theta}_1^2 \cos\theta_1 \\ l_2 \cos\theta_2\ddot{\theta}_2 &= l_2\dot{\theta}_2^2 \sin\theta_2 + l_1\ddot{\theta}_1 \cos\theta_1 - l_1\dot{\theta}_1^2 \sin\theta_1 \end{aligned} \tag{4.2.38}$$

which is identical to the system of equations in Eq. (4.2.35).

Simulation 4.1: The Slider-Crank

The velocity and acceleration equations just obtained for the slider-crank have been placed into the MATLAB® script slider_crank. The motion of the crankshaft is specified as $\theta_1(t) = \theta_{10} - 6t$ rad, where θ_{10} is the starting angle and t is the time in seconds which corresponds to having the crank driven at a constant clockwise angular velocity of 6 rad/s, as assumed for the slider-crank problem of Fig. 4.8. The positional equations of Eq. (4.2.30), the velocity equations of Eq. (4.2.31), and the acceleration equations of Eq. (4.2.35) are solved in the script over a time interval

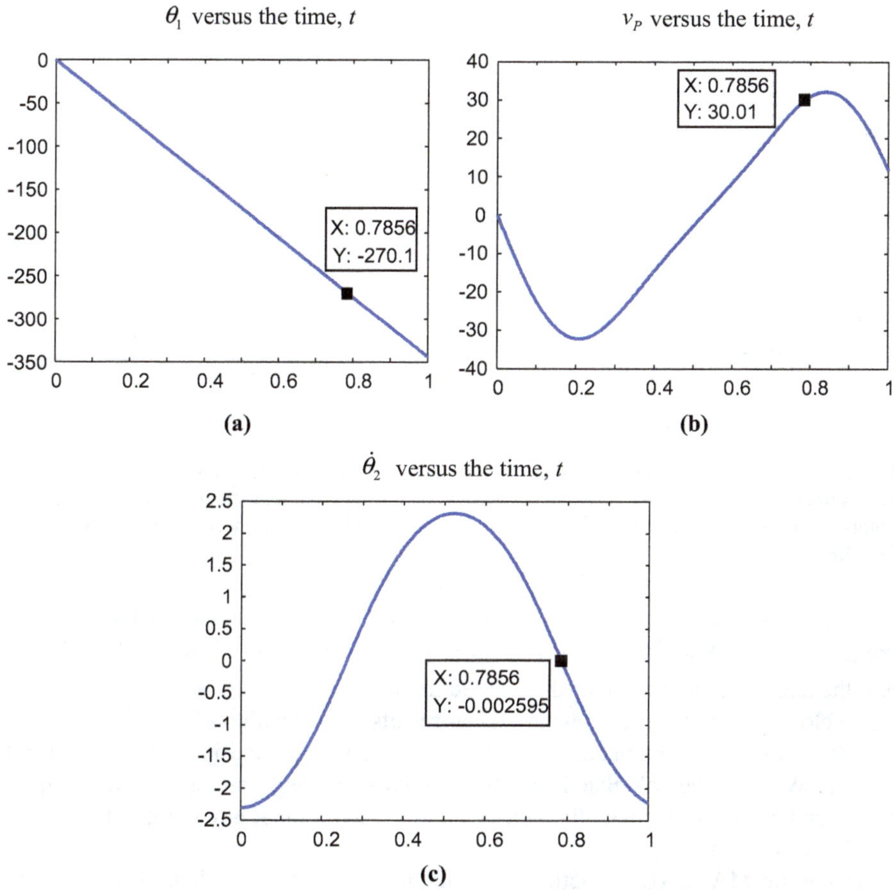

Fig. 4.12 Kinematic analysis of the crank-slider, showing (**a**) the angle of the crank (deg) versus time, (**b**) the velocity of the slider (in./s) versus time, and (**c**) the angular velocity (rad/s) of the connecting rod versus time. The marked points are the values approximately at the time when the position of the mechanism is as shown in Fig. 4.8

of 1 s, using $l_1 = 5$ in. and $l_2 = 13$ in., which again corresponds to the parameters present in Fig. 4.8. The script plots (1) the angular position, θ_1, of the crankshaft (in deg) versus time; (2) the velocity of the slider (piston), $\dot{X}_P(t) = v_P(t)$ in./s; (3) the angular velocity of the connecting rod, $\dot{\theta}_2(t)$ (rad/s); (4) the angular acceleration of the piston, $\ddot{X}_P(t) = a_P(t)$ in./s^2; and (5) the angular acceleration of the connecting rod, $\ddot{\theta}_2(t)$ rad/s^2. Figure 4.12a shows the angle of the crank versus time, where the crank starts out with $\theta_1 = 0$ ($\theta_{10} = 0$). At a time $t = 0.785$ s, the crank is at an angle of $-270.1°$ (Fig. 4.12a), which is very close to the case shown in Fig. 4.8. The velocity of the piston at this time is 30.01 in./s (Fig. 4.12b), and the angular velocity of the connecting rod at this time is a very small value of –0.0026 rad/s^2, approximately (Fig. 4.12c), values that are very close to what we obtained analytically. The corresponding acceleration results are shown in Fig. 4.13, where we can again see

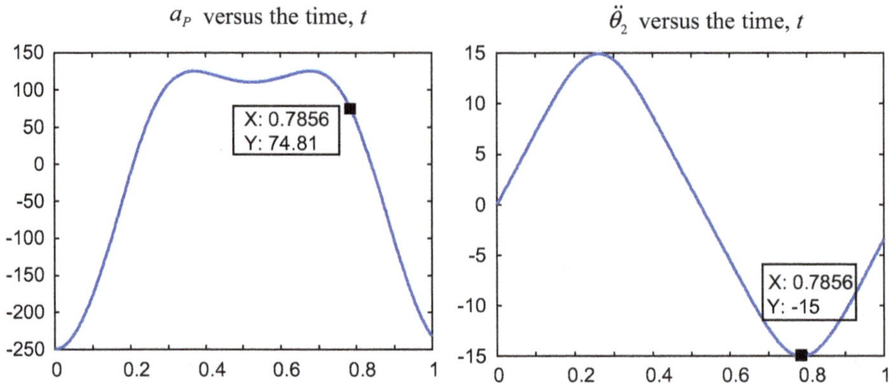

(a) (b)

Fig. 4.13 Kinematic analysis of the slider-crank, showing (**a**) the acceleration of the slider in in. /s^2 versus time, and (**b**) the angular acceleration of the connecting rod in rad/s^2 versus time. The marked points are values approximately at the time when the position of the mechanism is as shown in Fig. 4.8

that at the specific time of $t = 0.7856$ s, when the mechanism is close to the configuration of Fig. 4.8, the acceleration of the piston is $\ddot{X}_P = a_P = 74.81\ \text{in.}/\text{s}^2$ and the angular acceleration of the connecting rod is $\ddot{\theta}_2 = -15\ \text{rad}/\text{s}^2$, which again agree closely with our previous analytical results. The position of the piston, $X_P(t)$, and the angle of the connecting rod, $\theta_2(t)$, are also calculated in the script but not plotted. With all the calculated results the relative velocity and acceleration equations can be used to evaluate the velocity and acceleration at any point in the slider-crank mechanism.

Using the MATLAB® function animate_lines described in Chap. 3 it is easy to see the slider-crank motion explicitly for this problem (see Problem P4.17).

4.2.4 Singular Configurations

If a rigid body mechanism such as the slider-crank is not designed properly, there can be situations where the mechanism will "lock up" and a smooth progression of the motion ceases. To see this, let's obtain the explicit solution of the linear velocity equations of Eq. (4.2.31). We have

$$\begin{aligned}\dot{X}_P &= \frac{1}{l_2\cos\theta_2}\left(-l_1l_2\dot{\theta}_1\sin\theta_1\cos\theta_2 - l_1l_2\dot{\theta}_1\cos\theta_1\sin\theta_2\right)\\ &= \frac{1}{l_2\cos\theta_2}\left(-l_1l_2\dot{\theta}_1\sin\left(\theta_1+\theta_2\right)\right)\\ \dot{\theta}_2 &= \frac{1}{l_2\cos\theta_2}\left(l_1\dot{\theta}_1\cos\theta_1\right)\end{aligned} \tag{4.2.39}$$

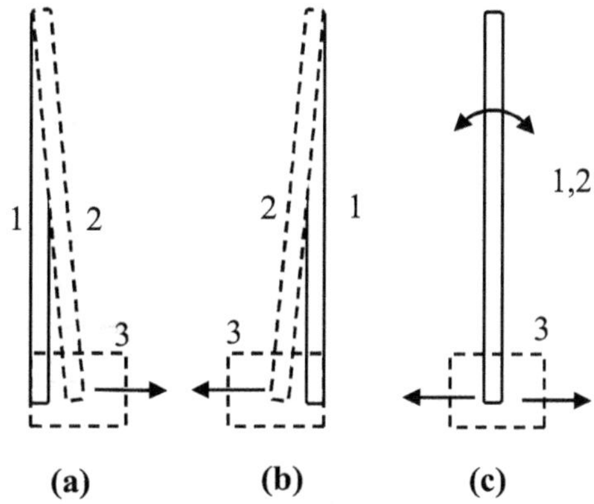

Fig. 4.14 The slider-crank near or at a "singular" position where multiple solutions are possible, such as (**a**) bar 2 moving with the piston to the right, or (**b**) where bar 2 moves with the piston to the left, or (**c**) where bars 1 and 1 are "locked" together in their motion and rotate as a unit to the left or right while the piston can move independently either to the left or to the right

where note that $\Delta = l_2 \cos\theta_2$ is the determinant of the matrix of coefficients in Eq. (4.2.31). If we let $l_1 = l_2 = l$, then the second positional constraint of Eq. (4.2.29) gives $\sin\theta_1 = \sin\theta_2$. If $\theta_1 = \pi/2$ then $\theta_2 = \pi/2$ and the determinant $\Delta = 0$. This physically corresponds to the configurations shown in Fig. 4.14 where there can be motion of bar 2 to the right or left together with the piston or where the two bars rotate to the left or right together as a unit while the piston moves independently to the left or right. We can of course design the mechanism so that such singular configurations do not occur, but in complex systems the detection of singular configurations may be more difficult since they also depend on the manner in which the system is driven. For example, consider the case where the motion of the slider (piston) in the slider-crank is specified instead of the crank angle. In this case we can rearrange the velocity equations of Eq. (4.2.31) or Eq. (4.2.34) so that we solve instead for $(\dot{\theta}_1, \dot{\theta}_2)$. We find

$$\begin{bmatrix} l_1 \sin\theta_1 & l_2 \sin\theta_2 \\ -l_1 \cos\theta_1 & l_2 \cos\theta_2 \end{bmatrix} \begin{Bmatrix} \dot{\theta}_1 \\ \dot{\theta}_2 \end{Bmatrix} = \begin{Bmatrix} -\dot{X}_P \\ 0 \end{Bmatrix} \tag{4.2.40}$$

with the solution

$$\begin{aligned} \dot{\theta}_1 &= \frac{1}{l_1 l_2 \sin(\theta_1 - \theta_2)} \left(-l_2 \cos\theta_2 \dot{X}_P\right) \\ \dot{\theta}_2 &= \frac{1}{l_1 l_2 \sin(\theta_1 - \theta_2)} \left(-l_1 \cos\theta_1 \dot{X}_P\right) \end{aligned} \tag{4.2.41}$$

where the determinant

$$\begin{aligned} \Delta &= l_1 l_2 \sin\theta_1 \cos\theta_2 - l_1 l_2 \sin\theta_2 \cos\theta_1 \\ &= l_1 l_2 \sin(\theta_1 - \theta_2) \end{aligned} \tag{4.2.42}$$

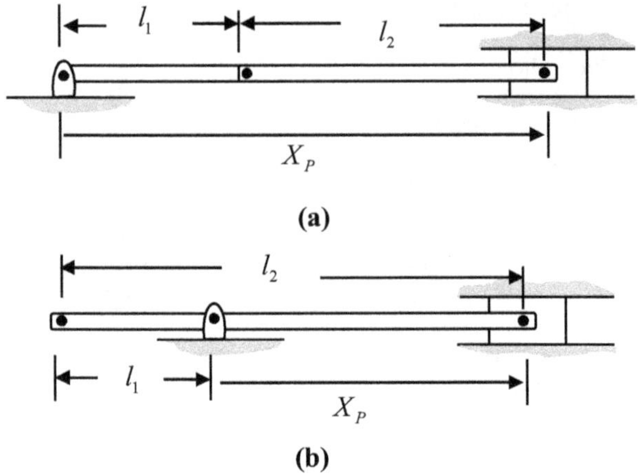

Fig. 4.15 The slider-crank in the case where the slider (piston) is being driven, i.e., $X_P(t)$ is specified, at "singular" positions where (**a**) $\theta_1 = \theta_2 = 0$, and (**b**) where $\theta_2 = 0$ and $\theta_1 = \pi$

where $\Delta = 0$, for example, if $\theta_1 = \theta_2 = 0$ corresponding to the singular configuration shown in Fig. 4.15a or for $\theta_1 = \pi$, $\theta_2 = 0$, as shown in Fig. 4.15b. Note, however, that these are not singular positions when the crankshaft is being driven.

4.2.5 Numerical Solution of the Position Equations

As seen in the slider-crank example, the relative velocity and acceleration expressions lead to a system of linear equations for the unknown velocities or accelerations. The constraint equations which relate the unknown generalized coordinates to the known "driving" coordinate(s), however, are not linear so that it may be difficult, as mentioned previously, to find a closed-form analytical solution as done in the slider-crank example. Thus, it is necessary to have a way to obtain this solution numerically. One way to do this is through the use of the *Newton-Raphson method*, which is a well-known method for solving systems of nonlinear algebraic equations. In this section we will discuss this method from a general standpoint where we have n nonlinear constraint equations for n generalized coordinates $\mathbf{q} = (q_1, q_2, \ldots, q_n)$ of the form

$$f_i(\mathbf{q}, t) = 0 \quad (i = 1, 2, \ldots, n) \tag{4.2.43}$$

that we want to solve for $\mathbf{q}$ as a function of the time, t. A kinematically driven holonomic system of constraints can always be placed in this form. For the slider-crank problem where the motion of the crankshaft is given, for example, we can write these constraint equations for $\mathbf{q} = (q_1, q_2, q_3) = (\theta_1, \theta_2, X_P)$ as

$$\begin{aligned}\theta_1 + 6t - \theta_{10} &= 0 \\ l_1\cos\theta_1 + l_2\cos\theta_2 - X_P &= 0 \\ l_1\sin\theta_1 - l_2\sin\theta_2 &= 0\end{aligned} \tag{4.2.44}$$

To solve the constraint equations, Eq. (4.2.43), we start with a guess for the solution as $\tilde{\mathbf{q}}^p$ at some fixed time, t_k. If we expand these functions in a Taylor series about this guess we have to first order

$$f_i(\mathbf{q}, t_k) = f_i(\tilde{\mathbf{q}}^p, t_k) + \sum_{m=1}^{n} \left.\frac{\partial f_i}{\partial q_m}\right|_{\mathbf{q}=\tilde{\mathbf{q}}^p, t=t_k} (q_m - \tilde{q}_m^p) \tag{4.2.45}$$

where we assume that $q_m - \tilde{q}_m^p$ is small enough so that we can neglect all higher order terms. The matrix of derivatives of the constraint functions is called the constraint *Jacobian matrix*, $\mathbf{f_q}(\tilde{\mathbf{q}}^p, t_k)$, where

$$\mathbf{f_q}(\tilde{\mathbf{q}}^p, t_k) = \begin{bmatrix} \partial f_1/\partial q_1 & \partial f_1/\partial q_2 & \dots & \partial f_1/\partial q_n \\ \partial f_2/\partial q_1 & \partial f_2/\partial q_2 & \dots & \partial f_2/\partial q_n \\ & & & \\ \partial f_n/\partial q_1 & \partial f_n/\partial q_2 & \dots & \partial f_n/\partial q_n \end{bmatrix} \tag{4.2.46}$$

We encountered a similar matrix in Chap. 2 when we examined constrained motion. Here, however, since we include the equations for the driving coordinates as well as the other coordinates that are dependent on those driving coordinates, the Jacobian matrix is always square. In vector-matrix form Eq. (4.2.45) can be written, using Eq. (4.2.43), as

$$\mathbf{f}(\mathbf{q}, t_k) = \mathbf{f}(\tilde{\mathbf{q}}^p, t_k) + \mathbf{f_q}(\tilde{\mathbf{q}}^p, t_k)(\mathbf{q} - \tilde{\mathbf{q}}^p) = 0 \tag{4.2.47}$$

If the constraint equations are independent, we generally expect the Jacobian matrix to be non-singular and have an inverse so that we can solve Eq. (4.2.47) to yield

$$\mathbf{q} = \tilde{\mathbf{q}}^p - \mathbf{f_q^{-1}}(\tilde{\mathbf{q}}^p, t_k)\mathbf{f}(\tilde{\mathbf{q}}^p, t_k) \tag{4.2.48}$$

and we can take $\mathbf{q}$ as an updated guess, $\tilde{\mathbf{q}}^{p+1}$, which then can be placed back into Eq. (4.2.48) to solve for the next updated set of values in the same manner. Thus, we successively perform the updates

$$\tilde{\mathbf{q}}^{p+1} = \tilde{\mathbf{q}}^p - \mathbf{f_q^{-1}}(\tilde{\mathbf{q}}^p, t_k)\mathbf{f}(\tilde{\mathbf{q}}^p, t_k) \quad (p = 1, 2, \dots) \tag{4.2.49}$$

until we reach some desired accuracy in the result. One way to specify this is to ensure that the magnitudes of the constraint functions themselves are sufficiently small, i.e.,

$$\left|\mathbf{f}\left(\tilde{\mathbf{q}}^{p}, t_k\right)\right| \leq \text{ftol} \tag{4.2.50}$$

where ftol is a user-specified tolerance. One can also specify an upper limit on the number of iterations, N_{iter}, i.e., $p \leq N_{\text{iter}}$, and a maximum allowable value, maxval, on the magnitude of the constraint functions, i.e., $\left|\mathbf{f}\left(\tilde{\mathbf{q}}^{p}, t_k\right)\right| \leq$ maxval, since if the initial guess is not close enough to the actual solution or the system is near a singular configuration the solution may diverge or converge too slowly. Note that to implement the Newton-Raphson approach we need to evaluate the constraint functions as well as the elements of the Jacobian matrix at the successive approximation values. For the slider-crank problem, therefore, we need to have the values of the constraint functions

$$\mathbf{f} = \begin{Bmatrix} f_1 \\ f_2 \\ f_3 \end{Bmatrix} = \begin{Bmatrix} \theta_1 + 6t - \theta_{10} \\ l_1 \cos\theta_1 + l_2 \cos\theta_2 - X_P \\ l_1 \sin\theta_1 - l_2 \sin\theta_2 \end{Bmatrix} \tag{4.2.51}$$

as well as values of the Jacobian

$$\mathbf{f_q} = \begin{bmatrix} 1 & 0 & 0 \\ -l_1 \sin\theta_1 & -l_2 \sin\theta_2 & -1 \\ l_1 \cos\theta_1 & -l_2 \cos\theta_2 & 0 \end{bmatrix} \tag{4.2.52}$$

Simulation 4.2: Numerical Solution of the Positional Constraints of a Kinematically Driven Rigid Body System

For the slider-crank problem the **f** and $\mathbf{f_q}$ functions have been placed into the MATLAB® functions slider_crank_f and slider_crank_J, respectively, which have the calling sequences

```
[f] = slider_crank_f (t, x,d1,d2);
[J] = slider_crank_J (t, x,d1,d2);
```

where the scalar t is the time, x is a column vector of generalized coordinates, $\{x\} = [\theta_1, \theta_2, X_P]^T$, and d1 and d2 are the lengths of the crankshaft and connecting rod. The output f is a column vector containing the values of the constraint functions and J is the matrix of derivative values for the constraints that define the Jacobian. A MATLAB® function, newton_raph, which implements the Newton-Raphson method, has the calling sequence

```
[x, iter] =newton_raph(@fun, @jac, t, x0, ftol, varargin);
```

where @fun and @jac are function handles for the constraint and Jacobian functions, which here are the functions slider_crank_f and slider_crank_J, respectively; the scalar t is the time; and x0 is a row vector containing an initial guess for the solution to this problem. The parameter, ftol, is a user-supplied error tolerance based on the magnitude of the constraint functions and varargin is an optional set of additional parameters that are input arguments for the constraint and Jacobian functions (besides the t and x). Here, varargin is replaced by d1, d2, and ang10, where ang10 = θ_{10}. Note that all of these optional parameters must be included as input arguments to both the constraint function and the Jacobian, even if not all of the parameters are used in both of those functions. The output argument, x, is a 3×1 column vector containing the approximate values of the three generalized coordinates, $(\theta_1, \theta_2, X_P)$, at the time, *t*, and iter is a scalar which contains the numbers of iterations needed at time, *t*, to find a solution. [Note: Within the newton_raph function there is an error tolerance variable jtol = e-10. If the determinant of the Jacobian matrix is less than this tolerance an error message is generated that the Jacobian is singular. There is also an internal variable maxval = 10,000. If any of the position functions exceed this value an error message is generated that the iterative procedure is diverging. Finally, there is an internal variable Niter = 100 that specifies the maximum allowable number of iterations. If this limit is reached an error message is generated that there is no convergence for the iterative process. To change any of these internal variables within newton_raph one has to edit that function and then save it.]

To illustrate the use of the Newton-Raphson scheme for the slider-crank problem a MATLAB® function slider_crank_sol2 was also written which has the calling sequence

```
[xout, numit] = slider_crank_sol2(tspan ,x0, d1, d2, ang10, ftol)
```

where tspan is a row vector of times at which we want to solve the position constraint equations. This function calls the MATLAB® Newton-Raphson function and the functions slider_crank_f and slider_crank_J at each of these times to generate the approximate solutions. The output argument, xout, contains all of the approximate values of the three generalized coordinates, $(\theta_1, \theta_2, X_P)$, in its three columns at the times specified in tspan and numit is a column vector which contains the numbers of iterations needed to find a solution at those times.

Finally, we have generated a MATLAB® script slider_crank_newton_sim2, which sets up default values of d1 = 5 in., d2 = 13 in., ang10 = 0, ftol = e-10, tspan = linspace(0, 1500), and x0 = [0, 0, 18] for the case we have been considering, where the initial guess contains the exact starting values at t = 0. All these values can be changed in the command window before executing the script to override the defaults. This script evaluates the exact analytical solutions (based on Eq. (4.2.30) and $\theta_1(t) = \theta_{10} - 6t$) at the specified times in tspan and then calls the slider_crank_sol2 function to obtain the approximate values of these variables at the same times with the Newton-Raphson method. The script then plots the exact and

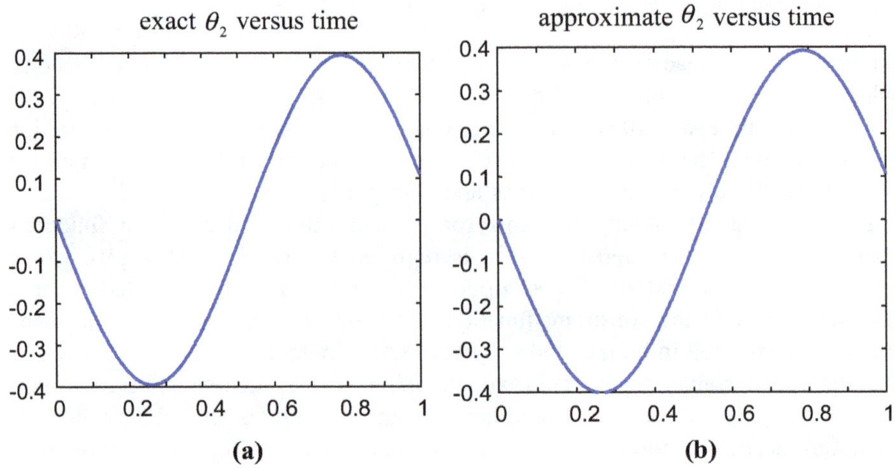

Fig. 4.16 The angle of the connecting rod versus time from (**a**) the exact analytical solution, and (**b**) the approximate numerical solution using the Newton-Raphson method

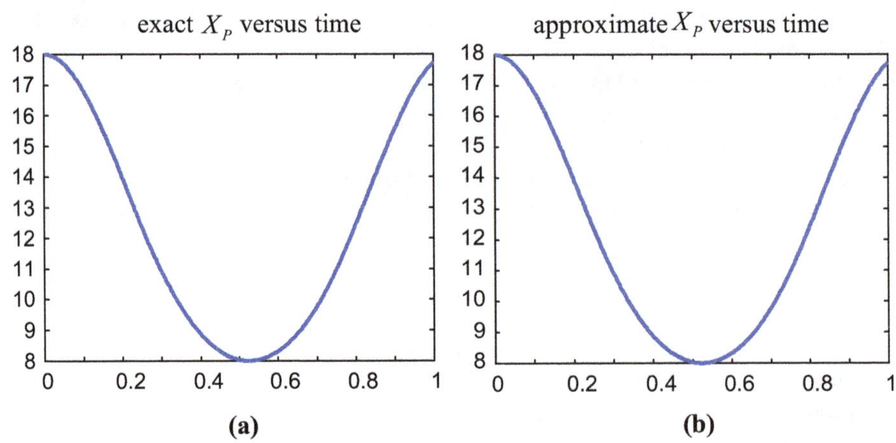

Fig. 4.17 The displacement of the slider versus time from (**a**) the exact analytical solution, and (**b**) the approximate numerical solution using the Newton-Raphson method

approximate values of θ_2 and X_P to demonstrate the accuracy of the numerical solution. These plots are shown in Figs. 4.16 and 4.17. It can be seen that the agreement between the numerical and analytical results is very good. If one evaluates the largest number of iterations for this example in MATLAB®, one finds

```
max(numit)
ans = 2
```

so that the convergence of the solution with the Newton-Raphson method was quite fast. If the initial guess only approximated the initial conditions, however, there will be some increase in the maximum number of iterations. This increase should not be large, however, as long as the initial guess is relatively close to the exact initial values. More iterations will also be needed, of course, if the time interval between solutions is increased.

4.2.6 Velocity and Acceleration Constraints

In the preceding example the default initial guesses for the solution at $t_1 = 0$ were the exact initial conditions while subsequent starting guesses at later times, t_k, were simply taken to be the final values of the Newton-Raphson iteration process coming from the previous time, t_{k-1}. This may only be effective, however, if we use very small time steps. But if we differentiate the position constraints on time we can use the resulting velocity and acceleration forms of the constraint equations to provide improved guesses at each instant.

If we differentiate the constraint equations on time, for example, we have

$$d\mathbf{f}/dt = \mathbf{f}_{\mathbf{q}}\dot{\mathbf{q}} + \mathbf{f}_t = 0 \tag{4.2.53}$$

where

$$\mathbf{f}_t = \begin{Bmatrix} \partial f_1/\partial t \\ \partial f_2/\partial t \\ \partial f_3/\partial t \end{Bmatrix} \tag{4.2.54}$$

and since we already have the Jacobian we have the linear system of equations

$$\mathbf{f}_{\mathbf{q}}\dot{\mathbf{q}} = -\mathbf{f}_t \tag{4.2.55}$$

which we can solve numerically for the generalized velocities. Taking another derivative on time gives

$$d^2\mathbf{f}/dt^2 = \mathbf{f}_{\mathbf{q}}\ddot{\mathbf{q}} + \left(\mathbf{f}_{\mathbf{q}}\dot{\mathbf{q}}\right)_{\mathbf{q}}\dot{\mathbf{q}} + 2\mathbf{f}_{\mathbf{q}t}\dot{\mathbf{q}} + \mathbf{f}_{tt} = 0 \tag{4.2.56}$$

where

$$\mathbf{f}_{tt} = \begin{Bmatrix} \partial^2 f_1/\partial t^2 \\ \partial^2 f_2/\partial t^2 \\ \\ \partial^2 f_n/\partial t^2 \end{Bmatrix} \tag{4.2.57a}$$

$$\mathbf{f}_{\mathbf{q}t} = \begin{bmatrix} \partial^2 f_1/\partial q_1 \partial t & \partial^2 f_1/\partial q_2 \partial t & \cdots & \partial^2 f_1/\partial q_n \partial t \\ \partial^2 f_2/\partial q_1 \partial t & \partial^2 f_2/\partial q_2 \partial t & \cdots & \partial^2 f_2/\partial q_n \partial t \\ \\ \partial^2 f_n/\partial q_1 \partial t & \partial^2 f_n/\partial q_2 \partial t & \cdots & \partial^2 f_n/\partial q_n \partial t \end{bmatrix} \tag{4.2.57b}$$

We can let $(\mathbf{f}_{\mathbf{q}}\dot{\mathbf{q}})_{\mathbf{q}} = \mathbf{F}_{\mathbf{q}}$ where $\mathbf{F} = \mathbf{f}_q\dot{\mathbf{q}}$ is given by the vector

$$\mathbf{F} = \begin{Bmatrix} F_1 \\ F_2 \\ \\ F_n \end{Bmatrix} = \begin{Bmatrix} (\partial f_1/\partial q_1)\dot{q}_1 + (\partial f_1/\partial q_2)\dot{q}_2 + \cdots + (\partial f_1/\partial q_n)\dot{q}_n \\ (\partial f_2/\partial q_1)\dot{q}_1 + (\partial f_2/\partial q_2)\dot{q}_2 + \cdots + (\partial f_2/\partial q_n)\dot{q}_n \\ \\ (\partial f_n/\partial q_1)\dot{q}_1 + (\partial f_n/\partial q_2)\dot{q}_2 + \cdots + (\partial f_n/\partial q_n)\dot{q}_n \end{Bmatrix} \tag{4.2.58a}$$

and so $\mathbf{F}_{\mathbf{q}}$ is the matrix

$$\mathbf{F}_{\mathbf{q}} = \begin{bmatrix} \partial F_1/\partial q_1 & \partial F_1/\partial q_2 & \cdots & \partial F_1/\partial q_n \\ \partial F_2/\partial q_1 & \partial F_2/\partial q_2 & \cdots & \partial F_2/\partial q_n \\ \\ \partial F_n/\partial q_1 & \partial F_n/\partial q_2 & \cdots & \partial F_n/\partial q_n \end{bmatrix} \tag{4.2.58b}$$

and, finally,

$$(\mathbf{f}_{\mathbf{q}}\dot{\mathbf{q}})\dot{\mathbf{q}} = \mathbf{F}_{\mathbf{q}}\dot{\mathbf{q}} = \begin{Bmatrix} \partial F_1/\partial q_1\dot{q}_1 + \partial F_1/\partial q_2\dot{q}_2 + \cdots + \partial F_1/\partial q_n\dot{q}_n \\ \partial F_2/\partial q_1\dot{q}_1 + \partial F_2/\partial q_2\dot{q}_2 + \cdots + \partial F_2/\partial q_n\dot{q}_n \\ \\ \partial F_n/\partial q_1\dot{q}_1 + \partial F_n/\partial q_2\dot{q}_2 + \cdots + \partial F_n/\partial q_n\dot{q}_n \end{Bmatrix} \tag{4.2.58c}$$

From Eq. (4.2.56) we obtain the system of linear equations for the accelerations

$$\mathbf{f}_{\mathbf{q}}\ddot{\mathbf{q}} = \mathbf{b}_v \tag{4.2.59}$$

where the vector $\mathbf{b}_v$ is

$$\mathbf{b}_v = -(\mathbf{f}_{\mathbf{q}}\dot{\mathbf{q}})_{\mathbf{q}}\dot{\mathbf{q}} - 2\mathbf{f}_{\mathbf{q}t}\dot{\mathbf{q}} - \mathbf{f}_{tt} \tag{4.2.60}$$

For the slider-crank problem we have for $\mathbf{f}_t$ and the velocity equations

$$\mathbf{f}_t = \begin{Bmatrix} -6 \\ 0 \\ 0 \end{Bmatrix}, \quad \begin{bmatrix} 1 & 0 & 0 \\ -l_1 \sin\theta_1 & -l_2 \sin\theta_2 & -1 \\ l_1 \cos\theta_1 & -l_2 \cos\theta_2 & 0 \end{bmatrix} \begin{Bmatrix} \dot{\theta}_1 \\ \dot{\theta}_2 \\ \dot{X}_P \end{Bmatrix} = \begin{Bmatrix} -6 \\ 0 \\ 0 \end{Bmatrix} \tag{4.2.61}$$

which is consistent with our previous results (see Eq. (4.2.31)).

Since $\mathbf{f}_{tt} = \mathbf{f}_{qt} = 0$ for the slider-crank problem, we need only to compute the term containing $\mathbf{F}$, where

$$\mathbf{F} = \begin{Bmatrix} \dot{\theta}_1 \\ -l_1 \sin\theta_1 \dot{\theta}_1 - l_2 \sin\theta_2 \dot{\theta}_2 - \dot{X}_P \\ l_1 \cos\theta_1 \dot{\theta}_1 - l_2 \cos\theta_2 \dot{\theta}_2 \end{Bmatrix} \tag{4.2.62}$$

and so

$$\mathbf{F}_\mathbf{q} = \begin{bmatrix} 0 & 0 & 0 \\ -l_1 \cos\theta_1 \dot{\theta}_1 & -l_2 \cos\theta_2 \dot{\theta}_2 & 0 \\ -l_1 \sin\theta_1 \dot{\theta}_1 & l_2 \sin\theta_2 \dot{\theta}_2 & 0 \end{bmatrix} \tag{4.2.63}$$

and

$$\mathbf{b}_v = -\mathbf{F}_\mathbf{q}\dot{\mathbf{q}} = \begin{Bmatrix} 0 \\ l_1 \cos\theta_1 \dot{\theta}_1^2 + l_2 \cos\theta_2 \dot{\theta}_2^2 \\ l_1 \sin\theta_1 \dot{\theta}_1^2 - l_2 \sin\theta_2 \dot{\theta}_2^2 \end{Bmatrix} \tag{4.2.64}$$

which gives, finally, the acceleration equations

$$\begin{aligned} &\begin{bmatrix} 1 & 0 & 0 \\ -l_1 \sin\theta_1 & -l_2 \sin\theta_2 & -1 \\ l_1 \cos\theta_1 & -l_2 \cos\theta_2 & 0 \end{bmatrix} \begin{Bmatrix} \ddot{\theta}_1 \\ \ddot{\theta}_2 \\ \ddot{X}_P \end{Bmatrix} \\ &= \begin{Bmatrix} 0 \\ l_1 \cos\theta_1 \dot{\theta}_1^2 + l_2 \cos\theta_2 \dot{\theta}_2^2 \\ l_1 \sin\theta_1 \dot{\theta}_1^2 - l_2 \sin\theta_2 \dot{\theta}_2^2 \end{Bmatrix} \end{aligned} \tag{4.2.65}$$

which is also consistent with our previous acceleration results (see Eq. (4.2.35)).

The way in which these velocity and acceleration values can be used in improving the positional analysis is that instead of using the values of $\tilde{\mathbf{q}}_k$ from the kth step as the initial guess for the $\tilde{\mathbf{q}}_{k+1}$ values at step (k+1), we can update instead, knowing the values of the velocities, $\dot{\mathbf{q}}_k$, and accelerations, $\ddot{\mathbf{q}}_k$, at the kth step, via

$$\tilde{\mathbf{q}}_{k+1} = \tilde{\mathbf{q}}_k + \dot{\mathbf{q}}_k \Delta t + \frac{1}{2} \ddot{\mathbf{q}}_k \Delta t^2 \tag{4.2.66}$$

where $\Delta t = t_{k+1} - t_k$ is the time step. To calculate the $(\dot{\mathbf{q}}, \ddot{\mathbf{q}})$ values we need a function that uses the Jacobian matrix, $\mathbf{f_q}$, and the vectors, $\mathbf{f}_t$ and $\mathbf{b}_v$, and returns those values at a given time, t. See Problem P4.2 which asks you to implement this improved updating by incorporating Eq. (4.2.66) into the Newton-Raphson method.

4.2.7 The Analysis of Kinematically Driven System: II

We can also perform a general computational analysis of kinematically driven rigid body systems where we solve for all the kinematic variables—positions, velocities, and accelerations—numerically. This is possible since as we have seen the positional, velocity, and acceleration relations for a kinematically driven rigid body system are given by

$$\begin{aligned} \mathbf{f} &= 0 \\ \mathbf{f_q}\dot{\mathbf{q}} &= -\mathbf{f}_t \\ \mathbf{f_q}\ddot{\mathbf{q}} &= \mathbf{b}_v \end{aligned} \tag{4.2.67}$$

In general, if the rigid body system is described in terms of n generalized coordinates but only has n_{dof} degrees of freedom, then those n generalized coordinates are related to each other by $n - n_{\text{dof}}$ kinematic (holonomic) constraints. However, if we partition the n coordinates into $n - n_{\text{dof}}$ (dependent) coordinates and explicitly specify the values of the n_{dof} (independent) coordinates as functions of time, then we always have n positional constraint equations for the n generalized coordinates. The specifications of the independent coordinates, as mentioned previously, are called driving constraints to distinguish them from the holonomic kinematic constraints. Thus, if we combine the kinematic and driving constraints together we can always write a total set of positional constraint equations as $\mathbf{f} = 0$, where there are n constraints (i.e., n functions in the vector $\mathbf{f}$) for the n generalized coordinates. As we have shown previously, the choice of dependent and independent coordinates is not fixed but determined by how we specify the rigid body system is driven. When we differentiate these n constraint functions once and twice on time, as just shown, we find that the velocities and accelerations satisfy the second and third equations in Eq. (4.2.67) where the Jacobian, $\mathbf{f_q}$, is a square matrix, and $\mathbf{f}_t$ and $\mathbf{b}_v$, are vectors. These (generally nonlinear) positional constraint equations can be solved numerically as a function of time for the generalized coordinates by the Newton-Raphson method, while the linear equations for the generalized velocities and accelerations can be solved directly (non-iteratively) as a function of time once all those generalized coordinate values are known, assuming that the rigid body system is designed so that singular configurations are not encountered during the motion. Thus, this computational approach numerically obtains values for the n coordinates, n velocities, and n accelerations that completely define the kinematic behavior of the rigid body system. Note that the velocity and acceleration equations in Eq. (4.2.67) can also be obtained from relative velocity and acceleration expressions for a rigid body system rather than through differentiation of the positional constraint equation, in the same manner as shown for the slider-crank problem.

Simulation 4.3: Complete Numerical Solution of a Kinematically Driven Rigid Body System

A MATLAB® script slider_crank_sim3 has been written which uses the positional analysis contained in the slider_crank_newton_sim2 and combines it with numerical solutions of the velocity equations (Eq. (4.2.61)) and acceleration equations (Eq. (4.2.65)), which are the explicit forms of the general solution for a kinematically driven rigid body system given by Eq. (4.2.67) for the slider-crank problem. The default values are the same as used in the script slider_crank_newton_sim2 so we can compare the outputs of the two scripts. The script slider_crank_sim3 plots the angular position, angular velocity, and angular acceleration of the connecting rod as well as position, velocity, and acceleration of the slider (piston). The results are shown in Figs. 4.18 and 4.19. Figure 4.18 shows the angle, angular velocity, and angular acceleration of the connecting rod versus time, which can be compared to the previous results for these same quantities in Figs. 4.17a, 4.12c, and 4.13b. Similarly, Fig. 4.19 shows the position, velocity, and acceleration versus time for the piston which can be compared to the results in Figs. 4.17a, 4.12b, and 4.13a. The difference between the present and previous studies is that here the position analysis as well as the velocity and acceleration analysis were done numerically. It can be seen that there is very good agreement between the simulations.

4.3 Motion on the Rotating Earth

Up to this point we have examined relative motion in moving coordinates from a purely kinematics perspective. However, relative motion also affects the relationship between measured accelerations and the forces appearing in Newton's second law. For example, we live on a moving earth so that when we make dynamics measurements in a coordinate system attached to the earth we are inherently using a set of moving coordinates. This means that the relative acceleration of a body that we measure in a coordinate system attached to the earth is only part of the total acceleration experienced by that body so that the other parts of the acceleration due to the motion of the earth itself will produce additional inertial force-like terms that affect the motion of the body. To see some specific examples, first consider the case where we drop a particle of mass m at rest from a height, h, above the surface of the rotating earth (Fig. 4.20). The (X, Y, Z) axes will be taken as a nonrotating inertial system whose origin O is at the center of the earth and the (x, y, z) system will be a

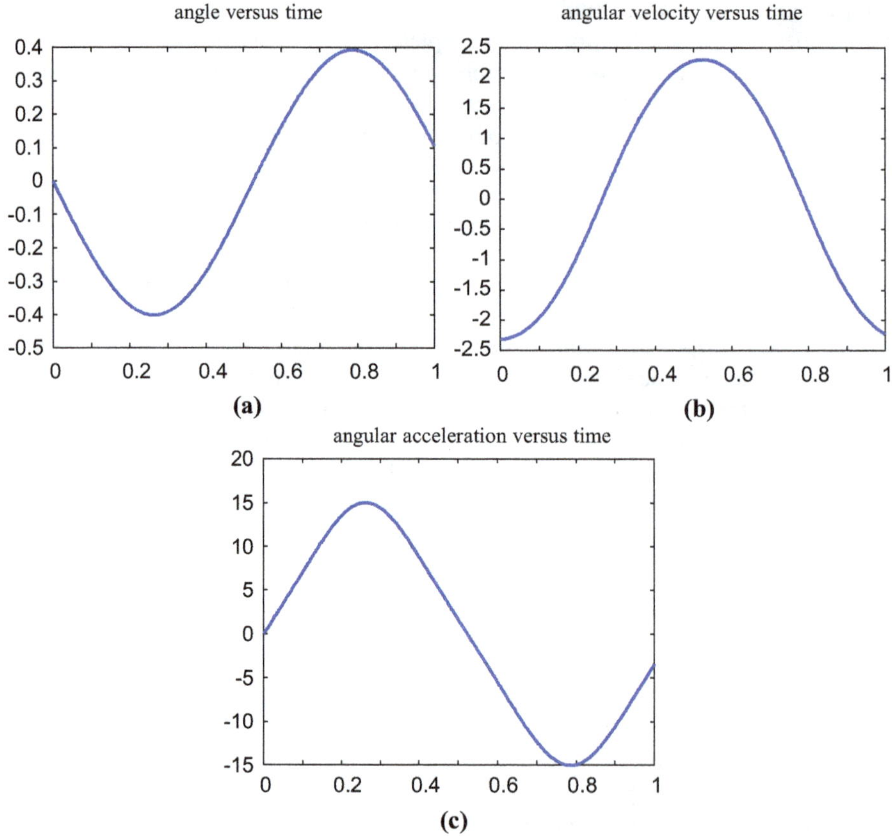

Fig. 4.18 Simulated kinematics of the slider_crank connecting rod for (**a**) its angular position in radians versus time, (**b**) its angular velocity in rad/s versus time, and (**c**) its angular acceleration in rad/s^2 versus time

coordinate system (called here the earth frame E) with origin at B and which rotates with the earth's angular velocity, $\boldsymbol{\Omega}$. We will assume that the height from which the mass is dropped is small enough so that the gravitational force is approximately its value at the earth's surface and we will also neglect any change in the angular velocity of the earth so that the angular acceleration of the earth is assumed to be zero. From Newton's second law and the relative acceleration expression of Eq. (4.2.19) we have

$$\begin{aligned} -mg\,\mathbf{e}_z &= m\mathbf{a}_{P/O} \\ &= m\Big(\mathbf{a}_{B/O} + \boldsymbol{\omega} \times \left(\boldsymbol{\omega} \times \mathbf{x}_{P/B}\right) + 2\boldsymbol{\omega} \times \left(\dot{\mathbf{x}}_{P/B}\right)_E + \left(\ddot{\mathbf{x}}_{P/B}\right)_E\Big) \end{aligned} \tag{4.3.1}$$

where $\boldsymbol{\omega} = \Omega \cos\theta\,\mathbf{e}_y + \Omega \sin\theta\,\mathbf{e}_z$ in terms of the latitude angle, θ, $\mathbf{x}_{P/B} = x\mathbf{e}_x + y\mathbf{e}_y + z\mathbf{e}_z$, $\left(\dot{\mathbf{x}}_{P/B}\right)_E = \dot{x}\,\mathbf{e}_x + \dot{y}\,\mathbf{e}_y + \dot{z}\,\mathbf{e}_z$, and $\left(\ddot{\mathbf{x}}_{P/B}\right)_E = \ddot{x}\mathbf{e}_x + \ddot{y}\mathbf{e}_y + \ddot{z}\mathbf{e}_z$. The magnitude

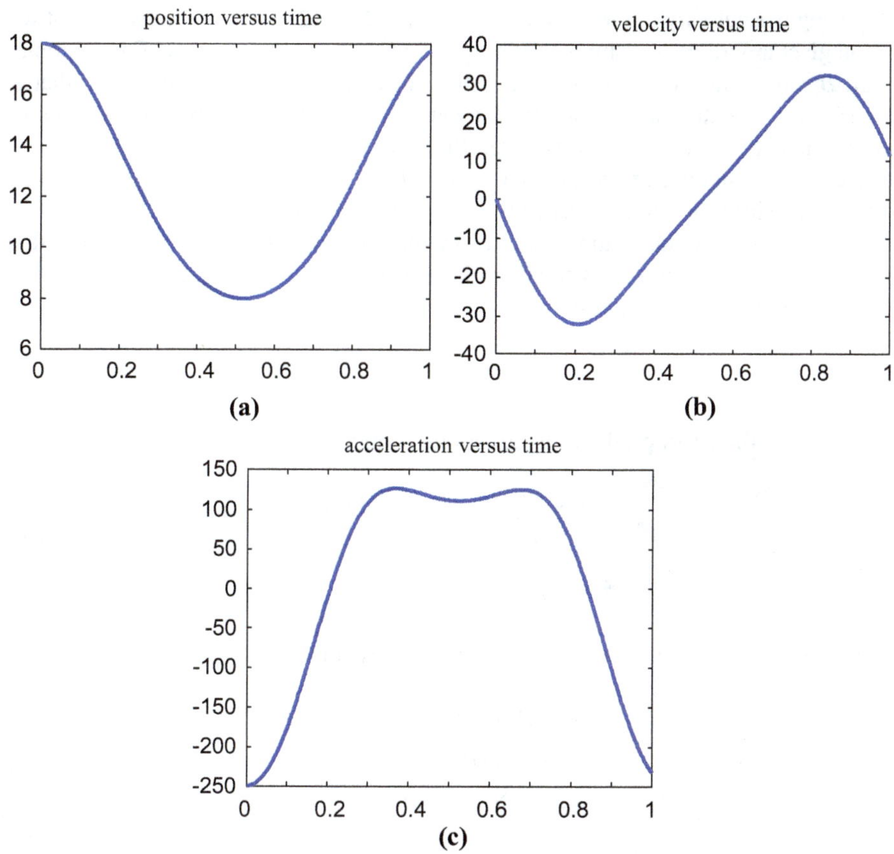

Fig. 4.19 Simulated kinematics of the slider-crank piston for (**a**) its position in inches versus time, (**b**) its velocity in in./s versus time, and (**c**) its acceleration in in./s^2 versus time

Fig. 4.20 The dropping of a ball of mass m from a height h above the surface of the earth, which is rotating with the constant angular velocity Ω. The angle θ is the latitude and the x-axis of the rotating (x, y, z) axes with origin at B is in the east direction

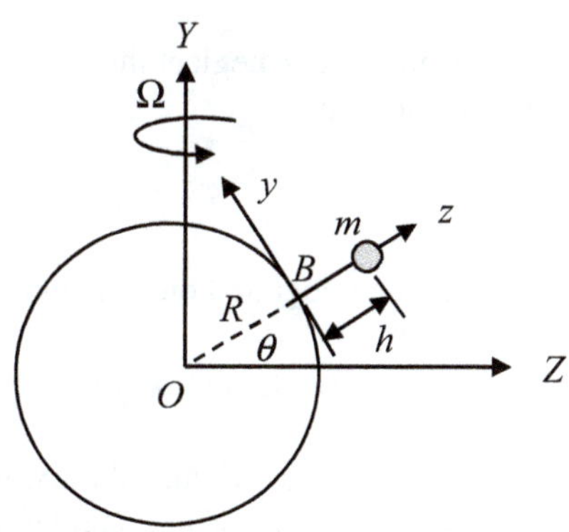

of the angular velocity of the earth is very small ($\Omega = 7.272 \times 10^{-5}$ rad/s) so that we can neglect any terms that involve the square of that angular velocity. For example, point B moves in a circular path of radius $R \cos \theta$ at a constant speed so that it experiences a centripetal acceleration towards the Y-axis of magnitude $R\Omega^2 \cos \theta$. However, even though R is very large this acceleration is much smaller than the acceleration, g, at the earth's surface (see Problem P1.1) so that the $\mathbf{a}_{B/O}$ term can be neglected. Similarly, the centripetal acceleration term $\boldsymbol{\omega} \times (\boldsymbol{\omega} \times \mathbf{x}_{P/B})$ has a magnitude proportional to the square of the angular velocity of the earth so it also can be neglected, leaving only the Coriolis acceleration and the relative acceleration as measured in the (x, y, z) system:

$$-g\mathbf{e}_z = 2\boldsymbol{\omega} \times \left(\dot{\mathbf{x}}_{P/B}\right)_E + \left(\ddot{\mathbf{x}}_{P/B}\right)_E \tag{4.3.2}$$

Carrying out the cross product the equations of motion become

$$\begin{aligned} \ddot{x} &= -2\dot{z}\ \Omega \cos\theta + 2\dot{y}\ \Omega \sin\theta \\ \ddot{y} &= -2\dot{x}\ \Omega \sin\theta \\ \ddot{z} &= -g + 2\dot{x}\ \Omega \cos\theta \end{aligned} \tag{4.3.3}$$

These equations can be integrated on time directly, giving

$$\begin{aligned} \dot{x} &= -2z\,\Omega \cos\theta + 2y\,\Omega \sin\theta + c_1 \\ \dot{y} &= -2x\Omega \sin\theta + c_2 \\ \dot{z} &= -gt + 2x\,\Omega \cos\theta + c_3 \end{aligned} \tag{4.3.4}$$

where (c_1, c_2, c_3) are constants of integration. If we place the expressions for $\dot{y}, \dot{z}$ into the equation for $\ddot{x}$ we find

$$\ddot{x} = -2\Omega \cos\theta(-gt + 2x\Omega \cos\theta + c_3) + 2\Omega \sin\theta(-2x\Omega \sin\theta + c_2) \tag{4.3.5}$$

where again we can neglect the terms involving the square of the angular velocity of the earth, so that

$$\ddot{x} = -2\Omega \cos\theta(c_3 - gt) + 2\Omega \sin\theta c_2 \tag{4.3.6}$$

which can be integrated on time twice to give

$$x = \Omega g \cos\theta t^3/3 - \Omega(c_3 \cos\theta - c_2 \sin\theta)t^2 + c_1 t + x_0 \tag{4.3.7}$$

If we place Eq. (4.3.7) into the expressions for $\dot{y}$ and $\dot{z}$, again neglect the terms involving the square of the angular velocity of the earth, and then integrate these expressions on time, we find

$$\begin{aligned} y &= -\Omega \sin\theta c_1 t^2 + (c_2 - 2\Omega \sin\theta x_0)t + y_0 \\ z &= (\Omega \cos\theta c_1 - g/2)t^2 + (c_3 + 2\Omega \cos\theta x_0)t + z_0 \end{aligned} \tag{4.3.8}$$

Since the mass is being dropped from a height h while at rest relative to the (x, y, z) coordinates, the initial conditions are $x_0 = y_0 = 0$, $z_0 = h$, $\dot{x}_0 = c_1 = 0$, $\dot{y}_0 = c_2 = 0$, and $\dot{z}_0 = c_3 = 0$. When the mass strikes the surface of the earth $z = 0$, so from the equation for z in Eq. (4.3.8) we find the time to drop is $t_d = \sqrt{2h/g}$. At this time we have $y = z = 0$ and

$$x = \Omega g (2h/g)^{3/2} \cos\theta/3 \tag{4.3.9}$$

so that the mass lands to the east of its initial position. This deflection is, of course, very small. If $h = 100$ m, for example, at a 45° latitude we find $x = 15$ mm, approximately.

The second example we will give of describing motion in a non-inertial coordinate system is the Foucault pendulum (Fig. 4.21a). The Foucault pendulum is famous because it is a simple dynamical system yet it can explicitly demonstrate through its motion that the earth is indeed rotating. It can be found in many science museums and planetariums around the world. It is simply a long pendulum which is started in a familiar back-and-forth oscillation in a given plane but where the earth's rotation causes that plane of motion to slowly change over time. In contrast to the falling mass (see Eq. (4.3.1)), Newton's second law for the pendulum is

$$\begin{aligned} \mathbf{T} - mg\,\mathbf{e}_z &= m\mathbf{a}_{P/O} \\ &= m\left(\mathbf{a}_{B/O} + \boldsymbol{\omega} \times \left(\boldsymbol{\omega} \times \mathbf{x}_{P/B}\right) + 2\boldsymbol{\omega} \times \left(\dot{\mathbf{x}}_{P/B}\right)_E + \left(\ddot{\mathbf{x}}_{P/B}\right)_E\right) \end{aligned} \tag{4.3.10}$$

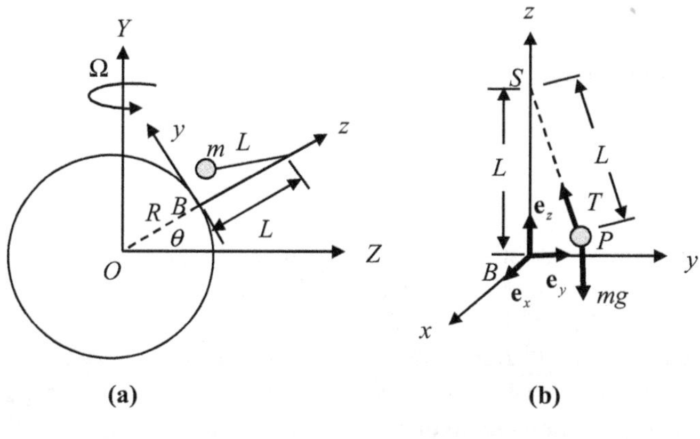

Fig. 4.21 (**a**) A pendulum of mass m and length L swinging in three-dimensional motion, suspended at a height L from the surface of the rotating earth, and (**b**) the free body diagram of the pendulum mass, showing the force of gravity and tension, T, in the supporting wire

where $\mathbf{T}$ is the vector tension in the wire supporting the mass, m (Fig. 4.21b). Like the falling mass problem we will neglect the acceleration $\mathbf{a}_{B/O}$ and the $\boldsymbol{\omega} \times (\boldsymbol{\omega} \times \mathbf{x}_{P/B})$ terms because of the very small angular velocity of the earth, again leaving only the Coriolis acceleration and the relative acceleration terms, which in this case gives

$$\mathbf{T} - mg\,\mathbf{e}_z = 2m\boldsymbol{\omega} \times \left(\dot{\mathbf{x}}_{P/B}\right)_E + m\left(\ddot{\mathbf{x}}_{P/B}\right)_E \tag{4.3.11}$$

Letting (x, y, z) be the coordinates of the mass at point P relative to the moving origin, B, we have $\mathbf{T} = T\mathbf{x}_{S/P}/|\mathbf{x}_{S/P}| = T\mathbf{x}_{S/P}/L$, where $\mathbf{x}_{S/P}$ is the position vector from the mass (point P) to the support point, S (Fig. 4.21b), and from the geometry $\mathbf{x}_{S/P} = \mathbf{x}_{S/B} - \mathbf{x}_{P/B}$ where we have $\mathbf{x}_{S/B} - \mathbf{x}_{P/B} = L\mathbf{e}_z - (x\mathbf{e}_x + y\mathbf{e}_y + z\mathbf{e}_z)$, giving

$$\mathbf{T} = -T\frac{x}{L}\mathbf{e}_x - T\frac{y}{L}\mathbf{e}_y + T\frac{L-z}{L}\mathbf{e}_z \tag{4.3.12}$$

Placing this expression into Eq. (4.3.10) and carrying out the cross product in that equation we find

$$\begin{aligned} m\ddot{x} &= -2m\dot{z}\;\Omega\cos\theta + 2m\dot{y}\;\Omega\sin\theta - Tx/L \\ m\ddot{y} &= -2m\dot{x}\;\Omega\sin\theta - Ty/\mathrm{L} \\ m\ddot{z} &= -mg + 2m\dot{x}\;\Omega\cos\theta + T(L-z)/L \end{aligned} \tag{4.3.13}$$

If, however, the length $L \gg z$ then the motion occurs primarily in the x–y plane so if we ignore the z-motion entirely by setting $z = \dot{z} = \ddot{z} = 0$ we obtain

$$\begin{aligned} m\ddot{x} &= 2m\dot{y}\;\Omega\sin\theta - Tx/L \\ m\ddot{y} &= -2m\dot{x}\;\Omega\sin\theta - Ty/L \\ T &= mg - 2m\dot{x}\;\Omega\cos\theta \end{aligned} \tag{4.3.14}$$

If we place the expression for T in Eq. (4.3.14) into the other two equations of motion and neglect any nonlinear (product) terms in those two equations, we arrive at the set of linear equations for the motion of the pendulum in the x–y plane:

$$\begin{aligned} \ddot{x} &= 2\dot{y}\;\Omega\sin\theta - gx/L \\ \ddot{y} &= -2\dot{x}\;\Omega\sin\theta - gy/L \end{aligned} \tag{4.3.15}$$

We have followed the steps found in most books to arrive at this system of equations since we can solve these equations directly for the motion. First, note that if $\Omega = 0$ then Eq. (4.3.15) is identical to decoupled small-amplitude motions of a simple pendulum of length L in both the x- and y-directions. These motions have a natural frequency $\omega_n = \sqrt{g/L}$ and a corresponding period $T_n = 2\pi/\omega_n$. However, there is another characteristic frequency parameter in Eq. (4.3.15) given by the "precession" frequency, $\omega_p = \Omega\,\sin\,\theta$, and a corresponding period $T_p = 2\pi/\omega_p$.

While ω_n, as we will see, is approximately the frequency of the ordinary back-and-forth motion of the mass in the x–y plane, ω_p is associated with changes of the plane of motion of the pendulum itself, i.e., it is the source of the precession of that plane that we readily observe in the behavior of a Foucault pendulum. This precession frequency is dependent on the latitude location of the pendulum, reaching a maximum magnitude at the poles and disappearing at the equator. Note, however, that these two characteristic frequencies and periods are vastly different. A very long pendulum of length 50 m, for example, will have a back-and-forth oscillation period of approximately 14 s, while the precession period at a latitude of 45° will be approximately 1.2×10^5 s or about 1.4 days!

To obtain the analytical solution of Eq. (4.3.15) we can define the complex variable $z = x+iy$ [note: z here is not the z-coordinate of the pendulum mass] so that we can write the two equations in Eq. (4.3.15) as one complex equation

$$\ddot{z} + 2i\Omega \sin\theta \dot{z} + gz/L = 0 \tag{4.3.16}$$

or, in terms of the natural frequencies,

$$\ddot{z} + 2i\omega_p \dot{z} + \omega_n^2 z = 0 \tag{4.3.17}$$

We can write an explicit solution to this equation. Consider, for example, if at time $t = 0$ we set $x = a$, $y = 0$, and $\dot{x} = \dot{y} = 0$. Then it is easy to verify that the solution for these initial conditions is just

$$z = \frac{a}{2\overline{\omega}_n}\left[(\overline{\omega}_n + \omega_p)\exp\{i(\overline{\omega}_n - \omega_p)t\} + (\overline{\omega}_n - \omega_p)\exp\{-i(\overline{\omega}_n + \omega_p)t\}\right] \tag{4.3.18a}$$

or, equivalently,

$$z = a\exp(-i\omega_p t)\left[\cos(\overline{\omega}_n t) + i\frac{\omega_p}{\overline{\omega}_n}\sin(\overline{\omega}_n t)\right] \tag{4.3.18b}$$

where $\overline{\omega}_n = \sqrt{\omega_n^2 + \omega_p^2}$. Since $\omega_n \gg \omega_p$ it follows that $\overline{\omega}_n \cong \omega_n$. The path followed by the pendulum mass is a curve called a hypocycloid. To put Eq. (4.3.18a) into the standard form for the hypocycloid we let

$$b = \frac{\overline{\omega}_n - \omega_p}{2\overline{\omega}_n}a, \quad \phi = \frac{2\overline{\omega}_n}{a}t \tag{4.3.19}$$

in which case Eq. (4.3.18a) becomes

$$z = (a - b)\exp[ib\phi] + b\exp[-i(a - b)\phi] \tag{4.3.20}$$

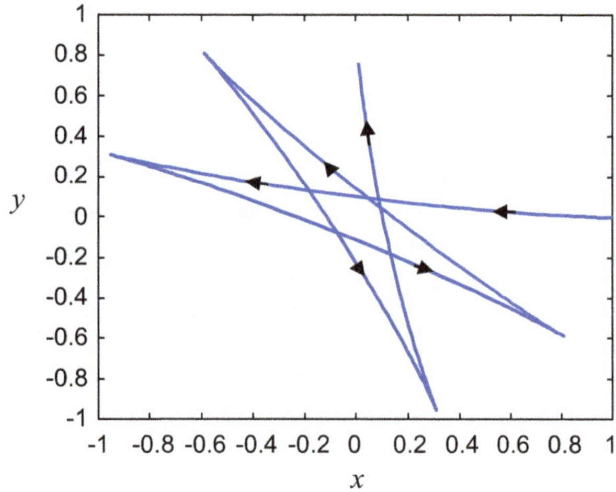

Fig. 4.22 Simulation of the hypocycloidal motion of the Foucault pendulum where the precession speed is greatly increased to show the details of the motion

which is the equation of the hypocycloid in complex form, equivalent to writing the real and imaginary parts separately as $x = f(a, b, \phi)$ and $y = g(a, b, \phi)$. If we want to see a typical hypocycloid path, let $a = 1, \overline{\omega}_n = 1$ and $\omega_p = 0.1$, and let ϕ range from 0 to 30. If we evaluate Eq. (4.3.20) under these conditions and plot the imaginary part of z versus the real part of z we see the hypocycloidal path shown in Fig. 4.22. For the real Foucault pendulum the precession angular velocity is extremely small and the arcs of the hypocycloid are nearly straight lines so that when we observe a Foucault pendulum in action the motion appears to be in a straight path. Thus, we have artificially increased the relative size of the precession angular velocity used to generate Fig. 4.22 so that the actual path of the motion can be seen clearly. In the northern hemisphere the precession frequency is positive and the plane of the pendulum rotates in a clockwise sense as seen in Fig. 4.22. In the southern hemisphere the rotation would instead be counterclockwise. The period for a complete to-and-fro swing of the pendulum is $\bar{T}_n = 2\pi/\overline{\omega}_n$ which is almost identical to the period $T_n = 2\pi/\omega_n$ when the rotation of the earth is absent. The value of z at $\bar{T}_n$ is $z = a \exp\left[-2\pi i\omega_p/\overline{\omega}_n\right]$ (from Eq. (4.3.18b)) so that the plane of the pendulum has rotated in a negative sense (clockwise) through an angle of $\Phi = 2\pi\omega_p/\overline{\omega}_n$ radians in the time $\bar{T}_n = 2\pi/\overline{\omega}_n$. Thus, the average angular velocity of the plane is just $\Phi/\bar{T}_n = \omega_p$, which justifies us calling ω_p the precession angular velocity.

Simulation 4.4: Foucault Pendulum

We can also solve the Foucault pendulum problem numerically but we should be aware of the fact that having two parameters with such different scales within the same problem can present some unique challenges. A system of equations is said to be "stiff" when such highly different timescales are present. The numerical solver used to solve such stiff systems must be able to resolve the very fast motions while maintaining accuracy over the timescale needed to demonstrate the slow motions. The adaptive time interval used for calculations in MATLAB®'s ode45 is very versatile and can often handle this problem but in some cases at the expense of very long solution times. To ensure the accuracy of the solution it is good practice to monitor the accuracy of the solution as a function of time for stiff systems (and for non-stiff systems as well). This is possible for the system of Eq. (4.3.15) since if we multiply the first equation by $\dot{x}$ and the second equation by $\dot{y}$ and add the two resulting equations the Coriolis terms cancel and we find

$$\ddot{x}\dot{x} + \ddot{y}\dot{y} + g(x\dot{x} + y\dot{y})/L = 0 \tag{4.3.21}$$

which we can write as a perfect differential, namely

$$\frac{d}{dt}\left(\frac{\dot{x}^2 + \dot{y}^2}{2} + g\frac{x^2 + y^2}{2L}\right) = 0 \tag{4.3.22}$$

to give a constant energy-like integral, E, where

$$E = \frac{\dot{x}^2 + \dot{y}^2}{2} + g\frac{x^2 + y^2}{2L} = \frac{\dot{x}_0^2 + \dot{y}_0^2}{2} + g\frac{x_0^2 + y_0^2}{2L} \tag{4.3.23}$$

Having an energy-like integral for this problem may seem to be at odds with our discussion in Chap. 2 of the motion of a particle on a rotating hoop where we saw that the total energy was not conserved but that instead there was an integral of the equations of motion called the Jacobi integral that was conserved. These properties are also true for the exact solution of Foucault pendulum but the linearized equations in Eq. (4.3.15) do exhibit an energy-like conservation relationship because a number of terms have been neglected in the approximations used. Note that when we are going to solve the Foucault pendulum problem numerically we do not have to use any approximations to the equations of motion to obtain the dynamic behavior of the pendulum. Here, however, we will use the linearized equations in Eq. (4.3.15) to form up a first-order system of equations of motion in the MATLAB® function f_pendulum. This function is nested in the MATLAB® function foucault which uses f_pendulum and the MATLAB® solver ode45 to obtain the numerical solution as done in our previous examples. The MATLAB® script foucault_sim sets up default input parameters and numerical tolerances for the problem which the user can override before executing the script, if desired. The x–y motion is plotted by the script as well as the normalized energy, $E(t)/E(0)$. For the simulation the parameters chosen are $L = 10$ m for the length of the pendulum and $\theta = 45°$ for the latitude, initial conditions are $x(0) = 0.01L$ and $y(0) = \dot{x}(0) = \dot{y}(0) = 0$, the time interval (in seconds) for calculations is tout = [0, 10,000], and error tolerances are rtol = 1e-10 and atol = 1e-10. These very small error tolerances were chosen to guarantee a high accuracy in energy conservation during the relatively long time interval. The

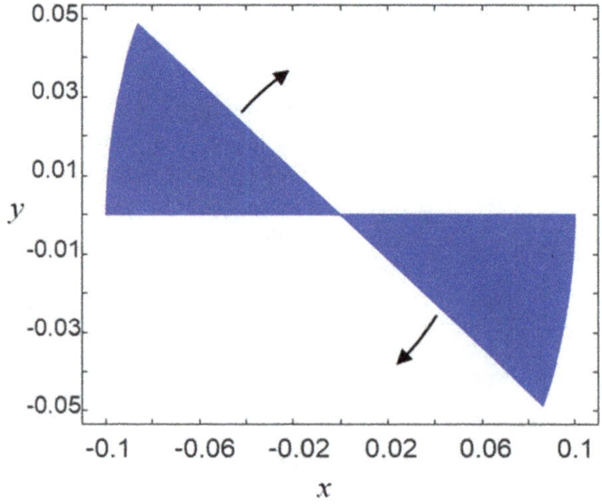

Fig. 4.23 The motion of the Foucault pendulum at a latitude of 45° showing the clockwise rotation of the plane of motion over a time interval of 10,000 s

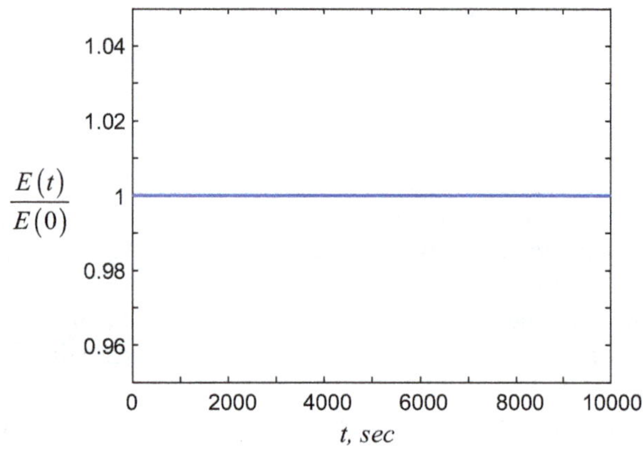

Fig. 4.24 The normalized energy versus time for the numerical Foucault pendulum problem

plot of the motion is shown in Fig. 4.23, where the individual oscillations are so close together that they appear as a continuously colored region, but where we can clearly see the clockwise motion of the oscillation plane which starts as approximately along the x-axis. The normalized energy as a function of time is shown in Fig. 4.24, which indeed gives us some confidence in the numerical solution. Note that with these very small error tolerances the solution with ode45 requires a computational time on the order of 10 s. This time can be substantially reduced by choosing larger tolerances but with a corresponding loss of some accuracy in the satisfaction of energy conservation.

4.4 Matrix Kinematics of Rigid Body Planar Motion

In Sect. 4.2.2 we developed the relative velocity and acceleration expressions relating the motion of two points in a rigid body, where the distance between the two points must remain fixed. Those expressions were derived in vector form, following the approach used in many dynamics texts. In this section we will obtain equivalent matrix-vector forms for relative position, velocity, and acceleration relations that will be used to describe the planar (two-dimensional) motion of rigid bodies. This will give us a very convenient form to analyze the motion of interconnected rigid bodies as found, for example, in many mechanical systems. In the next chapter, these results will be used to obtain and numerically solve the equations for such systems of rigid bodies in planar motion. Later, starting in Sect. 4.6, we will extend the matrix kinematics to cover more general three-dimension rigid body motions.

Since velocities and accelerations are vectors, we can simply write them as one-dimensional matrices. However, rigid body rotational motion like $\boldsymbol{\omega} \times \mathbf{r}$, $\boldsymbol{\alpha} \times \mathbf{r}$, and $\boldsymbol{\omega} \times (\boldsymbol{\omega} \times \mathbf{r})$ contain the cross product operation so that it would be useful to be able to write them instead in terms of the underlying angular rotations themselves. As we will see we can obtain such forms with matrices.

4.4.1 Positional Analysis

To begin, consider a planar rigid body that is translating and rotating in the *x*–*y* plane (Fig. 4.25). We will take point *A* as a reference point to define the translation of the

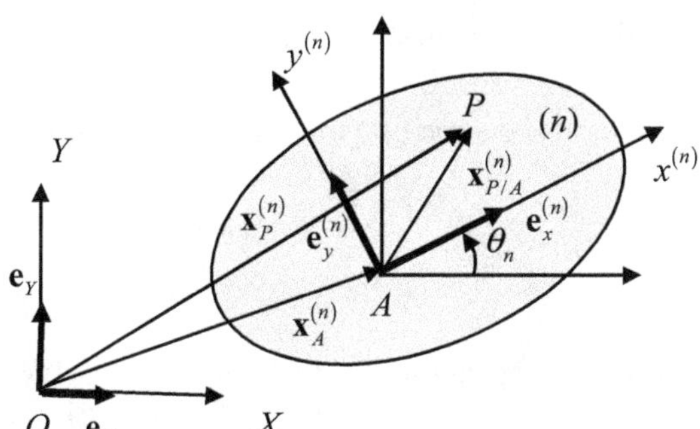

Fig. 4.25 A planar rigid body, labeled (n), translating and rotating in the *x*–*y* plane, where we can take point *A* as the reference point to define the translation and the angle θ_n defines the rotation of the body with respect to the *X*-axis of a set of fixed (*X*, *Y*) axes. The unit vectors $(\mathbf{e}_X, \mathbf{e}_Y)$ act along the fixed (X, Y) axes and the unit vectors $\left(\mathbf{e}_x^{(n)}, \mathbf{e}_y^{(n)}\right)$ act along a set of (x, y) axes that rotate with rigid body

body. The position of any other point, P, in that body is then related to the position of A via

$$\mathbf{x}_P^{(n)} = \mathbf{x}_A^{(n)} + \mathbf{x}_{P/A}^{(n)} \tag{4.4.1}$$

As done before we will use the superscript label (n) to denote the rigid body in a system so that we can distinguish relations for one rigid body from another. At a time, t, we will let the angle θ_n define orientation of body (n) with respect to the X-axis where the X–Y-axes are assumed to be fixed. We can define a set of unit vectors along the X–Y-axis as $(\mathbf{e}_X, \mathbf{e}_Y)$. We can also define a set of $(x^{(n)}, y^{(n)})$ axes that rotate with the rigid body (n) and a corresponding set of rotating unit vectors $\left(\mathbf{e}_x^{(n)}, \mathbf{e}_y^{(n)}\right)$ (see Fig. 4.25). From the geometry we can relate these two sets of unit vectors to each other:

$$\begin{aligned} \mathbf{e}_x^{(n)} &= \cos\theta_n\,\mathbf{e}_X + \sin\theta_n\,\mathbf{e}_Y \\ \mathbf{e}_y^{(n)} &= -\sin\theta_n\,\mathbf{e}_X + \cos\theta_n\,\mathbf{e}_Y \end{aligned} \tag{4.4.2}$$

This relationship can be written in a matrix-vector form as

$$\begin{Bmatrix} \mathbf{e}_x^{(n)} \\ \mathbf{e}_y^{(n)} \end{Bmatrix} = \begin{bmatrix} \cos\theta_n & \sin\theta_n \\ -\sin\theta_n & \cos\theta_n \end{bmatrix} \begin{Bmatrix} \mathbf{e}_X \\ \mathbf{e}_Y \end{Bmatrix} \tag{4.4.3}$$

From Eq. (4.4.2) it is easy to invert this relationship and instead write the unit vectors $(\mathbf{e}_X, \mathbf{e}_Y)$ in terms of $\left(\mathbf{e}_x^{(n)}, \mathbf{e}_y^{(n)}\right)$, which gives us the matrix form

$$\begin{Bmatrix} \mathbf{e}_X \\ \mathbf{e}_Y \end{Bmatrix} = \begin{bmatrix} \cos\theta_n & -\sin\theta_n \\ \sin\theta_n & \cos\theta_n \end{bmatrix} \begin{Bmatrix} \mathbf{e}_x^{(n)} \\ \mathbf{e}_y^{(n)} \end{Bmatrix} \tag{4.4.4}$$

which we will write in terms of a rotation matrix, $R^{(n)}(\theta_n)$, for body (n):

$$R^{(n)}(\theta_n) = \begin{bmatrix} \cos\theta_n & -\sin\theta_n \\ \sin\theta_n & \cos\theta_n \end{bmatrix} \tag{4.4.5}$$

Although a different matrix appears in Eq. (4.4.3), we can write it in terms of the matrix $R^{(n)}$ by noting that it is just the transpose of that matrix, $(R^{(n)})^T$. The matrix $R^{(n)}$ is a *proper orthogonal matrix*. Any proper orthogonal matrix, A, satisfies the following relations:

$$\begin{aligned} &\mathrm{AA}^T = \mathrm{A}^T\mathrm{A} = \mathrm{I} \\ &\det(\mathrm{A}) = 1 \\ &\mathrm{A}^{-1} = \mathrm{A}^T \end{aligned}$$

and it is easy to verify that $R^{(n)}$ has all of these properties. If we dot the two equations in Eq. (4.4.4) with the relative position vector, $\mathbf{x}_{P/A}^{(n)}$ (Fig. 4.25), and let

$$\begin{aligned}\mathbf{x}_{P/A}^{(n)} \cdot \mathbf{e}_X &= X_{P/A}^{(n)} \\ \mathbf{x}_{P/A}^{(n)} \cdot \mathbf{e}_Y &= Y_{P/A}^{(n)} \\ \mathbf{x}_{P/A}^{(n)} \cdot \mathbf{e}_x^{(n)} &= x_{P/A}^{(n)} \\ \mathbf{x}_{P/A}^{(n)} \cdot \mathbf{e}_y^{(n)} &= y_{P/A}^{(n)}\end{aligned} \tag{4.4.6}$$

then we find

$$\begin{Bmatrix} X_{P/A}^{(n)} \\ Y_{P/A}^{(n)} \end{Bmatrix} = \begin{bmatrix} \cos\theta_n & -\sin\theta_n \\ \sin\theta_n & \cos\theta_n \end{bmatrix} \begin{Bmatrix} x_{P/A}^{(n)} \\ y_{P/A}^{(n)} \end{Bmatrix} \tag{4.4.7}$$

If we write Eq. (4.4.1), written in terms of vector components with respect to the fixed (X, Y) axes, we have

$$\begin{Bmatrix} X_P^{(n)} \\ Y_P^{(n)} \end{Bmatrix} = \begin{Bmatrix} X_A^{(n)} \\ Y_A^{(n)} \end{Bmatrix} + \begin{Bmatrix} X_{P/A}^{(n)} \\ Y_{P/A}^{(n)} \end{Bmatrix} \tag{4.4.8}$$

or, using Eq. (4.4.7), we have

$$\begin{Bmatrix} X_P^{(n)} \\ Y_P^{(n)} \end{Bmatrix} = \begin{Bmatrix} X_A^{(n)} \\ Y_A^{(n)} \end{Bmatrix} + \begin{bmatrix} \cos\theta_n & -\sin\theta_n \\ \sin\theta_n & \cos\theta_n \end{bmatrix} \begin{Bmatrix} x_{P/A}^{(n)} \\ y_{P/A}^{(n)} \end{Bmatrix} \tag{4.4.9}$$

Equation (4.4.9) is the positional relationship between two points in a rigid body in a matrix-vector form. It involves components along both fixed (inertial) axes and components along a set of axes fixed in the moving body and a rotation matrix that defines the angular motion of the rigid body as well as any set of lines such as the line between points A and P that are fixed in that body. Thus, it is important to remember the underlying coordinates (frames) that define the terms in Eq. (4.4.9). We will also write Eq. (4.4.9) as

$$\begin{aligned}\mathbf{x}_P^{(n)} &= \mathbf{x}_A^{(n)} + \mathbf{x}_{P/A}^{(n)} \\ &= \mathbf{x}_A^{(n)} + \mathbf{R}^{(n)}(\theta_n)\bar{\mathbf{x}}_{P/A}^{(n)}\end{aligned} \tag{4.4.10}$$

where the "overbar" is used to explicitly indicate that the relative position vector $\bar{\mathbf{x}}_{P/A}^{(n)}$ is measured in a set of body coordinates. This means that for the $\left(x_{P/A}^{(n)}, y_{P/A}^{(n)}\right)$

coordinates seen in Eq. (4.4.9), for example, we have redefined those coordinates as $\left(x_{P/A}^{(n)}, y_{P/A}^{(n)}\right) \equiv \left(\bar{x}_{P/A}^{(n)}, \bar{y}_{P/A}^{(n)}\right)$ to indicate that those coordinates are those seen by an observer moving with the body. In contrast, all the other position vectors $\left(\mathbf{x}_P^{(n)}, \mathbf{x}_A^{(n)}, \mathbf{x}_{P/A}^{(n)}\right)$ appearing in Eq. (4.4.10) are measured in fixed (inertial) coordinates whose components are given in Eq. (4.4.8). The rotation matrix $\mathbf{R}^{(n)}$ transforms vectors like $\bar{\mathbf{x}}_{P/A}^{(n)}$ in body-fixed coordinates to the corresponding vectors in inertial coordinates. The meaning of the overbar in Eq. (4.4.10) is particularly important when differentiating $\mathbf{R}^{(n)}(\theta_n)\bar{\mathbf{x}}_{P/A}^{(n)}$ on time. Changes occur with respect to the components of the matrix $[R^{(n)}]$ seen in Eq. (4.4.9) but there are no changes with respect to the components $\left(\bar{x}_{P/A}^{(n)}, \bar{y}_{P/A}^{(n)}\right)$ since points P and A are always fixed relative to each other in the rigid body so these components are constants. In the matrix-vector form seen in Eq. (4.4.10) this means that upon differentiating we have

$$\dot{\mathbf{x}}_{P/A}^{(n)} = \dot{\mathbf{R}}^{(n)}(\theta_n)\bar{\mathbf{x}}_{P/A}^{(n)} \tag{4.4.11}$$

That is, $\bar{\mathbf{x}}_{P/A}^{(n)}$ is a constant vector during differentiation on time. We will use this relation in the next section when discussing the velocity relationship.

The particular forms of the positional relationship of two points A and P in a rigid body given by Eqs. (4.4.9) and (4.4.10) are most useful because we see that the motion of point P is decomposed into a motion of point A (a translational motion) plus a rotation of the rigid body as defined explicitly by the rotation matrix. Thus, we can differentiate these positional relations to obtain similar velocity and acceleration relationships in terms of translational and rotational terms.

4.4.2 Velocity Analysis

We can differentiate the positional equations of the last section on time to obtain a velocity relationship between points in a rigid body in a vector-matrix form. We have, from Eq. (4.4.10), using the more explicit notation discussed previously:

$$\left(\dot{\mathbf{x}}_P^{(n)}\right)_I = \left(\dot{\mathbf{x}}_A^{(n)}\right)_I + \dot{\mathbf{R}}^{(n)}(\theta_n)\bar{\mathbf{x}}_{P/A}^{(n)} \tag{4.4.12}$$

or, equivalently,

$$\left(\mathbf{v}_P^{(n)}\right)_I = \left(\mathbf{v}_A^{(n)}\right)_I + \dot{\mathbf{R}}^{(n)}\bar{\mathbf{x}}_{P/A}^{(n)} \tag{4.4.13}$$

which in terms of explicit components is

$$\begin{Bmatrix} \dot{X}_P^{(n)} \\ \dot{Y}_P^{(n)} \end{Bmatrix} = \begin{Bmatrix} \dot{X}_A^{(n)} \\ \dot{Y}_A^{(n)} \end{Bmatrix} + \left[\dot{R}^{(n)}(\theta_n)\right] \begin{Bmatrix} \bar{x}_{P/A}^{(n)} \\ \bar{y}_{P/A}^{(n)} \end{Bmatrix} \tag{4.4.14}$$

Now, considering the derivative of the rotation matrix we have

$$\left[\dot{R}^{(n)}\right] = \dot{\theta} \begin{bmatrix} -\sin\theta_n & -\cos\theta_n \\ \cos\theta_n & -\sin\theta_n \end{bmatrix} \tag{4.4.15}$$

which we will write as

$$\left[\dot{R}^{(n)}\right] = \dot{\theta}\left[R_\theta^{(n)}\right] \tag{4.4.16}$$

where

$$R_\theta^{(n)} = \begin{bmatrix} -\sin\theta_n & -\cos\theta_n \\ \cos\theta_n & -\sin\theta_n \end{bmatrix} \tag{4.4.17}$$

so our relative velocity equation, Eq. (4.4.13), becomes

$$\left(\mathbf{v}_P^{(n)}\right)_I = \left(\mathbf{v}_A^{(n)}\right)_I + \dot{\theta}\,\mathbf{R}_\theta^{(n)}\bar{\mathbf{x}}_{P/A}^{(n)} \tag{4.4.18}$$

which is the vector-matrix form of the relative velocity relationship we developed earlier in terms of vectors only (see, for example, Eq. (4.2.21a)). To see how these different forms are related, first note that we can rewrite Eq. (4.4.15) as

$$\left[\dot{R}^{(n)}\right] = \begin{bmatrix} 0 & -\dot{\theta}_n \\ \dot{\theta}_n & 0 \end{bmatrix} \begin{bmatrix} \cos\theta_n & -\sin\theta_n \\ \sin\theta_n & \cos\theta_n \end{bmatrix} \tag{4.4.19}$$

which we will write as

$$\left[\dot{R}^{(n)}\right] = \left[\Omega^{(n)}\right]\left[R^{(n)}\right] \tag{4.4.20}$$

where

$$\left[\Omega^{(n)}\right] = \begin{bmatrix} 0 & -\dot{\theta}_n \\ \dot{\theta}_n & 0 \end{bmatrix} \tag{4.4.21}$$

is a skew symmetrical angular velocity matrix. Note that if we post-multiply both sides of Eq. (4.4.20) by the transpose of the rotation matrix we obtain an expression for this skew symmetrical matrix in terms of the rotation matrix and its derivative as

$$\left[\Omega^{(n)}\right] = \left[\dot{R}^{(n)}\right]\left[R^{(n)}\right]^T \tag{4.4.22}$$

In Chap. 1 we indicated that we could replace a vector cross product of two vectors with an equivalent multiplication of a skew symmetrical matrix and a vector. That replacement was done in three dimensions in Chap. 1 but it can also be used for this two-dimensional case. First, we note that the angular velocity, ${}_I\boldsymbol{\omega}_B^{(n)}$, of our body (n), as measured in the inertial frame, is ${}_I\boldsymbol{\omega}_B^{(n)} = \dot{\theta}_n \mathbf{e}_Z$ since the rotational motion is about the fixed Z-axis. Now, consider placing Eq. (4.4.20) into the velocity relationship of Eq. (4.4.14). We find

$$\begin{aligned}
\begin{Bmatrix} \dot{X}_P^{(n)} \\ \dot{Y}_P^{(n)} \end{Bmatrix} &= \begin{Bmatrix} \dot{X}_A^{(n)} \\ \dot{Y}_A^{(n)} \end{Bmatrix} + \left[\Omega^{(n)}\right]\left[R^{(n)}\right]\begin{Bmatrix} \bar{x}_{P/A}^{(n)} \\ \bar{y}_{P/A}^{(n)} \end{Bmatrix} \\
&= \begin{Bmatrix} \dot{X}_A^{(n)} \\ \dot{Y}_A^{(n)} \end{Bmatrix} + \left[\Omega^{(n)}\right]\left[R^{(n)}\right]\left[R^{(n)}\right]^T\begin{Bmatrix} X_{P/A}^{(n)} \\ Y_{P/A}^{(n)} \end{Bmatrix} \\
&= \begin{Bmatrix} \dot{X}_A^{(n)} \\ \dot{Y}_A^{(n)} \end{Bmatrix} + \left[\Omega^{(n)}\right]\begin{Bmatrix} X_{P/A}^{(n)} \\ Y_{P/A}^{(n)} \end{Bmatrix}
\end{aligned} \tag{4.4.23}$$

which can be written more compactly as

$$\left(\mathbf{v}_P^{(n)}\right)_I = \left(\mathbf{v}_A^{(n)}\right)_I + {}_I\boldsymbol{\Omega}_B^{(n)}\mathbf{x}_{P/A}^{(n)} \tag{4.4.24}$$

where ${}_I\boldsymbol{\Omega}_B^{(n)} \equiv \boldsymbol{\Omega}^{(n)}$, i.e., we have emphasized that the angular velocity matrix is associated with the angular velocity of the body frame relative to the fixed inertia frame by writing it in the more explicit form ${}_I\boldsymbol{\Omega}_B^{(n)}$ just as we have done with the angular velocity vector. But

$$\begin{aligned}
{}_I\boldsymbol{\Omega}_B^{(n)}\mathbf{x}_{P/A}^{(n)} &= \begin{bmatrix} 0 & -\dot{\theta}_n \\ \dot{\theta}_n & 0 \end{bmatrix}\begin{Bmatrix} X_{P/A}^{(n)} \\ Y_{P/A}^{(n)} \end{Bmatrix} = \begin{Bmatrix} -\dot{\theta}_n Y_{P/A}^{(n)} \\ \dot{\theta}_n X_{P/A}^{(n)} \end{Bmatrix} \\
{}_I\boldsymbol{\Omega}_B^{(n)} \times \mathbf{x}_{P/A}^{(n)} &= \dot{\theta}_n \mathbf{e}_Z \times \left(X_{P/A}^{(n)}\mathbf{e}_X + Y_{P/A}^{(n)}\mathbf{e}_Y\right) = -\dot{\theta}_n Y_{P/A}^{(n)}\mathbf{e}_X + \dot{\theta}_n X_{P/A}^{(n)}\mathbf{e}_Y
\end{aligned} \tag{4.4.25}$$

which are just two different forms of the same result. Thus, the relative velocity relationship can be written in a number of equivalent forms:

$$\left(\mathbf{v}_P^{(n)}\right)_I = \left(\mathbf{v}_A^{(n)}\right)_I + \dot{\theta}_n \mathbf{R}_\theta^{(n)}\bar{\mathbf{x}}_{P/A}^{(n)} \tag{4.4.26a}$$

$$\left(\mathbf{v}_P^{(n)}\right)_I = \left(\mathbf{v}_A^{(n)}\right)_I + {}_I\boldsymbol{\Omega}_B^{(n)}\mathbf{x}_{P/A}^{(n)} \tag{4.4.26b}$$

$$\left(\mathbf{v}_P^{(n)}\right)_I = \left(\mathbf{v}_A^{(n)}\right)_I + \dot{\mathbf{R}}^{(n)}\left(\mathbf{R}^{(n)}\right)^T\mathbf{x}_{P/A}^{(n)} \tag{4.4.26c}$$

$$\left(\mathbf{v}_P^{(n)}\right)_I = \left(\mathbf{v}_A^{(n)}\right)_I + {}_I\boldsymbol{\omega}_B^{(n)} \times \mathbf{x}_{P/A}^{(n)} \tag{4.4.26d}$$

The mixed vector-matrix forms of Eqs. (4.4.26a) and (4.4.26b) are the ones we will typically use in our kinematic analysis of constrained rigid bodies in two-dimensional motion. The purely vector form of Eq. (4.4.26d) is the velocity relationship found in most introductory dynamics texts.

4.4.3 Acceleration Analysis

Consider the relative velocity relationship of Eq. (4.4.26a). If we differentiate this relationship on time we have

$$\left(\dot{\mathbf{v}}_P^{(n)}\right)_I = \left(\dot{\mathbf{v}}_A^{(n)}\right)_I + \ddot{\theta}_n\mathbf{R}_\theta^{(n)}\bar{\mathbf{x}}_{P/A}^{(n)} + \dot{\theta}_n\dot{\mathbf{R}}_\theta^{(n)}\bar{\mathbf{x}}_{P/A}^{(n)} \tag{4.4.27}$$

Now consider the matrix $\left[\dot{R}_\theta^{(n)}\right]$. We have, using Eq. (4.4.17),

$$\left[\dot{R}_\theta^{(n)}\right] = -\dot{\theta}_n\begin{bmatrix}\cos\theta_n & -\sin\theta_n\\ \sin\theta_n & \cos\theta_n\end{bmatrix} = -\dot{\theta}_n\left[R^{(n)}\right] \tag{4.4.28}$$

so we find the acceleration relationship in matrix-vector form:

$$\left(\mathbf{a}_P^{(n)}\right)_I = \left(\mathbf{a}_A^{(n)}\right)_I + \ddot{\theta}_n\mathbf{R}_\theta^{(n)}\bar{\mathbf{x}}_{P/A}^{(n)} - \left(\dot{\theta}_n\right)^2\mathbf{R}^{(n)}\bar{\mathbf{x}}_{P/A}^{(n)} \tag{4.4.29}$$

To compare this with the relative acceleration vector relationship developed earlier (see, for example, Eq. (4.2.21b)) first note that the angular acceleration vector, $\boldsymbol{\alpha}^{(n)}$, like the angular velocity, is along the fixed Z-axis, i.e., $\boldsymbol{\alpha}^{(n)} = \ddot{\theta}_n\mathbf{e}_Z$. We can rework the second terms on the right side of Eq. (4.4.29) in the same manner as done for the relative velocity term as

$$\ddot{\theta}_n\left[R_\theta^{(n)}\right]\begin{Bmatrix}\bar{x}_{P/A}^{(n)}\\ \bar{y}_{P/A}^{(n)}\end{Bmatrix} = \begin{bmatrix}0 & -\ddot{\theta}_n\\ \ddot{\theta}_n & 0\end{bmatrix}\left[R^{(n)}\right]\begin{Bmatrix}\bar{x}_{P/A}^{(n)}\\ \bar{y}_{P/A}^{(n)}\end{Bmatrix} = \left[\dot{\Omega}^{(n)}\right]\begin{Bmatrix}X_{P/A}^{(n)}\\ Y_{P/A}^{(n)}\end{Bmatrix} \tag{4.4.30}$$

where the skew symmetrical matrix $\dot{\boldsymbol{\Omega}}^{(n)}$ is given by

$$\left[\dot{\Omega}^{(n)}\right] = \begin{bmatrix} 0 & -\ddot{\theta}_n \\ \ddot{\theta}_n & 0 \end{bmatrix} \tag{4.4.31}$$

and this again is equivalent to a vector cross product so

$$\ddot{\theta}_n \left[R_\theta^{(n)}\right] \begin{Bmatrix} \bar{x}_{P/A}^{(n)} \\ \bar{y}_{P/A}^{(n)} \end{Bmatrix} = \left[\dot{\Omega}^{(n)}\right] \begin{Bmatrix} X_{P/A}^{(n)} \\ Y_{P/A}^{(n)} \end{Bmatrix} = {}_I\boldsymbol{\alpha}_B^{(n)} \times \mathbf{x}_{P/A}^{(n)} \tag{4.4.32}$$

For the third term on the right side of Eq. (4.4.29) it is easy to show that we have

$$-\left(\dot{\theta}_n\right)^2 \left[R^{(n)}\right] \begin{Bmatrix} \bar{x}_{P/A}^{(n)} \\ \bar{y}_{P/A}^{(n)} \end{Bmatrix} = -\left(\dot{\theta}_n\right)^2 \begin{Bmatrix} X_{P/A}^{(n)} \\ Y_{P/A}^{(n)} \end{Bmatrix} = {}_I\boldsymbol{\omega}_B^{(n)} \times \left({}_I\boldsymbol{\omega}_B^{(n)} \times \mathbf{x}_{P/A}^{(n)}\right) \tag{4.4.33}$$

From these results it follows that our relative acceleration expression has the equivalent forms:

$$\left(\mathbf{a}_P^{(n)}\right)_I = \left(\mathbf{a}_A^{(n)}\right)_I + \ddot{\theta}_n \mathbf{R}_\theta^{(n)} \bar{\mathbf{x}}_{P/A}^{(n)} - \left(\dot{\theta}_n\right)^2 \mathbf{R}^{(n)} \bar{\mathbf{x}}_{P/A}^{(n)} \tag{4.4.34a}$$

$$\left(\mathbf{a}_P^{(n)}\right)_I = \left(\mathbf{a}_A^{(n)}\right)_I + {}_I\dot{\boldsymbol{\Omega}}_B^{(n)} \mathbf{x}_{P/A}^{(n)} - \left(\dot{\theta}_n\right)^2 \mathbf{x}_{P/A}^{(n)} \tag{4.4.34b}$$

$$\left(\mathbf{a}_P^{(n)}\right)_I = \left(\mathbf{a}_A^{(n)}\right)_I + {}_I\boldsymbol{\alpha}_B^{(n)} \times \mathbf{x}_{P/A}^{(n)} + {}_I\boldsymbol{\omega}_B^{(n)} \times \left({}_I\boldsymbol{\omega}_B^{(n)} \times \mathbf{x}_{P/A}^{(n)}\right) \tag{4.4.34c}$$

where we again have emphasized that the skew symmetrical angular acceleration matrix is based on angular accelerations of the body B as measured in the inertial frame, I, by writing it as ${}_I\dot{\boldsymbol{\Omega}}_B^{(n)}$. Because we are dealing with the case of two-dimensional motion, the relative acceleration expression in terms of vectors only, Eq. (4.4.34c) can obviously also be written in the simpler form:

$$\left(\mathbf{a}_P^{(n)}\right)_I = \left(\mathbf{a}_A^{(n)}\right)_I + {}_I\boldsymbol{\alpha}_B^{(n)} \times \mathbf{x}_{P/A}^{(n)} - \left(\dot{\theta}_n\right)^2 \mathbf{x}_{P/A}^{(n)} \tag{4.4.35}$$

which is also equivalent to the form involving the tangential and normal components discussed previously. We will generally use the mixed forms of Eqs. (4.4.34a) and (4.4.34b) when writing the acceleration relationship in matrix-vector form. Equation (4.4.34c) or (4.4.35) is the customary form seen in dynamics texts when writing the relationship in terms of vectors only.

4.4.4 General Relative Velocity and Acceleration Relations

In rigid body motion problems, if we are relating the motion of two points in the same body then we can use the rigid body velocity and acceleration expressions

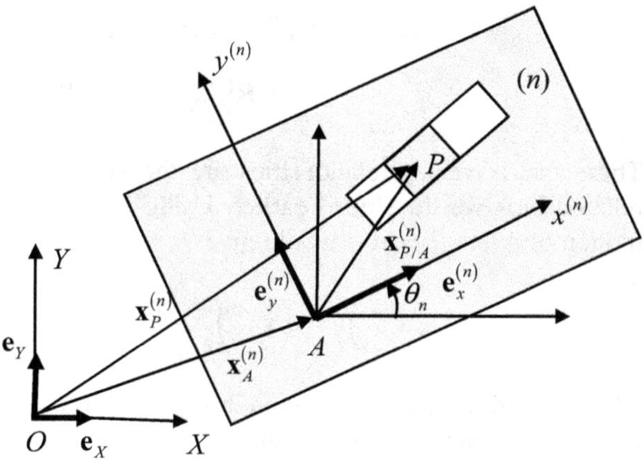

Fig. 4.26 The case where the motion of a point P is related to the motion of a point A in a rigid body where the distance between the two points is not fixed

developed in the last two subsections. However, we may need to consider the more general case where a point P moves relative to a point A that is fixed in a rigid body but where the point P itself may not be in the same rigid body. As a simple example, consider the case where we want to examine the motion of a point P on a block that moves within a slot of a rigid body, as shown in Fig. 4.26. In that case the positional relationship between points A and P can still be written as (Eq. (4.4.10))

$$\mathbf{x}_P^{(n)} = \mathbf{x}_A^{(n)} + \mathbf{R}^{(n)}(\theta_n)\bar{\mathbf{x}}_{P/A}^{(n)} \tag{4.4.36}$$

but the relative position vector of P with respect to A, as measured in body-fixed coordinates, is no longer a constant. Thus, when we differentiate Eq. (4.4.36) on time to obtain the relative velocity relationship, we have an additional term, i.e.,

$$\left(\mathbf{v}_P^{(n)}\right)_I = \left(\mathbf{v}_A^{(n)}\right)_I + \mathbf{R}^{(n)}\dot{\bar{\mathbf{x}}}_{P/A}^{(n)} + \dot{\theta}_n\mathbf{R}_\theta^{(n)}\bar{\mathbf{x}}_{P/A}^{(n)} \tag{4.4.37}$$

where $\dot{\bar{\mathbf{x}}}_{P/A}^{(n)}$ is the relative velocity of P with respect to A as seen by an observer who moves with the body, B. We can differentiate Eq. (4.4.37) once more to obtain, following the same steps as before,

$$\begin{aligned}\left(\mathbf{a}_P^{(n)}\right)_I = \left(\mathbf{a}_A^{(n)}\right)_I &+ \ddot{\theta}_n\mathbf{R}_\theta^{(n)}\bar{\mathbf{x}}_{P/A}^{(n)} - \left(\dot{\theta}_n\right)^2\mathbf{R}^{(n)}\bar{\mathbf{x}}_{P/A}^{(n)} \\ &+\mathbf{R}^{(n)}\ddot{\bar{\mathbf{x}}}_{P/A}^{(n)} + \dot{\theta}_n\mathbf{R}_\theta^{(n)}\dot{\bar{\mathbf{x}}}_{P/A}^{(n)} + \dot{\mathbf{R}}^{(n)}\dot{\bar{\mathbf{x}}}_{P/A}^{(n)}\end{aligned} \tag{4.4.38}$$

where $\ddot{\bar{\mathbf{x}}}_{P/A}^{(n)}$ is the relative acceleration of P with respect to A as seen by an observer who moves with the body, B. Using the fact that $\dot{\mathbf{R}}^{(n)} = \dot{\theta}_n\mathbf{R}_\theta^{(n)}$ (see Eq. (4.4.16)) we have, finally,

$$\begin{aligned}\left(\mathbf{a}_P^{(n)}\right)_I = \left(\mathbf{a}_A^{(n)}\right)_I + \ddot{\theta}_n \mathbf{R}_\theta^{(n)} \bar{\mathbf{x}}_{P/A}^{(n)} - \left(\dot{\theta}_n\right)^2 \mathbf{R}^{(n)} \bar{\mathbf{x}}_{P/A}^{(n)} \\ + \mathbf{R}^{(n)} \ddot{\bar{\mathbf{x}}}_{P/A}^{(n)} + 2\dot{\theta}_n \mathbf{R}_\theta^{(n)} \dot{\bar{\mathbf{x}}}_{P/A}^{(n)}\end{aligned} \tag{4.4.39}$$

These matrix-vector relationships are the equivalent forms to the purely vector relationships we developed earlier. Using Eq. (4.2.18), for example, to relate the motion of points P and A would give

$$\left(\dot{\mathbf{x}}_P^{(n)}\right)_I = \left(\dot{\mathbf{x}}_A^{(n)}\right)_I + \left(\dot{\mathbf{x}}_{P/A}^{(n)}\right)_B + {}_I\boldsymbol{\omega}_B^{(n)} \times \mathbf{x}_{P/A}^{(n)} \tag{4.4.40}$$

where I and B are inertial and body-fixed frames, respectively. The terms in Eq. (4.4.40) can be directly compared to those in Eq. (4.4.37). Similarly, from Eq. (4.2.19)

$$\begin{aligned}\left(\ddot{\mathbf{x}}_P^{(n)}\right)_I = \left(\ddot{\mathbf{x}}_A^{(n)}\right)_I + {}_I\boldsymbol{\alpha}_B^{(n)} \times \mathbf{x}_{P/A}^{(n)} + {}_I\boldsymbol{\omega}_B^{(n)} \times \left({}_I\boldsymbol{\omega}_B^{(n)} \times \mathbf{x}_{P/A}^{(n)}\right) \\ + \left(\ddot{\mathbf{x}}_{P/A}^{(n)}\right)_B + 2{}_I\boldsymbol{\omega}_B^{(n)} \times \left(\dot{\mathbf{x}}_{P/A}^{(n)}\right)_B\end{aligned} \tag{4.4.41}$$

whose terms can be directly compared to those in Eq. (4.4.39). In particular, you should note that the following terms are equal:

$$\left(\dot{\mathbf{x}}_{P/A}^{(n)}\right)_B = \mathbf{R}^{(n)} \dot{\bar{\mathbf{x}}}_{P/A}^{(n)}, \quad \left(\ddot{\mathbf{x}}_{P/A}^{(n)}\right)_B = \mathbf{R}^{(n)} \ddot{\bar{\mathbf{x}}}_{P/A}^{(n)} \tag{4.4.42}$$

Finally, we recall that while for the relative position vector we have

$$\mathbf{x}_{P/A}^{(n)} = \mathbf{R}^{(n)} \bar{\mathbf{x}}_{P/A}^{(n)} \tag{4.4.43}$$

for velocity and acceleration terms we have instead

$$\begin{aligned}\left(\dot{\mathbf{x}}_{P/A}^{(n)}\right)_I &= \mathbf{R}^{(n)} \dot{\bar{\mathbf{x}}}_{P/A}^{(n)} + {}_I\boldsymbol{\omega}_B^{(n)} \times \mathbf{x}_{P/A}^{(n)} \\ \left(\ddot{\mathbf{x}}_{P/A}^{(n)}\right)_I &= \mathbf{R}^{(n)} \ddot{\bar{\mathbf{x}}}_{P/A}^{(n)} + {}_I\boldsymbol{\alpha}_B^{(n)} \times \mathbf{x}_{P/A}^{(n)} + {}_I\boldsymbol{\omega}_B^{(n)} \times \left({}_I\boldsymbol{\omega}_B^{(n)} \times \mathbf{x}_{P/A}^{(n)}\right) \\ &\quad + 2{}_I\boldsymbol{\omega}_B^{(n)} \times \dot{\bar{\mathbf{x}}}_{P/A}^{(n)}\end{aligned} \tag{4.4.44}$$

But if we let the velocity and acceleration terms on the left side of Eq. (4.4.44) be

$$\begin{aligned}\left(\dot{\mathbf{x}}_{P/A}^{(n)}\right)_I &= \mathbf{v}_{P/A}^{(n)} \\ \left(\ddot{\mathbf{x}}_{P/A}^{(n)}\right)_I &= \mathbf{a}_{P/A}^{(n)}\end{aligned} \tag{4.4.45}$$

then we can also relate these relative velocity and acceleration vectors, as given in an inertial frame, to the same vectors in a body-fixed coordinates by the transformation given in Eq. (4.4.43), i.e.,

$$\begin{aligned}\mathbf{v}_{P/A}^{(n)} &= \mathbf{R}^{(n)}\overline{\mathbf{v}}_{P/A}^{(n)}\\ \mathbf{a}_{P/A}^{(n)} &= \mathbf{R}^{(n)}\overline{\mathbf{a}}_{P/A}^{(n)}\end{aligned} \tag{4.4.46}$$

However,

$$\begin{aligned}\overline{\mathbf{v}}_{P/A}^{(n)} &\neq \dot{\overline{\mathbf{x}}}_{P/A}^{(n)}\\ \overline{\mathbf{a}}_{P/A}^{(n)} &\neq \ddot{\overline{\mathbf{x}}}_{P/A}^{(n)}\end{aligned} \tag{4.4.47}$$

since the left-hand sides of Eq. (4.4.47) are the total relative velocity and acceleration terms as computed by an observer in an inertial frame and then transformed to a body-fixed frame while the right-hand sides are the relative velocities and accelerations as computed by an observer moving with the body-fixed frame. Indeed, by multiplying both sides of Eq. (4.4.44) by $(\mathbf{R}^{(n)})^T$ it is not difficult to show (see Problem P4.3) that

$$\begin{aligned}\overline{\mathbf{v}}_{P/A}^{(n)} &= \dot{\overline{\mathbf{x}}}_{P/A}^{(n)} + {}_I\overline{\boldsymbol{\omega}}_B^{(n)} \times \overline{\mathbf{x}}_{P/A}^{(n)}\\ \overline{\mathbf{a}}_{P/A}^{(n)} &= \ddot{\overline{\mathbf{x}}}_{P/A}^{(n)} + {}_I\overline{\boldsymbol{\alpha}}_B^{(n)} \times \overline{\mathbf{x}}_{P/A}^{(n)} + {}_I\overline{\boldsymbol{\omega}}_B^{(n)} \times \left({}_I\overline{\boldsymbol{\omega}}_B^{(n)} \times \overline{\mathbf{x}}_{P/A}^{(n)}\right)\\ &\quad + 2\,{}_I\overline{\boldsymbol{\omega}}_B^{(n)} \times \dot{\overline{\mathbf{x}}}_{P/A}^{(n)}\end{aligned} \tag{4.4.48}$$

where ${}_I\overline{\boldsymbol{\omega}}_B^{(n)} = \left(\mathbf{R}^{(n)}\right)^T {}_I\boldsymbol{\omega}_B^{(n)}$ and ${}_I\overline{\boldsymbol{\alpha}}_B^{(n)} = \left(\mathbf{R}^{(n)}\right)^T {}_I\boldsymbol{\alpha}_B^{(n)}$ are the angular velocity and angular acceleration vectors calculated in body-fixed coordinates. In writing Eq. (4.4.48) in these body-fixed coordinates obviously the rotation matrix $\mathbf{R}^{(n)}$ is absent, as it should be, since all the quantities are being computed in the body-fixed frame and there is no need to transform to any fixed coordinate system.

4.4.5 The Transport Theorem Revisited

Note that when we write the transport theorem in the form

$$\left(\frac{d\mathbf{p}}{dt}\right)_I = \left(\frac{d\mathbf{p}}{dt}\right)_B + {}_I\boldsymbol{\omega}_B \times \mathbf{p} \tag{4.4.49}$$

the notation $(\ )_I$ or $(\ )_B$ refers to the frame in which the time derivative is calculated and implicitly there must be another common frame present that we use to describe the components of all the terms present in the transport theorem after the derivatives are calculated. Some authors prefer to write the transport theorem as

$$\frac{{}^{I}d}{dt}(\mathbf{p}) = \frac{{}^{B}d}{dt}(\mathbf{p}) + {}_{I}\boldsymbol{\omega}_{B} \times \mathbf{p} \tag{4.4.50}$$

to indicate more clearly that I and B here refer to the frame in which the derivative is being taken. We could also bring in explicitly the underlying common frame we use to ultimately describe all the terms in the transport theorem by writing that theorem in a new notation as

$$\left[\frac{{}^{I}d}{dt}(\mathbf{p})\right]_{A} = \left[\frac{{}^{B}d}{dt}(\mathbf{p})\right]_{A} + [{}_{I}\boldsymbol{\omega}_{B} \times \mathbf{p}]_{A} \tag{4.4.51}$$

where $[\ ]_A$ indicates that components of all these vectors are being computed in a frame that we have labeled as A. Thus, for example, the term

$$\left[\frac{{}^{B}d}{dt}(\mathbf{p})\right]_{A} \tag{4.4.52}$$

then means that (1) we are taking the derivative of $\mathbf{p}$ when $\mathbf{p}$ is expressed in the B frame and (2) we are expressing that differentiated vector in the A frame. The A frame can be an inertial frame, a body-fixed frame, or any other frame we choose. For example, in this more explicit notation we could write Eq. (4.4.50) when evaluated in the inertial I frame as

$$\left[\frac{{}^{I}d}{dt}(\mathbf{p})\right]_{I} = \left[\frac{{}^{B}d}{dt}(\mathbf{p})\right]_{I} + [{}_{I}\boldsymbol{\omega}_{B} \times \mathbf{p}]_{I} \tag{4.4.53}$$

and we have

$$\left[\frac{{}^{B}d}{dt}(\mathbf{p})\right]_{I} = \mathbf{R}^{(n)}\left[\frac{{}^{B}d}{dt}(\mathbf{p})\right]_{B} = \mathbf{R}^{(n)}\dot{\bar{\mathbf{p}}} \tag{4.4.54}$$

where $\dot{\bar{\mathbf{p}}}$ is the vector $\mathbf{p}$ expressed in body-fixed coordinates whose time derivative indicated by the "dot" is taken relative to that coordinate system, i.e., holding the unit vectors along the B frame coordinates fixed. Thus, we can also write the transport theorem in this case as

$$\left[\frac{{}^{I}d}{dt}(\mathbf{p})\right]_{I} = \mathbf{R}^{(n)}\dot{\bar{\mathbf{p}}} + \left[{}_{I}\boldsymbol{\omega}_{B}^{(n)} \times \mathbf{p}\right]_{I} \tag{4.4.55}$$

where the rotation matrix, $\mathbf{R}^{(n)}$, which transforms $\dot{\mathbf{p}}$ from the B frame to the I frame, is now explicitly present.

Note, however, that in this new notation $[\ ]_A$ *refers to the frame A in which the expression in the square brackets is being evaluated while in our previous notation* $(\ )_A$ *referred to the frame A in which the time derivative is being calculated. Thus, while the notations look similar the meanings are quite different.*

To illustrate how these notations are related, consider Eq. (4.4.42), which can be written in the equivalent forms (assuming we are evaluating the end results in the I frame):

$$\begin{aligned}\left(\dot{\mathbf{x}}_{P/A}^{(n)}\right)_B &= \left[\frac{{}^B d\,\mathbf{x}_{P/A}}{dt}\right]_I = \mathbf{R}^{(n)}\dot{\bar{\mathbf{x}}}_{P/A}^{(n)} \\ \left(\ddot{\mathbf{x}}_{P/A}^{(n)}\right)_B &= \left[\frac{{}^B d^2\mathbf{x}_{P/A}}{dt^2}\right]_I = \mathbf{R}^{(n)}\ddot{\bar{\mathbf{x}}}_{P/A}^{(n)}\end{aligned} \tag{4.4.56}$$

where only the middle form shows the I frame explicitly. This new notation is a bit more cumbersome but we will use it when it is necessary to be more explicit.

4.5 Matrix-Vector Kinematics of Constraints and Kinematically Driven Systems

We have presented a matrix-vector formulation of the positional, velocity, and acceleration relationships since these provide an alternative to the traditional vector approach to formulating constraints and solving kinematically driven systems that we have discussed in the previous sections. The matrix-vector approach is able to deal with highly complex systems of interconnected and interacting rigid bodies since we can use it to describe the kinematics of each body in terms of exactly the same type of generalized coordinates. To see this, consider the slider-crank problem again, as shown in Fig. 4.27a. We previously used the generalized coordinates $(\tilde{\theta}_1, \tilde{\theta}_2, X_P)$ to define the geometry of this problem. These generalized coordinates were related to each other by two constraints since this was a one degree of freedom problem. We could continue to use this same approach for other problems, i.e., each time we would choose a set of generalized coordinates suitable for the given problem and then define the constraints those coordinates must meet. But, we could also treat the three rigid bodies in this problem by using the inertial coordinates of the center of mass, $(X^{(n)}, Y^{(n)})$, for each body and the counterclockwise angle that the body makes with respect to the positive X-axis, θ_n, as our generalized coordinates (Fig. 4.27b). We have distinguished the angles $(\tilde{\theta}_1, \tilde{\theta}_2)$ used previously from the current angles (θ_1, θ_2) because in the previous approach the angles were chosen to suit the problem at hand. Thus, while the angle of body (1) was taken to be a counterclockwise angle as measured from the positive X-axis, the angle of body (2) was taken as a clockwise angle with respect to the negative X-axis. These choices are natural ones for the

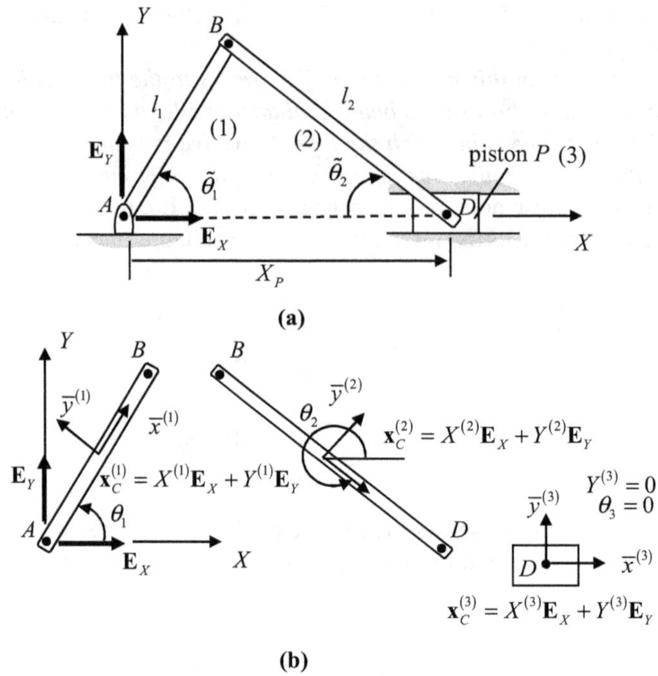

Fig. 4.27 (**a**) The geometry of the slider-crank as defined by the generalized coordinates $(\tilde{\theta}_1, \tilde{\theta}_2, X_P)$ and (**b**) the same elements defined by their absolute coordinates $(X^{(1)}, Y^{(1)}, \theta_1)$, $(X^{(2)}, Y^{(2)}, \theta_2)$, and $(X^{(3)}, Y^{(3)}, \theta_3)$. Note that the pin on the slider block has been relabeled as point D

problem being considered but they will typically change, depending on the problem. In contrast, *we will always take the angle of a body being considered by our matrix-vector approach to be the angle to the body-fixed x-axis as measured from the positive inertial coordinate X-axis*. This choice must not be changed since the rotation matrix used to describe the motion also makes this assumption. The angle θ_2 in Fig. 4.27b follows this convention and is related to the angle $\tilde{\theta}_2$ used previously by $\theta_2 = 2\pi - \tilde{\theta}_2$. For the connecting rod as shown in Fig. 4.27b we also could take its orientation as being at the negative angle $\theta_2 = -\tilde{\theta}_2$, which only differs by a factor of 2π from the angle θ_2 shown in Fig. 4.27b so that both choices describe the same orientation of the rod. Using $(X^{(n)}, Y^{(n)}, \theta_n)$ as our generalized coordinates means, of course, that we have to use more coordinates and more constraints on those coordinates to solve the problem than with other coordinate choices. But if we are going to obtain the solution numerically by computer anyway, this disadvantage can be outweighed by the fact that in a complex problem with many bodies the motion of each body is always treated in exactly the same fashion. Also, each type of constraint is also described in terms of these same coordinates so that the constraints also can be treated in a uniform manner for complex problems. For the slider-crank problem we have two types of constraints—constraints between bodies that are pinned to one

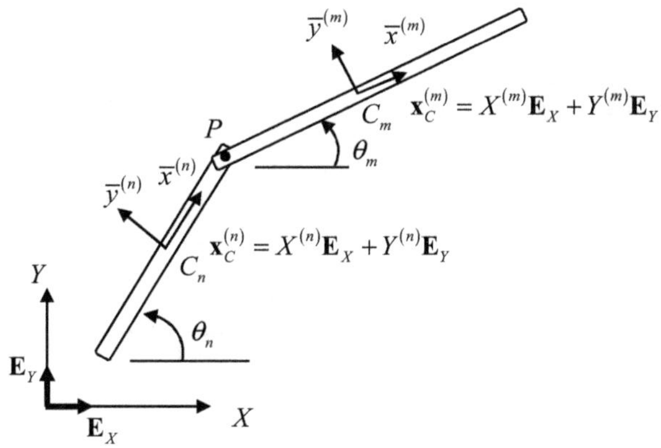

Fig. 4.28 The geometry for two rigid bodies (n) and (m), shown separated, that are pinned together at point P

another (or to the fixed ground) and the constraint of the slider to only translate along the X-axis. The slider constraint is easy to formulate here as

$$\begin{aligned} Y^{(n)} &= 0 \\ \theta_n &= 0 \end{aligned} \tag{4.5.1}$$

where the slider is labeled as body three ($n = 3$). For the pinned constraint (also called a *revolute joint* constraint), consider two rigid bodies (n) and (m) that are pinned at point P, as shown in Fig. 4.28. Point P is common to both bodies and is fixed in distance from all other points in both bodies so that we have

$$\begin{gathered} \mathbf{x}_P^{(n)} = \mathbf{x}_P^{(m)} \\ \text{or} \\ \mathbf{x}_C^{(n)} + \mathbf{R}^{(n)}(\theta_n)\bar{\mathbf{x}}_P^{(n)} = \mathbf{x}_C^{(m)} + \mathbf{R}^{(m)}(\theta_m)\bar{\mathbf{x}}_P^{(m)} \end{gathered} \tag{4.5.2}$$

with $\mathbf{x}_C^{(n)} = X^{(n)}\mathbf{E}_X + Y^{(n)}\mathbf{E}_Y$ and $\mathbf{x}_C^{(m)} = X^{(m)}\mathbf{E}_X + Y^{(m)}\mathbf{E}_Y$. Strictly speaking, to be consistent with our previous notation we should write $\bar{\mathbf{x}}_P^{(n)} = \bar{\mathbf{x}}_{P/C_n}^{(n)}$ and $\bar{\mathbf{x}}_P^{(m)} = \bar{\mathbf{x}}_{P/C_m}^{(m)}$, i.e., these vectors are position vectors in body-fixed coordinates relative to the different centers of mass, C_n and C_m (see Fig. 4.28), but we will drop the explicit reference to those centers for economy of notation with the understanding that these body-fixed relative position vectors always will be relative to the center of mass locations of the body under consideration unless otherwise stated. If we want to solve a kinematically driven slider-crank problem then we also have to include a driving constraint. If the crank is being driven we can define this driving constraint as

$$\theta_1 = g(t) \tag{4.5.3}$$

With the use of all these constraints we can then write the constraint equations for the kinematically driven slider-crank problem as

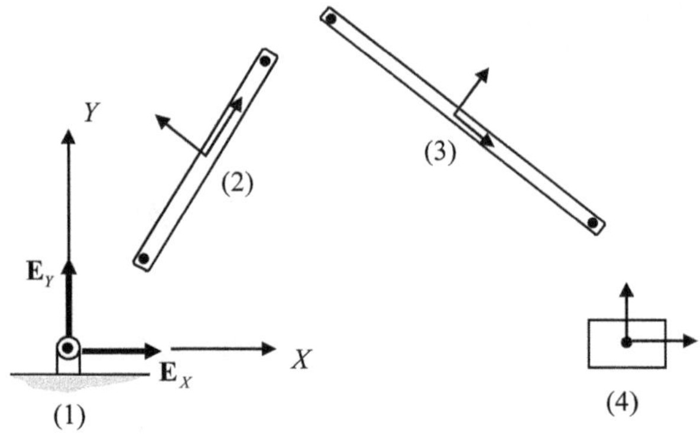

Fig. 4.29 The slider-crank problem treated as a four-body problem, where the ground is considered to be a completely fixed body

$$\mathbf{f}(\mathbf{q},t) = \begin{bmatrix} \mathbf{x}_C^{(1)} + \mathbf{R}^{(1)}\bar{\mathbf{x}}_A^{(1)} \\ \mathbf{x}_C^{(2)} + \mathbf{R}^{(2)}\bar{\mathbf{x}}_B^{(2)} - \mathbf{x}_C^{(1)} - \mathbf{R}^{(1)}\bar{\mathbf{x}}_B^{(1)} \\ \mathbf{x}_C^{(3)} + \mathbf{R}^{(3)}\bar{\mathbf{x}}_D^{(3)} - \mathbf{x}_C^{(2)} - \mathbf{R}^{(2)}\bar{\mathbf{x}}_D^{(2)} \\ Y^{(3)} \\ \theta_3 \\ \theta_1 - g(t) \end{bmatrix} = \begin{bmatrix} \mathbf{0} \\ \mathbf{0} \\ \mathbf{0} \\ 0 \\ 0 \\ 0 \end{bmatrix} \tag{4.5.4}$$

where the generalized coordinates are $\mathbf{q} = [X^{(1)}, Y^{(1)}, \theta_1, X^{(2)}, Y^{(2)}, \theta_2, X^{(3)}, Y^{(3)}, \theta_3]^T$. Note that it is common practice (see [1], for example) to consider the fixed ground as a fourth body in this problem (see Fig. 4.29). This allows the pin constraint at the ground (point A) to be written in the same fashion as the other pins and to write the fixity constraints of the ground explicitly, which is important for more complex problems. In this case, the explicit constraint for the ground is (Fig. 4.29)

$$\begin{aligned} X^{(1)} &= 0 \\ Y^{(1)} &= 0 \\ \theta_1 &= 0 \end{aligned} \tag{4.5.5}$$

and the constraints would then be written as

$$\mathbf{f}(\mathbf{q},t) = \begin{bmatrix} X^{(1)} \\ Y^{(1)} \\ \theta_1 \\ \mathbf{x}_C^{(2)} + \mathbf{R}^{(2)}\bar{\mathbf{x}}_A^{(2)} - \mathbf{x}_C^{(1)} - \mathbf{R}^{(1)}\bar{\mathbf{x}}_A^{(1)} \\ \mathbf{x}_C^{(3)} + \mathbf{R}^{(3)}\bar{\mathbf{x}}_B^{(3)} - \mathbf{x}_C^{(2)} - \mathbf{R}^{(2)}\bar{\mathbf{x}}_B^{(2)} \\ \mathbf{x}_C^{(4)} + \mathbf{R}^{(4)}\bar{\mathbf{x}}_D^{(4)} - \mathbf{x}_C^{(3)} - \mathbf{R}^{(3)}\bar{\mathbf{x}}_D^{(3)} \\ Y^{(4)} \\ \theta_4 \\ \theta_2 - g(t) \end{bmatrix} = \begin{bmatrix} 0 \\ 0 \\ 0 \\ \mathbf{0} \\ \mathbf{0} \\ \mathbf{0} \\ 0 \\ 0 \\ 0 \end{bmatrix} \tag{4.5.6}$$

where each body (and its coordinates) now has a different label number. We will use the simpler form of the constraints given in Eq. (4.5.4) as in this book we will not need to use the generality afforded by Eq. (4.5.6) and both forms lead to exactly the same solution. We can write the constraints of Eq. (4.5.4) in terms of the generalized coordinates directly as

$$\mathbf{f}(\mathbf{q},t)=\begin{bmatrix} X^{(1)}-l_1\cos\theta_1/2 \\ Y^{(1)}-l_1\sin\theta_1/2 \\ X^{(2)}-l_2\cos\theta_2/2-X^{(1)}-l_1\cos\theta_1/2 \\ Y^{(2)}-l_2\sin\theta_2/2-Y^{(1)}-l_1\sin\theta_1/2 \\ X^{(3)}-X^{(2)}-l_2\cos\theta_2/2 \\ Y^{(3)}-Y^{(2)}-l_2\sin\theta_2/2 \\ Y^{(3)} \\ \theta_3 \\ \theta_1-g(t) \end{bmatrix}=\begin{bmatrix}0\\0\\0\\0\\0\\0\\0\\0\\0\end{bmatrix} \tag{4.5.7}$$

Equation (4.5.7) is a nonlinear system of nine equations for the generalized coordinates which we can solve with the Newton-Raphson method as done before. We also need to have the Jacobian matrix $\mathbf{f}_\mathbf{q}$, which in this case is a 9×9 matrix given by

$$\mathbf{f}_\mathbf{q}=\begin{bmatrix} 1 & 0 & l_1\sin\theta_1/2 & 0 & 0 & 0 & 0 & 0 & 0 \\ 0 & 1 & -l_1\cos\theta_1/2 & 0 & 0 & 0 & 0 & 0 & 0 \\ -1 & 0 & l_1\sin\theta_1/2 & 1 & 0 & l_2\sin\theta_2/2 & 0 & 0 & 0 \\ 0 & -1 & -l_1\cos\theta_1/2 & 0 & 1 & -l_2\cos\theta_2/2 & 0 & 0 & 0 \\ 0 & 0 & 0 & -1 & 0 & l_2\sin\theta_2/2 & 1 & 0 & 0 \\ 0 & 0 & 0 & 0 & -1 & -l_2\cos\theta_2/2 & 0 & 1 & 0 \\ 0 & 0 & 0 & 0 & 0 & 0 & 0 & 1 & 0 \\ 0 & 0 & 0 & 0 & 0 & 0 & 0 & 0 & 1 \\ 0 & 0 & 1 & 0 & 0 & 0 & 0 & 0 & 0 \end{bmatrix} \tag{4.5.8}$$

To obtain the constraints written in terms of the velocity, recall $\dot{\mathbf{f}}=\mathbf{f}_\mathbf{q}\dot{\mathbf{q}}+\mathbf{f}_t=0$ so that we obtain a set of linear velocity equations given as $\mathbf{f}_\mathbf{q}\dot{\mathbf{q}}=-\mathbf{f}_t$ and formally we can write these terms as

$$\mathbf{f}_\mathbf{q}\dot{\mathbf{q}}=\begin{bmatrix} \dot{\mathbf{x}}_C^{(1)}+\dot{\theta}_1\mathbf{R}_\theta^{(1)}\bar{\mathbf{x}}_A^{(1)} \\ \dot{\mathbf{x}}_C^{(2)}+\dot{\theta}_2\mathbf{R}_\theta^{(2)}\bar{\mathbf{x}}_B^{(2)}-\dot{\mathbf{x}}_C^{(1)}-\dot{\theta}_1\mathbf{R}_\theta^{(1)}\bar{\mathbf{x}}_B^{(1)} \\ \dot{\mathbf{x}}_C^{(3)}+\dot{\theta}_3\mathbf{R}_\theta^{(3)}\bar{\mathbf{x}}_D^{(3)}-\dot{\mathbf{x}}_C^{(2)}-\dot{\theta}_2\mathbf{R}_\theta^{(2)}\bar{\mathbf{x}}_D^{(2)} \\ \dot{Y}^{(3)} \\ \dot{\theta}_3 \\ \dot{\theta}_1 \end{bmatrix},\quad -\mathbf{f}_t=\begin{bmatrix}\mathbf{0}\\\mathbf{0}\\\mathbf{0}\\0\\0\\dg/dt\end{bmatrix} \tag{4.5.9}$$

Since $\mathbf{f}_{\mathbf{q}t} = 0$ for these pinned joint constraints, differentiating the velocity constraint on time once more gives us the acceleration relationship $d(\mathbf{f}_{\mathbf{q}}\dot{\mathbf{q}})/dt + \mathbf{f}_{tt} = 0$, or, equivalently, $d(\mathbf{f}_{\mathbf{q}}\dot{\mathbf{q}})/dt = -\mathbf{f}_{tt}$, where formally we have

$$\frac{d(\mathbf{f}_{\mathbf{q}}\dot{\mathbf{q}})}{dt} = \begin{bmatrix} \ddot{\mathbf{x}}_C^{(1)} + \ddot{\theta}_1 \mathbf{R}_\theta^{(1)} \bar{\mathbf{x}}_A^{(1)} \\ \ddot{\mathbf{x}}_C^{(2)} + \ddot{\theta}_2 \mathbf{R}_\theta^{(2)} \bar{\mathbf{x}}_B^{(2)} - \ddot{\mathbf{x}}_C^{(1)} - \ddot{\theta}_1 \mathbf{R}_\theta^{(1)} \bar{\mathbf{x}}_B^{(1)} \\ \ddot{\mathbf{x}}_C^{(3)} + \ddot{\theta}_3 \mathbf{R}_\theta^{(3)} \bar{\mathbf{x}}_D^{(3)} - \ddot{\mathbf{x}}_C^{(2)} - \ddot{\theta}_2 \mathbf{R}_\theta^{(2)} \bar{\mathbf{x}}_D^{(2)} \\ \ddot{Y}^{(3)} \\ \ddot{\theta}_3 \\ \ddot{\theta}_1 \end{bmatrix} + \begin{bmatrix} \dot{\theta}_1 \dot{\mathbf{R}}_\theta^{(1)} \bar{\mathbf{x}}_A^{(1)} \\ \dot{\theta}_2 \dot{\mathbf{R}}_\theta^{(2)} \bar{\mathbf{x}}_B^{(2)} - \dot{\theta}_1 \dot{\mathbf{R}}_\theta^{(1)} \bar{\mathbf{x}}_B^{(1)} \\ \dot{\theta}_3 \dot{\mathbf{R}}_\theta^{(3)} \bar{\mathbf{x}}_D^{(3)} - \dot{\theta}_2 \dot{\mathbf{R}}_\theta^{(2)} \bar{\mathbf{x}}_D^{(2)} \\ 0 \\ 0 \\ 0 \end{bmatrix} \tag{4.5.10}$$

and

$$-\mathbf{f}_{tt} = \begin{bmatrix} \mathbf{0} \\ \mathbf{0} \\ \mathbf{0} \\ 0 \\ 0 \\ d^2 g/dt^2 \end{bmatrix} \tag{4.5.11}$$

Comparing Eq. (4.5.10) with Eq. (4.5.9) and using Eq. (4.4.28) for $\dot{\mathbf{R}}_\theta^{(n)}$ we see that Eq. (4.5.10) is of the form

$$\frac{d(\mathbf{f}_{\mathbf{q}}\dot{\mathbf{q}})}{dt} = \mathbf{f}_{\mathbf{q}}\ddot{\mathbf{q}} + \begin{bmatrix} -(\dot{\theta}_1)^2 \mathbf{R}^{(1)} \bar{\mathbf{x}}_A^{(1)} \\ -(\dot{\theta}_2)^2 \mathbf{R}^{(2)} \bar{\mathbf{x}}_B^{(2)} + (\dot{\theta}_1)^2 \mathbf{R}^{(1)} \bar{\mathbf{x}}_B^{(1)} \\ -(\dot{\theta}_3)^2 \mathbf{R}^{(3)} \bar{\mathbf{x}}_D^{(3)} + (\dot{\theta}_2)^2 \mathbf{R}^{(2)} \bar{\mathbf{x}}_D^{(2)} \\ 0 \\ 0 \\ 0 \end{bmatrix} \tag{4.5.12}$$

so that if we write the acceleration constraint equation as $\mathbf{f}_{\mathbf{q}}\ddot{\mathbf{q}} = \mathbf{b}_v$, we have

$$\mathbf{b}_v = \begin{bmatrix} (\dot{\theta}_1)^2 \mathbf{R}^{(1)} \bar{\mathbf{x}}_A^{(1)} \\ (\dot{\theta}_2)^2 \mathbf{R}^{(2)} \bar{\mathbf{x}}_B^{(2)} - (\dot{\theta}_1)^2 \mathbf{R}^{(1)} \bar{\mathbf{x}}_B^{(1)} \\ (\dot{\theta}_3)^2 \mathbf{R}^{(3)} \bar{\mathbf{x}}_D^{(3)} - (\dot{\theta}_2)^2 \mathbf{R}^{(2)} \bar{\mathbf{x}}_D^{(2)} \\ 0 \\ 0 \\ d^2 g/dt^2 \end{bmatrix} \tag{4.5.13}$$

which becomes explicitly

$$\mathbf{b}_v = \begin{bmatrix} -(\dot{\theta}_1)^2 l_1 \cos\theta_1/2 \\ -(\dot{\theta}_1)^2 l_1 \sin\theta_1/2 \\ -(\dot{\theta}_2)^2 l_2 \cos\theta_2/2 - (\dot{\theta}_1)^2 l_1 \cos\theta_1/2 \\ -(\dot{\theta}_2)^2 l_2 \sin\theta_2/2 - (\dot{\theta}_1)^2 l_1 \sin\theta_1/2 \\ -(\dot{\theta}_2)^2 l_2 \cos\theta_2/2 \\ -(\dot{\theta}_2)^2 l_2 \sin\theta_2/2 \\ 0 \\ 0 \\ d^2 g/dt^2 \end{bmatrix} \tag{4.5.14}$$

Simulation 4.5: Matrix Kinematics of a Slider-Crank
The complete analysis of a kinematically driven slider-crank with a matrix approach is again governed by the positional, velocity, and acceleration equations:

$$\begin{aligned} \mathbf{f} &= 0 \\ \mathbf{f}_\mathbf{q}\dot{\mathbf{q}} &= -\mathbf{f}_t \\ \mathbf{f}_\mathbf{q}\ddot{\mathbf{q}} &= \mathbf{b}_v \end{aligned} \tag{4.5.15}$$

The positional equations are solved via the Newton-Raphson method, while the linear velocity and acceleration equations are solved with the system of linear equations solver built into MATLAB®. A script, slider_crank_matrix_sim, follows the same structure of the other slider-crank scripts. Default values are same as those used in the previous slider-crank scripts so that we can compare results with those formulations. The script uses two functions, slider_crank_f_matrix and slider_crank_J_matrix, that implement the positional constraints of Eq. (4.5.7) and the Jacobian of Eq. (4.5.8), respectively. The script calls the function slider_crank_matrix_sol which uses the Newton-Raphson method to solve for $[X^{(1)}, Y^{(1)}, \theta_1, X^{(2)}, Y^{(2)}, \theta_2, X^{(3)}, Y^{(3)}, \theta_3]^T$, the nine generalized coordinates used in the matrix approach. These values are then used in the script to solve the velocity and acceleration equations. To compare with our previous results the position, velocity, and acceleration of the connecting rod are shown as a function of time in Fig. 4.30 and the same parameters of the slider are shown in Fig. 4.31. These can be compared to the same results shown previously in Figs. 4.18 and 4.19, respectively. Other than a change in sign between Figs. 4.18 and 4.30 the two figures are in excellent agreement. (The change of sign occurs because the angle θ_2 used here is the

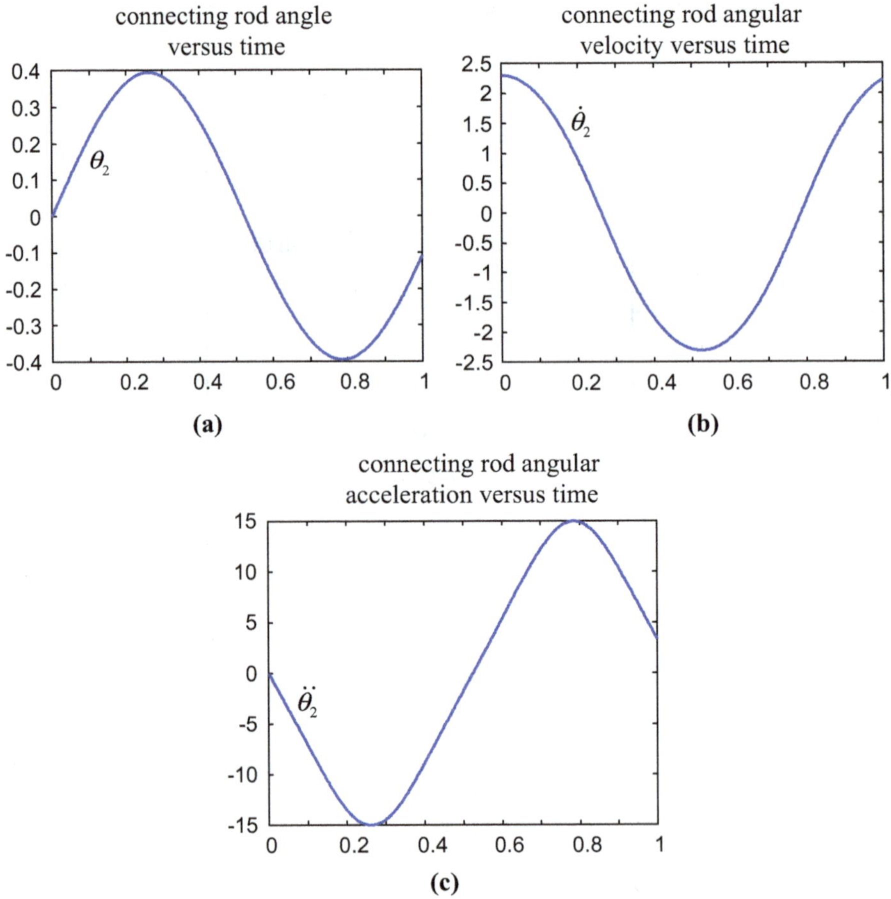

Fig. 4.30 Simulated kinematics of the slider-crank connecting rod using a matrix approach for (**a**) its angular position in radians versus time, (**b**) its angular velocity in rad/s versus time, and (**c**) its angular acceleration in rad/s^2 versus time. These results can be compared with Fig. 4.18 where the behavior is identical except for a change in sign due to the difference in definitions of the angles used in the two figures

negative of the angle $\tilde{\theta}_2$ used previously. This is, however, not the $\theta_2 = 2\pi - \tilde{\theta}_2$ angle shown in Fig. 4.27b. To obtain the θ_2 shown in Fig. 4.27b we would have had to set the initial value of θ_2 at $t = 0$ to be $\theta_{20} = 2\pi$ rather than the $\theta_{20} = 0$ initial value used in the script. Either choice, of course, results in identical behavior.) Similarly, the results shown in Figs. 4.31 and 4.19 agree completely for the dynamics of the slider (piston).

From the data generated in the MATLAB® workspace by the script you can also examine the kinematics of the center of mass for the crank and the connecting rod. These results are shown in Figs. 4.32 and 4.33, respectively. Recall that we also solved analytically for the kinematics of the slider-crank for the specific position of

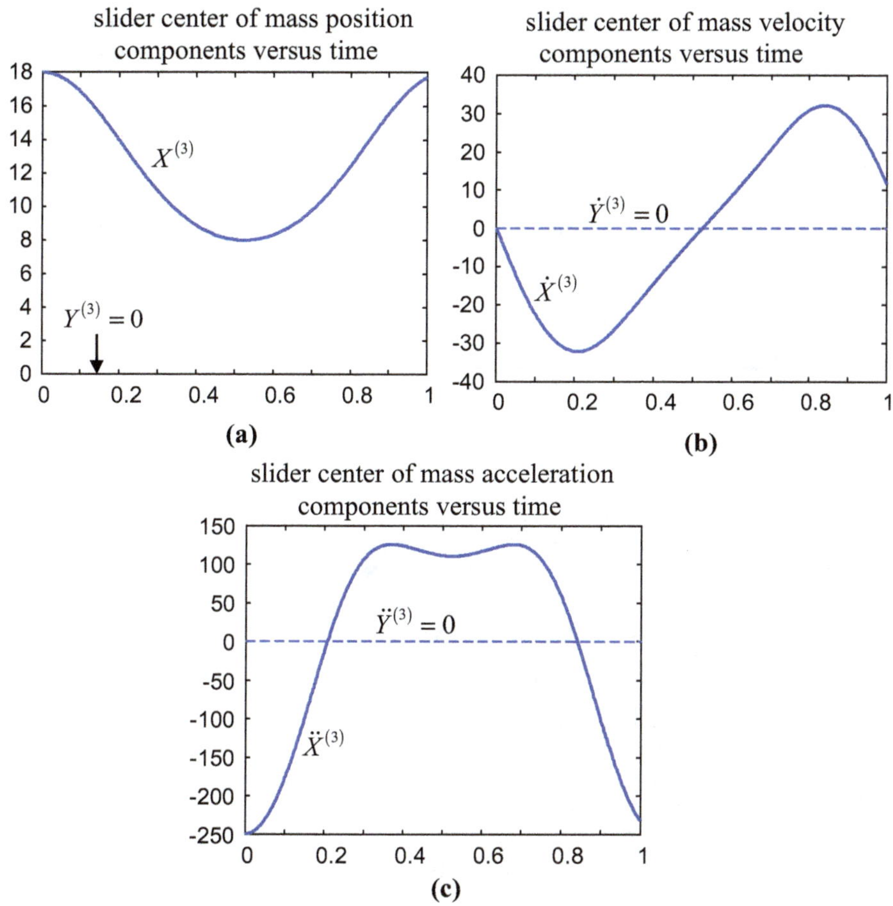

Fig. 4.31 Simulated kinematics of the slider-crank slider (piston) with a matrix approach for (**a**) its position in inches versus time, (**b**) its velocity in in./s versus time, and (**c**) its acceleration in in./s^2 versus time. These results can be compared to Fig. 4.19

the mechanism shown in Fig. 4.7. At this position, the velocities and accelerations of the three components can be easily obtained with the relative velocity and acceleration expressions. The results are shown in Fig. 4.34. Recall also that the mechanism reaches the position shown in Fig. 4.34 at $t = 0.7865$ s, approximately. It can be seen that the curves in Figs. 4.31–4.33 are consistent with those analytical values at that time. Of course the curves can give us a complete picture of the kinematics throughout the entire motion of the mechanism, something that is generally not practical to obtain by hand with analytical, fixed time calculations.

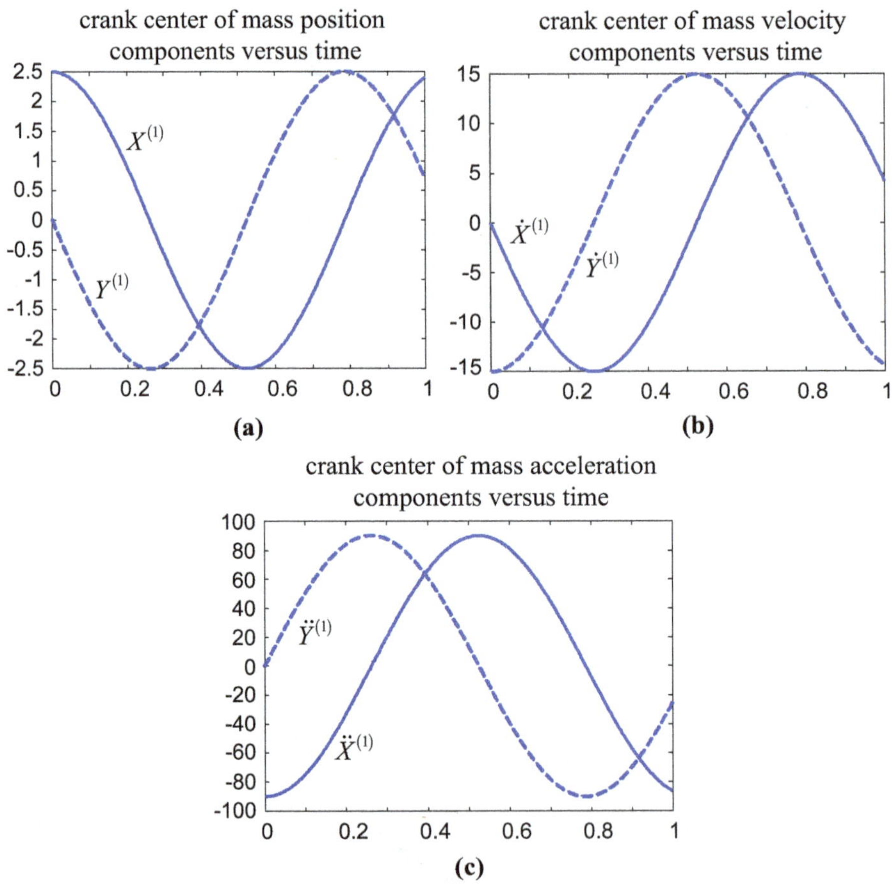

Fig. 4.32 Kinematics of the crank in the slider-crank mechanism for (**a**) the position components of the center of mass (inches), (**b**) the velocity components of the center of mass (in./s), and (**c**) the acceleration components of the center of mass (in./s^2)

4.6 Three-Dimension Motion: Finite Rotations and Relative Position

In dealing with the three-dimensional motion of rigid bodies one of the major complicating factors is their behavior during rotations. In two-dimensional (planar) motions the body only experiences a rotation about a single inertial Cartesian axis, usually taken as the Z-axis. However, in three-dimensional motions rotations can occur about all three Cartesian axes. This may not seem to be a major change but when one tries to combine rotations one soon finds that the motion of a rigid body depends on the order of the rotations. For example, consider the block shown in Fig. 4.35. If the block is rotated 90 degrees about the Y-axis and then 90 degrees

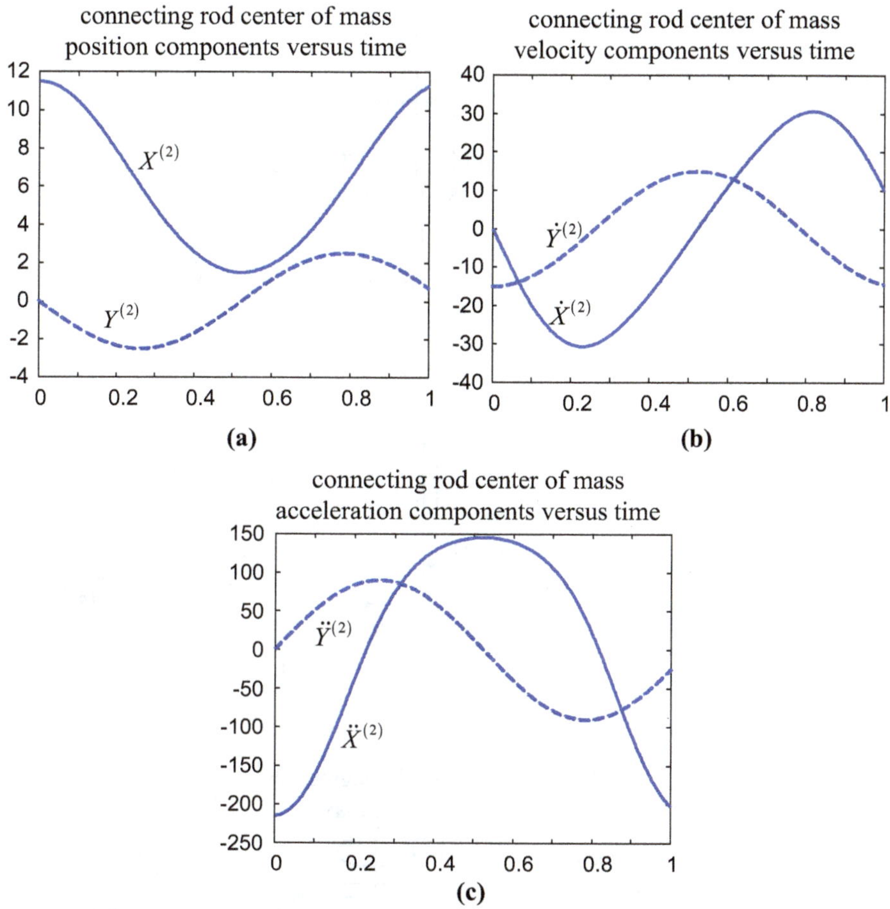

Fig. 4.33 Kinematics of the connecting rod in the slider-crank mechanism for (**a**) the position components of the center of mass (inches), (**b**) the velocity components of the center of mass (in./s), and (**c**) the acceleration components of the center of mass (in./s^2)

about the Z-axis, the resulting orientation of the block is different from when the Z-axis rotation occurs first, followed by the Y-rotation. If we call $(\theta_X, \theta_Y, \theta_Z)$ the rotations about the (X, Y, Z) axes, respectively, then these rotations cannot be characterized as components of a vector since adding vector components is not order dependent. However, if we differentiate these coordinates on time, the derivatives $\left(\dot{\theta}_X, \dot{\theta}_Y, \dot{\theta}_Z\right)$ are just the angular velocity components $(\omega_X, \omega_Y, \omega_Z)$ about the Cartesian inertial axes, which are components of an angular velocity vector $\boldsymbol{\omega} = \omega_X\mathbf{E}_X+\omega_Y\mathbf{E}_Y+\omega_Z\mathbf{E}_Z$. The Newton-Euler moment equations involve the angular velocities and their derivatives so they cannot be integrated to yield the rotations $(\theta_X, \theta_Y, \theta_Z)$ except in special cases like planar motion where there is only one rotation about a fixed axis. Since we cannot integrate the angular velocities directly we must find a set of generalized coordinates and their derivatives, $(\mathbf{q}, \dot{\mathbf{q}})$, in terms of which

Fig. 4.34 (**a**) The positions and velocities of the center of masses for the slider-crank elements at the instant shown and (**b**) the corresponding accelerations

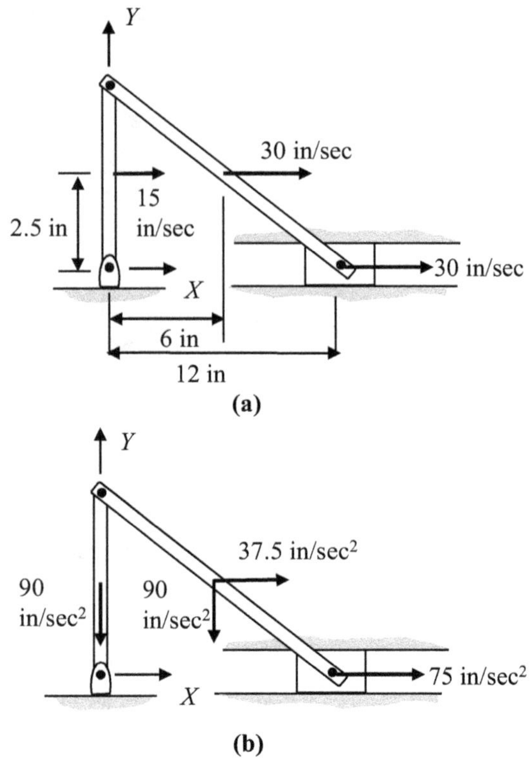

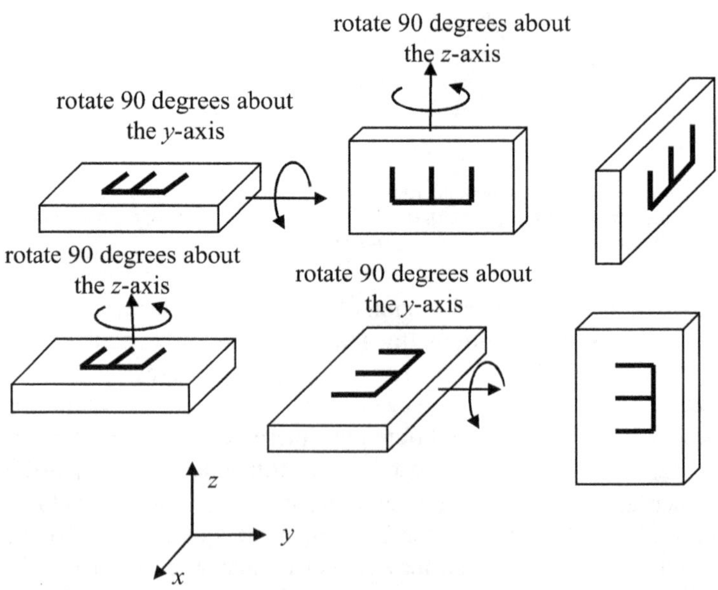

Fig. 4.35 An illustration of the order dependency of rotations, showing that the rotation of a block about the y-axis and then the z-axis is not equivalent to the rotation about the z-axis followed by a rotation about the y-axis

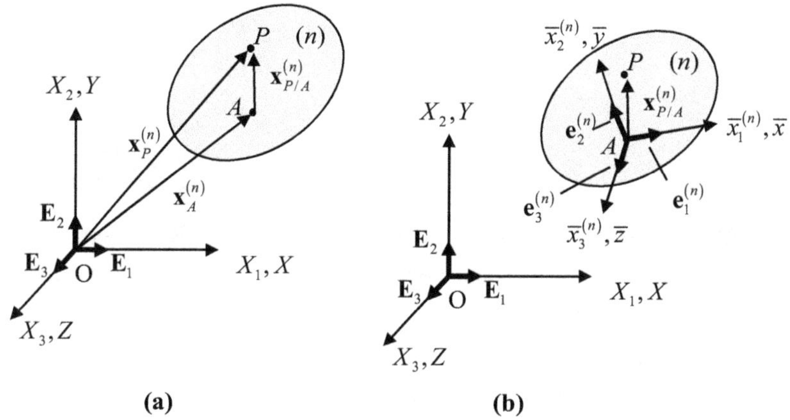

Fig. 4.36 (**a**) Geometry relating two points A and P that are both in the same rigid body, and (**b**) body-fixed coordinates $\left(\bar{x}_1^{(n)}, \bar{x}_2^{(n)}, \bar{x}_3^{(n)}\right) = (\bar{x}, \bar{y}, \bar{z})$ and unit vectors $\left(\mathbf{e}_1^{(n)}, \mathbf{e}_2^{(n)}, \mathbf{e}_3^{(n)}\right)$

we can express the angular velocities in the form $\boldsymbol{\omega} = \boldsymbol{\omega}(\mathbf{q}, \dot{\mathbf{q}})$ and in terms of which we can carry out the integrations to obtain the rotational motions. The need to define these intermediate generalized coordinates and their derivatives often makes the solution of three-dimensional dynamics problems complex. Because of the significant difference between planar motions and 3-D motions of rigid bodies we will have to generalize many of the kinematic results discussed earlier in this chapter. Thus, bear with me while we do that generalization.

Although rotations are not vectors, we can represent them in terms of matrices. For example, consider the rigid body shown in Fig. 4.36a. As we have already seen motion of this rigid body can be decomposed into a translation and a rotation. We will take point A as a reference point moving with the body to define the translation of that body. The position of any other point, P, in that body is then related to the position of A via

$$\mathbf{x}_P^{(n)} = \mathbf{x}_A^{(n)} + \mathbf{x}_{P/A}^{(n)} \tag{4.6.1}$$

where $\mathbf{x}_{P/A}^{(n)}$ is the vector defining the relative position of P with respect to A. Again we will use the superscript label (n) to denote the rigid body so that we can distinguish relations for one rigid body from another. Now, define a set of unit vectors along the fixed (inertial) $(X_1, X_2, X_3) = (X, Y, Z)$ Cartesian axis as $(\mathbf{E}_1, \mathbf{E}_2, \mathbf{E}_3)$. We will also define a set of $\left(\bar{x}_1^{(n)}, \bar{x}_2^{(n)}, \bar{x}_3^{(n)}\right) = \left(\bar{x}^{(n)}, \bar{y}^{(n)}, \bar{z}^{(n)}\right)$ Cartesian axes with origin at A that rotate with the rigid body (n) and a corresponding set of rotating unit vectors $\left(\mathbf{e}_1^{(n)}, \mathbf{e}_2^{(n)}, \mathbf{e}_3^{(n)}\right)$ (see Fig. 4.36b). We can relate these two sets of unit vectors to each other through a *rotation matrix*, $[R^{(n)}]$, where

$$\begin{Bmatrix} \mathbf{E}_1 \\ \mathbf{E}_2 \\ \mathbf{E}_3 \end{Bmatrix} = \begin{bmatrix} R_{11}^{(n)} & R_{12}^{(n)} & R_{13}^{(n)} \\ R_{21}^{(n)} & R_{22}^{(n)} & R_{23}^{(n)} \\ R_{31}^{(n)} & R_{32}^{(n)} & R_{33}^{(n)} \end{bmatrix} \begin{Bmatrix} \mathbf{e}_1^{(n)} \\ \mathbf{e}_2^{(n)} \\ \mathbf{e}_3^{(n)} \end{Bmatrix} \tag{4.6.2}$$

If we use the fact that the components $R_{ij}^{(n)} = \mathbf{E}_i \cdot \mathbf{e}_j^{(n)} = \cos\left(\theta_{ij}^{(n)}\right)$ this rotation matrix is called a *direction cosine matrix* where $\theta_{ij}^{(n)}$ is the angle between the $\mathbf{E}_i$ unit vector (or X_i axis) and the $\mathbf{e}_j^{(n)}$ unit vector (or the $\bar{x}_j^{(n)}$ axis). We can write these equations as explicit sums

$$\mathbf{E}_i = \sum_{j=1}^{3} R_{ij}^{(n)} \mathbf{e}_j^{(n)} \quad (i = 1, 2, 3) \tag{4.6.3}$$

But these are unit vectors and they are orthogonal to each other so that we have

$$\begin{aligned} \mathbf{E}_i \cdot \mathbf{E}_i &= \sum_{j=1}^{3} R_{ij}^{(n)} R_{ij}^{(n)} = 1 \quad (i = 1, 2, 3) \\ \mathbf{E}_i \cdot \mathbf{E}_k &= \sum_{j=1}^{3} R_{ij}^{(n)} R_{kj}^{(n)} = 0 \quad (i \neq k) \end{aligned} \tag{4.6.4}$$

which can also be written as

$$\sum_{j=1}^{3} R_{ij}^{(n)} R_{kj}^{(n)} = \delta_{ik} \quad (i, k = 1, 2, 3) \tag{4.6.5}$$

where the Kronecker delta, δ_{ij}, is defined as

$$\delta_{ij} = \begin{cases} 1 & i = j \\ 0 & i \neq j \end{cases} \tag{4.6.6}$$

Equation (4.6.5) can also be written in matrix form as

$$\mathbf{R}^{(n)} \left(\mathbf{R}^{(n)}\right)^T = \mathbf{I} \tag{4.6.7}$$

where $\mathbf{I}$ is the 3×3 identity matrix. Since there are six relationships between the nine direction cosines, there are only three independent parameters needed to define a general 3-D rotation. Equation (4.6.7) also demonstrates that the rotation matrix $[R^{(n)}]$ is an orthogonal matrix. This rotation matrix can also be shown to be a proper orthogonal matrix where $\det([R^{(n)}]) = 1$.

Now, let us take the relative position vector $\mathbf{x}_{P/A}^{(n)}$ in Eq. (4.6.1) and form up the dot products

$$\begin{aligned}
\mathbf{x}_{P/A}^{(n)} \cdot \mathbf{E}_1 &= X_{P/A}^{(n)} \\
\mathbf{x}_{P/A}^{(n)} \cdot \mathbf{E}_2 &= Y_{P/A}^{(n)} \\
\mathbf{x}_{P/A}^{(n)} \cdot \mathbf{E}_3 &= Z_{P/A}^{(n)} \\
\mathbf{x}_{P/A}^{(n)} \cdot \mathbf{e}_1^{(n)} &= \bar{x}_{P/A}^{(n)} \\
\mathbf{x}_{P/A}^{(n)} \cdot \mathbf{e}_2^{(n)} &= \bar{y}_{P/A}^{(n)} \\
\mathbf{x}_{P/A}^{(n)} \cdot \mathbf{e}_3^{(n)} &= \bar{z}_{P/A}^{(n)}
\end{aligned} \tag{4.6.8}$$

where $\left(X_{P/A}^{(n)}, Y_{P/A}^{(n)}, Z_{P/A}^{(n)}\right)$ and $\left(\bar{x}_{P/A}^{(n)}, \bar{y}_{P/A}^{(n)}, \bar{z}_{P/A}^{(n)}\right)$ are the components of the relative position vector as measured in the inertial and body-fixed coordinate systems, respectively. Then we have

$$\begin{Bmatrix} X_{P/A}^{(n)} \\ Y_{P/A}^{(n)} \\ Z_{P/A}^{(n)} \end{Bmatrix} = \begin{bmatrix} R_{11}^{(n)} & R_{12}^{(n)} & R_{13}^{(n)} \\ R_{21}^{(n)} & R_{22}^{(n)} & R_{23}^{(n)} \\ R_{31}^{(n)} & R_{32}^{(n)} & R_{33}^{(n)} \end{bmatrix} \begin{Bmatrix} \bar{x}_{P/A}^{(n)} \\ \bar{y}_{P/A}^{(n)} \\ \bar{z}_{P/A}^{(n)} \end{Bmatrix} \tag{4.6.9}$$

and we can write Eq. (4.6.1) in inertia coordinates as

$$\begin{Bmatrix} X_{P}^{(n)} \\ Y_{P}^{(n)} \\ Z_{P}^{(n)} \end{Bmatrix} = \begin{Bmatrix} X_{A}^{(n)} \\ Y_{A}^{(n)} \\ Z_{A}^{(n)} \end{Bmatrix} + \begin{bmatrix} R_{11}^{(n)} & R_{12}^{(n)} & R_{13}^{(n)} \\ R_{21}^{(n)} & R_{22}^{(n)} & R_{23}^{(n)} \\ R_{31}^{(n)} & R_{32}^{(n)} & R_{33}^{(n)} \end{bmatrix} \begin{Bmatrix} \bar{x}_{P/A}^{(n)} \\ \bar{y}_{P/A}^{(n)} \\ \bar{z}_{P/A}^{(n)} \end{Bmatrix} \tag{4.6.10}$$

or, in matrix-vector form,

$$\mathbf{x}_P^{(n)} = \mathbf{x}_A^{(n)} + \mathbf{R}^{(n)} \bar{\mathbf{x}}_{P/A}^{(n)} \tag{4.6.11}$$

where we have again placed a bar over the relative displacement term in Eq. (4.6.11) to indicate that it is measured in the body-fixed axes, i.e., $\bar{\mathbf{x}}_{P/A}^{(n)} = \left(\bar{x}_{P/A}^{(n)}, \bar{y}_{P/A}^{(n)}, \bar{z}_{P/A}^{(n)}\right)$. Equation (4.6.11) is the three-dimensional relationship equivalent to the planar case (see Eq. (4.4.10)). The rotation matrix for rotations about a single axis is much simpler than the general case. For example, if a rotation, $\theta_Z^{(n)}$, as measured counter-clockwise from the positive X-axis, occurs around the Z-axis only (see Fig. 4.37) then

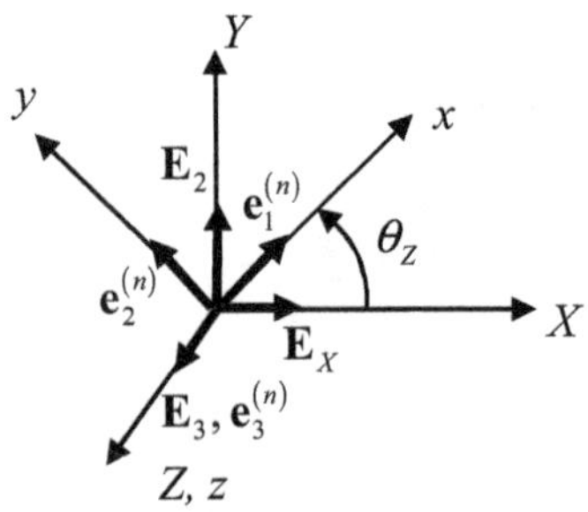

Fig. 4.37 Rotation about the *Z*-axis

$$\left[R^{(n)}\right] = \begin{bmatrix} \cos\theta_Z^{(n)} & -\sin\theta_Z^{(n)} & 0 \\ \sin\theta_Z^{(n)} & \cos\theta_Z^{(n)} & 0 \\ 0 & 0 & 1 \end{bmatrix} \tag{4.6.12}$$

since

$$\begin{aligned} R_{11}^{(n)} &= \cos\left(\theta_{11}^{(n)}\right) = \cos\left(\theta_Z^{(n)}\right), \\ R_{12}^{(n)} &= \cos\left(\theta_{12}^{(n)}\right) = \cos\left(\theta_Z^{(n)} + \pi/2\right) = -\sin\left(\theta_Z^{(n)}\right), \\ &\text{etc.} \end{aligned}$$

Equation (4.6.12) is the 3-D counterpart of the rotation matrix used for planar problems. If we have a rotation through an angle $\theta_X^{(n)}$ about the *X*-axis, then similar to Eq. (4.6.12) we find

$$\left[R^{(n)}\right] = \begin{bmatrix} 1 & 0 & 0 \\ 0 & \cos\theta_X^{(n)} & -\sin\theta_X^{(n)} \\ 0 & \sin\theta_X^{(n)} & \cos\theta_X^{(n)} \end{bmatrix} \tag{4.6.13}$$

and for a rotation $\theta_Y^{(n)}$ about the *Y*-axis

$$\left[R^{(n)}\right] = \begin{bmatrix} \cos\theta_Y^{(n)} & 0 & \sin\theta_Y^{(n)} \\ 0 & 1 & 0 \\ -\sin\theta_Y^{(n)} & 0 & \cos\theta_Y^{(n)} \end{bmatrix} \tag{4.6.14}$$

(Note the change in signs on the sine terms for the *Y*-axis rotation versus the *X*- and *Z*-axis rotations.)

You may wonder why we chose to define the rotation matrix as a transformation from body-fixed axis quantities to fixed (inertial) axis quantities (see Eqs. (4.6.2) and (4.6.9)) instead of vice versa. To see why, consider the rotation matrix for rotation about the *Z*-axis (Eq. (4.6.12)). The 2-D rotation submatrix that relates the (*X*, *Y*) coordinates to the (*x*, *y*) coordinates gives

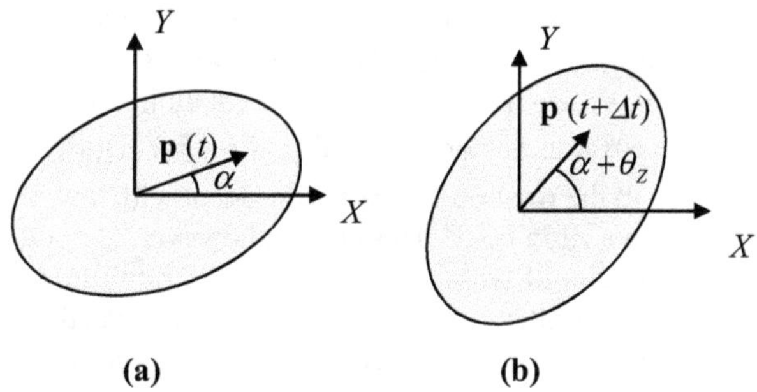

Fig. 4.38 (**a**) A planar rigid body and a vector **p** of fixed length that rotates with the body, and (**b**) the same body and vector at a later time where the body and vector have rotated through an angle θ_Z about the Z-axis

$$\begin{Bmatrix} X^{(n)}_{P/A} \\ Y^{(n)}_{P/A} \end{Bmatrix} = \begin{bmatrix} \cos\theta^{(n)}_Z & -\sin\theta^{(n)}_Z \\ \sin\theta^{(n)}_Z & \cos\theta^{(n)}_Z \end{bmatrix} \begin{Bmatrix} \bar{x}^{(n)}_{P/A} \\ \bar{y}^{(n)}_{P/A} \end{Bmatrix} \tag{4.6.15a}$$

while the inverse relationship gives

$$\begin{Bmatrix} \bar{x}^{(n)}_{P/A} \\ \bar{y}^{(n)}_{P/A} \end{Bmatrix} = \begin{bmatrix} \cos\theta^{(n)}_Z & \sin\theta^{(n)}_Z \\ -\sin\theta^{(n)}_Z & \cos\theta^{(n)}_Z \end{bmatrix} \begin{Bmatrix} X^{(n)}_{P/A} \\ Y^{(n)}_{P/A} \end{Bmatrix} \tag{4.6.15b}$$

Now, consider a rigid body that rotates about the Z-axis and consider a vector **p** in the X–Y plane between two points in the rigid body that rotates with that body as shown in Fig. 4.38. The length of the vector is fixed so we have $\mathbf{p}(t+\Delta t) = (p\cos(\alpha+\theta_Z), p\sin(\alpha+\theta_Z))$, and $\mathbf{p}(t) = (p\cos\alpha, p\sin\alpha)$. Thus the components along the fixed axes at time $t + \Delta t$ are

$$\begin{aligned} p_X(t+\Delta t) &= p\cos(\alpha+\theta_Z) \\ &= p\cos\alpha\cos\theta_Z - p\sin\alpha\sin\theta_Z \\ &= p_X(t)\cos\theta_Z - p_Y(t)\sin\theta_Z \end{aligned} \tag{4.6.16}$$

and

$$\begin{aligned} p_Y(t+\Delta t) &= p\sin(\alpha+\theta_Z) \\ &= p\sin\alpha\cos\theta_Z + p\sin\theta_Z\cos\alpha \\ &= p_Y(t)\cos\theta_Z + p_X(t)\sin\theta_Z \end{aligned} \tag{4.6.17}$$

which we can write in terms of a rotation matrix as

$$\begin{Bmatrix} p_X(t+\Delta t) \\ p_Y(t+\Delta t) \end{Bmatrix} = \begin{bmatrix} \cos\theta_Z & -\sin\theta_z \\ \sin\theta_Z & \cos\theta_Z \end{bmatrix} \begin{Bmatrix} p_X(t) \\ p_Y(t) \end{Bmatrix} \tag{4.6.18}$$

Thus, the rotation of $\mathbf{p}$ (or any line in the body) through an angle θ_Z is characterized by the rotation matrix of Eq. (4.6.18). Comparing with Eqs. (4.6.15a) and (4.6.15b) we see that the comparable rotation matrix relating the components of the relative position vector $\mathbf{x}_{P/A}^{(n)}$ is Eq. (4.6.15a). Thus, our choice to call the matrix of Eq. (4.6.9) the rotation matrix is consistent with how that matrix affects the rotations of lines in a rigid body. This choice, however, is not a universal one and so in other texts you should be careful to determine how the rotation matrix is defined. We should also point out that even if other texts define the rotation matrix consistent with our choice they may instead define the direction cosine matrix $[D]$ through the transformation

$$\begin{Bmatrix} \bar{x}_{P/A}^{(n)} \\ \bar{y}_{P/A}^{(n)} \\ \bar{z}_{P/A}^{(n)} \end{Bmatrix} = \begin{bmatrix} D_{11}^{(n)} & D_{12}^{(n)} & D_{13}^{(n)} \\ D_{21}^{(n)} & D_{22}^{(n)} & D_{23}^{(n)} \\ D_{31}^{(n)} & D_{32}^{(n)} & D_{33}^{(n)} \end{bmatrix} \begin{Bmatrix} X_{P/A}^{(n)} \\ Y_{P/A}^{(n)} \\ Z_{P/A}^{(n)} \end{Bmatrix} \tag{4.6.19}$$

where $D_{ij}^{(n)} = \mathbf{e}_i^{(n)} \cdot \mathbf{E}_j$. In this case we see $D_{ij}^{(n)} = R_{ji}^{(n)}$ or $[D] = [R]^T$ so the direction cosine matrix and the rotation matrix are then the transposes of each other.

4.7 Angular Velocity and Relative Velocity Relations

Equation (4.6.11) gives us the basic matrix-vector expression for the motion of an arbitrary point P in a rigid body as a combination of translation of a reference point A and a rotation, $\mathbf{R}^{(n)}$, about that point:

$$\mathbf{x}_P^{(n)} = \mathbf{x}_A^{(n)} + \mathbf{R}^{(n)}\bar{\mathbf{x}}_{P/A}^{(n)} \tag{4.7.1}$$

The velocity of point P can be obtained by differentiation to obtain

$$\dot{\mathbf{x}}_P^{(n)} = \dot{\mathbf{x}}_A^{(n)} + \dot{\mathbf{R}}^{(n)}\bar{\mathbf{x}}_{P/A}^{(n)} \tag{4.7.2}$$

But the rotation matrix is orthogonal so that

$$\mathbf{R}^{(n)}\left(\mathbf{R}^{(n)}\right)^T = \mathbf{I} \tag{4.7.3}$$

Differentiating Eq. (4.7.3) gives

$$\dot{\mathbf{R}}^{(n)}\left(\mathbf{R}^{(n)}\right)^T = -\mathbf{R}^{(n)}\left(\dot{\mathbf{R}}^{(n)}\right)^T \tag{4.7.4}$$

which can be written as

$$\dot{\mathbf{R}}^{(n)}\left(\mathbf{R}^{(n)}\right)^T = -\left[\dot{\mathbf{R}}^{(n)}\left(\mathbf{R}^{(n)}\right)^T\right]^T \tag{4.7.5}$$

A matrix $\mathbf{W}$ that satisfies $\mathbf{W} = -\mathbf{W}^T$ is a skew symmetric matrix so that from Eq. (4.7.5) we have

$$\dot{\mathbf{R}}^{(n)}\left(\mathbf{R}^{(n)}\right)^T = \boldsymbol{\Omega}^{(n)} \tag{4.7.6}$$

which we will call the skew symmetric *angular velocity matrix*, $\boldsymbol{\Omega}^{(n)}$. This matrix has the form

$$\boldsymbol{\Omega}^{(n)} = \begin{bmatrix} 0 & -\omega_3^{(n)} & \omega_2^{(n)} \\ \omega_3^{(n)} & 0 & -\omega_1^{(n)} \\ -\omega_2^{(n)} & \omega_1^{(n)} & 0 \end{bmatrix} \tag{4.7.7}$$

and where the components of $\boldsymbol{\Omega}^{(n)}$ define the angular velocity vector $\boldsymbol{\omega}^{(n)} = \left(\omega_1^{(n)}, \omega_2^{(n)}, \omega_3^{(n)}\right)$ as measured in inertial coordinates. If we post-multiply both sides of Eq. (4.7.6) by $\mathbf{R}^{(n)}$ we find an expression for the derivative of the rotation matrix as

$$\dot{\mathbf{R}}^{(n)} = \boldsymbol{\Omega}^{(n)}\mathbf{R}^{(n)} \tag{4.7.8}$$

and if we place this expression into Eq. (4.7.2) we have

$$\dot{\mathbf{x}}_P^{(n)} = \dot{\mathbf{x}}_A^{(n)} + \boldsymbol{\Omega}^{(n)}\mathbf{R}^{(n)}\bar{\mathbf{x}}_{P/A}^{(n)} \tag{4.7.9}$$

which we can also write as

$$\dot{\mathbf{x}}_P^{(n)} = \dot{\mathbf{x}}_A^{(n)} + \boldsymbol{\Omega}^{(n)}\mathbf{x}_{P/A}^{(n)} \tag{4.7.10}$$

Equations (4.7.9) and (4.7.10) are equivalent matrix-vector expressions for the relative velocity relationship of a rigid body. They are the three-dimensional counterparts of the relations we found for planar motion. We can also write the relative velocity relationship in purely vector form by noting that $\boldsymbol{\Omega}^{(n)}\mathbf{x}_{P/A}^{(n)} = \boldsymbol{\omega}^{(n)} \times \mathbf{x}_{P/A}^{(n)}$ so that

$$\dot{\mathbf{x}}_P^{(n)} = \dot{\mathbf{x}}_A^{(n)} + \boldsymbol{\omega}^{(n)} \times \mathbf{x}_{P/A}^{(n)} \tag{4.7.11}$$

As in planar case there are a number of ways to write the relative velocity relationship (see Eqs. (4.4.26a)–(4.4.26d)). Here we have the equivalent forms

$$\left(\mathbf{v}_P^{(n)}\right)_I = \left(\mathbf{v}_A^{(n)}\right)_I + {}_I\boldsymbol{\Omega}_B^{(n)}\mathbf{R}^{(n)}\bar{\mathbf{x}}_{P/A}^{(n)} \tag{4.7.12a}$$

$$\left(\mathbf{v}_P^{(n)}\right)_I = \left(\mathbf{v}_A^{(n)}\right)_I + {}_I\boldsymbol{\Omega}_B^{(n)}\mathbf{x}_{P/A}^{(n)} \tag{4.7.12b}$$

$$\left(\mathbf{v}_P^{(n)}\right)_I = \left(\mathbf{v}_A^{(n)}\right)_I + \dot{\mathbf{R}}^{(n)}\left(\mathbf{R}^{(n)}\right)^T\mathbf{x}_{P/A}^{(n)} \tag{4.7.12c}$$

$$\left(\mathbf{v}_P^{(n)}\right)_I = \left(\mathbf{v}_A^{(n)}\right)_I + {}_I\boldsymbol{\omega}_B^{(n)} \times \mathbf{x}_{P/A}^{(n)} \tag{4.7.12d}$$

where we have set $\dot{\mathbf{x}}_P^{(n)} = \left(\mathbf{v}_P^{(n)}\right)_I, \dot{\mathbf{x}}_A^{(n)} = \left(\mathbf{v}_A^{(n)}\right)_I$ to emphasize that these are velocities calculated by taking derivatives in an inertial frame, I. Similarly, we have written the angular velocity matrix and angular velocity vector as ${}_I\boldsymbol{\Omega}_B^{(n)}$ and ${}_I\boldsymbol{\omega}_B^{(n)}$, respectively, to emphasize that these are the angular velocity matrix and angular velocity vector of the body with respect to the inertial frame. It is also useful to represent the relative velocity of P with respect to A in body-fixed coordinates. Consider the relative velocity term as measured in the inertial frame

$$\left(\mathbf{v}_{P/A}^{(n)}\right)_I = {}_I\boldsymbol{\Omega}_B^{(n)}\mathbf{x}_{P/A}^{(n)} \tag{4.7.13}$$

where we made the fact that this is the relative velocity calculated in the inertial frame explicit by including an I subscript. If we express all of the terms in Eq. (4.7.13) in body-fixed coordinates we can write that equation as

$$\bar{\mathbf{v}}_{P/A}^{(n)} = {}_I\overline{\boldsymbol{\Omega}}_B^{(n)}\bar{\mathbf{x}}_{P/A}^{(n)} \tag{4.7.14}$$

where $\bar{\mathbf{v}}_{P/A}^{(n)}$ is the relative velocity $\left(\mathbf{v}_{P/A}^{(n)}\right)_I$ as expressed in body-fixed coordinates so they are related through the rotation matrix, which transforms body-fixed components into inertial frame components:

$$\left(\mathbf{v}_{P/A}^{(n)}\right)_I = \mathbf{R}^{(n)}\bar{\mathbf{v}}_{P/A}^{(n)} \tag{4.7.15}$$

Thus, we have

$$\left(\mathbf{v}_{P/A}^{(n)}\right)_I = \mathbf{R}^{(n)}{}_I\overline{\boldsymbol{\Omega}}_B^{(n)}\left(\mathbf{R}^{(n)}\right)^T\mathbf{x}_{P/A}^{(n)} \tag{4.7.16}$$

which, comparing with Eq. (4.7.13), gives the relationship between the angular velocity matrix as measured in the inertial frame and the matrix measured in the body-fixed frame as

$$ {}_I\mathbf{\Omega}_B^{(n)} = \mathbf{R}^{(n)}\,{}_I\overline{\mathbf{\Omega}}_B^{(n)}\left(\mathbf{R}^{(n)}\right)^T \tag{4.7.17a} $$

which can be easily inverted by pre- and post-multiplying both sides of Eq. (4.7.17a) by the rotation matrices $(\mathbf{R}^{(n)})^T$ and $\mathbf{R}^{(n)}$, respectively, to obtain

$$ {}_I\overline{\mathbf{\Omega}}_B^{(n)} = \left(\mathbf{R}^{(n)}\right)^T\,{}_I\mathbf{\Omega}_B^{(n)}\mathbf{R}^{(n)} \tag{4.7.17b} $$

It is easy to show that the angular velocity matrix ${}_I\overline{\mathbf{\Omega}}_B^{(n)}$ in body-fixed coordinates is a skew symmetric matrix with components given as

$$ {}_I\overline{\mathbf{\Omega}}_B^{(n)} = \begin{bmatrix} 0 & -\overline{\omega}_3^{(n)} & \overline{\omega}_2^{(n)} \\ \overline{\omega}_3^{(n)} & 0 & -\overline{\omega}_1^{(n)} \\ -\overline{\omega}_2^{(n)} & \overline{\omega}_1^{(n)} & 0 \end{bmatrix} \tag{4.7.18} $$

where the angular velocity vector in body-fixed coordinates is ${}_I\overline{\boldsymbol{\omega}}_B^{(n)} = \left(\overline{\omega}_1^{(n)}, \overline{\omega}_2^{(n)}, \overline{\omega}_3^{(n)}\right)$. This angular velocity in body-fixed coordinates is related to the same vector in inertial coordinates and vice versa through

$$ \begin{aligned} {}_I\overline{\boldsymbol{\omega}}_B^{(n)} &= \left(\mathbf{R}^{(n)}\right)^T\,{}_I\boldsymbol{\omega}_B^{(n)} \\ {}_I\boldsymbol{\omega}_B^{(n)} &= \mathbf{R}^{(n)}\,{}_I\overline{\boldsymbol{\omega}}_B^{(n)} \end{aligned} \tag{4.7.19} $$

Thus, we can also write the rigid body relative velocity relationship in the matrix-vector forms

$$ \left(\mathbf{v}_P^{(n)}\right)_I = \left(\mathbf{v}_A^{(n)}\right)_I + \mathbf{R}^{(n)}\,{}_I\overline{\mathbf{\Omega}}_B^{(n)}\overline{\mathbf{x}}_{P/A}^{(n)} \tag{4.7.20a} $$

$$ \left(\mathbf{v}_P^{(n)}\right)_I = \left(\mathbf{v}_A^{(n)}\right)_I + \mathbf{R}^{(n)}\,{}_I\overline{\boldsymbol{\omega}}_B^{(n)} \times \overline{\mathbf{x}}_{P/A}^{(n)} \tag{4.7.20b} $$

We can also multiply these equations by the transpose of $\mathbf{R}^{(n)}$ to produce equations written entirely in terms of body-fixed coordinates, i.e.,

$$ \begin{aligned} \overline{\mathbf{v}}_P^{(n)} &= \overline{\mathbf{v}}_A^{(n)} + {}_I\overline{\mathbf{\Omega}}_B^{(n)}\overline{\mathbf{x}}_{P/A}^{(n)} \\ \overline{\mathbf{v}}_P^{(n)} &= \overline{\mathbf{v}}_A^{(n)} + {}_I\overline{\boldsymbol{\omega}}_B^{(n)} \times \overline{\mathbf{x}}_{P/A}^{(n)} \end{aligned} \tag{4.7.20c} $$

4.8 Angular Velocity and Euler Angles

From the discussion of the last section you can see that the angular velocity components were related to elements of the skew symmetric matrix $\dot{\mathbf{R}}^{(n)}\left(\mathbf{R}^{(n)}\right)^T$ but were not explicitly derivatives of any angular rotation variables. We need to define the rotation matrix in terms of a set of angular rotations so that we can relate the angular velocity components to those rotations and their derivatives. As seen by our discussion of direction cosines we need three independent rotations. One choice is to use a set of angular rotations, called Euler angles, taken in a specified order about three non-orthogonal directions. As we will see, there are a number of different combinations (sets) of Euler angles possible, but we will describe here a set which is called the 3-1-3 set. This set, shown in Fig. 4.39, is used in many texts to describe rigid body problems such as tops and gyroscopes. Starting with a set of inertial (X, Y, Z) coordinates it consists of rotation about the Z-axis, followed by a rotation about the x'-axis and then about the z''-axis, as shown in Fig. 4.39, to produce a final set of rotated $(\bar{x}, \bar{y}, \bar{z})$ body-fixed axes. Since the rotation occurs about the third, first, and third coordinates of the (X, Y, Z), (x', y', z'), and (x'', y'', z'') systems, respectively, it is called a 3-1-3 set of Euler angles. The angles (ϕ, θ, ψ) are called the *precession*, *nutation*, and *spin* angles because for a spinning disk on an axle that rotates about a fixed point, as shown in Fig. 4.40, the rotation ϕ of the axle and disk assembly around the Z-axis is called precession, while the "bobbing" or "nodding" up and down of the assembly, corresponding to changes in the angle θ, is called nutation, and the rotation of the disk about the axle produces changes in the spin angle, ψ. The rotation matrix for the $\phi = \phi^{(n)}$ rotation of a rigid body (n) is

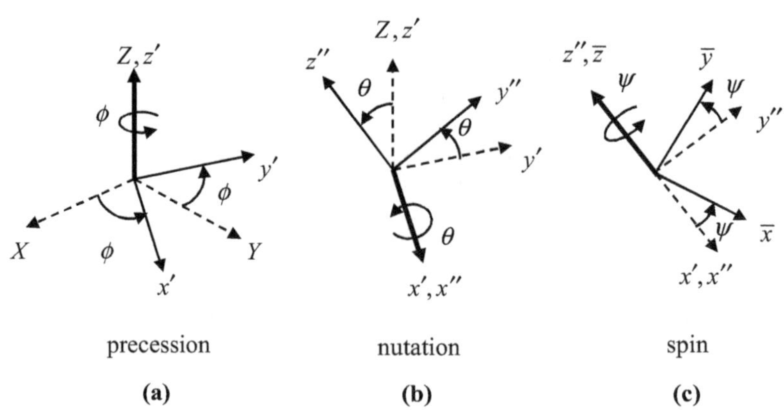

Fig. 4.39 The definition of the Euler angles consisting of (**a**) a first rotation, ϕ, about the Z-axis of an (X, Y, Z) coordinate system to produce a set of (x', y', z') axes; (**b**) a second rotation, θ, about the x'-axis to produce a set of (x'', y'', z'') axes; and (**c**) a third rotation about the z''-axis to produce the final set of rotated $(\bar{x}, \bar{y}, \bar{z})$ axes

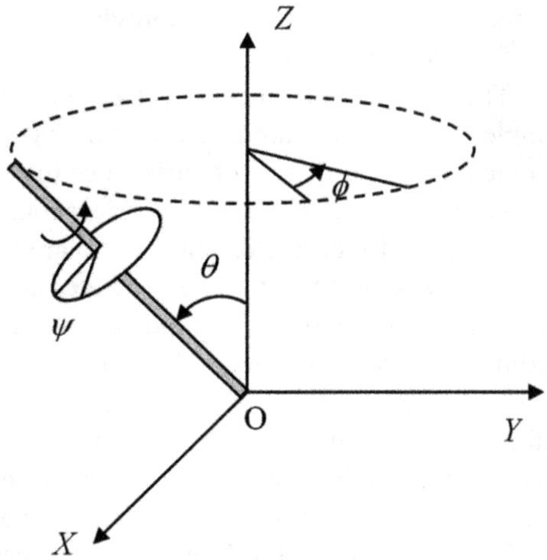

Fig. 4.40 The motion of a spinning disk on an axle that rotates about a fixed point, O, which exhibits precessional (ϕ), nutational (θ), and spinning (ψ) rotations

$$\mathbf{R}_1^{(n)} = \begin{bmatrix} \cos\phi^{(n)} & -\sin\phi^{(n)} & 0 \\ \sin\phi^{(n)} & \cos\phi^{(n)} & 0 \\ 0 & 0 & 1 \end{bmatrix} \tag{4.8.1}$$

and the $\theta = \theta^{(n)}$ rotation yields the rotation matrix

$$\mathbf{R}_2^{(n)} = \begin{bmatrix} 1 & 0 & 0 \\ 0 & \cos\theta^{(n)} & -\sin\theta^{(n)} \\ 0 & \sin\theta^{(n)} & \cos\theta^{(n)} \end{bmatrix} \tag{4.8.2}$$

Finally, the $\psi = \psi^{(n)}$ rotation gives

$$\mathbf{R}_3^{(n)} = \begin{bmatrix} \cos\psi^{(n)} & -\sin\psi^{(n)} & 0 \\ \sin\psi^{(n)} & \cos\psi^{(n)} & 0 \\ 0 & 0 & 1 \end{bmatrix} \tag{4.8.3}$$

so the total rotation matrix that relates the inertial (X, Y, Z) coordinates to the body-fixed coordinates (x, y, z) is

$$\begin{aligned} \mathbf{R}^{(n)} &= \mathbf{R}_1^{(n)}\mathbf{R}_2^{(n)}\mathbf{R}_3^{(n)} \\ &= \begin{bmatrix} c\psi^{(n)}c\phi^{(n)} - c\theta^{(n)}s\phi^{(n)}s\psi^{(n)} & -s\psi^{(n)}c\phi^{(n)} - c\theta^{(n)}s\phi^{(n)}c\psi^{(n)} & s\theta^{(n)}s\phi^{(n)} \\ c\psi^{(n)}s\phi^{(n)} + c\theta^{(n)}c\phi^{(n)}s\psi^{(n)} & -s\psi^{(n)}s\phi^{(n)} + c\theta^{(n)}c\phi^{(n)}c\psi^{(n)} & -s\theta^{(n)}c\phi^{(n)} \\ s\theta^{(n)}s\psi^{(n)} & s\theta^{(n)}c\psi^{(n)} & c\theta^{(n)} \end{bmatrix} \end{aligned} \tag{4.8.4}$$

where we have used an abbreviated notation where $c\psi^{(n)} = \cos(\psi^{(n)})$ and $s\psi^{(n)} = \sin(\psi^{(n)})$, etc.

The 3-1-3 set of Euler angles is only one possibility. In choosing the first Euler angle rotation there are three rotation axes to choose from but for the second Euler angle there are only two choices since making the choice of axis the same as the first choice would not give an independent rotation. Similarly, there are only two choices for the third Euler angle. Thus, there are a total of $3 \times 2 \times 2 = 12$ combination of choices we could make to define Euler angles. Three of those choices, the 3-1-3 set we have described above, and the 3-2-1, and the 3-2-3 sets, are the ones most commonly used. Although Euler angles are popular choices to represent rotations, they have some restrictions. For the 3-1-3 set we see that if the $\bar{z}$ and the Z-axes are aligned ($\theta = 0$) or in opposite alignment ($\theta = \pi$) then the rotations about those two axes are not independent and the Euler angles cannot represent a general rotation. Such singularities also occur at certain values of the rotations for all of the other Euler angle sets as well. In practice, one can choose a set where the singularities are not encountered during the normal motion of the system. Or one can resort to a different set of parameters to define the motion that are not singular. A common choice of this type is the choice of the Euler parameters discussed in Sect. 4.10.

To obtain the angular velocity we can differentiate the rotation matrix and then use Eq. (4.7.6) to find the angular velocity components from the components of the angular velocity matrix. However, we can also use the relative angular velocity expression. If we let the inertial (X, Y, Z) frame be the frame I, the (x', y', z') frame be the frame A', the (x'', y'', z'') frame be the frame A'', and the frame $(\bar{x}, \bar{y}, \bar{z})$ be the body frame B, then the angular velocity of a body B as measured in an inertial system is

$$\begin{aligned} {}_I\boldsymbol{\omega}_B^{(n)} &= {}_I\boldsymbol{\omega}_{A'}^{(n)} + {}_{A'}\boldsymbol{\omega}_{A''}^{(n)} + {}_{A''}\boldsymbol{\omega}_B^{(n)} \\ &= \dot{\phi}^{(n)}\mathbf{E}_Z + \dot{\theta}^{(n)}\mathbf{e}'_x + \dot{\psi}^{(n)}\mathbf{e}''_z \end{aligned} \tag{4.8.5}$$

where $(\mathbf{E}_Z, \mathbf{e}'_x, \mathbf{e}''_z)$ are unit vectors along their respective axes as shown in Fig. 4.41. Note that while we cannot simply add vector components that are resolved in different frames we can add them if they are all resolved in the same frame. We can, for example, write all these unit vectors in the inertial system to find (see Fig. 4.41).

$$\begin{aligned} {}_I\boldsymbol{\omega}_B^{(n)} &= \dot{\phi}^{(n)}\mathbf{E}_Z + \dot{\theta}^{(n)}\left(\cos\phi^{(n)}\mathbf{E}_X + \sin\phi^{(n)}\mathbf{E}_Y\right) + \dot{\psi}^{(n)}\left(\cos\theta^{(n)}\mathbf{E}_Z - \sin\theta^{(n)}\mathbf{e}'_y\right) \\ &= \dot{\theta}^{(n)}\cos\phi^{(n)}\mathbf{E}_X + \dot{\theta}^{(n)}\sin\phi^{(n)}\mathbf{E}_Y + \left(\dot{\phi}^{(n)} + \dot{\psi}^{(n)}\cos\theta^{(n)}\right)\mathbf{E}_Z \\ &\quad - \dot{\psi}^{(n)}\sin\theta^{(n)}\left(-\sin\phi^{(n)}\mathbf{E}_X + \cos\phi^{(n)}\mathbf{E}_Y\right) \\ &= \left(\dot{\theta}^{(n)}\cos\phi^{(n)} + \dot{\psi}^{(n)}\sin\theta^{(n)}\sin\phi^{(n)}\right)\mathbf{E}_X + \left(\dot{\theta}^{(n)}\sin\phi^{(n)} - \dot{\psi}^{(n)}\sin\theta^{(n)}\cos\phi^{(n)}\right)\mathbf{E}_Y \\ &\quad + \left(\dot{\phi}^{(n)} + \dot{\psi}^{(n)}\cos\theta^{(n)}\right)\mathbf{E}_Z \end{aligned} \tag{4.8.6}$$

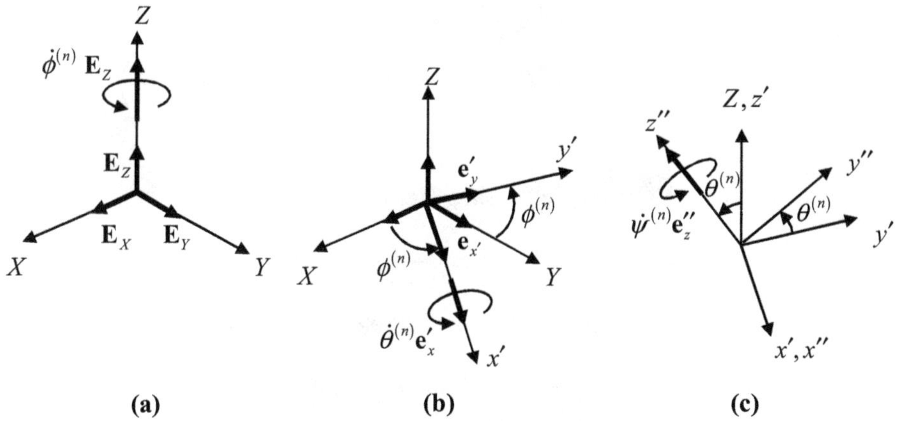

Fig. 4.41 The angular velocities about (**a**) the *Z*-axis, (**b**) the x' axis, and (**c**) the z'' axis

which gives the inertial frame components

$$\begin{aligned}
\left[{}_I\boldsymbol{\omega}_B^{(n)}\right]_X &= \dot{\theta}^{(n)}\cos\phi^{(n)} + \dot{\psi}^{(n)}\sin\theta^{(n)}\sin\phi^{(n)} \\
\left[{}_I\boldsymbol{\omega}_B^{(n)}\right]_Y &= \dot{\theta}^{(n)}\sin\phi^{(n)} - \dot{\psi}^{(n)}\sin\theta^{(n)}\cos\phi^{(n)} \\
\left[{}_I\boldsymbol{\omega}_B^{(n)}\right]_Z &= \dot{\phi}^{(n)} + \dot{\psi}^{n}\cos\theta^{(n)}
\end{aligned} \tag{4.8.7}$$

which can also be written in matrix-vector form as

$${}_I\boldsymbol{\omega}_B^{(n)} = \mathbf{H}^{(n)}\dot{\boldsymbol{\theta}}^{(n)} \tag{4.8.8}$$

where

$$\mathbf{H}^{(n)} = \begin{bmatrix} 0 & \cos\phi^{(n)} & \sin\theta^{n}\sin\phi^{(n)} \\ 0 & \sin\phi^{(n)} & -\sin\theta^{(n)}\cos\phi^{(n)} \\ 1 & 0 & \cos\theta^{(n)} \end{bmatrix} \tag{4.8.9}$$

and $\dot{\boldsymbol{\theta}}^{(n)}$ is the time derivative of $\boldsymbol{\theta}^{(n)} = [\phi^{(n)}, \theta^{(n)}, \psi^{(n)}]^T$, a column vector of the Euler angles. The columns of the $\mathbf{H}^{(n)}$ can be recognized as the (X, Y, Z) components of the unit vectors acting along the axes about which the Euler angle rotations $(\phi^{(n)}, \theta^{(n)}, \psi^{(n)})$ are taken. We can also find the angular velocity components for a set of $(\bar{x}, \bar{y}, \bar{z})$ body-fixed axes in the same fashion. We have

$$
\begin{aligned}
{}_I\bar{\boldsymbol{\omega}}_B^{(n)} &= \dot{\phi}^{(n)}\mathbf{E}_Z + \dot{\theta}^{(n)}\left(\cos\psi^{(n)}\mathbf{e}_x - \sin\psi^{(n)}\mathbf{e}_y\right) + \dot{\psi}^{(n)}\mathbf{e}_z \\
&= \dot{\phi}^{(n)}\left(\cos\theta^{(n)}\mathbf{e}_z + \sin\theta^{(n)}\mathbf{e}_y''\right) + \dot{\theta}^{(n)}\left(\cos\psi^{(n)}\mathbf{e}_x - \sin\psi^{(n)}\mathbf{e}_y\right) \\
&\quad + \dot{\psi}^{(n)}\mathbf{e}_z \\
&= \dot{\phi}^{(n)}\sin\theta^{(n)}\left(\cos\psi^{(n)}\mathbf{e}_y + \sin\psi^{(n)}\mathbf{e}_x\right) \\
&\quad + \dot{\theta}^{(n)}\left(\cos\psi^{(n)}\mathbf{e}_x - \sin\psi^{(n)}\mathbf{e}_y\right) + \left(\dot{\phi}^{(n)}\cos\theta^{(n)} + \dot{\psi}^{(n)}\right)\mathbf{e}_z \\
&= \left(\dot{\phi}^{(n)}\sin\theta^{(n)}\sin\psi^{(n)} + \dot{\theta}^{(n)}\cos\psi^{(n)}\right)\mathbf{e}_x \\
&\quad + \left(\dot{\phi}^{(n)}\sin\theta^{(n)}\cos\psi^{(n)} - \dot{\theta}^{(n)}\sin\psi^{(n)}\right)\mathbf{e}_y \\
&\quad + \left(\dot{\phi}^{(n)}\cos\theta^{(n)} + \dot{\psi}^{(n)}\right)\mathbf{e}_z
\end{aligned}
\tag{4.8.10}
$$

where $(\mathbf{e}_x, \mathbf{e}_y, \mathbf{e}_z)$ are unit vectors along the $\left(\bar{x}, \bar{y}, \bar{z}\right)$ axes. Resolving all these components in the body-fixed frame, the $\left(\bar{x}, \bar{y}, \bar{z}\right)$ components are

$$
\begin{aligned}
\left[{}_I\bar{\boldsymbol{\omega}}_B^{(n)}\right]_{\bar{x}} &= \dot{\phi}^{(n)}\sin\theta^{(n)}\sin\psi^{(n)} + \dot{\theta}^{(n)}\cos\psi^{(n)} \\
\left[{}_I\bar{\boldsymbol{\omega}}_B^{(n)}\right]_{\bar{y}} &= \dot{\phi}^{(n)}\sin\theta^{(n)}\cos\psi^{(n)} - \dot{\theta}^{(n)}\sin\psi^{(n)} \\
\left[{}_I\bar{\boldsymbol{\omega}}_B^{(n)}\right]_{\bar{z}} &= \dot{\phi}^{(n)}\cos\theta^{(n)} + \dot{\psi}^{(n)}
\end{aligned}
\tag{4.8.11}
$$

which can also be written as

$$
{}_I\bar{\boldsymbol{\omega}}_B^{(n)} = \bar{\mathbf{H}}^{(n)}\dot{\boldsymbol{\theta}}^{(n)} \tag{4.8.12}
$$

where

$$
\bar{\mathbf{H}}^{(n)} = \begin{bmatrix} \sin\theta^{(n)}\sin\psi^{(n)} & \cos\psi^{(n)} & 0 \\ \sin\theta^{(n)}\cos\psi^{(n)} & -\sin\psi^{(n)} & 0 \\ \cos\theta^{(n)} & 0 & 1 \end{bmatrix} \tag{4.8.13}
$$

The columns of the $\bar{\mathbf{H}}^{(n)}$ can similarly be recognized as the $\left(\bar{x}, \bar{y}, \bar{z}\right)$ components of the unit vectors along the axes about which the Euler angle rotations $(\phi^{(n)}, \theta^{(n)}, \psi^{(n)})$ are taken.

Having these angular velocity relations, we now can write the relative velocity term directly in terms of derivatives of the Euler angles. From Eqs. (4.7.12d) and (4.7.20b) we have

$$\left(\mathbf{v}_P^{(n)}\right)_I = \left(\mathbf{v}_A^{(n)}\right)_I - \mathbf{x}_{P/A}^{(n)} \times {}_I\boldsymbol{\omega}_B^{(n)} \tag{4.8.14a}$$

$$\left(\mathbf{v}_P^{(n)}\right)_I = \left(\mathbf{v}_A^{(n)}\right)_I - \mathbf{R}^{(n)}\bar{\mathbf{x}}_{P/A}^{(n)} \times {}_I\bar{\boldsymbol{\omega}}_B^{(n)} \tag{4.8.14b}$$

where we have changed the order in the cross products. We now can replace these cross products with multiplications by skew symmetric matrices

$$\mathbf{X}_{P/A}^{(n)} = \begin{bmatrix} 0 & -Z_{P/A}^{(n)} & Y_{P/A}^{(n)} \\ Z_{P/A}^{(n)} & 0 & -X_{P/A}^{(n)} \\ -Y_{P/A}^{(n)} & X_{P/A}^{(n)} & 0 \end{bmatrix}, \ \bar{\mathbf{X}}_{P/A}^{(n)} = \begin{bmatrix} 0 & -\bar{z}_{P/A}^{(n)} & \bar{y}_{P/A}^{(n)} \\ \bar{z}_{P/A}^{(n)} & 0 & -\bar{x}_{P/A}^{(n)} \\ -\bar{y}_{P/A}^{(n)} & \bar{x}_{P/A}^{(n)} & 0 \end{bmatrix} \tag{4.8.15}$$

to give, in terms of the angular velocity vector,

$$\left(\mathbf{v}_P^{(n)}\right)_I = \left(\mathbf{v}_A^{(n)}\right)_I - \mathbf{X}_{P/A}^{(n)}{}_I\boldsymbol{\omega}_B^{(n)} \tag{4.8.16a}$$

$$\left(\mathbf{v}_P^{(n)}\right)_I = \left(\mathbf{v}_A^{(n)}\right)_I - \mathbf{R}^{(n)}\bar{\mathbf{X}}_{P/A}^{(n)}{}_I\bar{\boldsymbol{\omega}}_B^{(n)} \tag{4.8.16b}$$

or, equivalently, in terms of the derivatives of the Euler angles

$$\left(\mathbf{v}_P^{(n)}\right)_I = \left(\mathbf{v}_A^{(n)}\right)_I - \mathbf{X}_{P/A}^{(n)}\mathbf{H}^{(n)}\dot{\boldsymbol{\theta}}^{(n)} \tag{4.8.17a}$$

$$\left(\mathbf{v}_P^{(n)}\right)_I = \left(\mathbf{v}_A^{(n)}\right)_I - \mathbf{R}^{(n)}\bar{\mathbf{X}}_{P/A}^{(n)}\bar{\mathbf{H}}^{(n)}\dot{\boldsymbol{\theta}}^{(n)} \tag{4.8.17b}$$

4.9 Acceleration Relations

If we differentiate the velocity relation of Eq. (4.7.2) on time we find the acceleration relation

$$\ddot{\mathbf{x}}_P^{(n)} = \ddot{\mathbf{x}}_A^{(n)} + \ddot{\mathbf{R}}^{(n)}\bar{\mathbf{x}}_{P/A}^{(n)} \tag{4.9.1}$$

From Eq. (4.7.8)

$$\ddot{\mathbf{R}}^{(n)} = {}_I\dot{\boldsymbol{\Omega}}_B^{(n)}\mathbf{R}^{(n)} + {}_I\boldsymbol{\Omega}_B^{(n)}\dot{\mathbf{R}}^{(n)} \tag{4.9.2}$$

where we can use Eq. (4.7.8) again to obtain

$$\ddot{\mathbf{R}}^{(n)} = {}_I\dot{\boldsymbol{\Omega}}_B^{(n)}\mathbf{R}^{(n)} + {}_I\boldsymbol{\Omega}_B^{(n)}{}_I\boldsymbol{\Omega}_B^{(n)}\mathbf{R}^{(n)} \tag{4.9.3}$$

where ${}_I\dot{\boldsymbol{\Omega}}_B^{(n)}$ is the skew symmetric angular acceleration matrix

$${}_I\dot{\boldsymbol{\Omega}}_B^{(n)} = \begin{bmatrix} 0 & -\dot{\omega}_3^{(n)} & \dot{\omega}_2^{(n)} \\ \dot{\omega}_3^{(n)} & 0 & -\dot{\omega}_1^{(n)} \\ -\dot{\omega}_2^{(n)} & \dot{\omega}_1^{(n)} & 0 \end{bmatrix} = \begin{bmatrix} 0 & -\alpha_3^{(n)} & \alpha_2^{(n)} \\ \alpha_3^{(n)} & 0 & -\alpha_1^{(n)} \\ -\alpha_2^{(n)} & \alpha_1^{(n)} & 0 \end{bmatrix} \tag{4.9.4}$$

as measured in inertial coordinates. Thus, the acceleration relation becomes

$$\begin{aligned} \ddot{\mathbf{x}}_P^{(n)} &= \ddot{\mathbf{x}}_A^{(n)} + {}_I\dot{\boldsymbol{\Omega}}_B^{(n)}\mathbf{R}^{(n)}\bar{\mathbf{x}}_{P/A}^{(n)} + {}_I\boldsymbol{\Omega}_B^{(n)}{}_I\boldsymbol{\Omega}_B^{(n)}\mathbf{R}^{(n)}\bar{\mathbf{x}}_{P/A}^{(n)} \\ &= \ddot{\mathbf{x}}_A^{(n)} + {}_I\dot{\boldsymbol{\Omega}}_B^{(n)}\mathbf{x}_{P/A}^{(n)} + {}_I\boldsymbol{\Omega}_B^{(n)}{}_I\boldsymbol{\Omega}_B^{(n)}\mathbf{x}_{P/A}^{(n)} \end{aligned} \tag{4.9.5}$$

which we can also write in terms of vectors and vector cross products as

$$\ddot{\mathbf{x}}_P^{(n)} = \ddot{\mathbf{x}}_A^{(n)} + {}_I\boldsymbol{\alpha}_B^{(n)} \times \mathbf{x}_{P/A}^{(n)} + {}_I\boldsymbol{\omega}_B^{(n)} \times \left({}_I\boldsymbol{\omega}_B^{(n)} \times \mathbf{x}_{P/A}^{(n)}\right) \tag{4.9.6}$$

From our earlier results in this chapter we recognize the ${}_I\boldsymbol{\alpha}_B^{(n)} \times \mathbf{x}_{P/A}^{(n)}$ term as the tangential component of the acceleration and ${}_I\boldsymbol{\omega}_B^{(n)} \times \left({}_I\boldsymbol{\omega}_B^{(n)} \times \mathbf{x}_{P/A}^{(n)}\right)$ as the normal component. As with the velocity components we can express the relative acceleration in terms of body-fixed components as

$$\ddot{\mathbf{x}}_P^{(n)} = \ddot{\mathbf{x}}_A^{(n)} + \mathbf{R}^{(n)}{}_I\overline{\dot{\boldsymbol{\Omega}}}_B^{(n)}\bar{\mathbf{x}}_{P/A}^{(n)} + \mathbf{R}^{(n)}{}_I\overline{\boldsymbol{\Omega}}_B^{(n)}{}_I\overline{\boldsymbol{\Omega}}_B^{(n)}\bar{\mathbf{x}}_{P/A}^{(n)} \tag{4.9.7a}$$

or, in terms of vector cross products,

$$\ddot{\mathbf{x}}_P^{(n)} = \ddot{\mathbf{x}}_A^{(n)} + \mathbf{R}^{(n)}\left({}_I\overline{\boldsymbol{\alpha}}_B^{(n)} \times \bar{\mathbf{x}}_{P/A}^{(n)}\right) + \mathbf{R}^{(n)}\left[{}_I\overline{\boldsymbol{\omega}}_B^{(n)} \times \left({}_I\overline{\boldsymbol{\omega}}_B^{(n)} \times \bar{\mathbf{x}}_{P/A}^{(n)}\right)\right] \tag{4.9.7b}$$

It is useful to recast the matrix-vector forms in terms of the angular accelerations. For Eq. (4.9.5) we have in inertial coordinates

$$\ddot{\mathbf{x}}_P^{(n)} = \ddot{\mathbf{x}}_A^{(n)} - \mathbf{X}_{P/A}^{(n)}{}_I\boldsymbol{\alpha}_B^{(n)} + {}_I\boldsymbol{\Omega}_B^{(n)}{}_I\boldsymbol{\Omega}_B^{(n)}\mathbf{x}_{P/A}^{(n)} \tag{4.9.8}$$

which can be transformed to

$$\ddot{\mathbf{x}}_P^{(n)} = \ddot{\mathbf{x}}_A^{(n)} - \mathbf{R}^{(n)}\bar{\mathbf{X}}_{P/A}^{(n)}{}_I\bar{\boldsymbol{\alpha}}_B^{(n)} + \mathbf{R}^{(n)}{}_I\bar{\boldsymbol{\Omega}}_B^{(n)}{}_I\bar{\boldsymbol{\Omega}}_B^{(n)}\bar{\mathbf{x}}_{P/A}^{(n)} \tag{4.9.9}$$

In terms of Euler angles since from Eqs. (4.8.8) and (4.8.12)

$${}_I\boldsymbol{\omega}_B^{(n)} = \mathbf{H}^{(n)}\dot{\boldsymbol{\theta}}^{(n)}, \quad {}_I\bar{\boldsymbol{\omega}}_B^{(n)} = \bar{\mathbf{H}}^{(n)}\dot{\boldsymbol{\theta}}^{(n)}$$

we have

$$\begin{aligned} {}_I\boldsymbol{\alpha}_B^{(n)} &= {}_I\dot{\boldsymbol{\omega}}_B^{(n)} = \mathbf{H}^{(n)}\ddot{\boldsymbol{\theta}}^{(n)} + \dot{\mathbf{H}}^{(n)}\dot{\boldsymbol{\theta}}^{(n)} \\ {}_I\bar{\boldsymbol{\alpha}}_B^{(n)} &= {}_I\dot{\bar{\boldsymbol{\omega}}}_B^{(n)} = \bar{\mathbf{H}}^{(n)}\ddot{\boldsymbol{\theta}}^{(n)} + \dot{\bar{\mathbf{H}}}^{(n)}\dot{\boldsymbol{\theta}}^{(n)} \end{aligned} \tag{4.9.10}$$

so that the rigid body acceleration relations can be expressed as

$$\ddot{\mathbf{x}}_P^{(n)} = \ddot{\mathbf{x}}_A^{(n)} - \mathbf{X}_{P/A}^{(n)}\mathbf{H}^{(n)}\ddot{\boldsymbol{\theta}}^{(n)} - \mathbf{X}_{P/A}^{(n)}\dot{\mathbf{H}}^{(n)}\dot{\boldsymbol{\theta}}^{(n)} + {}_I\boldsymbol{\Omega}_B^{(n)}{}_I\boldsymbol{\Omega}_B^{(n)}\mathbf{x}_{P/A}^{(n)} \tag{4.9.11}$$

or, in a form where the relative position terms are in body-fixed coordinates and hence constants,

$$\begin{aligned} \ddot{\mathbf{x}}_P^{(n)} = \ddot{\mathbf{x}}_A^{(n)} &- \mathbf{R}^{(n)}\bar{\mathbf{X}}_{P/A}^{(n)}\bar{\mathbf{H}}^{(n)}\ddot{\boldsymbol{\theta}}^{(n)} \\ &- \mathbf{R}^{(n)}\bar{\mathbf{X}}_{P/A}^{(n)}\dot{\bar{\mathbf{H}}}^{(n)}\dot{\boldsymbol{\theta}}^{(n)} + \mathbf{R}^{(n)}{}_I\bar{\boldsymbol{\Omega}}_B^{(n)}{}_I\bar{\boldsymbol{\Omega}}_B^{(n)}\bar{\mathbf{x}}_{P/A}^{(n)} \end{aligned} \tag{4.9.12}$$

4.10 Euler Parameters

Since all the Euler angles have singular configurations where they cannot uniquely represent a general rotation they are not a good choice in general for doing numerical calculations, particularly in complex situations where it cannot be guaranteed that the singular configurations will not occur. Euler parameters are often used as a replacement for Euler angles since they do not have such singular behavior. Before introducing the Euler parameters, we need to examine the rotation matrix in more detail. According to Euler's theorem, the general rotation of a rigid body about a fixed point can always be expressed as a single rotation about a given axis, called the axis of rotation. Thus, consider a position vector, $\bar{\mathbf{x}}$, that goes from O to the point $\bar{P}$ that is fixed in a body where the body rotates about a fixed point O through an angle Θ in a plane perpendicular to the axis of rotation, OC, whose direction is given by a unit vector, $\mathbf{n}$ (Fig. 4.42a). [In this section we will drop the (n) superscript to denote

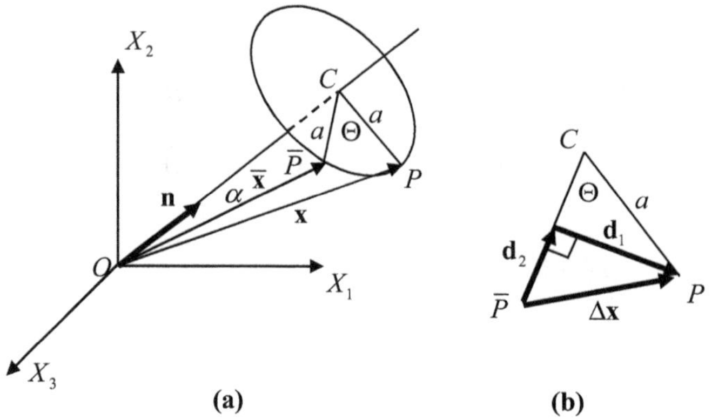

Fig. 4.42 (**a**) Rotation of a rigid body about an axis of rotation defined by the unit vector, **n**, that transforms the vector $\bar{\mathbf{x}}$ to **x** and (**b**) the geometry relating $\Delta\mathbf{x}$ to the other parameters

the body for economy of notation.] This rotation will cause the body-fixed vector $\bar{\mathbf{x}}$ to change by an amount $\Delta\mathbf{x}$, producing a new position vector, **x**, where

$$\mathbf{x} = \bar{\mathbf{x}} + \Delta\mathbf{x} \tag{4.10.1}$$

The vector $\Delta\mathbf{x}$ lies in a plane perpendicular to **n**, as shown in Fig. 4.42b, and can be written as

$$\Delta\mathbf{x} = \mathbf{d}_1 + \mathbf{d}_2 \tag{4.10.2}$$

where $\mathbf{d}_1$ is perpendicular to the plane $OC\bar{P}$ so that it is in the direction $\mathbf{n}\times\bar{\mathbf{x}}$. Thus, $\mathbf{d}_1$ has the magnitude $|\mathbf{d}_1| = a \sin \Theta$ (see Fig. 4.42b). But $a = |\bar{\mathbf{x}}|\sin\alpha = |\mathbf{n}\times\mathbf{x}|$ where α is the angle between **n** and $\bar{\mathbf{x}}$ so that

$$\mathbf{d}_1 = a\sin\Theta\frac{\mathbf{n}\times\bar{\mathbf{x}}}{|\mathbf{n}\times\bar{\mathbf{x}}|} = (\mathbf{n}\times\bar{\mathbf{x}})\sin\Theta \tag{4.10.3}$$

Now consider the vector, $\mathbf{d}_2$, which has a magnitude $|\mathbf{d}_2| = (1 - \cos\Theta)a = 2a \sin(\Theta/2)$. This vector is perpendicular to **n** and also is perpendicular to the unit vector $\mathbf{n}\times\bar{\mathbf{x}}/a$, which defines the direction of $\mathbf{d}_1$ so we have

$$\mathbf{d}_2 = 2a\sin^2(\Theta/2)\frac{\mathbf{n}\times(\mathbf{n}\times\bar{\mathbf{x}})}{a} = 2[\mathbf{n}\times(\mathbf{n}\times\bar{\mathbf{x}})]\sin^2(\Theta/2) \tag{4.10.4}$$

Thus, from Eqs. (4.10.4), (4.10.3), and (4.10.2), Eq. (4.10.1) becomes

$$\mathbf{x} = \bar{\mathbf{x}} + (\mathbf{n}\times\bar{\mathbf{x}})\sin\Theta + 2[\mathbf{n}\times(\mathbf{n}\times\bar{\mathbf{x}})]\sin^2(\Theta/2) \tag{4.10.5}$$

But if the cross products are written in terms of the skew symmetric matrix

$$\mathbf{N} = \begin{bmatrix} 0 & -n_3 & n_2 \\ n_3 & 0 & -n_1 \\ -n_2 & n_1 & 0 \end{bmatrix} \tag{4.10.6}$$

we can write Eq. (4.10.5) as

$$\mathbf{x} = \left[\mathbf{I} + \mathbf{N}\sin\Theta + 2(\mathbf{N})^2 \sin^2(\Theta/2)\right]\bar{\mathbf{x}} \tag{4.10.7}$$

where $\mathbf{I}$ is the 3×3 identity matrix and the matrix product $(\mathbf{N})^2 = \mathbf{NN}$. This equation can also be written as

$$\mathbf{x} = \mathbf{R}\bar{\mathbf{x}} \tag{4.10.8}$$

where the rotation matrix for the body is

$$\begin{aligned} \mathbf{R} &= \left[\mathbf{I} + \mathbf{N}\sin\Theta + 2(\mathbf{N})^2 \sin^2(\Theta/2)\right] \\ &= \left[\mathbf{I} + 2\mathbf{N}\sin(\Theta/2)\cos(\Theta/2) + 2(\mathbf{N})^2 \sin^2(\Theta/2)\right] \end{aligned} \tag{4.10.9}$$

which is also called the *Rodrigues' formula.* Since $\mathbf{R} = \mathbf{R}(\mathbf{n}, \Theta)$ and $\mathbf{n}$ is a unit vector, the rotation matrix is characterized by three independent parameters.

To avoid the singularities associated with the writing of the rotation matrix in terms of three Euler angles, one can define four parameters, $\boldsymbol{\theta} = (\theta_0, \theta_1, \theta_2, \theta_3)$, called the *Euler parameters*, as

$$\begin{aligned} &\theta_0 = \cos(\Theta/2), \quad \theta_1 = n_1 \sin(\Theta/2), \quad \theta_2 = n_2 \sin(\Theta/2), \\ &\theta_3 = n_3 \sin(\Theta/2) \end{aligned} \tag{4.10.10a}$$

which obviously satisfy

$$\boldsymbol{\theta}^T\boldsymbol{\theta} = 1 \tag{4.10.10b}$$

Using these relations the rotation matrix can be written in terms of these parameters in the form

$$\mathbf{R} = \begin{bmatrix} 1 - 2\theta_2^2 - 2\theta_3^2 & 2(\theta_1\theta_2 - \theta_0\theta_3) & 2(\theta_1\theta_3 + \theta_0\theta_2) \\ 2(\theta_1\theta_2 + \theta_0\theta_3) & 1 - 2\theta_1^2 - 2\theta_3^2 & 2(\theta_2\theta_3 - \theta_0\theta_1) \\ 2(\theta_1\theta_3 - \theta_0\theta_2) & 2(\theta_2\theta_3 + \theta_0\theta_1) & 1 - 2\theta_1^2 - 2\theta_2^2 \end{bmatrix} \tag{4.10.11}$$

We can see that the rotation matrix depends on products of the Euler parameters. We can define this matrix as the product of two 3 × 4 matrices $(\mathbf{E}, \bar{\mathbf{E}})$ that only depend linearly on the Euler parameters as

$$\mathbf{R} = \mathbf{E}\bar{\mathbf{E}}^T \tag{4.10.12}$$

with

$$\mathbf{E} = \begin{bmatrix} -\theta_1 & \theta_0 & -\theta_3 & \theta_2 \\ -\theta_2 & \theta_3 & \theta_0 & -\theta_1 \\ -\theta_3 & -\theta_2 & \theta_1 & \theta_0 \end{bmatrix}, \quad \bar{\mathbf{E}} = \begin{bmatrix} -\theta_1 & \theta_0 & \theta_3 & -\theta_2 \\ -\theta_2 & -\theta_3 & \theta_0 & \theta_1 \\ -\theta_3 & \theta_2 & -\theta_1 & \theta_0 \end{bmatrix} \tag{4.10.13}$$

Some relations that $\mathbf{E}, \bar{\mathbf{E}}$ can be shown to satisfy are [1, 2]

$$\begin{aligned} &\mathbf{E}\mathbf{E}^T = \bar{\mathbf{E}}\bar{\mathbf{E}}^T = \mathbf{I}, \;\; \mathbf{E}^T\mathbf{E} = \bar{\mathbf{E}}^T\bar{\mathbf{E}} = \mathbf{I}_4 - \boldsymbol{\theta}\boldsymbol{\theta}^T \\ &\mathbf{E}\boldsymbol{\theta} = \bar{\mathbf{E}}\boldsymbol{\theta} = 0, \;\; \dot{\mathbf{E}}\dot{\boldsymbol{\theta}} = \dot{\bar{\mathbf{E}}}\dot{\boldsymbol{\theta}} = 0, \;\; \mathbf{E}\dot{\bar{\mathbf{E}}}^T = \dot{\mathbf{E}}\bar{\mathbf{E}}^T \end{aligned} \tag{4.10.14}$$

where $\mathbf{I}$ is the 3 × 3 identity matrix and $\mathbf{I}_4$ is the 4 × 4 identity matrix. In terms of the rotation matrix and its derivative the skew symmetric angular velocity matrix, $\boldsymbol{\Omega}$, is $\boldsymbol{\Omega} = \dot{\mathbf{R}}\mathbf{R}^T$, which contains the angular velocity components of the body relative to an inertial frame. Similarly, $\overline{\boldsymbol{\Omega}} = \mathbf{R}^T\boldsymbol{\Omega}\mathbf{R} = \mathbf{R}^T\dot{\mathbf{R}}$ is in terms of the angular velocity components measured in a body-fixed frame. Thus, using Eq. (4.10.12) and the properties of $\mathbf{E}, \bar{\mathbf{E}}$ we find

$$\boldsymbol{\Omega} = \begin{bmatrix} 0 & -\omega_3 & \omega_2 \\ \omega_3 & 0 & -\omega_1 \\ -\omega_2 & \omega_1 & 0 \end{bmatrix} = \dot{\mathbf{R}}\mathbf{R}^T = 2\dot{\mathbf{E}}\mathbf{E}^T \tag{4.10.15}$$

which gives the following relationship between the angular velocity components and the Euler parameters:

$$\begin{aligned} \omega_1 &= 2\left(\dot{\theta}_3\theta_2 - \dot{\theta}_2\theta_3 + \dot{\theta}_1\theta_0 - \dot{\theta}_0\theta_1\right) \\ \omega_2 &= 2\left(\dot{\theta}_1\theta_3 - \dot{\theta}_0\theta_2 - \dot{\theta}_3\theta_1 + \dot{\theta}_2\theta_0\right) \\ \omega_3 &= 2\left(\dot{\theta}_2\theta_1 + \dot{\theta}_3\theta_0 - \dot{\theta}_0\theta_3 - \dot{\theta}_1\theta_2\right) \end{aligned} \tag{4.10.16}$$

which can also be written in the matrix-vector forms

$$\boldsymbol{\omega} = 2\mathbf{E}\dot{\boldsymbol{\theta}} = -2\dot{\mathbf{E}}\boldsymbol{\theta} \tag{4.10.17}$$

In the body-fixed frame

$$\overline{\boldsymbol{\Omega}} = \begin{bmatrix} 0 & -\overline{\omega}_3 & \overline{\omega}_2 \\ \overline{\omega}_3 & 0 & -\overline{\omega}_1 \\ -\overline{\omega}_2 & \overline{\omega}_1 & 0 \end{bmatrix} = \mathbf{R}^T\dot{\mathbf{R}} = 2\bar{\mathbf{E}}\dot{\bar{\mathbf{E}}}^T \tag{4.10.18}$$

giving

$$\begin{aligned} \overline{\omega}_1 &= 2\left(\theta_3\dot{\theta}_2 - \theta_2\dot{\theta}_3 - \theta_1\dot{\theta}_0 + \theta_0\dot{\theta}_1\right) \\ \overline{\omega}_2 &= 2\left(\theta_1\dot{\theta}_3 + \theta_0\dot{\theta}_2 - \theta_3\dot{\theta}_1 - \theta_2\dot{\theta}_0\right) \\ \overline{\omega}_3 &= 2\left(\theta_2\dot{\theta}_1 - \theta_3\dot{\theta}_0 + \theta_0\dot{\theta}_3 - \theta_1\dot{\theta}_2\right) \end{aligned} \tag{4.10.19}$$

or in matrix-vector forms we have

$$\overline{\boldsymbol{\omega}} = 2\bar{\mathbf{E}}\dot{\boldsymbol{\theta}} = -2\dot{\bar{\mathbf{E}}}\boldsymbol{\theta} \tag{4.10.20}$$

In obtaining the equations of motion in body-fixed coordinates it is also convenient to find the inverse relationship of Eq. (4.10.20), i.e., we want to write the derivatives of the Euler parameters in terms of the angular velocity components in body-fixed coordinates and the Euler parameters.

If we multiply the first equation in Eq. (4.10.20) by $\bar{\mathbf{E}}^T$ we obtain the inverse relationship since

$$\bar{\mathbf{E}}^T\overline{\boldsymbol{\omega}} = 2\bar{\mathbf{E}}^T\bar{\mathbf{E}}\dot{\boldsymbol{\theta}} = 2\left(\mathbf{I}_4 - \boldsymbol{\theta}^T\right)\dot{\boldsymbol{\theta}} = 2\dot{\boldsymbol{\theta}} \tag{4.10.21}$$

or

$$\dot{\boldsymbol{\theta}} = \frac{1}{2}\bar{\mathbf{E}}^T\overline{\boldsymbol{\omega}} \tag{4.10.22}$$

where $\boldsymbol{\theta}^{\mathbf{T}}\dot{\boldsymbol{\theta}} = 0$ since $\boldsymbol{\theta}^T\boldsymbol{\theta} = 1$. More explicitly, Eq. (4.10.22) is

$$\begin{Bmatrix} \dot{\theta}_0 \\ \dot{\theta}_1 \\ \dot{\theta}_2 \\ \dot{\theta}_3 \end{Bmatrix} = \frac{1}{2}\begin{bmatrix} -\theta_1 & -\theta_2 & -\theta_3 \\ \theta_0 & -\theta_3 & \theta_2 \\ \theta_3 & \theta_0 & -\theta_1 \\ -\theta_2 & \theta_1 & \theta_0 \end{bmatrix}\begin{Bmatrix} \omega_1 \\ \omega_2 \\ \omega_3 \end{Bmatrix} \tag{4.10.23}$$

which can be rearranged to also give

$$\begin{Bmatrix} \dot{\theta}_0 \\ \dot{\theta}_1 \\ \dot{\theta}_2 \\ \dot{\theta}_3 \end{Bmatrix} = \frac{1}{2} \begin{bmatrix} 0 & -\omega_1 & -\omega_2 & -\omega_3 \\ \omega_1 & 0 & \omega_3 & -\omega_2 \\ \omega_2 & -\omega_3 & 0 & \omega_1 \\ \omega_3 & \omega_2 & -\omega_1 & 0 \end{bmatrix} \begin{Bmatrix} \theta_0 \\ \theta_1 \\ \theta_2 \\ \theta_3 \end{Bmatrix} \tag{4.10.24}$$

where we see that the 4 × 4 matrix in Eq. (4.10.24) is skew symmetric.

In the case of Euler angles we wrote the relationship between the angular velocity vector and the derivative of the Euler angles as

$${}_I\boldsymbol{\omega}_B^{(n)} = \mathbf{H}^{(n)}\dot{\boldsymbol{\theta}}, \quad {}_I\overline{\boldsymbol{\omega}}_B^{(n)} = \bar{\mathbf{H}}^{(n)}\dot{\boldsymbol{\theta}} \tag{4.10.25}$$

Here, we can use the same matrix-vector forms where $\dot{\boldsymbol{\theta}} = [\dot{\theta}_0, \dot{\theta}_1, \dot{\theta}_2, \dot{\theta}_3]$ and

$$\mathbf{H} = 2\mathbf{E}, \quad \bar{\mathbf{H}} = 2\bar{\mathbf{E}} \tag{4.10.26}$$

In terms of these matrices and the Euler parameters we have the following relations:

$$\begin{aligned} \dot{\mathbf{H}}\mathbf{H}^T &= -\mathbf{H}\dot{\mathbf{H}}^T = 2\,{}_I\boldsymbol{\Omega}_B \\ \bar{\mathbf{H}}\dot{\bar{\mathbf{H}}}^T &= -\dot{\bar{\mathbf{H}}}\bar{\mathbf{H}}^T = 2\,{}_I\overline{\boldsymbol{\Omega}}_B \\ \mathbf{H}\dot{\boldsymbol{\theta}} &= -\dot{\mathbf{H}}\boldsymbol{\theta}, \quad \bar{\mathbf{H}}\dot{\boldsymbol{\theta}} = -\dot{\bar{\mathbf{H}}}\boldsymbol{\theta} \end{aligned} \tag{4.10.27a}$$

in addition to the relations we listed earlier for $\mathbf{E}, \bar{\mathbf{E}}$ (Eq. (4.10.14)), which in terms of $\mathbf{H}, \bar{\mathbf{H}}$ are

$$\begin{aligned} &\mathbf{H}\mathbf{H}^T = \bar{\mathbf{H}}\bar{\mathbf{H}}^T = 4\mathbf{I}, \;\; \mathbf{H}^T\mathbf{H} = \bar{\mathbf{H}}^T\bar{\mathbf{H}} = 4(\mathbf{I}_4 - \boldsymbol{\theta}\boldsymbol{\theta}^T) \\ &\mathbf{H}\boldsymbol{\theta} = \bar{\mathbf{H}}\boldsymbol{\theta} = 0, \;\; \dot{\mathbf{H}}\dot{\boldsymbol{\theta}} = \dot{\bar{\mathbf{H}}}\dot{\boldsymbol{\theta}} = 0, \;\; \mathbf{H}\dot{\bar{\mathbf{H}}}^T = \dot{\mathbf{H}}\bar{\mathbf{H}}^T \end{aligned} \tag{4.10.27b}$$

The relations for the angular accelerations in inertial or body-fixed coordinates for Euler parameters are simpler than for Euler angles (see Eq. (4.9.9)) since

$${}_I\boldsymbol{\alpha}_B^{(n)} = {}_I\dot{\boldsymbol{\omega}}_B^{(n)} = \mathbf{H}^{(n)}\ddot{\boldsymbol{\theta}}, \quad {}_I\overline{\boldsymbol{\alpha}}_B^{(n)} = {}_I\dot{\overline{\boldsymbol{\omega}}}_B^{(n)} = \bar{\mathbf{H}}^{(n)}\ddot{\boldsymbol{\theta}} \tag{4.10.28}$$

where we have used the relations $\dot{\mathbf{H}}\dot{\boldsymbol{\theta}} = \dot{\bar{\mathbf{H}}}\dot{\boldsymbol{\theta}} = 0$ from Eq. (4.10.27b). Note that we also can determine the angular acceleration in body-fixed coordinates from the angular acceleration in inertial coordinates, giving the same result, i.e.,

$$\begin{aligned} {}_I\overline{\boldsymbol{\alpha}}_B^{(n)} &= \mathbf{R}^T{}_I\boldsymbol{\alpha}_B^{(n)} = \frac{1}{4}\bar{\mathbf{H}}^{(n)}\left(\mathbf{H}^{(n)}\right)^T\mathbf{H}^{(n)}\ddot{\boldsymbol{\theta}} \\ &= \bar{\mathbf{H}}^{(n)}\ddot{\boldsymbol{\theta}} - (\bar{\mathbf{H}}^{(n)}\boldsymbol{\theta})\boldsymbol{\theta}^T\ddot{\boldsymbol{\theta}} = \bar{\mathbf{H}}^{(n)}\ddot{\boldsymbol{\theta}} \end{aligned} \tag{4.10.29}$$

Table 4.1 Euler parameters in terms of Euler angles for the common Euler angle sets

Euler angle set:	3-1-3	3-2-3	3-2-1
Euler angles:	ϕ,θ,ψ	ϕ,θ,ψ	ψ,θ,ϕ
θ_0	$\cos\left(\frac{\phi+\psi}{2}\right)\cos\left(\frac{\theta}{2}\right)$	$\cos\left(\frac{\phi+\psi}{2}\right)\cos\left(\frac{\theta}{2}\right)$	$\cos\left(\frac{\psi}{2}\right)\cos\left(\frac{\theta}{2}\right)\cos\left(\frac{\phi}{2}\right)$ $+\sin\left(\frac{\psi}{2}\right)\sin\left(\frac{\theta}{2}\right)\sin\left(\frac{\phi}{2}\right)$
θ_1	$\cos\left(\frac{\phi-\psi}{2}\right)\sin\left(\frac{\theta}{2}\right)$	$\sin\left(\frac{\psi-\phi}{2}\right)\sin\left(\frac{\theta}{2}\right)$	$\cos\left(\frac{\psi}{2}\right)\cos\left(\frac{\theta}{2}\right)\sin\left(\frac{\phi}{2}\right)$ $-\sin\left(\frac{\psi}{2}\right)\sin\left(\frac{\theta}{2}\right)\cos\left(\frac{\phi}{2}\right)$
θ_2	$\sin\left(\frac{\phi-\psi}{2}\right)\sin\left(\frac{\theta}{2}\right)$	$\cos\left(\frac{\psi-\phi}{2}\right)\sin\left(\frac{\theta}{2}\right)$	$\cos\left(\frac{\psi}{2}\right)\sin\left(\frac{\theta}{2}\right)\cos\left(\frac{\phi}{2}\right)$ $+\sin\left(\frac{\psi}{2}\right)\cos\left(\frac{\theta}{2}\right)\sin\left(\frac{\phi}{2}\right)$
θ_3	$\sin\left(\frac{\phi+\psi}{2}\right)\cos\left(\frac{\theta}{2}\right)$	$\sin\left(\frac{\phi+\psi}{2}\right)\cos\left(\frac{\theta}{2}\right)$	$\sin\left(\frac{\psi}{2}\right)\cos\left(\frac{\theta}{2}\right)\cos\left(\frac{\phi}{2}\right)$ $-\cos\left(\frac{\psi}{2}\right)\sin\left(\frac{\theta}{2}\right)\sin\left(\frac{\phi}{2}\right)$

so that we can get the angular acceleration in the body-fixed frame, ${}_I\overline{\boldsymbol{\alpha}}_B^{(n)}$, either by differentiating the angular velocity components in the body-fixed frame (Eq. (4.10.28)) or, as shown in Eq. (4.10.29), by taking the time derivative of the angular velocity in an inertial frame and then transforming those derivatives to the body-fixed frame.

When using the Euler parameters, it is often convenient to specify the body orientation in terms of Euler angles and then convert them to equivalent Euler parameter values. Specifying initial conditions, for example, is a case where this conversion is useful. We can obtain the Euler parameters in terms of the Euler angles by examining the rotation matrix in terms of either the Euler angles or Euler parameters and then equating components or combinations of components of the matrix. The process is rather tedious so in Table 4.1 we simply list the relationships for different Euler angle choices [3].

4.11 The Commonly Used Euler Angle Sets

We mentioned previously that of the 12 possible Euler angle sets, 3 sets were those most commonly used—the 3-1-3 set, the 3-2-3 set, and the 3-2-1 set. In this section we will give the rotation matrix, **R**, associated with these sets where, recall, this matrix transforms body-fixed vector components such as $\overline{\mathbf{x}} = (\overline{x}, \overline{y}, \overline{z})$ into inertial components such as $\mathbf{X} = (X, Y, Z)$, i.e.,

$$\mathbf{X} = \mathbf{R}\bar{\mathbf{x}} \tag{4.11.1}$$

To write these matrices succinctly we will use the notation $c\theta_i = \cos(\theta_i)$ and $s\theta_i = \sin(\theta_i)$ where the angles in $\boldsymbol{\theta} = (\theta_1, \theta_2, \theta_3)$ are the angles for the first, second, and third rotations, respectively, in any of these sets. We will also give the matrices that transform the angular velocity, measured in terms of body-fixed components, and the Euler angle derivatives, i.e.,

$$\bar{\boldsymbol{\omega}} = \bar{\mathbf{H}}\dot{\boldsymbol{\theta}}, \quad \dot{\boldsymbol{\theta}} = \bar{\mathbf{H}}^{-1}\bar{\boldsymbol{\omega}} \tag{4.11.2}$$

4.11.1 3-1-3 Set

$$\mathbf{R} = \begin{bmatrix} c\theta_3 c\theta_1 - s\theta_3 c\theta_2 s\theta_1 & -s\theta_3 c\theta_1 - c\theta_3 c\theta_2 s\theta_1 & s\theta_2 s\theta_1 \\ c\theta_3 s\theta_1 + s\theta_3 c\theta_2 c\theta_1 & -s\theta_3 s\theta_1 + c\theta_3 c\theta_2 c\theta_1 & -s\theta_2 c\theta_1 \\ s\theta_3 s\theta_2 & c\theta_3 s\theta_2 & c\theta_2 \end{bmatrix} \tag{4.11.3}$$

$$\bar{\mathbf{H}} = \begin{bmatrix} s\theta_3 s\theta_2 & c\theta_3 & 0 \\ s\theta_2 c\theta_3 & -s\theta_3 & 0 \\ c\theta_2 & 0 & 1 \end{bmatrix}, \quad \bar{\mathbf{H}}^{-1} = \frac{1}{s\theta_2} \begin{bmatrix} s\theta_3 & c\theta_3 & 0 \\ s\theta_2 c\theta_3 & -s\theta_2 s\theta_3 & 0 \\ -c\theta_2 s\theta_3 & -c\theta_2 c\theta_3 & s\theta_2 \end{bmatrix} \tag{4.11.4}$$

4.11.2 3-2-3 Set

$$\mathbf{R} = \begin{bmatrix} c\theta_3 c\theta_2 c\theta_1 - s\theta_3 s\theta_1 & -s\theta_3 c\theta_2 c\theta_1 - c\theta_3 s\theta_1 & s\theta_2 c\theta_1 \\ c\theta_3 c\theta_2 s\theta_1 + s\theta_3 c\theta_1 & -s\theta_3 c\theta_2 s\theta_1 + c\theta_3 c\theta_1 & s\theta_2 s\theta_1 \\ -c\theta_3 s\theta_2 & s\theta_3 s\theta_2 & c\theta_2 \end{bmatrix} \tag{4.11.5}$$

$$\bar{\mathbf{H}} = \begin{bmatrix} -s\theta_2 c\theta_3 & s\theta_3 & 0 \\ s\theta_2 s\theta_3 & c\theta_3 & 0 \\ c\theta_2 & 0 & 1 \end{bmatrix}, \quad \bar{\mathbf{H}}^{-1} = \frac{1}{s\theta_2} \begin{bmatrix} -c\theta_3 & s\theta_3 & 0 \\ s\theta_2 s\theta_3 & s\theta_2 c\theta_3 & 0 \\ c\theta_2 c\theta_3 & -c\theta_2 s\theta_3 & s\theta_2 \end{bmatrix} \tag{4.11.6}$$

4.11.3 3-2-1 Set

$$\mathbf{R} = \begin{bmatrix} c\theta_2 c\theta_1 & s\theta_3 s\theta_2 c\theta_1 - c\theta_3 s\theta_1 & c\theta_3 s\theta_2 c\theta_1 + s\theta_3 s\theta_1 \\ c\theta_2 s\theta_1 & s\theta_3 s\theta_2 s\theta_1 + c\theta_3 c\theta_1 & c\theta_3 s\theta_2 s\theta_1 - s\theta_3 c\theta_1 \\ -s\theta_2 & s\theta_3 c\theta_2 & c\theta_3 c\theta_2 \end{bmatrix} \tag{4.11.7}$$

$$\bar{\mathbf{H}} = \begin{bmatrix} -s\theta_2 & 0 & 1 \\ c\theta_2 s\theta_3 & c\theta_3 & 0 \\ c\theta_2 c\theta_3 & -s\theta_3 & 0 \end{bmatrix}, \quad \bar{\mathbf{H}}^{-1} = \frac{1}{c\theta_2} \begin{bmatrix} 0 & s\theta_3 & c\theta_3 \\ 0 & c\theta_2 c\theta_3 & -c\theta_2 s\theta_3 \\ c\theta_2 & s\theta_2 s\theta_3 & s\theta_2 c\theta_3 \end{bmatrix} \tag{4.11.8}$$

4.12 Combination Rotations with Body-Fixed and Space-Fixed Frames

The Euler angle sets are examples of the use of frames that are fixed to the body as it undergoes a combination of rotations about a set of axes. For example, if we have three rotation matrices $(\mathbf{R}_1, \mathbf{R}_2, \mathbf{R}_3)$ that define the rotations of a body relating the initial $\mathbf{X} = [X_1, X_2, X_3]^T$ axes to the final $\mathbf{x} = [x, y, z]^T$ rotated axes then we can consider the rotations as applied successively as

$$\begin{aligned} \mathbf{X} &= \mathbf{R}_1 \mathbf{x}_1 \\ \mathbf{x}_1 &= \mathbf{R}_2 \mathbf{x}_2 \\ \mathbf{x}_2 &= \mathbf{R}_3 \mathbf{x} \end{aligned} \tag{4.12.1}$$

where $\mathbf{x}_1 = [x_1, y_1, z_1]^T$ and $\mathbf{x}_2 = [x_2, y_2, z_2]^T$. The total composite rotation $\mathbf{X} = \mathbf{R}\mathbf{x}$ is then given by

$$\mathbf{X} = \mathbf{R}_1 \mathbf{R}_2 \mathbf{R}_3 \mathbf{x} \tag{4.12.2}$$

For example, a 3-1-3 set of Euler angle rotations then corresponds to where $\mathbf{R}_1$ is a rotation about the Z_1 axis, $\mathbf{R}_2$ is a rotation about the x_1 axis, and $\mathbf{R}_3$ is a rotation about the z_2 axis. This result can be easily generalized to n rotations, where for a sequence of body-fixed rotations relating an initial frame, $\mathbf{X}$, to a final frame, $\mathbf{x}$, we have

$$\mathbf{X} = \mathbf{R}_1 \mathbf{R}_2 \ldots \mathbf{R}_n \mathbf{x} \tag{4.12.3}$$

Note that for body-fixed frames, therefore, we need to multiply the rotations sequentially going from left to right. When dealing with a set of body-fixed rotations in this manner the orientations of the axes of rotation are constantly changing. We could, however, when relating the frame $\mathbf{X}$ to the final frame $\mathbf{x}$, define a set of rotations that are all taken about the initial space-fixed (inertial) frame, $\mathbf{X}$. For three rotations, for example, we would have

$$\begin{aligned}\mathbf{X} &= \mathbf{R}_1\mathbf{x}_1\\ \mathbf{X} &= \mathbf{R}_2\mathbf{x}_2\\ \mathbf{X} &= \mathbf{R}_3\mathbf{x}_3\end{aligned} \tag{4.12.4}$$

In this case, to get the total effects of all the rotations we need to define an intermediate frame, which we will call $\mathbf{x}_f$, and the final frame, $\mathbf{x}$, in addition to the frames shown in Eq. (4.12.4). Relating the $\mathbf{X}$ frame to the $\mathbf{x}_1$ frame is simple since we can just use the first equation in Eq. (4.12.4). Let the $\mathbf{x}_f$ frame be the result of applying both the $\mathbf{R}_1$ and $\mathbf{R}_2$ rotations. Since both the $\mathbf{x}_f$ frame and the $\mathbf{x}_2$ frame share the same rotation, $\mathbf{R}_2$, they must only differ by the rotation $\mathbf{R}_1$ so we have

$$\mathbf{x}_2 = \mathbf{R}_1\mathbf{x}_f \tag{4.12.5}$$

and then it follows that

$$\mathbf{X} = \mathbf{R}_2\mathbf{R}_1\mathbf{x}_f \tag{4.12.6}$$

The final frame, $\mathbf{x}$, which is a result of applying all three rotations, shares the same $\mathbf{R}_3$ rotation with the frame $\mathbf{x}_3$ so they must only differ by the first two rotations, giving

$$\mathbf{x}_3 = \mathbf{R}_2\mathbf{R}_1\mathbf{x} \tag{4.12.7}$$

which leads, finally, to

$$\mathbf{X} = \mathbf{R}_3\mathbf{R}_2\mathbf{R}_1\mathbf{x} \tag{4.12.8}$$

For rotations involving space-fixed frames, therefore, we need to multiply the rotations sequentially from right to left, which is opposite to what was found with the use of body-fixed frames. We can again generalize these results for a sequence of n space-fixed rotations relating an initial frame, $\mathbf{X}$, to a final frame, $\mathbf{x}$, as

$$\mathbf{X} = \mathbf{R}_n \ldots \mathbf{R}_2\mathbf{R}_1\mathbf{x} \tag{4.12.9}$$

Note, however, that these orderings depend on our definition of the rotation matrix. If we define instead, as some books do [4], rotation matrices $\mathbf{R}' = \mathbf{R}^T$ where $\mathbf{x} = \mathbf{R}'\mathbf{X}$ then we find for body-fixed axes

$$\mathbf{x} = \mathbf{R}_n' \ldots \mathbf{R}_2'\mathbf{R}_1'\mathbf{X} \tag{4.12.10}$$

and for space-fixed axes

$$\mathbf{x} = \mathbf{R}_1'\mathbf{R}_2' \ldots \mathbf{R}_n'\mathbf{X} \tag{4.12.11}$$

But in this book we will stick with our original definition of the rotation matrix as presented in Sect. 4.6 (see, for example, Eq. (4.6.9)). One can combine body-fixed and space-fixed rotations as long as one adheres to these orderings. For example if we perform rotations $\mathbf{R}_1$ and $\mathbf{R}_2$ about body-fixed axes but then perform a rotation $\mathbf{R}_3$ about a space-fixed axis, we have $\mathbf{X} = \mathbf{R}_3\mathbf{R}_1\mathbf{R}_2\mathbf{x}$. Body-fixed rotations are normally preferred over space-fixed rotations since angular velocities and accelerations are usually more conveniently expressed in terms of axes attached to the body.

4.13 Problems

P4.1. Prove the general form of the transport theorem, Eq. (4.2.7), by using Eq. (4.2.6) twice (I to R_1, I to R_2) and the relative angular velocity equation, Eq. (4.2.11).

P4.2. Use the updating formula Eq. (4.2.66), to improve the initial guess used in the Newton-Raphson method. Write a MATLAB® script similar to Simulation 4.2 except where the Newton-Raphson method is modified to use the updating of Eq. (4.2.66). Compare the performance of the Newton-Raphson method with and without this improvement on the problem considered in Simulation 4.2.

P4.3. Show that the relative velocity and acceleration terms expressed in body-fixed axes are given by Eq. (4.4.48).

P4.4. Consider the slider-crank mechanism shown in Fig. P4.1 where the slider is in a smooth inclined slot that makes an angle α with respect to the X-axis. If the crankshaft AB rotates at a constant angular velocity of 6 rad/s and $\alpha = 30°$, determine the angular velocity and angular acceleration of the connecting rod BC and the velocity of the slider at the given instant of time when the mechanism is in the position shown in Fig. P4.1. Use the relative velocity and acceleration expressions as done in Sect. 4.2.2.

P4.5. The slider-crank mechanism of Fig. P4.2a is in a general configuration defined by the angles ϕ_1 and ϕ_2 and the position of the slider along the inclined slot, D_P, which makes an angle α with respect to the X-axis. Alternatively, we can describe the system in the inclined coordinates by the parameters θ_1, θ_2, and X_P of Fig. P4.2b which are similar to those used in the slider-crank mechanism considered previously in this chapter. (a) Show that in terms of the inclined coordinates of Fig. P4.2b the constraint equations are

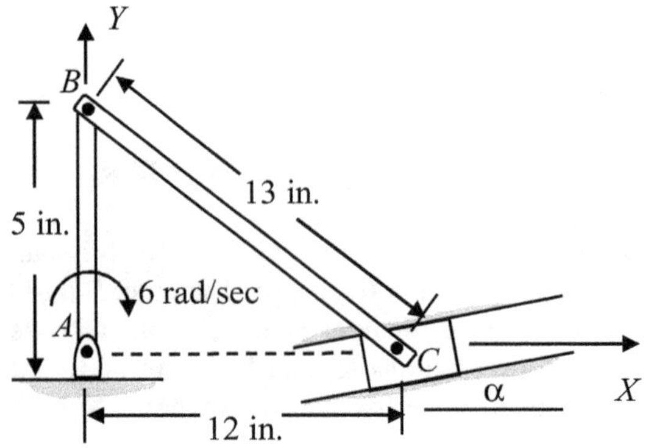

Fig. P4.1 A slider-crank mechanism at a given instant of time where the slider rides in a smooth inclined slot

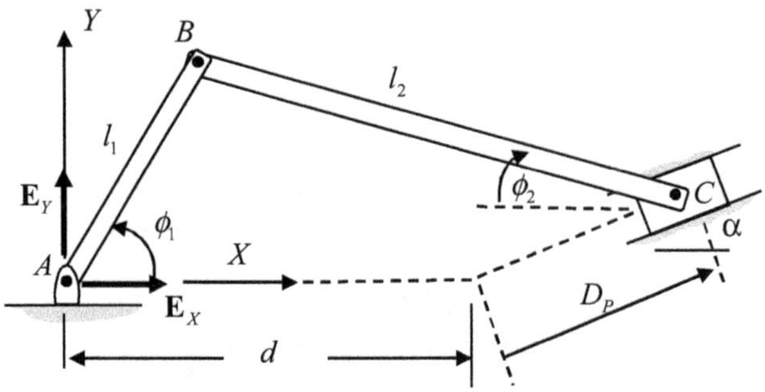

(a)

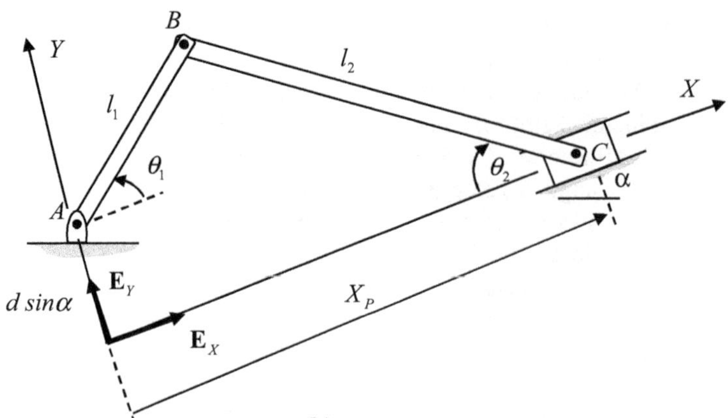

(b)

Fig. P4.2 (**a**) A slider-crank mechanism in a general positional configuration where the slider (piston) rides in a smooth inclined slot. (**b**) The slider-crank described in a set of *X*–*Y* coordinates along and perpendicular to the smooth slot. Using the X_P coordinate of the piston and the angles ϕ_1,ϕ_2 this problem is very similar to the slider-crank problem of Fig. 4.10 but where there is an offset, $d \sin \alpha$, between the slot and point *A* in the *Y*-direction

$$
\begin{aligned}
X_P &= l_1 \cos\theta_1 + l_2 \cos\theta_2 \\
l_2 \sin\theta_2 &= l_1 \sin\theta_1 + d \sin\alpha
\end{aligned}
\tag{P4.1}
$$

These are identical to Eq. (4.2.28) when $\alpha = 0$ so that except for the solution of these positional equations the determination of the velocity and acceleration constraints for this problem proceeds in an entirely similar manner to the $\alpha = 0$ case. In this case the solution for θ_2 gives

$$
\theta_2 = \sin^{-1}(l_1 \sin\theta_1 / l_2 + d \sin\alpha / l_2) \tag{P4.2}
$$

which will always have a real solution if $l_1 + d \sin\alpha \le l_2$. Take $l_1 = 5$ in., $l_2 = 13$ in., $d = 12$ in., and $\alpha = 30^\circ$ which will satisfy this inequality. Following similar steps as done in Simulation 4.1, use the constraint equations in Eq. (P4.1) and their derivatives to write a MATLAB® script that solves for velocities and accelerations and plots them versus time for the slider-crank mechanism of Fig. P4.2b when the crankshaft is being kinematically driven, e.g., $\theta_1(t) = \theta_{10} - 6t$. Use the same default geometry parameters found in Simulation 4.1 and take $d = 12$ in. and $\alpha = 30°$. Note, however, that the starting initial conditions here are different.

P4.6. Consider the slider-crank mechanism of Fig. P4.2a in a general configuration as defined by the angles ϕ_1 and ϕ_2 and the position of the slider along the inclined slot, D_P, which makes an angle α with respect to the X-axis. Show that the constraint equations become

$$
\begin{aligned}
l_1 \sin\phi_1 &= l_2 \sin\phi_2 + D_P \sin\alpha \\
l_1 \cos\phi_1 + l_2 \cos\phi_2 &= d + D_P \cos\alpha
\end{aligned}
\tag{P4.3}
$$

Numerically solve the positional constraints of Eq. (P4.3), using the Newton-Raphson method. Write a MATLAB® script following steps similar to those found in Simulation 4.3. Use the same default geometry parameters of that script as well as $d = 12$ in. and $\alpha = 30°$. Note, however, that the starting initial conditions here are different.

P4.7. Solve the problem of the kinematically driven slider-crank of Fig. P4.2a using the vector-matrix approach of Sect. 4.5 and following similar steps to those found Simulation 4.5. Use the same default geometry parameters of that script as well as $d = 12$ in. and $\alpha = 30°$. Note, however, that the starting initial conditions here are different.

P4.8. Figure P4.3 shows two bars that are pinned at point B and where bar (2) has a smooth slot which rides in a fixed pin at D. (a) If the crank (bar (1)) rotates with a constant clockwise angular velocity of 6 rad/s as shown, at the instant when the crank is vertical, determine the angular velocity and angular acceleration of the slotted bar, BD, using the relative velocity and acceleration relations. Also, determine the velocity and acceleration of the center of masses of the two uniform bars at this instant. Let $l_1 = 5$ in., $d = 12$ in., and $l_2 = 18$ in.

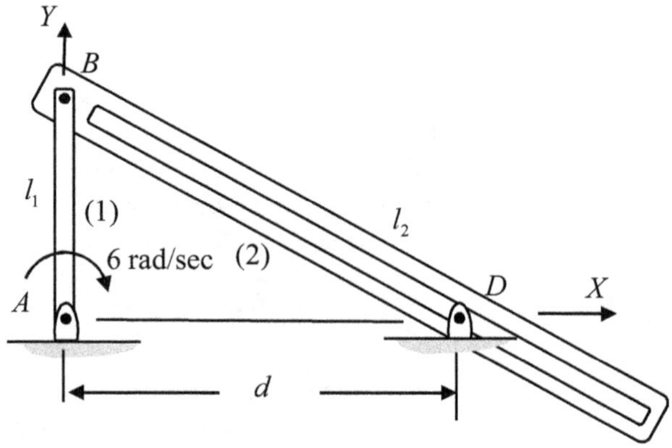

Fig. P4.3 A mechanism where a crank (body (1)) is pinned to a slotted bar (body (2)) which rides in a smooth fixed pin at D. The crank rotates at a constant angular velocity of 6 rad/s in the clockwise direction. At the instant shown the crank is vertical

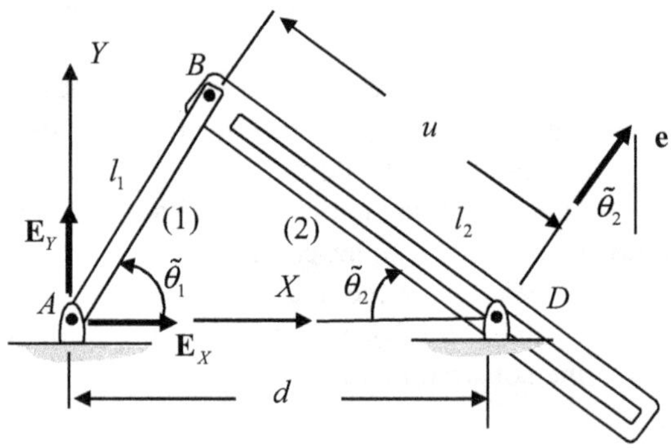

Fig. P4.4 A mechanism at an arbitrary time where a crank (body (1)) is pinned to a slotted bar (body (2)) which rides in a smooth fixed pin at D

P4.9. Figure P4.4 shows the crank-slotted bar mechanism in a general position as defined by the angles $\left(\tilde{\theta}_1, \tilde{\theta}_2\right)$. This is a one degree of freedom system so we need to find a single positional constraint between the two angles. This is most easily done by noting that the position vector from B to D is always perpendicular to the unit vector, $\mathbf{e}$, which is normal to the slotted bar, which can be written as the constraint $f\left(\tilde{\theta}_1, \tilde{\theta}_2\right) = \left(\mathbf{x}_{B/A} - d\,\mathbf{E}_X\right) \cdot \mathbf{e} = 0$.

(a) Show that this relationship leads to the constraint equation in the form

$$d \sin \tilde{\theta}_2 - l_1 \sin\left(\tilde{\theta}_1 + \tilde{\theta}_2\right) = 0 \tag{P4.4}$$

which is a simple but nonlinear relationship between the two angles which is difficult to solve analytically for $\tilde{\theta}_2$. However, from the geometry of Fig. P4.4 we also see that

$$u = \sqrt{l_1^2 + d^2 - 2l_1 d \cos\tilde{\theta}_1}$$
$$\sin\tilde{\theta}_2/l_1 = \sin\tilde{\theta}_1/u \tag{P4.5}$$

which can be solved analytically for $\tilde{\theta}_2$ as

$$\tilde{\theta}_2 = \sin^{-1}\left(\frac{l_1 \sin\tilde{\theta}_1}{\sqrt{l_1^2 + d^2 - 2l_1 d \cos\tilde{\theta}_1}}\right) \tag{P4.6}$$

(b) Differentiate Eq. (P4.4) once and twice to get relationships for the angular velocities and angular accelerations. Let $\tilde{\theta}_1 = -6t$, $l_1 = 5$ in., $d = 12$ in., and $l_2 = 18$ in. and obtain the angular velocity and angular acceleration of the slotted bar by solving for $\tilde{\theta}_2$ and then using that solution to find the angular velocities and accelerations from $t = 0$ to $t = 1$ s. Show that your results agree with Problem P4.8 at that specific instant of time.

P4.10. Repeat part (b) of Problem P4.9 except solve the nonlinear Eq. (P4.4) numerically using the Newton-Raphson method in conjunction with the velocity and acceleration expressions.

P4.11. The kinematics of the mechanism of Fig. P4.4 can also be described in a general matrix-vector form, as considered previously for the slider-crank problem by using the variables $(X^{(1)}, Y^{(1)}, \theta_1, X^{(2)}, Y^{(2)}, \theta_2)$, where $(X^{(n)}, Y^{(n)})$ are the coordinates of the center of mass of bar (n) and θ_n is the angle that bar (n) makes with respect to the positive X-axis. Note that $\theta_2 \neq \tilde{\theta}_2$ was also true for the slider-crank problem.

(a) Obtain the five positional constraints. Four of these come from the conditions at the pins A and B and at pin D we have the fifth constraint $g(\theta_1, \theta_2) = (\mathbf{X}^{(2)} - d\,\mathbf{E}_X) \cdot \mathbf{e} = 0$, which again shows that this is a one degree of freedom system.

(b) Differentiate the positional constraints once and twice to get the velocity and acceleration relations and solve them numerically in conjunction with the positional equations (using Newton-Raphson) for the same parameters given in Problem P4.10. Show that your results agree with those for the angular velocities and accelerations and velocities and accelerations of the center of mass obtained in Problem P4.8.

P4.12. Figure P4.5 shows a four-bar mechanism which is a configuration commonly used in practice. It is called a four-bar mechanism since the distance, l_1, can be considered to represent a fixed, fourth bar.

(a) If one uses the three angles $(\theta_2, \theta_3, \theta_4)$ to define the positions of the bars, show that they are related through the two positional constraints:

Fig. P4.5 A four-bar mechanism

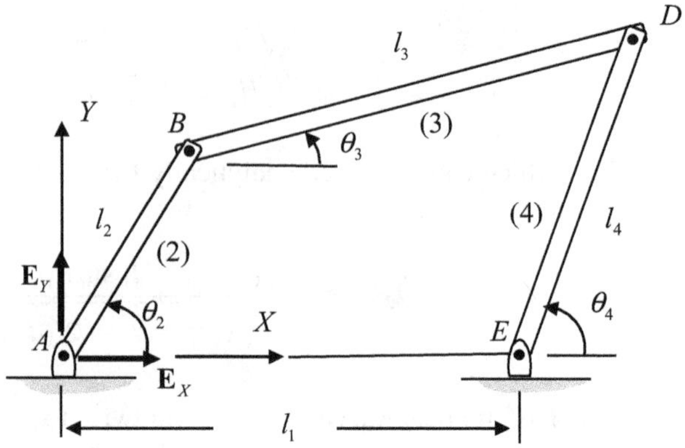

$$\begin{aligned} & l_2 \cos\theta_2 + l_3 \cos\theta_3 - l_4 \cos\theta_4 - l_1 = 0 \\ & l_2 \sin\theta_2 + l_3 \sin\theta_3 - l_4 \sin\theta_4 = 0 \end{aligned} \quad \text{(P4.7)}$$

(b) Differentiate these equations to obtain the velocity and acceleration constraints in matrix-vector form.

P4.13. By a particular choice of the lengths, a four-bar mechanism can simulate different types of motion. Figure P4.6, for example, shows two cases where we assume that θ_2 is the driving degree of freedom such that it describes a continuous full revolution motion. In case (a) where $s+l < p+q$, and where s denotes the length of the shortest bar and l denotes the length of the longest bar (and p and q are the other two bar lengths), then the angle θ_4 also describes a continuous, full revolution. In case (b), where again $s+l < p+q$ even though θ_2 describes a continuous full revolution type of motion the angle θ_4 exhibits a back-and-forth rocking motion between a smallest and a largest angle, as shown:

(a) Choose a set of values for the bar lengths so that the motion of the mechanism is of the double-crank type shown in Fig. P4.6a. Prove this by solving the positional equations numerically with the Newton-Raphson method for $0 \leq \theta_2 \leq 2\pi$ and plot the corresponding angles (θ_3, θ_4).
(b) Using the bar lengths chosen in part (a) and the velocity and acceleration expressions obtained in Problem P4.12, take $\theta_2(t) = 5\ t$ rad, where t is the time in seconds, and determine and plot the angular velocities and accelerations over a time interval sufficient for the driving angle, θ_2, to complete at least one revolution.

P4.14. Repeat Problem P4.13 for the case where the four-bar mechanism is of the crank-rocker type shown in Fig. P4.6b and the time interval is sufficiently

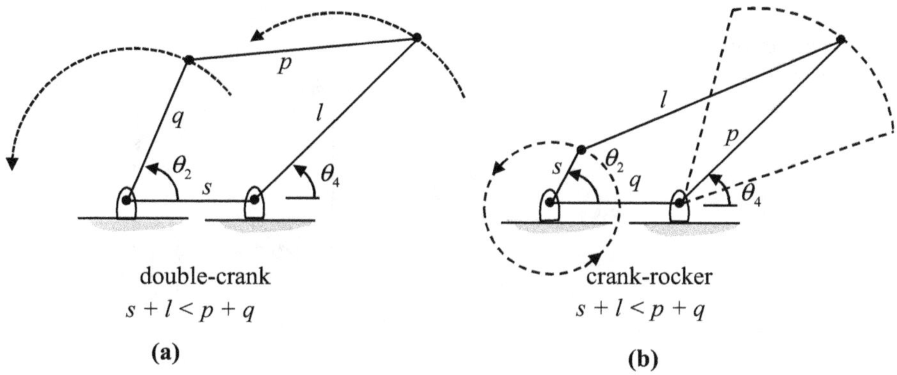

Fig. P4.6 Four-bar mechanisms where the driving angle, θ_2, executes a continuous full revolution type of motion and (**a**) where the angle θ_4 also exhibits a continuous, full revolution motion, and (**b**) where the angle θ_4 exhibits a rocking motion back and forth between two extreme values. The label s denotes the length of the shortest bar and the label l denotes the length of the largest bar. In both cases one must have $s+l < p+q$ satisfied, where p and q are the lengths of the bars shown

long so that the angle θ_4 reaches maximum and minimum values several times.

P4.15. (a) Define the eight positional constraints for the four-bar mechanism of Fig. P4.5 by using as unknowns $(X^{(2)}, Y^{(2)}, \theta_2, X^{(3)}, Y^{(3)}, \theta_3, X^{(4)}, Y^{(4)}, \theta_4)$, where $(X^{(n)}, Y^{(n)})$ are the X- and Y-coordinates of the center of mass of bar (n) and θ_n is the angle of bar (n) with respect to the positive X-axis. Differentiate these positional constraints to obtain the velocity and acceleration constraints.
(b) Letting $\theta_2(t) = 5\ t$ rad, where t is the time in seconds, determine and plot the center of mass coordinates and angles versus time, using the same bar lengths as chosen in Problem P4.13 so that the mechanism acts as the double-crank mechanism of Fig. P4.6a.

P4.16. Repeat Problem P4.15 using the same bar lengths as chosen in Problem P4.14 so that the mechanism acts as the crank-rocker mechanism of Fig. P4.6b.

P4.17. Write a MATLAB® script that uses the function animate_lines to animate the motion of the slider_crank problem of Simulation 4.1, evaluating the motion at 1000 times over the interval [0, 10] s.

P4.18. The 4-in.-long arm OA in the final position shown in Fig. P4.7 is initially along the X-axis and is free to rotate about point O. (a) Show that the final position of the arm can be reached through space-fixed rotations of 30° about the ***negative*** Y-axis and then 45° about the positive Z-axis or by body-fixed rotations of 45° about the Z-axis and then 30° about the ***negative*** y-axis (not shown) obtained after the Z-axis rotation. Find the individual and total rotation matrices explicitly. (b) Determine the (X, Y, Z) coordinates of point A on the arm after the rotations.

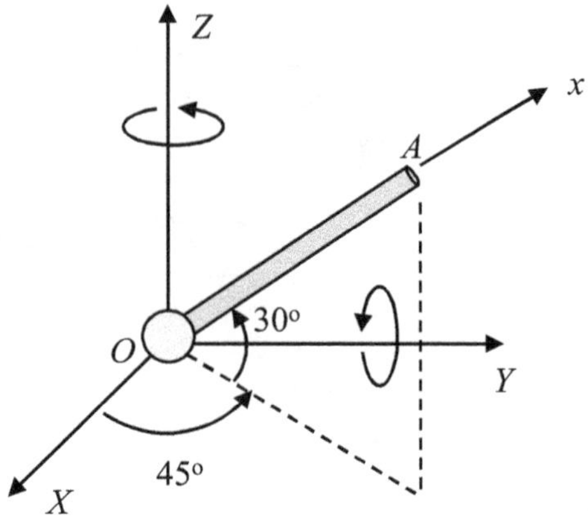

Fig. P4.7 Three-dimensional rotation of a 4-in.-long arm *OA*

P4.19. A series of connected arms form a robotic linkage as shown in Fig. P4.8a where the lengths are $OA = 1.5$ m, $AB = 1.0$ m, and $BC = 0.25$ m. Arm *OA* can rotate about the *Z*-axis, arm *AB* can rotate about its own axis, and arm *BC* is pinned to *AB* with the pin for rotation initially horizontal. Initially the linkage lies in in the *X–Z* plane with the variable angles $\alpha = \beta = \gamma = 0$. The $35°$ angle is fixed. The assembly is given a set of rotations $\alpha = 45°, \beta = 30°$, and $\gamma = 25°$. The objective is to determine the (X, Y, Z) coordinates of the endpoint *C* after these rotations, using the coordinate systems (frames) shown in Fig. P4.8b. The (x_1, y_1, z_1) frame is attached to *OA*, the (x_2, y_2, z_2) frame is attached to arm *AB*, and the (x_3, y_3, z_3) frame is attached to arm *BC*. In this problem, use the relative position relation

$$\mathbf{x}_{C/O} = \mathbf{x}_{A/O} + \mathbf{x}_{B/A} + \mathbf{x}_{C/B} \tag{P4.8}$$

and the appropriate rotation matrices to determine the coordinates $(X_{C/O}, Y_{C/O}, Z_{C/O})$ of $\mathbf{x}_{C/O}$ by writing Eq. (P4.8) in matrix form with

$$\mathbf{x}_{C/O} = \left[X_{C/O}, Y_{C/O}, Z_{C/O}\right]\begin{Bmatrix}\mathbf{E}_X\\ \mathbf{E}_Y\\ \mathbf{E}_Z\end{Bmatrix}, \quad \mathbf{x}_{A/O} = \left[x_{A/O}, y_{A/O}, z_{A/O}\right]\begin{Bmatrix}\mathbf{e}_{x_1}\\ \mathbf{e}_{y_1}\\ \mathbf{e}_{z_1}\end{Bmatrix}$$
$$\mathbf{x}_{B/A} = \left[x_{B/A}, y_{B/A}, z_{B/A}\right]\begin{Bmatrix}\mathbf{e}_{x_2}\\ \mathbf{e}_{y_2}\\ \mathbf{e}_{z_2}\end{Bmatrix}, \quad \mathbf{x}_{C/B} = \left[x_{C/B}, y_{C/B}, z_{C/B}\right]\begin{Bmatrix}\mathbf{e}_{x_3}\\ \mathbf{e}_{y_3}\\ \mathbf{e}_{z_3}\end{Bmatrix} \tag{P4.9}$$

where $(x_{A/O}, y_{A/O}, z_{A/O})$ are the coordinates of point *A* as measured with respect to point *O* in the (x_1, y_1, z_1) frame, with similar definitions for the *B/A* and *C/B* terms as measured in the other two frames. The vectors

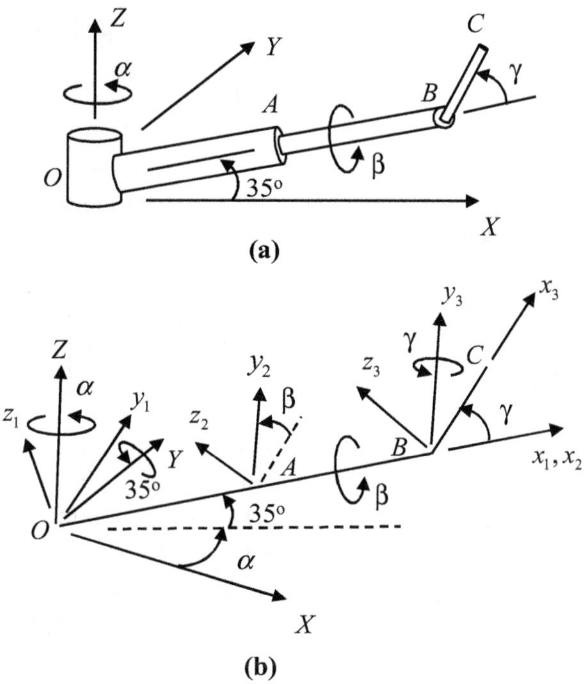

Fig. P4.8 (**a**) A robotic arm assembly, and (**b**) the coordinate systems and rotations used to define the system

$\left(\mathbf{e}_{x_j}, \mathbf{e}_{y_j}, \mathbf{e}_{z_j}\right)$ $(j = 1, 2, 3)$ are the coordinate unit vectors associated with the (x_j, y_j, z_j) axes and $(\mathbf{E}_X, \mathbf{E}_Y, \mathbf{E}_Z)$ are unit vectors along the axes of the fixed (X, Y, Z) frame. From our definition of rotation matrices and their relationship to these unit vectors we can write

$$\begin{Bmatrix} \mathbf{e}_{x_1} \\ \mathbf{e}_{y_1} \\ \mathbf{e}_{z_1} \end{Bmatrix} = [R_1]^T \begin{Bmatrix} \mathbf{E}_X \\ \mathbf{E}_Y \\ \mathbf{E}_Z \end{Bmatrix}, \quad \begin{Bmatrix} \mathbf{e}_{x_2} \\ \mathbf{e}_{y_2} \\ \mathbf{e}_{z_2} \end{Bmatrix} = [R_2]^T \begin{Bmatrix} \mathbf{e}_{x_1} \\ \mathbf{e}_{y_1} \\ \mathbf{e}_{z_1} \end{Bmatrix} = [R_2]^T [R_1]^T \begin{Bmatrix} \mathbf{E}_X \\ \mathbf{E}_Y \\ \mathbf{E}_Z \end{Bmatrix}$$

$$\begin{Bmatrix} \mathbf{e}_{x_3} \\ \mathbf{e}_{y_3} \\ \mathbf{e}_{z_3} \end{Bmatrix} = [R_3]^T \begin{Bmatrix} \mathbf{e}_{x_2} \\ \mathbf{e}_{y_2} \\ \mathbf{e}_{z_2} \end{Bmatrix} = [R_3]^T [R_2]^T [R_1]^T \begin{Bmatrix} \mathbf{E}_X \\ \mathbf{E}_Y \\ \mathbf{E}_Z \end{Bmatrix} \tag{P4.10}$$

(Why are the transposes present in Eq. (P4.10)?) Substituting these unit vector expressions into Eq. (P4.9) and cancelling the common (X, Y, Z) unit vectors in Eq. (P4.8) give

$$\left[X_{C/O}, Y_{C/O}, Z_{C/O}\right] = \left[x_{A/O}, y_{A/O}, z_{A/O}\right][R_1]^T + \left[x_{B/A}, y_{B/A}, z_{B/A}\right][R_2]^T[R_1]^T$$
$$+ \left[x_{C/B}, y_{C/B}, z_{C/B}\right][R_3]^T[R_2]^T[R_1]^T \tag{P4.11}$$

and taking the transpose of Eq. (P4.11) leads to the relationship in terms of matrices and column vectors as

$$\begin{Bmatrix} X_{C/O} \\ Y_{C/O} \\ Z_{C/O} \end{Bmatrix} = [R_1]\begin{Bmatrix} x_{A/O} \\ y_{A/O} \\ z_{A/O} \end{Bmatrix} + [R_1][R_2]\begin{Bmatrix} x_{B/A} \\ y_{B/A} \\ z_{B/A} \end{Bmatrix} + [R_1][R_2][R_3]\begin{Bmatrix} x_{C/B} \\ y_{C/B} \\ z_{C/B} \end{Bmatrix} \tag{P4.12}$$

Equation (P4.12) will solve our problem once we have obtained the necessary rotation matrices. Note that the only nonzero coordinates on the right side of Eq. (P4.12) are $x_{A/O} = 1.5$ m, $x_{B/A} = 1.0$ m, and $x_{C/B} = 0.25$ m from the data specified for the problem.

(a) The rotation matrix $[R_1]$ can be obtained from a space-fixed frame rotation about the ***negative*** Y-axis followed by a rotation about the space-fixed (and body-fixed) Z-axis, i.e., $[R_1] = [R_Z][R_Y]$ (see Fig. P4.8b) or, equivalently, by a rotation about the Z-axis followed by a rotation about the body-fixed ***negative*** y_1-axis, i.e., $[R_1] = [R_Z]\left[R_{y_1}\right]$. Determine the total matrix $[R_1]$ numerically.

(b) Similarly obtain the matrices $[R_2]$ and $[R_3]$ numerically, where $[R_2]$ is a rotation about the x_2 (or x_1) axis and $[R_3]$ is a rotation about the ***negative*** y_3 (or y_2) axis.

(c) Using the rotation matrices from (a) and (b) and Eq. (P4.12) obtain the $(X_{C/O}, Y_{C/O}, Z_{C/O})$ coordinates of point C in the fixed (X, Y, Z) frame.

Note that since we can write all the rotations in terms of rotations about body-fixed axes and we know the sequence in which such rotation matrices should be multiplied from the results in Sect. 4.12, we could have written down the terms appearing in Eq. (P4.12) directly, thus bypassing the rather lengthy derivation of that equation.

References

1. A.A. Shabana, *Computational Dynamics*, 3rd edn. (John Wiley, Chichester, 2010)
2. A.A. Shabana, *Dynamics of Multibody Systems*, 4th edn. (Cambridge University Press, New York, 2013)
3. H. Baruh, *Analytical Dynamics* (McGraw-Hill, New York, 1999)
4. J.H. Ginsberg, *Advanced Engineering Dynamics*, 2nd edn. (Cambridge University Press, New York, 1995)

Chapter 5
Planar Dynamics of Rigid Bodies

This chapter examines the dynamics of rigid bodies in planar motion and the forces/moments that act upon them. It will be shown how the equations of motion can be generated using a classical Newton-Euler approach as well as with Lagrange's equations. The augmented approach considered for single particles and systems of particles is also discussed. The matrix-vector kinematics developed in Chap. 4 is used throughout this chapter and the emphasis is on obtaining complete numerical solutions.

5.1 Governing Equations for a Rigid Body in Plane Motion

A rigid body whose mass is continuously distributed throughout its volume can be considered to be the limit of an infinitely large collection of rigidly constrained particles. In Chap. 3 we saw that the equations governing the system of particles consisted of the force and moment equations (see Eqs. (3.2.5) and (3.2.14)):

$$\sum_{i=1}^{N} m_i \ddot{\mathbf{x}}_i = \mathbf{F}^{\text{sys}} \tag{5.1.1}$$

$$\sum_{i=1}^{N} \mathbf{x}_{Bi} \times m_i \ddot{\mathbf{x}}_i = \mathbf{M}_B^{\text{sys}} \tag{5.1.2}$$

where $\mathbf{F}^{\text{sys}}$ is the net force acting on the system from the known external forces and external forces of constraint and $\mathbf{M}_B^{\text{sys}}$ is the corresponding net external moment with respect to an arbitrary point B (see Fig. 5.1a). As shown in Chap. 3 the internal constraint forces and their moments do not appear in Eqs. (5.1.1) and (5.1.2) because

L. W. Schmerr, *Engineering Dynamics 2.0*, Solid Mechanics and Its Applications 254,
https://doi.org/10.1007/978-3-319-98470-4_5

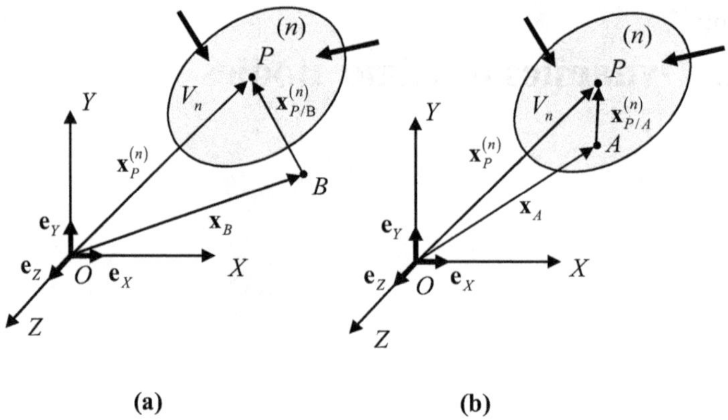

Fig. 5.1 (**a**) Geometry for describing the motion of a continuous rigid body. Point P is an arbitrary point in the body and reference point B can be an arbitrary (fixed or moving) point. (**b**) The same geometry but where reference point A is a point moving with the rigid body

those rigid constraints satisfy Newton's third law. If we let $m_i = \rho\,\Delta V_i$ where ρ is the mass density (mass/volume) of the particle and ΔV_i is a small volume, then in the limit as the number of particles becomes infinitely large we can replace the sums in Eqs. (5.1.1) and (5.1.2) by integrals over the entire volume, V_n, of the continuous rigid body, which we will label as (n), yielding

$$\begin{aligned}
&\int\limits_{V_n} \rho \ddot{\mathbf{x}}_P^{(n)} dV = \mathbf{F}^{(n)} \\
&\int\limits_{V_n} \rho \left[\mathbf{x}_{P/B}^{(n)} \times \left(\ddot{\mathbf{x}}_P^{(n)}\right)\right] dV = \mathbf{M}_B^{(n)}
\end{aligned} \tag{5.1.3}$$

where $\ddot{\mathbf{x}}_P^{(n)}$ is the acceleration (in an inertial system) of an arbitrary point P in the rigid body (n), $\mathbf{x}_{P/B}^{(n)}$ is the position of P relative to a reference point B, and $\mathbf{F}^{(n)}$ and $\mathbf{M}_B^{(n)}$ are the net force and moment on the rigid body. In Chap. 3 we showed that a rigidly constrained system of particles moving in 3-D had six degrees of freedom. The same is true for a continuous rigid body so the six Newton-Euler equations in Eq. (5.1.3) are sufficient to determine the motion of a rigid body. The governing equations in Eq. (5.1.3) are valid for any rigid body and the reference point B can be taken as an arbitrary point for each body (n). In this chapter, however, we want to consider only planar motion problems. As done previously, when discussing planar motion kinematics, we will take the translational motion to be in the X–Y plane and the rotational motion about the Z-axis (Fig. 5.2). *We will also take the reference point to be a point A that is fixed in the same rigid body as P* (Fig. 5.1b). [We will come back to the more general case of Fig. 5.1a where point B is arbitrary in a later section.] In the

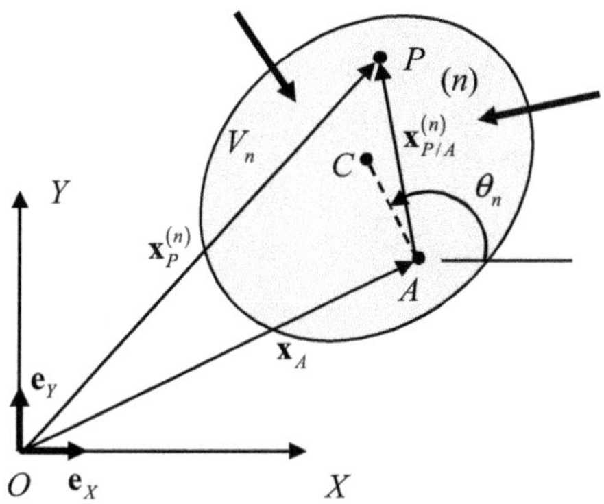

Fig. 5.2 A rigid body in plane motion where all the motion is in the X–Y plane and the rotation is about the Z-axis (in the $\mathbf{e}_Z$ direction). The rotation angle θ_n is shown as measured to the line from A to the center of mass C but it could be measured from the positive X-axis to any fixed line in the body

case shown in Fig. 5.1b we can write the position vector for P, $\mathbf{x}_P^{(n)}$, in matrix-vector form as (see Eq. (4.4.10))

$$\begin{aligned}\mathbf{x}_P^{(n)} &= \mathbf{x}_A^{(n)} + \mathbf{x}_{P/A}^{(n)} \\ &= \mathbf{x}_A^{(n)} + \mathbf{R}^{(n)}(\theta_n)\bar{\mathbf{x}}_{P/A}^{(n)}\end{aligned} \tag{5.1.4}$$

Note that if we differentiate on time this relationship $\bar{\mathbf{x}}_{P/A}^{(n)}$ is a constant since both A and P move with the rigid body and $\bar{\mathbf{x}}_{P/A}^{(n)}$ is the relative position of point P with respect to point A as measured in a set of body-fixed coordinates. Performing that differentiation twice we can relate the accelerations of P and A as (see Eq. (4.4.29))

$$\ddot{\mathbf{x}}_P^{(n)} = \ddot{\mathbf{x}}_A^{(n)} + \ddot{\theta}_n \mathbf{R}_\theta^{(n)} \bar{\mathbf{x}}_{P/A}^{(n)} - \left(\dot{\theta}_n\right)^2 \mathbf{R}^{(n)} \bar{\mathbf{x}}_{P/A}^{(n)} \tag{5.1.5}$$

so that Eq. (5.1.3) becomes

$$\begin{aligned}&\ddot{\mathbf{x}}_A^{(n)} \int_{V_n} \rho dV + \ddot{\theta}_n \mathbf{R}_\theta^{(n)} \int_{V_n} \rho \bar{\mathbf{x}}_{P/A}^{(n)} dV - \left(\dot{\theta}_n\right)^2 \mathbf{R}^{(n)} \int_{V_n} \rho \bar{\mathbf{x}}_{P/A}^{(n)} dV = \mathbf{F}^{(n)} \\ &\left(\int_{V_n} \rho \mathbf{x}_{P/A}^{(n)} dV\right) \times \ddot{\mathbf{x}}_A^{(n)} + \ddot{\theta}_n \int_{V_n} \rho \mathbf{x}_{P/A}^{(n)} \times \mathbf{R}_\theta^{(n)} \bar{\mathbf{x}}_{P/A}^{(n)} dV = \mathbf{M}_A^{(n)}\end{aligned} \tag{5.1.6}$$

where we have used the fact that

$$-\left(\dot{\theta}_n\right)^2 \int_{V_n} \rho \mathbf{x}_{P/A}^{(n)} \times R^{(n)} \bar{\mathbf{x}}_{P/A}^{(n)} dV = -\left(\dot{\theta}_n\right)^2 \int_{V_n} \rho \left(\mathbf{x}_{P/A}^{(n)} \times \mathbf{x}_{P/A}^{(n)}\right) dV = 0 \tag{5.1.7}$$

Because of the cross product terms Eq. (5.1.6) is still in a "mixed" form with both matrix and vector operations. But we can put it in a purely matrix-vector form by noting that in 2-D problems the cross product of any two 2-D vectors **a** and **b** can be written as

$$\begin{aligned}\mathbf{a}\times\mathbf{b} &= \left(a_x\mathbf{e}_x + a_y\mathbf{e}_y\right)\times\left(b_x\mathbf{e}_x + b_y\mathbf{e}_y\right)\\ &= \left(a_x b_y - a_y b_x\right)\mathbf{e}_z\\ &= \left(\widehat{\mathbf{a}}^T\mathbf{b}\right)\mathbf{e}_z = -\left(\widehat{\mathbf{b}}^T\right)\mathbf{e}_z\end{aligned} \tag{5.1.8}$$

where, for any 2-D vector $\mathbf{v} = [v_x, v_y]^T$, we have $\widehat{\mathbf{v}} = \left[-v_y, v_x\right]^T$. Using these results and noting that $\mathbf{M}_A^{(n)} = M_{Az}^{(n)}\mathbf{e}_Z$ in the planar case, we can write Eq. (5.1.6) as

$$\begin{aligned}&\ddot{\mathbf{x}}_A^{(n)}\int_{V_n}\rho dV + \ddot{\theta}_n\int_{V_n}\rho\mathbf{R}_\theta^{(n)}\,\bar{\mathbf{x}}_{P/A}^{(n)}dV - \left(\dot{\theta}_n\right)^2\mathbf{R}^{(n)}\int_{V_n}\rho\bar{\mathbf{x}}_{P/A}^{(n)}dV = \mathbf{F}^{(n)}\\ &\left(\int_{V_n}\rho\left(\widehat{\mathbf{x}}_{P/A}^{(n)}\right)^T dV\right)\ddot{\mathbf{x}}_A^{(n)} + \ddot{\theta}_n\int_{V_n}\rho\left(\widehat{\mathbf{x}}_{P/A}^{(n)}\right)^T\mathbf{R}_\theta^{(n)}\bar{\mathbf{x}}_{P/A}^{(n)}dV = M_{Az}^{(n)}\end{aligned} \tag{5.1.9}$$

We can write Eq. (5.1.9) in a more symmetrical form by examining the coefficient of $\ddot{\theta}_n$ in the force equation:

$$\begin{aligned}\int_{V_n}\rho\mathbf{R}_\theta^{(n)}\,\bar{\mathbf{x}}_{P/A}^{(n)}dV &= \int_{V_n}\rho\mathbf{R}_\theta^{(n)}\left(\mathbf{R}^{(n)}\right)^T\mathbf{x}_{P/A}^{(n)}dV\\ &= \int_{V_n}\rho\begin{bmatrix}-\sin\theta & -\cos\theta\\ \cos\theta & -\sin\theta\end{bmatrix}\begin{bmatrix}\cos\theta & \sin\theta\\ -\sin\theta & \cos\theta\end{bmatrix}\mathbf{x}_{P/A}^{(n)}dV\\ &= \int_{V_n}\rho\begin{bmatrix}0 & -1\\ 1 & 0\end{bmatrix}\mathbf{x}_{P/A}^{(n)}dV = \int_{V_n}\rho\,\widehat{\mathbf{x}}_{P/A}^{(n)}dV\end{aligned} \tag{5.1.10}$$

so that Eq. (5.1.9) becomes

$$\begin{aligned}&\ddot{\mathbf{x}}_A^{(n)}\int_{V_n}\rho dV + \ddot{\theta}_n\int_{V_n}\rho\,\widehat{\mathbf{x}}_{P/A}^{(n)}dV - \left(\dot{\theta}_n\right)^2\int_{V_n}\rho\mathbf{x}_{P/A}^{(n)}dV = \mathbf{F}^{(n)}\\ &\left(\int_{V_n}\rho\left(\widehat{\mathbf{x}}_{P/A}^{(n)}\right)^T dV\right)\ddot{\mathbf{x}}_A^{(n)} + \ddot{\theta}_n\int_{V_n}\rho\left(\widehat{\mathbf{x}}_{P/A}^{(n)}\right)^T\mathbf{R}_\theta^{(n)}\bar{\mathbf{x}}_{P/A}^{(n)}dV = M_{Az}^{(n)}\end{aligned} \tag{5.1.11}$$

where we have also used $\mathbf{R}^{(n)}\bar{\mathbf{x}}_{P/A}^{(n)} = \mathbf{x}_{P/A}^{(n)}$. If we choose the generalized coordinates $\mathbf{q}^{(n)} = \left(\mathbf{x}_A^{(n)}, \theta_n\right)$ to describe the motion of rigid body (n) then we see that the Newton-Euler equations can be written in matrix-vector form as

$$\mathbf{M}^{(n)}\ddot{\mathbf{q}}^{(n)} = \mathbf{Q}^{in;(n)} + \mathbf{Q}^{(n)} \tag{5.1.12}$$

where there is a 3 × 3 "mass" matrix, $\mathbf{M}^{(n)}$, given by

$$\mathbf{M}^{(n)} = \begin{bmatrix} \mathbf{m}_{XX}^{(n)} & \mathbf{m}_{X\theta}^{(n)} \\ \mathbf{m}_{\theta X}^{(n)} & m_{\theta\theta}^{(n)} \end{bmatrix} \tag{5.1.13}$$

whose components are

$$\begin{aligned} \mathbf{m}_{XX}^{(n)} &= \left(\int_{V_n} \rho dV\right)\mathbf{I} = m^{(n)}\mathbf{I} \\ \mathbf{m}_{X\theta}^{(n)} &= \left(\mathbf{m}_{\theta X}^{(n)}\right)^T = \int_{V_n} \rho\, \widehat{\mathbf{x}}_{P/A}^{(n)} dV \\ m_{\theta\theta}^{(n)} &= \int_{V_n} \rho \left(\widehat{\mathbf{x}}_{P/A}^{(n)}\right)^T \mathbf{R}_\theta^{(n)} \bar{\mathbf{x}}_{P/A}^{(n)} dV \end{aligned} \tag{5.1.14}$$

where $m^{(n)}$ is the mass of the body, $\mathbf{I}$ is the 2 × 2 identity matrix, $\mathbf{m}_{X\theta}^{(n)}$ is a 2 × 1 vector, $\mathbf{m}_{\theta X}^{(n)}$ is a 1 × 2 vector, and $m_{\theta\theta}^{(n)}$ is a scalar. There is also a term in Eq. (5.1.11) that does not involve the accelerations. This has been placed in a generalized 3 × 1 inertial force vector, $\mathbf{Q}^{in;(n)}$, while the other 3 × 1 generalized force vector, $\mathbf{Q}^{(n)}$, contains the actual forces and moments, i.e.,

$$\mathbf{Q}^{in;(n)} = \left\{ \begin{array}{c} \left(\dot{\theta}_n\right)^2 \int_{V_n} \rho \mathbf{x}_{P/A}^{(n)} dV \\ 0 \end{array} \right\}, \quad \mathbf{Q}^{(n)} = \left\{ \begin{array}{c} \mathbf{F}^{(n)} \\ M_{AZ}^{(n)} \end{array} \right\} \tag{5.1.15}$$

Choosing the point A to be an arbitrary point in the rigid body causes the Newton-Euler equations to be rather complex. However, choosing A to be C, the center of mass point for the rigid body results in a significant simplification since

$$\int_{V_n} \rho \mathbf{x}_{P/C}^{(n)} dV = \int_{V_n} \rho\, \widehat{\mathbf{x}}_{P/C}^{(n)} dV = \int_{V_n} \rho \left(\widehat{\mathbf{x}}_{P/C}^{(n)}\right)^T dV = 0 \tag{5.1.16}$$

which, when placed into Eq. (5.1.6), gives $\mathbf{Q}^{in;\,(n)} = 0$ and

$$\begin{aligned} m^{(n)}\ddot{\mathbf{x}}_C^{(n)} &= \mathbf{F}^{(n)} \\ \ddot{\theta}_n \int_{V_n} \rho \left(\widehat{\mathbf{x}}_{P/C}^{(n)}\right)^T \mathbf{R}_\theta^{(n)} \bar{\mathbf{x}}_{P/C}^{(n)} dV &= M_{CZ}^{(n)} \end{aligned} \tag{5.1.17}$$

so that the mass matrix becomes

$$\mathbf{M}^{(n)} = \begin{bmatrix} \mathbf{m}_{XX}^{(n)} & \mathbf{0} \\ \mathbf{0} & m_{\theta\theta}^{(n)} \end{bmatrix} = \begin{bmatrix} m^{(n)} & 0 & 0 \\ 0 & m^{(n)} & 0 \\ 0 & 0 & m_{\theta\theta}^{(n)} \end{bmatrix} \tag{5.1.18}$$

Thus, the mass matrix is a diagonal matrix and all coupling terms between the accelerations of point A and the rotational motion are eliminated. We see that the generalized inertial force also vanishes. Now consider $m_{\theta\theta}^{(n)}$. We find

$$\begin{aligned} m_{\theta\theta}^{(n)} &= \int_{V_n} \rho \left(\widehat{\mathbf{x}}_{P/C}^{(n)}\right)^T \mathbf{R}_\theta^{(n)} \bar{\mathbf{x}}_{P/C}^{(n)} dV = \int_{V_n} \rho \left(\widehat{\mathbf{x}}_{P/C}^{(n)}\right)^T \mathbf{R}_\theta^{(n)} \left(\mathbf{R}^{(n)}\right)^T \mathbf{x}_{P/C}^{(n)} dV \\ &= \int_{V_n} \rho \left(\widehat{\mathbf{x}}_{P/C}^{(n)}\right)^T \begin{bmatrix} 0 & -1 \\ 1 & 0 \end{bmatrix} \mathbf{x}_{P/C}^{(n)} dV = \int_{V_n} \rho \left(\widehat{\mathbf{x}}_{P/C}^{(n)}\right)^T \widehat{\mathbf{x}}_{P/C}^{(n)} dV \\ &= \int_{V_n} \rho \left[\left(X_{P/C}^{(n)}\right)^2 + \left(Y_{P/C}^{(n)}\right)^2\right] dV = \int_{V_n} \rho \left[\left(\bar{x}_{P/C}^{(n)}\right)^2 + \left(\bar{y}_{P/C}^{(n)}\right)^2\right] dV \\ &= I_C^{(n)} \end{aligned} \tag{5.1.19}$$

since

$$\begin{aligned} \left(\widehat{\mathbf{x}}_{P/C}^{(n)}\right)^T \widehat{\mathbf{x}}_{P/C}^{(n)} &= \begin{bmatrix} -Y_{P/C}^{(n)} & X_{P/C}^{(n)} \end{bmatrix} \begin{bmatrix} -Y_{P/C}^{(n)} \\ X_{P/C}^{(n)} \end{bmatrix} \\ &= \left[\left(X_{P/C}^{(n)}\right)^2 + \left(Y_{P/C}^{(n)}\right)^2\right] \end{aligned}$$

is just the distance squared between P and C in terms of inertial coordinates $\mathbf{x}_{p/C}^{(n)} = \left(X_{P/C}^{(n)}, Y_{P/C}^{(n)}\right)$ which also can be written as

$$\begin{aligned}\left\{\mathbf{x}_{P/C}^{(n)}\right\}^T\left\{\mathbf{x}_{P/C}^{(n)}\right\} &= \left\{\mathbf{R}^{(n)}\bar{\mathbf{x}}_{P/C}^{(n)}\right\}^T\left\{\mathbf{R}^{(n)}\bar{\mathbf{x}}_{P/C}^{(n)}\right\} \\ &= \left\{\bar{\mathbf{x}}_{P/C}^{(n)}\right\}^T\left(\mathbf{R}^{(n)}\right)^T\mathbf{R}^{(n)}\left\{\bar{\mathbf{x}}_{P/C}^{(n)}\right\} = \left\{\bar{\mathbf{x}}_{P/C}^{(n)}\right\}^T\left\{\bar{\mathbf{x}}_{P/C}^{(n)}\right\} \\ &= \left[\left(\bar{x}_{P/C}^{(n)}\right)^2 + \left(\bar{y}_{P/C}^{(n)}\right)^2\right]\end{aligned}$$

in terms of the body-fixed coordinates $\bar{\mathbf{x}}_{P/C}^{(n)} = \left(\bar{x}_{P/C}^{(n)}, \bar{y}_{P/C}^{(n)}\right)$. In fact, the distance squared between P and C is the same, regardless of the planar (x, y) coordinates used. The moment equation is therefore $I_C^{(n)}\ddot{\theta}_n = M_{CZ}^{(n)}$ and

$$I_C^{(n)} = \int_{V_n} \rho\left[\left(X_{P/C}^{(n)}\right)^2 + \left(Y_{P/C}^{(n)}\right)^2\right]dV = \int_{V_n} \rho\left[\left(\bar{x}_{P/C}^{(n)}\right)^2 + \left(\bar{y}_{P/C}^{(n)}\right)^2\right]dV \tag{5.1.20}$$

is a mass moment of inertia about the center of mass.

The governing equations for the motion of the rigid body with respect to the center of mass, therefore, have the final simple form

$$\begin{aligned} m^{(n)}\ddot{\mathbf{x}}_C^{(n)} &= \mathbf{F}^{(n)} \\ I_C^{(n)}\ddot{\theta}_n &= M_{CZ}^{(n)} \end{aligned} \tag{5.1.21}$$

In terms of inertial components we have

$$\begin{aligned} m^{(n)}\ddot{X}_C^{(n)} &= F_X^{(n)} \\ m^{(n)}\ddot{Y}_C^{(n)} &= F_Y^{(n)} \\ I_C^{(n)}\ddot{\theta}_n &= M_{CZ}^{(n)} \end{aligned} \tag{5.1.22}$$

These three equations are sufficient to determine the motion of a rigid body, since like the plane motion of a rigidly constrained system of particles discussed in Chap. 3, an unconstrained continuous rigid body in plane motion only has three degrees of freedom.

In terms of inertial components the Newton-Euler equations can also be written as

$$\begin{bmatrix} m^{(n)} & 0 & 0 \\ 0 & m^{(n)} & 0 \\ 0 & 0 & I_C^{(n)} \end{bmatrix}\begin{Bmatrix} \ddot{X}_C^{(n)} \\ \ddot{Y}_C^{(n)} \\ \ddot{\theta}_n \end{Bmatrix} = \begin{Bmatrix} F_X^{(n)} \\ F_Y^{(n)} \\ M_{CZ}^{(n)} \end{Bmatrix} \tag{5.1.23a}$$

or, equivalently, in matrix-vector form

$$\mathbf{M}^{(n)}\ddot{\mathbf{q}} = \mathbf{Q}^{(n)} \tag{5.1.23b}$$

where the inertial force is absent and $\mathbf{q} = \left[X_C^{(n)}, Y_C^{(n)}, \theta_n\right]^T$ and $\mathbf{Q}^{(n)} = \left[F_X^{(n)}, F_Y^{(n)}, M_{CZ}^{(n)}\right]^T$.

We derived the equations of planar motion for a rigid body, Eq. (5.1.22), using a matrix-vector approach. In most dynamics texts you will likely find instead the derivation using only vectors. You may be more familiar with the vector version so for completeness we will give it here so you can compare it to the matrix-vector approach. As shown in Chap. 4 for planar motion the acceleration of a point P in a rigid body relative to a point A in the same rigid body can be written in several equivalent vector forms:

$$\begin{aligned}\ddot{\mathbf{x}}_P^{(n)} &= \ddot{\mathbf{x}}_A^{(n)} + r_{P/A}^{(n)}\ddot{\theta}_n\mathbf{e}_t - r_{P/A}^{(n)}\dot{\theta}_n^2\mathbf{e}_n \\ &= \ddot{\mathbf{x}}_A^{(n)} + \boldsymbol{\alpha}^{(n)} \times \mathbf{x}_{P/A}^{(n)} - r_{P/A}^{(n)}\dot{\theta}_n^2\mathbf{e}_n \\ &= \ddot{\mathbf{x}}_A^{(n)} + \boldsymbol{\alpha}^{(n)} \times \mathbf{x}_{P/A}^{(n)} + \boldsymbol{\omega}^{(n)} \times \left(\boldsymbol{\omega}^{(n)} \times \mathbf{x}_{P/A}^{(n)}\right)\end{aligned} \tag{5.1.24}$$

where $\boldsymbol{\omega}^{(n)} = \dot{\theta}_n\mathbf{e}_Z$ and $\boldsymbol{\alpha}^{(n)} = \ddot{\theta}_n\mathbf{e}_Z$ and the distance $r_{P/A}^{(n)}$ and normal and tangential unit vectors $(\mathbf{e}_n, \mathbf{e}_t)$ are shown in Fig. 5.3a. If we use elements of these forms in Eq. (5.1.3) we find

$$\begin{aligned}&\ddot{\mathbf{x}}_A^{(n)}\int_{V_n}\rho dV + \boldsymbol{\alpha}^{(n)} \times \int_{V_n}\rho\mathbf{x}_{P/A}^{(n)}dV - \dot{\theta}_n^2\int_{V_n}\rho\mathbf{x}_{P/A}^{(n)}dV = \mathbf{F}^{(n)} \\ &\int_{V_n}\rho\mathbf{x}_{P/A}^{(n)}dV \times \left(\ddot{\mathbf{x}}_A^{(n)}\right) + \ddot{\theta}_n\int_{V_n}\rho\mathbf{x}_{P/A}^{(n)} \times r_{P/A}^{(n)}\mathbf{e}_t dV - \dot{\theta}_n^2\int_{V_n}\rho\mathbf{x}_{P/A}^{(n)} \times r_{P/A}^{(n)}\mathbf{e}_n dV = \mathbf{M}_A^{(n)}\end{aligned} \tag{5.1.25}$$

But we have $\mathbf{x}_{P/A}^{(n)} = -r_{P/A}^{(n)}\mathbf{e}_n$ and $\mathbf{e}_n \times \mathbf{e}_t = -\mathbf{e}_Z$ so that these equations become

$$\begin{aligned}&\ddot{\mathbf{x}}_A^{(n)}\int_{V_n}\rho dV + \boldsymbol{\alpha}^{(n)} \times \int_{V_n}\rho\mathbf{x}_{P/A}^{(n)}dV - \dot{\theta}_n^2\int_{V_n}\rho\mathbf{x}_{P/A}^{(n)}dV = \mathbf{F}^{(n)} \\ &\int_{V_n}\rho\mathbf{x}_{P/A}^{(n)}dV \times \left(\ddot{\mathbf{x}}_A^{(n)}\right) + \ddot{\theta}_n\mathbf{e}_Z\int_{V_n}\rho\left(r_{P/A}^{(n)}\right)^2 dV = \mathbf{M}_A^{(n)}\end{aligned} \tag{5.1.26}$$

If we make the choice of point A to be the center of mass, C, we find

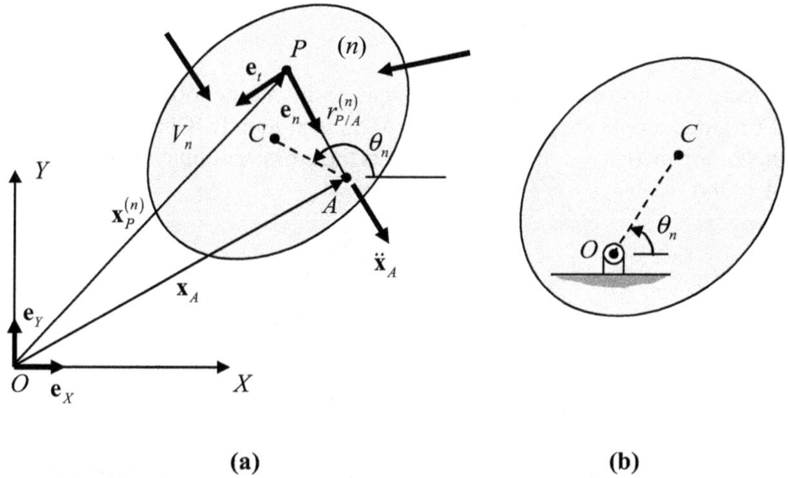

Fig. 5.3 (**a**) The same geometry of Fig. 5.2 showing the acceleration of the reference point A and the normal and tangential unit vectors used to express the relative acceleration of point P with respect to A. (**b**) The special case where the rigid body is rotating about a fixed point, O. In both cases the angle θ_n is shown as measured from the positive X-axis to the line drawn from the reference point to the center of mass point C but it could also be measured from the positive X-axis to any fixed line in the body

$$\begin{aligned} m\ddot{\mathbf{x}}_C^{(n)} &= \mathbf{F}^{(n)} \\ \left(\int_{V_n} \rho\left(r_{P/C}^{(n)}\right)^2 dV\right)\ddot{\theta}_n\mathbf{e}_Z &= \mathbf{M}_C^{(n)} \end{aligned} \tag{5.1.27}$$

which again is Eq. (5.1.21) since we recognize the integral in Eq. (5.1.27) as just $I_C^{(n)}$. Although the moment equation has a simple form when the center of mass point is used as the reference point to take moments, the same simplification occurs if the rigid body is rotating about a fixed point O in the body as shown in Fig. 5.3b and we use that fixed point as our reference point. In that case, since $\ddot{\mathbf{x}}_A = \ddot{\mathbf{x}}_O = 0$ and $r_{P/A}^{(n)} = r_{P/O}^{(n)}$ in the moment equation of Eq. (5.1.26) we have

$$\begin{aligned} &\ddot{\theta}_n\mathbf{e}_Z \int_{V_n} \rho\left(r_{P/O}^{(n)}\right)^2 dV = \mathbf{M}_O^{(n)} \\ &or \\ &I_O^{(n)}\ddot{\theta}_n\mathbf{e}_Z = \mathbf{M}_O^{(n)} \end{aligned} \tag{5.1.28}$$

In the matrix-vector approach Eq. (5.1.28) also follows directly from the moment equation in Eq. (5.1.11).

5.2 Moment of Inertia

The moment equation requires that we calculate the moment of inertia for the rigid body. For some simple shapes this is easy to do analytically. Consider, for example, the thin, homogeneous rod of Fig. 5.4a. If A is the cross-sectional area of the rod then $dV = Adx$ and

$$\begin{aligned} I_C &= \int_{-L/2}^{L/2} \rho x^2 A dx = \rho A \frac{x^3}{3}\bigg|_{-L/2}^{L/2} \\ &= \frac{1}{12}\rho A L^3 = \frac{1}{12} m L^2 \end{aligned} \tag{5.2.1}$$

where $m = \rho AL$ is the total mass of the rod. Similarly, for the homogeneous disk of Fig. 5.4b we can take $dV = 2\pi R\, t\, dR$, where t is the thickness, and

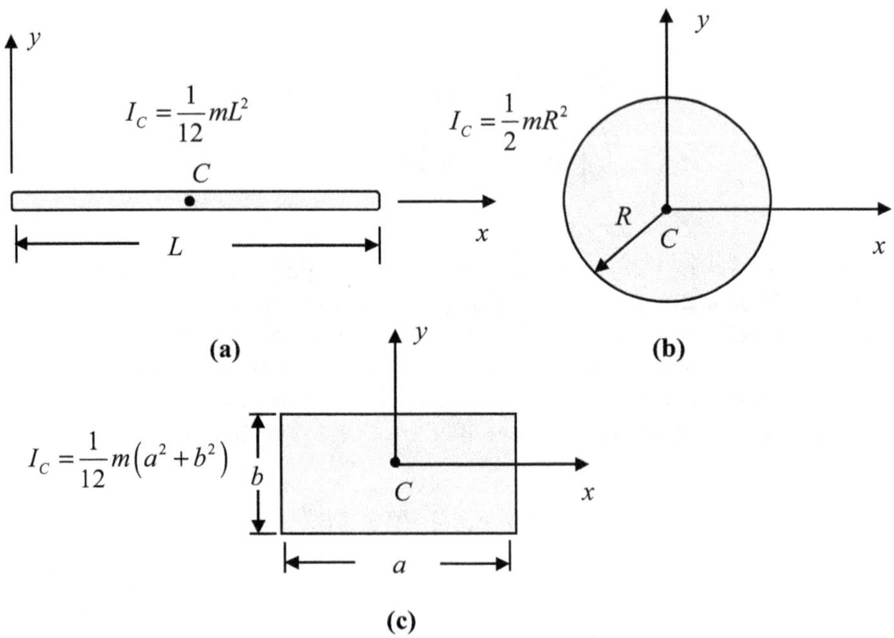

Fig. 5.4 Moments of inertia about the center of mass for (**a**) a homogeneous rod, (**b**) a homogeneous disk, and (**c**) a homogeneous rectangle

$$
\begin{aligned}
I_C &= \int_0^R \rho R^2 (2\pi t R) dR = 2\pi\rho t \frac{R^4}{4}\bigg|_0^R \\
&= \rho\left(\pi R^2 t\right)\frac{R^2}{2} = \frac{mR^2}{2}
\end{aligned}
\tag{5.2.2}
$$

where the mass of the disk $m = \rho\pi R^2 t$. Finally, for the rectangle of Fig. 5.4c $dV = t\,dx\,dy$, where t is the thickness, and

$$
\begin{aligned}
I_C &= \int_{-b/2}^{b/2}\int_{-a/2}^{a/2} \rho\left(x^2 + y^2\right) t dx dy \\
&= \rho t \int_{-b/2}^{b/2} \left.\frac{x^3}{3} + y^2 x\right|_{-a/2}^{a/2} dy = \frac{\rho t a^3 b}{12} + \rho t a \frac{y^3}{3}\bigg|_{-b/2}^{b/2} \\
&= \frac{\rho t a^3 b}{12} + \frac{\rho t a b^3}{12} = \frac{1}{12} m\left(a^2 + b^2\right)
\end{aligned}
\tag{5.2.3}
$$

where $m = \rho tab$ is the total mass.

These results can also be used for composite bodies that are combinations of these (or other) simple shapes. For example, consider the two bodies shown in Fig. 5.5. To find the moment of inertia about the center of mass for the composite body we need to know the location of the center of mass, $\mathbf{x}_C = (x_C, y_C)$. By definition, this location is computed as a weighted average, which in integral form for a continuously distributed mass is

$$
m\mathbf{x}_C = \int_V \rho \mathbf{x} dV = \int_{V_1} \rho \mathbf{x} dV + \int_{V_2} \rho \mathbf{x} dV
\tag{5.2.4}
$$

where $\mathbf{x}$ is the location of an arbitrary point in the composite body and $m = m_1 + m_2$ is the total mass. However, by breaking the integration over the composite body into

Fig. 5.5 A composite body

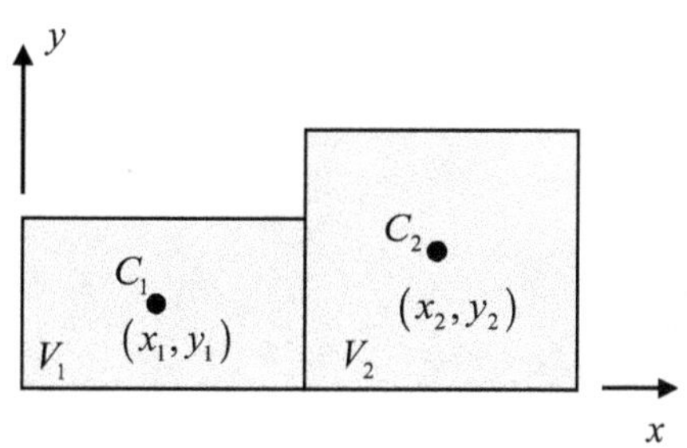

integrations over the components, we can also recognize those components are integrals that define the location of the center of masses of the individual components,$\mathbf{x}_{C_1}$ and $\mathbf{x}_{C_2}$, respectively, so that

$$m\mathbf{x}_C = m_1\mathbf{x}_{C_1} + m_2\mathbf{x}_{C_2} \tag{5.2.5}$$

or, in terms of components (see Fig. 5.5),

$$\begin{aligned} x_C &= \frac{m_1x_1 + m_2x_2}{m_1 + m_2} \\ y_C &= \frac{m_1y_1 + m_2y_2}{m_1 + m_2} \end{aligned} \tag{5.2.6}$$

Thus, if we know the locations of the centers of mass of the components, we can use Eq. (5.2.6) to calculate the center of mass location for the composite body. Obviously, this same process can be used for more than two components, if necessary. We can also determine the moment of inertia for the composite body about this center of mass location. This requires that we know how the moment of inertia values about the center of masses of the individual components change when we want to calculate them about a different point. Another way to view this is that we want to shift the moment of inertia values from a z-axis located at the center of mass for each component to a different (but parallel) z-axis located at another point (in this case the center of mass of the entire composite body). There is a general expression for making that shift called the parallel axis theorem which we will now derive. Consider, for example, a body with a center of mass location $C = (x_{C/O}, y_{C/O})$, as shown in Fig. 5.6. If we want to calculate the moment of inertia with respect to point O, we have

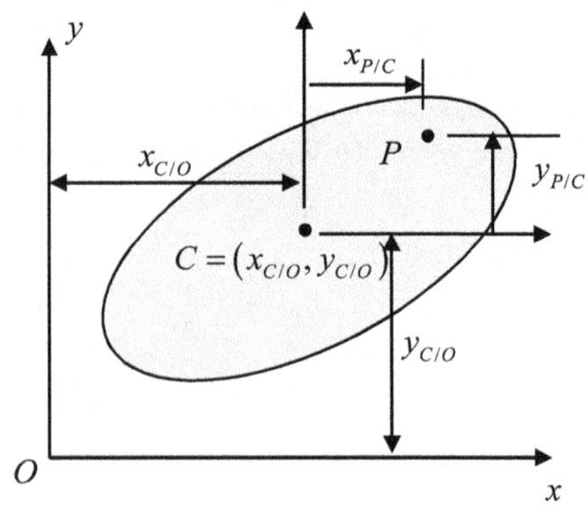

Fig. 5.6 Geometry for use in the parallel axis theorem

$$
\begin{aligned}
I_O &= \int_V \rho\left(x^2 + y^2\right)dV \\
&= \int_V \rho\left[\left(x_{C/O} + x_{P/C}\right)^2 + \left(y_{C/O} + y_{P/C}\right)^2\right]dV \\
&= \left(x_{C/O}^2 + y_{C/O}^2\right)\int_V \rho dV + 2x_{C/O}\int_V \rho x_{P/C} dV \\
&\quad + 2y_{C/O}\int_V \rho y_{P/C} dV + \int_V \rho\left[\left(x_{P/C}\right)^2 + \left(y_{P/C}\right)^2\right]dV
\end{aligned}
\tag{5.2.7}
$$

But we have

$$
\begin{aligned}
m &= \int_V \rho dV \\
\int_V \rho x_{P/C} dV &= \int_V \rho y_{P/C} dV = 0 \\
I_C &= \int_V \rho\left[\left(x_{P/C}\right)^2 + \left(y_{P/C}\right)^2\right]dV
\end{aligned}
\tag{5.2.8}
$$

where we have used the definitions of the mass and the moment of inertia about the center of mass. The two "first-moment" integrals in Eq. (5.2.8) vanish since the distances in those integrals are measured from the center of mass so the center of mass location in those coordinates would be at $(0, 0)$. Using these results in Eq. (5.2.7) we find

$$
I_O = I_C + m\left(x_{C/O}^2 + y_{C/O}^2\right) \tag{5.2.9}
$$

which is the *parallel axis theorem*, where $x_{C/O}^2 + y_{C/O}^2 = r_{C/O}^2$ is just the distance squared between a z-axis located at O and a z-axis located at C.

We can apply the parallel axis theorem to our composite body problem of Fig. 5.5. We have

$$
\begin{aligned}
I_C &= I_{C1} + m_1\left[(x_C - x_1)^2 + (y_C - y_1)^2\right] \\
&\quad + I_{C2} + m_2\left[(x_C - x_2)^2 + (y_C - y_2)^2\right]
\end{aligned}
\tag{5.2.10}
$$

where (I_{C1}, I_{C2}) are the mass moments of inertia about the center of masses of the individual components.

In planar motion problems only one mass moment of inertia appears in the governing equations but for 3-D rigid body problems there are more moments of inertia that need to be defined and calculated. This should not be surprising because we saw such multiple mass moments when we considered the 3-D motion of a rigidly constrained system of particles in Chap. 3. For more information on rigid body mass moments see Appendix B.

5.3 Planar Problems and Constraint Forces

In Chaps. 2 and 3 we saw that we can obtain the equations of motion and constraint forces for individual particles or systems of particles by a number of different methods. The same options are available when we solve for the motion of continuous rigid bodies. In this section we want to examine some of the approaches available for solving continuous rigid body problems.

5.3.1 Newton-Euler Equations

As a first example of solving a planar dynamics problem, consider the motion of the pendulum shown in Fig. 5.7 where a thin, rigid rod of length L and mass m is supported by a smooth pin at O and swings under the action of its own weight. This is an example of a *physical pendulum* where the body has distributed inertial properties that are less idealized than that of the point mass of the simple pendulum. Let us examine this problem initially by using the Newton-Euler equations directly. The free body diagram of the pendulum is shown in Fig. 5.7b. From that diagram the Newton-Euler equations are

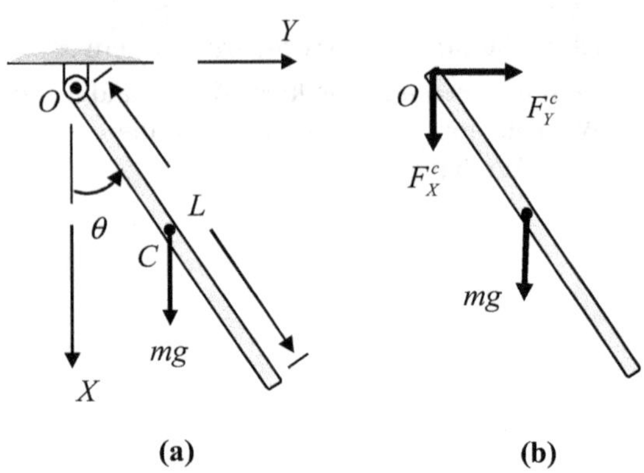

Fig. 5.7 (**a**) A physical pendulum and (**b**) its free body diagram

$$\begin{aligned} m\ddot{X}_C &= mg + F_X^c \\ m\ddot{Y}_C &= F_Y^c \\ \frac{1}{12}mL^2\ddot{\theta} &= F_X^c L\sin\theta/2 - F_Y^c L\cos\theta/2 \end{aligned} \tag{5.3.1}$$

where (F_X^c, F_Y^c) are the constraint forces acting at the smooth pin at point O (a smooth pin cannot support a moment). At the fixed pin we must have $\mathbf{x}_O = \mathbf{x}_C + \mathbf{x}_{O/C} = 0$ which gives

$$\begin{aligned} X_C - L\cos\theta/2 &= 0 \\ Y_C - L\sin\theta/2 &= 0 \end{aligned} \tag{5.3.2}$$

Taking one derivative on time of these constraints we find the velocity constraints

$$\begin{aligned} \dot{X}_C + L\sin\theta\dot{\theta}/2 &= 0 \\ \dot{Y}_C - L\cos\theta\dot{\theta}/2 &= 0 \end{aligned} \tag{5.3.3}$$

and taking another derivative gives the acceleration constraints

$$\begin{aligned} \ddot{X}_C + L\sin\theta\ddot{\theta}/2 &= -L\dot{\theta}^2\cos\theta/2 \\ \ddot{Y}_C - L\cos\theta\ddot{\theta}/2 &= -L\dot{\theta}^2\sin\theta/2 \end{aligned} \tag{5.3.4}$$

Placing these acceleration constraints into the first two equations in Eq. (5.3.1) we find

$$\begin{aligned} F_X^c &= -mg - \frac{mL}{2}\left(\sin\theta\ddot{\theta} + \dot{\theta}^2\cos\theta\right) \\ F_Y^c &= \frac{mL}{2}\left(\cos\theta\ddot{\theta} - \dot{\theta}^2\sin\theta\right) \end{aligned} \tag{5.3.5}$$

which, when placed into the moment equation of Eq. (5.3.1), gives

$$\ddot{\theta} = -\frac{3g}{2L}\sin\theta \tag{5.3.6}$$

Using this equation of motion in Eq. (5.3.5) the constraint forces become

$$\begin{aligned} F_X^c &= -mg\left(1 - \frac{3}{4}\sin^2\theta\right) - \frac{mL}{2}\dot{\theta}^2\cos\theta \\ F_Y^c &= -\frac{3}{4}mg\sin\theta\cos\theta - \frac{mL}{2}\dot{\theta}^2\sin\theta \end{aligned} \tag{5.3.7}$$

These constraint forces are the only forces acting at O since the pin is assumed to be smooth. However, if we calculate the equivalent constraints around other points we must also include a constraint moment. Note that the constraint moment about the center of mass, C, generated by these constraint forces, $M_{CZ}^{c} = F_{X}^{c} L \sin\theta/2 - F_{Y}^{c} L \cos\theta/2$, can be obtained from these constraint forces and Eq. (5.3.1) as

$$M_{CZ}^{c} = -\frac{1}{8} mgL \sin\theta \tag{5.3.8}$$

Finally, the other two equations of motion, from Eq. (5.3.1), are

$$\begin{aligned} \ddot{X}_C &= \frac{3}{4} g \sin^2\theta - \frac{L}{2} \dot{\theta}^2 \cos\theta \\ \ddot{Y}_C &= -\frac{3}{4} g \sin\theta \cos\theta - \frac{L}{2} \dot{\theta}^2 \sin\theta \end{aligned} \tag{5.3.9}$$

Since we have three degrees of freedom and two holonomic constraints this is a one degree of freedom problem. If we take θ as the independent coordinate and (X_C, Y_C) as the dependent coordinates then we can solve the problem completely by only solving Eq. (5.3.6) for $\theta = \theta(t)$. We could have obtained Eq. (5.3.6) much quicker, of course, by simply using the moment equation about the fixed point O since

$$I_O \ddot{\theta} = -\frac{1}{2} mgL \sin\theta \tag{5.3.10}$$

The moment inertia about O we can get from the center of mass value through the parallel axis theorem, which gives

$$\begin{aligned} I_O &= I_C + m\left(\frac{L}{2}\right)^2 = \frac{1}{12} mL^2 + \frac{1}{4} mL^2 \\ &= \frac{1}{3} mL^2 \end{aligned} \tag{5.3.11}$$

which, when placed into Eq. (5.3.10), gives Eq. (5.3.6).

Although we solved this problem for a thin bar, in fact we can obtain the solution for a body of any shape that has the X–Y plane as a plane of symmetry. In that case the moment equation about the Z-axis (Fig. 5.8) is

$$I_O \ddot{\theta} = -mgh \sin\theta \tag{5.3.12}$$

where h is the distance from O to the center of mass and $I_O = I_C + mh^2$ in terms of the center of mass moment of inertia, I_C, for the given shape. While Eq. (5.3.12) is still true for an unsymmetrical body, in that case there will also be out-of-plane moments acting on the pin at O so that it is really a three-dimensional problem.

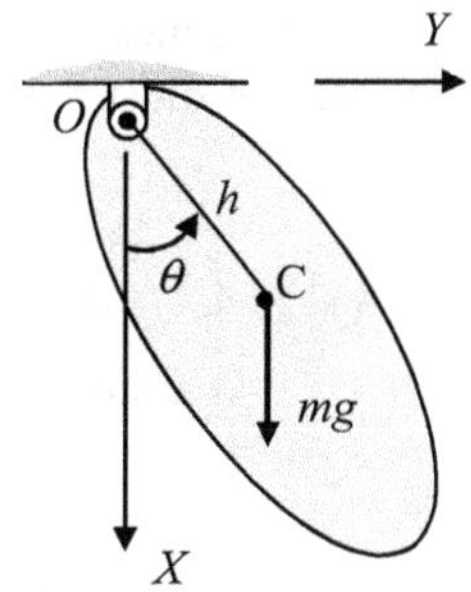

Fig. 5.8 A physical pendulum

5.3.2 An Augmented Approach

While solving the Newton-Euler equations and differentiated constraints by hand can be used in principle on any rigid body problem, as we have seen in our discussion of particles there are other approaches that can be more effective and are more suitable for computer-based implementation. The augmented approach, for example, was one method that yielded explicitly both the equations of motion and generalized constraint forces. This same approach can be used for continuous rigid bodies. Consider the case we have been discussing where we use the generalized coordinates $\mathbf{q} = (X_C, Y_C, \theta)$ to describe the motion of a rigid body. The governing equations for planar motion can be written as

$$\mathbf{M}\ddot{\mathbf{q}} = \mathbf{Q}^e + \mathbf{Q}^c \tag{5.3.13}$$

where, for our pendulum example, the mass matrix is

$$\mathbf{M} = \begin{bmatrix} m & 0 & 0 \\ 0 & m & 0 \\ 0 & 0 & mL^2/12 \end{bmatrix} \tag{5.3.14}$$

and the generalized external and constraint forces are

$$\mathbf{Q}^e = \begin{Bmatrix} mg \\ 0 \\ 0 \end{Bmatrix}, \quad \mathbf{Q}^c = \begin{Bmatrix} F_X^c \\ F_Y^c \\ M_{CZ}^c \end{Bmatrix} \tag{5.3.15}$$

The constraint equations, written in terms of the accelerations as $\mathbf{A}\ddot{\mathbf{q}} = \mathbf{b}$, give

$$\mathbf{A} = \begin{bmatrix} 1 & 0 & L\sin\theta/2 \\ 0 & 1 & -L\cos\theta/2 \end{bmatrix}, \mathbf{b} = \begin{Bmatrix} -L\dot{\theta}^2\cos\theta/2 \\ -L\dot{\theta}^2\sin\theta/2 \end{Bmatrix} \tag{5.3.16}$$

Are the generalized constraint forces at the smooth pin ideal constraints? If they are then we must have $\mathbf{Q}^c = \mathbf{A}^T\boldsymbol{\lambda}$, where $\boldsymbol{\lambda}$ is a 2×1 vector. For our case

$$\mathbf{Q}^c = \begin{bmatrix} 1 & 0 \\ 0 & 1 \\ L\sin\theta/2 & -L\cos\theta/2 \end{bmatrix} \begin{Bmatrix} \lambda_1 \\ \lambda_2 \end{Bmatrix} = \begin{Bmatrix} \lambda_1 \\ \lambda_2 \\ L\lambda_1 \sin\theta/2 - L\lambda_2 \cos\theta/2 \end{Bmatrix} \tag{5.3.17}$$

which we see is the same as Eq. (5.3.15) with $\lambda_1 = F^c_X, \lambda_2 = F^c_Y$ and M^c_{CZ} (see Eq. (5.3.1)) so the constraints are indeed ideal. We see that the generalized force vector contains both constraint force components and a moment generated about the reference point C used in the governing equations (see Fig. 5.9a). These are obviously not the actual constraint forces and moments at the pin (point O), which have been labeled (O^c_X, O^c_Y, M^c_{OZ}) in Fig. 5.9b. To obtain the pin constraint forces we use the fact that the two force systems (I, II) shown in Fig. 5.9 must be equivalent (also called "*equipollent*"), i.e., they both must produce the same total force, $\mathbf{F}$, and total moment, $\mathbf{M}_P$, about any reference point P:

$$\begin{aligned} (\mathbf{F})_I &= (\mathbf{F})_{II} \\ (\mathbf{M}_P)_I &= (\mathbf{M}_P)_{II} \end{aligned} \tag{5.3.18}$$

For our pendulum problem this means that if we take the reference point P to be, say, C we must have

$$\begin{aligned} F^c_X &= O^c_X \\ F^c_Y &= O^c_Y \\ M^c_{CZ}\mathbf{e}_z &= M^c_{OZ}\mathbf{e}_z + \mathbf{x}_{O/C} \times \mathbf{O}^c \\ &= M^c_{OZ}\mathbf{e}_z + \left(O^c_X L\sin\theta/2 - O^c_Y L\cos\theta/2\right)\mathbf{e}_Z \end{aligned} \tag{5.3.19}$$

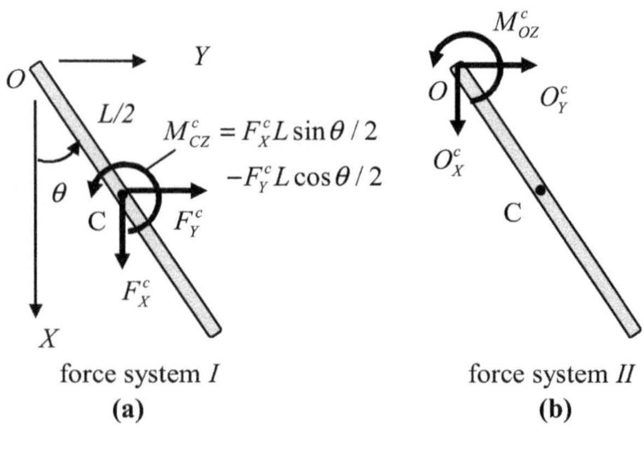

Fig. 5.9 (**a**) The components of the generalized constraint force (force system I) and (**b**) components of the actual constraint force at the pin O (force system II)

which gives $\left(O_X^c, O_Y^c, M_{OZ}^c\right) = \left(F_X^c, F_Y^c, 0\right)$. These are now the true constraint forces and zero moment which act at the smooth pin. Thus, in rigid body problems we must typically use the condition of equipollence to transform the generalized constraint forces solved for at the center of mass into the constraint forces and moments acting at the point where the constraints are actually applied. Note, however, that for choices of generalized coordinates other than the $\mathbf{q} = (X_C, Y_C, \theta)$ coordinates used here or for other types of constraints the $\boldsymbol{\lambda}$ vector may not represent the actual constraint forces (see Chap. 2 for some examples).

Since the pin constraints are ideal constraints, we need to find the solution of the system

$$\begin{aligned} \mathbf{M}\ddot{\mathbf{q}} &= \mathbf{Q}^e + \mathbf{A}^T\boldsymbol{\lambda} \\ \mathbf{A}\ddot{\mathbf{q}} &= \mathbf{b} \end{aligned} \tag{5.3.20}$$

which is in exactly the same form as found in an augmented approach for particles. Thus, if the constraints are independent we can immediately write the solution down explicitly for the equations of motion and generalized constraint force as

$$\begin{aligned} \mathbf{Q}^c &= \mathbf{A}^T\left(\mathbf{A}\mathbf{M}^{-1}\mathbf{A}^T\right)^{-1}\left(\mathbf{b} - \mathbf{A}\mathbf{M}^{-1}\mathbf{Q}^e\right) \\ \ddot{\mathbf{q}} &= \mathbf{M}^{-1}\mathbf{Q}^e + \mathbf{M}^{-1}\mathbf{A}^T\left(\mathbf{A}\mathbf{M}^{-1}\mathbf{A}^T\right)^{-1}\left(\mathbf{b} - \mathbf{A}\mathbf{M}^{-1}\mathbf{Q}^e\right) \end{aligned} \tag{5.3.21}$$

Since we have explicit algebraic expressions for ($\mathbf{A}$, $\mathbf{b}$, $\mathbf{M}$, $\mathbf{Q}^e$) with the appropriate MATLAB® matrix operations we can find all the terms needed in Eq. (5.3.21) numerically without obtaining explicit algebraic expressions for the end results. However, we can also perform the operations symbolically to obtain explicit expressions. MATLAB® can help make this process much less time consuming through its symbolic math toolbox so we will show an example of such MATLAB® symbolic manipulations for this problem.

To begin, we need to declare the variables we will use as symbolic. For this case we have the variables m, L, θ, $\dot{\theta}$, and g which we will call the symbolic variables m, L, ang, angd, and g in MATLAB®:

```
syms m L ang angd g
```

We now can define symbolically the matrices and vectors $\mathbf{A}$, $\mathbf{M}$, $\mathbf{b}$, and $\mathbf{Q}^e$ in MATLAB® as

```
A = [ 1 , 0 , L*sin(ang)/2; 0 , 1 , -L*cos(ang)/2];
b = [ -L*angd^2*cos(ang)/2 ; -L*angd^2*sin(ang)/2];
Qe = [ m*g ; 0 ; 0];
M = [ m , 0 , 0 ; 0 , m , 0; 0 , 0 , m*L^2/12];
```

We need to calculate the matrix $\mathbf{V} = (\mathbf{AM}^{-1}\mathbf{A}^T)^{-1}$, which is

```
V = inv(A*inv(M)*A.');
```

Note that in computing the transpose in MATLAB® we used the operation A.' rather than A' since in general a matrix A could be complex. In that case the MATLAB® operation A' produces the complex conjugate of the transpose of matrix A, which is still correct here as everything is real, but gives a more complex-looking symbolic expression. The operation A.' simply gives the transpose, which is what we want. If you look at V explicitly in MATLAB® you will see that it is not in a compact form. However, we can simplify this expression and get a more readable version with the MATLAB® function simplified, which yields the simplified version of V, Vs, as

```
Vs = simplify(V)

Vs = [ -(m*(3*sin(ang)^2 - 4))/4 ,          (3*m*sin(2*ang))/8]
     [    (3*m*sin(2*ang))/8     ,    m*((3*sin(ang)^2)/4 + 1/4)]
```

We now define the vector $\mathbf{B} = \mathbf{b} - \mathbf{AM}^{-1}\mathbf{Q}^e$ and the generalized constraint force, $\mathbf{Q}^c$:

```
B = b - A*inv(M)*Qe;
Qc = A.'*Vs*B;
```

If we simplify Qc we find the simplified version, Qcs:

```
Qcs = simplify(Qc)

Qcs =

 -(m*(g + 3*g*cos(ang)^2 + 2*L*angd^2*cos(ang)))/4
   -(m*(4*L*sin(ang)*angd^2 + 3*g*sin(2*ang)))/8
                    -(L*g*m*sin(ang))/8
```

and using this generalized constraint force we can obtain the equations of motion for $\ddot{\mathbf{q}}$ as

```
qdd = inv(M)*Qe + inv(M)*Qcs

qdd =

 (3*g)/4 - (3*g*cos(ang)^2)/4 - (L*angd^2*cos(ang))/2
     - (3*g*sin(2*ang))/8 - (L*angd^2*sin(ang))/2
                   -(3*g*sin(ang))/(2*L)
```

If we rewrite these MATLAB® expressions in terms of our original variables we find (with some minor rearrangement for better readability)

$$\begin{aligned}\mathbf{Q}^c &= \begin{Bmatrix} F_X^c \\ F_Y^c \\ M_{CZ}^c \end{Bmatrix} = \begin{Bmatrix} -mg(1+3\cos^2\theta)/4 - mL\dot{\theta}^2\cos\theta/2 \\ -3mg\sin 2\theta/8 - mL\dot{\theta}^2\sin\theta/2 \\ -mgL\sin\theta/8 \end{Bmatrix} \\ \ddot{\mathbf{q}} &= \begin{Bmatrix} \ddot{X}_C \\ \ddot{Y}_C \\ \ddot{\theta} \end{Bmatrix} = \begin{Bmatrix} 3g/4 - 3g\cos^2\theta/4 - L\dot{\theta}^2\cos\theta/2 \\ -3g\sin 2\theta/8 - L\dot{\theta}^2\sin\theta/2 \\ -3g\sin\theta/2L \end{Bmatrix}\end{aligned} \tag{5.3.22}$$

which are the same results derived previously in Eqs. (5.3.6), (5.3.7), (5.3.8), and (5.3.9) except that the $\sin^2\theta$ terms have been written in Eq. (5.3.22) in terms of $\cos^2\theta$ and the $\sin\theta\cos\theta$ terms have been written in Eq. (5.3.22) using the double-angle formula $\sin 2\theta = 2\sin\theta\cos\theta$.

We now have all the symbolic expressions we need to set up the equations of motion and evaluate the constraint force. We could use those functions to generate by hand MATLAB® functions that solve the problem numerically, but MATLAB® can also do that for us with the built-in function called matlabFunction. Let's see an example of this by solving Eq. (5.3.6) for the angle θ. First, we put this equation of motion in the form of a first-order system by defining the symbolic column vector xdot:

```
xdot = [ angd; qdd(3)]
xdot =        angd
     -(3*g*sin(ang))/(2*L)
```

The vector xdot is exactly the vector needed to be calculated by ode45 but it is still in symbolic form. Since the function that ode45 uses also has to have the time t in its input arguments, we need to define a symbolic value for that time:

```
syms t
```

Now we can use the matlabFunction to generate an ordinary MATLAB® function file that ode45 can use from the symbolic vector xdot. The form of the generated function is

```
function [OUT1, OUT2, ...] = NAME(IN1, IN2, ...)
```

and the form of the call to matlabFunction is

```
matlabFunction(xdot, 'File', 'NAME', 'Vars', {IN1, IN2, ...},
'Outputs', {OUT1,OUT2,...});
```

which will generate a function m-file whose name is the string 'NAME'. The input variable 'Vars' contains the input variable names IN1, IN2,... in a cell array, and 'Outputs' contains the output variable names in the strings 'OUT1', 'OUT2', .. in a cell array. For our example we will call the file name phys_pend2, the input variables will be { t, [ang; angd], g, L}, and the output of the function will be called dxdt. The call to matlabFunction is

```
matlabFunction(xdot,'File','phys_pend2', 'Vars', {t, [ang;angd] ,g, L},
'Outputs',{'dxdt'});
```

We can see the function that matlabFunction generates with the command type:

```
type phys_pend2
```

which gives, as we can see below, a function of exactly the type needed by ode45 for the physical pendulum:

```
function dxdt = phys_pend2(t,in2,g,L)
%PHYS_PEND2
%   DXDT = PHYS_PEND2(T,IN2,G,L)
%   This function was generated by the Symbolic Math Toolbox
%    version 7.0.
%   04-Jan-2017 15:13:48

ang = in2(1,:);
angd = in2(2,:);
dxdt = [angd;(g.*sin(ang).*(-3.0./2.0))./L];
```

Now, we can simply use this function in ode45 in the normal manner, by giving a time interval tspan = [0 10] and initial conditions for $\theta(0)$ and $\dot{\theta}(0)$ as $\pi/2$ radians and zero, respectively, and provide values for (g, L) as (9.8, 10):

```
[tout, xout] =ode45(@phys_pend2, [0 10], [pi/2, 0],[],9.8, 10);
```

which uses, for course, the default error tolerances. We can then plot the angle versus time

```
plot(tout,xout(:,1))
```

whose plot is shown in Fig. 5.10. We can also generate a file for calculating the constraint forces from the symbolic vector Qcs generated previously. The call to matlabFunction in this case is

```
matlabFunction(Qcs,'File','pend_con', 'Vars',{[ang; angd] ,m, g, L},
'Outputs',{'Qcon'});
```

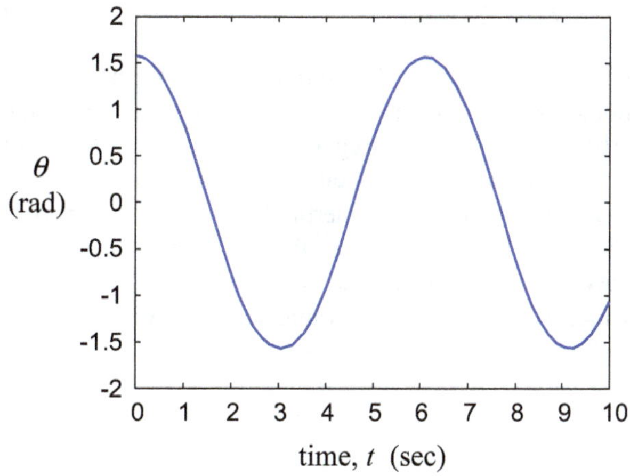

Fig. 5.10 Angular motion of the physical pendulum

where we have called the pendulum constrain function pend_con. Note that we need the mass here also as an input variable and the generalized constraint force vector is in the output vector Qcon. We can see the function file generated again with the command type

```
type pend_con
function Qcon = pend_con(in1,m,g,L)
%PEND_CON
%   QCON = PEND_CON(IN1,M,G,L)

%   This function was generated by the Symbolic Math Toolbox
%   version 7.0.
%   04-Jan-2017 18:13:36
ang = in1(1,:);
angd = in1(2,:);
t2 = cos(ang);
t3 = angd.^2;
t4 = sin(ang);
Qcon = [m.*(g + g.*t2.^2.*3.0 + L.*t2.*t3.*2.0).*(- 1.0./4.0);
m.*(g.*sin(ang.*2.0).*3.0 + L.*t3.*t4.*4.0).*(-1.0./8.0);
L.*g.*m.*t4.*(-1.0./8.0)];
```

We can use the outputs of ode45 as the inputs to this function to generate the generalized constraint force vector as a function of time:

```
Qvect = pend_con(xout(:,1)', xout(:,2)', 1, 9.8, 10);
```

Note that we have set the mass $m = 1$ here and we have changed the xout variables from column to row vectors to be compatible with the forms seen in the pend_con function. We can then plot the first generalized force component, F_X^c, versus time as

```
plot(tout', Qvect(1,:));
```

which gives the plot shown in Fig. 5.11. From that figure we see that the curves are not smooth so the default error tolerances are likely too large. We can change those error tolerance values to get a more accurate result, but we will not do that here. At $t = 0$ the rod is horizontal and has zero velocity. At that time all the forces and moments (including the inertial force $-m\ddot{X}_C$and inertial moment $-I_C\ddot{\theta}$) are shown in Fig. 5.12. We see that these forces and moments are in "dynamic equilibrium", i.e., they balance, and that the constraint force component $F_X^c = -mg/4 = -2.45\ N$, which is also the starting value seen in Fig. 5.11.

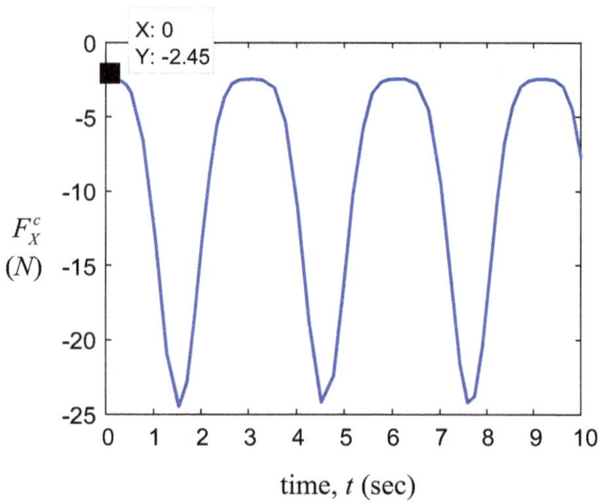

Fig. 5.11 The X-component of the constraint force at the pin for the physical pendulum versus time. Also shown is the starting value for the force at time $t = 0$

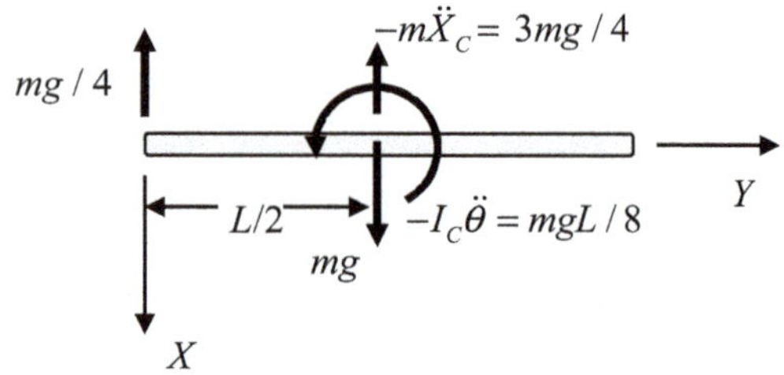

Fig. 5.12 All the forces and moments acting on the physical pendulum at the starting time $t = 0$, including the inertial forces and moment

5.3.3 Rolling Without Slipping

One type of rigid body constraint that occurs frequently in dynamics is the constraint of rolling without slipping. In planar motion problems where the rotational motion is always relative to a fixed axis this constraint is a holonomic constraint but for three-dimensional problems rolling without slipping is non-holonomic. This should not be surprising since in Chap. 4 we saw that the rolling axle problem was non-holonomic. In that case while the motion was planar motion, the axle (and hence the axis of rotation of the wheels) was changing.

Consider now a circular disk of radius r that is rolling on a horizontal plane (Fig. 5.13a). We will assume that the surface is rough so that the surface can exert both a normal force, N, and a tangential force, F. The normal force as we have seen is a force of constraint that prevents motion of the disk at point D in the Y-direction. Similarly, if the disk is rolling without slipping the friction force will be a constraint force that maintains that condition. No slipping means that there must be no relative velocity of the disk at point D, where it touches the plane, with respect to the plane. But the plane itself is motionless so that the velocity of the disk at D must be zero. Thus, we could express the constraint of no slipping in terms of the velocity as simply $\mathbf{v}_D = 0$. However, we said that the constraint is holonomic so that we should also be able to express the constraint in geometrical terms. As shown in Fig. 5.14, since there is no relative motion between the points on the edge of the disk and the plane, within one complete revolution of the disk there should be a unique correspondence between points on the disk edge and corresponding points on the plane where both points were in contact. Some such points are labeled points 1–4 in Fig. 5.14 for both the disk and the plane. But this also means that the distance between points 1 and 4 along the disk edge must be the same distance between points 1 and 4 along the plane. (Imagine rolling the disk back to its original position when the two points labeled 1 were in contact. During the rolling all successive points 4, 3, 2, and then 1 will coincide so the distance between any of these points laid out along the edge must be the same distance along the plane.) The distance along the edge is just $r\alpha$ while the distance moved along the plane is X_C, which is also the distance

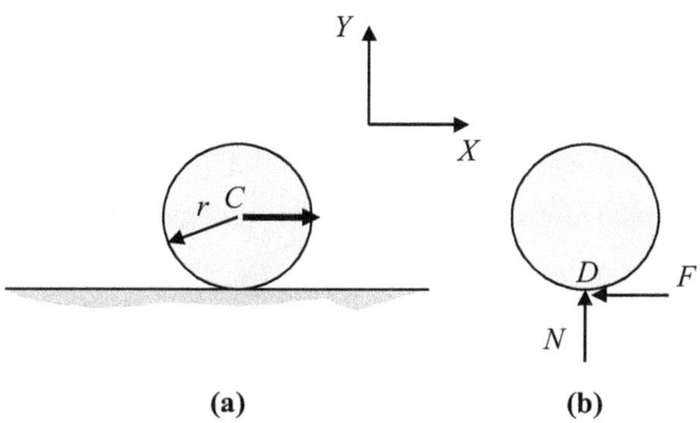

Fig. 5.13 (**a**) A disk rolling without slipping to the right on a horizontal plane and (**b**) the constraint forces normal and tangential to the surface

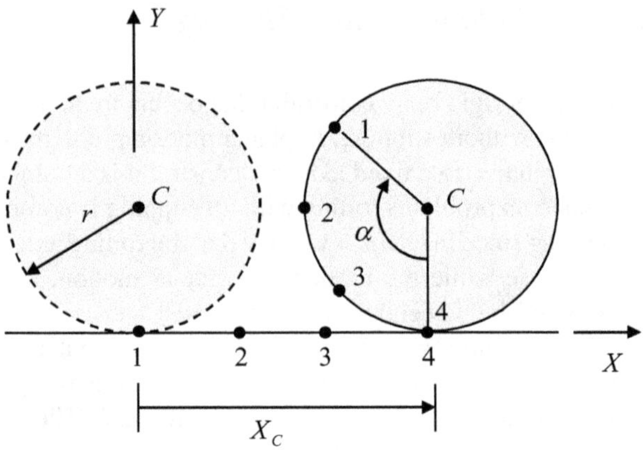

Fig. 5.14 An illustration of the conditions of rolling without slipping

moved by the center of mass in the X-direction. In the Y-direction, $Y_C = r =$ constant. Note, however, that the angle α (Fig. 5.14) is measured clockwise. If we take $\mathbf{q} = (X_C, Y_C, \theta)$ as our generalized coordinates for the disk where, recall, θ was the rotation of the rigid body as measured positive counterclockwise from a fixed positive X-axis, then our constraints must be written as

$$\begin{aligned} X_C &= -r\theta \\ Y_C &= r \end{aligned} \tag{5.3.23}$$

Differentiating once the velocity constraints are

$$\begin{aligned} \dot{X}_C + r\dot{\theta} &= 0 \\ \dot{Y}_C &= 0 \end{aligned} \tag{5.3.24}$$

and differentiating again the acceleration constraints are

$$\begin{aligned} \ddot{X}_C + r\ddot{\theta} &= 0 \\ \ddot{Y}_C &= 0 \end{aligned} \tag{5.3.25}$$

which written as $\mathbf{A}\ddot{\mathbf{q}} = \mathbf{b}$ gives

$$\mathbf{A} = \begin{bmatrix} 1 & 0 & r \\ 0 & 1 & 0 \end{bmatrix} \quad \mathbf{b} = 0 \tag{5.3.26}$$

An ideal generalized constraint force $\mathbf{Q}^c = \mathbf{A}^T\boldsymbol{\lambda}$, therefore, would be

$$\mathbf{Q}^c = \begin{Bmatrix} F_X^c \\ F_Y^c \\ M_{CZ}^c \end{Bmatrix} = \begin{bmatrix} 1 & 0 \\ 0 & 1 \\ r & 0 \end{bmatrix} \begin{Bmatrix} \lambda_1 \\ \lambda_2 \end{Bmatrix} = \begin{Bmatrix} \lambda_1 \\ \lambda_2 \\ r\lambda_1 \end{Bmatrix} \tag{5.3.27}$$

which, as discussed before, is just the force system from the constraints at the plane as measured at the center of mass point C (Fig. 5.15a) which is equipollent to the force system at the point of contact D with the plane (Fig. 5.15b). Thus, we recognize λ_2 as the normal force and λ_1 as the friction force, which is pointing in the plus X-direction since if the rotation, θ, of the disk is positive (counterclockwise) the disk will be rolling to the left. Note that if we use the constraint equations (Eq. (5.3.24) in the relative velocity relationship between points C and D we have

$$\begin{aligned} \mathbf{v}_D &= \mathbf{v}_C + \boldsymbol{\omega} \times \mathbf{x}_{D/C} \\ &= \mathbf{v}_C + \dot{\theta}\,\mathbf{e}_Z \times (-r\mathbf{e}_Y) \\ &= \begin{Bmatrix} \dot{X}_C \\ \dot{Y}_C \end{Bmatrix} + \begin{Bmatrix} r\dot{\theta} \\ 0 \end{Bmatrix} = \begin{Bmatrix} 0 \\ 0 \end{Bmatrix} \end{aligned} \tag{5.3.28}$$

so that indeed rolling without slipping is equivalent to the velocity constraint $\mathbf{v}_D = 0$. However, even though the velocity of point D on the disk vanishes, its acceleration is not zero. In fact using the acceleration constraints, Eq. (5.3.25), in the relative acceleration relationship between points C and D we have

$$\begin{aligned} \mathbf{a}_D &= \mathbf{a}_C + \boldsymbol{\alpha} \times \mathbf{x}_{D/C} - \dot{\theta}^2\mathbf{x}_{D/C} \\ &= \mathbf{a}_C + \ddot{\theta}\mathbf{e}_Z \times (-r\mathbf{e}_Y) + r\dot{\theta}^2\mathbf{e}_Y \\ &= \begin{Bmatrix} \ddot{X}_C \\ \ddot{Y}_C \end{Bmatrix} + \begin{Bmatrix} r\ddot{\theta} \\ 0 \end{Bmatrix} + \begin{Bmatrix} 0 \\ r\dot{\theta}^2 \end{Bmatrix} = \begin{Bmatrix} 0 \\ r\dot{\theta}^2 \end{Bmatrix} \end{aligned} \tag{5.3.29}$$

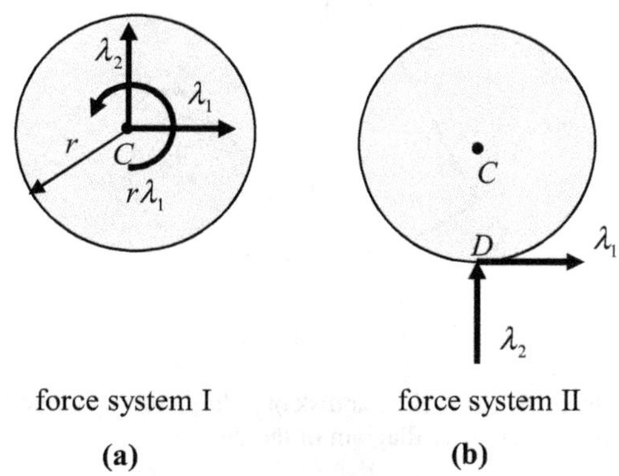

Fig. 5.15 (**a**) The generalized constraint force system acting at the center of mass, C, for a disk rolling without slipping, and (**b**) the equivalent system acting at the contact point D

so that the point D on the disk that is in contact with the plane has a Y-component (vertical) acceleration of $r\dot{\theta}^2$. Note that if in Eq. (5.1.26) we take point A in the moment equation to be the contact point D in a problem of rolling without slipping we find

$$\int_{V_n} \rho \mathbf{x}_{P/D}^{(n)} dV \times \left(\ddot{\mathbf{x}}_D^{(n)}\right) + \ddot{\theta}_n \mathbf{e}_Z \int_{V_n} \rho \left(r_{P/D}^{(n)}\right)^2 dV = \mathbf{M}_D^{(n)}$$

$$or \qquad (5.3.30)$$

$$m\mathbf{x}_{C/D}^{(n)} \times \ddot{\mathbf{x}}_D^{(n)} + I_D^{(n)} \ddot{\theta}_n \mathbf{e}_Z = \mathbf{M}_D^{(n)}$$

But $\mathbf{x}_{C/D}^{(n)}$ is parallel to $\ddot{\mathbf{x}}_D^{(n)}$ so the cross product is zero and the moment equation about the contact point D of no slipping has just the same simple form as about the center of mass or a fixed point about which a rigid body rotates:

$$I_D^{(n)} \ddot{\theta}_n \mathbf{e}_Z = \mathbf{M}_D^{(n)} \qquad (5.3.31)$$

We also obtain the same result from the moment equation in Eq. (5.1.9) by noting that since $\ddot{\mathbf{x}}_D^{(n)} = \tau \mathbf{x}_{C/D}^{(n)}$, where τ is a proportionality factor, the term $\left(\widehat{\mathbf{x}}_{C/D}^{(n)}\right)^T \ddot{\mathbf{x}}_D^{(n)}$ $= \tau \left(\widehat{\mathbf{x}}_{C/D}^{(n)}\right)^T \mathbf{x}_{C/D}^{(n)} = 0$ because $\widehat{\mathbf{a}}^T \mathbf{a} = 0$ for any vector $\mathbf{a}$.

Equation (5.3.30) is often very useful in rolling without slipping problems since by taking moments about D the constraint forces are automatically eliminated. As an example, consider a homogeneous disk of radius r and mass m rolling without slipping on a circular track of radius R, as shown in Fig. 5.16a. If we take moments about point D, then from the free body diagram of Fig. 5.16b we have

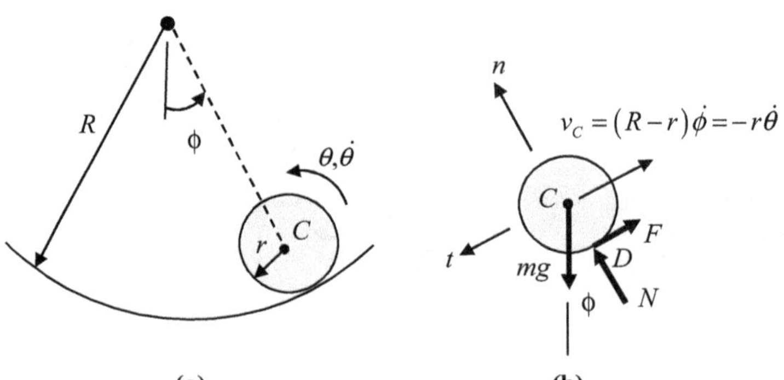

Fig. 5.16 (**a**) A circular disk of radius r rolling without slipping on a circular track of radius R, and (**b**) the free body diagram of the disk

$$mgr\sin\phi = I_D\ddot{\theta} \tag{5.3.32}$$

where, using the parallel axis theorem, $I_D = mr^2/2+mr^2 = 3mr^2/2$. Since the center of mass point C is moving in a circular path of radius $(R - r)$ it has a speed tangent to the path given as $v_C = (R-r)\dot{\phi}$. But the rolling without slipping constraint also gives $v_C = -r\dot{\theta}$, where the minus sign is due to the fact that θ and $\dot{\theta}$ are taken as positive in the counterclockwise direction (see Fig. 5.16b), so that we have $\dot{\theta} = -(R-r)\dot{\phi}/r$ and hence $\ddot{\theta} = -(R-r)\ddot{\phi}/r$. Placing this relationship into Eq. (5.3.31) gives the equation of motion in terms of ϕ as

$$\ddot{\phi} + \frac{2g}{3(R-r)}\sin\phi = 0 \tag{5.3.33}$$

This equation of motion is identical in general form to that of the simple pendulum. However, one difference is that Eq. (5.3.32) is only valid if the disk rolls without slipping. The frictional force, F, is the force that maintains this constraint. But this frictional force must satisfy $|F| < \mu_s N$ so that if it becomes too large (i.e., $|F| \geq \mu_s N$) the disk will slip and we have to solve a rolling-with-slipping dynamics problem where $|F| = \mu_k N$ and the direction of the friction force is opposite to the velocity of the point D on the disk. The coefficients μ_s and μ_k are the static and dynamic coefficients of friction, where generally μ_k is somewhat smaller than μ_s. Thus, when solving Eq. (5.3.32) we need to check to see if $|F| < \mu_s N$ is satisfied. This is a good example of a problem where even though we used a very efficient method (the moment equation about D) to generate the equation of motion, we still do need to obtain the constraint forces to validate the solution. This is not difficult to do here using the free body diagram of Fig. 5.16b and the force equations for the motion of the center of mass in the n- and t-directions. We have

$$\begin{aligned} mg\sin\phi - F &= ma_{Ct} = mr\ddot{\theta} = -m(R-r)\ddot{\phi} \\ N - mg\cos\phi &= ma_{Cn} = m\frac{v_C^2}{(R-r)} = m(R-r)\dot{\phi}^2 \end{aligned} \tag{5.3.34}$$

where we have used the constraint of rolling without slipping to write the acceleration of the center of mass in the t-direction as $a_{Ct} = r\ddot{\theta}$ and we have also used the fact that the center of mass is moving in a circular path of radius $\rho = (R - r)$ about the center of the track so that it experiences an acceleration in the n-direction, $a_{Cn} = v_C^2/\rho$. Solving for F and N we have

$$\begin{aligned} F &= mg\sin\phi/3 \\ N &= mg\left(\cos\phi + (R-r)\dot{\phi}^2/g\right) \end{aligned} \tag{5.3.35}$$

which leads to the requirement that for rolling without slipping we must have

$$|F|/N = \frac{|\sin\phi|}{3\left(\cos\phi + (R-r)\dot{\phi}^2/g\right)} \leq \mu_s \tag{5.3.36}$$

[Note: We have assumed that N is always positive so we are not considering the possibility that $N = 0$ and the disk can leave the circular track.]

Simulation 5.1: Rolling Without Slipping

A MATLAB® function rolling_disk has been written which contains a nested function disk_eqs that implements the equation of motion in Eq. (5.3.32) as a first-order system of ordinary differential equations and uses ode45 with that nested function to obtain a solution for the problem of the disk rolling without slipping on the circular track. The calling sequence for the rolling_disk function is

```
[tout, xout] = rolling_disk(tspan, x0, rtol, atol, g, R, r);
```

where tspan and x0 are the time interval for the solution and the initial conditions $x0 = \left[\phi(0), \dot{\phi}(0)\right]$, and g, R, and r are the acceleration of gravity (in m/s^2), the radius of the track (in m), and the radius of the disk (in m), respectively. The rolling_disk function returns the times at which the solution was found in tout and the values of $\left(\phi, \dot{\phi}\right)$ in the columns of the xout matrix. A script rolling_disk_sim sets up defaults for all the parameters, calls the function rolling_disk with those default parameters, and obtains the solution. The script then plots the solution for the angle vesus time, $\phi(t)$, and on a second plot plots both $|F|$ and $\mu_s N$ (see Fig. 5.17). The default values

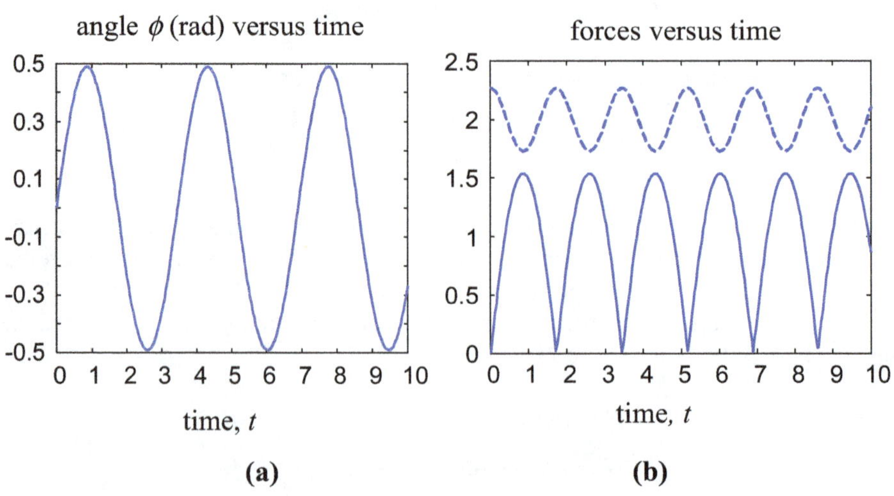

Fig. 5.17 (**a**) The angle ϕ versus time for the disk rolling without slipping on a circular track, and (**b**) the magnitude of the friction force, $|F|$ (solid line), and the maximum allowable friction force for rolling without slipping, $\mu_s N$ (dashed line)

used in the script are m = 1, g = 9.81, R = 2, r = 0.1, $\mu_s = 0.2$, tspan = [0, 10], and x0 = [0, 0.9] which can be changed by entering new values in the command window before executing the script. It can be seen from Fig. 5.17b that $|F| < \mu_s N$ is satisfied for the motion so that the condition of rolling without slipping is satisfied but you can verify that for a somewhat higher initial angular velocity such as $\dot{\phi}\,(0) = 1.2$ rad/s slipping does occur. We will not consider such cases here.

5.4 Kinetic Energy and Work-Energy

5.4.1 Kinetic Energy of a Rigid Body in Plane Motion

A system of N particles has a total kinetic energy:

$$T = \frac{1}{2}\sum_{i=1}^{N} m_i \dot{\mathbf{x}}_i \cdot \dot{\mathbf{x}}_i \tag{5.4.1}$$

If again we let $m_i = \rho \Delta V_i$ where ρ is the mass density (mass/volume) of the particle and ΔV_i is a small volume, then in the limit as the number of particles becomes infinitely large we can write the kinetic energy for a continuous body (n) as

$$T^{(n)} = \frac{1}{2}\int_{V_n} \rho \dot{\mathbf{x}}_P^{(n)} \cdot \dot{\mathbf{x}}_P^{(n)} dV \tag{5.4.2}$$

where $\dot{\mathbf{x}}_P^{(n)}$ is the velocity at an arbitrary point P in the body. If the body is a rigid body we can express the position vector to P in terms of the position vector to a reference point A in the body and a rotation of the body:

$$\mathbf{x}_P^{(n)} = \mathbf{x}_A^{(n)} + R^{(n)}\bar{\mathbf{x}}_{P/A}^{(n)} \tag{5.4.3}$$

where $\bar{\mathbf{x}}_{P/A}^{(n)}$ is constant since it is in body-fixed coordinates. Differentiating this position vector we find

$$\dot{\mathbf{x}}_P^{(n)} = \dot{\mathbf{x}}_A^{(n)} + \dot{\theta}_n \mathbf{R}_\theta^{(n)}\bar{\mathbf{x}}_{P/A}^{(n)} \tag{5.4.4}$$

which we can put in the alternate matrix-vector form as the product of a 2×3 matrix and a 3×1 vector, namely

$$\dot{\mathbf{x}}_P^{(n)} = \left[\mathbf{I}\ \mathbf{R}_\theta^{(n)}\bar{\mathbf{x}}_{P/A}^{(n)}\right]\begin{Bmatrix} \dot{\mathbf{x}}_A^{(n)} \\ \dot{\theta}_n \end{Bmatrix} \tag{5.4.5}$$

where **I** is the 2 × 2 identity matrix. The kinetic energy then becomes

$$\begin{aligned} T^{(n)} &= \frac{1}{2}\int\limits_{V_n} \rho \begin{Bmatrix} \dot{\mathbf{x}}_A^{(n)} \\ \dot{\theta}_n \end{Bmatrix}^T \begin{bmatrix} \mathbf{I} \\ \left(\bar{\mathbf{x}}_{P/A}^{(n)}\right)^T \left(\mathbf{R}_\theta^{(n)}\right)^T \end{bmatrix} \begin{bmatrix} \mathbf{I} & \mathbf{R}_\theta^{(n)} \bar{\mathbf{x}}_{P/A}^{(n)} \end{bmatrix} \begin{Bmatrix} \dot{\mathbf{x}}_A^{(n)} \\ \dot{\theta}_n \end{Bmatrix} dV \\ &= \frac{1}{2} \begin{Bmatrix} \dot{\mathbf{x}}_A^{(n)} \\ \dot{\theta}_n \end{Bmatrix}^T \left(\int\limits_{V_n} \rho \begin{bmatrix} \mathbf{I} & \mathbf{R}_\theta^{(n)} \bar{\mathbf{x}}_{P/A}^{(n)} \\ \left(\bar{\mathbf{x}}_{P/A}^{(n)}\right)^T \left(\mathbf{R}_\theta^{(n)}\right)^T & \left(\bar{\mathbf{x}}_{P/A}^{(n)}\right)^T \bar{\mathbf{x}}_{P/A}^{(n)} \end{bmatrix} dV \right) \begin{Bmatrix} \dot{\mathbf{x}}_A^{(n)} \\ \dot{\theta}_n \end{Bmatrix} \end{aligned} \tag{5.4.6}$$

where we have used the fact that $\left(\mathbf{R}_\theta^{(n)}\right)^T \mathbf{R}_\theta^{(n)} = \mathbf{I}$. Equation (5.4.6) can be written as

$$T^{(n)} = \frac{1}{2}\left(\dot{\mathbf{q}}^{(n)}\right)^T \mathbf{M}^{(n)} \dot{\mathbf{q}}^{(n)} \tag{5.4.7}$$

where $\mathbf{q}^{(n)} = \begin{bmatrix} \mathbf{x}_A^{(n)} & \theta_n \end{bmatrix}^T$ and the 3 × 3 mass matrix $\mathbf{M}^{(n)}$ is again given by (see Eqs. (5.1.13) and (5.1.14))

$$\mathbf{M}^{(n)} = \begin{bmatrix} \mathbf{m}_{XX}^{(n)} & \mathbf{m}_{X\theta}^{(n)} \\ \mathbf{m}_{\theta X}^{(n)} & m_{\theta\theta}^{(n)} \end{bmatrix} \tag{5.4.8}$$

with

$$\begin{aligned} \mathbf{m}_{XX}^{(n)} &= \left(\int\limits_{V_n} \rho dV\right)\mathbf{I} = m^{(n)}\mathbf{I} \\ \mathbf{m}_{X\theta}^{(n)} &= \left(\mathbf{m}_{\theta X}^{(n)}\right)^T = \mathbf{R}_\theta^{(n)} \int\limits_{V} \rho \bar{\mathbf{x}}_{P/A}^{(n)} dV = \int\limits_{V_n} \rho \, \widehat{\mathbf{x}}_{P/A}^{(n)} dV \\ m_{\theta\theta}^{(n)} &= \int\limits_{V_n} \rho \left(\bar{\mathbf{x}}_{P/A}^{(n)}\right)^T \bar{\mathbf{x}}_{P/A}^{(n)} dV = I_{AZ}^{(n)} \end{aligned} \tag{5.4.9}$$

One important special case is when point A is at the center of mass, C. In that case

$$\mathbf{M}^{(n)} = \begin{bmatrix} m^{(n)}\mathbf{I} & 0 \\ 0 & I_{CZ}^{(n)} \end{bmatrix} \tag{5.4.10}$$

so that

$$T^{(n)} = \frac{1}{2} m^{(n)} \mathbf{v}_C^{(n)} \cdot \mathbf{v}_C^{(n)} + \frac{1}{2} I_{CZ}^{(n)} \dot{\theta}_n^2 \tag{5.4.11}$$

If instead point A is a fixed point O about which the body is rotating we have $\dot{\mathbf{x}}_A^{(n)} = 0$ and $T^{(n)}$ becomes

$$T^{(n)} = \frac{1}{2} I_{OZ}^{(n)} \dot{\theta}_n^2 \tag{5.4.12}$$

5.4.2 *Work-Energy Principle for a Rigid Body in Plane Motion*

The work-energy theorem for a single particle says that the differential work done by the forces acting on the particle is equal to the differential change of the kinetic energy of the particle (see Eq. (3.2.40)):

$$dW_i = dT_i \tag{5.4.13}$$

If we sum over all N particles then this is also true for the entire system of particles, i.e.,

$$dW = \sum_{i=1}^{N} dW_i = dT = \sum_{i=1}^{N} dT_i \tag{5.4.14}$$

For a rigidly constrained system of particles, however, as shown in Chap. 3 the internal forces do no net work so that the total work is only due to the external applied forces, $\mathbf{F}_i^a$. Therefore, we have

$$dW = \sum_{i=1}^{N} \mathbf{F}_i^a \cdot d\mathbf{x}_i = dT \tag{5.4.15}$$

Suppose now that we have a rigidly constrained system of particles in planar motion (Fig. 5.18a). Then if we have a point A which moves with the rigid system we can write

$$\begin{aligned} \mathbf{x}_i &= \mathbf{x}_A + \mathbf{x}_{i/A} \\ &= \mathbf{x}_A + \mathbf{R}(\theta)\bar{\mathbf{x}}_{i/A} \end{aligned} \tag{5.4.16}$$

where $\bar{\mathbf{x}}_{i/A}$ is the position vector of the ith particle with respect to point A as measured in a set of body-fixed coordinates and $\mathbf{R}$ is the rotation of the rigid system. The same relationship is also true for a continuous rigid body (Fig. 5.18b) if we let $\mathbf{x}_i$ be the

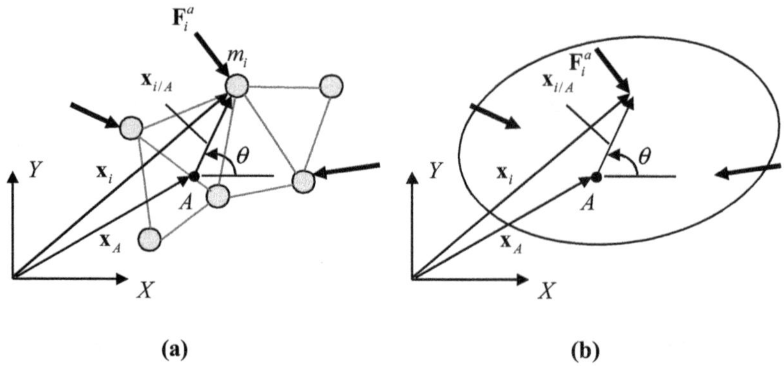

Fig. 5.18 (**a**) A system of forces acting on rigidly connected particles in planar motion and a point A that moves with the rigid system. The angle θ is shown as measured to the line from A to the ith particle but it could be the angle from the positive X-axis to any line that moves with the rigid system. (**b**) The same set of forces acting on a continuous rigid body in planar motion where $\mathbf{x}_i$ is a point on the rigid body where the ith force acts. The angle θ is shown as measured to the line from A to $\mathbf{x}_i$ but it could be the angle from the positive X-axis to any line that moves with the rigid system

point on the rigid body where F_i^a acts and point A is an arbitrary point that moves with the rigid body. Thus, Eq. (5.4.16) is applicable to either a rigidly constrained system of particles or a continuous rigid body. In a continuous rigid body we could also have forces which act over an area or a line on the body surface or are distributed throughout the body (such as gravity). Those forces could be included in our discussion of work but for simplicity we will treat all forces as discrete forces as found in the system of particles case. Upon differentiation of Eq. (5.4.16) we find

$$d\mathbf{x}_i = d\mathbf{x}_A + d\theta \mathbf{R}_\theta(\theta)\bar{\mathbf{x}}_{i/A} \tag{5.4.17}$$

Thus,

$$\begin{aligned} dW = \sum_{i=1}^{N} \mathbf{F}_i^a \cdot d\mathbf{x}_i &= \left(\sum_{i=1}^{N} \mathbf{F}_i^a\right) \cdot d\mathbf{x}_A + d\theta \sum_{i=1}^{N} \mathbf{R}_\theta \bar{\mathbf{x}}_{i/A} \cdot \mathbf{F}_i^a \\ &= \mathbf{F}^e \cdot d\mathbf{x}_A + d\theta \sum_{i=1}^{N} \left(\mathbf{R}_\theta \bar{\mathbf{x}}_{i/A}\right) \cdot \mathbf{F}_i^a \end{aligned} \tag{5.4.18}$$

where $\mathbf{F}^e$ is the total external force acting on the rigid body. Now, consider the term

$$\begin{aligned} \left(\mathbf{R}_\theta \bar{\mathbf{x}}_{i/A}\right) \cdot \mathbf{F}_i^a &= \left(\mathbf{R}_\theta \bar{\mathbf{x}}_{i/A}\right)^T \mathbf{F}_i^a \\ &= \left(\bar{\mathbf{x}}_{i/A}\right)^T \mathbf{R}_\theta^T \mathbf{F}_i^a \end{aligned} \tag{5.4.19}$$

To evaluate this term we will show that for any vector, **A**, that lies in the *X*–*Y* plane we have

$$\left(\mathbf{x}_{i/A} \times \mathbf{A}\right) \cdot \mathbf{e}_Z = \left(\bar{\mathbf{x}}_{i/A}\right)^T (\mathbf{R}_\theta)^T \mathbf{A} \tag{5.4.20}$$

To prove this, we first note that from the definition of the cross product in terms of components we have for the left side of Eq. (5.4.20)

$$\left(\mathbf{x}_{i/A} \times \mathbf{A}\right) \cdot \mathbf{e}_Z = X_{i/A} A_Y - Y_{i/A} A_X \tag{5.4.21}$$

We also have

$$\begin{aligned}\left(\bar{\mathbf{x}}_{i/A}\right)^T (\mathbf{R}_\theta)^T \mathbf{A} &= \left((\mathbf{R})^T \mathbf{x}_{i/A}\right)^T (\mathbf{R}_\theta)^T \mathbf{A} \\ &= \left(\mathbf{x}_{i/A}\right)^T \mathbf{R} (\mathbf{R}_\theta)^T \mathbf{A}\end{aligned}$$

But it is easy to show that

$$\mathbf{R}(\mathbf{R}_\theta)^T = \begin{bmatrix} 0 & 1 \\ -1 & 0 \end{bmatrix}$$

so that in terms of components

$$\begin{aligned}\left(\bar{\mathbf{x}}_{i/A}\right)^T (\mathbf{R}_\theta)^T \mathbf{A} &= \begin{bmatrix} X_{i/A} & Y_{i/A} \end{bmatrix} \begin{bmatrix} 0 & 1 \\ -1 & 0 \end{bmatrix} \begin{Bmatrix} A_X \\ A_Y \end{Bmatrix} \\ &= X_{i/A} A_Y - Y_{i/A} A_X\end{aligned}$$

which, from Eq. (5.4.21), is identical to the left-hand side of Eq. (5.4.20) so that Eq. (5.4.20) has been proven. If we let $\mathbf{A} = \mathbf{F}_i^a$ then Eq. (5.4.18) can be written as

$$dW = \mathbf{F}^e \cdot d\mathbf{x}_A + d\theta \sum_{i=1}^{N} \left(\mathbf{x}_{i/A} \times \mathbf{F}_i^a\right) \cdot \mathbf{e}_Z \tag{5.4.22}$$

But

$$\sum_{i=1}^{N} \left(\mathbf{x}_{i/A} \times \mathbf{F}_i^a\right) \cdot \mathbf{e}_Z = M_{AZ}^e \tag{5.4.23}$$

where M_{AZ}^e is the total external moment about point *A* in the *Z*-direction, so we have finally

$$dW = \mathbf{F}^e \cdot d\mathbf{x}_A + M^e_{AZ} d\theta \quad (5.4.24)$$

and the work-energy theorem for the rigid body in plane motion becomes

$$dW = \mathbf{F}^e \cdot d\mathbf{x}_A + M^e_{AZ} d\theta = dT \quad (5.4.25)$$

If the rigid body goes through a motion where $(\mathbf{x}_A, \theta)$ changes from a starting position at $(\mathbf{x}_{As}, \theta_s)$ where the kinetic energy is T_s to a final position at $(\mathbf{x}_{Af}, \theta_f)$ where the kinetic energy is T_f then we have for the total work, W,

$$W = \int_{\mathbf{x}_{As}}^{\mathbf{x}_{Af}} \mathbf{F}^e \cdot d\mathbf{x}_A + \int_{\theta_s}^{\theta_f} M^e_{AZ} d\theta = T_f - T_s = \Delta T \quad (5.4.26)$$

where the integrals will depend on the actual motion. We can also divide both sides of Eq. (5.4.24) by the differential time dt and relate it to the kinetic energy to obtain a power relationship:

$$\begin{aligned} P = \frac{dW}{dt} &= \mathbf{F}^e \cdot \frac{d\mathbf{x}_A}{dt} + M^e_{AZ} \frac{d\theta}{dt} \\ &= \mathbf{F}^e \cdot \mathbf{v}_A + M^e_{AZ} \omega_Z = \frac{dT}{dt} \end{aligned} \quad (5.4.27)$$

where P is the power (rate at which work is done), $\mathbf{v}_A$ is the velocity of point A, and $\omega_Z = \dot{\theta}$ is the angular velocity about the Z-axis. We can then integrate Eq. (5.4.27) on time to obtain

$$W = \int_{t_s}^{t_f} P dt = \int_{t_s}^{t_f} \mathbf{F}^e \cdot \mathbf{v}_A dt + \int_{t_s}^{t_f} M^e_{AZ} \omega_Z dt = T(t_f) - T(t_s) = \Delta T \quad (5.4.28)$$

If all the external forces and moments are derivable from potential energy functions then $dW = -dV$ where V is the total potential energy and we have $d(T+V) = 0$ so that the total energy is conserved for a rigid body. But note that the total external force on a rigid body may come from constraints as well as known applied forces so that all of the forces must be derivable from potentials or be workless for the total energy to be conserved. Although ideal constraint forces do no virtual work it is not true in general that ideal constraint forces are workless. For a set of constraints $\mathbf{f}(\mathbf{q}, t) = 0$ an ideal generalized constraint force $\mathbf{Q}^c = \mathbf{f_q}\boldsymbol{\lambda}$. The differential work done by that force is $dW = d\mathbf{q}^T\mathbf{Q}^c$ so that $dW = d\mathbf{q}^T\mathbf{f_q}^T\boldsymbol{\lambda} = (\mathbf{f_q}d\mathbf{q})^T\boldsymbol{\lambda}$. But from differentiating the constraints we have $\mathbf{f_q}d\mathbf{q}+\mathbf{f}_t = 0$ so $dW = -(\mathbf{f}_t)^T\boldsymbol{\lambda}$ which is not zero unless $\mathbf{f}_t = 0$, i.e., the constraints will be workless if the constraints are not explicit functions of time.

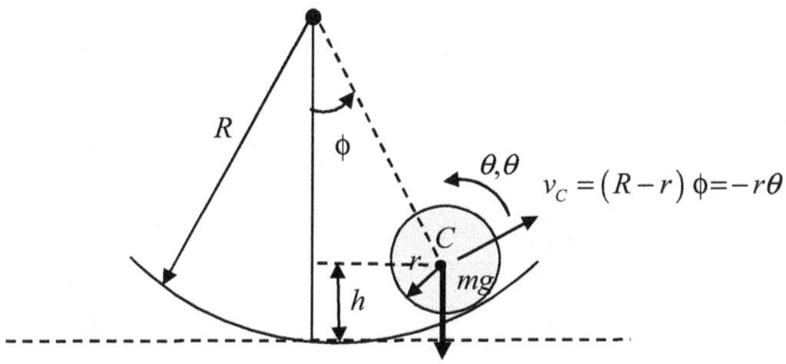

Fig. 5.19 Geometric parameters and velocities for a disk rolling without slipping on a circular track and the only force acting on the disk (its own weight) that does work

The total energy must be conserved for the problem of a disk rolling on a circular track considered in the last section since the no-slipping constraints are not explicitly functions of time and the only other force is the force of gravity, which is derivable from a potential. Thus, the total energy in that problem is (see Fig. 5.19)

$$\begin{aligned} E = T + V &= \frac{1}{2}mv_C^2 + \frac{1}{2}I_C\dot{\theta}^2 + mgh \\ &= \frac{1}{2}m(R-r)^2\dot{\phi}^2 + \frac{1}{2}\left(\frac{1}{2}mr^2\right)\left(\frac{R-r}{r}\right)^2\dot{\phi}^2 + mgh \\ &= \frac{3}{4}m(R-r)^2\dot{\phi}^2 + mg[R-(R-r)\cos\phi] \end{aligned} \tag{5.4.29}$$

If the MATLAB® script rolling_disk_sim is executed, all the terms needed to find E are available in the MATLAB® workspace as a function of the time, t, where phi $= \phi$ and phidot $= \dot{\phi}$. Thus if after running rolling_disk_sim you enter

```
E = 3*m*(R-r)^2*phidot.^2/4 +m*g*(R-(R-r)*cos(phi));
plot(t, E/E(1))
axis([ 0 10 0.95 1.05])
```

you will see the plot of $E(t)/E(0)$ shown in Fig. 5.20, where obviously the total energy of the numerical solution has been conserved (you can refine the vertical scale if you want to see explicitly the very small changes present).

5.5 Angular Momentum and the Moment Equation

5.5.1 *Motion Relative to a Point that Moves with a Rigid Body*

One of the differences between the motion of a particle and the motion of a rigid body is that the motion of the rigid body is governed by both force and moment

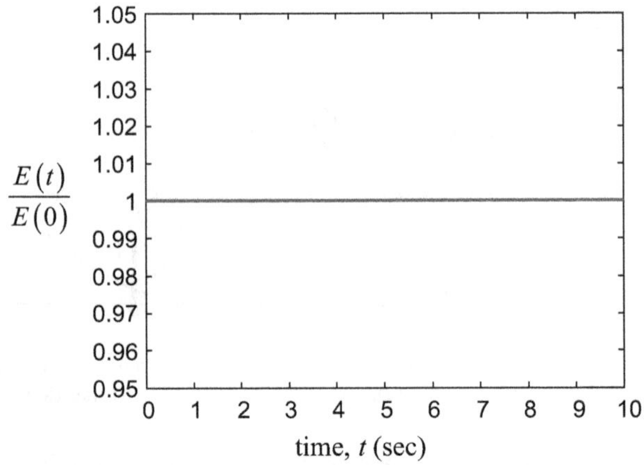

Fig. 5.20 The normalized total energy versus time for the problem of a disk rolling without slipping on a circular track

equations while the particle only requires the force equations of Newton's second law. The moments acting on a rigid body are related to changes in the angular momentum so that we must examine that relationship in more depth. Even for planar motion problems the relationship can be rather involved.

For a system of N rigidly constrained particles, the angular momentum about a point A that moves with the rigid system is

$$\mathbf{L}_A = \sum_{i=1}^{N} \mathbf{x}_{i/\mathrm{A}} \times m_i \dot{\mathbf{x}}_i \tag{5.5.1}$$

where $\mathbf{x}_{i/A}$ is the position vector of the ith particle with respect to point A. If, as before, we let $m_i = \rho \Delta V_i$ where ρ is the mass density (mass/volume) of the particle and ΔV_i is a small volume, then in the limit as the number of particles becomes infinitely large the angular momentum about point A for a continuous rigid body (n) is

$$\mathbf{L}_A^{(n)} = \int_{V_n} \rho \mathbf{x}_{P/A}^{(n)} \times \dot{\mathbf{x}}_P^{(n)} dV \tag{5.5.2}$$

But we can relate the velocity of P to the velocity of A as

$$\dot{\mathbf{x}}_P^{(n)} = \dot{\mathbf{x}}_A^{(n)} + \dot{\theta}_n R_\theta^{(n)} \bar{\mathbf{x}}_{P/A}^{(n)} \tag{5.5.3}$$

so

$$\begin{aligned}\mathbf{L}_A^{(n)} &= \int\limits_{V_n} \rho \mathbf{x}_{P/A}^{(n)} dV \times \dot{\mathbf{x}}_A^{(n)} + \int\limits_{V_n} \rho \mathbf{x}_{P/A}^{(n)} \times \dot{\theta}_n \mathbf{R}_\theta^{(n)} \bar{\mathbf{x}}_{P/A}^{(n)} dV \\ &= m^{(n)} \mathbf{x}_{C/A}^{(n)} \times \mathbf{v}_A^{(n)} + \dot{\theta}_n \mathbf{e}_Z \int\limits_{V_n} \rho \left(\widehat{\mathbf{x}}_{P/A}^{(n)} \right)^T \mathbf{R}_\theta^{(n)} \bar{\mathbf{x}}_{P/A}^{(n)} dV \\ &= m^{(n)} \mathbf{x}_{C/A}^{(n)} \times \mathbf{v}_A^{(n)} + I_{AZ}^{(n)} \dot{\theta}_n \mathbf{e}_Z\end{aligned} \tag{5.5.4}$$

where we have used Eqs. (5.1.8) and (5.1.19) to rewrite the cross product of the second term on the right-hand side of Eq. (5.5.4), first in purely matrix-vector form and then as the moment of inertia. Two important special cases are the following:

1. Point A is at the center of mass point, C. In that case $\mathbf{x}_{C/A}^{(n)} = 0$ so

$$\mathbf{L}_C^{(n)} = I_{CZ}^{(n)} \dot{\theta}_n \mathbf{e}_Z \tag{5.5.5}$$

2. Point A is at a fixed point O in the rigid body about which the body rotates. In that case $\mathbf{v}_A^{(n)} = 0$ so

$$\mathbf{L}_O^{(n)} = I_{OZ}^{(n)} \dot{\theta}_n \mathbf{e}_Z \tag{5.5.6}$$

We have already examined the moment equation for these two special cases in Eqs. (5.1.22) and (5.1.28). Thus, in terms of the angular momentum these moment equations become

$$\mathbf{M}_C^{(n)} = \dot{\mathbf{L}}_C^{(n)} = I_C^{(n)} \ddot{\theta}_n \mathbf{e}_Z \tag{5.5.7a}$$

and

$$\mathbf{M}_O^{(n)} = \dot{\mathbf{L}}_O^{(n)} = I_O^{(n)} \ddot{\theta}_n \mathbf{e}_Z \tag{5.5.7b}$$

5.5.2 Motion Relative to a General Point

It is also instructive to examine the angular momentum and the moment equation when we use a general reference point B (see Fig. 5.1a). In that case the angular momentum about B is

$$\mathbf{L}_B^{(n)} = \int\limits_{V_n} \rho \mathbf{x}_{P/B}^{(n)} \times \dot{\mathbf{x}}_P^{(n)} dV \tag{5.5.8}$$

For the terms in Eq. (5.5.8) we can write $\mathbf{x}_{P/B}^{(n)} = \mathbf{x}_{C/B}^{(n)} + \mathbf{x}_{P/C}^{(n)}$ and $\dot{\mathbf{x}}_{P}^{(n)} = \mathbf{v}_{P}^{(n)} = \mathbf{v}_{C}^{(n)} + \mathbf{v}_{P/C}^{(n)}$ where C is the center of mass so using these relations Eq. (5.5.8) becomes

$$\begin{aligned}\mathbf{L}_B^{(n)} = & \left(\int\limits_{V_n} \rho dV\right) \mathbf{x}_{C/B}^{(n)} \times \mathbf{v}_C^{(n)} + \left(\int\limits_{V} \rho \mathbf{x}_{P/C}^{(n)} dV\right) \times \mathbf{v}_C^{(n)} \\ & + \mathbf{x}_{C/B}^{(n)} \times \left(\int\limits_{V_n} \rho \mathbf{v}_{P/C}^{(n)} dV\right) + \int\limits_{V_n} \rho \mathbf{x}_{P/C}^{(n)} \times \mathbf{v}_{P/C}^{(n)} dV\end{aligned} \tag{5.5.9}$$

The first term in brackets on the right side of Eq. (5.5.9) is just the mass $m^{(n)}$ while the other two terms in brackets vanish, leaving

$$\mathbf{L}_B^{(n)} = m^{(n)} \mathbf{x}_{C/B}^{(n)} \times \mathbf{v}_C^{(n)} + \int\limits_{V_n} \rho \mathbf{x}_{P/C}^{(n)} \times \mathbf{v}_{P/C}^{(n)} dV \tag{5.5.10}$$

But since P and C are both points in the same rigid body $\mathbf{v}_{P/C}^{(n)} = \dot{\theta}_n \mathbf{R}_\theta^{(n)} \bar{\mathbf{x}}_{P/C}^{(n)}$ and we have, following the same steps that we used when considering the relations for a body-fixed point A,

$$\begin{aligned}\mathbf{L}_B^{(n)} &= m^{(n)} \mathbf{x}_{C/B}^{(n)} \times \mathbf{v}_C^{(n)} + \int\limits_{V_n} \rho \mathbf{x}_{P/C}^{(n)} \times \dot{\theta}_n \mathbf{R}_\theta^{(n)} \bar{\mathbf{x}}_{P/C}^{(n)} dV \\ &= m^{(n)} \mathbf{x}_{C/B}^{(n)} \times \mathbf{v}_C^{(n)} + \dot{\theta}_n \mathbf{e}_Z \int\limits_{V_n} \rho\, \widehat{\mathbf{x}}_{P/C}^{(n)} \mathbf{R}_\theta^{(n)} \bar{\mathbf{x}}_{P/C}^{(n)} dV \\ &= m^{(n)} \mathbf{x}_{C/B}^{(n)} \times \mathbf{v}_C^{(n)} + I_{CZ}^{(n)} \dot{\theta}_n \mathbf{e}_Z\end{aligned} \tag{5.5.11}$$

If we differentiate this angular momentum expression on time we have

$$\dot{\mathbf{L}}_B^{(n)} = m^{(n)} \mathbf{v}_{C/B}^{(n)} \times \mathbf{v}_C^{(n)} + m^{(n)} \mathbf{x}_{C/B}^{(n)} \times \mathbf{a}_C^{(n)} + I_{CZ}^{(n)} \ddot{\theta}_n \mathbf{e}_z \tag{5.5.12}$$

Now, consider the moment equation about the general point B from Eq. (5.1.3):

$$\mathbf{M}_B^{(n)} = \int\limits_{V_n} \rho \left[\mathbf{x}_{P/B}^{(n)} \times \left(\ddot{\mathbf{x}}_P^{(n)}\right)\right] dV \tag{5.5.13}$$

Letting $\mathbf{x}_{P/B}^{(n)} = \mathbf{x}_{C/B}^{(n)} + \mathbf{x}_{P/C}^{(n)}$ again we have

$$\mathbf{M}_B = \mathbf{x}_{C/B}^{(n)} \times \left(\int_{V_n} \rho \ddot{\mathbf{x}}_P^{(n)} dV \right) + \int_{V_n} \rho \mathbf{x}_{P/C}^{(n)} \times \ddot{\mathbf{x}}_P^{(n)} dV \tag{5.5.14}$$

which we can write as

$$\mathbf{M}_B = m^{(n)} \mathbf{x}_{C/B}^{(n)} \times \mathbf{a}_C^{(n)} + \int_{V_n} \rho \mathbf{x}_{P/C}^{(n)} \times \ddot{\mathbf{x}}_P^{(n)} dV \tag{5.5.15}$$

where $\mathbf{a}_C^{(n)}$ is the acceleration of the center of mass point C. The integral term is identical in form to the one considered previously when considering a body-fixed point A so we will briefly repeat the pertinent steps here in its evaluation:

$$\begin{aligned}
&\int_{V_n} \rho \mathbf{x}_{P/C}^{(n)} \times \ddot{\mathbf{x}}_P^{(n)} dV = \int_{V_n} \rho \mathbf{x}_{P/C}^{(n)} \times \left(\ddot{\mathbf{x}}_C^{(n)} + \ddot{\theta}_n \mathbf{R}_\theta^{(n)} \bar{\mathbf{x}}_{P/C}^{(n)} - (\dot{\theta}_n)^2 \mathbf{R}^{(n)} \bar{\mathbf{x}}_{P/C}^{(n)} \right) dV \\
&= \left(\int_{V_n} \rho \mathbf{x}_{P/C}^{(n)} \right) \times \mathbf{a}_C^{(n)} + \ddot{\theta}_n \mathbf{e}_Z \int_{V_n} \left(\widehat{\mathbf{x}}_{P/C}^{(n)} \right)^T R_\theta^{(n)} \bar{\mathbf{x}}_{P/C}^{(n)} dV - \dot{\theta}_n^2 \mathbf{e}_Z \int_{V_n} \rho \left(\widehat{\mathbf{x}}_{P/C}^{(n)} \right)^T \mathbf{x}_{P/C}^{(n)} dV \\
&= 0 + I_{CZ}^{(n)} \ddot{\theta}_n \mathbf{e}_Z + 0
\end{aligned} \tag{5.5.16}$$

giving

$$\mathbf{M}_B = m^{(n)} \mathbf{x}_{C/B}^{(n)} \times \mathbf{a}_C^{(n)} + I_{CZ}^{(n)} \ddot{\theta}_n \mathbf{e}_Z \tag{5.5.17}$$

If we compare Eqs. (5.5.17) and (5.5.12) we see that

$$\mathbf{M}_B^{(n)} = \dot{\mathbf{L}}_B^{(n)} - m^{(n)} \mathbf{v}_{C/B}^{(n)} \times \mathbf{v}_C^{(n)} \tag{5.5.18a}$$

or, using $\mathbf{v}_{C/B}^{(n)} = \mathbf{v}_C^{(n)} - \mathbf{v}_B^{(n)}$,

$$\mathbf{M}_B^{(n)} = \dot{\mathbf{L}}_B^{(n)} + m^{(n)} \mathbf{v}_B^{(n)} \times \mathbf{v}_C^{(n)} \tag{5.5.18b}$$

Equations (5.5.18a) and (5.5.18b) are the most general forms of the relationship between the moment and the angular momentum. Although we derived these results assuming planar motion, Eqs. (5.5.18a) and (5.5.18b) are in fact also valid for general 3-D problems as we will see in Chap. 8. Since point B is an arbitrary point we can recover all of the previous special cases considered from Eqs. (5.5.18a) and (5.5.18b). These include the following:

1. Point B is at the center of mass, C. Since $\mathbf{v}_B \times \mathbf{v}_C^{(n)} = \mathbf{v}_C^{(n)} \times \mathbf{v}_C^{(n)} = 0$

$$\mathbf{M}_C^{(n)} = \dot{\mathbf{L}}_C^{(n)} \tag{5.5.19a}$$

   Furthermore, it follows from Eq. (5.5.5) that $\mathbf{L}_C^{(n)} = I_{CZ}^{(n)}\dot{\theta}_n\mathbf{e}_Z$ and hence $\dot{\mathbf{L}}_C^{(n)} = I_{CZ}^{(n)}\ddot{\theta}_n\mathbf{e}_Z$.
2. Point B is a fixed point O in the body about which the body is rotating. Since $\mathbf{v}_B = \mathbf{v}_O = 0$

$$\mathbf{M}_O^{(n)} = \dot{\mathbf{L}}_O^{(n)} \tag{5.5.19b}$$

   Furthermore, it follows from Eq. (5.5.6) that $\mathbf{L}_O^{(n)} = I_{OZ}^{(n)}\dot{\theta}_n\mathbf{e}_Z$ and hence $\dot{\mathbf{L}}_O^{(n)} = I_{OZ}^{(n)}\ddot{\theta}_n\mathbf{e}_Z$.
3. Point B is a point D in the body where $\mathbf{x}_{C/D}^{(n)}$ is parallel to $\mathbf{a}_D$. Using Eq. (5.5.18a) and letting point A be point D in Eq. (5.5.4) we find

$$\begin{aligned}\mathbf{M}_D^{(n)} &= \dot{\mathbf{L}}_D^{(n)} - m^{(n)}\mathbf{v}_{C/D}^{(n)} \times \mathbf{v}_C^{(n)} \\ &= m^{(n)}\mathbf{x}_{C/D}^{(n)} \times \mathbf{a}_D^{(n)} + I_{DZ}^{(n)}\ddot{\theta}_n\mathbf{e}_Z \\ &= I_{DZ}^{(n)}\ddot{\theta}_n\mathbf{e}_Z\end{aligned} \tag{5.5.19c}$$

   The case of rolling without slipping of a cylinder on a surface where point D is the point of contact with the surface and the center of mass is at the center of the cylinder is an example of this type of case that we have seen before in Sect. 5.3.3. However note that if the center of mass is offset from the center of the cylinder then $\dot{\mathbf{L}}_D^{(n)} \neq I_{DZ}^{(n)}\ddot{\theta}_n\mathbf{e}_Z$ in general.
4. Point B is a fixed (inertial) point Q where Q is not a point moving with the rigid body. Since $\mathbf{v}_Q = 0$

$$\mathbf{M}_Q^{(n)} = \dot{\mathbf{L}}_Q^{(n)} \tag{5.5.19d}$$

   but we cannot relate the angular momentum about Q to a mass moment of inertia so that this choice of a point is of limited use for rigid body problems.

5.6 Solving Systems of Rigid Bodies in Plane Motion

In this chapter we have given detailed discussions of the general forms of the equations that govern planar rigid body motion. In principle, we could use any of those forms to solve a particular problem. However, in practice there are several forms and approaches that are often preferred. In Sect. 5.1, for example, we saw that we can describe the equations of plane motion of a rigid body (n) in terms of Cartesian center of mass coordinates $\left(X_C^{(n)}, Y_C^{(n)}\right)$ and the angle θ_n that a line in the body makes with respect to the positive X-axis. In general, that body will have a net external force acting it whose Cartesian components are $\left(F_X^{(n);e}, F_Y^{(n);e}\right)$ and net external moment about the center of mass is $M_{CZ}^{(n);e}$. Similarly, there may be constraints acting on the body generating constraint force components $\left(F_X^{(n);c}, F_Y^{(n);c}\right)$ and a constraint moment about the center of mass, $M_{CZ}^{(n);c}$. As we discussed, the equations of motion in these coordinates can be written in matrix-vector form as

$$\mathbf{M}^{(n)}\ddot{\mathbf{q}}^{(n)} = \mathbf{Q}^{(n);e} + \mathbf{Q}^{(n);c} \tag{5.6.1}$$

in terms of a mass matrix, $\mathbf{M}^{(n)}$, and generalized forces which are separated into externally applied forces, $\mathbf{Q}^{(n;\,e)}$, and constraint forces, $\mathbf{Q}^{(n);\,c}$:

$$\mathbf{M}^{(n)} = \begin{bmatrix} m^{(n)} & 0 & 0 \\ 0 & m^{(n)} & 0 \\ 0 & 0 & I_C^{(n)} \end{bmatrix}, \quad \mathbf{Q}^{(n);e} = \begin{Bmatrix} F_X^{(n);e} \\ F_Y^{(n);e} \\ M_{CZ}^{(n);e} \end{Bmatrix}, \quad \mathbf{Q}^{(n);c} = \begin{Bmatrix} F_X^{(n);c} \\ F_Y^{(n);c} \\ M_{CZ}^{(n);c} \end{Bmatrix} \tag{5.6.2}$$

and where $m^{(n)}$ is the mass of body (n) and $I_C^{(n)}$ is its moment of inertia with respect to the center of mass in the Z-direction. For a system of N_b interconnected rigid bodies in these coordinates we can collect all of the equations of motion in a similar matrix-vector form as

$$\mathbf{M}\ddot{\mathbf{q}} = \mathbf{Q}^e + \mathbf{Q}^c \tag{5.6.3}$$

or, in more explicit terms,

$$\begin{bmatrix} \mathbf{M}^{(1)} & 0 & \cdots & 0 \\ 0 & \mathbf{M}^{(2)} & & \vdots \\ \vdots & & \ddots & 0 \\ 0 & \cdots & 0 & \mathbf{M}^{(N_b)} \end{bmatrix} \begin{Bmatrix} \ddot{\mathbf{q}}^{(1)} \\ \ddot{\mathbf{q}}^{(2)} \\ \vdots \\ \ddot{\mathbf{q}}^{(N_b)} \end{Bmatrix} = \begin{Bmatrix} \mathbf{Q}^{(1);e} \\ \mathbf{Q}^{(2);e} \\ \vdots \\ \mathbf{Q}^{(N_b);e} \end{Bmatrix} + \begin{Bmatrix} \mathbf{Q}^{(1);c} \\ \mathbf{Q}^{(2);c} \\ \vdots \\ \mathbf{Q}^{(N_b);c} \end{Bmatrix} \tag{5.6.4}$$

where $\mathbf{M},\mathbf{q},\mathbf{Q}^e,\mathbf{Q}^c$ are collections of all the individual rigid body terms in the system. If the constraint forces are ideal constraints then we can also write them in the standard form we have seen before, namely, $\mathbf{Q}^c = \mathbf{A}^T\boldsymbol{\lambda}$, where the $\mathbf{A}$ matrix appears in the constraint equations written as

$$\mathbf{A}\ddot{\mathbf{q}} = \mathbf{b} \tag{5.6.5}$$

and, assuming that the constraints are independent, we can solve the augmented system of Eqs. (5.6.4) and (5.6.5) for the equations of motion and constraint forces as

$$\begin{aligned}\ddot{\mathbf{q}} &= \mathbf{M}^{-1}\mathbf{Q}^e + \mathbf{M}^{-1}\mathbf{A}^T\left(\mathbf{A}\mathbf{M}^{-1}\mathbf{A}^T\right)^{-1}\left(\mathbf{b} - \mathbf{A}\mathbf{M}^{-1}\mathbf{Q}^e\right)\\ \mathbf{Q}^c &= \mathbf{A}^T\left(\mathbf{A}\mathbf{M}^{-1}\mathbf{A}^T\right)^{-1}\left(\mathbf{b} - \mathbf{A}\mathbf{M}^{-1}\mathbf{Q}^e\right)\end{aligned} \tag{5.6.6}$$

Making this choice of coordinates and using an augmented approach means that for N_b rigid bodies we must solve $3N_b$ equations. This may be too large of a system to solve by hand, but it is ideally set up for a numerical computer-based solution. Some of the complexities of forming up the large vectors and matrices with this augmented approach can also be handled with the use of computer-based symbolic algebra. Because of the importance of this approach, in many of the problems at the end of this chapter we have asked that the solution be obtained with this augmented approach method. Of course, as mentioned earlier, we must be aware of the problem of constraint drift and ensure that we have an accurate numerical solution with this method.

5.6.1 Lagrange's Equations

If one wants to use generalized coordinates other than the center of mass Cartesian coordinates and angles of rotation for each rigid body to describe the motion then it is often convenient to obtain the equations of motion using Lagrange's equations. For a continuous rigid body Lagrange's equations will appear in much the same forms we have seen in the particle or multiple-particle cases but there are some differences in the underlying expressions so we will examine the derivation of Lagrange's equations here for planar motion. To keep the discussion simple, consider a single unconstrained rigid body in planar motion. Let the rigid body be described by the generalized coordinates $\mathbf{q} = (X_C, Y_C, \theta)$ and suppose we transform to a new set of generalized $\mathbf{u}$-coordinates through the transformation $\mathbf{q} = \mathbf{q}(\mathbf{u}, t)$. Then the kinetic energy is

$$T = \frac{1}{2}m\mathbf{v}_C\left(\mathbf{u}, \dot{\mathbf{u}}\right) \cdot \mathbf{v}_C\left(\mathbf{u}, \dot{\mathbf{u}}\right) + \frac{1}{2}I_C\boldsymbol{\omega}\left(\mathbf{u}, \dot{\mathbf{u}}\right) \cdot \boldsymbol{\omega}\left(\mathbf{u}, \dot{\mathbf{u}}\right) \tag{5.6.7}$$

Now, consider

$$\begin{aligned}\frac{\partial T}{\partial \dot{u}_k} &= m\mathbf{v}_C \cdot \frac{\partial \mathbf{v}_C}{\partial \dot{u}_k} + I_C \boldsymbol{\omega} \cdot \frac{\partial \boldsymbol{\omega}}{\partial \dot{u}_k} \\ &= m\mathbf{v}_C \cdot \mathbf{c}_k + I_C \boldsymbol{\omega} \cdot \mathbf{d}_k \\ \frac{\partial T}{\partial u_k} &= m\mathbf{v}_C \cdot \frac{\partial \mathbf{v}_C}{\partial u_k} + I_C \boldsymbol{\omega} \cdot \frac{\partial \boldsymbol{\omega}}{\partial u_k}\end{aligned} \tag{5.6.8}$$

where we have let

$$\mathbf{c}_k = \frac{\partial \mathbf{v}_C}{\partial \dot{u}_k}, \quad \mathbf{d}_k = \frac{\partial \boldsymbol{\omega}}{\partial \dot{u}_k} \tag{5.6.9}$$

Thus,

$$\begin{aligned}\frac{d}{dt}\left(\frac{\partial T}{\partial \dot{u}_k}\right) &= \frac{d(m\mathbf{v}_C)}{dt} \cdot \mathbf{c}_k + \frac{d(I_C \boldsymbol{\omega})}{dt} \cdot \mathbf{d}_k + m\mathbf{v}_C \cdot \dot{\mathbf{c}}_k + I_C \boldsymbol{\omega} \cdot \dot{\mathbf{d}}_k \\ &= \mathbf{F}^e \cdot \mathbf{c}_k + \mathbf{M}_C^e \cdot \mathbf{d}_k + m\mathbf{v}_C \cdot \dot{\mathbf{c}}_k + I_C \boldsymbol{\omega} \cdot \dot{\mathbf{d}}_k\end{aligned} \tag{5.6.10}$$

However, similar to the particle case we have

$$\mathbf{v}_C = \frac{d\mathbf{X}_C}{dt} = \sum_k \frac{\partial \mathbf{X}_C}{\partial u_k} \dot{u}_k + \frac{\partial \mathbf{X}_C}{\partial t} \tag{5.6.11}$$

and

$$\boldsymbol{\omega} = \frac{d\boldsymbol{\theta}}{dt} = \sum_k \frac{\partial \boldsymbol{\theta}}{\partial u_k} \dot{u}_k + \frac{\partial \boldsymbol{\theta}}{\partial t} \tag{5.6.12}$$

where $\mathbf{X}_C = (X_C, Y_C)$ and $\boldsymbol{\theta} = \theta \mathbf{e}_Z$, which show that

$$\mathbf{c}_k = \frac{\partial \mathbf{v}_C}{\partial \dot{u}_k} = \frac{\partial \mathbf{X}_C}{\partial u_k}, \quad \mathbf{d}_k = \frac{\partial \boldsymbol{\omega}}{\partial \dot{u}_k} = \frac{\partial \boldsymbol{\theta}}{\partial u_k} \tag{5.6.13}$$

Using these results we see

$$\begin{aligned}\frac{\partial \mathbf{v}_C}{\partial u_i} &= \frac{\partial}{\partial u_i}\left(\sum_k \frac{\partial \mathbf{X}_C}{\partial u_k} \dot{u}_k + \frac{\partial \mathbf{X}_C}{\partial t}\right) = \sum_i \dot{u}_k \frac{\partial \mathbf{c}_k}{\partial u_i} \\ &= \sum_k \dot{u}_k \frac{\partial^2 \mathbf{X}_C}{\partial u_i \partial u_k} = \sum_k \dot{u}_k \frac{\partial}{\partial u_k}\left(\frac{\partial \mathbf{X}_C}{\partial u_i}\right) \\ &= \sum_k \dot{u}_k \frac{\partial \mathbf{c}_i}{\partial u_k} = \dot{\mathbf{c}}_i\end{aligned} \tag{5.6.14}$$

and in an entirely similar manner we can show

$$\frac{\partial \boldsymbol{\omega}}{\partial u_i} = \dot{\mathbf{d}}_i \tag{5.6.15}$$

Using Eqs. (5.6.14) and (5.6.15) into Eq. (5.6.10), and using Eq. (5.6.8), Eq. (5.6.10) becomes

$$\frac{d}{dt}\left(\frac{\partial T}{\partial \dot{u}_k}\right) - \frac{\partial T}{\partial u_k} = \tilde{Q}_k^e \tag{5.6.16}$$

which is Lagrange's equation with a generalized external force, $\tilde{Q}_k^e$, given by

$$\begin{aligned} \tilde{Q}_k^e &= \mathbf{F}^e \cdot \mathbf{c}_k + \mathbf{M}_C^e \cdot \mathbf{d}_k \\ &= \mathbf{F}^e \cdot \frac{\partial \mathbf{v}_C}{\partial \dot{u}_k} + \mathbf{M}_C^e \cdot \frac{\partial \boldsymbol{\omega}}{\partial \dot{u}_k} \end{aligned} \tag{5.6.17}$$

If there are constraints acting on the rigid body then there will also be generalized constraint forces, whose components, $\tilde{Q}_k^c$, are given in terms of Cartesian components:

$$\begin{aligned} \tilde{Q}_k^c &= \mathbf{F}^c \cdot \mathbf{c}_k + \mathbf{M}_C^c \cdot \mathbf{d}_k \\ &= \sum_{i=1}^{2}\left(F_i^c \frac{\partial \dot{X}_{Ci}}{\partial \dot{u}_k} + M_{CZ}^c \frac{\partial \dot{\theta}}{\partial \dot{u}_k}\right) \end{aligned} \tag{5.6.18}$$

where we have let $(X_C, Y_C, \theta) = (X_{C1}, X_{C2}, \theta)$. In the original **q**-coordinates the generalized forces were $\left(Q_1^c, Q_2^c, Q_3^c\right) = \left(F_1^c, F_2^c, M_{Cz}^c\right)$ so from Eq. (5.6.18) we can write the new generalized constraint force, $\tilde{Q}_k^c$, as

$$\tilde{Q}_k^c = \sum_{i=1}^{3} Q_i^c \frac{\partial \dot{q}_i}{\partial \dot{u}_k} \tag{5.6.19}$$

Now suppose there are p constraint equations ($p = 1$ or 2) in terms of the original **q**-variables in the form we have seen many times before:

$$\sum_{i=1}^{3} A_{ki}\ddot{q}_i = b_k \quad (k = 1, \ldots, p) \tag{5.6.20}$$

and that these constraints generate ideal constraint "forces." Then from the Lagrangian prescription we have

$$Q_i^c = \sum_{k=1}^{p} A_{ki}\lambda_k \tag{5.6.21}$$

But if we have $q_i = q_i(\mathbf{u}, t)$ then

$$\begin{aligned}\dot{q}_i &= \sum_{m=1}^{3} \frac{\partial q_i}{\partial u_m}\dot{u}_m + \frac{\partial q_i}{\partial t} \\ \ddot{q}_i &= \sum_{m=1}^{3} \frac{\partial q_i}{\partial u_m}\ddot{u}_m + \sum_{m=1}^{3} \dot{u}_m \frac{d}{dt}\left(\frac{\partial q_i}{\partial u_m}\right) + \frac{\partial^2 q_i}{\partial t^2}\end{aligned} \tag{5.6.22}$$

which shows that we have

$$\frac{\partial \dot{q}_i}{\partial \dot{u}_m} = \frac{\partial \ddot{q}_i}{\partial \ddot{u}_m} = \frac{\partial q_i}{\partial u_m} \tag{5.6.23}$$

which is a relationship called the "cancelling of the dots" for the obvious reason. If we place the acceleration expression in Eq. (5.6.22) into the constraint equations, Eq. (5.6.20), then the constraints can be written as

$$\sum_{i=1}^{3} \sum_{m=1}^{3} A_{ki} \frac{\partial q_i}{\partial u_m}\ddot{u}_m = \tilde{b}_k \quad (k = 1, \ldots, p) \tag{5.6.24}$$

or, equivalently,

$$\sum_{m=1}^{3} \tilde{A}_{km}\ddot{u}_m = \tilde{b}_k \quad (k = 1, \ldots, p) \tag{5.6.25}$$

where

$$\tilde{A}_{km} = \sum_{i=1}^{3} A_{ki} \frac{\partial q_i}{\partial u_m} = \sum_{i=1}^{3} A_{ki} \frac{\partial \dot{q}_i}{\partial \dot{u}_m} \tag{5.6.26}$$

Now, let us return to the generalized constraint force, Eq. (5.6.19), and use Eqs. (5.6.21) and (5.6.26). We find

$$\begin{aligned}\tilde{Q}_m^c &= \sum_{i=1}^{3}\sum_{k=1}^{p} A_{ki}\frac{\partial \dot{q}_i}{\partial \dot{u}_m}\lambda_k \\ &= \sum_{k=1}^{p}\tilde{A}_{km}\lambda_k\end{aligned} \tag{5.6.27}$$

which shows that the generalized constraint force in the new generalized **u**-coordinates remains an ideal constraint, i.e., it satisfies the Lagrange prescription.

In summary, the motion of a constrained rigid body can be written, using Lagrange's equations and the constraint equations in the new set of generalized (and constrained) coordinates, as

$$\begin{aligned}&\frac{d}{dt}\left(\frac{\partial T}{\partial \dot{u}_m}\right) - \frac{\partial T}{\partial u_m} = \tilde{Q}_m^e + \sum_{k=1}^{p}\tilde{A}_{km}\lambda_k \\ &\sum_{m=1}^{3}\tilde{A}_{km}\ddot{u}_m = \tilde{b}_k \quad (k = 1, \ldots, p)\end{aligned} \tag{5.6.28}$$

Also, it is important to note that we have shown that *the form of Eq. (5.6.28) for the equations of motion and the acceleration-based constraint equations for a constrained rigid body with ideal constraint forces is the same for any generalized coordinates we may want to use.*

Once the kinetic energy terms in Lagrange's equations are evaluated explicitly we can identify the $\mathbf{M}\ddot{\mathbf{u}}$ term and the inertial force term, $\tilde{\mathbf{Q}}^{in}$ (which was missing in the original **q**-coordinates), and solve for the equations of motion and generalized constraint force with the standard augmented approach in the generalized **u**-coordinates. We can use Lagrange's equations in exactly the same manner for a system of rigid bodies, but we will not show that generalization here. A number of the problems in the next section use this Lagrangian approach to generate the equations of motion.

5.7 Problems

P5.1. The two ends of a homogeneous thin bar of mass m and length l ride in smooth slots as shown in Fig. P5.1. (a) Derive the equation of motion for this one degree of freedom problem using a direct Newton-Euler approach, i.e., draw a free body diagram and apply the force and moment equations of motion. Using the constraint equations, eliminate the unknown constraint forces and obtain a single differential equation for the angle, θ. Solve this equation numerically. Let $m = 4$ kg, $l = 1.5$ m, and $\theta(0) = 3\pi/4$ rad, $\dot{\theta}(0) = 0$ and take the solution time to be long enough to see a complete sequence

Fig. P5.1 A rigid bar of mass m and length l rides on rollers in a smooth slot. Neglect the masses of the rollers

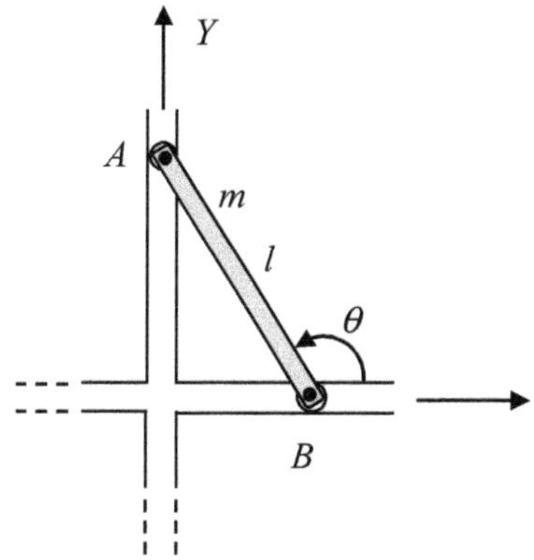

of up-and-down and back-and-forth motions for the bar. (b) Animate the motion using the MATLAB® function animate_lines described in Chap. 3.

P5.2. Use an augmented approach to solve for the motion of the bar in Fig. P5.1 in terms of the coordinates (X_C, Y_C, θ) where (X_C, Y_C) are the X- and Y-coordinates of the center of mass of the bar and θ is the angle shown in Fig. P5.1 Solve the equations of motion numerically for the bar in conjunction with the twice-differentiated constraint equations. Let $m = 4$ kg, $l = 1.5$ m, and $\theta(0) = 3\pi/4$ rad, $\dot{\theta}(0) = 0$ and take the solution time to be long enough to see a complete sequence of up-and-down and back-and-forth motions for the bar. Compare with the solution obtained in problem P5.1.

P5.3. Using the numerical solution of problem P5.2 determine the constraint forces acting at A and B on the bar.

P5.4. The motion of the bar shown in Fig. P5.1 is a conservative problem so show that the numerical solution obtained in problem P5.2 satisfies the conservation of energy during the motion. What about the satisfaction of the positional constraint equation and its first derivative during the motion?

P5.5. Consider the case where a thin homogeneous bar is connected to two sliding masses as shown in Fig. P5.2. Use a Lagrangian approach to obtain the equation of motion of this single degree of freedom problem for the angle, θ. Let $m_1 = m_2 = 2$ kg, $m_3 = 3$ kg, $l_2 = 1.5$ m, and $\theta_2(0) = 3\pi/4$, $\dot{\theta}_2(0) = 0$. Solve this equation numerically. Take the solution time to be long enough to see a complete sequence of up-and-down and back-and-forth motions for the bar.

P5.6. Consider the system shown in Fig. P5.2. Use an augmented approach to solve for the motion of the bar and the masses in terms of the coordinates $(Y^{(1)}, X^{(2)}, Y^{(2)}, \theta_2, X^{(3)})$ where $(X^{(n)}, Y^{(n)})$ are the X- and Y-coordinates of the center of mass of body (n) and let $m_1 = m_2 = 2$ kg, $m_3 = 3$ kg, $l_2 = 1.5$ m, and $\theta_2(0) = 3\pi/4$, $\dot{\theta}_2(0) = 0$. Again take the solution time to be long enough to see a complete sequence of up-and-down and back-and-forth motions for the bar. Compare this solution with the solution of problem P5.5.

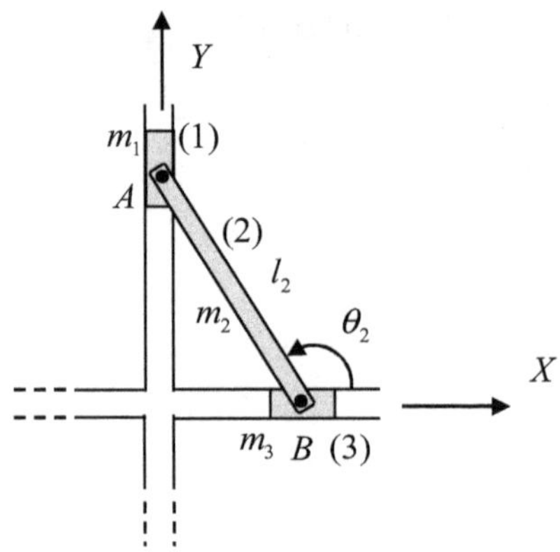

Fig. P5.2 A rigid bar of length l_2 and mass m_2 is pinned to two masses (m_1, m_3) that slide in smooth slots

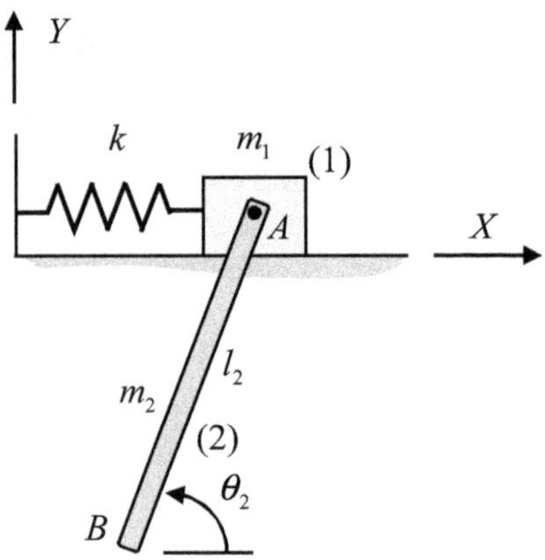

Fig. P5.3 A mass m_1 attached to a linear spring with spring constant k and pinned to a thin, homogeneous bar of mass m_2 and length l_2

P5.7. Using the numerical solution of problem P5.6 determine the constraint forces acting at A and B on the bar.

P5.8. P5.8 The motion of the bar shown in Fig. P5.2 is a conservative problem so show that the numerical solution obtained in problem P5.6 satisfies the conservation of energy during the motion. What about the satisfaction of the positional constraint equation and its first derivative during the motion?

P5.9. A block of mass m_1 slides on a horizontal surface and is attached to a linear spring with spring constant k. The block is pinned to a thin homogeneous rod of mass m_2 and length l_2, as shown in Fig. P5.3. (a) Use a Lagrangian approach to obtain the equations of motion for this two degree of freedom system in terms of $(X^{(1)}, \theta_2)$ where $X^{(1)}$ is the horizontal position of the block. Solve this system numerically. Assume that the spring is unstretched when $X^{(1)} = 0$. Let $m_1 = 2$ kg, $m_2 = 3$ kg, $l_2 = 2$ m, and $k = 15$ N/m and take the rod to be vertical at time $t = 0$ when the spring has a stretch of 1 m and the system is released from rest. Let the time interval for the solution be 15 s. (b) Animate the motion using the MATLAB® function animate_lines described in Chap. 3.

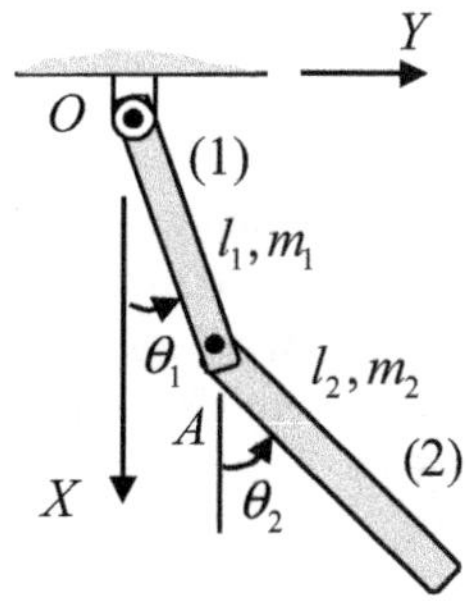

Fig. P5.4 A two-bar pendulum

P5.10. Consider the system shown in Fig. P5.3. Use an augmented approach to solve for the motion of this two degree of freedom system using the generalized coordinates $(X^{(1)}, X^{(2)}, Y^{(2)}, \theta_2)$, where $(X^{(n)}, Y^{(n)})$ are the coordinates in the X- and Y-directions of the center of mass of body (n). As in problem P5.9 assume that the spring is unstretched when $X^{(1)} = 0$. Let $m_1 = 2$ kg, $m_2 = 3$ kg, $l_2 = 2$ m, and $k = 15$ N/m and take the rod to be vertical at time $t = 0$ when the spring has a stretch of 1 m and the system is released from rest. Let the time interval for the solution be 15 s. Compare your solution with the solution of problem P5.9.

P5.11. Using the numerical solution of problem P5.10 determine the constraint forces acting at A on the rod.

P5.12. The motion of the system in Fig. P5.3 is a conservative problem so show that the numerical solution obtained in problem P5.10 satisfies the conservation of energy during the motion. What about the satisfaction of the positional constraint equation and its first derivative during the motion?

P5.13. The double pendulum of Fig. P5.4 consists of two thin, homogeneous bars where bar (1) has a mass m_1 and a length l_1 and bar (2) has a mass m_2 and a length l_2. (b) Use a Lagrangian approach to obtain the equations of motion for this two degree of freedom system in terms of (θ_1, θ_2). Let $m_1 = 2$ kg, $l_1 = 1.5$ m, $m_2 = 3$ kg, and $l_2 = 2$ m. At time $t = 0$ take $\theta_1(0) = \pi/4$ and $\theta_2(0) = \pi/4$ when the system is released from rest and solve for the motion over a time interval of 10 s. (b) Animate the motion using the MATLAB® function animate_lines described in Chap. 3.

P5.14. Consider the double pendulum shown in Fig. P5.4. Use an augmented approach to solve for the motion of this two degree of freedom system using the generalized coordinates $(X^{(1)}, Y^{(1)}, \theta_1, X^{(2)}, Y^{(2)}, \theta_2)$, where $(X^{(n)}, Y^{(n)})$ are the coordinates in the X- and Y-directions of the center of mass of bar (n). As in problem P5.13 let $m_1 = 2$ kg, $l_1 = 1.5$ m, $m_2 = 3$ kg, and $l_2 =$ 2 m. At time $t = 0$ take $\theta_1(0) = \pi/4$ and $\theta_2(0) = \pi/4$ when the system is released from rest and solve for the motion over a time interval of 10 s. Compare to the solution obtained previously in problem P5.13. Animate the motion using the MATLAB® function animate_lines described in Chap. 3.

P5.15. Using the numerical solution of problem P5.14 determine the constraint forces acting at O and A on bar (1) and the constraint forces acting at A on bar (2).

P5.16. The motion of the system in Fig. P5.4 is a conservative problem so show that the numerical solution obtained in problem P5.14 satisfies the conservation of energy during the motion. What about the satisfaction of the positional constraint equation and its first derivative during the motion?

P5.17. Consider the crank-slider system shown in Fig. P5.5 where the crank AB is being driven by an external moment, $M(t)$. (a) Use a Lagrangian approach to obtain the equations of motion of this system in terms of the coordinates $\left(X_P, \tilde{\theta}_1, \tilde{\theta}_2\right)$. Incorporate the constraint equations as ideal constraint forces, which are given explicitly in terms of the second-order time derivatives of the constraint equations in an augmented approach. Let $m_1 g = 2$ lb., $m_2 g = 1$ lb., $m_3 g = 3$ lb., $l_1 = 0.5$ ft., and $l_2 = 1.0$ ft., and assume that the driving moment is a constant given by $M = 2$ ft.-lb. At time $t = 0$ assume that the system is released from rest and $\tilde{\theta}_1 = 30°$. Solve numerically for the motion over at least three revolutions of the crank. (b) Animate the motion using the MATLAB® function animate_lines described in Chap. 3.

P5.18. Solve the problem of the crank-slider shown in Fig. P5.5 using an augmented approach with the coordinates $(X^{(1)}, Y^{(1)}, \theta_1, X^{(2)}, Y^{(2)}, \theta_2, X^{(3)})$ where $(X^{(n)}, Y^{(n)})$ are the coordinates of the center of mass of body (n) and (θ_1, θ_2) are angles measured from the positive X-axis (note the change of θ_2 from the angle $\tilde{\theta}_2$ shown in Fig. P5.5). Let $m_1 g = 2$ lb., $m_2 g = 1$ lb., $m_3 g = 3$ lb., $l_1 = 0.5$ ft., and $l_2 = 1.0$ ft., and assume that the driving moment is a constant given by $M = 2$ ft.-lb. At time $t = 0$ assume that the system is released from rest and $\tilde{\theta}_1 = 30°$. Solve numerically for the motion over at least three revolutions of the crank. Compare your solution with the solution obtained in problem P5.17. Animate the motion using the MATLAB® function animate_lines described in Chap. 3.

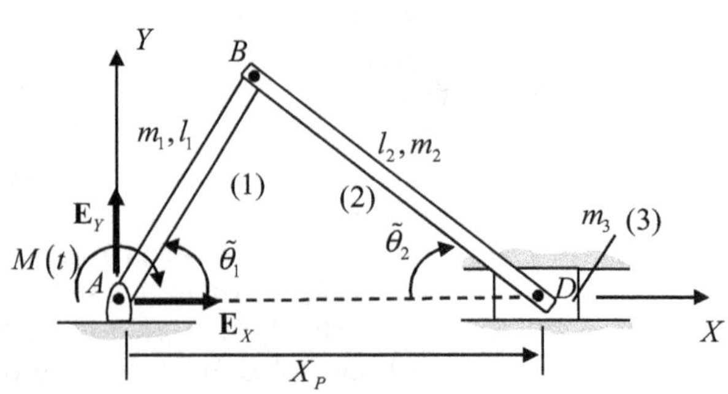

Fig. P5.5 Motion of a crank-slider mechanism

Fig. P5.6 A two-bar mechanism where the slot along the bar *BE* rides in a smooth, fixed pin at *D*

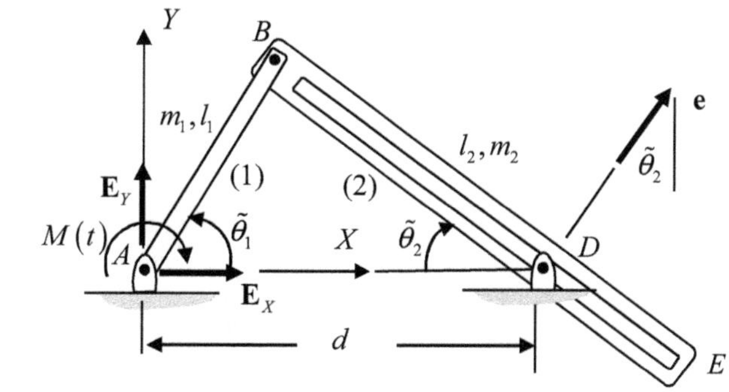

P5.19. Using the numerical solution of problem P5.18 determine the constraint forces acting at *A* and *B* on bar (1) and the forces at *B* and *D* acting on bar (2).

P5.20. Examine the satisfaction of the constraint equation and its time derivative for the numerical solution of problem P5.18.

P5.21. Consider the two-bar mechanism shown in Fig. P5.6 where bar *AB* is being driven by an external moment, *M*(*t*), and bar *BE* has a smooth slot which rides in a fixed pin at *D*. (a) Use the Lagrangian for this system to obtain the equations of motion in terms of the coordinates $(\tilde{\theta}_1, \tilde{\theta}_2)$ and use an augmented approach to solve these two equations of motion in conjunction with the constraint equation for this one degree of freedom system. Note that the constraint equation can be obtained from the geometry of Fig. P5.6 as $f(\tilde{\theta}_1, \tilde{\theta}_2) = (\mathbf{x}_{B/A} - d\,\mathbf{E}_X) \cdot \mathbf{e} = 0$, where **e** is a unit vector normal to bar *BE* at point *D*. Let $m_1 g = 2$ lb., $m_2 g = 3$ lb., $l_1 = 1.0$ ft., and $l_2 = 2.0$ ft., and assume that the driving moment is a constant given by $M = 3$ ft.-lb. At time $t = 0$ assume that the system is released from rest and $\tilde{\theta}_1 = 90°$. Solve numerically for the motion over at least three revolutions of the crank. (b) Animate the motion using the MATLAB® function animate_lines described in Chap. 3.

P5.22. Solve the problem of the two-bar mechanism shown in Fig. P5.6, where bar *AB* is being driven by an external moment, *M*(*t*), and bar *BE* has a smooth slot which rides in a fixed pin at *D*. (a) Use an augmented approach with the coordinates $(X^{(1)}, Y^{(1)}, \theta_1, X^{(2)}, Y^{(2)}, \theta_2)$ where $(X^{(n)}, Y^{(n)})$ are the coordinates of the center of mass of body (*n*) and θ_n are angles of body (*n*) as measured from the positive *X*-axis (note the change of θ_2 from the angle $\tilde{\theta}_2$ shown in Fig. P5.6). The constraint equation at *D* can be obtained from the geometry of Fig. P5.6 as $g(\theta_1, \theta_2) = (\mathbf{x}_{B/A} - d\,\mathbf{E}_X) \cdot \mathbf{e} = 0$, where **e** is a unit vector normal to bar *BE* at point *D*. Let $m_1 g = 2$ lb., $m_2 g = 3$ lb., $l_1 = 1.0$ ft., and $l_2 = 2.0$ ft., and assume that the driving moment is a constant given by $M = 3$ ft.-lb. At time $t = 0$ assume that the system is released from rest and $\tilde{\theta}_1 = 90°$. Solve numerically for the motion over at least three revolutions of the crank. Compare your solution with the solution obtained in problem P5.21.

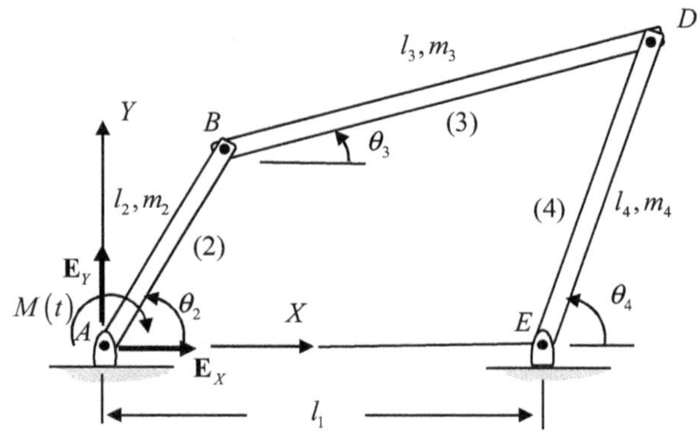

Fig. P5.7 A four-bar mechanism

P5.23. Using the numerical solution of problem P5.22 determine the constraint forces acting on bar (1) at *A* and *B* and the constraint forces acting on bar (2) at *B* and *D*.

P5.24. Examine the satisfaction of the constraint equation and its time derivative for the numerical solution of problem P5.22.

P5.25. Consider the four-bar mechanism shown in Fig. P5.7 where bar *AB* is being driven by an external moment, *M*(*t*). (a) Use the Lagrangian for this system to obtain the equations of motion in terms of the coordinates $(\theta_2, \theta_3, \theta_4)$ and use an augmented approach to solve these three equations of motion in conjunction with the two constraint equations for this one degree of freedom system. Let $m_2 = 3$ kg, $m_3 = 3$ kg, $m_4 = 3$ kg, $l_1 = 1.0$ m, $l_2 = 0.4$ m, $l_3 = 0.8$ m, and $l_4 = 0.7$ m, and assume that the driving moment is $M(t) = 2t$ N-m, where *t* is the time in seconds. At time $t = 0$ the system is released from rest and $\theta_2 = \pi/2$ rad. In this mechanism, while bar *AB* can execute continuous revolutions, bar *DE* performs a rocking back-and-forth motion. Solve numerically for the motion over a time frame long enough to see four complete back-and-forth motions of *DE*. (b) Animate the motion using the MATLAB® function animate_lines described in Chap. 3.

P5.26. Solve the problem of the four-bar mechanism of Fig. P5.7 where bar AB is being driven by an external moment, *M*(*t*). (a) Use an augmented approach to solve this problem with the coordinates $(X^{(2)}, Y^{(2)}, \theta_2, X^{(3)}, Y^{(3)}, \theta_3, X^{(4)}, Y^{(4)}, \theta_4)$ where $(X^{(n)}, Y^{(n)})$ are the coordinates of the center of mass of body (*n*) and θ_n are angles of body (*n*) as measured from the positive *X*-axis. Here there are nine equations of motion and eight constraint equations for this one degree of freedom problem. Let $m_2 = 3$ kg, $m_3 = 3$ kg, $m_4 = 3$ kg, $l_1 = 1.0$ m, $l_2 = 0.4$ m, $l_3 = 0.8$ m, and $l_4 = 0.7$ m, and assume that the driving moment is $M(t) = 2t$ N-m, where *t* is the time in seconds. At time $t = 0$ the system is released from rest and $\theta_2 = \pi/2$ rad. In this mechanism, while bar *AB* can execute continuous revolutions, bar *DE* performs a rocking back-and-forth motion. Solve numerically for the motion over a time frame long enough to see four complete back-and-forth motions of *DE*. Compare your solution with the solution obtained in problem P5.25. (b) Animate the motion using the MATLAB® function animate_lines described in Chap. 3.

P5.27. Use the numerical solution of problem P5.26 to obtain the constraint forces at *A* and *B* acting on bar (2), the constraint forces at *B* and *D* acting on bar (3), and the constraint forces at *D* and *E* acting on bar DE.

P5.28. Examine the satisfaction of the constraint equation and its time derivative for the numerical solution of problem P5.26.

Chapter 6
Dynamic and Static Stability

The motion of dynamical systems is strongly dependent on whether their behavior is stable or unstable. This chapter obtains the conditions which determine stability for certain types of systems that we have analyzed in previous chapters, using a direct method that involves the properties of the potential energy or the dynamic potential energy. We also discuss an indirect method that first requires the linearization of the equations of motion, but which is applicable to a wider class of systems. Since the direct dynamic stability method relies on the behavior of the potential energy, a quantity that in most dynamical systems is independent of time, this stability criterion is also applicable to static systems, i.e., systems that are designed to inherently be in equilibrium. Stability of static systems is rarely treated in any depth in traditional statics or dynamics courses so we will also analyze some of the important ways in which such static systems can lose stability.

6.1 Dynamic Stability

There are a number of ways to define dynamic stability. In this section we outline in physical terms one common stability criterion using the graphical depictions of Rosenberg [1]. Consider, for example, a two degrees of freedom motion of a dynamical system where we can define the motion in terms of the two generalized coordinates (q_1, q_2) (Fig. 6.1). Under a given set of initial conditions we can plot the motion of the system as a path followed in the (q_1, q_2, t) coordinates, where t is the time. The solid line, labeled "original motion" in Fig. 6.1, is such a path. If we perturb the initial conditions slightly then the motion will occur along a different path, shown as the dashed line in Fig. 6.1. If at some time this different path is close to the unperturbed path and remains close to it for all later times, we say that the motion is stable. We can visualize this criterion by saying that the path of motion must always lie in a region which surrounds the unperturbed path and does not grow with time such as the tube shown in Fig. 6.1. This "definition" of stability can be

L. W. Schmerr, *Engineering Dynamics 2.0*, Solid Mechanics and Its Applications 254,
https://doi.org/10.1007/978-3-319-98470-4_6

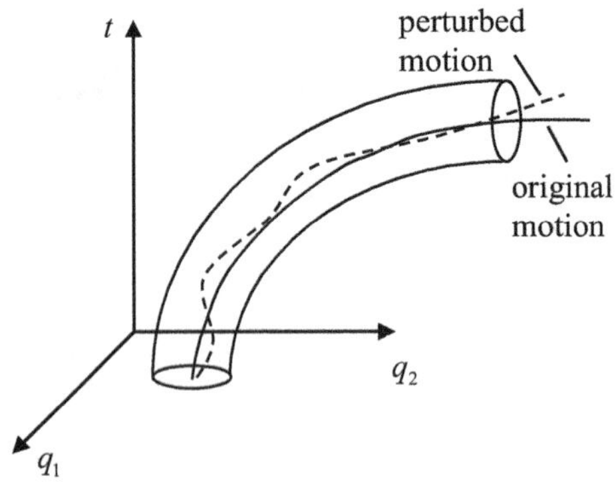

Fig. 6.1 Representation of Lyapunov stability of a system where the perturbed motion (dashed line) always stays near the original unperturbed motion (solid line)

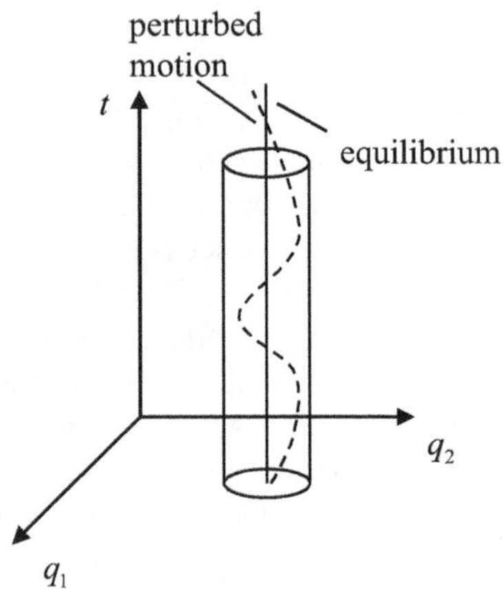

Fig. 6.2 Representation of Lyapunov stability about an equilibrium position

given in a mathematically more precise form called stability in the sense of Lyapunov. Stability of dynamical systems about equilibrium points is a special case of this definition of stability as shown in Fig. 6.2. If an equilibrium point at $t = 0$ has the values (q_{10}, q_{20}) then the unperturbed system always has these same equilibrium values, which is the vertical solid line shown in Fig. 6.2. A stable perturbed solution will always remain within a region such as the vertical tube seen in that figure. Since dynamical systems often exhibit significant nonlinear behavior, as we have seen in many of the example problems we have already considered, directly proving that a specific system is Lyapunov stable or not is often not possible. One can, of course, run a series of numerical experiments with different initial conditions to examine the behavior of the system to see if any unstable behavior is manifestly present. While such experiments cannot prove that a system is stable, they can provide some indication of the behavior of the system under "normal" conditions. There is, however, one important special case, which we will examine in the following section, where we can examine stability in a more direct and complete manner.

6.2 Stability of a Natural, Conservative System Near Equilibrium

Consider the case of a natural system with p degrees of freedom. Recall from Chap. 2 that for natural systems when we transform from Cartesian to p independent generalized coordinates the time, t, is not explicitly present in that transformation so that the kinetic energy, T, is just quadratic in the generalized velocities, i.e. (see Eq. (2.7.14)),

$$T = T_2 = \frac{1}{2}\sum_{i=1}^{p}\sum_{j=1}^{p} M_{ij}(\mathbf{q})\dot{q}_i\dot{q}_j \tag{6.2.1}$$

where M_{ij} is a positive definite mass matrix. If in addition all the forces are derivable from potential energy functions, then the total energy, $E = T+V = T_2+V$, is conserved, where V is the total potential energy of all the forces present. For such natural, conservative systems one has Lagrange-Dirichlet stability theorem which says:

Lagrange-Dirichlet Stability Theorem
An equilibrium position of a natural, conservative dynamical system with p degrees of freedom is (Lyapunov) stable if the potential energy, $V(q_1, q_2, \ldots, q_p)$, which is stationary at an equilibrium point $\mathbf{q}_0 = (q_{10}, q_{20}, \ldots, q_{p0})$, also has a minimum value at that equilibrium point relative to all neighboring points.

According to this theorem an equilibrium state is *stable* if when the system is given any small change in the generalized coordinates from the equilibrium state, then the change of the potential energy is positive, i.e., $\Delta V > 0$. If, for some small change of one or more coordinates, $\Delta V < 0$, then the equilibrium state is *unstable*, while if $\Delta V = 0$ for some small change of one or more coordinates but $\Delta V > 0$ for all other coordinate changes then we will say that the equilibrium state is *neutrally stable*. These conditions make sense from a physical standpoint since conservation of the total energy and the kinetic energy being given by Eq. (6.2.1) means that a positive change of the potential energy requires that the kinetic energy must decrease as a system moves away from the equilibrium position, thus limiting how far from equilibrium the system can go. Likewise, a negative change of the potential energy means that the kinetic energy will actually increase as the system moves to a nearby position and the system will tend to be driven away from the equilibrium position. A zero change of potential energy means that there is no tendency to either keep the system close to equilibrium or drive it away from equilibrium. As a simple example of this reasoning, consider the case of a one degree of freedom system such as the ball of mass m constrained to move in the x–y plane on a cylindrical surface as shown in Fig. 6.3. Here the total potential energy of the system is just that of gravity, i.e., $V = mgy$. For an equilibrium position at the bottom of a cylindrical trough (Fig. 6.3a), we see that y must increase for any small displacement either to the

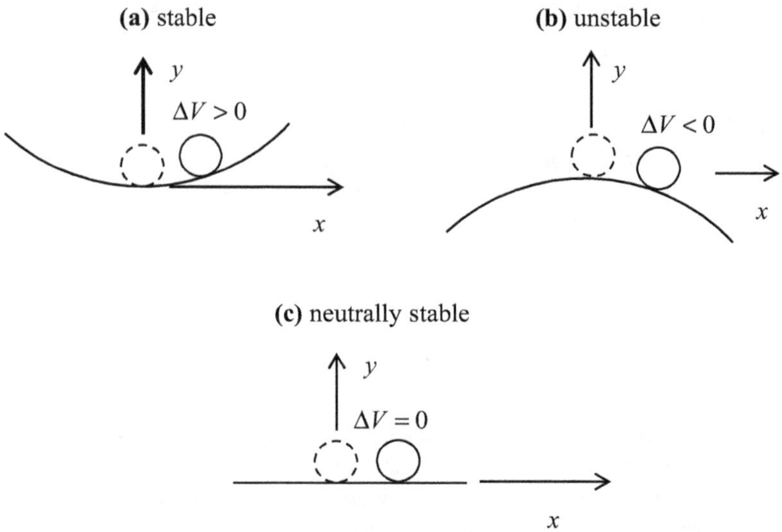

Fig. 6.3 Examples of (**a**) a stable equilibrium position, (**b**) an unstable equilibrium position, and (**c**) a neutrally stable position

right or to the left so that $\Delta V > 0$ always and the equilibrium state is stable. For the cylindrical hill of Fig. 6.3b, however, we see that $\Delta V < 0$ away from equilibrium at the top of the hill so that the equilibrium state is unstable. Finally, for the case of the ball on the plane surface of Fig. 6.3c it is clear that y is a constant for any small displacement from an equilibrium position so $\Delta V = 0$ and the equilibrium is by definition neutrally stable.

The fact that the potential energy is stationary at equilibrium follows from Lagrange's equations for conservative systems at equilibrium (where the kinetic energy is identically zero):

$$\left.\frac{\partial V}{\partial q_k}\right|_{\mathbf{q}=\mathbf{q}_0} = 0 \quad (k, = 1, 2, \ldots, p) \tag{6.2.2}$$

Determining if the potential energy has a minimum is generally done by examining the second derivatives of the potential energy near an equilibrium position. For example, if we expand the potential energy in a Taylor series about an equilibrium point we have

$$V(\mathbf{q}) = V(\mathbf{q}_0) + \sum_{k=1}^{p} \left.\frac{\partial V}{\partial q_k}\right|_{\mathbf{q}=\mathbf{q}_0} \Delta q_k + \frac{1}{2} \sum_{k=1}^{p} \sum_{m=1}^{p} \left.\frac{\partial^2 V}{\partial q_k \partial q_m}\right|_{\mathbf{q}=\mathbf{q}_0} \Delta q_k \Delta q_m + H.O.T \tag{6.2.3}$$

$$|A| = \left| \begin{array}{ccc} \partial^2 V / q_1^2 & & \\ \vdots & & \\ \partial^2 V / \partial q_{N-1}\partial q_1 & \partial^2 V / \partial q_1 \partial q_{N-1} \;\; \partial^2 V / \partial q_{N-1}^2 & \partial^2 V / \partial q_1 \partial q_N \\ \partial^2 V / \partial q_N \partial q_1 & \cdots & \partial^2 V / q_N^2 \end{array} \right| \quad D_{N-1}, \; D_N$$

Fig. 6.4 Determinant of the matrix of second derivatives of the total potential energy and of the principal minors of that matrix

where *H.O.T.* indicates "higher order terms." Since the first derivatives vanish at equilibrium we have $\Delta V = V(\mathbf{q}) - V(\mathbf{q}_o) > 0$ for all small changes Δq_k if the matrix of second derivatives, A_{km}, is positive definite, where $A_{km} = \partial^2 V/\partial q_k \partial q_m$. This assumes, of course, that all the second derivatives do not vanish, since if this occurs one must then consider the higher order derivatives in the *H.O.T.* that contribute to ΔV. A positive definite matrix has the property that its determinant must be positive and the determinant of all its principal minors must also be positive, where successive principal minors are obtained by deleting a row and column from the remaining matrix. Figure 6.4, for example, shows the matrix, A, of second derivatives of a N degree of freedom system where for stability one must have the successive determinants $D_N > 0, D_{N-1} > 0, \ldots, D_1 > 0$. For a system with a large number of degrees of freedom it becomes computationally expensive to perform an exhaustive test of all these determinants in order to establish stability but for a system with one or two degrees of freedom it is easy to determine stability in this manner. In fact for a one degree of freedom system we simply have

$$\frac{d^2V}{dq^2} \begin{cases} > 0 & \text{stable} \\ < 0 & \text{unstable} \\ = 0 & \text{indeterminate} \end{cases} \tag{6.2.4}$$

The indeterminate case, as stated earlier, requires an examination of higher order derivatives to classify the stability.

We have already commented on stability for a number of problems that were natural, conservative systems, often stating stability results without proof. Here, we want to examine the stability in those problems in more detail. For example, in Chap. 2, Sect. 2.3.2, we considered a roller-coaster problem (Fig. 6.5a) where a particle was launched and forced to move on a sinusoidal path $y = A[1 - \cos(\pi x/L)]/2$ and the potential energy was $V = mgy$. For this one degree of freedom system setting $dV/dx = 0$ gives $\sin(\pi x/L) = 0$ so that the possible equilibrium points are at $x = \pm nL \quad (n = 0, 1, 2, \ldots)$. For d^2V/dx^2 we have

$$\frac{d^2V}{dx^2} = \frac{mgA\pi^2}{2L^2} \cos\left(\frac{\pi x}{L}\right) \tag{6.2.5}$$

Fig. 6.5 (**a**) Particle moving on sinusoidal "roller-coaster" track and (**b**) a simple pendulum

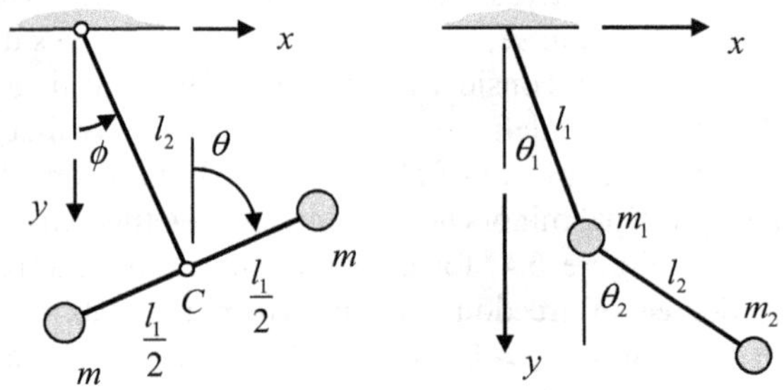

Fig. 6.6 (**a**) A multiparticle pendulum, and (**b**) a double pendulum

which shows that the second derivative is positive at the equilibrium points at the bottom of the coaster track, $x = \pm 2mL \quad (m = 0, 1, 2, \ldots)$, so that the positions are stable, and at the high points $x = \pm(2m{+}1)L \quad (m = 0, 1, 2, \ldots)$ the second derivative is negative so that those positions are unstable. These results, of course, correspond to our expectations from the example discussed in Fig. 6.3.

The simple pendulum that we discussed in Chap. 2, Sect. 2.5.3, was also a natural, conservative, one-dimensional system (Fig. 6.5b). In that case the potential energy $V = -mgL\cos\theta$ so that $\partial V/\partial\theta = mgL\sin\theta = 0$ gives the equilibrium positions $\theta = \pm n\pi \quad (n = 0, 1, 2, \ldots)$ and so for $\partial^2 V/\partial\theta^2 = mgL\cos\theta$ we see that the second derivative is positive at equilibrium positions $\theta = \pm 2m\pi \quad (m = 0, 1, 2, \ldots)$ and negative when $\theta = \pm(2m{+}1)\pi \quad (m = 0, 1, 2, \ldots)$ which correspond to the pendulum hanging down and up vertically, as we would expect. Note that another way to find the equilibrium positions is directly from the equation of motion $\ddot{\theta} + g\sin\theta/L = 0$ by setting all time derivatives present equal to zero, which again leads to $\sin\theta = 0$ at equilibrium.

The multiparticle pendulum of Chap. 3, Sect. 3.3 (Fig. 6.6a), without dampers present would appear to be an example of a two degrees of freedom stability problem but if we examine the potential energy we find that adding the potential energy of the two equal masses gives

$$\begin{aligned} V &= -mgl_2\cos\phi + mgl_1\cos\theta/2 - mgl_2\cos\phi - mgl_1\cos\theta/2 \\ &= -2mgl_2\cos\phi \end{aligned} \tag{6.2.6}$$

which is independent of the angle θ; this makes sense from a physical standpoint since the two masses can balance at any fixed angle θ about the pivot point C shown in Fig. 3.12a. Recall, however, that the final θ equilibrium position of this system, which had rotational dampers that are not shown in Fig. 6.5a, was not arbitrary but was determined by the initial conditions. The equilibrium condition stability criteria for this problem (with dampers absent) are thus identical to a simple pendulum of mass 2 m and length l_2, giving stable equilibrium positions for $\phi = \pm 2m\pi$ $(m = 0, 1, 2, \ldots)$ and unstable positions for $\phi = \pm (2m+1)\pi$ $(m = 0, 1, 2, \ldots)$.

The double pendulum of Chap. 3, Sect. 3.4.1 (Fig. 6.6b), is a system with true two degrees of freedom stability criteria. We have for the potential energy $V = -m_1gl_1\cos\theta_1 - m_2g(l_1\cos\theta_1 + l_2\cos\theta_2)$ so that setting $\partial V/\partial\theta_1 = 0$ and $\partial V/\partial\theta_2 = 0$ gives the equilibrium conditions $\sin\theta_1 = 0$ and $\sin\theta_2 = 0$. This results in the four possible equilibrium positions shown in Fig. 6.7 which are $\theta_1 = 0$, $\theta_2 = 0$ (Fig. 6.7a), $\theta_1 = 0$, $\theta_2 = \pi$ (Fig. 6.7b), $\theta_1 = \pi$, $\theta_2 = 0$ (Fig. 6.7c), and $\theta_1 = \pi$, $\theta_1 = \pi$ (Fig. 6.7d). There are, of course, additional positions that are multiples of $\pm 2\pi$, as given in the previous examples, but which are not shown here. The second derivatives of the potential energy are

$$\begin{aligned} \frac{\partial^2 V}{\partial\theta_1^2} &= (m_1 + m_2)gl_1\cos\theta_1 \\ \frac{\partial^2 V}{\partial\theta_2^2} &= m_2gl_2\cos\theta_2 \\ \frac{\partial^2 V}{\partial\theta_1\partial\theta_2} &= 0 \end{aligned} \tag{6.2.7}$$

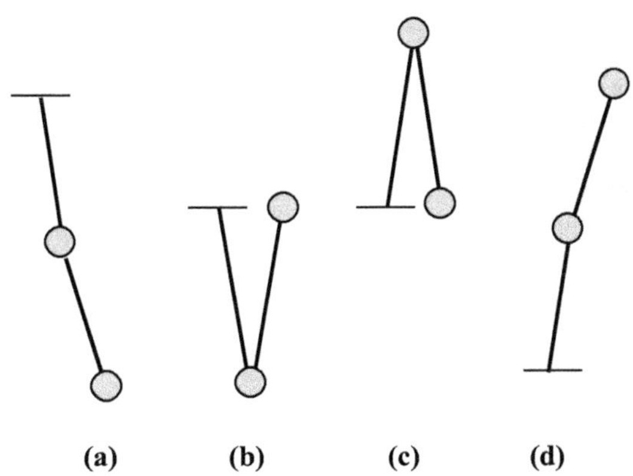

Fig. 6.7 The double pendulum near each of its four possible equilibrium positions (shown for $l_1 = l_2$)

The stability requirements in terms of the determinants of the matrix of second derivatives of the total potential energy and of the principal minors are

$$\frac{\partial^2 V}{\partial\theta_1^2}\frac{\partial^2 V}{\partial\theta_2^2} - \left(\frac{\partial^2 V}{\partial\theta_1\partial\theta_2}\right)^2 > 0, \quad \frac{\partial^2 V}{\partial\theta_1^2} > 0 \tag{6.2.8}$$

which, because the mixed second derivative vanishes, can be reduced to the equivalent stability criteria

$$\frac{\partial^2 V}{\partial\theta_1^2} > 0, \quad \frac{\partial^2 V}{\partial\theta_2^2} > 0 \tag{6.2.9}$$

Thus, placing the four possible equilibrium states into these conditions gives (a) stable, (b) unstable, (c) unstable, and (d) unstable. Again, these results coincide with what we intuitively expect to be the behavior of the double pendulum from our discussion of Fig. 6.3.

6.3 Stability of a Nonnatural System Near Equilibrium

All of the examples discussed have been for natural, conservative systems. However, recall our discussion of Lagrange's equations and the particle on a rotating hoop example in Chap. 2, Sect. 2.7 (see Fig. 6.8). There we showed that, for nonnatural systems where all the forces are derivable from potential energy functions and the Lagrangian is not an explicit function of time (i.e., $\partial L/\partial t = 0$), the Jacobi integral $h = T_2 + V - T_o$ is a constant. This conservation law was also written in the alternate form $h = T_2 + U =$ constant where $U = V - T_0$ is called the dynamic potential. For nonnatural dynamical systems where the Jacobi integral is a constant and where $T_1 = 0$ we can again use the stability theorem of Lagrange-Dirichlet where we

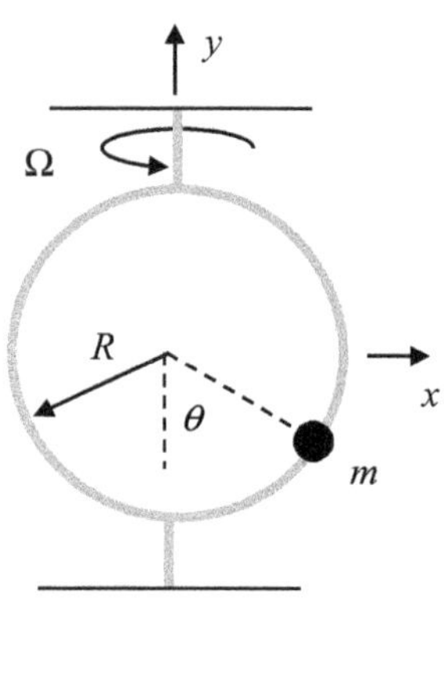

Fig. 6.8 A particle moving on a smooth hoop under the action of gravity, where the hoop is rotating at a constant angular velocity, Ω, about the y-axis:

simply replace the stability criteria on the total potential energy by the same criteria on the dynamic potential. The particle on the rotating hoop problem is such a problem so that for this one degree of freedom system we have $\partial U/\partial q = 0$ at equilibrium and the stability conditions are

$$\frac{d^2U}{dq^2}\begin{cases} > 0 & \text{stable} \\ < 0 & \text{unstable} \\ = 0 & \text{indeterminate} \end{cases} \tag{6.3.1}$$

In the rotating hoop problem $T_2 = \frac{1}{2}mR^2\dot{\theta}^2$ and

$$U = -\frac{1}{2}m(R\sin\theta)^2\Omega^2 - mgR\cos\theta \tag{6.3.2}$$

At equilibrium

$$\frac{\partial U}{\partial \theta} = mR\sin\theta\left(g - R\Omega^2\cos\theta\right) = 0 \tag{6.3.3}$$

which gives the same conditions as found in Eq. (2.7.46) whose solutions (neglecting multiples of $\pm 2\pi$) are $\theta_1 = 0$, $\theta_2 = \pm\pi$, and $\theta_3 = \pm\cos^{-1}(g/R\Omega^2)$, as found previously in Chap. 2. The second derivative of the dynamic potential is

$$\begin{aligned}\frac{\partial^2 U}{\partial \theta^2} &= mR\cos\theta\left(g - R\Omega^2\cos\theta\right) + mR^2\Omega^2\sin^2\theta \\ &= mgR\cos\theta - mR^2\Omega^2\cos 2\theta\end{aligned} \tag{6.3.4}$$

From this result it follows that the $\theta_2 = \pm\pi$ positions are always unstable since the second derivative is negative, as can be easily seen from the expression for the second derivative in the second line of Eq. (6.3.4). At $\theta_1 = 0$ we have $\partial^2U/\partial\theta^2 = mR(g - R\Omega^2)$ so this position is stable if $\Omega^2 < g/R$ and unstable if $\Omega^2 > g/R$. However, if $\Omega^2 < g/R$ there is no real equilibrium value at θ_3 (since $\cos\theta_3 = g/R\Omega^2$ would have to be greater than one in that case) and for $\Omega^2 > g/R$ there are real θ_3 equilibrium positions that are stable since the first line in Eq. (6.3.4) gives $\partial^2U/\partial\theta^2 = mR^2\Omega^2\sin^2\theta_3 > 0$. All of these results agree with those stated without proof in Chap. 2.

Use of the dynamic potential in this fashion requires that $T_1 = 0$ for the nonnatural system since the T_1 kinetic energy term typically comes from dynamic "gyroscopic" effects which can produce stability or instability of the system.

6.4 Stability Analysis Through Linearization

The direct methods discussed in the previous sections are applicable only when the potential energy or dynamic potential energy controls their behavior. Systems that are nonconservative and dissipative thus cannot be treated with those methods. Although there are other stability methods available to examine a wider class of systems they are rather involved and best treated in a dedicated course on stability. There is, however, an important "indirect" approach to stability that we will consider. This approach relies on first linearizing the equations of motion about the possible equilibrium positions. These linear equations can then be analyzed and used to infer the behavior of the original nonlinear system from which they came.

The equations of motion of a system in terms of n independent generalized coordinates can be written in the general case as

$$\ddot{\mathbf{q}}(t) = \mathbf{f}(\mathbf{q}(t), \dot{\mathbf{q}}(t), t) \tag{6.4.1}$$

When solving these equations numerically we placed them in the form of a system of first-order differential equations by defining $2n$ variables $\mathbf{x}(t) = (\mathbf{q}(t), \dot{\mathbf{q}}(t))$ and writing the system as

$$\dot{\mathbf{x}}(t) = \mathbf{g}(\mathbf{x}(t), t) \tag{6.4.2}$$

Here, we will consider the case when the system has the form

$$\dot{\mathbf{x}}(t) = \mathbf{g}(\mathbf{x}(t)) \tag{6.4.3}$$

In this case the system of equations is said to be *autonomous*. Thus, our analysis will be limited to autonomous systems but this still includes a wide variety of commonly occurring problems. An equilibrium position of the autonomous system of Eq. (6.4.3), $\mathbf{x}_e = (\mathbf{q}_e, \mathbf{0})$, must satisfy

$$\mathbf{g}(\mathbf{x}_e) = 0 \tag{6.4.4}$$

since $\dot{\mathbf{q}} = \ddot{\mathbf{q}} = 0$ at equilibrium so that $\dot{\mathbf{x}} = (\dot{\mathbf{q}}, \ddot{\mathbf{q}}) = 0$ as well. Now, let us assume that we expand the motion about this equilibrium position by letting $\mathbf{x}(t) = \mathbf{x}_e + \boldsymbol{\varepsilon}(t)$, where $\boldsymbol{\varepsilon}(t)$ is assumed to be small. Then neglecting all terms except the zeroth- and first-order ones we have

$$\begin{aligned}\dot{\boldsymbol{\varepsilon}}(t) &= \mathbf{g}(\mathbf{x}_e) + \left.\frac{\partial \mathbf{g}}{\partial \mathbf{x}}\right|_{\mathbf{x}=\mathbf{x}_e} \boldsymbol{\varepsilon}(t) \\ &= \mathbf{B}\boldsymbol{\varepsilon}(t)\end{aligned} \tag{6.4.5}$$

where $\mathbf{B}$ is a constant matrix of the derivatives of $\mathbf{g}$ at equilibrium whose components are $B_{ij} = \partial g_i / \partial x_j$, and in Eq. (6.4.5) we have used the equilibrium conditions of

Eq. (6.4.4) to eliminate the zeroth-order term. Thus, our original nonlinear system of Eq. (6.4.3) has been reduced to the linear system

$$\dot{\boldsymbol{\varepsilon}}(t) = \mathbf{B}\,\boldsymbol{\varepsilon}(t) \tag{6.4.6}$$

for small motions about an equilibrium position. To study stability we let the small motion be given as $\boldsymbol{\varepsilon} = \mathbf{E}\exp(\lambda t)$. This will be a solution of Eq. (6.4.6) if we can find a solution of the homogeneous set of linear equations

$$[\mathbf{B} - \lambda\mathbf{I}]\mathbf{E} = 0 \tag{6.4.7}$$

where $\mathbf{I}$ is the identity matrix. But this is only possible if we set the determinate of coefficients of the vector $\mathbf{E}$ equal to zero, i.e.,

$$\det[\mathbf{B} - \lambda\mathbf{I}] = 0 \tag{6.4.8}$$

This is a polynomial equation in λ, called the *characteristic equation*, whose roots we must find to satisfy Eq. (6.4.8). This is also a standard eigenvalue problem where the roots λ_k are the eigenvalues and the corresponding amplitude values, $\mathbf{E}_k$, are the eigenvectors. As we have seen before, we can easily solve such eigenvalues problems with MATLAB® using the eig function. The roots of the characteristic equation may be complex in which case they will occur in complex conjugate pairs such as $\lambda_k = \alpha_k + i\omega_k$ and $\lambda_{k+1} = \alpha_k - i\omega_k$. Then the solution for such a pair will be given by

$$\varepsilon = \exp(\alpha_k t)[E_k \exp(i\omega_k t) + E_{k+1}\exp(-i\omega_k t)] \tag{6.4.9}$$

or, equivalently, by

$$\varepsilon = \exp(\alpha_k t)[(E_k + E_{k+1})\cos(\omega_k t) + i(E_k - E_{k+1})\sin(\omega_k t)] \tag{6.4.10}$$

These solutions can be used to illustrate why the real parts of the roots, α_k, determine stability, as stated in the following conditions:

Dynamic Stability via a Linearized System Analysis

1. When all the roots, λ_k, have negative real parts, α_k, the solution of the linearized system is asymptotically stable.
2. When one or more than one root, λ_k, has a positive real part, α_k, the solution of the linearized system is unstable.
3. When some roots have zero real parts while the remaining ones, if any, have negative real parts, the solution of the linearized system is critically stable.

Furthermore, the nonlinearities present in the original system, Eq. (6.4.2), do not change the conclusion on stability based on the linearized equations of Eq. (6.4.6) when the linearized system is either asymptotically stable or unstable. For the critical

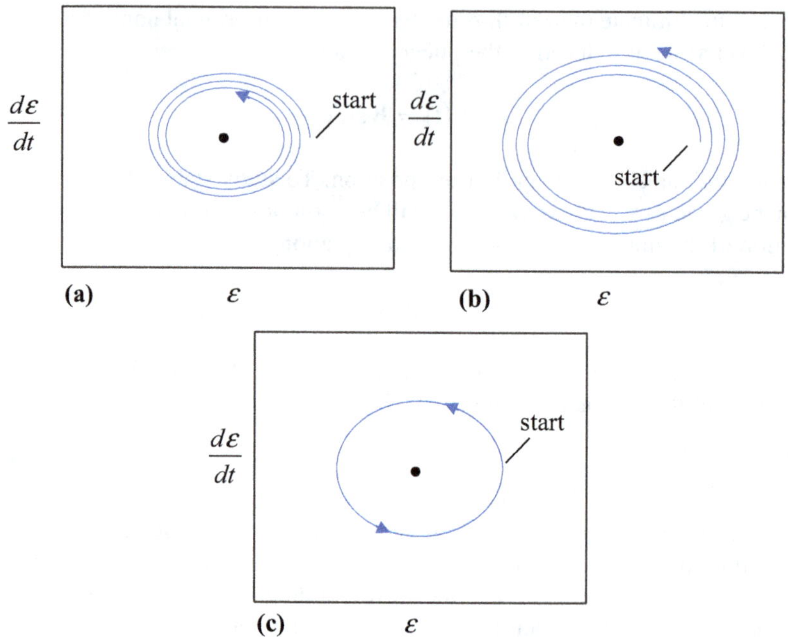

Fig. 6.9 Phase plots of $\dot{\varepsilon}$ versus ε about the equilibrium position $(\varepsilon, \dot{\varepsilon}) = (0, 0)$ (the black dots shown in the figures) for (**a**) the case when the real part of a pair of complex conjugate roots of the characteristic equation is less than zero and the linear system is asymptotically stable, (**b**) when the real part of the complex conjugate roots is greater than zero and the system is unstable, and (**c**) when the real part of the complex conjugate roots is equal to zero and the system is critically stable. The points labeled "start" are where the motion began

stability case, additional other tests are needed to determine if the original system of Eq. (6.4.2) is stable or unstable.

Case (1) is called asymptotically stable because when all the roots have negative real parts the linearized solution is composed of a combination of decaying exponentials which all go to zero as $t \to \infty$. This behavior can be seen from the phase plot of a motion where a pair of complex conjugate roots yield the motion of Fig. 6.9a where the system spirals into the stable equilibrium position $(\varepsilon, \dot{\varepsilon}) = (0, 0)$. Likewise, the system is unstable for Case (2) since if one or more roots have a positive real part the solution will grow infinitely large as $t \to \infty$, as shown in Fig. 6.9b, where the pair of complex conjugate solutions produce a motion that spirals outward from the unstable equilibrium position (0, 0). In the critical stability Case (3), no conclusion can be reached as to stability of the original nonlinear system (based on the linearized analysis) where the pair of complex conjugate solutions produce a stable orbit of the linear system about the equilibrium position (Fig. 6.9c) but the nonlinearities are crucial and can cause the original nonlinear system to be either stable or unstable depending on the nature of the nonlinearities present. In that case one could perform other stability tests or do some numerical studies of the behavior of the solutions of the original nonlinear equations of motion.

As an example of these linearized system stability tests, consider a simple pendulum where we include a linear angular velocity term so that the equation of motion is

$$\ddot{\theta} + c\dot{\theta} + g\sin\theta/L = 0 \tag{6.4.11}$$

If we let $x_1 = \theta$ and $x_2 = \dot{\theta}$ then the equations of motion become

$$\begin{aligned} \dot{x}_1 &= x_2 \\ \dot{x}_2 &= -\frac{g\sin x_1}{L} - cx_2 \end{aligned} \tag{6.4.12}$$

The addition of the linear angular velocity term does not affect the equilibrium positions of the pendulum which are $\mathbf{x}_{e1} = (0, 0)$ and $\mathbf{x}_{e2} = (\pi, 0)$. If we consider small motions around the first equilibrium point, $\mathbf{x}_{e1}$, by setting $\mathbf{x} = \mathbf{x}_{e1} + \boldsymbol{\varepsilon}$ then the linearized equations of motion are

$$\begin{Bmatrix} \dot{\varepsilon}_1 \\ \dot{\varepsilon}_2 \end{Bmatrix} = \begin{bmatrix} 0 & 1 \\ -g/L & -c \end{bmatrix} \begin{Bmatrix} \varepsilon_1 \\ \varepsilon_2 \end{Bmatrix} \tag{6.4.13}$$

since $\sin x_1 = \sin\varepsilon_1 \cong \varepsilon_1$. The characteristic equation is then

$$\det\begin{bmatrix} -\lambda & 1 \\ -g/L & -c-\lambda \end{bmatrix} = \lambda^2 + c\lambda + g/L = 0 \tag{6.4.14}$$

whose roots are

$$\lambda_{1,2} = -c/2 \pm \sqrt{(c/2)^2 - g/L} \tag{6.4.15}$$

First, consider the case when $|c/2| < g/L$. In that case the roots are the complex conjugate pair $\lambda_{1,2} = -c/2 \pm i\sqrt{g/L - (c/2)^2}$. If $c > 0$ the real parts of the roots are both negative so this equilibrium position is asymptotically stable while if $c < 0$ both roots have a positive real part and the position is unstable. This makes sense physically since the $c > 0$ case corresponds to having an angular damper present which causes the motion to decay around the equilibrium position while the $c < 0$ case would correspond to a "negative" damping which would reinforce the motion, causing it to grow. In the $c = 0$ case the roots are $\lambda_{1,2} = \pm i\sqrt{g/L}$ so the system is critically stable. In that case the system is a natural, conservative system so that our previous direct analysis tells us that the total potential energy is a minimum and the equilibrium position is indeed stable for the original nonlinear system. Now, examine the case when $|c/2| \geq g/L$. In this case the two roots are real. If $c > 0$ both roots are negative so the system is again asymptotically stable. If $c < 0$ both roots are positive so the system is again unstable.

Next, if we expand the motion about the second equilibrium position $\mathbf{x}_{e2} = (\pi, 0)$ then we have $\sin x_1 = \sin(\pi+\varepsilon_1) \cong -\varepsilon_1$ so the characteristic equation is

$$\det\begin{bmatrix} -\lambda & 1 \\ g/L & -c-\lambda \end{bmatrix} = \lambda^2 + c\lambda - g/L = 0 \tag{6.4.16}$$

and the roots are

$$\lambda_{1,2} = -c/2 \pm \sqrt{(c/2)^2 + g/L} \tag{6.4.17}$$

so there is always at least one positive real root for all values of c, including $c = 0$, and this equilibrium position is unstable.

As an example of the use of the linearization approach for a multi-degree of freedom problem consider the double pendulum. If we let the two lengths of the pendulum be equal as well as the masses, i.e., $l_1 = l_2 = l$ and $m_1 = m_2 = m$, then the equations of motion can be written as (see Eq. (3.4.30))

$$\begin{aligned} &2\ddot{\theta}_1 + \cos(\theta_1 - \theta_2)\ddot{\theta}_2 + \sin(\theta_1 - \theta_2)\dot{\theta}_2^2 + 2g\sin\theta_1/l = 0 \\ &\cos(\theta_1 - \theta_2)\ddot{\theta}_1 + \ddot{\theta}_2 - \sin(\theta_1 - \theta_2)\dot{\theta}_1^2 + g\sin\theta_2/l = 0 \end{aligned} \tag{6.4.18}$$

We have already found the four possible equilibrium positions: (1) $\theta_1 = 0, \theta_2 = 0$, (2) $\theta_1 = 0, \theta_2 = \pi$, (3) $\theta_1 = \pi, \theta_2 = 0$, and (4) $\theta_1 = \pi, \theta_1 = \pi$. Let $\mathbf{x} = (x_1, x_2, x_3, x_4) = (\theta_1, \dot{\theta}_1, \theta_2, \dot{\theta}_2)$. Then we have

$$\begin{aligned} &\dot{x}_1 = x_2 \\ &2\dot{x}_2 + \cos(x_1 - x_3)\dot{x}_4 + \sin(x_1 - x_3)x_4^2 + 2g\sin x_1/l = 0 \\ &\dot{x}_3 = x_4 \\ &\cos(x_1 - x_3)\dot{x}_2 + \dot{x}_4 - \sin(x_1 - x_3)x_2^2 + g\sin x_3/l = 0 \end{aligned} \tag{6.4.19}$$

For an equilibrium position $\mathbf{x}_e = (x_{e1}, 0, x_{e3}, 0)$ and motions about that position, $\mathbf{x} = \mathbf{x}_e + \boldsymbol{\varepsilon}$ Eq. (6.4.19) becomes

$$\begin{aligned} &\dot{\varepsilon}_1 = \varepsilon_2 \\ &2\dot{\varepsilon}_2 + \cos(x_{e1} + \varepsilon_1 - x_{e3} - \varepsilon_3)\dot{\varepsilon}_4 + \sin(x_{e1} + \varepsilon_1 - x_{e3} - \varepsilon_3)\varepsilon_4^2 + 2g\sin(x_{e1} + \varepsilon_1)/l = 0 \\ &\dot{\varepsilon}_3 = \varepsilon_4 \\ &\cos(x_{e1} + \varepsilon_1 - x_{e3} - \varepsilon_3)\dot{\varepsilon}_2 + \dot{\varepsilon}_4 - \sin(x_{e1} + \varepsilon_1 - x_{e3} - \varepsilon_3)\varepsilon_2^2 + g\sin(x_{e3} + \varepsilon_3)/l = 0 \end{aligned} \tag{6.4.20}$$

since $\dot{\mathbf{x}} = \dot{\boldsymbol{\varepsilon}}$ and $x_2 = \varepsilon_2$, $x_4 = \varepsilon_4$. We can now use Eq. (6.4.20) to obtain the linearized equations about any of the equilibrium positions. Consider the first position $(x_{e1}, x_{e3}) = (0, 0)$. Then Eq. (6.4.20) becomes

$$\begin{aligned}
&\dot{\varepsilon}_1 = \varepsilon_2\\
&2\dot{\varepsilon}_2 + \cos(\varepsilon_1 - \varepsilon_3)\dot{\varepsilon}_4 + \sin(\varepsilon_1 - \varepsilon_3)\varepsilon_4^2 + 2g\sin(\varepsilon_1)/l = 0\\
&\dot{\varepsilon}_3 = \varepsilon_4\\
&\cos(\varepsilon_1 - \varepsilon_3)\dot{\varepsilon}_2 + \dot{\varepsilon}_4 - \sin(\varepsilon_1 - \varepsilon_3)\varepsilon_2^2 + g\sin(\varepsilon_3)/l = 0
\end{aligned} \tag{6.4.21}$$

and using the approximations $\cos u \cong 1$ and $\sin u \cong u$ for u small our equations for small motions about the equilibrium position to first order are

$$\begin{aligned}
&\dot{\varepsilon}_1 = \varepsilon_2\\
&2\dot{\varepsilon}_2 + \dot{\varepsilon}_4 + 2g\varepsilon_1/l = 0\\
&\dot{\varepsilon}_3 = \varepsilon_4\\
&\dot{\varepsilon}_2 + \dot{\varepsilon}_4 + g\varepsilon_3/l = 0
\end{aligned} \tag{6.4.22}$$

which can easily be reduced to the first-order system in the form we want, i.e.,

$$\begin{aligned}
&\dot{\varepsilon}_1 = \varepsilon_2\\
&\dot{\varepsilon}_2 = (g\varepsilon_3/l - 2g\varepsilon_1/l)\\
&\dot{\varepsilon}_3 = \varepsilon_4\\
&\dot{\varepsilon}_4 = (2g\varepsilon_1/l - 2g\varepsilon_3/l)
\end{aligned} \tag{6.4.23}$$

by combining the second and fourth equations in Eq. (6.4.22) appropriately (the difference of the second and fourth equations in Eq. (6.4.22) just gives us the second equation in Eq. (6.4.23) and placing this result into the fourth equation in Eq. (6.4.22) gives us the fourth equation in Eq. (6.4.23)). This result can also be written in matrix form as

$$\begin{Bmatrix} \dot{\varepsilon}_1 \\ \dot{\varepsilon}_2 \\ \dot{\varepsilon}_3 \\ \dot{\varepsilon}_4 \end{Bmatrix} = \begin{bmatrix} 0 & 1 & 0 & 0 \\ -2g/l & 0 & g/l & 0 \\ 0 & 0 & 0 & 1 \\ 2g/l & 0 & -2g/l & 0 \end{bmatrix} \begin{Bmatrix} \varepsilon_1 \\ \varepsilon_1 \\ \varepsilon_3 \\ \varepsilon_4 \end{Bmatrix} \tag{6.4.24}$$

and the characteristic equation is

$$\det\begin{bmatrix} -\lambda & 1 & 0 & 0 \\ -2g/l & -\lambda & g/l & 0 \\ 0 & 0 & -\lambda & 1 \\ 2g/l & 0 & -2g/l & -\lambda \end{bmatrix} = \lambda^4 + 4(g/l)\lambda^2 + 2(g/l)^2 = 0 \tag{6.4.25}$$

Solving for λ^2 gives

$$\lambda_{1,2}^2 = \left(-2 \pm \sqrt{2}\right) g/l \tag{6.4.26}$$

which are both real and negative so that taking the square roots gives the four purely imaginary roots

$$\lambda_{1,2,3,4} = \pm i\sqrt{\left(2 \mp \sqrt{2}\right) g/l} \tag{6.4.27}$$

and the linearized analysis shows that this position is critically stable. The direct analysis, of course, showed that the original system is stable here.

When we considered the double-pendulum problem in Chap. 3 we also examined the solution for small oscillations about this stable equilibrium position. We found that there were two natural frequencies of vibration $\omega_{1,2} = \sqrt{2 \mp \sqrt{2}}\,\omega_0$ where $\omega_0 = \sqrt{g/l}$ is the natural frequency of small oscillations for a simple pendulum of length l. Thus, the four roots of the characteristic equation are just $\lambda_{1,2} = \pm i\omega_1$, $\lambda_{3,4} = \pm i\omega_2$. Corresponding to the natural frequencies ω_1,ω_2 we also found the normalized normal mode shapes

$$\mathbf{e}^{(1)} = \begin{Bmatrix} e_1^{(1)} \\ e_2^{(1)} \end{Bmatrix} = \begin{Bmatrix} \sqrt{1/3} \\ \sqrt{2/3} \end{Bmatrix}, \quad \mathbf{e}^{(2)} = \begin{Bmatrix} e_1^{(2)} \\ e_2^{(2)} \end{Bmatrix} = \begin{Bmatrix} \sqrt{1/3} \\ -\sqrt{2/3} \end{Bmatrix}$$

respectively, in Chap. 3. The eigenvectors of the $\mathbf{B}$ matrix are also related to these normal mode shapes. For the $\lambda_1 = +i\omega_1$ eigenvalue the corresponding eigenvector is of the form $\mathbf{E}^{(1)} = \left[Ce_1^{(1)}, i\omega_1 Ce_1^{(1)}, Ce_2^{(1)}, i\omega_1 Ce_2^{(1)}\right]^T$ while for the $\lambda_2 = -i\omega_1$ eigenvalue the eigenvector is of the form $\mathbf{E}^{(2)} = \left[Ce_1^{(1)}, -i\omega_1 Ce_1^{(1)}, Ce_2^{(1)}, -i\omega_1 Ce_2^{(1)}\right]^T$, where C is an arbitrary (complex) constant that can be chosen differently for each eigenvector (note: these particular forms arise because the elements of these eigenvectors corresponding to values of the variables $\left[\theta_1, \dot{\theta}_1, \theta_2, \dot{\theta}_2\right]^T$). For the $\lambda_3 = +i\omega_2$ eigenvalue the eigenvector is of the form $\mathbf{E}^{(3)} = \left[Ce_1^{(2)}, i\omega_2 Ce_1^{(2)}, Ce_2^{(2)}, i\omega_2 Ce_2^{(2)}\right]^T$ and for $\lambda_4 = -i\omega_2$ the eigenvector is of the form $\mathbf{E}^{(4)} = \left[Ce_1^{(2)}, -i\omega_2 Ce_1^{(2)}, Ce_2^{(2)}, -i\omega_2 Ce_2^{(2)}\right]^T$. For a stability analysis these eigenvectors are not needed but we have shown them here to connect this linearized analysis with our previous linearized small vibrations solutions.

Now, consider the second equilibrium position where $(x_{e1}, x_{e3}) = (0, \pi)$. Then Eq. (6.4.20) becomes

$$\begin{aligned}
&\dot{\varepsilon}_1 = \varepsilon_2 \\
&2\dot{\varepsilon}_2 + \cos(\varepsilon_1 - \varepsilon_3 - \pi)\dot{\varepsilon}_4 + \sin(\varepsilon_1 - \varepsilon_3 - \pi)\varepsilon_4^2 + 2g\sin(\varepsilon_1)/l = 0 \\
&\dot{\varepsilon}_3 = \varepsilon_4 \\
&\cos(\varepsilon_1 - \varepsilon_3 - \pi)\dot{\varepsilon}_2 + \dot{\varepsilon}_4 - \sin(\varepsilon_1 - \varepsilon_3 - \pi)\varepsilon_2^2 + g\sin(\pi + \varepsilon_3)/l = 0
\end{aligned} \tag{6.4.28}$$

or, equivalently,

$$\begin{aligned}
&\dot{\varepsilon}_1 = \varepsilon_2 \\
&2\dot{\varepsilon}_2 - \cos(\varepsilon_1 - \varepsilon_3)\dot{\varepsilon}_4 - \sin(\varepsilon_1 - \varepsilon_3)\varepsilon_4^2 + 2g\sin(\varepsilon_1)/l = 0 \\
&\dot{\varepsilon}_3 = \varepsilon_4 \\
&-\cos(\varepsilon_1 - \varepsilon_3)\dot{\varepsilon}_2 + \dot{\varepsilon}_4 + \sin(\varepsilon_1 - \varepsilon_3)\varepsilon_2^2 - g\sin(\varepsilon_3)/l = 0
\end{aligned} \tag{6.4.29}$$

and for small motions we find to first order

$$\begin{aligned}
&\dot{\varepsilon}_1 = \varepsilon_2 \\
&2\dot{\varepsilon}_2 - \dot{\varepsilon}_4 = -2\varepsilon_1 g/l \\
&\dot{\varepsilon}_3 = \varepsilon_4 \\
&-\dot{\varepsilon}_2 + \dot{\varepsilon}_4 = \varepsilon_3 g/l
\end{aligned} \tag{6.4.30}$$

which can be rewritten as

$$\begin{aligned}
&\dot{\varepsilon}_1 = \varepsilon_2 \\
&\dot{\varepsilon}_2 = \varepsilon_3 g/l - 2\varepsilon_1 g/l \\
&\dot{\varepsilon}_3 = \varepsilon_4 \\
&\dot{\varepsilon}_4 = 2\varepsilon_3 g/l - 2\varepsilon_1 g/l
\end{aligned} \tag{6.4.31}$$

or, in matrix form,

$$\begin{Bmatrix} \dot{\varepsilon}_1 \\ \dot{\varepsilon}_2 \\ \dot{\varepsilon}_3 \\ \dot{\varepsilon}_4 \end{Bmatrix} = \begin{bmatrix} 0 & 1 & 0 & 0 \\ -2g/l & 0 & g/l & 0 \\ 0 & 0 & 0 & 1 \\ -2g/l & 0 & 2g/l & 0 \end{bmatrix} \begin{Bmatrix} \varepsilon_1 \\ \varepsilon_1 \\ \varepsilon_3 \\ \varepsilon_4 \end{Bmatrix} \tag{6.4.32}$$

so that the characteristic equation is

$$\det\begin{bmatrix} -\lambda & 1 & 0 & 0 \\ -2g/l & -\lambda & g/l & 0 \\ 0 & 0 & -\lambda & 1 \\ -2g/l & 0 & 2g/l & -\lambda \end{bmatrix} = \lambda^4 - 2(g/l)^2 = 0 \tag{6.4.33}$$

which has the four roots

$$\lambda_{1,2,3,4} = \pm\sqrt{\pm\sqrt{2}g/l} \tag{6.4.34}$$

The root with both plus signs obviously is positive and so this position is unstable, as also found with the direct analysis. You can examine the other two cases in the same manner and show that they also are unstable via the linearized equations.

Examining systems with only a few degrees of freedom can be easily done by examining all the roots of the characteristic equation but this can become burdensome for more complex systems. Fortunately, there is a way to examine the stability without explicitly finding all the roots. The characteristic equation yields a polynomial whose coefficients can be used to derive algebraic criteria for the determination of stability called the *Routh-Hurwitz criteria*. We will not examine those criteria here but you can see some examples of their use in [2].

6.5 Static Stability

The minimization of the potential energy to determine the dynamic stability of natural, conservative systems near equilibrium does not require us to consider any details of the motion of a system about the equilibrium positions. Thus, we can use the minimization of the potential energy criterion also as a criterion to discuss the stability of static systems (systems in equilibrium) where the system is designed to be used without motion being present. Although many statics engineering texts include a discussion of static stability, the coverage is typically very brief. We will try to provide a more comprehensive static analysis that discusses some of the important types of static instabilities such as bifurcation instability, limit load instability, and snap-through instability.

Most treatments of static stability start from the principle of virtual work, so let us first examine that principle in light of our previous discussions of dynamical systems. Recall, for example, the D'Alembert-Lagrange principle which was examined in Chap.3 for a system of N particles. For a system described by n independent generalized coordinates that principle was given as (see Eq. (3.4.4))

$$\sum_{k=1}^{n}\left[\sum_{i=1}^{N} m_i \ddot{\mathbf{x}}_i \cdot \frac{\partial \mathbf{x}_i}{\partial q_k} - Q_k^e + \frac{\partial V^{\text{int}}}{\partial q_k}\right]\delta q_k = 0 \tag{6.5.1}$$

where Q_k^e were the components of the external forces, V^{int} was the potential energy of the internal forces (such as springs), and δq_k were virtual changes of the generalized displacements. For static equilibrium problems, therefore, we have

$$\sum_{k=1}^{n}\left[-Q_k^e + \frac{\partial V^{\text{int}}}{\partial q_k}\right]\delta q_k = 0 \tag{6.5.2}$$

However,

$$\delta W^e = \sum_{k=1}^{n} Q_k^e \delta q_k \tag{6.5.3}$$

where δW^e is the virtual work of the external generalized force, Q_k^e, due to the virtual displacements, δq_k, and

$$\delta V^{\text{int}} = \sum_{k=1}^{n} \frac{\partial V^{\text{int}}}{\partial q_k}\delta q_k \tag{6.5.4}$$

where δV^{int} is just the virtual change of the internal potential energy of the system. Thus, at equilibrium the D'Alembert-Lagrange principle becomes the *principle of virtual work* given as

$$\delta W^e = \delta V^{\text{int}} \tag{6.5.5}$$

i.e., the virtual work done by the external loads, δW^e, is equal to the virtual change of the internal potential energy, δV^{int}, of the system. Although here we have obtained the principle of virtual work from the D'Alembert-Lagrange principle, historically the principle of virtual work preceded the D'Alembert-Lagrange principle. In fact the principle of virtual work for static problems is what inspired D'Alembert and Lagrange to extend this principle to dynamical systems. Although we used a system of particles to obtain this principle, it is also directly applicable to systems of rigid bodies.

In calculating the virtual work of the external generalized force, the generalized displacement is varied but not the generalized force itself. This means that the generalized external forces Q_k^e can be treated as constants in calculating δW^e. As seen in Chap. 2 constant forces are derivable from a potential energy, V^e, and the potential energy is in that case of the form { – ("force") times ("displacement")} so that here we could let $V^e = -\sum_{k=1}^{n} Q_k^e q_e$ (V^e is often called the "force potential") and we can write

$$\delta W^e = -\sum_{k=1}^{n} \frac{\partial V^e}{\partial q_k} \delta q_k = -\delta V^e \tag{6.5.6}$$

where δV^e is the virtual change of the force potential. Thus, the principle of virtual work becomes

$$\delta V^{\mathrm{tot}} = 0 \tag{6.5.7}$$

where $V^{\mathrm{tot}} = V^e + V^{\mathrm{int}}$ is the total potential energy of the internal and external forces. For forces derivable from a potential we can also relate the total work done by the external force, W^e, to the change in potential energy through $W^e = -\Delta V^e$. But the potential energy is always determined up to an irrelevant constant anyway so we can replace the change in potential energy with just the potential energy itself and simply let the work be equal to the negative of the potential energy, i.e., $W^e = -V^e$. The total potential energy of the system is then

$$V^{\mathrm{tot}} = V^{\mathrm{int}} - W^e \tag{6.5.8}$$

[Note that in this form the work is computed as just {("force") times ("displacement")}, i.e., $W^e = \sum_{k=1}^{n} Q_k^e q_k$, so that there is no minus sign present in obtaining W^e but W^e is multiplied by a minus sign in Eq. (6.5.8) in the calculation of the total potential energy.]

Since the total potential energy is completely defined by the set of generalized coordinates, $(q_1, q_2, \ldots, q_n)$ the principle of virtual work becomes

$$\delta V^{\mathrm{tot}} = \sum_{k=1}^{n} \frac{\partial V^{\mathrm{tot}}}{\partial q_k} \delta q_k = 0 \tag{6.5.9}$$

which, since the generalized coordinates are assumed to be independent, gives the conditions of equilibrium as

$$\frac{\partial V^{\mathrm{tot}}}{\partial q_k} = 0 \quad (k = 1, 2, \ldots, n) \tag{6.5.10}$$

[Note: We earlier obtained this same result directly from Lagrange's equations by simply setting all the kinetic energy terms to zero for a conservative system. We derived this result from the principle of virtual work to illustrate the explicit use of that principle, following the traditional way in which this result is often obtained.]

Static stability criteria for conservative systems follow the same reasoning used for dynamic systems. To guarantee stability we must have any virtual change of the total potential energy, δV^{tot}, to be positive in a region about an equilibrium position. We can write this condition by considering first- and second-order virtual coordinate changes which causes the change in the total potential energy to be

$$\begin{aligned} \delta V^{\text{tot}} &= \sum_{k=1}^{n} \frac{\partial V^{ot}}{\partial q_k} \delta q_k + \frac{1}{2} \sum_{k=1}^{n} \sum_{m=1}^{n} \frac{\partial^2 V^{\text{tot}}}{\partial q_k \partial q_m} \delta q_k \delta q_m + \quad H.O.T. \\ &= \frac{1}{2} \sum_{k=1}^{n} \sum_{m=1}^{n} \frac{\partial^2 V^{\text{tot}}}{\partial q_k \partial q_m} \delta q_k \delta q_m + \quad H.O.T. \end{aligned} \tag{6.5.11}$$

since the linear terms in the first line of Eq. (6.5.11) vanish by virtue of the equilibrium conditions of Eq. (6.5.10) and the higher order terms *(H.O.T.)* have been neglected. Thus, as found previously, in a small region about an equilibrium position the conditions for stability depend on the potential energy through the matrix of its second derivatives. For multiple degree of freedom systems for static stability we need to again have the matrix of second derivative terms in Eq. (6.5.11) to be a positive definite matrix, leading to the conditions on the principal determinants of that matrix, as discussed earlier. For the commonly occurring one degree of freedom case we have, as before, the static stability criteria:

$$\frac{d^2 V^{\text{tot}}}{dq^2} \begin{cases} > 0 & \text{stable} \\ < 0 & \text{unstable} \\ = 0 & \text{indeterminate} \end{cases} \tag{6.5.12}$$

where the indeterminate case requires an examination of higher order derivatives to classify the stability.

6.6 Bifurcations and Buckling

Consider a one degree of freedom system in equilibrium. While many dynamic stability studies examine the consequences of changing the initial conditions to see their effects on the subsequent motion, most static stability analyses instead examine stability due to changes of the external load on the system. As the external load is slowly increased on this system and the system deforms, a plot of the load versus position (or angle) at a collection of equilibrium positions is called a load path. One common way that static stability is lost is through the occurrence of the possibility of more than one load path at a given value of the load. The load at which this bifurcation (splitting) of loading paths occurs is often called the *critical load* or the *buckling load*. This situation is shown schematically in Fig. 6.10. If one or more

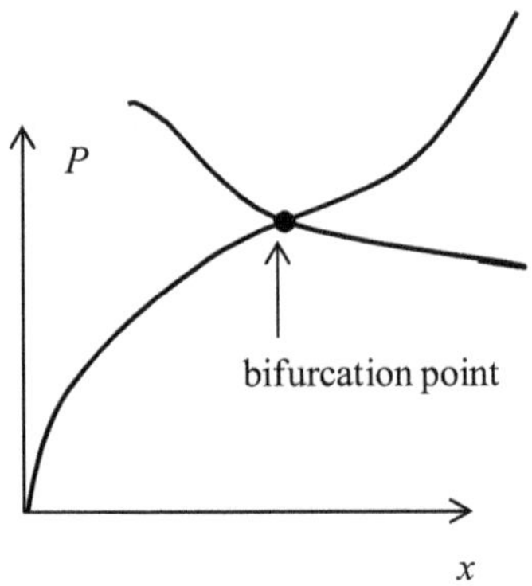

Fig. 6.10 A system where there are multiple potential loading paths that intersect at a bifurcation point

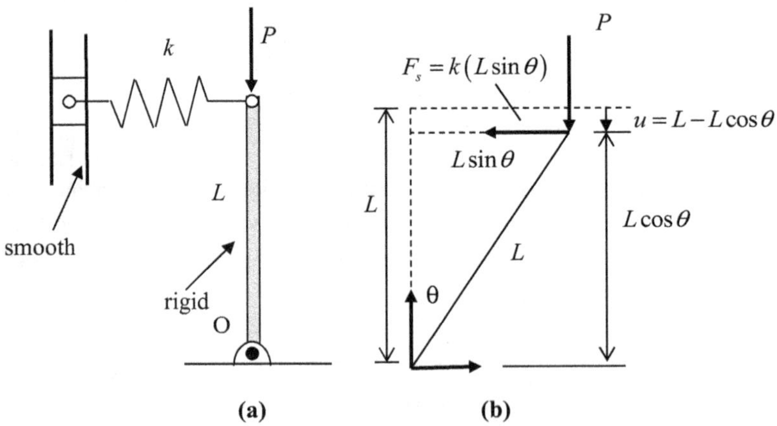

Fig. 6.11 (**a**) A one degree of freedom system that has a critical load at which the system loses stability and (**b**) the geometry and loads of the system in a deflected position

of the possible loading paths are unstable then when the critical load is reached the system may move to a new equilibrium position far from the bifurcation point, resulting in collapse.

To see this type of instability in an explicit example, consider the bar-spring system of Fig. 6.11a. The bar is assumed to be rigid and weightless and one end of the linear spring is assumed to ride in a smooth horizontal slot so that the spring always remains horizontal. It is also assumed that the spring is initially not stretched when the bar is vertical. The work, W^e, done by this external force can be calculated, from our previous discussion, as simply

$$W^e = Pu = PL(1 - \cos\theta) \tag{6.6.1}$$

where u is the displacement in the direction of the load (Fig. 6.11b). The internal (potential) energy of the linear spring is also

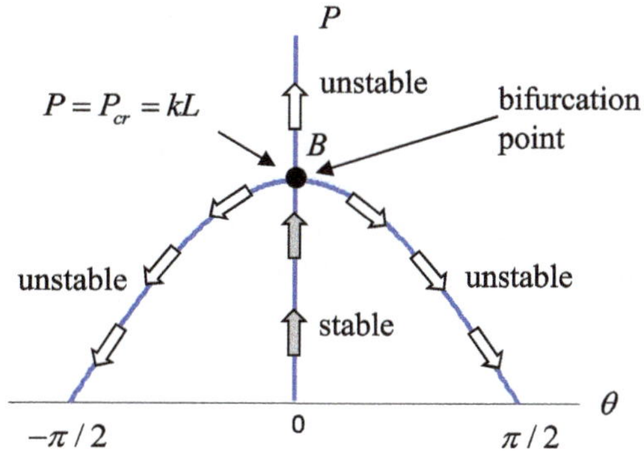

Fig. 6.12 Loading paths for the system of Fig. 6.11 and their stability characteristics

$$V^{\text{int}} = \frac{1}{2} k s^2 = \frac{1}{2} k (L \sin \theta)^2 \tag{6.6.2}$$

where k is the spring constant (force/unit length) and s is the stretch in the spring (Fig. 6.11b) so the total potential energy is

$$V^{\text{tot}}(\theta) = -PL(1 - \cos \theta) + \frac{1}{2} k (L \sin \theta)^2 \tag{6.6.3}$$

and for equilibrium we must have

$$\frac{dV^{\text{tot}}}{d\theta} = (-P + kL \cos \theta) L \sin \theta = 0 \tag{6.6.4}$$

Equation (6.6.4) has two solutions for $-\pi/2 < \theta < \pi/2$. One solution is with the angle θ equal to zero. Then the load P is arbitrary and equilibrium is possible anywhere along the vertical loading path line of Fig. 6.12, corresponding to the bar remaining vertical and not deflected while the load increases. However, if θ is not zero, then we can divide out the $\sin\theta$ term in Eq. (6.6.4) and we obtain the second load path

$$P = kL \cos \theta \tag{6.6.5}$$

which is the curved loading path shown in Fig. 6.12. We see the two loading paths intersect at the bifurcation point B where the load $P = P_{cr} = kL$ is called the critical load or the buckling load. To examine the conditions of stability along these loading paths, we compute the second derivative of the potential energy to find

$$\frac{d^2 V^{\text{tot}}}{d\theta^2} = -PL \cos \theta + kL^2 \left(\cos^2\theta - \sin^2\theta \right) \tag{6.6.6}$$

Along the vertical loading path where $\theta = 0$ we have

$$\frac{d^2 V^{\mathrm{tot}}}{d\theta^2} = kL^2 - PL \tag{6.6.7}$$

which is positive when $P < kL$, and negative for $P > kL$, resulting in the stability labels shown in Fig. 6.12 on the primary load path. For the other equilibrium solution of Eq. (6.6.4) we find

$$\frac{d^2 V^{\mathrm{tot}}}{d\theta^2} = -kL^2 \sin^2\theta \tag{6.6.8}$$

showing that the curved loading paths in Fig. 6.12 are always unstable for $-\pi/2 < \theta < \pi/2$. These results demonstrate that the bar can be stably loaded with no angular deflection for all loads up to the critical load, but to continue along this path beyond the critical load is unstable so any small disturbance will cause some angular deflection. However, no angular position along the curved path is stable so the bar will collapse to the floor at $\theta = \pm\pi/2$. Thus, we say that the post-critical load behavior of this system is unstable. In one degree of freedom systems such as this one we can also plot the total potential energy to show the stability conditions explicitly. Figure 6.13a–c shows the normalized total potential energy when the load is one-half the critical load, at the critical load, and 1.2 times the critical load, respectively, showing the change from stable to unstable conditions on the $\theta = 0$ path and the unstable nature of the other equilibrium solutions along the curved loading paths. [Note that in these plots of the potential energy and in all subsequent plots we will omit the vertical axis values, as the potential energy is always defined only to within an arbitrary constant value so the exact values on these plots are unimportant and can be a distraction from focusing on the important changing behavior of these potential energy curves.]

From this example we see that we could fully characterize the stability of this simple system by considering both first- and second-order virtual variations of the total potential energy. Setting the first variation equal to zero, $\delta V^{\mathrm{tot}} = (\partial V^{\mathrm{tot}}/\partial\theta)\delta\theta = 0$, (1) guaranteed that equilibrium was satisfied, (2) gave us the possible equilibrium load paths, and (3) gave us the value of the critical load at the bifurcation point. Examining the sign of the second-order variation $\delta^2 V^{\mathrm{tot}} = (1/2)(\partial^2 V^{\mathrm{tot}}/\partial\theta^2)(\delta\theta)^2$ gave us information on the stability of the post-critical load behavior. We see from Eq. (6.6.7) that we could also have found the value of the critical load by determining the load when $\delta^2 V^{\mathrm{tot}} = 0$, since this condition requires $P = P_{cr} = kL$. This criterion makes sense from a physical standpoint since we can define the critical load to be the load where we go from a stable equilibrium location on the load path where $\delta^2 V^{\mathrm{tot}} > \theta$ to an unstable location where $\delta^2 V^{\mathrm{tot}} < 0$. At the boundary of these stable/unstable conditions $\delta^2 V^{\mathrm{tot}} = 0$.

In more complex multiple degree of freedom stability problems, performing a full stability characterization of this type is often not possible because of the complicated

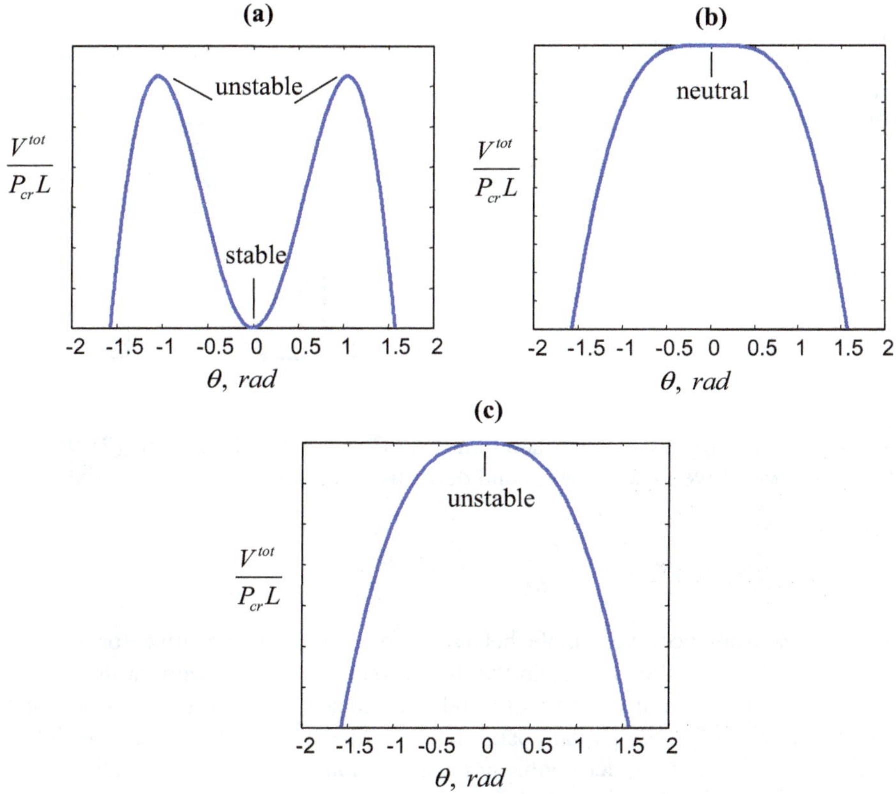

Fig. 6.13 The normalized total potential energy, $V^{tot}/P_{cr}L$, versus θ for the problem of Fig. 6.11a and the nature of the equilibrium positions for (**a**) the load $P = 0.5P_{cr}$, (**b**) the load $P = 1.0P_{cr}$, and (**c**) the load $P = 1.2P_{cr}$

nonlinear behavior of the system during post-buckling. However, consider what happens if we linearize this problem for very small deflections to first order about the non-deflected ($\theta = 0$) equilibrium position. In this case if we let $\sin\theta \simeq \theta$, $\cos\theta \simeq 1 - \theta^2/2$, the potential energy is

$$V^{\text{tot}}(\theta) \cong -\frac{1}{2}PL\theta^2 + \frac{1}{2}k(L\theta)^2 \tag{6.6.9}$$

and the equilibrium equation becomes

$$\frac{dV^{\text{tot}}}{d\theta} = (-P + kL)L\theta = 0 \tag{6.6.10}$$

Note that the equilibrium equation is indeed linear in θ, but that in the linearized analysis we need to keep second-order (quadratic) terms in the total energy. This

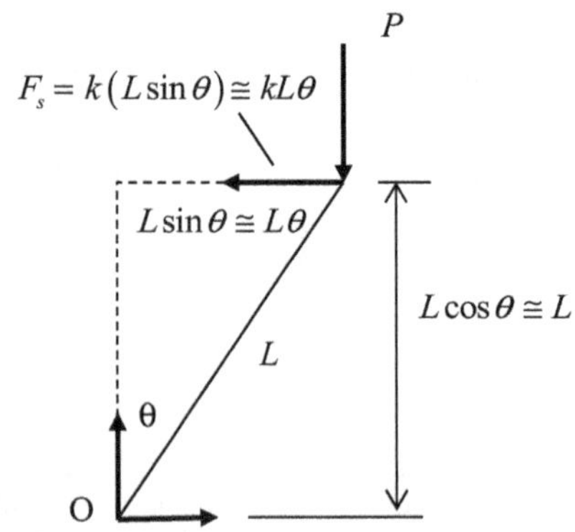

Fig. 6.14 Free body diagram for the one degree of freedom system of Fig. 6.11 for small (first order) deflections of the system

linearized problem has the two solutions: (1) $\theta = 0$, P arbitrary and (2) $\theta \neq 0$, $P = P_{cr} = kL$. If we look at the second derivative, we have

$$\frac{d^2 V^{\text{tot}}}{d\theta^2} = (-P + kL)L \tag{6.6.11}$$

which goes from positive (stable behavior) for $P < kL$ to negative for $P > kL$ (unstable behavior), showing again that the second derivative is zero at the critical load. Thus, linearization allows us to find the critical load. Linearization does not allow us to analyze in detail the post-critical load behavior of the system, which is controlled by its higher order nonlinear behavior along other possible loading paths besides the primary ($\theta = 0$) path. Even so, having the critical load is an important result from a safety standpoint and a linearized analysis can often be performed on very complex systems using numerical methods such as finite elements to obtain that critical load. Linearization only requires that we solve the equilibrium conditions for small changes about a non-deflected position. Note that we can establish linearized equilibrium conditions such as Eq. (6.6.10) either with energy methods, as done here, or directly from the Newton-Euler equations at equilibrium. For example, Fig. 6.14 shows a free body diagram of the system for small (linearized) deflections. Equilibrium of moments about point O then gives, for positive counterclockwise moments,

$$\begin{aligned} &\sum M_z = 0 \\ &(kL\theta)(L) - (P)(L\theta) = 0 \end{aligned} \tag{6.6.12}$$

which is identical with Eq. (6.6.10). Solving for critical loads with either linearized energy or Newton-Euler methods is sometimes more precisely called a *linearized precritical load analysis* to emphasize the fact that after linearization no information is obtained on post-critical behavior.

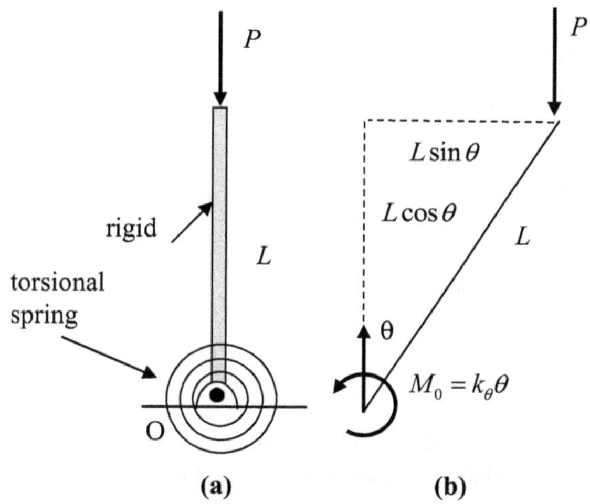

Fig. 6.15 (**a**) A one degree of freedom system that has a bifurcation behavior and (**b**) the geometry and loads of the system in a deflected position

Knowing post-critical behavior is also important since the existence of a bifurcation load does not by itself imply that the system will have some catastrophic change. For example, consider the simple system of a weightless bar and a linear torsional spring (Fig. 6.15a) which supplies a moment, M_0, about point O that is proportional to the angle of rotation, θ, i.e., $M_0 = k_\theta \theta$. In this case the work of the load is again given by Eq. (6.6.1) and the internal energy of the torsional spring is just $k_\theta \theta^2/2$ so the total potential energy is

$$V^{\text{tot}}(\theta) = -PL(1 - \cos\theta) + \frac{1}{2} k_\theta \theta^2 \tag{6.6.13}$$

At equilibrium we have

$$\frac{dV^{\text{tot}}}{d\theta} = -PL\sin\theta + k_\theta \theta = 0 \tag{6.6.14}$$

which has two solutions: (1) $\theta = 0$, P arbitrary, and (2) $\theta \neq 0$, $P = k_\theta \theta / L \sin\theta$. Again, these two loading paths have an intersection (bifurcation point) at the critical load $P = P_{cr} = k_\theta / L$. If we examine the second derivative of the potential energy we find

$$\frac{d^2 V^{\text{tot}}}{d\theta^2} = -PL\cos\theta + k_\theta \tag{6.6.15}$$

so that for the $\theta = 0$ path we have

$$\frac{d^2 V^{\text{tot}}}{d\theta^2} = -PL + k_\theta \tag{6.6.16}$$

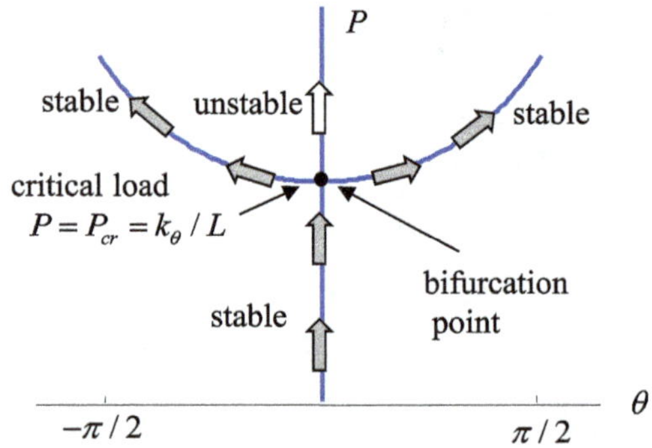

Fig. 6.16 Loading paths for the system of Fig. 6.15a and their stability characteristics

and the second derivative is positive for $P < P_{cr}$ and negative for $P > P_{cr}$, consistent with the labels shown in Fig. 6.16. For the $\theta \neq 0$ path from Eqs. (6.6.14) and (6.6.15) we have

$$\frac{d^2 V^{\text{tot}}}{d\theta^2} = LP_{\text{cr}}\left(1 - \frac{\theta \cos\theta}{\sin\theta}\right) \tag{6.6.17}$$

which shows that the second derivative is always positive along this path since $\theta/\tan\theta < 1$. Thus, the initial primary loading path is stable, as shown in Fig. 6.16. Once the critical load is reached for this system, continuing along the $\theta = 0$ path is unstable, but the system can continue along the stable post-critical curved load path. In this case the spring supports the bar in a deflected position but without any catastrophic behavior. This behavior can also be seen by examining the total potential energy of this system, as shown in Fig. 6.17. Below the bifurcation load the system is stable in the no-deflection position and above the bifurcation load the no-deflection load path is unstable but the other deflected paths are both stable.

If one linearizes the problem then the potential energy is given as

$$V^{\text{tot}}(\theta) \cong -PL\frac{\theta^2}{2} + \frac{1}{2}k_\theta \theta^2 \tag{6.6.18}$$

and the equilibrium equation is

$$\frac{dV^{\text{tot}}}{d\theta} = (-PL + k_\theta)\theta = 0 \tag{6.6.19}$$

which again indicates that either $\theta = 0$ and P is arbitrary or $\theta \neq 0$ and $P = P_{cr}$. The second derivative becomes

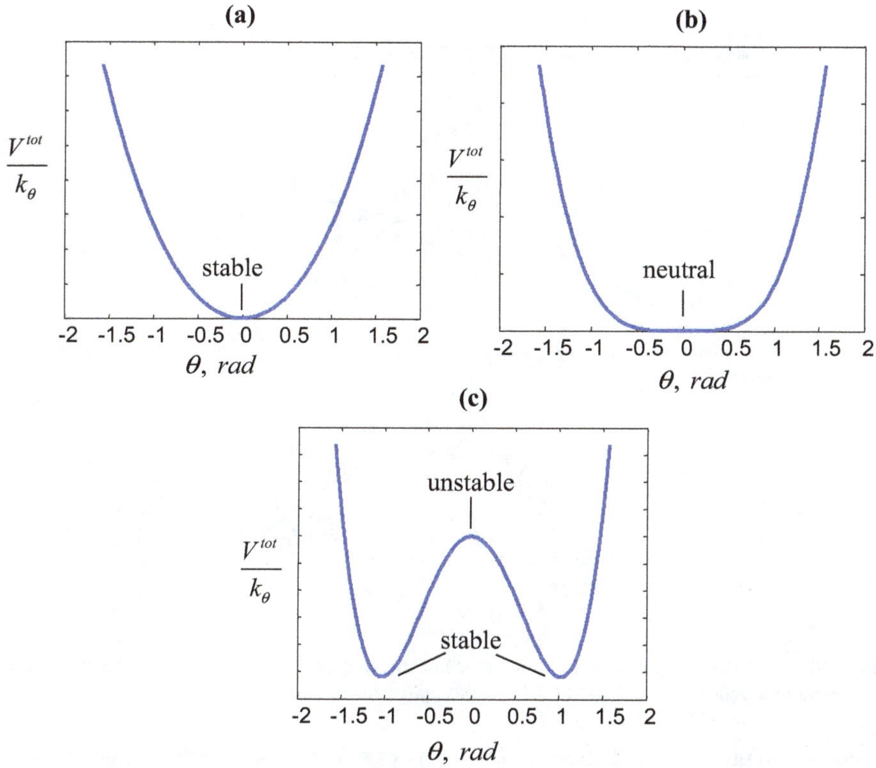

Fig. 6.17 The normalized total potential energy, V^{tot}/k_θ, versus θ for the problem of Fig. 6.15a and the nature of the equilibrium positions for (**a**) the load $P = 0.5P_{cr}$, (**b**) the load $P = 1.0P_{cr}$, and (**c**) the load $P = 1.2P_{cr}$

$$\frac{d^2V^{\text{tot}}}{d\theta^2} = -PL + k_\theta \tag{6.6.20}$$

which shows that we go from stable to unstable behavior at the critical load, which is certainly true along the primary ($\theta = 0$) loading path (Fig. 6.16) but this does not describe the nature of the post-critical load behavior. You can verify that we also get Eq. (6.6.19) by a Newton-Euler approach, i.e., guaranteeing equilibrium of moments about point O for the linearized case of small deflections.

These one degree of freedom examples show us many of the key features associated with a bifurcation type of instability. Simple two degrees of freedom systems, however, can give us a clearer window into the behavior of more complex systems. Consider, for example, the system of Fig. 6.18a where three rigid bars are pinned together and support an axial load, P. It is assumed that there are linear torsional springs with spring constants (k_1, k_2), respectively, at the two center pins. The left-hand-side pin is fixed while the right-hand-side pin is allowed to move horizontally. We will first examine this system with a linearized Newton-Euler type

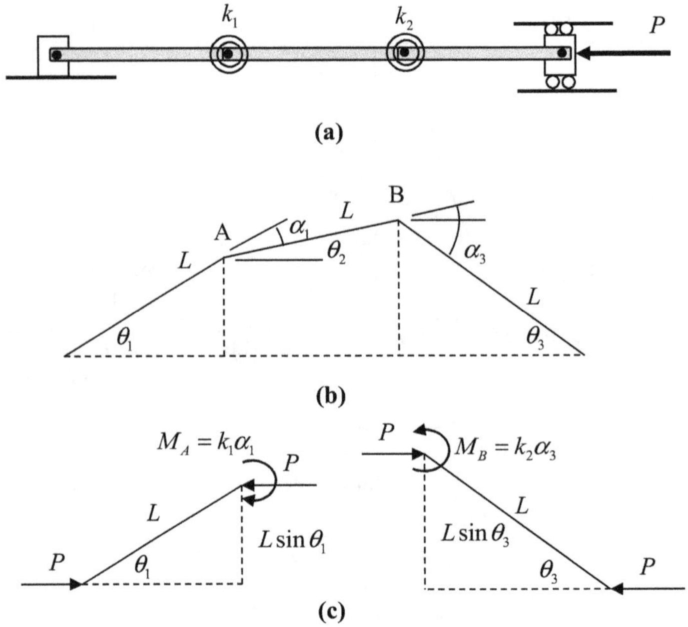

Fig. 6.18 (**a**) A two degrees of freedom mechanism, and (**b**) the definition of the geometric parameters in a deflected shape, and (**c**) free body diagrams of two of the three bars

of analysis. This is a two degrees of freedom system because a deflected position of the system, as shown in Fig. 6.18b, can be completely described by the three angles $(\theta_1, \theta_2, \theta_3)$ but these are not independent since if we examine the vertical location of pin B, for example, we have from the geometry

$$L\sin\theta_1 + L\sin\theta_2 = L\sin\theta_3 \tag{6.6.21}$$

which, for small angles in a linearized analysis, gives

$$\theta_1 + \theta_2 = \theta_3 \tag{6.6.22}$$

Thus, we will take the angles (θ_1, θ_3) as the two independent degrees of freedom. We will assume that the torsional springs are both not stretched in the configuration of Fig. 6.18a. In the deformed position of Fig. 6.18b the moments generated by the springs are proportional to the angles (α_1, α_3), which are the changes of the angles of the two bars attached to the springs from their not-stretched positions. From the geometry of Fig. 6.18b and Eq. (6.6.22) we see that

$$\begin{aligned} \alpha_1 &= \theta_1 - \theta_2 = 2\theta_1 - \theta_3 \\ \alpha_3 &= \theta_2 + \theta_3 = 2\theta_3 - \theta_1 \end{aligned} \tag{6.6.23}$$

From moment equilibrium of the two bars whose free body diagrams are shown in Fig. 6.18c we have

$$\begin{aligned} k_1(2\theta_1 - \theta_3) - PL\sin\theta_1 &= 0 \\ k_2(2\theta_3 - \theta_1) - PL\sin\theta_3 &= 0 \end{aligned} \tag{6.6.24}$$

You can verify that moment equilibrium for the other (center) bar is satisfied identically by the equations of Eq. (6.6.24) so we do not have to also examine that bar. Since we are doing a linearized analysis, to first order these equilibrium equations become

$$\begin{aligned} (2k_1 - PL)\theta_1 - k_1\theta_3 &= 0 \\ -k_2\theta_1 + (2k_2 - PL)\theta_3 &= 0 \end{aligned} \tag{6.6.25}$$

Equation (6.6.25) is a system of homogeneous linear equations for the two angles (θ_1, θ_3). The term homogeneous means that there are no constant terms present in these equations and so the right-hand-side terms of these equations are identically zero. As in the one degree of freedom case there are two possibilities. The first is that $\theta_1 = \theta_3 = 0$ and P is arbitrary, which is just the case of no deflection as the system is loaded. However, if the angles are not zero then a solution of these two homogeneous equation is only possible if the determinant of the coefficients of (θ_1, θ_3) is equal to zero, i.e.,

$$\begin{vmatrix} (2k_1 - PL) & -k_1 \\ -k_2 & (2k_2 - PL) \end{vmatrix} = 0 \tag{6.6.26}$$

If we evaluate this determinant, we find a quadratic equation for P:

$$P^2 - 2\frac{(k_1 + k_2)}{L}P + 3\frac{k_1}{L}\frac{k_2}{L} = 0 \tag{6.6.27}$$

Thus, there are two roots, each of which represents a critical load for the system, given by

$$\begin{aligned} P_1 &= \frac{(k_1 + k_2)}{L} - \sqrt{\frac{(k_1 + k_2)^2}{L^2} - 3\frac{k_1}{L}\frac{k_2}{L}} \\ P_2 &= \frac{(k_1 + k_2)}{L} + \sqrt{\frac{(k_1 + k_2)^2}{L^2} - 3\frac{k_1}{L}\frac{k_2}{L}} \end{aligned} \tag{6.6.28}$$

which can also be written as

$$
\begin{aligned}
P_1 &= \frac{(k_1 + k_2)}{L} - \sqrt{\frac{(k_1 - k_2)^2}{L^2} + \frac{k_1}{L}\frac{k_2}{L}} \\
P_2 &= \frac{(k_1 + k_2)}{L} + \sqrt{\frac{(k_1 - k_2)^2}{L^2} + \frac{k_1}{L}\frac{k_2}{L}}
\end{aligned}
\tag{6.6.29}
$$

From the forms shown in Eq. (6.6.29) we see that the terms in the square roots are always positive so the two critical loads are both real. From Eq. (6.6.28) we see that the square root terms are also always less than that of the terms outside the square roots so that the critical loads are both positive and $P_1 < P_2$. Thus, the smallest critical load is P_1. However, in this two degrees of freedom case, the linearized analysis tells us more than just the critical loads. It also tells us the nature of the deflected shape (also called the mode shape or buckling mode). To keep the algebra simple in examining these modes, let's consider the case when $k_1 = k_2 = k$ so the two torsional springs are identical. In this case the two critical loads are

$$
P_1 = \frac{k}{L}, \quad P_2 = \frac{3k}{L} \tag{6.6.30}
$$

Setting the determinant of the two equations in Eq. (6.6.25) equal to zero means that at the critical loads these two equations are not independent; that is, they really only represent one set of conditions for the two unknowns (θ_1, θ_3). However, from either one of these equations we can determine a relationship between these angles at the critical load, which gives a system configuration called a mode shape. For example, the first critical load of Eqs. (6.6.30) and (6.6.25) gives $\theta_1 = \theta_3$, the mode shape shown in Fig. 6.19a. In contrast, at the second critical load we find, from Eq. (6.6.25), $\theta_1 = -\theta_3$, the mode shape of Fig. 6.19b. Note that we can only find the relationship between the angles here, not their actual values, so specific amplitudes of the deflections shown in Fig. 6.19a, b are not determined.

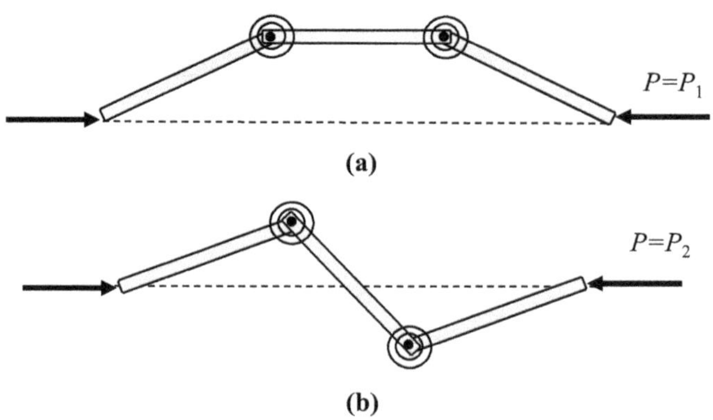

Fig. 6.19 (**a**) The system at the first critical load in a symmetric buckling pattern, and (**b**) the system at the second critical load in an antisymmetrical buckling pattern

We can write these results for this two degrees of freedom system in matrix form that gives us some valuable new ways to look at this problem. First, we note that the relationship between the moments of the torsional springs and the angles (θ_1, θ_3) for the $k_1 = k_2 = k$ case is

$$\begin{Bmatrix} M_A \\ M_B \end{Bmatrix} = \begin{bmatrix} 2k & -k \\ -k & 2k \end{bmatrix} \begin{Bmatrix} \theta_1 \\ \theta_3 \end{Bmatrix} \tag{6.6.31}$$

or, more symbolically, in matrix notation,

$$\{M\} = [K]\{\theta\} \tag{6.6.32}$$

where the matrix of coefficients, $[K]$, is called a stiffness matrix. This stiffness matrix also appears when we write the equilibrium equations, Eq. (6.6.25), in matrix-vector form as

$$\begin{bmatrix} 2k & -k \\ -k & 2k \end{bmatrix} \begin{Bmatrix} \theta_1 \\ \theta_3 \end{Bmatrix} = PL \begin{Bmatrix} \theta_1 \\ \theta_3 \end{Bmatrix} \tag{6.6.33}$$

This is a matrix problem in the form of the eigenvalue problem:

$$[K]\{\theta\} = \lambda\{\theta\} \tag{6.6.34}$$

which, as we have seen before in this book in a number of places, is a common problem. In this case we see that the values of λ that satisfy this equation (the eigenvalues) are just proportional to the critical loads, and the corresponding values for the vectors $\{\theta\}$ at these eigenvalues (the eigenvectors) are the mode shapes. Again the MATLAB® function eig can be used to obtain both the eigenvalues and eigenvectors. For example, if we normalize the stiffness matrix by dividing by k, our problem becomes the completely equivalent problem in nondimensional form

$$\begin{bmatrix} 2 & -1 \\ -1 & 2 \end{bmatrix} \begin{Bmatrix} \theta_1 \\ \theta_3 \end{Bmatrix} = \frac{PL}{k} \begin{Bmatrix} \theta_1 \\ \theta_3 \end{Bmatrix} \tag{6.6.35}$$

Now consider a MATLAB® solution of this eigenvalue problem. First we form up the nondimensional stiffness matrix:

```
K = [2 -1 ; -1 2]

K =  2   -1
    -1    2
```

If we use this matrix as an input argument to the MATLAB® function eig, the modes are stored as vectors in the columns of a matrix we have called modes and the eigenvalues of (1, 3) are placed on the diagonal of a matrix we have called vals:

```
[modes, vals] = eig(K)

modes = -0.7071  -0.7071
        -0.7071   0.7071

vals = 1   0
       0   3
```

Here the eigenvalue of $\lambda = 1$ ($P_1L/k = 1$) is associated with the mode $\{-0.7071 \quad -0.7071\}^T$ and the eigenvalue of $\lambda = 3$ ($P_2L/k = 3$) corresponds to the mode $\{-0.7071 \quad 0.7071\}^T$. Both of these mode vectors have been normalized so that they are of unit magnitude.

The value of looking at the critical load determination with matrices is that it can be generalized to very complex structures. Let us return to Eq. (6.6.33) and rewrite it as

$$\begin{bmatrix} 2k & -k \\ -k & 2k \end{bmatrix} \begin{Bmatrix} \theta_1 \\ \theta_3 \end{Bmatrix} = P \begin{bmatrix} L & 0 \\ 0 & L \end{bmatrix} \begin{Bmatrix} \theta_1 \\ \theta_3 \end{Bmatrix} \tag{6.6.36}$$

or, in symbolic form,

$$[K_s]\{\theta\} = P[K_g]\{\theta\} \tag{6.6.37}$$

where $[K_s]$ is the stiffness matrix and $[K_g]$ is called the geometric matrix. In complicated structures we can break up the structure into small elements and use potential energy methods to determine the stiffness matrix and geometric matrix for each element. If these elements are then all assembled into a representation of the entire structure then we can form up the "global" stiffness and geometric matrices associated with the entire structure and find an equation of the form of Eq. (6.6.37) but where one may be dealing with thousands if not millions of degrees of freedom. Solving this eigenvalue problem is then done just for a few of the smallest eigenvalues since those are associated with the most important critical loads that first occur. Finite element software packages can implement this approach in an efficient manner and give an engineer an important tool for obtaining critical loads at bifurcation.

Let us now examine this two degrees of freedom problem from a linearized potential energy standpoint. For the system of Fig. 6.18a, for example, the internal energy of the two springs can be written, for small angles, as approximately

$$\begin{aligned} U &= \frac{1}{2}k_1\alpha_1^2 + \frac{1}{2}k_2\alpha_3^2 \\ &\cong \frac{1}{2}k_1(2\theta_1 - \theta_3)^2 + \frac{1}{2}(2\theta_3 - \theta_1)^2 \end{aligned} \tag{6.6.38}$$

The work done by the load, P, is also approximately

$$\begin{aligned} W^e &= Pu = P[3L - L\cos\theta_1 - L\cos\theta_2 - L\cos\theta_3] \\ &= PL\left[3 - \left(1 - \frac{\theta_1^2}{2} + \ldots\right) - \left(1 - \frac{\theta_2^2}{2} + \ldots\right) + \left(1 - \frac{\theta_3^2}{2} + \ldots\right)\right] \\ &\cong \frac{PL\left[\theta_1^2 + \theta_2^2 + \theta_3^2\right]}{2} = \frac{PL\left[\theta_1^2 + (\theta_3 - \theta_1)^2 + \theta_3^2\right]}{2} \end{aligned} \tag{6.6.39}$$

The total potential energy is then approximately given by

$$V^{\text{tot}} = \frac{1}{2}k_1(2\theta_1 - \theta_3)^2 + \frac{1}{2}k_2(2\theta_3 - \theta_1)^2 - \frac{PL}{2}\left[\theta_1^2 + (\theta_3 - \theta_1)^2 + \theta_3^2\right] \tag{6.6.40}$$

so that at equilibrium we have

$$\begin{aligned} \frac{\partial V^{\text{tot}}}{\partial\theta_1} &= k_1(2\theta_1 - \theta_3)(2) + k_2(2\theta_3 - \theta_1)(-1) - PL[\theta_1 + (\theta_3 - \theta_1)(-1)] \\ &= (4k_1 + k_2)\theta_1 - (2k_2 + 2k_2)\theta_3 - PL(2\theta_1 - \theta_3) = 0 \\ \frac{\partial V^{\text{tot}}}{\partial\theta_3} &= k_1(2\theta_1 - \theta_3)(-1) + k_2(2\theta_3 - \theta_1)(2) - PL[\theta_3 + (\theta_3 - \theta_1)] \\ &= -(2k_1 + 2k_2)\theta_1 + (k_1 + 4k_2)\theta_3 - PL(2\theta_3 - \theta_1) = 0 \end{aligned} \tag{6.6.41}$$

For the $k_1 = k_2 = k$ case, these equations can be written in matrix form as

$$\begin{bmatrix} 5 & -4 \\ -4 & 5 \end{bmatrix}\begin{Bmatrix} \theta_1 \\ \theta_3 \end{Bmatrix} - \frac{PL}{k}\begin{bmatrix} 2 & -1 \\ -1 & 2 \end{bmatrix}\begin{Bmatrix} \theta_1 \\ \theta_3 \end{Bmatrix} = 0 \tag{6.6.42}$$

These equilibrium equations look quite different from those that we obtained with the Newtonian-Euler approach (Eq. (6.6.36)) but if we compute the inverse of the second matrix, given by

$$\begin{bmatrix} 2 & -1 \\ -1 & 2 \end{bmatrix}^{-1} = (1/3)\begin{bmatrix} 2 & 1 \\ 1 & 2 \end{bmatrix}$$

and pre-multiply Eq. (6.6.42) by this inverse, we recover Eq. (6.6.36) so both methods do give the same matrices and hence the same critical loads and modes.

However, this multiplication is not necessary. We can directly solve the generalized eigenvalue problem of Eq. (6.6.42) in MATLAB® with the eig function. We have

```
Ks= [5 -4; -4 5];
Kg = [2 -1 ; -1 2];
[modes, vals] = eig(Ks, Kg)
modes=  -0.7071  -0.4082
        -0.7071   0.4082
vals =    1     0
          0     3
```

which are the same values obtained before if we normalize the eigenvectors so they are both unit vectors.

For the second derivatives of the potential energy we have

$$\begin{aligned}\frac{\partial^2 V^{tot}}{\partial \theta_1^2} &= (4k_1 + k_2) - 2PL = 5k - 2PL \\ \frac{\partial^2 V^{tot}}{\partial \theta_3^2} &= (k_1 + 4k_2) - 2PL = 5k - 2PL \\ \frac{\partial^2 V^{tot}}{\partial \theta_1 \partial \theta_3} &= -(2k_1 + 2k_2) + PL = -4k + PL\end{aligned} \tag{6.6.43}$$

From our discussion of dynamic stability we know that we must keep the change of the total potential positive for any virtual change of the independent variables (θ_1, θ_3), which is also the requirement for static stability. Here this change is to second order

$$\delta V^{\text{tot}} = \frac{1}{2}\left(\frac{\partial^2 V^{\text{tot}}}{\partial \theta_1^2}\delta\theta_1^2 + \frac{\partial^2 V^{\text{tot}}}{\partial \theta_3^2}\delta\theta_3^2 + \frac{\partial^2 V^{\text{tot}}}{\partial \theta_1 \partial \theta_3}\delta\theta_1\delta\theta_3 + \frac{\partial^2 V^{\text{tot}}}{\partial \theta_3 \partial \theta_1}\delta\theta_1\delta\theta_3\right) \tag{6.6.44}$$

so that the change of the potential energy is characterized by its second derivatives, which we can again place in the form of a matrix as

$$A = \begin{bmatrix} \partial^2 V^{\text{tot}}/\partial\theta_1^2 & \partial^2 V^{\text{tot}}/\partial\theta_1\partial\theta_3 \\ \partial^2 V^{\text{tot}}/\partial\theta_1\partial\theta_3 & \partial^2 V^{\text{tot}}/\partial\theta_3^2 \end{bmatrix} \tag{6.6.45}$$

Since for static stability we must have $\delta V^{\text{tot}} > 0$ for all changes $(\delta\theta_1, \delta\theta_3)$ the matrix A must be positive definite. Recall that a positive definite matrix has the property that its determinant must be positive and the determinant of all its principal minors also must be positive, where successive principal minors are obtained by deleting a row and column from the remaining matrix. However, Langhaar [3] has shown that if the elements of A, the matrix of second derivatives, are sufficiently well-behaved functions of the load P then we can determine the critical load to be the smallest value, P_{cr}, where the determinant of the entire matrix A goes to zero and where below that value A is positive definite and for at least some finite interval above P_{cr} the matrix A is no longer positive definite. For example, in this case we must have for stability the following two determinants as positive:

$$D_2 = \begin{vmatrix} \partial^2 V^{\text{tot}}/\partial\theta_1^2 & \partial^2 V^{\text{tot}}/\partial\theta_1\partial\theta_3 \\ \partial^2 V^{\text{tot}}/\partial\theta_1\partial\theta_3 & \partial^2 V^{\text{tot}}/\partial\theta_3^2 \end{vmatrix} > 0, \quad D_1 = \left|\frac{\partial^2 V^{\text{tot}}}{\partial\theta_1^2}\right| > 0 \quad (6.6.46)$$

or, equivalently,

$$\begin{aligned} D_2 &= \left(\partial^2 V^{\text{tot}}/\partial\theta_1^2\right)\left(\partial^2 V^{\text{tot}}/\partial\theta_3^2\right) - \left(\partial^2 V^{\text{tot}}/\partial\theta_1\partial\theta_3\right)^2 > 0, \\ D_1 &= \partial^2 V^{\text{tot}}/\partial\theta_1^2 > 0 \end{aligned} \quad (6.6.47)$$

In our problem these determinants are

$$D_2 = (5k - 2Pl)^2 - (Pl - 4k)^2, \quad D_1 = 5k - 2Pl \quad (6.6.48)$$

By expanding the determinant D_2 and collecting terms it is easy to see that the determinants can be rewritten as

$$\begin{aligned} D_2 &= 3L^2(k/L - P)(3k/L - P) \\ D_1 &= 2L(5k/2L - P) \end{aligned} \quad (6.6.49)$$

If one examines the expression for D_2 in Eq. (6.6.49) we see that the smallest value of P where this determinant goes to zero is $P = k/L$. Below this load Eq. (6.6.49) shows that $D_1 > 0, D_2 > 0$ so the system is stable and Eq. (6.6.49) also shows that there are values immediately above this load where $D_2 < 0$ and the system is not stable, so $P_{cr} = k/L$. Note that there are other values where $D_2 = 0$ ($P = 3k/L$ is such a value in this example) but the system becomes unstable before such loads and their corresponding modes are ever reached. However, if one constrains the system so that the lower critical modes cannot occur, then these higher loads may become the new critical loads that govern the stability of the system. For example, if we place some rollers at the center of this system, as shown in Fig. 6.20, then we must have

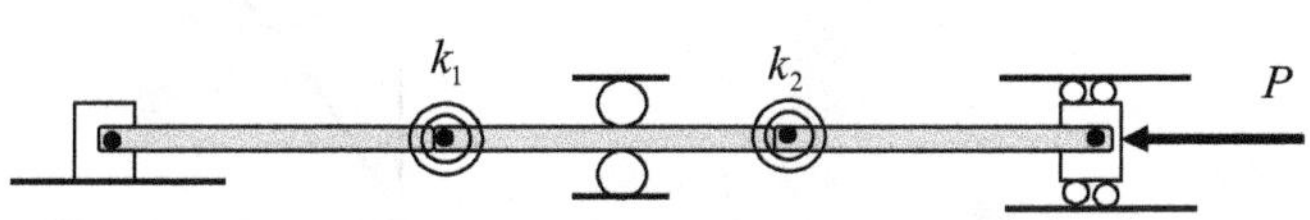

Fig. 6.20 The system of Fig. 6.9a supported so as to prevent the first critical mode from occurring

$\theta_1 = -\theta_3$. Placing this constraint into our potential energy expression, Eq. (6.6.40), with $k_1 = k_2 = k$ gives

$$V^{\text{tot}} = 9k\theta_1^2 - 3PL\theta_1^2 \tag{6.6.50}$$

which leads to the equilibrium equation

$$\frac{\partial V^{\text{tot}}}{\partial \theta_1} = (18k - 6PL)\theta_1 = 0 \tag{6.6.51}$$

and the second derivative

$$\frac{\partial^2 V^{tot}}{\partial \theta_1^2} = (18k - 6PL) \tag{6.6.52}$$

showing that the critical load is now $P_{cr} = 3k/L$. We see that knowing the mode of deformation of the first critical load, and supporting the system so as to prevent the buckling behavior of that first mode, allows us to increase the critical load by a factor of three, a substantial increase that leads to a much safer system.

6.7 Limit Load Instability

In the case of a bifurcation type of instability multiple load paths exist at the critical loads. However, even when a system follows a single load path, if a maximum or minimum limit load is reached the system may become unstable. This is illustrated schematically in Fig. 6.21 for a plot of a load P versus a deflection, Δ, where along the load path up to the maximum load the system is stable, while past that limiting load the system is unstable. We can modify the same simple one degree of freedom systems discussed earlier to exhibit this limit load behavior. First consider the problem shown in Fig. 6.22a, where the load of Fig. 6.11a is now applied to the rigid bar at a distance, e, along an extension. It is assumed that the load still remains constant in magnitude and direction and the spring is not stretched when the bar is

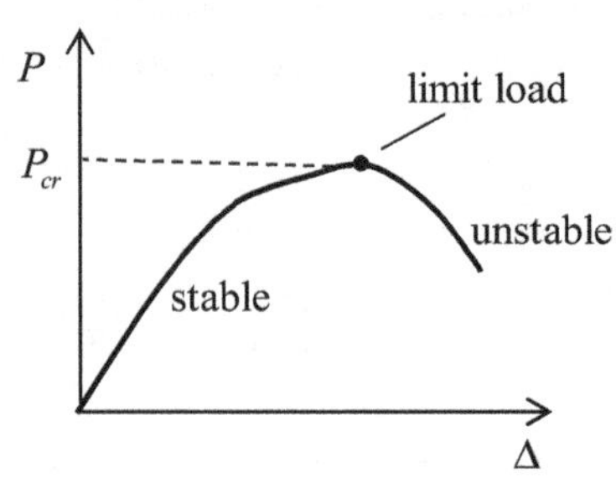

Fig. 6.21 A limit load type of instability

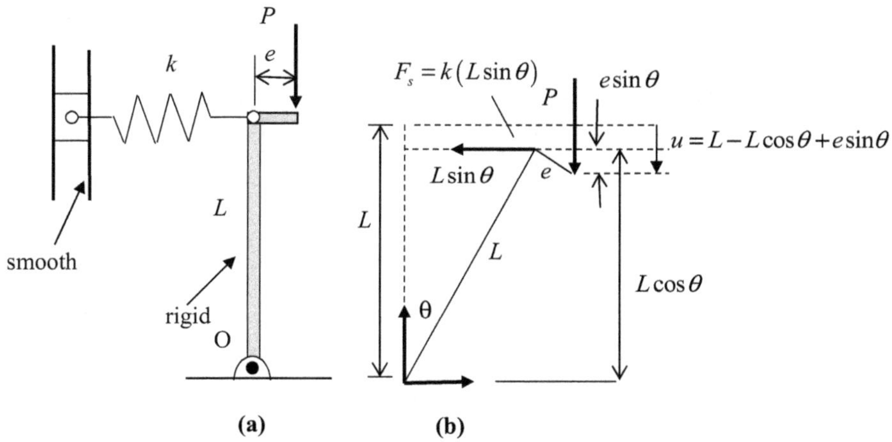

Fig. 6.22 (**a**) A one degree of freedom system (slightly modified from Fig. 6.10a) in its undeformed position and (**b**) the geometry and loads of the system in a deformed position

vertical. The stretch of the spring in the deformed position (Fig. 6.22b) is still $L \sin \theta$ but the vertical displacement of the load is now $u = L - L \cos \theta + e \sin \theta$. Thus, the total potential energy is

$$V^{\text{tot}} = \frac{1}{2}k(L\sin\theta)^2 - P(L - L\cos\theta + e\sin\theta) \tag{6.7.1}$$

At equilibrium

$$\frac{\partial V^{\text{tot}}}{\partial \theta} = kL^2 \sin\theta\cos\theta - PL(\sin\theta + e\cos\theta/L) = 0 \tag{6.7.2}$$

which we can solve for the load P versus θ as

$$P = \frac{kL\sin\theta\cos\theta}{\sin\theta + e\cos\theta/L} \tag{6.7.3}$$

Figure 6.23 plots the normalized load, P/kL, versus θ for various values of the parameter e/L, which is called the *eccentricity*. It can be seen that the eccentricity of the loading has destroyed the symmetrical nature of the problem. As P increases from zero, θ is also a positive, increasing angle until the load reaches a maximum limiting load (that is a function of e/L). Beyond the limit load the load decreases with increasing angle, showing an unstable behavior. As the eccentricity becomes very small, the limit load approaches the critical bifurcation load of the $e/L = 0$ case. Note that there are no stable equilibrium positions when the angle θ is negative (Fig. 6.23). These conditions can be confirmed by examining $\partial^2 V^{\text{tot}}/\partial\theta^2$ or by simply plotting

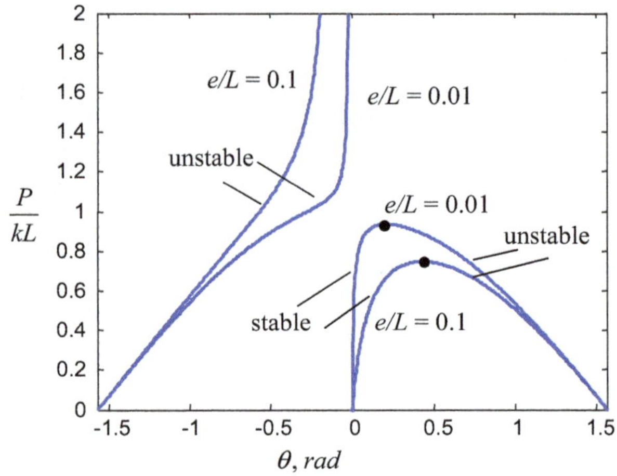

Fig. 6.23 Load curves showing limit load behavior for different eccentricities of loading for the system of Fig. 6.22a

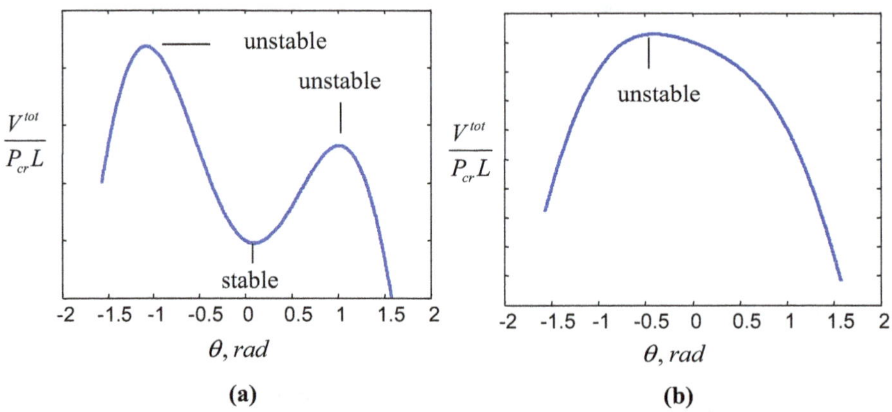

Fig. 6.24 The normalized total potential energy, $V^{tot}/P_{cr}L$, versus θ for the problem of Fig. 6.22a and the nature of the equilibrium positions for $e/L = 0.1$ and (**a**) $P = 0.5P_{cr}$, and (**b**) $P = 1.2P_{cr}$

the total potential energy (at a given eccentricity) versus θ for different loads. Figure 6.24, for example, shows the total potential energy for $e/L = 0.1$ at loads both below and above the limit load. We see both stable and unstable solutions for $\theta > 0$ but only unstable solutions for $\theta < 0$.

Some texts study bifurcation stability problems using a linearized approach when introducing eccentricities to the load. For example, in this case for small angles θ we can let $\sin\theta \cong \theta$ and $\cos\theta \cong 1$ and the equation for the load curves becomes

$$P = \frac{kL\theta}{\theta + e/L} \tag{6.7.4}$$

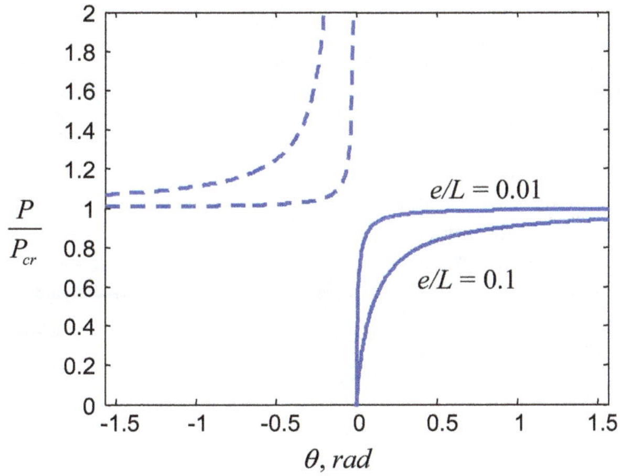

Fig. 6.25 Load curves for a linearized eccentricity analysis of the problem of Fig. 6.15a

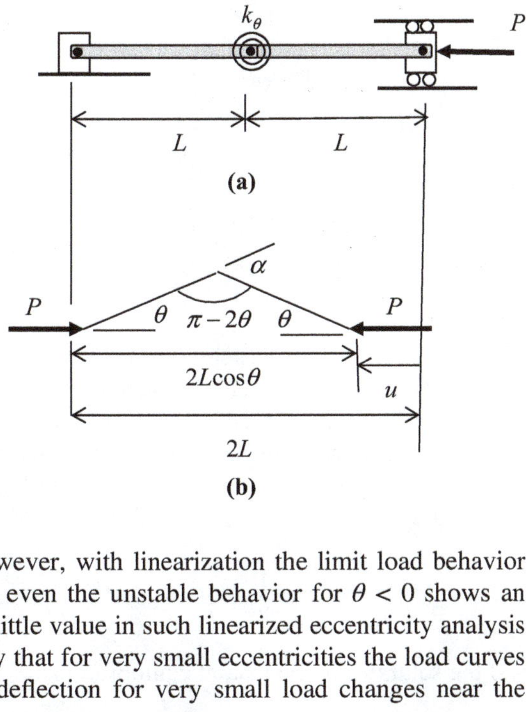

Fig. 6.26 (**a**) An axially loaded system of two rigid bars pinned together and acted upon by a linear torsional spring, and (**b**) the geometry in a deformed configuration

which is shown in Fig. 6.25. However, with linearization the limit load behavior disappears entirely for $\theta > 0$ and even the unstable behavior for $\theta < 0$ shows an incorrect behavior. Thus, there is little value in such linearized eccentricity analysis except to demonstrate qualitatively that for very small eccentricities the load curves exhibit a very rapid increase in deflection for very small load changes near the critical bifurcation load.

To see another way in which a system with a bifurcation behavior changes into a limit load behavior, consider the system shown in Fig. 6.26a where a linear torsional spring acts on two rigid bars that are pinned together and at their ends. It is assumed that the spring is not stretched when the bars are collinear, as shown in Fig. 6.26a. Figure 6.26b shows the system in a deflected position, where from the geometry we

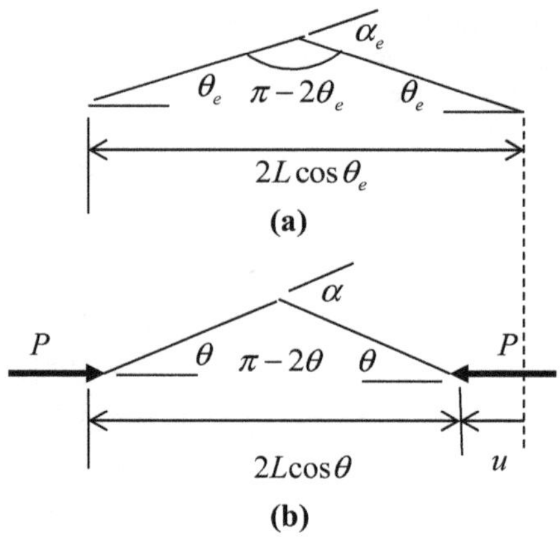

Fig. 6.27 (**a**) The system of Fig. 6.26a when there is a small initial misalignment of the bars shown when (**a**) the system is in its original unloaded position and the torsional spring is not stretched in this position, and (**b**) the system is in a loaded and deformed position

see that the angle $\alpha = 2\theta$. In the deformed configuration the total potential energy is given by

$$\begin{aligned} V^{\text{tot}} &= \frac{1}{2}k_\theta \alpha^2 - Pu \\ &= \frac{1}{2}k_\theta (2\theta)^2 - P(2L - 2L\cos\theta) \end{aligned} \tag{6.7.5}$$

and at equilibrium $\partial V^{\text{tot}}/\partial\theta = 0$, giving

$$4k_\theta \theta - 2PL\sin\theta = 0 \tag{6.7.6}$$

There are two equilibrium solutions: (1) $\theta = 0$, P arbitrary, and (2) $\theta \neq 0$, $P = 2k_\theta\theta/L\sin\theta$. This leads to a stable bifurcation behavior almost identical to the one bar system of Fig. 6.15a, as seen in Fig. 6.16, except the critical load is now $P_{cr} = 2k_\theta/L$. However, let us now introduce an eccentricity into this system by assuming that there is initially a small angle, θ_e, when the load is absent and the spring is not stretched. In this case from Fig. 6.27a, b we see that the total potential energy

$$V^{\text{tot}} = \frac{1}{2}k_\theta (2\theta - 2\theta_e)^2 - P(2L\cos\theta_e - 2L\cos\theta) \tag{6.7.7}$$

and the solution for the load at equilibrium is

$$P = \frac{2k_\theta(\theta - \theta_e)}{L\sin\theta} \tag{6.7.8}$$

Figure 6.28 shows the loading paths for this system where there is a small angular misalignment $\theta_e = 0.07$. We see that as the load is increased from zero, the angle θ increases with the load along a stable equilibrium loading path with no critical angle

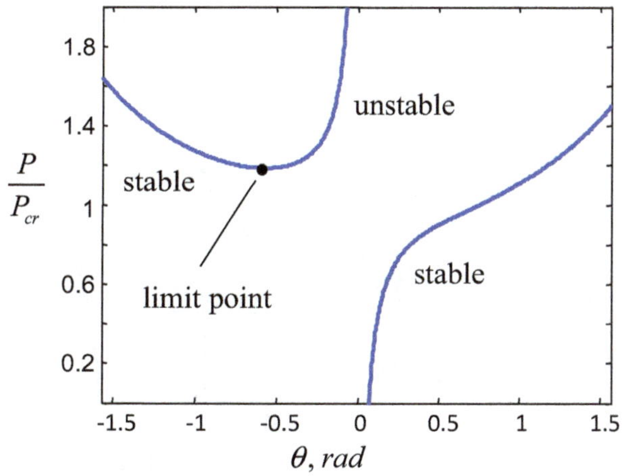

Fig. 6.28 Loading paths for the misaligned case (see Fig. 6.27) of the system shown in Fig. 6.26a where $\theta_e = 0.07$ rad

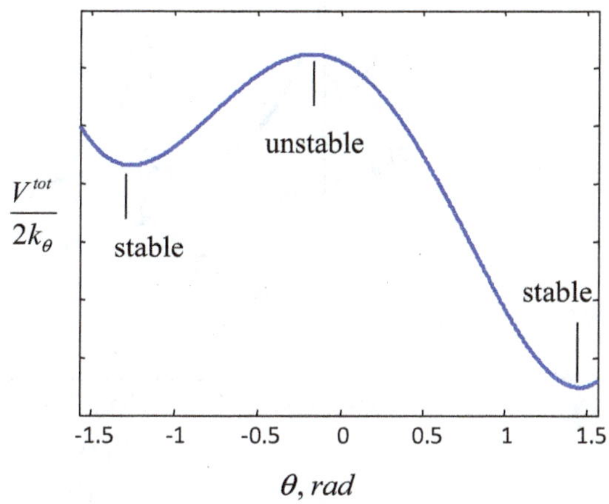

Fig. 6.29 The total potential energy for the misaligned case (see Fig. 6.26) where $\theta_e = 0.07$ rad and $P/P_{cr} = 1.4$

or limit load behavior along this primary load path. However, now there is also a set of stable (and unstable) equilibrium positions for negative angles as well. This can be seen by plotting the total potential energy at a load $P/P_{cr} = 1.4$ as shown in Fig. 6.29, where there are stable potential "wells" for both positive and negative angles, as well as an unstable potential "hill" at a small negative angle, agreeing with the labels shown in Fig. 6.28. Thus, if the bars were forced to one of these stable negative angle positions we could hold it there in stable equilibrium with the appropriate force. If we then started to reduce that force, we would follow the arrows along the stable unloading path of Fig. 6.30a until a minimum limit point is reached at a load of approximately $P/P_{cr} = 1.186$ at point A. The total potential energy curve at this limit load value is shown in Fig. 6.30b. If the load was reduced slightly from this limit load value there is no stable position for the system to go to for negative angles so the bars "snap through" to a new equilibrium position B at a positive angle, as shown in

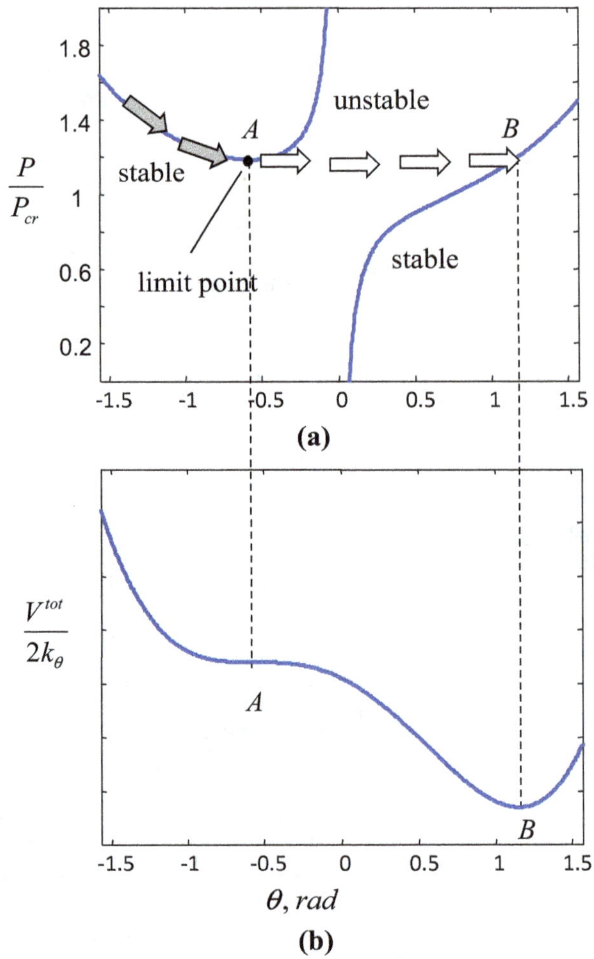

Fig. 6.30 (**a**) Following a stable unloading path to a limit point where "snap-through" behavior occurs for the misaligned case (see Fig. 6.26), and (**b**) the total potential energy curve at the limit point where $P/P_{\mathrm{cr}} = 1.186$. Again $\theta_e = 0.07$ rad in this example

Fig. 6.30a, corresponding to a motion from the neutral equilibrium position of point A to the stable equilibrium position of position B on the potential energy curve of Fig. 6.30b. At B the system could then continue along a stable unloading or loading path. Snap-through instability is another type of instability behavior that is possible in some systems. Abrupt snap-through behavior must be introduced somewhat artificially in this example by starting the system out from a particular loaded position, but as we will see in the next section snap-through behavior can also occur along a primary loading path when a limit load is reached along that path and the system jumps to a new stable position.

6.8 Snap-Through Instability

To see how snap-through instabilities can occur directly in a simple case, consider the single degree of freedom bar and spring system of Fig. 6.31a, where θ (Fig. 6.31b) is the angle of the bar at a load P. It is assumed that the spring is initially not stretched and the bar (whose weight is neglected) is unloaded when $\theta = \theta_0$. From Fig. 6.31b we see that total potential energy of the system is

$$V^{\text{tot}} = \frac{1}{2}k(L\sin\theta - L\sin\theta_0)^2 - P(L\cos\theta_0 - L\cos\theta) \tag{6.8.1}$$

and at equilibrium $\partial V^{\text{tot}}/\partial\theta = 0$ gives

$$P = \frac{kL(\sin\theta - \sin\theta_0)\cos\theta}{\sin\theta} \tag{6.8.2}$$

The nondimensional load, P/kL, is plotted versus θ in Fig. 6.32 for $\theta_0 = 9^\circ$. We see a stable loading path up to a limit load $P/kL = 0.598$ at point A where $\theta = 32.5^\circ$. If the load is increased slightly above this limit load the only stable position is at B so the system snaps through to that point where $\theta = 174^\circ$. Note that the bar can remain in equilibrium when it is horizontal ($\theta = 90^\circ$) under no load but this is an unstable

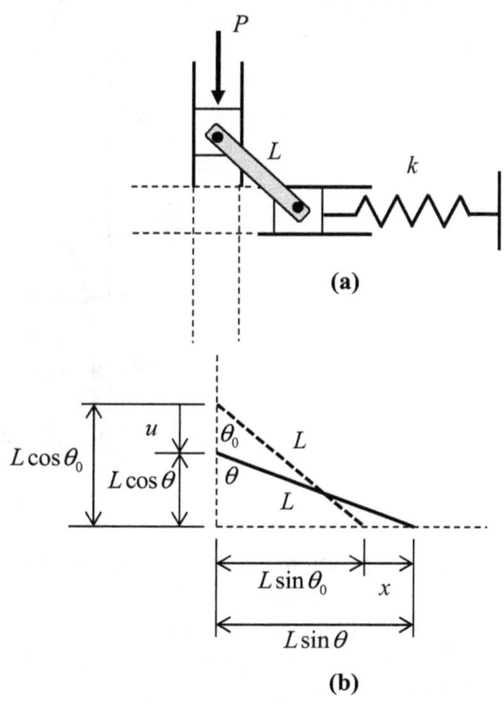

Fig. 6.31 (**a**) A bar-spring system and (**b**) the deflection of the system by the load P (from its unloaded position where $\theta = \theta_0$ and the spring is not stretched) to its loaded position

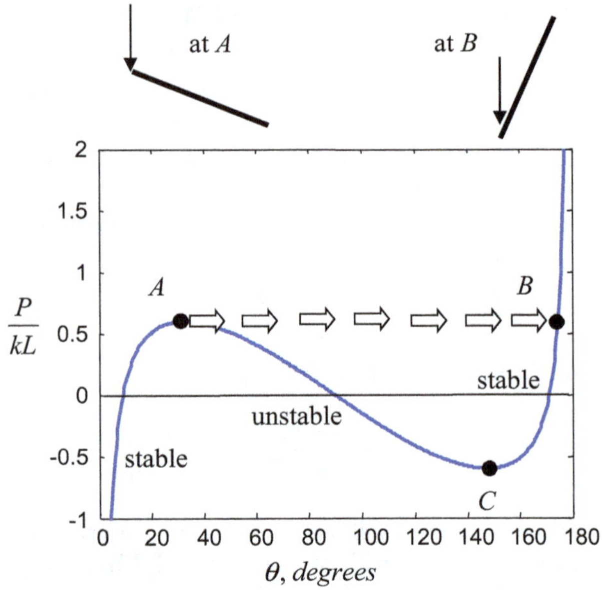

Fig. 6.32 Snap-through behavior of the bar-spring of Fig. 6.31a

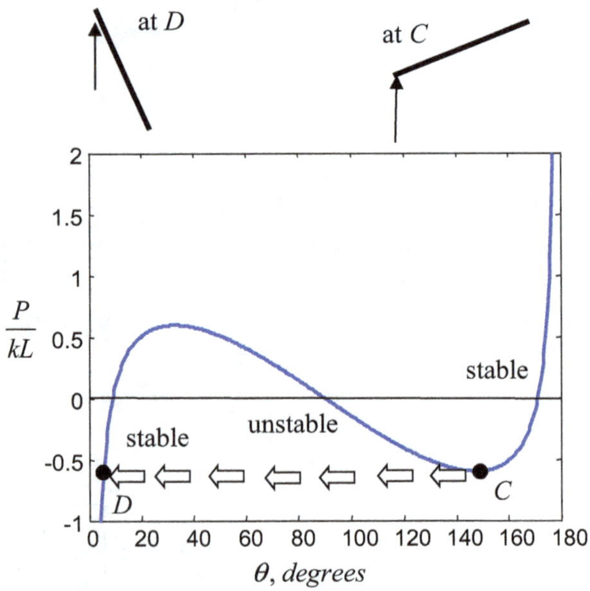

Fig. 6.33 Snap-through behavior returning the bar-spring of Fig. 6.31a to a small positive angle

equilibrium position and the bar snaps through at a much smaller angle before this position is ever reached. At B the load could continue to be stably increased or it could also be decreased to a new negative limit load at point C (see Fig. 6.33) where the bar then snaps through back to a stable position at point D with a positive angle under the action of the negative load.

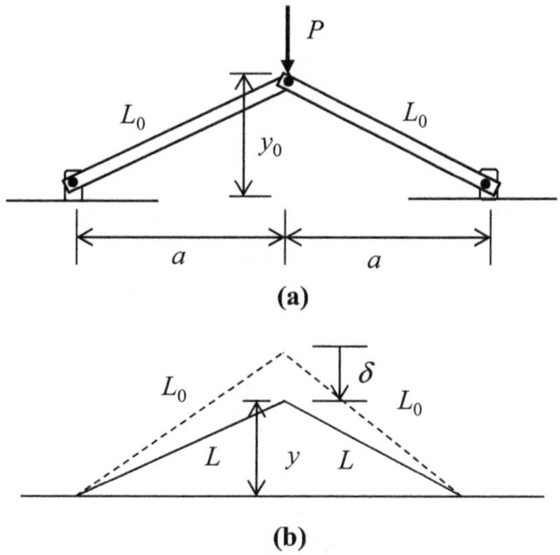

Fig. 6.34 (**a**) A shallow truss whose initial unloaded dimensions are as shown and (**b**) its deformation under the load P

A structure that can exhibit snap-through buckling and is often used to illustrate this behavior is the truss shown in Fig. 6.34a. It is assumed that the truss is shallow, i.e., $y_0 << a$, and that we have a linear elastic (springlike) deformation of the truss members under the load, P. The spring constant of each truss member is $k = EA/L_0$ where A is the cross-sectional area of a member, E is Young's modulus (a material constant), and L_0 is the member's initial length when $P = 0$. Application of the load causes this length to change a small amount and become the new length, L, while the initial height, y_0, changes to a height, y, under the load P, as shown in Fig. 6.34b. Since both members act as springs the total potential energy of this truss is

$$\begin{aligned} V^{tot} &= \frac{1}{2}k(L - L_0)^2 \times 2 - P\delta \\ &= k\left[\sqrt{y^2 + a^2} - \sqrt{y_0^2 + a^2}\right]^2 - P\delta \\ &= k\left[\sqrt{(y_0 - \delta)^2 + a^2} - \sqrt{y_0^2 + a^2}\right]^2 - P\delta \end{aligned} \tag{6.8.3}$$

where $\delta = y_0 - y$ is the distance moved by the load, P. In this one degree of freedom system $V^{tot} = V^{tot}(\delta)$ so at equilibrium we must have

$$\frac{dV^{tot}}{d\delta} = 2k\left[\sqrt{(y_0 - \delta)^2 + a^2} - \sqrt{y_0^2 + a^2}\right]\left[\frac{\delta - y_0}{\sqrt{(y_0 - \delta)^2 + a^2}}\right] - P = 0 \tag{6.8.4}$$

which gives the load-deflection relationship, which we can write in nondimensional form as

$$\frac{P}{ky_0} = 2\left[\sqrt{(1-\delta/y_0)^2 + (a/y_0)^2} - \sqrt{1+(a/y_0)^2}\right]\left[\frac{\delta/y_0 - 1}{\sqrt{(1-\delta/y_0)^2 + a^2/y_0^2}}\right] \tag{6.8.5}$$

Equation (6.8.5) looks rather complex, but we can simplify it, using the fact that $a/y_0 >> 1$, which gives, after some manipulation and collecting terms,

$$\begin{aligned}\frac{P}{ky_0} &\cong 2\left[\sqrt{\frac{(1-\delta/y_0)^2}{(a/y_0)^2}+1} - \sqrt{\frac{1}{(a/y_0)^2}+1}\right][\delta/y_0 - 1]\\ &\cong 2\left[\frac{(1-\delta/y_0)^2}{2(a/y_0)^2} + 1 - 1 - \frac{1}{2(a/y_0)^2}\right][\delta/y_0 - 1]\\ &= \frac{y_0^2}{a^2}\left[-2\delta/y_0 + \delta^2/y_0^2\right][\delta/y_0 - 1]\\ &= \frac{y_0^2}{a^2}\left[\frac{\delta}{y_0}\right]\left[1-\frac{\delta}{y_0}\right]\left[2-\frac{\delta}{y_0}\right]\end{aligned} \tag{6.8.6}$$

Equation (6.8.6) shows that the approximate load versus deflection curve is a cubic curve which we can write finally as

$$\frac{P}{P_0} = \left[\frac{\delta}{y_0}\right]\left[1-\frac{\delta}{y_0}\right]\left[2-\frac{\delta}{y_0}\right] \tag{6.8.7}$$

where $P_0 = ky_0^3/a^2$. If we plot the nondimensional load of Eq. (6.8.7) versus nondimensional deflection we see the typical loading curve characteristic of snap-through buckling (Fig. 6.35), which shows that we can load the truss stably up to a

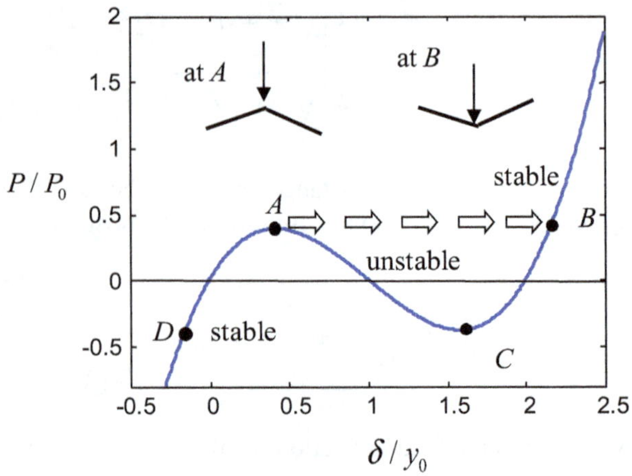

Fig. 6.35 Normalized load versus normalized deflection for the shallow truss of Fig. 6.34a, showing its snap-through behavior

limit point A (at $P/P_0 = 0.385$ with $\delta/y_0 = 0.423$) where the truss snaps through to a new stable equilibrium point at B. As in the previous example it is also possible to snap the truss back from point C to point D. Eq. (6.8.7) shows that the load is zero when $\delta/y_0 = (0, 1, 2)$, corresponding, respectively, to the original unloaded position, truss members both horizontal (which requires no load to maintain this unstable equilibrium position), and the truss in a snapped-through position, mirroring the original unloaded position. Although all these results are based on Eq. (6.8.7), which is an approximate expression, you can plot the original expression (Eq. (6.8.5)) and verify that Eq. (6.8.7) captures the snap-through behavior very well.

Snap-through behavior is sometimes called "oil canning" since the slightly outwardly curved bottom of an oil can will often snap through to an inwardly curved position when pushed on by a finger. Snap-through behavior can also occur in large, shallow domes or arch structures.

6.9 Problems

P6.1. Plot the behavior of the normalized dynamic potential, U/mgR, for the problem of a particle sliding on rotating hoop (Eq. (6.3.2)) for $\Omega^2 R/g = 0.5$ and $\Omega^2 R/g = 2.0$ and identify the stable and unstable points.

P6.2. One end of a simple pendulum is attached to an assembly that rotates at a constant angular velocity, Ω, about the z-axis as shown in Fig. P6.1. Determine the equilibrium positions and the stability of those positions.

P6.3. A mass, m, is attached to a rigid arm of length L and negligible weight and to a linear spring with spring constant, k (Fig. P6.2). The spring rides in a smooth horizontal track so that it always remains horizontal and the spring is unstretched when the arm is vertical. This dynamic problem is analogous to the static stability problem discussed in Sect. 6.6 (see Fig. 6.11) where the load P in the static problem is now the weight mg of the moving mass. (a) Determine the equation of motion for the system in terms of the angle, θ, and find the equilibrium positions of the mass for $mg < kL$ and $mg > kL$. Are

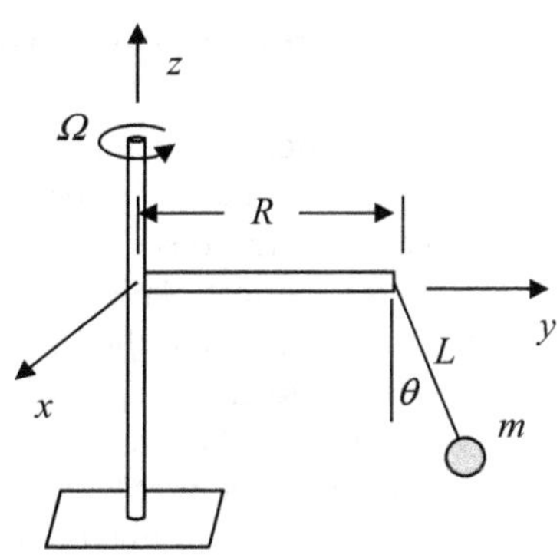

Fig. P6.1 A simple pendulum on a rotating arm (Prob. P6.2)

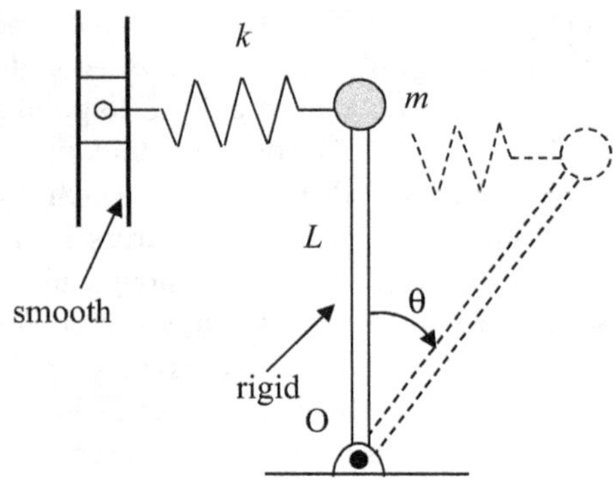

Fig. P6.2 Mass attached to a rigid arm and a linear spring (Prob. P 6.3)

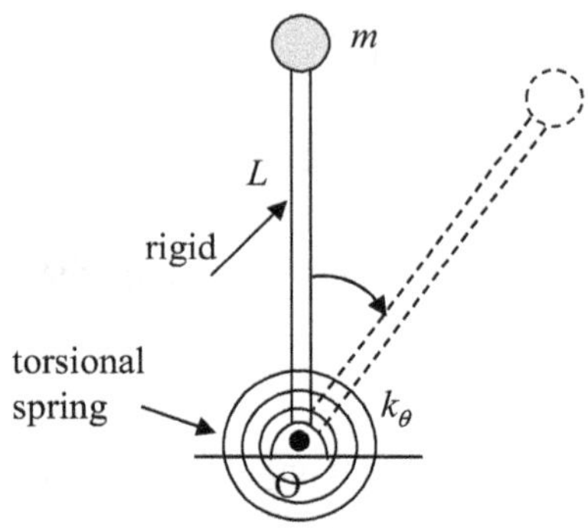

Fig. P6.3 Mass attached to a rigid arm and a torsional spring (Prob. P6.4)

these equilibrium positions stable or unstable? Plot the total potential energy for $-5\pi/4 < \theta < 5\pi/4$ rad to verify your results. (b) This is a conservative system so that the total energy, $E = T+V$, is a constant. Obtain a phase plot of the motion by finding the constant energy contours of the normalized energy, E/mgL, for $mg/kL = 0.7$ and $L/g = 0.1$ sec^2, plotting 20 contours over the region $-5\pi/4 < \theta < 5\pi/4$ rad and $-5 < \dot{\theta} < 5$ rad/s (use at least 500 points for both θ and $\dot{\theta}$ to generate the contour plot). Identify the stable and unstable equilibrium positions on this phase plot. Based on the phase plot, describe the nature of the motion you would expect if the initial conditions are $\theta = \theta_0$, $\dot{\theta} = 0$ for $\theta_0 = 30°$, and $\theta_0 = 60°$. How does the phase plot and your description of the motion change if $mg/kL = 1.2$? (c) Solve the equation of motion numerically for different initial condition values θ_0 and verify that your descriptions of the motion deduced from the phase plots in parts (a) and (b) are correct. Animate the motion using the MATLAB® animate_lines.

P6.4. A mass, m, is attached to a rigid arm of length L and negligible weight and to a linear torsional spring that applies a resisting moment $M = k_\theta \theta$ about the pivot point O (Fig. P6.3). The spring is unstretched when the arm is vertical. This dynamic problem is analogous to the static stability problem discussed in Sect. 6.6 (see Fig. 6.15) where the load P in the static problem is now the weight mg of the moving mass. (a) Determine the equation of motion for the system in terms of the angle, θ, and find the equilibrium positions of the mass for $mg < k_\theta/L$ and $mg > k_\theta/L$. Are these equilibrium positions stable or unstable? Plot the total potential energy for $-5\pi/4 < \theta < 5\pi/4$ rad to verify your results. (b) This is a conservative system so that the total energy, $E = T + V$, is a constant. Obtain a phase plot of the motion by finding the constant energy contours of the normalized energy, E/mgL, for $mgL/k_\theta = 0.7$ and $L/g = 0.1\ \text{sec}^2$, plotting 20 contours over the region $-5\pi/4 < \theta < 5\pi/4$ rad and $-5 < \dot{\theta} < 5$ rad/s (use at least 500 points for both θ and $\dot{\theta}$ to generate the contour plot). Identify the equilibrium positions on this phase plot. Based on the phase plot, describe the nature of the motion you would expect if the initial conditions are $\theta = 0$ and $\dot{\theta} = 1.5$ rad/s. How does the phase plot and your description of the motion change if $mgL/k_\theta = 1.5$? (c) Solve the equation of motion numerically for different initial conditions and verify that your descriptions of the motion deduced from the phase plots in parts (a) and (b) are correct. Animate the motion using the MATLAB® animate_lines.

P6.5. Two masses, each of mass m, are connected by a weightless rigid bar, as shown in Fig. P6.4 and ride in smooth horizontal and vertical slots. The mass in the horizontal slot is also attached to a linear spring with spring constant, k. The spring is unstretched when the angle $\theta = \theta_0$. [Note: This is not the initial starting angle or equilibrium angle of the system.] This dynamic problem is analogous to the static problem discussed in Sect. 6.8 (see Fig. 6.31) where the load P in the static problem is now the weight of the moving mass in the vertical slot. (a) Let $\theta_0 = 9^\circ$. Determine the equation of motion for the system in terms of the angle, θ, and determine the equilibrium positions of the system for $mg/kL = 0.4$ and $mg/kL = 0.7$. Are these equilibrium positions stable or unstable? Plot the total potential energy for $-5\pi/4 < \theta < 5\pi/4$ rad to verify your results. (b) This is a conservative system so that the total energy, $E = T + V$, is a constant. Obtain a phase plot of the motion by finding the constant energy contours of the normalized energy, E/mgL, for $mg/kL = 0.4$ and $L/g = 0.1$, plotting 40 contours over the region $-5\pi/4 < \theta < 5\pi/4$ rad and $-5 < \dot{\theta} < 5$ rad/s (use at least 1000 points for both θ and $\dot{\theta}$ to generate the contour plot). Identify the equilibrium positions on this phase plot. Based on

Fig. P6.4 Mass-spring system (Prob. P6.5)

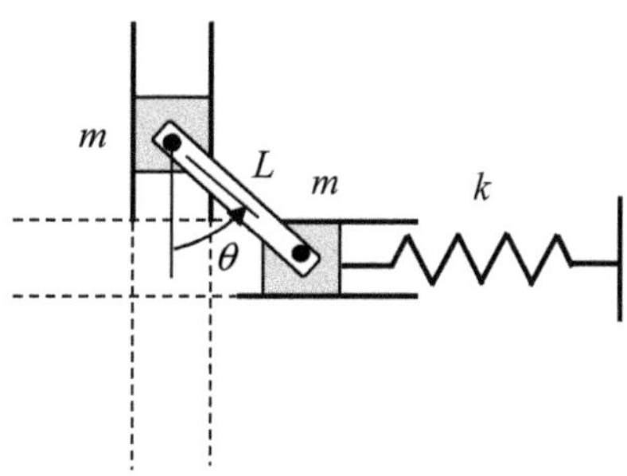

the phase plot, describe the nature of the motion you would expect if the initial conditions are $\theta = 30°$, $\dot{\theta} = 0$, or if $\theta = 65°$ $\dot{\theta} = 0$. Can you connect this behavior to the static analysis in Sect. 6.8? How does the phase plot and your description of the motion change if $mg/kL = 0.7$? (c) Solve the equation of motion numerically for different initial starting angles (and zero velocity) and verify that your descriptions of the motion deduced from the phase plots in parts (a) and (b) are correct. Animate the motion using the MATLAB® animate_lines.

P6.6. A single rigid bar of length L (whose weight can be neglected) is supported by two springs which ride in a smooth vertical track as shown in Fig. P6.5. Both springs are unstretched when the bar is vertical. Let the spring constants be $k_1 = k,\ k_2 = 2k$. (a) Determine the critical load and the post-critical load behavior for large deflections. (b) Perform a linearized analysis of this problem and show that the critical buckling load of part (a) is obtained from this analysis.

P6.7. A rigid bar of length L (whose weight can be neglected) is supported by a spring of length ηL which is attached to a fixed support as shown in Fig. P6.6. The spring is unstretched when the bar is in a vertical position. (a) Determine the critical load and the post-critical load behavior for large deflections for $\eta = 0.5, 1.0, 5.0$. How do these results differ from the case considered in the text when the spring rides in a smooth slot? (b) Perform a linearized analysis of this problem and show that the critical loads of part (a) are obtained from this analysis. Does the linearized analysis of this problem differ from the case considered in the text when the spring rides in a smooth slot?

P6.8. Two bars, each of length L (whose weight can be neglected), are pinned to each other and supported by a spring as shown in Fig. P6.7. The spring is unstretched when the bars are vertical. The top bar rides in a smooth slot and the lower bar is pinned at one end. Perform a linearized analysis of this problem and determine the critical load.

Fig. P6.5 (see Prob. P6.6)

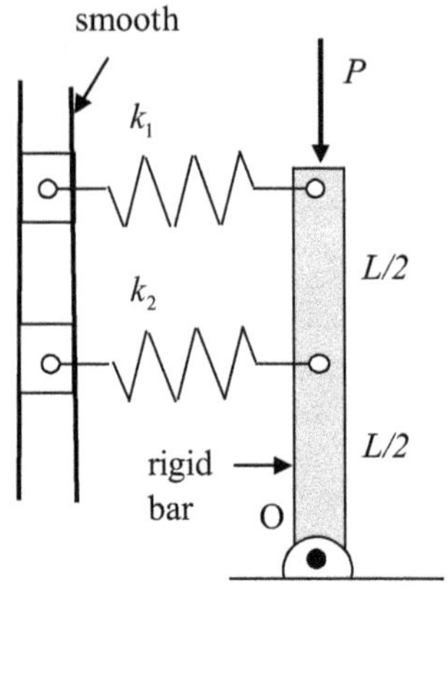

Fig. P6.6 (see Prob. P6.7)

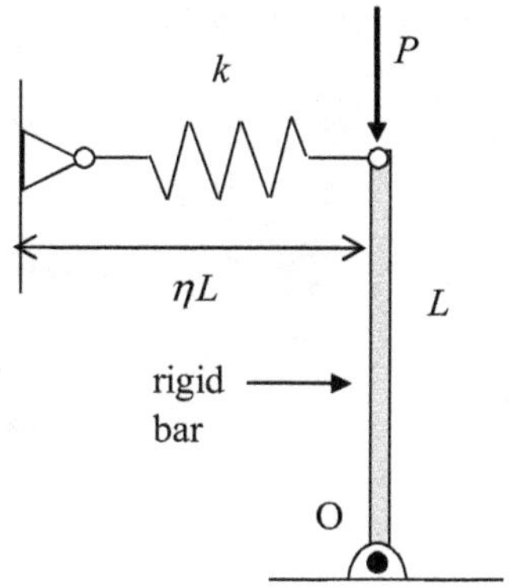

Fig. P6.7 (see Prob. P6.8)

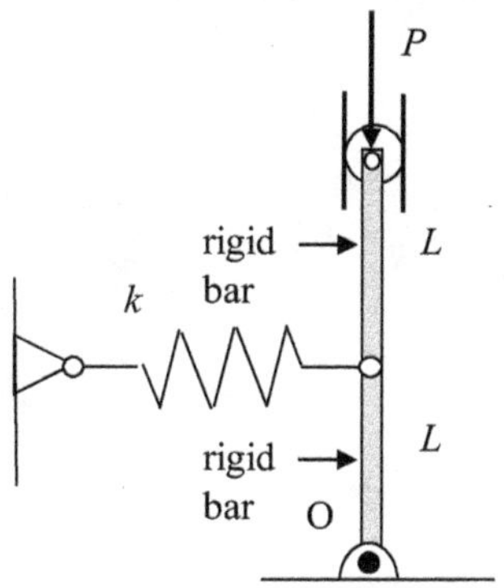

P6.9. Two rigid bars, each of length L (whose weight can be neglected), are pinned to each other and supported by springs as shown in Fig. P6.8. The springs are unstretched when the bars are vertical. Let the spring constants be $k_1 = k$, $k_2 = 3k$. Using a linearized analysis, determine the critical loads and mode shapes for this problem.

P6.10. Three rigid bars, each of length L (whose weight can be neglected), are pinned to each other and supported by springs as shown in Fig. P6.9. The springs are unstretched when the bars are vertical. The top bar rides in a smooth slot and the lowest bar is pinned to a fixed support at its lowest end. Let the spring constants be $k_1 = k$, $k_2 = 2k$. Using a linearized analysis, determine the critical loads and mode shapes for this problem.

Fig. P6.8 (see Prob. P6.9)

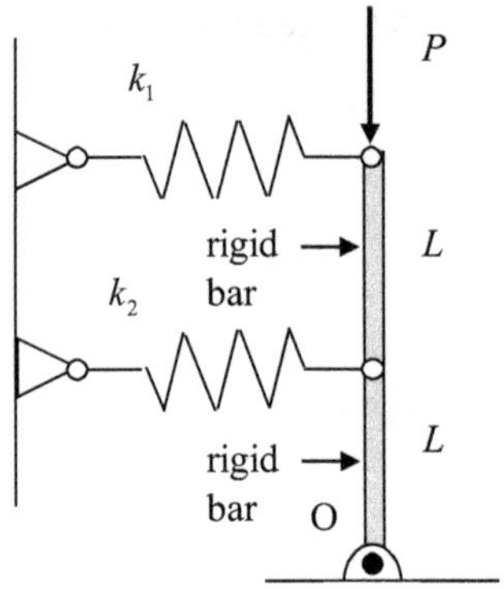

Fig. P6.9 (see Prob. P6.10)

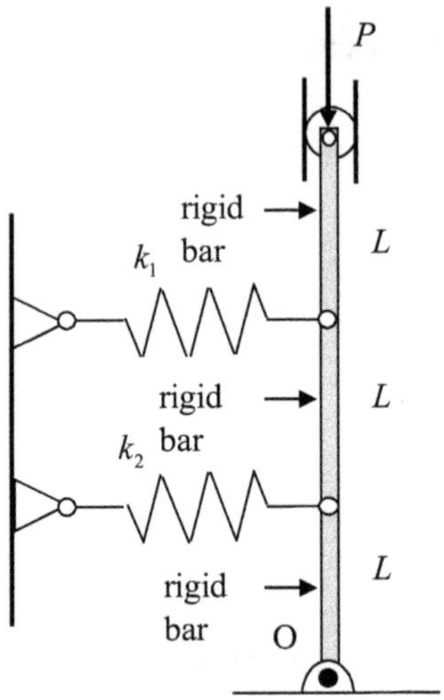

References

1. R.M. Rosenberg, *Analytical Dynamics of Discrete Systems* (Plenum Press, New York, 1977)
2. M.R.M. Crespo Da Silva, *Fundamentals of Dynamics and Analysis of Motion* (Dover Publications, New York, 2016)
3. H.L. Langhaar, *Energy Methods in Applied Mechanics* (John Wiley, New York, 1962)

Chapter 7
Vibrations of Dynamical Systems

Engineers must often analyze and design systems that vibrate. In many cases the vibrations are of very small amplitude so that the equations governing them are linear. This allows one to use a variety of analytical tools to solve for the motion and forces. This chapter examines vibrating systems with multiple degrees of freedom where matrix methods can be used to great advantage. The vibration of single degree of freedom systems is covered in Appendix D.

7.1 An Overview of Linearized Vibrating Systems

We have already analyzed a number of dynamical systems where the motion was a repetitive oscillatory motion. Many of those systems were one degree of freedom systems such as the bead on a parabolic wire; the simple or physical pendulum; the particle sliding on a rotating, circular hoop; the mass connected to a spring-dashpot and moving on an inclined plane; and others. In general the motion was governed by nonlinear equations having different forms but when the motion is assumed to be a small-amplitude motion around an equilibrium position, all of these problems are governed by the same type of linear equations. Below we list the equations of motion, under the small-amplitude assumption, for the problems mentioned above (as well as the original nonlinear equation, if any, shown in parentheses).

1. Bead on a parabolic wire:

$$\ddot{x} + 2agx = 0 \quad \left(\ddot{x} + \frac{2ax\left(g + 2a\dot{x}^2\right)}{1 + 4a^2x^2} = 0\right) \tag{7.1.1a}$$

L. W. Schmerr, *Engineering Dynamics 2.0*, Solid Mechanics and Its Applications 254,
https://doi.org/10.1007/978-3-319-98470-4_7

2. Simple pendulum:

$$\ddot{\theta} + \frac{g}{l}\theta \quad \left(\ddot{\theta} + \frac{g}{l}\sin\theta = 0\right) \tag{7.1.1b}$$

3. Physical pendulum:

$$\ddot{\theta} + \frac{mgh}{I_0}\theta = 0 \quad \left(I_O\ddot{\theta} + mgh\sin\theta = 0\right) \tag{7.1.1c}$$

4. Particle on a rotating hoop:

$$\ddot{\theta} + \left(\frac{g}{R} - \Omega^2\right)\theta = 0 \quad \left(\ddot{\theta} + \left(\frac{g}{R} - \Omega^2\cos\theta\right)\sin\theta = 0\right) \tag{7.1.1d}$$

5. Mass connected to a spring and dashpot sliding on a moving inclined plane:

$$m\ddot{d} + c\dot{d} + kd = -m\ddot{u}\cos\alpha \tag{7.1.1e}$$

The equation of motion for all of these problems is of the general form

$$m\ddot{q} + c\dot{q} + kq = F(t) \tag{7.1.2}$$

which is also the equation of motion of a block of mass m sliding on a smooth, horizontal plane where the mass is connected to a linear spring, whose spring constant is k, and a linear dashpot whose viscous damping coefficient is c. The block is driven by a force, $F(t)$ (see Fig. 7.1a). Using the free body diagram of the mass shown in Fig. 7.1b and Newton's second law gives Eq. (7.1.2) directly. This equation is a linear differential equation with constant coefficients so it can be solved analytically in many cases. Some of those solutions and their properties are given in Appendix D. Here we will examine vibration problems for systems with multiple degrees of freedom. Remarkably, the generalization of Eq. (7.1.2) in a matrix-vector form given by

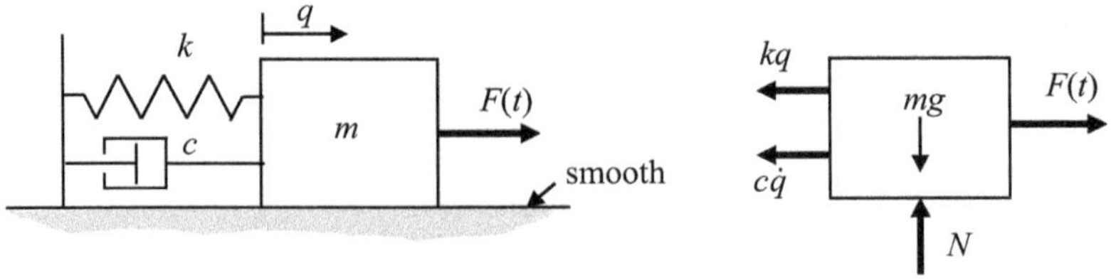

Fig. 7.1 (**a**) Model of a 1-D vibrating system and (**b**) its free body diagram

$$\mathbf{M}\ddot{\mathbf{q}} + \mathbf{C}\dot{\mathbf{q}} + \mathbf{K}\mathbf{q} = \mathbf{Q} \tag{7.1.3}$$

governs the small-amplitude motion around equilibrium of many multiple degree of freedom dynamical systems, where **M** is a mass matrix, **C** is a damping matrix, **K** is a stiffness matrix, and **Q** is a generalized driving force term. We have already seen several examples of this type of equation of motion in the small-amplitude motion of the spring pendulum and the double pendulum:

1. Spring pendulum:

$$\begin{bmatrix} 1 & 0 \\ 0 & 1 \end{bmatrix} \begin{Bmatrix} \ddot{r}_n \\ \ddot{\theta} \end{Bmatrix} + \begin{bmatrix} 1 & 0 \\ 0 & mg/\mathrm{kL}_e \end{bmatrix} \begin{Bmatrix} r_n \\ \theta \end{Bmatrix} = 0 \tag{7.1.4a}$$

where here the dot indicated a derivative on a normalized time, $\tau = t\sqrt{k/m}$.

2. Double pendulum:

$$\begin{bmatrix} 1 & 0 \\ 0 & 1 \end{bmatrix} \begin{Bmatrix} \ddot{\theta}_1 \\ \ddot{\theta}_2 \end{Bmatrix} + \begin{bmatrix} \mu g/l_1 & (\mu - 1)g/l_1 \\ -\mu g/l_2 & \mu g/l_2 \end{bmatrix} \begin{Bmatrix} \theta_1 \\ \theta_2 \end{Bmatrix} = 0 \tag{7.1.4b}$$

In the next section we will see why Eq. (7.1.3) covers a wide range of systems.

7.2 Linearized Motion Near Equilibrium

To see how Eq. (7.1.3) arises in many dynamics problems we will examine the behavior of the kinetic and potential energies for small-amplitude motion near an equilibrium position and derive the equations using Lagrange's equations. Recall Lagrange's equations for a system where the forces derivable from a potential energy function, V, are

$$\frac{d}{dt}\left(\frac{\partial T}{\partial \dot{q}_k}\right) - \frac{\partial T}{\partial q_k} + \frac{\partial V}{\partial q_k} = 0 \tag{7.2.1}$$

We have seen before that the kinetic energy can be generally expressed in terms of three separate terms (T_2, T_1, T_0) as

$$\begin{aligned} T &= \frac{1}{2}\sum_{i=1}^{n}\sum_{j=1}^{n} M_{ij}\dot{q}_i\dot{q}_j + \sum_{i=1}^{n}\mu_i\dot{q}_i + T_0 \\ &= T_2 + T_1 + T_0 \end{aligned} \tag{7.2.2}$$

We can combine the potential energy function together with the T_0 part of the kinetic energy into the dynamic potential $U = V - T_0$ to write Lagrange's equations as

$$\frac{d}{dt}\left(\frac{\partial \widehat{T}}{\partial \dot{q}_k}\right) - \frac{\partial \widehat{T}}{\partial q_k} + \frac{\partial U}{\partial q_k} = 0 \tag{7.2.3}$$

where now $\widehat{T} = T_2 + T_1$. If we let J represent either the dynamic potential or one of the remaining kinetic energy terms (T_2 or T_1), then assuming any explicit time dependency is not present we have in general $J = J(\mathbf{q}, \dot{\mathbf{q}})$. If we want to examine Lagrange's equations near an equilibrium position $(\mathbf{q}^e, \dot{\mathbf{q}}^e = 0)$ and obtain the first-order (linear) behavior of the system we need to keep at most quadratic terms in the J functions since we have to take a derivative of these functions on q or $\dot{q}$, as seen in Eq. (7.2.3), to obtain the equations of motion. Note also that we can also redefine the generalized coordinates so that equilibrium is at $\mathbf{q} = \mathbf{q}^e = \mathbf{0}$. This means we need to examine a Taylor series expansion for J to second order given by

$$\begin{aligned} J(\mathbf{q}, \dot{\mathbf{q}}) = J(\mathbf{0}, \mathbf{0}) &+ \sum_{i=1}^{n} \frac{\partial J}{\partial q_i}\bigg|_{(\mathbf{0},\mathbf{0})} q_i + \sum_{i=1}^{n} \frac{\partial J}{\partial \dot{q}_i}\bigg|_{(\mathbf{0},\mathbf{0})} \dot{q}_i + \frac{1}{2}\sum_{i=1}^{n}\sum_{j=1}^{n} \frac{\partial^2 J}{\partial q_i \partial q_j}\bigg|_{(\mathbf{0},\mathbf{0})} q_i q_j \\ &+ \sum_{i=1}^{n}\sum_{j=1}^{n} \frac{\partial^2 J}{\partial q_i \partial \dot{q}_j}\bigg|_{(\mathbf{0},\mathbf{0})} q_i \dot{q}_j + \frac{1}{2}\sum_{i=1}^{n}\sum_{j=1}^{n} \frac{\partial^2 J}{\partial \dot{q}_i \partial \dot{q}_j}\bigg|_{(\mathbf{0},\mathbf{0})} \dot{q}_i \dot{q}_j \end{aligned} \tag{7.2.4}$$

First, consider the dynamic potential and let $J = U$. Since this term is only a function of the generalized coordinates we can approximate U as

$$U(\mathbf{q}) = U(0) + \sum_{i=1}^{n} \frac{\partial U}{\partial q_i}\bigg|_{(\mathbf{0})} q_i + + \frac{1}{2}\sum_{i=1}^{n}\sum_{j=1}^{n} \frac{\partial^2 U}{\partial q_i \partial q_j}\bigg|_{(\mathbf{0})} q_i q_j \tag{7.2.5}$$

But $\partial U / \partial q_i = 0$ at equilibrium, so that

$$\begin{aligned} U(\mathbf{q}) &= U(0) + \frac{1}{2}\sum_{i=1}^{n}\sum_{j=1}^{n} \frac{\partial^2 U}{\partial q_i \partial q_j}\bigg|_{(\mathbf{0})} q_i q_j \\ &= U(0) + \frac{1}{2}\sum_{i=1}^{n}\sum_{j=1}^{n} K_{ij}^e q_i q_j \end{aligned} \tag{7.2.6}$$

where

$$K_{ij}^{e} = \left.\frac{\partial^{2} U}{\partial q_{i} \partial q_{j}}\right|_{(\mathbf{0})} \tag{7.2.7}$$

are the components of the constant *stiffness matrix*, $\mathbf{K}^e$, which are the second derivatives of the dynamic potential evaluated at the equilibrium position. It is called the stiffness matrix since if we have a linear spring with a stretch, x, we have $U = V = kx^2/2$ and $k = \partial^2 V/\partial x^2$ is the spring stiffness so that K_{ij}^e plays the same role for a multiple degree of freedom problem as that spring constant. Next, consider the case where we let $J = T_2$. If we examine all the terms in Eq. (7.2.4) we see only the last term to be nonzero and we find

$$\begin{aligned} T_2(\mathbf{q}, \dot{\mathbf{q}}) &= \frac{1}{2}\sum_{i=1}^{n}\sum_{j=1}^{n} \left.\frac{\partial^2 T_2}{\partial \dot{q}_i \partial \dot{q}_j}\right|_{(\mathbf{0},\mathbf{0})} \dot{q}_i \dot{q}_j \\ &= \frac{1}{2}\sum_{i=1}^{n}\sum_{j=1}^{n} M_{ij}^e \dot{q}_i \dot{q}_j \end{aligned} \tag{7.2.8}$$

where $\mathbf{M}^e$ is the constant mass matrix at equilibrium. Finally, let $J = T_1$. In this case, the only nonzero terms on the right side of Eq. (7.2.4) are the third and the last terms, giving

$$\begin{aligned} T_1(\mathbf{q}, \dot{\mathbf{q}}) &= \sum_{i=1}^{n} \left.\frac{\partial J}{\partial \dot{q}_i}\right|_{(\mathbf{0},\mathbf{0})} \dot{q}_i + \sum_{i=1}^{n}\sum_{j=1}^{n} \left.\frac{\partial^2 J}{\partial q_i \partial \dot{q}_j}\right|_{(\mathbf{0},\mathbf{0})} q_i \dot{q}_j \\ &= \sum_{i=1}^{n} \mu_i^e \dot{q}_i + \sum_{i=1}^{n}\sum_{j=1}^{n} B_{ij}^e q_i \dot{q}_j \end{aligned} \tag{7.2.9}$$

where $B_{ij}^e = \left.\dfrac{\partial \mu_i}{\partial q_j}\right|_{(\mathbf{0},\mathbf{0})}$. With all these results we see that

$$\begin{aligned} \frac{\partial \widehat{T}}{\partial \dot{q}_k} &= \sum_{j=1}^{n} M_{kj}^e \dot{q}_j + \mu_k^e + \sum_{j=1}^{n} B_{jk}^e q_j \\ \frac{\partial \widehat{T}}{\partial q_k} &= \sum_{j=1}^{n} B_{kj}^e \dot{q}_j \\ \frac{\partial U}{\partial q_k} &= \sum_{j=1}^{n} K_{kj}^e q_j \end{aligned} \tag{7.2.10}$$

which, when placed into Lagrange's equations, gives

$$\sum_{j=1}^{n} M_{kj}^{e}\ddot{q}_j + \sum_{j=1}^{n}\left(B_{jk}^{e} - B_{kj}^{e}\right)\dot{q}_j + \sum_{j=1}^{n} K_{kj}^{e}q_j = 0 \quad (7.12.11)$$

or in matrix form

$$[M^e]\{\ddot{q}\} + [G^e]\{\dot{q}\} + [K^e]\{q\} = 0 \quad (7.2.12)$$

where the *gyroscopic matrix* $[G^e] = [B^e]^T - [B^e]$. If the system is not conservative so that there are other generalized forces $\mathbf{Q}^{nc}$ acting, then the equations of motion become

$$[M^e]\{\ddot{q}\} + [G^e]\{\dot{q}\} + [K^e]\{q\} = \{Q^{nc}\} \quad (7.2.13)$$

Equation (7.2.13) is in the form of Eq. (7.1.3) but here the gyroscopic matrix is related to Coriolis effects and is not associated with dissipation. The matrix $[G^e]$ can produce stabilizing or destabilizing effects on the motion about the equilibrium position. Note that the stiffness matrix comes from the dynamic potential $U = V - T_0$, where the T_0 term often arises in problems when the motion of the system takes place in a non-inertial system. In that case, T_0 can contain centripetal acceleration effects which may also be destabilizing. We saw this in the case of a particle on a rotating hoop where the stiffness was $k = (g/R - \Omega^2)$ in terms of the acceleration of gravity, g; the radius of the hoop, R; and the angular velocity of the hoop, Ω (see Eq. (7.1.1d)). As long as $\Omega < \sqrt{g/R}$ the stiffness is positive and the particle can oscillate near the $\theta = 0$ equilibrium position but when $\Omega > \sqrt{g/R}$ the stiffness is negative and the particle is driven away from $\theta = 0$. We saw examples of that type of behavior in Chap. 2.

In Chap. 6 we showed that the linearized equations of motion could give us important information about the stability of a system. There we expressed the equations of motion as a system of first-order differential equations. The second-order system of Eq. (7.2.13) can be placed into that first-order form or we can use Eq. (7.2.13) directly since the behavior of this second-order system of linear differential equations is also easy to examine.

If the dynamical system is a natural system then $T_1 = T_0 = 0$ and the equations of motion become

$$[M^e]\{\ddot{q}\} + [K^e]\{q\} = \{Q^{nc}\} \quad (7.2.14)$$

which is in the form of Eq. (7.1.3) but with no damping matrix present.

If there are losses due to linear viscous damping types of forces, then they are easy to include in a Lagrangian formulation of a dynamical system. We have seen a 1-D example of this in the behavior of a linear dashpot, where the force of the dashpot acting on a particle in motion along the x-direction, for example, is given by $F = -c\dot{x}$. This force can be expressed in terms of potential-like function called the Rayleigh dissipation function, D, where $D = c\dot{x}^2/2$. Clearly the force, F, is then given by $F = -\partial D/\partial \dot{x}$ which is similar to the relationship between the force of a spring and its potential energy function. For more general systems we can include multiple viscous damping coefficients and write a Rayleigh dissipation function as

$$D = \frac{1}{2}\sum_{i=1}^{n}\sum_{j=1}^{n} C_{ij}\dot{q}_i\dot{q}_j \tag{7.2.15}$$

in terms of the generalized velocities and a damping matrix, $[C]$. In the 1-D case we must have $c > 0$ in order to produce damping of the motion. Otherwise, the included term will actually cause the motion to grow. Similarly, the matrix $[C]$ is assumed to be a symmetrical, positive definite matrix to ensure that physically it does produce a damping of the motion. For a natural system Lagrange's equations in the presence of these viscous damping forces, therefore, become

$$\frac{d}{dt}\left(\frac{\partial T_2}{\partial \dot{q}_k}\right) + \frac{\partial V}{\partial q_k} + \frac{\partial D}{\partial \dot{q}_k} = Q_k^{nc} \tag{7.2.16}$$

and these equations, when linearized about an equilibrium position, become

$$[M^e]\{\ddot{q}\} + [C^e]\{\dot{q}\} + [K^e]\{q\} = \{Q^{nc}\} \tag{7.2.17}$$

where $[C^e]$ is the *damping matrix*, $[C]$, evaluated at the equilibrium position. Equation (7.2.17) is now in exactly the form seen in Eq. (7.1.3).

Finally, we will mention that the Rayleigh function can also be used to describe another type of force called circulatory forces. Circulatory forces can arise, for example, from the electromagnetic forces in electric motors. In this case the Rayleigh dissipation function can be written as

$$D = \frac{1}{2}\sum_{i=1}^{n}\sum_{j=1}^{n} C_{ij}\dot{q}_i\dot{q}_j + \sum_{i=1}^{n}\sum_{j=1}^{n} H_{ij}\dot{q}_i q_j \tag{7.2.18}$$

If this dissipation function is included in a nonnatural dynamical system then the equations of motion linearized about an equilibrium position become

$$[M^e]\{\ddot{q}\} + ([C^e] + [G^e])\{\dot{q}\} + ([K^e] + [H^e])\{q\} = \{Q^{nc}\} \tag{7.2.19}$$

where $[H^e]$ is the *circulatory matrix*, $[H]$, evaluated at the equilibrium position.

Equation (7.2.19) covers the general case when damping, gyroscopic, and circulatory matrices are all present in addition to the usual mass and stiffness matrices. In this chapter we will examine vibrations of structural systems where gyroscopic and circulatory forces are absent.

7.3 Free Vibrations Without Damping

Consider the small-amplitude motion of a natural system near equilibrium where there are no damping or circulatory forces present and there is no applied force, **Q**(t). In this case the system can only be excited by giving it some initial generalized displacements or velocities and it will execute free (i.e., unforced) vibrations. The governing equations in this case are

$$[M^e]\{\ddot{q}\} + [K^e]\{q\} = 0 \tag{7.3.1}$$

To examine the vibratory motion of the system we will express the solution in the form of sinusoidal motions as

$$\{q\} = \{u\}\exp(i\omega t) \tag{7.3.2}$$

where $\{u\}$ is a vector amplitude and we have written the sinusoidal response in complex exponential terms where ω is a frequency of oscillation in rad/s. Using this solution, Eq. (7.3.2) becomes

$$\left([K^e] - \omega^2[M^e]\right)\{u\} = 0 \tag{7.3.3}$$

which are a set of homogeneous algebraic equations. We want to find values for ω and $\{u\}$ that satisfy these equations. This type of problem, as we have seen before, is an *eigenvalue problem*, where solutions for ω^2 are called *eigenvalues* and the corresponding solutions for $\{u\}$ are called *eigenvectors*. Physically, the eigenvalues represent the squares of the *natural frequencies* at which this system freely vibrates and the eigenvectors represent the amplitudes the system coordinates have during the motion at these natural frequencies. The eigenvectors are usually called the *natural modes* or *normal modes* of the system. One can show that there are only nontrivial solutions of Eq. (7.3.3) if the determinant of the coefficient matrix in Eq. (7.3.3) is zero:

$$\det\left([K^e] - \omega^2[M^e]\right) = 0 \tag{7.3.4}$$

Equation (7.3.4) is called the *characteristic equation*. For a system described by n degrees of freedom it represents an nth-order polynomial in ω^2 that has n roots which we will call ω_p^2 $(p = 1, 2,\ldots, n)$ so corresponding to each root we will have the natural frequencies $\omega_p = \pm\sqrt{\omega_p^2}$. The values for the eigenvectors at each of these roots we will call $\{u_p\}$ $(p = 1, 2,\ldots, n)$. Since the system of equations in Eq. (7.3.3) is homogeneous, the solutions for the eigenvectors can only be found up to an arbitrary multiplicative constant. This makes sense since if $\{u_p\}$ is a solution to

Table 7.1 Properties of positive definite and positive semi-definite matrices

For a symmetric matrix that is also:	Positive definite	Positive semi-definite
Diagonal elements	All diagonal elements >0	All diagonal elements ≥0
Minor determinants	All minor determinants >0	All minor determinants ≥0
Main determinant	Main determinant >0	Main determinant ≥0
Eigenvalues	All eigenvalues real and >0	All eigenvalues real and ≥0
Eigenvectors	All eigenvectors are real	All eigenvectors are real

Eq. (7.3.3) then $C_p\{u_p\}$ will also be a solution, where C_p is a constant. The properties of roots are determined by the properties of the stiffness matrix, $[K^e]$. The mass matrix $[M^e]$ is generally a symmetrical, positive definite matrix but the stiffness matrix may be a symmetrical positive definite or positive semi-definite matrix when we are examining stable oscillatory motion about an equilibrium position. Since these matrix types have important characteristics that determine the behavior of the system, we list in Table 7.1 some of their mathematical properties [1]:

In order for a matrix to be positive definite or positive semi-definite it must be a symmetrical matrix. As indicated in Table 7.1 when the stiffness matrix is positive definite then the eigenvalues and eigenvectors are all real and the eigenvalues are all positive so that taking the square root there are real positive and negative natural frequencies given by $\omega_p = \pm\sqrt{\omega_p^2}$. A general solution for the free response, $\{q\}$, will be a superposition of all these individual solutions, therefore, having the form

$$\{q\} = \sum_{p=1}^{n} \{u_p\}\left(C_p \exp\left(i\omega_p t\right) + D_p \exp\left(-i\omega_p t\right)\right) \tag{7.3.5}$$

where (C_p, D_p) are arbitrary (complex) constants. Alternatively, we could write the free response in real form as

$$\{q\} = \sum_{p=1}^{n} A_p\{u_p\} \cos\left(\omega_p t - \phi_p\right) \tag{7.3.6}$$

where (A_p, ϕ_p) are real constants. The values of the constants (C_p, D_p) or (A_p, ϕ_p) must be chosen to satisfy the initial conditions.

We previously solved an eigenvalue problem of the type we have been discussing here for the double pendulum in Chap. 3 and also saw that an eigenvalue problem arose in Chap. 6 when considering the stability of the small-amplitude solutions of the double pendulum about an equilibrium position. In that simple two degree of freedom double-pendulum problem we solved for the eigenvalues and eigenvectors analytically but we also pointed out that MATLAB® has a built-in function, eig, that can be used to solve eigenvalue problems. That MATLAB® function can also be used effectively here. Consider, for example, the three degree of freedom system

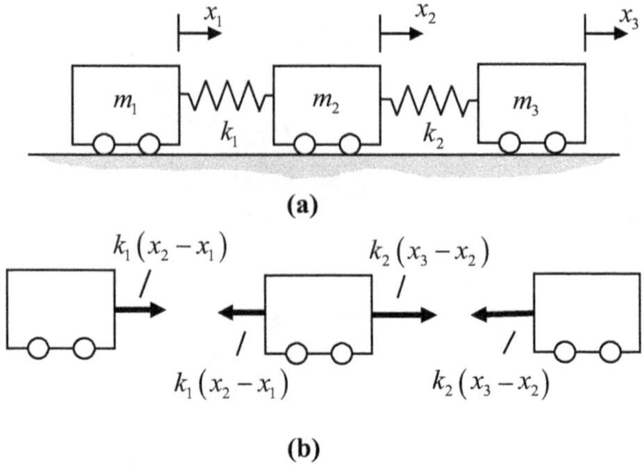

Fig. 7.2 (**a**) A multiple degree of freedom system and (**b**) the free body diagram showing only the horizontal forces acting on its elements

shown in Fig. 7.2a where we have three carts connected together by two springs. Using the free body diagram of Fig. 7.2b Newton's second law gives us directly

$$\begin{aligned} m_1\ddot{x}_1 &= k_1(x_2 - x_1) \\ m_2\ddot{x}_2 &= k_2(x_3 - x_2) - k_1(x_2 - x_1) \\ m\ddot{x}_3 &= -k_2(x_3 - x_2) \end{aligned} \tag{7.3.7}$$

which can be put in the form of Eq. (7.3.1) where

$$[M^e] = \begin{bmatrix} m_1 & 0 & 0 \\ 0 & m_2 & 0 \\ 0 & 0 & m_3 \end{bmatrix}, \quad [K^e] = \begin{bmatrix} k_1 & -k_1 & 0 \\ -k_1 & k_1 + k_2 & -k_2 \\ 0 & -k_2 & k_2 \end{bmatrix} \tag{7.3.8}$$

MATLAB® can solve the eigenvalue problem

$$([K] - \lambda[M])\{u\} = 0 \tag{7.3.9}$$

by using the eig function in the following form:

```
[V, D] = eig(K, M)
```

where D is a matrix having the eigenvalues along its diagonal (the other elements of D are zeros) and V is a matrix whose columns contain the corresponding eigenvectors. For example, suppose we take simply $m_1 = m_2 = m_3 = 2$ kg and $k_1 = k_2 = k_3 = 2$ N/m. Then setting this problem up in MATLAB® we have

```
M = [ 2 0 0; 0 2 0; 0 0 2]

M = 2   0   0
    0   2   0
    0   0   2

K = [ 2 -2 0; -2 4 -2; 0 -2 2]

K =  2  -2   0
    -2   4  -2
     0  -2   2

[V, D] = eig(K, M)

V =  -0.4082  -0.5000   0.2887
     -0.4082   0.0000  -0.5774
     -0.4082   0.5000   0.2887

D = -0.0000         0        0
          0    1.0000        0
          0         0   3.0000
```

Thus the values of $\lambda = \omega^2$ are $(\lambda_1, \lambda_2, \lambda_3) = (0, 1, 3)$ corresponding to the natural frequencies $\omega_1 = 0$, $\omega_2 = \pm 1$, and $\omega_3 = \pm\sqrt{3}$ rad/s, respectively. The corresponding eigenvectors are in the columns of V. But since we can multiply those eigenvectors by arbitrary constants we could write them more succinctly as the values $\{u_1\} = [1, \;\; 1, \;\; 1]^T$, $\{u_2\} = [-1, \;\; 0, \;\; 1]^T$, and $\{u_3\} = [1, -2, \;\; 1]^T$. The fact that one of the eigenvalues is zero indicates that the stiffness matrix is only positive semi-definite. Physically, this case corresponds to a rigid body motion where all the masses have the same displacement so that the springs remain unchanged in length (see Fig. 7.3). This rigid body motion is possible since none of the springs are attached to a fixed support. In more general mechanical systems zero eigenvalues will also indicate the existence of rigid body motions.

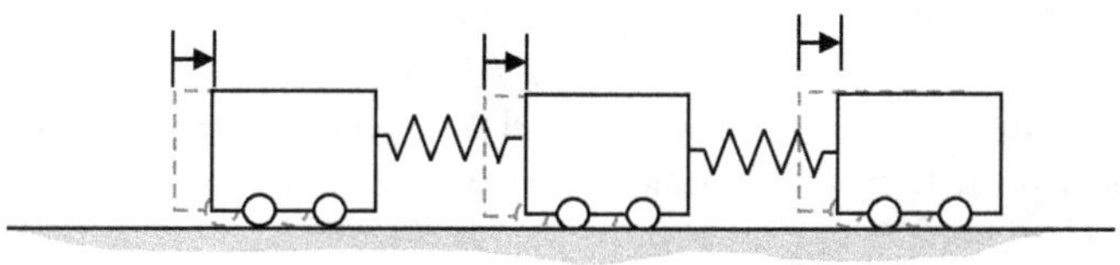

Fig. 7.3 Motion of the system at a zero eigenvalue (rigid body motion)

The natural modes (eigenvectors) have an important property called orthogonality. Consider two modes corresponding to two different natural frequencies ω_p,ω_q $(p \neq q)$. Then

$$\begin{aligned}\left([K^e] - \omega_p^2[M^e]\right)\{u_p\} = 0\\ \left([K^e] - \omega_q^2[M^e]\right)\{u_q\} = 0\end{aligned} \tag{7.3.10}$$

If we multiply the first equation by $\{u_q\}^T$ and the second equation by $\{u_p\}^T$ and subtract the two resulting equations we find

$$\left(\omega_q^2 - \omega_p^2\right)\{u_q\}^T[M^e]\{u_p\} = 0 \tag{7.3.11}$$

where we have used the fact that since both $[M^e]$ and $[K^e]$ are symmetrical matrices it follows that

$$\begin{aligned}\{u_q\}^T[K^e]\{u_p\} = \{u_p\}^T[K^e]\{u_q\}\\ \{u_q\}^T[M^e]\{u_p\} = \{u_p\}^T[M^e]\{u_q\}\end{aligned} \tag{7.3.12}$$

But since $\left(\omega_q^2 - \omega_p^2\right)$ is not zero Eq. (7.3.11) gives

$$\{u_q\}^T[M^e]\{u_p\} = 0 \tag{7.3.13a}$$

If we multiply the first equation in Eq. (7.3.10) by $\{u_q\}^T$ again and use Eq. (7.3.13a) then we also find

$$\{u_q\}^T[K^e]\{u_p\} = 0 \tag{7.3.13b}$$

These orthogonality relations are analogous to the ordinary orthogonality described for vectors where two vectors **u** and **v** are orthogonal if their dot product is zero, i.e., $\mathbf{u} \cdot \mathbf{v} = 0$, which can also be written as $\{u\}^T\{v\} = 0$. The eigenvector orthogonality relations of Eqs. (7.3.13a) and (7.3.13b) are similar but also involve the weighting factors of the mass or stiffness matrices in the orthogonality relations. We have shown that orthogonality is satisfied for any two natural frequencies that are not equal. If there are multiple repeated natural frequencies one can show that their natural modes, while they are not unique, can also be selected so that Eqs. (7.3.13a) and (7.3.13b) still hold. We will not prove that fact here but you can find the proof in many linear algebra texts. When we use the MATLAB® function eig the natural eigenvectors, including those for zero eigenvalues, will be orthogonal in the sense of Eq. (7.3.13a).

Because the mass matrix is always positive definite, it follows that for any natural mode we have

$$\{u_p\}^T[M^e]\{u_p\} > 0$$

As we have seen the natural modes are always defined to within constant values so it is useful to normalize them by consistently choosing those constants. There are a number of ways to make that choice but here we will normalize them by requiring that

$$\{u_p\}^T[M^e]\{u_p\} = 1 \tag{7.3.14a}$$

Again, this is similar to the usual normalization seen for vectors where we can make the magnitude of a vector unity by requiring that $\mathbf{u} \cdot \mathbf{u} = 1$ or, equivalently, $\{u\}^T\{u\} = 1$. Equation (7.3.14a) is similar but also uses the mass matrix as a weighting factor, so this normalization for the eigenvectors does not make them unit vectors. If the eigenvectors are normalized according to Eq. (7.3.14a) then from Eq. (7.3.10) we also have

$$\{u_p\}^T[K^e]\{u_p\} = \omega_p^2 \tag{7.3.14b}$$

We can combine Eqs. (7.3.13a) and (7.3.13b) with Eqs. (7.3.14a) and (7.3.14b) to give the *orthonormality relations*:

$$\{u_p\}^T[M^e]\{u_q\} = \delta_{pq}, \quad \{u_p\}^T[K^e]\{u_q\} = \omega_p^2\delta_{pq} \tag{7.3.15a}$$

or, in terms of the matrix V that is produced by the MATLAB® eig function and which has all the eigenvectors in its columns,

$$[V]^T[M^e][V] = [I], \quad [V]^T[K^e][V] = [\Omega] \tag{7.3.15b}$$

where $[I]$ and $[\Omega]$ are diagonal matrices with ones and the squares of the natural frequencies, respectively, along their diagonals and where δ_{pq} is the Kronecker delta, defined as

$$\delta_{pq} = \begin{cases} 1 & p = q \\ 0 & p \neq q \end{cases} \tag{7.3.16}$$

In solving for the eigenvalues and eigenvectors with the MATLAB® eig function the eigenvectors obtained are orthogonal, as mentioned previously, and they satisfy the normalization of Eqs. (7.3.14a) and (7.3.14b) as long as the stiffness matrix is symmetric and the mass matrix is positive definite which is always the case for dynamics problems. We can see this for the cart problem just considered by calculating the matrix $[N] = [V]^T[M^e][V]$:

```
N = V.'*M*V

N =  1.0000  -0.0000  -0.0000
    -0.0000   1.0000   0.0000
    -0.0000   0.0000   1.0000
```

which verifies the orthonormality relations of Eq. (7.3.15b) for the mass matrix. For the stiffness matrix we have

```
N2 = V.'*K*V
N2 = 0.0000  -0.0000  -0.0000
    -0.0000   1.0000   0.0000
    -0.0000     0      3.0000
```

so the stiffness matrix orthonormality relations of Eq. (7.3.15b) are also satisfied. [Note that because the eigenvalues and eigenvectors are computed numerically they are only approximate values and so it is possible for the eigenvalues and eigenvectors to have very small imaginary components and for the eigenvectors to be not exactly orthogonal. Thus we may have to remove any such small unwanted components before we do further calculations with them. Also note that other eigenvalue problem solvers may not normalize the eigenvectors according to Eq. (7.3.14a). In that case, to make the eigenvectors have the normalization of Eq. (7.3.14a) we must compute

$$\{u_p\}^T[M^e]\{u_p\} = \alpha_p$$

and then form the normalized eigenvectors, $\{u_p\}_{\text{norm}}$, as $\{u_p\}_{\text{norm}} = \sqrt{\alpha_p}\{u_p\}$].

The orthogonality of the natural modes implies that they are an independent set of vectors and that we can express any n-dimensional vector as a linear combination of these natural modes multiplied by appropriate coefficients. This very important result is called the *expansion theorem*. In particular, we can express our n generalized coordinates, $\{q(t)\}$, for the free vibration problem as

$$\{q(t)\} = \sum_{p=1}^{n} z_p(t)\{u_p\} \tag{7.3.17a}$$

Since the MATLAB® eig function generates the V matrix having all the eigenvectors in its columns, we can also write Eq. (7.3.17a) as

$$\{q(t)\} = [V]\{z(t)\} \tag{7.3.17b}$$

where $\{z\} = [z_1, z_2, \ldots, z_n]^T$. Similarly, by multiplying both sides of Eq. (7.3.17a) by $\{u_q\}^T[M^e]$ and using the orthonormality relation we find

$$z_q(t) = \{u_q\}^T[M^e]\{q(t)\} \tag{7.3.18a}$$

or

$$\{z(t)\} = [V]^T[M^e]\{q(t)\} \tag{7.3.18b}$$

The $z_p(t)$ coefficients in Eq. (7.3.17a) are called *modal coordinates* or *principal coordinates*. This transformation from generalized coordinates to modal coordinates makes the solution to our free vibration problem very easy. To see this, we place Eq. (7.3.17a) into Eq. (7.3.1) and multiply the resulting equation by $\{u_q\}^T$ to find

$$\sum_{p=1}^{n}\{u_q\}^T[M^e]\{u_p\}\ddot{z}_p(t) + \sum_{p=1}^{n}\{u_q\}^T[K^e]\{u_p\}z_p(t) = 0 \tag{7.3.19}$$

But by the orthonormality relations and the properties of the Kronecker delta these equations reduce to simply

$$\ddot{z}_q(t) + \omega_q^2 z_q(t) = 0 \tag{7.3.20}$$

Thus, in the modal coordinates we have n uncoupled equations of motion, each of which is identical to a 1-D free, undamped vibration problem. As shown in Appendix D the solution for each of these modal coordinates in terms of their initial values is just

$$z_q(t) = z_q(0)\cos(\omega_q t) + \frac{\dot{z}_q(0)}{\omega_q}\sin(\omega_q t) \tag{7.3.21}$$

which is well behaved for $\omega_q \neq 0$. For problems with rigid body modes where $\omega_q = 0$ the solution becomes

$$z_q(t) = z_q(0) + \dot{z}_q(0)t \tag{7.3.22}$$

We can get this solution for $\omega_q = 0$ from Eq. (7.3.21) by taking the limit of that equation as $\omega_q \to 0$ but since ω_q is present in both the denominator and the numerator of the second term in Eq. (7.3.21) we formally have a 0/0 value. L' Hopital's rule gives us the correct limit and hence Eq. (7.3.22). When we try to evaluate such terms numerically in MATLAB® we will get an error if we try to do a straight division with $\omega_q = 0$. A work-around to this problem in MATLAB® is to simply replace ω_q in Eq. (7.3.21) by ω_q + eps, where eps is a very small number that loosely speaking is on the order of the computer machine precision. With that replacement we then can also use Eq. (7.3.21) for $\omega_q = 0$.

By using Eq. (7.3.18a) and its derivative we can get the initial values of the modal coordinates from the initial values of the original generalized coordinate as

$$\begin{aligned} z_q(0) &= \{u_q\}^T[M^e]\{q(0)\} \\ \dot{z}_q(0) &= \{u_q\}^T[M^e]\{\dot{q}(0)\} \end{aligned} \tag{7.3.23a}$$

which can also be written in matrix-vector form

$$\begin{aligned} \{z(0)\} &= [V]^T[M^e]\{q(0)\} \\ \{\dot{z}(0)\} &= [V]^T[M^e]\{\dot{q}(0)\} \end{aligned} \tag{7.3.23b}$$

where V is the matrix of eigenvectors generated by the MATLAB® eig function with the eigenvectors in its columns. From these initial values and the modal coordinate solutions of Eq. (7.3.21) we can write the solution for the generalized coordinates, Eq. (7.3.17b).

Simulation 7.1: Vibration of Masses on a Stretched String

As an example of a multi-degree of freedom undamped system consider the problem shown in Fig. 7.4a where an array of n masses is held by a stretched elastic string, where T_e is the tension in the string (Fig. 7.4b). We want to examine the free, small vertical vibrations of this system. We will take the generalized coordinates to be the vertical displacements $(q_1, q_2, \ldots, q_n)$ of the masses, as shown in Fig. 7.5a, and assume that the two ends of the string are fixed. During the vertical motion the distance between masses k and $(k + 1)$ (see Fig. 7.5b) is approximately (for small displacements, i.e., $q_k << d$)

$$\left[d^2 + (q_{k+1} - q_k)^2\right]^{1/2} \cong d + \frac{1}{2d}(q_{k+1} - q_k)^2 + \ldots$$

where d is the distance between particles at equilibrium. The elongation in the elastic string, Δl, is therefore

$$\Delta l = \frac{1}{2d}(q_{k+1} - q_k)^2$$

The work done by the internal tension T_e during this elongation, $T_e\Delta l$, must be equal to the change in the internal potential energy, ΔV, of the string so that we have

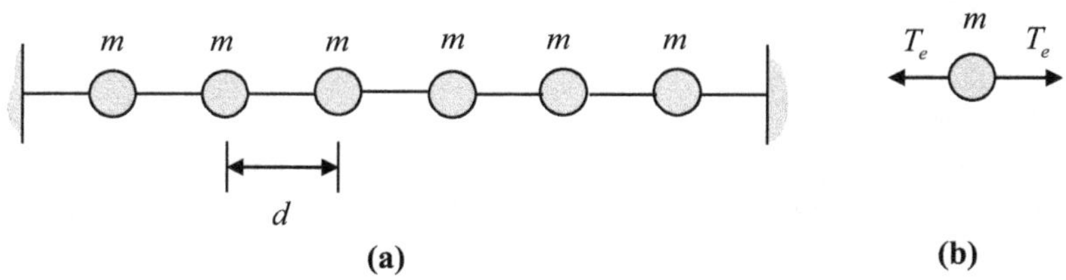

Fig. 7.4 (**a**) A linear array of n equally spaced masses on a stretched string. (**b**) The free body diagram of an individual mass

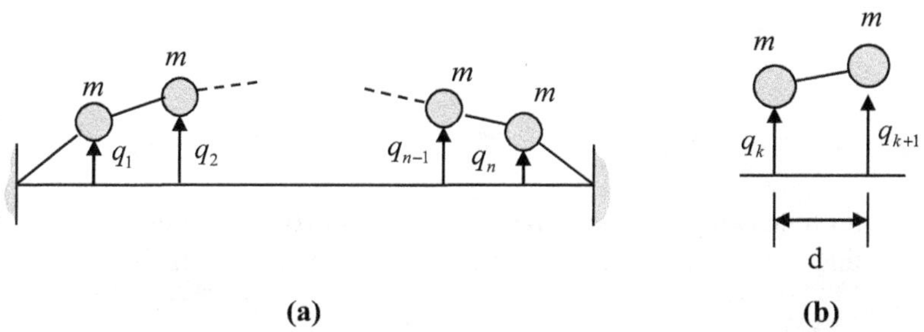

Fig. 7.5 (**a**) The masses on the string when it is in motion, and (**b**) the vertical displacements of a generic pair of masses

$$T_e \Delta l = \Delta V = \frac{T_e}{2d}\left(q_{k+1} - q_k\right)^2$$

and the total potential energy of the string (neglecting any irrelevant constant value) is

$$V = \frac{1}{2}\sum_{k=0}^{n+1} k_e\left(q_{k+1} - q_k\right)^2 \tag{7.3.24}$$

where we have included the parts of the springs from the first and last masses to the fixed ends by summing from $k = 0$ to $k = n + 1$. At those ends $q_0 = q_{n+1} = 0$. The constant $k_e = T_e/d$ acts as a stiffness constant for this system. The kinetic energy of the system is just

$$T = \frac{1}{2}\sum_{k=1}^{n} m\dot{q}_k^2 \tag{7.3.25}$$

Since in this case $\partial T/\partial q_p = 0$ Lagrange's equations (Eq. (7.2.1)) are just

$$\frac{d}{dt}\left(\frac{\partial T}{\partial \dot{q}_p}\right) + \frac{\partial V}{\partial q_p} = 0 \quad (p = 1, 2, \ldots, n) \tag{7.3.26}$$

In the potential energy expression when we take a derivative on q_p we will obtain contributions from two terms at $k = p$ and $k = p - 1$ since those terms both contain q_p. That is,

$$V = \quad \ldots + \frac{1}{2}k_e\left(q_p - q_{p-1}\right)^2 + \frac{1}{2}k_e\left(q_{p+1} - q_p\right)^2 + \ldots$$

so that

$$\frac{\partial V}{\partial q_p} = k_e\left(q_p - q_{p-1}\right) - k_e\left(q_{p+1} - q_p\right) \tag{7.3.27}$$

Note that this holds for $p = 1$ to $p = n$ with $q_0 = q_{n+1} = 0$. Using this relationship and differentiating the kinetic energy we find the equations of motion as

$$m\ddot{q}_p + k_e\left(q_p - q_{p-1}\right) - k_e\left(q_{p+1} - q_p\right) = 0 \quad (p = 1, 2, \ldots, n) \tag{7.3.28}$$

which can also be placed in matrix-vector form as

$$[M]\{\ddot{q}\} + [K]\{q\} = 0 \tag{7.3.29}$$

where the mass and stiffness matrices are

$$[M] = \begin{bmatrix} m & 0 & 0 & \cdots & 0 \\ 0 & m & 0 & \cdots & 0 \\ 0 & 0 & m & \cdots & 0 \\ \cdots & \cdots & \cdots & \cdots & \cdots \\ 0 & 0 & 0 & \cdots & m \end{bmatrix}, [K] = \begin{bmatrix} 2k_e & -k_e & 0 & \cdots & 0 \\ -k_e & 2k_e & -k_e & \cdots & 0 \\ 0 & -k_e & 2k_e & \cdots & 0 \\ \cdots & \cdots & \cdots & \cdots & \cdots \\ 0 & 0 & 0 & \cdots & 2k_e \end{bmatrix} \tag{7.3.30}$$

Now, consider the problem of the free vibration of a string with five masses with $m = 2$ kg and $k_e = 10$ N/m. We could generate these matrices by hand but we can let MATLAB® do it for us. The mass matrix is a diagonal matrix. If we generate a vector with the masses in them and use the MATLAB® function diag we will obtain the diagonal mass matrix:

```
mv = ones(1,5)*2;
M = diag(mv)
M =  2   0   0   0   0
     0   2   0   0   0
     0   0   2   0   0
```

```
0   0   0   2   0
0   0   0   0   2
```

The stiffness matrix is a tridiagonal matrix with elements on the main diagonal (which is labeled as the K = 0 diagonal) and on the diagonal immediately below it (labeled as the K = −1 diagonal) and on the diagonal immediately above it (labeled as the K = 1 diagonal). The MATLAB® function diag(v, K) generates a diagonal matrix with the elements of the vector v on the Kth diagonal so if we generate vectors (vd, vu, vl) with the stiffnesses on the K = 0, K = 1 and K = −1 diagonals, respectively, we can then use the diag function to obtain the tridiagonal matrix that contains those values:

```
vd = ones(1,5)*20;
vu = ones(1,4)*(-10);
vl = vu;
K = diag(vd) +diag(vu,1) +diag(vl, -1)

K =  20  -10    0    0    0
    -10   20  -10    0    0
      0  -10   20  -10    0
      0    0  -10   20  -10
      0    0    0  -10   20
```

where diag(vd) as used above is equivalent to diag(vd, 0). Notice that the diagonal and off-diagonal vectors have different dimensions. There are other ways in MATLAB® to generate a tridiagonal matrix but the above method works fine for our purposes. Having the mass and stiffness matrices we can obtain the eigenvalues and normalized eigenvectors by using the eig function:

```
[V, D] = eig(K, M)

V = -0.2041  -0.3536  -0.4082  -0.3536   0.2041
    -0.3536  -0.3536   0.0000   0.3536  -0.3536
    -0.4082   0.0000   0.4082  -0.0000   0.4082
    -0.3536   0.3536   0.0000  -0.3536  -0.3536
    -0.2041   0.3536  -0.4082   0.3536   0.2041

D =  1.3397      0         0         0         0
     0        5.0000       0         0         0
     0           0      10.0000      0         0
     0           0         0      15.0000      0
     0           0         0         0      18.6603
```

Besides the eigenvalues and eigenvectors, the only additional quantities needed are the initial conditions for the string and then one can solve for the motion. In this case we will give the string the initial shape shown in Fig. 7.6a with zero velocities for all the masses. The generalized displacements and velocities are then $\mathbf{q}(0) = [1,\ 1.6,\ 2.0,\ 1.6,\ 1]$, $\dot{\mathbf{q}}(0) = [0,\ 0,\ 0,\ 0,\ 0]$. The MATLAB® script vibrating_string_anim contains all the steps listed above and sets up the default parameters previously

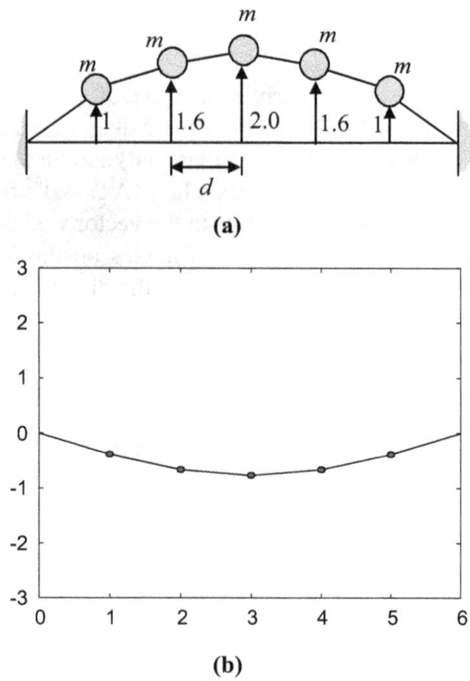

Fig. 7.6 (**a**) The initial displacements of the masses at $t = 0$. The initial velocities are all zero. These displacements are shown exaggerated since they are assumed to be small relative to the spacing, d, between the masses. (**b**) The position of the string at the final time of the animation generated by the MATLAB® script vibrating_string_anim

described. The script then uses Eq. (7.3.23b) to get the initial values for the modal coordinates. The solution for modal coordinates is found from Eq. (7.3.21) as a function of time over the interval from $t = 0$ to $t = t_{\max} = 20$ s at times spaced apart by a time interval dt $= 0.1$ s. These modal coordinates and Eq. (7.3.17b) are then used to evaluate the corresponding coordinates of the masses over that time interval and to display those coordinates as an animation of the motion. The final position of the string from that animation is shown in Fig. 7.6b. As can be seen in the entire animation the string vibrates primarily in the form of the first mode which has a natural frequency of approximately $\omega_1 = 1.16$ rad/s. This occurs since the initial string displacements are close to the first eigenvector which, when normalized so that its maximum component is 2, is given by $\{u_1\} = [1,\ \ 1.73,\ \ 2,\ \ 1.73,\ \ 1]^T$, a result that is very close to the initial displacements shown in Fig. 7.6a.

Solving the vibrating string problem numerically is an effective approach but we can gain a more complete understanding of this problem by conducting some of the analysis in analytical form. Consider, for example, the equations of motion given in the form of Eq. (7.3.28) which we write again here as

$$m\ddot{q}_p + k_e(q_p - q_{p-1}) - k_e(q_{p+1} - q_p) = 0, \qquad q_0 = q_{n+1} = 0 \tag{7.3.31}$$

If we assume harmonic solutions $q_p = a_p \cos(\omega t)$ then this is equivalent to taking the real part of the vector form we used before, namely $\{q\} = \{u\} \exp(i\omega t)$, where the components of $\{u\}$ are now the a_p, i.e., $\{u\} = [a_1, a_2, \ldots, a_n]^T$. These components satisfy

$$-m\omega^2 a_p = k_e(a_{p+1} - 2a_p + a_{p-1}), \qquad a_0 = a_{n+1} = 0 \tag{7.3.32}$$

To have nontrivial solutions to these homogeneous equations we must have the determinant of the coefficients equal to zero which leads to the characteristic equation

$$\begin{vmatrix} 2k_e - m\omega^2 & -k_e & 0 & \cdots & 0 \\ -k_e & 2k_e - m\omega^2 & -k_e & \cdots & 0 \\ 0 & -k_e & 2k_e - m\omega^2 & \cdots & 0 \\ \cdots & \cdots & \cdots & \cdots & \cdots \\ 0 & 0 & 0 & \cdots & 2k_e - m\omega^2 \end{vmatrix} = 0 \tag{7.3.33}$$

whose n roots are the eigenvalues obtained previously. However, we can find those roots and the eigenvectors by working with the recursion relationship of Eq. (7.3.32) directly. Let us write the a_p in the form

$$a_p = A \sin(p\phi) \tag{7.3.34}$$

Substituting Eq. (7.3.34) into Eq. (7.3.32) gives

$$-m\omega^2 A \sin(p\phi) = k_e A[\sin(p\phi - \phi) - 2\sin(p\phi) + \sin(p\phi + \phi)] \tag{7.3.35}$$

Using the trigonometric relations $\sin(B \pm C) = \sin B \cos C \pm \sin C \cos B$ Eq. (7.3.35) reduces to

$$m\omega^2 = k_e(2 - 2\cos\phi) = 4k_e \sin^2\left(\frac{\phi}{2}\right) \tag{7.3.36}$$

which yields

$$\omega = 2\omega_0 \sin\left(\frac{\phi}{2}\right) \tag{7.3.37}$$

where $\omega_0 = \sqrt{k_e/m}$. To find the parameter ϕ we note that Eq. (7.3.34) satisfies the one end condition $a_0 = 0$ but not the other one. Requiring $a_{n+1} = A \sin[(n+1)\phi] = 0$ means that we can find n values for the angle ϕ, ϕ_s $(s = 1, 2, \ldots, n)$ given by

$(n{+}1)\phi_s = s\pi$ that satisfies that end condition. Each of these s values gives us one of the natural frequencies and eigenvectors. Thus Eq. (7.3.37) gives us the n natural frequencies:

$$\omega_s = 2\omega_0 \sin\left(\frac{s\pi}{2(n+1)}\right) \quad (s = 1, 2, \ldots, n) \tag{7.3.38}$$

and the components of the sth eigenvector, $(a_p)_s$, from Eq. (7.3.34) are

$$\left(a_p\right)_s = A_s \sin\left(\frac{ps\pi}{n+1}\right) \quad (p = 1, 2, \ldots, n) \quad (s = 1, 2, \ldots, n) \tag{7.3.39}$$

Equations (7.3.38) and (7.3.39) are explicit analytical solutions for the natural frequencies and natural modes. Notice, however, that the natural modes are not normalized here. When the system is vibrating only in the sth mode we then have

$$\left(q_p\right)_s = \left(a_p\right)_s \cos\left(\omega_s t - \psi_s\right) = A_s \sin\left(\frac{\pi ps}{n+1}\right) \cos\left(\omega_s t - \psi_s\right) \tag{7.3.40}$$

where we have included arbitrary phase terms, ψ_s. A general expression for the motion can then be formed as a superposition of all the natural modes:

$$q_p = \sum_{s=1}^{n} \left(q_p\right)_s = \sum_{s=1}^{n} A_s \sin\left(\frac{\pi ps}{n+1}\right) \cos\left(\omega_s t - \psi_s\right) \tag{7.3.41}$$

where (A_s, ψ_s) can be determined from the initial conditions. We can compare these analytical solutions with the numerical results we obtained in Simulation 7.1 where m = 2, $k_e = 10$, n = 5. Then evaluating Eq. (7.3.38) in MATLAB® and comparing with the square root of the D matrix generated by the eig function give

```
s =1:5;
wexact = 2*sqrt(ke/m)*sin(s*pi/12) % exact eigenvalues
wexact =   1.1575   2.2361   3.1623   3.8730   4.3198
sqrt(D)  % eigenvalues calculated from eig

ans =  1.1575      0         0        0       0
          0   2.2361         0        0       0
          0        0    3.1623        0       0
          0        0         0   3.8730       0
          0        0         0        0  4.3198
```

which are the same natural frequencies. In comparing the eigenvectors we have to remember that the analytical values of Eq. (7.3.39) are not normalized so we have to account for that fact. For example, the first mode in the V matrix generated by eig can be renormalized so its max value is one and we have the same maximum value for the exact first mode given by Eq. (7.3.39) if we set $A_1 = 1$. These steps in MATLAB® yield

```
p = 1:5;
u1 = V(:,1)'/min(V(:,1)) % first mode eigenvector calculated with eig
re-normalized
u1 = 0.5000   0.8660   1.0000   0.8660   0.5000
u1exact = sin(p*pi/6)   % exact first mode eigenvector values
u1exact =  0.5000   0.8660   1.0000   0.8660   0.5000
```

so the numerical and exact first eigenvector values agree. Similar agreement is found for the other eigenvectors.

The fact that the analytical values for the eigenvector components are given by a sine function has a physical significance. If the particles are all separated by a distance d then $x = pd$ is the distance of the pth particle from the left end of the string, and we can write Eq. (7.3.39) as

$$\left(a_p\right)_s = A_s \sin\left(\frac{s\pi pd}{(n+1)d}\right) \quad = A_s \sin\left(\frac{s\pi x}{L}\right) \tag{7.3.42}$$

where $L = (n+1)d$ is the total length of the string. When the number of masses is large and the distance d is small, the system should closely model a real string where the mass is continuously distributed along its length and Eq. (7.3.42) should model the eigenvectors for vibrations of that continuous string where the variable x is a continuous function. If we plot the first three exact eigenvectors of Eq. (7.3.39) and the continuous functions of x given by Eq. (7.3.42) for $A_s = 1$ we find the plots shown in Fig. 7.7. Thus, the modal coordinates of the discrete masses lie along these continuous curves and as the number of masses in a given string length increases the modal coordinates of the masses "fill in" these curves at more and more points. We also see clearly from these plots that as the mode number increases the number of nulls (zeros) in the mode shapes increases.

Finally, consider the expression for the natural frequencies, Eq. (7.3.38), when the number of masses, n, is large and d is small so that we are modeling a string with continuously distributed mass. In that case let us examine the lowest natural frequencies where $s < < n$. Then these frequencies are given approximately as

$$\omega_s = 2\omega_0 \sin\left(\frac{s\pi}{2(n+1)}\right) \cong \omega_0 d\frac{s\pi}{L} \quad (s << n) \tag{7.3.43}$$

where $\omega_0 d = \sqrt{Td/m} = \sqrt{T/\rho_l}$, and where ρ_l is the linear mass density (mass/unit length) for the string. Thus, we see that the lower natural frequencies of the continuous string are approximately multiples of the first (fundamental) frequency, ω_1, i.e.,

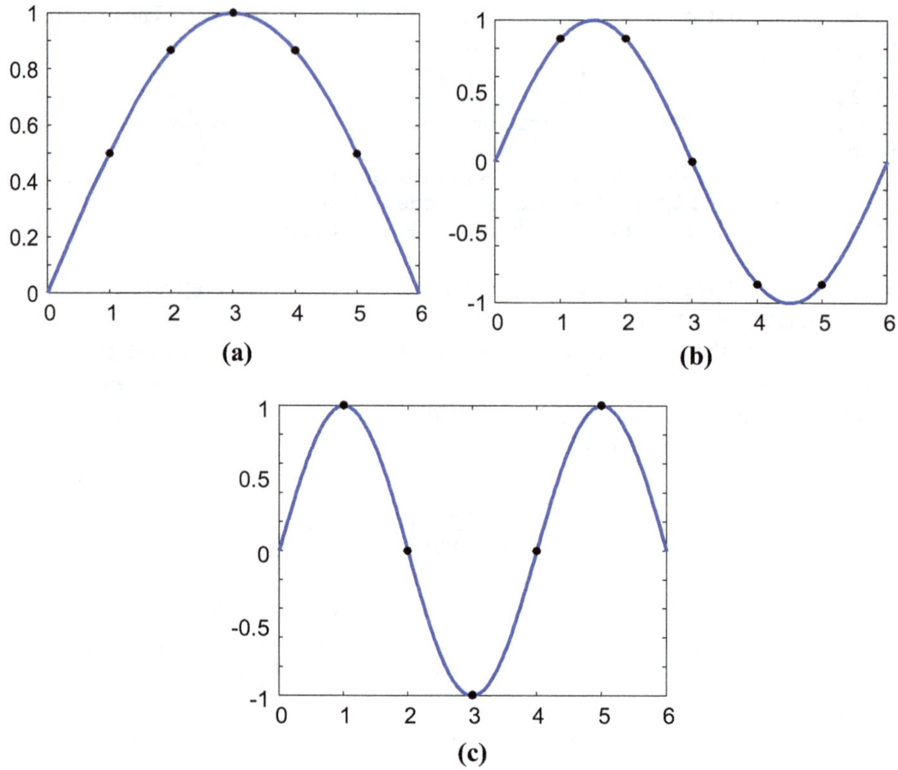

Fig. 7.7 The first three mode shapes for a string with a continuously distributed mass (solid line) and the modal coordinates for a string with five discrete masses (dots) for (**a**) the first mode, (**b**) the second mode, and (**c**) the third mode

$$\omega_s = s\omega_1 \quad (s << n), \quad \omega_1 = \frac{\pi}{L}\sqrt{\frac{T}{\rho_l}} \tag{7.3.44}$$

In Chap. 9 we will consider the motion of continuous deformable bodies and examine the axial motion of an elastic rod, where the continuous rod is modeled as the limit of a large number of masses connected by linear springs. The continuous rod is shown to satisfy the 1-D wave equation for the axial displacement, u, given by

$$\frac{1}{c^2}\frac{\partial^2 u}{\partial t^2} = \frac{\partial^2 u}{\partial x^2} \tag{7.3.45}$$

where $c = \sqrt{E/\rho}$ is the wave speed for waves traveling in the rod. The constant E is a material constant called Young's modulus and ρ is the mass density (mass/unit volume). The rod model comes from a discrete model that is identical in form to our

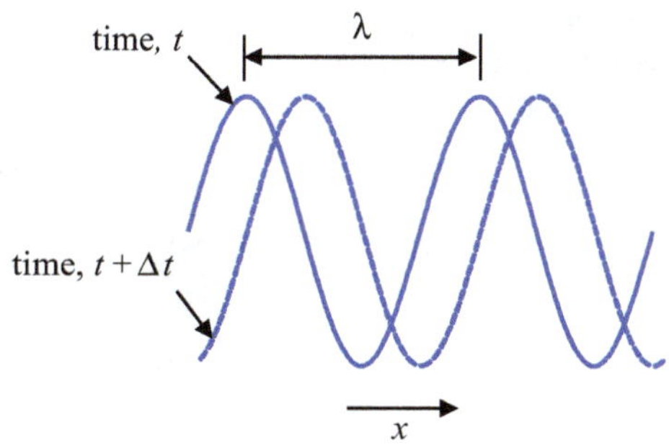

Fig. 7.8 A sinusoidal wave traveling in the positive x-direction as seen at two successive times

model for the transverse motions of a string if we let $c = \sqrt{T_e/\rho_l}$ instead so the vertical q-displacement in a continuous string also satisfies

$$\frac{1}{c^2}\frac{\partial^2 q}{\partial t^2} = \frac{\partial^2 q}{\partial x^2} \tag{7.3.46}$$

As seen in Chap. 9 there are traveling wave solutions to the 1-D wave equation such as $q = f(x - ct) + g(x + ct)$, where $f(x - ct)$ represents a wave traveling in the positive x-direction and $g(x + ct)$ represents a wave traveling in the negative x-direction. Suppose we examine the special case of a combination of rightward and leftward propagating sinusoidal waves in the string of identical form where

$$q = A\sin\left[\frac{2\pi}{\lambda}(x - ct)\right] + A\sin\left[\frac{2\pi}{\lambda}(x + ct)\right] \tag{7.3.47}$$

The parameter λ is called the wavelength of the sinusoid; that is, it is the distance between two successive maxima (or minima) in the sinusoid (see Fig. 7.8). The wavelength is related to the circular frequency of oscillations of the wave, ω, in rad/s, by $\lambda = 2\pi c/\omega$. Figure 7.8 plots the function $\sin[2\pi(x - ct)/\lambda]$ at two times t and $t + \Delta t$, showing that indeed the sinusoid is moving in the positive x-direction. Similarly, $\sin[2\pi(x + ct)/\lambda]$ can be shown to be a sinusoidal wave moving in the negative x-direction. If we again use the trigonometric relations $\sin(B \pm C) = \sin B \cos C \pm \sin C \cos B$ we can write Eq. (7.3.47) as

$$q = 2A\sin\left(\frac{2\pi x}{\lambda}\right)\cos(\omega t) \tag{7.3.48}$$

In comparison, a continuous string vibrating in its lowest (fundamental) mode satisfies

$$q = A_1 \sin\left(\frac{\pi x}{L}\right)\cos(\omega_1 t) = A_1 \sin\left(\frac{\pi x}{L}\right)\cos\left(\frac{\pi c}{L}t\right) \tag{7.3.49}$$

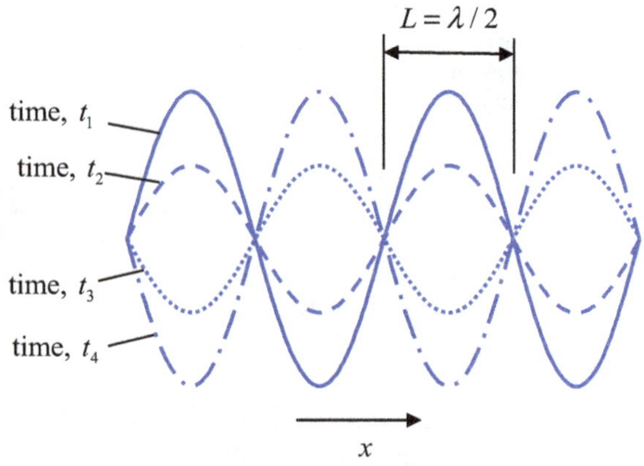

Fig. 7.9 A standing wave seen at four successive times

Comparing Eq. (7.3.48) with Eq. (7.3.49) we see that they are identical if $A_1 = 2A$ and $\lambda = 2L$. Thus, we see that this vibrational motion of the string, which is also called a *standing wave* since the motion is purely an up-and-down motion that does not move to the right or left (see Fig. 7.9 which plots Eq. (7.3.49) at several times), can be thought of as a superposition of rightward and leftward *propagating* sinusoidal waves. The condition $\lambda = 2L$ guarantees that the boundary conditions $q(0) = q(L) = 0$ are satisfied. A similar comparison can be done for the higher modes as well so that we see that there is a duality between the vibrational motions of dynamical systems and waves that propagate back and forth in those systems.

7.4 Forced Vibrations Without Damping

The last section showed that the use of modal coordinates makes the solution of free vibration problems relatively simple. The use of modal coordinates makes the solution of many undamped forced multi-degree of freedom problems easy as well. Consider again the small-amplitude motion of a natural system near equilibrium where there are no damping or circulatory forces present but where now there is an applied generalized force, $\mathbf{Q}(t)$. Then the equations of motion for the generalized coordinates are

$$[M^e]\{\ddot{q}\} + [K^e]\{q\} = \{Q\} \tag{7.4.1}$$

We again place the modal expansion of Eq. (7.3.17a) into Eq. (7.4.1) and multiply the resulting equation by $\{u_q\}^T$ to find

$$\sum_{p=1}^{n} \{u_q\}^T [M^e] \{u_p\} \ddot{z}_p(t) + \sum_{p=1}^{n} \{u_q\}^T [K^e] \{u_p\} z_p(t) = \{u_q\}^T \{Q(t)\} \quad (7.4.2)$$

and the orthonormality relations now give

$$\ddot{z}_q(t) + \omega_q^2 z_q(t) = N_q(t) \quad (7.4.3)$$

where $N_q(t) = \{u_q\}^T \{Q(t)\}$. Since the modal coordinates are uncoupled we can solve all of the 1-D forced vibration problems in Eq. (7.4.3) independently.

As shown in Appendix D there are two ways in which we can write the solutions of Eq. (7.4.3). One way, which is valid for any force $Q(t)$, to combine the free vibration solution with a convolution integral expression that provides a particular solution that accounts for the force. In this case we have for $\omega_q \neq 0$

$$z_q(t) = z_q(0) \cos(\omega_q t) + \frac{\dot{z}_q(0)}{\omega_q} \sin(\omega_q t) + \int_0^t N_q(\tau) \frac{\sin(\omega_q (t - \tau))}{\omega_q} d\tau \quad (7.4.4)$$

and for $\omega_q = 0$

$$z_q(t) = z_q(0) + \dot{z}_q(0) t + \int_0^t N_q(\tau)(t - \tau) d\tau \quad (7.4.5)$$

Again, we can use Eq. (7.4.4) in numerical evaluations in MATLAB® for the case $\omega_q = 0$ if we use the same work-around described in the last section.

A second way to obtain the solution is to write it as a sum of a homogeneous solution and a particular solution, i.e., $z_q(t) = z_q^{\text{hom}}(t) + z_q^{\text{part}}(t)$, where the homogeneous solution is a general solution to the free vibration problem which we could write in real form as

$$z_q^{\text{hom}}(t) = C_q \sin(\omega_q t) + D_q \cos(\omega_q t) \quad (7.4.6)$$

and the particular solution is any explicit solution of Eq. (7.4.3) when the force is present. The convolutional integral in Eq. (7.4.4) is such a particular solution but it is also a particular solution that satisfies zero initial conditions for the modal coordinates, i.e., $z_q(0) = \dot{z}_q(0) = 0$ (see Appendix D). Thus, the homogeneous solution of Eq. (7.4.3) in that case must satisfy by itself the initial conditions (so that it is identical with the free vibration solution), leading to the form seen in Eq. (7.4.4) where $C_q = \dot{z}_q(0)/\omega_q$ and $D_q = z_q(0)$. If the particular solution, however, is simply chosen to satisfy Eq. (7.4.3) then the initial conditions on the modal coordinates must be satisfied by the combination of the homogeneous and particular solutions:

$$\begin{aligned} z_q(0) &= z_q^{\text{hom}}(0) + z_q^{\text{part}}(0) \\ \dot{z}_q(0) &= \dot{z}_q^{\text{hom}}(0) + \dot{z}_q^{\text{part}}(0) \end{aligned} \tag{7.4.7}$$

which is a set of equations to determine (C_q, D_q) since the particular solution is an explicit solution without arbitrary constants. Again, the initial conditions for the modal coordinates are obtained from the initial conditions on the original generalized coordinates by Eq. (7.3.23b). This second approach is most useful when the form of the force function is simple so that one can easily obtain an analytical particular solution. We show an example of such a case for a 1-D system in Appendix D. Here, multiple degree of freedom problems can be solved in the same fashion.

7.4.1 Harmonic Driving Forces

Harmonic or periodic driving forces are often encountered in engineering systems. These forces can excite a system into large and dangerous vibrations so that it is important to examine the vibratory response. The models we have examined so far have all neglected damping but we will see later that if there is any damping present the free vibration or homogeneous solution of a damped system decays with time so that if we have a continuous harmonic driving force the particular solution for the motion will be all that remains at large times. This particular solution is called the *steady-state solution*. The steady-state solution must satisfy the governing equations but it does not necessarily satisfy the initial conditions. We can find steady-state solutions directly from the equations of motion in the generalized coordinates which are given by

$$[M^e]\{\ddot{q}(t)\} + [K^e]\{q(t)\} = \{Q(t)\} \tag{7.4.8}$$

If we have harmonic driving forces we can write them in complex form as

$$\{Q(t)\} = \{Q_0\}\exp(i\omega t) \tag{7.4.9}$$

where $\{Q_0\}$ is a vector amplitude and ω is the driving frequency (rad/s). To obtain a steady-state solution, $\{q_{ss}(t)\}$, we can assume a particular solution of Eq. (7.4.8) as

$$\{q_{ss}(t)\} = \{A\}\exp(i\omega t) \tag{7.4.10}$$

Placing Eq. (7.4.10) into Eq. (7.4.8) and dividing out a common $\exp(i\omega t)$ factor, we find

$$\left([K^e] - \omega^2[M^e]\right)\{A\} = \{Q_0\} \tag{7.4.11}$$

which is a system of linear equations for the unknown steady-state amplitudes, $\{A\}$. As long as $\omega \neq \omega_q$ these equations have a solution given formally as

$$\{A\} = \left([K^e] - \omega^2[M^e]\right)^{-1}\{Q_0\} \tag{7.4.12}$$

This is analogous to the solution found by simple division for a one degree of freedom system in Appendix D (see Eq. (D.1.33) with no damping). It is also instructive to examine the steady-state response using the modal coordinates where

$$\ddot{z}_q(t) + \omega_q^2 z_q(t) = \{u_q\}^T\{Q(t)\} = \{u_q\}^T\{Q_0\}\exp(i\omega t) \tag{7.4.13}$$

Letting $z_q(t) = Z_q \exp(i\omega t)$ we can solve for each mode response as

$$z_q(t) = \frac{\{u_q\}^T\{Q_0\}}{\omega_q^2}\frac{1}{1-\omega^2/\omega_q^2}\exp(i\omega t) \tag{7.4.14}$$

and the total steady-state response is

$$\{q_{ss}(t)\} = \sum_{q=1}^{n}\{u_q\}z_q(t) = \left[\sum_{q=1}^{n}\frac{\{u_q\}^T\{Q_0\}}{\omega_q^2}\frac{1}{1-\omega^2/\omega_q^2}\{u_q\}\right]\exp(i\omega t) \tag{7.4.15}$$

Equation (7.4.15) shows that the steady-state response becomes infinitely large at the natural frequencies (resonances) of the system, due to the fact that we have ignored damping. Each mode is multiplied by a factor of $1/\omega_q^2$ so the relative contribution to the steady response is larger for the lower frequency modes.

7.5 Free Vibrations with Damping

Consider a multi-degree of freedom system with linear viscous damping but with no gyroscopic or circulatory forces and no driving forces. Then the small-amplitude motions around equilibrium satisfy

$$[M^e]\{\ddot{q}\} + [C^e]\{\dot{q}\} + [K^e]\{q\} = 0 \tag{7.5.1}$$

If we examine solutions of the form

$$\{q\} = \{u\}\exp(\lambda t) \tag{7.5.2}$$

we will obtain the homogeneous equations

$$\left([K^e] + \lambda[C^e] + \lambda^2[M^e]\right)\{u\} = 0 \tag{7.5.3}$$

This system of equations again only has a nontrivial solution if the determinant of the coefficients is equal to zero which leads to the characteristic equation

$$\det\left([K^e] + \lambda[C^e] + \lambda^2[M^e]\right) = 0 \tag{7.5.4}$$

Since the coefficients in the stiffness, damping, and mass matrices are all real, one can show that for an n degree of freedom system the $2n$ roots of this equation must either be real or occur in complex conjugate pairs. For a dissipative system, i.e., one that has no energy-creating sources, the real parts of all the roots will be negative. For each root λ_p there will be a corresponding solution for the vectors $\{u_p\}$. If the root is real the corresponding vector will be real but for a pair of roots that are complex conjugates of each other the modal vectors will also be complex conjugates of each other. For an underdamped system (see Appendix D for the 1-D underdamped case) without rigid body modes we expect all the roots to be distinct and complex so the homogeneous solution will be of the form

$$\{q\} = \sum_{p=1}^{2n} C_p\{u_p\}\exp\left(\lambda_p t\right) \tag{7.5.5}$$

where the $2n$ coefficients C_p will occur in complex conjugate pairs so in a free vibration problem they can be determined by the $2n$ initial conditions on $\{q(0)\}$ and $\{\dot{q}(0)\}$.

Equation (7.5.3) is not in a form where we can directly use the MATLAB® eig function but we can calculate the eigenvalues and eigenvectors with eig by reducing the second-order equations of Eq. (7.5.1) to a system of first-order differential equations. Recall in Chap. 6 that we also used a first-order system to study stability via linearization. Letting $\{x\} = \left[\{q\}^T, \{\dot{q}\}^T\right]^T$ the system of first-order equations are

$$\begin{bmatrix} I & 0 \\ 0 & M^e \end{bmatrix}\{\dot{x}\} = \begin{bmatrix} 0 & I \\ -K^e & -C^e \end{bmatrix}\{x\} \tag{7.5.6}$$

where I is an $n \times n$ identity matrix and the zero matrix entry shown is also an $n \times n$ matrix. Equation (7.5.6) is in the form

$$[A]\{\dot{x}\} = [B] \tag{7.5.7}$$

so if we let

$$\{x\} = \{a\}\exp(\lambda t) \tag{7.5.8}$$

we have

$$([B] - \lambda[A])\{a\} = 0 \tag{7.5.9}$$

which is in the form of a standard eigenvalue problem. For an underdamped system where all the roots are distinct and complex we can write the homogeneous solution as

$$\{x\} = \sum_{k=1}^{2n} C_k\{a_k\}\exp(\lambda_k t) \tag{7.5.10}$$

and we can solve for the unknown constants C_k from the $2n$ initial conditions $\{x_0\}$ since

$$\{x_0\} = \sum_{k=1}^{2n} C_k\{a_k\} \tag{7.5.11}$$

or in matrix-vector form where the eigenvectors are in the columns of the matrix [V]:

$$[V]\{C\} = \{x_0\} \tag{7.5.12}$$

This is a linear system of equations that we can solve for the constants. Then we can write the free vibration solution also in matrix-vector form by defining the diagonal matrix, Λ, as

$$[\Lambda(t)] = \begin{bmatrix} \exp(\lambda_1 t) & 0 & \cdots & 0 \\ 0 & \exp(\lambda_2 t) & \cdots & 0 \\ \cdots & \cdots & \cdots & \cdots \\ 0 & 0 & \cdots & \exp(\lambda_{2n} t) \end{bmatrix} \tag{7.5.13}$$

and writing the solution as

$$\{x(t)\} = [V][\Lambda(t)]\{C\} \tag{7.5.14}$$

Although this solution is valid only for underdamped systems without any rigid body modes, this is typically the case found in the vibrations of many structural systems since the damping is normally small and rigid body motions are usually not allowed. This book will not cover more general damped free vibration solutions. For those cases, see any text on differential equations that covers the eigenvalue problem for a first-order system of ordinary differential equations in depth.

Simulation 7.2: Free Vibrations of an Underdamped Two Degree of Freedom System

As an example of free vibrations of an underdamped system, consider the two degree of freedom system shown in Fig. 7.10a. From the free body diagram in Fig. 7.10b we see that the equations of motion are

$$\begin{aligned} m_1\ddot{x}_1 &= k_2(x_2 - x_1) - k_1x_1 - c_1\dot{x}_1 + c_2(\dot{x}_2 - \dot{x}_1) \\ m_2\ddot{x}_2 &= -k_3x_2 - k_2(x_2 - x_1) - c_3\dot{x}_2 - c_2(\dot{x}_2 - \dot{x}_1) \end{aligned} \tag{7.5.15}$$

which is of the form of Eq. (7.5.1) where

$$[M^e] = \begin{bmatrix} m_1 & 0 \\ 0 & m_2 \end{bmatrix}, \quad [K^e] = \begin{bmatrix} (k_1 + k_2) & -k_2 \\ -k_2 & (k_2 + k_3) \end{bmatrix}, \quad [C^e] = \begin{bmatrix} (c_1 + c_2) & -c_2 \\ -c_2 & (c_2 + c_3) \end{bmatrix} \tag{7.5.16}$$

The first-order system will then be in the form of Eq. (7.5.6). The script free_vibs_2D sets up the eigenvalue problem for this first-order system and solves it with the MATLAB® eig function. These eigenvalues and eigenvectors are used to find the constants in the homogeneous solution that satisfy the initial conditions and the solution is then given by Eq. (7.5.14). The script plots the motion of mass m_1 from $t = 0$ to $t = 50$ s. The default parameters used are $m_1 = 2$ kg, $m_2 = 4$ kg, $k_1 = k_3 = 2$ N/m, $k_2 = 3$ N/m, and $c_1 = c_3 = 0.25$ N-s/m, $c_2 = 0.5$ N-s/m. The initial conditions are $x0 = [x_1(0), x_2(0), \dot{x}_1(0), \dot{x}_2(0)] = [1, 0, 0, 0]$. The eigenvalues are $\lambda_{1,2} = -0.2406 \pm 1.7459\ i$ and $\lambda_{3,4} = -0.0407 \pm 0.8014\ i$ so the eigenvalues are all distinct and occur in complex conjugate pairs with negative real parts, as discussed previously. Thus, all our solutions are of the form seen in Eq. (7.5.10)

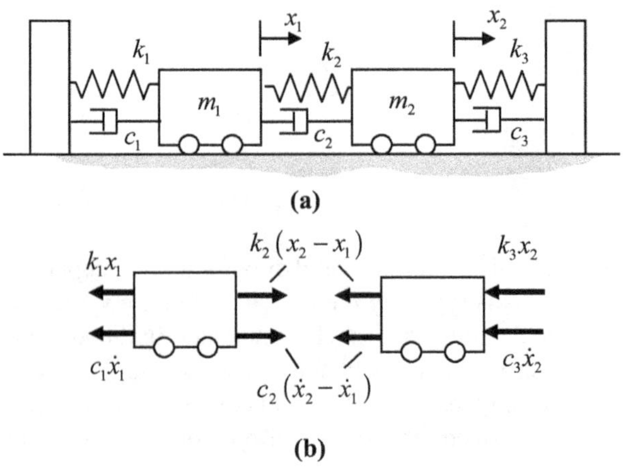

Fig. 7.10 (**a**) A two degree of freedom damped system and (**b**) the free body diagram of the horizontal forces acting on the system

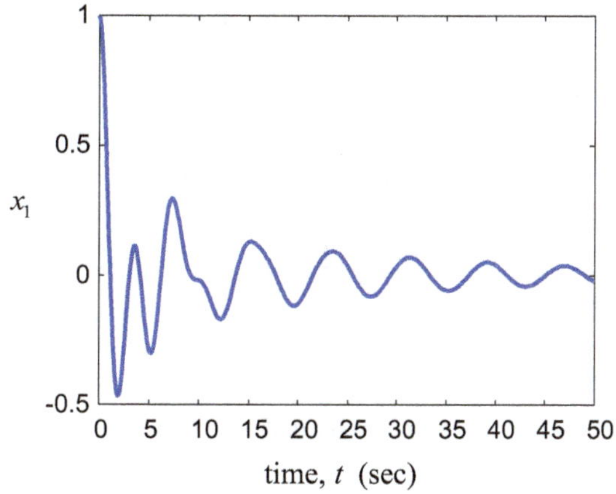

Fig. 7.11 The free vibrations of mass m_1 in the two degree of freedom system shown in Fig. 7.10a

so we are justified in obtaining the solution in the script by following the same steps previously outlined, leading to the underdamped, free vibration solution of Eq. (7.5.14). Figure 7.11 shows a plot of the displacement of mass m_1 versus time. We can see that the initial behavior starts from the given initial condition and is rather complex but settles down to a decaying simple oscillation of the mass.

Some engineering texts also discuss an approximate approach to the underdamped vibration problem which does not require going to a system of first-order equations. This approach uses the natural frequencies, ω_q, and natural modes, $\{u_q\}$, of the *undamped* system and makes the same expansion of the solution in terms of those modes, Eq. (7.3.17a). Placing that expansion into our equation for a damped system and using the orthonormality relations as before, Eq. (7.5.1) gives

$$\ddot{z}_q(t) + \sum_{p=1}^{n} \tilde{c}_{qp}\dot{z}_p(t) + \omega_q^2 z_q(t) = 0 \tag{7.5.17}$$

where the matrix components are

$$\tilde{c}_{qp} = \{u_q\}^T[C]\{u_p\} \quad (q, p = 1, 2, \ldots, n) \tag{7.5.18a}$$

or equivalently in matrix-vector form

$$[\tilde{c}] = [V]^T[C][V] \tag{7.5.18b}$$

It can be seen that in the damped case the use of the undamped modal coordinates is not particularly useful since the equations remain coupled. However, if it is possible to ignore the off-diagonal components of the $\tilde{c}_{qp}$ matrix, and if we define a damping factor $\varsigma_q = \dfrac{\tilde{c}_{qq}}{2\omega_q}$, the modal equations do decouple and we have

$$\ddot{z}_q(t) + 2\varsigma_q\omega_q\dot{z}(t) + \omega_q^2 z_q(t) = 0 \tag{7.5.19}$$

which is identical to the equation for a one degree of freedom damped system. Using the solution for the underdamped free vibration problem obtained in Appendix D the solution for each mode can be written as

$$z_q(t) = \exp(-\varsigma_q\omega_q t)\left[z_q(0)\cos(\omega_{qd}t) + \frac{\dot{z}_q(0) + z_q(0)\varsigma_q\omega_q}{\omega_{qd}}\sin(\omega_{qd}t)\right] \tag{7.5.20}$$

where $\omega_{qd} = \omega_q\sqrt{1-\varsigma_q^2}$ is the damped natural frequency. We can write Eq. (7.5.20) in matrix-vector form by defining two diagonal matrices, (Λ, Λ^d), where the diagonal elements are

$$\begin{aligned}\Lambda_{qq}(t) &= \exp(-\varsigma_q\omega_q t)\left[\cos(\omega_{qd}t) + \frac{\varsigma_q\omega_q}{\omega_{qd}}\sin(\omega_{qd}t)\right] \quad (q = 1, 2, \ldots, n)\\ \Lambda_{qq}^d(t) &= \exp(-\varsigma_q\omega_q t)\sin(\omega_{qd}t)/\omega_{qd}\end{aligned} \tag{7.5.21}$$

so that we have

$$\{z(t)\} = [\Lambda(t)]\{z(0)\} + [\Lambda^d(t)]\{\dot{z}(0)\} \tag{7.5.22}$$

and in terms of the initial conditions of the generalized coordinates

$$\{z(t)\} = [\Lambda(t)][V]^T[M^e]\{q(0)\} + [\Lambda^d(t)][V]^T[M^e]\{\dot{q}(0)\} \tag{7.5.23}$$

so the final free vibration solution is

$$\{q(t)\} = [V][\Lambda(t)][V]^T[M^e]\{q(0)\} + [V][\Lambda^d(t)][V]^T[M^e]\{\dot{q}(0)\} \tag{7.5.24}$$

Because we ignore the off-diagonal terms in the $[\tilde{c}]$ matrix this is only an approximate solution. There is no rigorous mathematical or physical justification of this approximation other than to note that when the damping is relatively light, as it is for most structural vibration problems, the omission of the off-diagonal terms is likely to lead to errors that are not larger than the errors made in estimating the damping terms in the original $[C]$ matrix since the accurate characterization of the damping in real multiple degree of freedom systems is notoriously difficult.

Simulation 7.3: Approximate Free Vibrations of Underdamped Systems
The MATLAB® script free_vibs_2D_approx implements the approximate formulation just described for the two degree of freedom system of Fig. 7.10a considered previously in the script free_vibs_2D. The natural undamped frequencies used in the

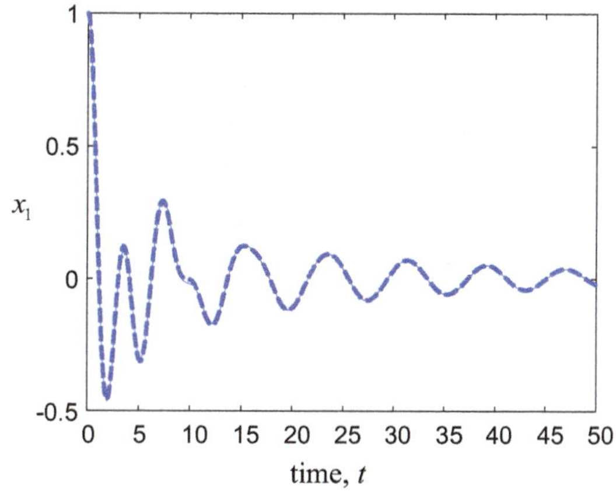

Fig. 7.12 The free vibrations of mass m_1 in the two degree of freedom system shown in Fig. 7.10a under the same initial conditions used for Fig. 7.11. Solid line—exact analysis, dashed line—approximate analysis

approximate solution were $\omega_1 = 0.8024$ rad/s and $\omega_2 = 1.7624$ rad/s. The diagonal damping factors were $\varsigma_1 = 0.0507$ and $\varsigma_2 = 0.1365$. In the exact analysis recall that the eigenvalues obtained were $\lambda_{1,2} = -0.2406 \pm 1.7459\ i$ and $\lambda_{3,4} = -0.0407 \pm 0.8014\ i$. From Appendix D we saw that the eigenvalues of a one degree of freedom system were related to the damping factor and the undamped natural frequency by (see Eq. (D.1.12))$\lambda = -\varsigma\omega_n \pm i\omega_n\sqrt{1-\varsigma^2}$. Thus, if we let $\lambda = -a \pm ib$ we have

$$\varsigma\omega_n = a, \quad \omega_n\sqrt{1-\varsigma^2} = b \tag{7.5.25}$$

which can be solved for ω_n and ς as

$$\omega_n = \sqrt{a^2+b^2}, \quad \varsigma = a/\omega_n \tag{7.5.26}$$

Using the eigenvalues from the exact analysis and Eq. (7.5.26) we find the same natural undamped frequencies and damping factors (to four decimal places) used in the approximate analysis so that one expects that the approximate solution will be quite accurate in this case. This is verified by plotting the two solutions for the motion of mass m_1 on the same plot, as shown in Fig. 7.12, where there is very little difference, even in the early time response. There is, of course, no guarantee that such agreement will always be found but this comparison does show that the approximate analysis can lead to good results.

7.6 Forced Vibrations with Damping

7.6.1 Harmonic Driving Forces

In a damped multi-degree of freedom system with driving forces governed by the equations

$$[M^e]\{\ddot{q}(t)\} + [C^e]\{\dot{q}(t)\} + [K^e]\{q(t)\} = \{Q(t)\} \tag{7.6.1}$$

we can again consider the case of harmonic driving forces $\{Q(t)\} = \{Q_0\} \exp(i\omega t)$ and examine steady-state harmonic solutions of the form $\{q_{ss}(t)\} = \{A\} \exp(i\omega t)$ which gives us the resulting equations

$$\left([K^e] + i\omega[C^e] - \omega^2[M^e]\right)\{A\} = \{Q_0\} \tag{7.6.2}$$

which can be solved for the amplitude vector $\{A\}$ as

$$\{A\} = [G]\{Q_0\} \tag{7.6.3}$$

where the *transfer matrix* $[G]$ (note: this is not the gyroscopic matrix considered earlier) is given by

$$[G] = \left([K^e] + i\omega[C^e] - \omega^2[M^e]\right)^{-1} \tag{7.6.4}$$

The elements of $[G]$, $G_{ij}(\omega)$, are complex numbers that represent the amplitude and phase of the steady-state response, $(q_{ss})_i$, in terms of the q_i coordinate due to a force component of unit amplitude, $Q_{0j} = 1$, with all the other components equal to zero. This transfer matrix is symmetric because the mass, stiffness, and damping matrix are symmetric. The general steady-state response is thus

$$\{q_{ss}\} = [G(\omega)]\{Q_0\}\exp(i\omega t) \tag{7.6.5a}$$

or, in component form for an n degree of freedom system,

$$(q_{ss})_i = \sum_{j=1}^{n} G_{ij}(\omega)Q_{0j}\exp(i\omega t) \tag{7.6.5b}$$

The transfer matrix is the direct multi-degree of freedom generalization of the scalar transfer function found for a one degree of freedom system in Appendix D. It is a matrix of particular solutions of Eq. (7.6.1) for harmonic driving forces. The steady-state solution here is also the generalization of the steady-state solution found in Appendix D for a single degree of freedom system. In general, this matrix of solutions is the response of the system after all the transient solutions have died out and so it does not satisfy the initial conditions.

7.6.2 System Impulse Response

In Appendix D we show that the transfer function for 1-D systems is closely related to a system impulse response. We have a similar relationship for multi-degree of

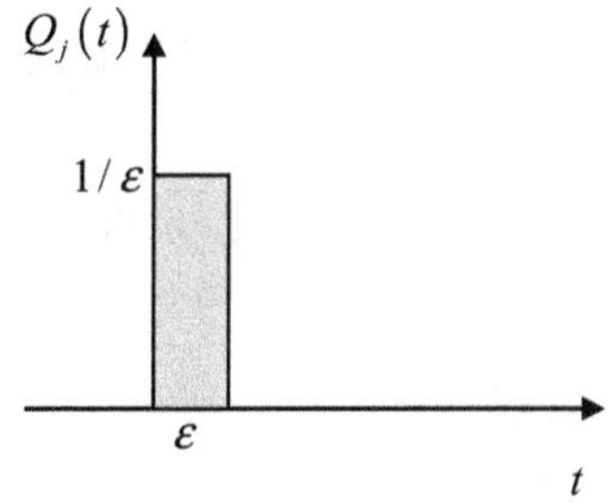

Fig. 7.13 A generalized force that produces a unit impulse (unit area under the force curve) at time $t = 0$

freedom systems. Consider, for example a generalized force that has only one nonzero component, $Q_j(t)$, with

$$Q_j(t) = \begin{cases} 1/\varepsilon & 0 < t < \varepsilon \\ 0 & \text{otherwise} \end{cases} \tag{7.6.6}$$

where ε is a constant. The graph of this force is shown in Fig. 7.13. If this constant is a small number then this force component is a large force acting over a small time but where the integral of the force, called the impulse, is unity. The complete force vector is then $\{Q(t)\} = [0, 0, \ldots, Q_j(t), 0, 0, \ldots]^T$. Now suppose we integrate Eq. (7.6.1) from $t = 0$ to $t = \varepsilon$ and assume zero initial conditions. If we let the ith component of generalized displacement due to a unit impulse force with only a k-component be $g_{ik}(t)$, then in terms of these components we have

$$\sum_{i=1}^{n} M^e_{ji} \dot{g}_{ik}\big|_{t=0}^{t=\varepsilon} + \sum_{i=1}^{n} C^e_{ji} \int_{t=0}^{t=\varepsilon} \dot{g}_{ik} dt + \sum_{i=1}^{n} K^e_{ji} \int_{t=0}^{t=\varepsilon} g_{ik} dt = \delta_{jk} \quad (j, k = 1, 2, \ldots, n) \tag{7.6.7}$$

where

$$\delta_{jk} = \begin{cases} 1 & j = k \\ 0 & \text{otherwise} \end{cases} \tag{7.6.8}$$

is called the *Kronecker delta*. These Kronecker delta components are just the components of the identity matrix, **I**, which has ones on its diagonal and zeros elsewhere. If we have zero initial conditions the lower limit seen in Eq. (7.6.7) vanishes and in the limit as $\varepsilon \to 0$ the two integrals vanish since both the velocity and displacement integrands must be bounded in that limit, leaving

$$\sum_{i=1}^{n} M_{ji}^{e}\dot{g}_{ik}(0^{+}) = \delta_{jk} \quad (j,k = 1,2,\ldots,n) \tag{7.6.9}$$

where $\dot{g}_{jk}(0^{+})$ means that these velocities are to be calculated just after the unit impulse is applied at $t = 0$. We can write Eq. (7.6.9) in matrix-vector form as

$$[M^{e}][\dot{g}(0^{+})] = [I] \tag{7.6.10}$$

or multiplying by the inverse of the mass matrix we have

$$[\dot{g}(0^{+})] = [M^{e}]^{-1} \tag{7.6.11a}$$

We also have

$$[g(0^{+})] = 0 \tag{7.6.11b}$$

(recall from physics that an impulsive force can produce an instantaneous change in velocity but it cannot produce an instantaneous change in position) so we see the matrix of unit impulse response solutions, $[g(t)]$, are just the solutions to the initial value problems specified by Eqs. (7.6.11a) and (7.6.11b). These are just generalizations, for a multiple degree of freedom system, of the impulse response function for a one degree of freedom system considered in Appendix D.

7.6.3 Convolution Integrals

Now, consider the problem where one of the external force components, Q_j, is a specified function of time, i.e., $Q_j = Q_j(t)$. We can consider a small portion of that force at time $t = \tau$, as shown in Fig. 7.14. That portion acts as an impulsive force of strength $Q_j(\tau)\Delta\tau$ at time $t = \tau$. Following the steps outlined in Appendix D for the one degree of freedom case, since our multi-degree of freedom system is a linear

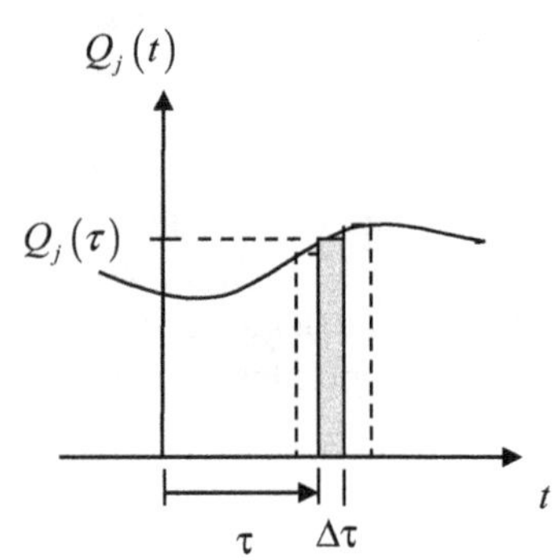

Fig. 7.14 An element of an applied force (shaded area) which produces an impulse of strength $Q_j(\tau)d\tau$ acting at time $t = \tau$

time-shift invariant system, we can write the response at the generalized coordinate q_i at time t due to the impulsive force Q_j at time τ as $q_i(t, \tau)$ where

$$q_i(t,\tau) = g_{ij}(t-\tau)Q_j(\tau)\Delta\tau \tag{7.6.12}$$

The total response due to all the force components and impulse-like elements of the total force curve shown in Fig. 7.14 is therefore

$$q_i(t) = \sum_{\tau}\sum_{j=1}^{n} g_{ij}(t-\tau)Q_j(\tau)\Delta\tau \tag{7.6.13}$$

or in the limit as $\Delta\tau \to 0$ we have an integral over all times τ as

$$q_i(t) = \sum_{j=1}^{n}\int_{-\infty}^{+\infty} g_{ij}(t-\tau)Q_j(\tau)d\tau \tag{7.6.14}$$

In matrix-vector form, Eq. (7.6.14) can also be written compactly as

$$\{q(t)\} = \int_{-\infty}^{+\infty} [g(t-\tau)]\{Q(\tau)\}d\tau \tag{7.6.15}$$

This is called a convolution integral. As in the 1-D case we can also use the properties of the impulse responses and the forces to express the convolution in alternate forms. For example, for $t < 0$ $g_{ij}(t) = 0$ and we normally have $\{Q(t)\} = 0$ for $t < 0$; we can write the convolution as

$$\{q(t)\} = \int_{0}^{t} [g(t-\tau)]\{Q(\tau)\}d\tau \tag{7.6.16}$$

and by a simple change of variables it can also be written as

$$\{q(t)\} = \int_{0}^{t} [g(\tau)]\{Q(t-\tau)\}d\tau \tag{7.6.17}$$

Following the same steps as used in Appendix D for the one degree of freedom case we can show that these convolution integrals represent particular solutions that satisfy zero initial conditions. Thus, we merely have to add the free vibration (homogeneous) solution to the convolution integral to obtain a complete solution to a forced vibration problem with given initial conditions. For example, if we use the approximate free vibration solution developed earlier for an underdamped system with no repeated natural frequencies the complete solution in matrix-vector form is

$$\begin{aligned}\{q(t)\} = {} & [V][\Lambda(t)][V]^T[M^e]\{q(0)\} + [V]\left[\Lambda^d(t)\right][V]^T[M^e]\{\dot{q}\,(0)\} \\ & + \int_0^t [g(\tau)]\{Q(t-\tau)\}d\tau\end{aligned} \tag{7.6.18}$$

In Appendix D it is also shown how the steady-state solution is related to the impulse response function. In the multi-degree of freedom case, if we have $\{Q(t)\} = \{Q_0\}\exp(i\omega t)$ for all times from $t = -\infty$ to the present, since the damped homogeneous solution becomes small, we have

$$\{q(t)\} = \exp(i\omega t)\int_{-\infty}^{t}[g(\tau)]\{Q_0\}\exp(-i\omega\tau)d\tau \tag{7.6.19}$$

and at very large times the solution must approach the steady-state solution so from Eq. (7.6.5a)

$$[G(\omega)]\{Q_0\}\exp(i\omega t) = \left(\int_{-\infty}^{+\infty}[g(\tau)]\exp(-i\omega\tau)d\tau\right)\{Q_0\}\exp(i\omega t)$$

which gives

$$[G(\omega)] = \int_{-\infty}^{+\infty}[g(\tau)]\exp(-i\omega\tau)d\tau \tag{7.6.20}$$

Equation (7.6.20) shows that the steady-state transfer matrix is the Fourier transform of the matrix of impulse responses. From the inverse Fourier transform then we also have

$$[g(\tau)] = \frac{1}{2\pi}\int_{-\infty}^{+\infty}[G(\omega)]\exp(i\omega\tau)d\tau \tag{7.6.21}$$

Appendix E discusses these Fourier transforms and shows how they can be used to numerically determine the impulse response from the transfer function (and vice versa) for the one degree of freedom case considered in Appendix D.

7.7 Problems

P7.1. Consider the cart-rod system shown in Fig. P7.1 where a cart of mass m is attacked to the vertical wall by a linear spring with spring constant k and a linear dashpot with damping constant, c. The cart is also pinned to a thin rod of mass M and length L. A driving force, $F(t)$, also acts on the cart. The spring is unstretched when $x = 0$.

(a) Determine the equations of motion for this two degree of freedom system.
(b) Linearize the equations of motion for small vibrations about the equilibrium position $x = 0, \theta = 0$ and obtain the mass, damping, and stiffness matrices that define this system.
(c) Let $m = 2$ kg, $M = 4$ kg, $L = 1$ m, $k = 10$ N/m, and $c = 0$ and set $F(t) = 0$. Determine the natural frequencies and natural modes (eigenvectors) for this system. Plot the values of these modal vectors and show graphically the motion system when it is vibrating in these natural modes.
(d) Show that the matrix V that contains the eigenvectors in its columns satisfies the orthonormality relations of Eq. (7.3.15b), or, if you have used an eigenvalue solver that does not normalize the eigenvectors in this form, normalize your eigenvectors so that they do satisfy these orthonormality relations.

P7.2. The three carts in Fig. P7.2 are connected via linear springs and linear dashpots as shown.

(a) Determine the equations of motion for this three degree of freedom system and obtain the mass, damping, and stiffness matrices that define this system.
(b) Let $m_1 = 1$ kg, $m_2 = 2$ kg, $m_3 = 4$ kg, $k_1 = 20$ N/m, $k_2 = 40$ N/m, $k_3 = 10$ N/m, $k_4 = 20$ N/m,, $c_1 = c_2 = c_3 = c_4 = 0$, and $F_1 = F_2 = F_3 = 0$.

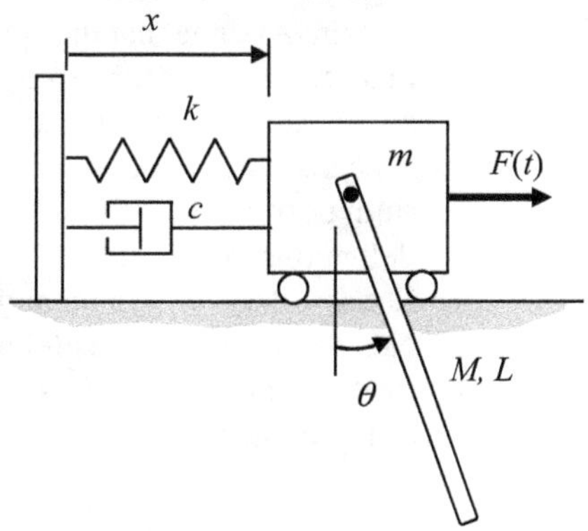

Fig. P7.1 A cart-rod two degree of freedom damped system

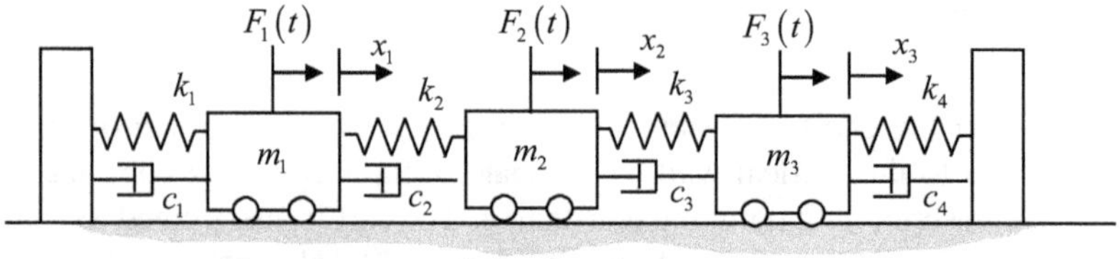

Fig. P7.2 A three degree of freedom system

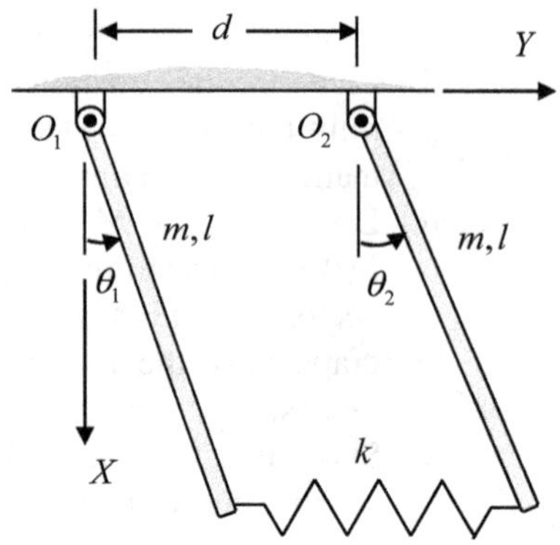

Fig. P7.3 Two identical thin rods connected through a linear spring

Determine the natural frequencies and natural modes (eigenvectors) for this system. Plot the values of these modal vectors.

(c) Show that the matrix V that contains the eigenvectors in its columns satisfies the orthonormality relations of Eq. (7.3.15b), or, if you have used an eigenvalue solver that does not normalize the eigenvectors in this form, normalize your eigenvectors so that they do satisfy these orthonormality relations.

P7.3. Two identical thin rods, each of mass m and length l, are connected by a linear spring with spring constant k, as shown in Fig. P7.3.

(a) Determine the equations of motion for this two degree of freedom system. Assume that the spring is unstretched when the bars are vertical; that is, the unstretched spring length is the distance d. Note that if the angles are not small the spring is not horizontal.

(b) Linearize the equations of motion for small vibrations about the equilibrium position $\theta_1 = \theta_2 = 0$ and obtain the mass and stiffness matrices that define this system.

(c) Let $m = 2$ kg, $l = 1.5$ m, and $k = 10$ N/m. Determine the natural frequencies and natural modes (eigenvectors) for this system. Plot the values of these modal vectors and show graphically the system when it is vibrating in these natural modes.

(d) Show that the matrix V that contains the eigenvectors in its columns satisfies the orthonormality relations of Eq. (7.3.15b), or, if you have used an eigenvalue solver that does not normalize the eigenvectors in this form, normalize your eigenvectors so that they do satisfy these orthonormality relations.

P7.4. Solve for the free undamped small vibrations of cart-rod system of Fig. P7.1 using modal coordinates for $m = 2$ kg, $M = 4$ kg, $L = 1$ m, $c = 0$, and $F(t) = 0$. At time $t = 0$ take $x(0) = 0, \dot{x}(0) = 0, \theta(0) = 0$, and $\dot{\theta}(0) = 2$ rad/s.

(a) Use the analysis of problem P7.1 to obtain the natural frequencies and natural modes of the system [be sure to eliminate any small imaginary components from these numerical results]. From the initial conditions, determine the corresponding initial conditions for the modal coordinates, $\{z(0)\}$ and $\{\dot{z}(0)\}$(see Eq. (7.3.23b)).
(b) From the initial conditions on the modal coordinates, obtain the undamped free vibration solutions for the $\{z(t)\}$ (see Eq. (7.3.21)) for a time interval from $t = 0$ to $t = 15$ s.
(c) Use the solutions for the modal coordinates obtained in part (b) to obtain the corresponding solutions for the undamped free vibrations in terms of the original coordinates $(x(t), \theta(t))$ and plot these solutions for the original coordinates versus time.

P7.5. Solve for the free undamped vibrations of the three degree of freedom system shown in Fig. P7.2. Let $m_1 = 1$ kg, $m_2 = 2$ kg, $m_3 = 4$ kg, $k_1 = 20$ N/m, $k_2 = 40$ N/m, $k_3 = 10$ N/m, $k_4 = 20$ N/m, $c_1 = c_2 = c_3 = c_4 = 0$, and $F_1 = F_2 = F_3 = 0$. At time $t = 0$ take $x_1(0) = 0, \dot{x}_1(0) = 0, x_2(0) = 0, \dot{x}_2(0) = 0, x_3(0) = 0$, and $\dot{x}_3(0) = 2$ m/s.

(a) Use the analysis of problem P7.2 to obtain the natural frequencies and natural modes of the system [be sure to eliminate any small imaginary components from these numerical results]. From the initial conditions, determine the corresponding initial conditions for the modal coordinates, $\{z(0)\}$, $\{\dot{z}(0)\}$(see Eq. (7.3.23b)).
(b) From the initial conditions on the modal coordinates, obtain the undamped free vibration solutions for the $\{z(t)\}$ (see Eq. (7.3.21)) for a time interval from $t = 0$ to $t = 10$ s.
(c) Use the solutions for the modal coordinates obtained in part (b) to obtain the corresponding solutions for the undamped free vibrations in terms of the original coordinates $(x(t), \theta(t))$ and plot these solutions for the original coordinates versus time.

P7.6. Solve for the free undamped small vibrations of the two-pendulum system shown in Fig. P7.3. Let $m = 2$ kg, $l = 1.5$ m, and $k = 10$ N/m. At time $t = 0$, take $\theta_1(0) = 0, \dot{\theta}_1(0) = 0, \theta_2(0) = 0.2$ rad, and $\dot{\theta}_2(0) = 0$.

(a) Use the analysis of problem P7.3 to obtain the natural frequencies and natural modes of the system [be sure to eliminate any small imaginary components from these numerical results]. From the initial conditions, determine the corresponding initial conditions for the modal coordinates, $\{z(0)\}$, $\{\dot{z}(0)\}$(see Eq. (7.3.23b)).

(b) From the initial conditions on the modal coordinates, obtain the undamped free vibration solutions for the $\{z(t)\}$ (see Eq. (7.3.21)) for a time interval from $t = 0$ to $t = 20$ s.

(c) Use the solutions for the modal coordinates obtained in part (b) to obtain the corresponding solutions for the undamped free vibrations in terms of the original coordinates $(\theta_1(t), \theta_2(t))$ and plot these solutions for the original coordinates versus time.

P7.7. Consider the forced undamped vibrations of the cart-rod system of Fig. P7.1 where there is a harmonic driving force $F(t) = F_0 \exp(i\omega t)$. Since the force is always present any small damping in a real system will result eventually in a steady-state motion of the system given by $\{q_{ss}(t)\} = \{A\} \exp(i\omega t)$, where $\{q_{ss}(t)\} = [x_{ss}(t), \theta_{ss}(t)]^T$ and $\{A\}$ is the steady-state amplitude. We can write this steady-state amplitude in terms of the natural frequencies and natural modes as a function of the driving frequency, ω (see Eq. (7.4.15)). Taking $F_0 = 1$ and using the other parameters of problem P7.1, plot the components of the steady-state amplitude versus frequency for a frequency from zero to 1.5 times the largest natural frequency. Note that this steady-state amplitude will go to infinity at the natural frequencies since there is no damping.

P7.8. Consider the forced undamped vibrations of the three degree of freedom system shown in Fig. P7.2 where there are harmonic driving forces $F_n(t) = F_{0n} \exp(i\omega t)$ $(n = 1, 2, 3)$. Since these forces are always present any small damping in this system will result eventually in a steady-state motion of the system given by $\{q_{ss}(t)\} = \{A\} \exp(i\omega t)$, where $\{q(t)\} = [x_{1;ss}(t), x_{2;ss}(t), x_{3;ss}]^T$ and $\{A\}$ is the steady-state amplitude. We can write this steady-state amplitude in terms of the natural frequencies and natural modes as a function of the driving frequency, ω (see Eq. (7.4.15)). Taking $F_{01} = 0, F_{02} = 1, F_{03} = 0$ and using the other parameters of problem P7.2, plot the components of the steady-state amplitude versus frequency for a frequency from zero to 1.5 times the largest natural frequency. Note that this steady-state amplitude will go to infinity at the natural frequencies since there is no damping.

P7.9. Show that if the spring and dashpot of the cart-rod system of Fig. P7.1 are removed, then this system can have a natural frequency of zero. Use the corresponding zero-frequency eigenvector to describe the motion of the system with the same parameters for m, M, and L $(k = 0,\ c = 0,\ F(t) = 0)$ found in problem P7.1. Show graphically the system in this rigid body mode.

P7.10. Three identical thin rods, each of mass m and length l, are connected by two identical linear springs with spring constant k, as shown in Fig. P7.4.

Fig. P7.4 Three identical rods connected through two identical springs

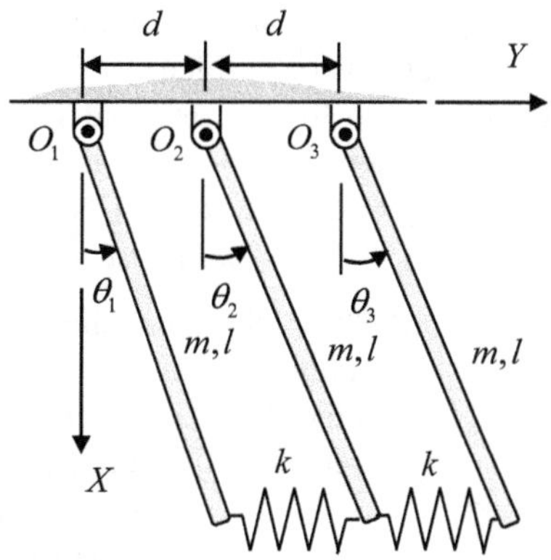

(a) Determine the linearized equations of motion for this two degree of freedom system about the equilibrium position $\theta_1 = \theta_2 = 0$ and obtain the mass and stiffness matrices that define this system. Assume that the springs are unstretched when the bars are vertical; that is, both unstretched spring lengths are the distance d.

(b) Let $m = 2$ kg, $l = 1.5$ m, and $k = 10$ N/m. Determine the natural frequencies and natural modes (eigenvectors) for this system. Plot the values of these modal vectors and show graphically the system when it is vibrating in these natural modes.

(c) Show that the matrix V that contains the eigenvectors in its columns satisfies the orthonormality relations of Eq. (7.3.15b), or, if you have used an eigenvalue solver that does not normalize the eigenvectors in this form, normalize your eigenvectors so that they do satisfy these orthonormality relations.

P7.11. Consider the forced undamped vibrations of the cart-rod system of Fig. P7.1 when there is a transient driving force, $F(t)$. Use the convolution integral solution of Eq. (7.4.4) to obtain the solution for $(x(t), \theta(t))$. Let the force be given as

$$F(t) = \begin{cases} F_0 & 0 \le t \le t_0 \\ 0 & \text{otherwise} \end{cases} \tag{P7.1}$$

and take $F_0 = 1$ N, $t_0 = 2$ s. Note that for this force we can do the convolution integral analytically. Use the other parameters and initial conditions of problem P7.4 and plot the solution over a time span $0 \le t \le 5t_0$.

P7.12. Consider the forced undamped vibrations of the three degree of freedom system shown in Fig. P7.2 when there is a transient driving force, $F_3(t)$, given by the same form of Eq. (P7.1) and the other forces $F_1 = F_2 = 0$. Use the convolution integral solution of Eq. (7.4.4) with $F_0 = 1$ N, $t_0 = 2$ s to obtain the solution for $(x_1(t), x_2(t), x_3(t))$. Note that for this force we can do the convolution integral analytically. Use the other parameters and initial conditions of problem P7.5 and plot the solution over a time span $0 \leq t \leq 5t_0$.

P7.13. Solve for the free underdamped vibrations of the three degree of freedom system shown in Fig. P7.2. Let $m_1 = 3$ kg, $m_2 = 2$ kg, $m_3 = 3$ kg, $k_1 = 2$ N/m, $k_2 = 4$ N/m, $k_3 = 4$ N/m, $k_2 = 2$ N/m, $c_1 = 1.0$ N-s/m, $c_2 = 0.5$ N-s/m, $c_3 = 2$ N-s/m, and $c_4 = 1.0$ N-s/m.

(a) Show that the eigenvalues for the system of first-order differential equations that describe this problem demonstrate that this is an underdamped system with three distinct natural frequencies.
(b) Let the initial conditions be $x_1(0)$= 1.0 m, $x_2(0) = x_3(0) = 0$, and $\dot{x}_1(0) = \dot{x}_2(0) = \dot{x}_3(0) = 0$ and solve for the free vibrations of this system following the same steps outlined in Simulation 7.2. Plot the motion of mass m_1 from $t = 0$ to $t = 40$ s.

P7.14. Solve for the free underdamped vibrations of the three degree of freedom system shown in Fig. P7.2 by using the approximate analysis of Simulation 7.3. Use the same parameters given in problem P7.13 and compare the motion of mass m_1 obtained in that problem with the approximate solution obtained here.

Reference

1. H. Baruh, *Applied Dynamics* (CRC press, Boca Raton, 2015)

Chapter 8
General Spatial Dynamics of Rigid Bodies

The planar problems examined in Chap. 5 do not describe a number of the dynamical behaviors found in 3-D problems such as gyroscopic effects. In this chapter we describe the spatial dynamics of rigid bodies and the numerical solutions of the equations of motion using both Euler angles and Euler parameters. We again examine both a Newton-Euler approach and Lagrange's equations. There are many other methods and issues that could be considered for 3-D problems that we will leave to more advanced treatments of dynamics.

8.1 Angular Momentum

The equations of motion of rigid bodies for general three-dimensional motion are more complex than the planar motion cases considered in Chap. 5. As seen in Chap. 4, most of this complexity arises because of the nature of general rotations. The rotational motion part of the Newton-Euler equations involves changes in angular momentum so in this section we obtain expressions for the angular momentum that are generalizations of the system of particles case considered in Chap. 3.

8.1.1 Angular Momentum About a Body-Fixed Point

If we have a point A that moves with a continuous rigid body (Fig. 8.1) the angular momentum, $d\mathbf{L}_A^{(n)}$, about that point for a mass element, $dm = \rho dV$, where ρ is the mass density and dV is a volume element, is

L. W. Schmerr, *Engineering Dynamics 2.0*, Solid Mechanics and Its Applications 254,
https://doi.org/10.1007/978-3-319-98470-4_8

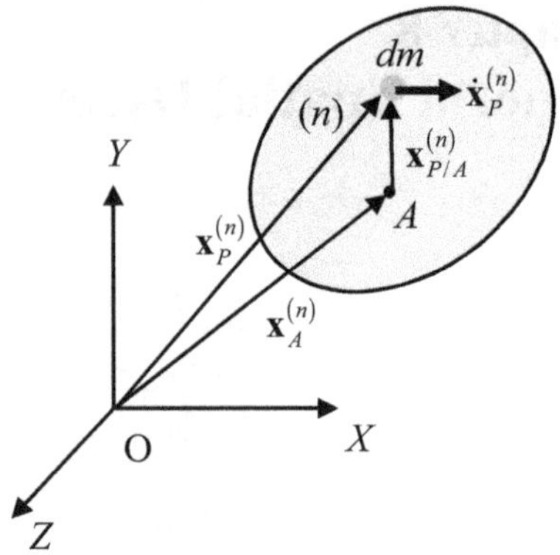

Fig. 8.1 Geometry of a rigid body for determining its angular momentum about a point A moving with the body

$$d\mathbf{L}_A^{(n)} = \mathbf{x}_{P/A}^{(n)} \times \dot{\mathbf{x}}_P^{(n)} \rho dV$$

so the total angular momentum is

$$\mathbf{L}_A^{(n)} = \int_{V_n} \rho \mathbf{x}_{P/A}^{(n)} \times \dot{\mathbf{x}}_P^{(n)} dV \tag{8.1.1}$$

Using the matrix-vector expression for the velocity developed in Chap. 4, Eq. (4.8.16a), and replacing the cross product in Eq. (8.1.1) with multiplication by a skew symmetric matrix, $\mathbf{X}_{P/A}^{(n)}$, we have

$$\mathbf{L}_A^{(n)} = \left(\int_{V_n} \rho \mathrm{X}_{P/A}^{(n)} dV \right) \mathbf{v}_A^{(n)} + \left(\int_{V_n} \rho \mathrm{X}_{P/A}^{(n)} \left(\mathrm{X}_{P/A}^{(n)} \right)^T dV \right) {}_I\boldsymbol{\omega}_B^{(n)} \tag{8.1.2}$$

The 3×3 matrix $\mathbf{I}_A^{(n)}$ given by

$$\mathbf{I}_A^{(n)} = \int_{V_n} \rho \mathrm{X}_{P/A}^{(n)} \left(\mathrm{X}_{P/A}^{(n)} \right)^T dV \tag{8.1.3}$$

is the moment of inertia matrix with components relative to the inertial (X, Y, Z) axes given by

$$\mathbf{I}_A^{(n)} = \begin{bmatrix} I_{XX}^{(n)} & I_{XY}^{(n)} & I_{XZ}^{(n)} \\ I_{YX}^{(n)} & I_{YY}^{(n)} & I_{YZ}^{(n)} \\ I_{ZX}^{(n)} & I_{ZY}^{(n)} & I_{ZZ}^{(n)} \end{bmatrix} \tag{8.1.4}$$

where

$$
\begin{aligned}
I_{XX}^{(n)} &= \int\limits_{V_n} \rho\left[\left(Y_{P/A}^{(n)}\right)^2 + \left(Z_{P/A}^{(n)}\right)^2\right] dV \\
I_{YY}^{(n)} &= \int\limits_{V_n} \rho\left[\left(X_{P/A}^{(n)}\right)^2 + \left(Z_{P/A}^{(n)}\right)^2\right] dV \\
I_{ZZ}^{(n)} &= \int\limits_{V_n} \rho\left[\left(X_{P/A}^{(n)}\right)^2 + \left(Y_{P/A}^{(n)}\right)^2\right] dV \\
I_{XY}^{(n)} &= I_{YX}^{(n)} = -\int\limits_{V_n} \rho X_{P/A}^{(n)} Y_{P/A}^{(n)} dV \\
I_{XZ}^{(n)} &= I_{ZX}^{(n)} = -\int\limits_{V_n} \rho X_{P/A}^{(n)} Z_{P/A}^{(n)} dV \\
I_{YZ}^{(n)} &= I_{ZY}^{(n)} = -\int\limits_{V_n} \rho Y_{P/A}^{(n)} Z_{P/A}^{(n)} dV
\end{aligned}
\tag{8.1.5}
$$

These moments of inertia integrals are the counterparts of the sums seen for a rigidly constrained system of particles in Chap. 3. Appendix B discusses moments and products of inertia and gives a table of their values for some simple shapes. In terms of the moment of inertia matrix the angular momentum expressions become

$$
\mathbf{L}_A^{(n)} = \left(\int\limits_{V_n} \rho \mathbf{X}_{P/A}^{(n)} dV\right) \mathbf{v}_A^{(n)} + \mathbf{I}_A^{(n)}\,{}_I\boldsymbol{\omega}_B^{(n)} = m^{(n)} \mathbf{x}_{C/A}^{(n)} \times \mathbf{v}_A^{(n)} + \mathbf{I}_A^{(n)}\,{}_I\boldsymbol{\omega}_B^{(n)} \tag{8.1.6}
$$

If point A is chosen to be the center of mass, C, then the cross product term vanishes so that the angular momentum is instead

$$
\mathbf{L}_C^{(n)} = \mathbf{I}_C^{(n)}\,{}_I\boldsymbol{\omega}_B^{(n)} \tag{8.1.7}
$$

Because the moment of inertia components are measured in a fixed inertial system their values will change during the motion. However, if we compute the angular momentum in body-fixed coordinates, $\bar{\mathbf{L}}_A^{(n)} = \left(\mathbf{R}^{(n)}\right)^T \mathbf{L}_A^{(n)}$, using Eq. (4.8.16b) we find

$$
\bar{\mathbf{L}}_A^{(n)} = \left(\int\limits_{V_n} \rho \overline{\mathbf{X}}_{P/A}^{(n)} dV\right) \bar{\mathbf{v}}_A^{(n)} + \left(\int\limits_{V_n} \rho \overline{\mathbf{X}}_{P/A}^{(n)} \left(\overline{\mathbf{X}}_{P/A}^{(n)}\right)^T dV\right) {}_I\overline{\boldsymbol{\omega}}_B^{(n)} \tag{8.1.8}
$$

where now the moment of inertia matrix $\bar{\mathbf{I}}_A^{(n)} = \int\limits_{V_n} \rho \overline{\mathbf{X}}_{P/A}^{(n)} \left(\overline{\mathbf{X}}_{P/A}^{(n)}\right)^T dV$ has the components

$$
\begin{aligned}
\bar{I}_{xx}^{(n)} &= \int\limits_{V_n} \rho\left[\left(\bar{y}_{P/A}^{(n)}\right)^2 + \left(\bar{z}_{P/A}^{(n)}\right)^2\right] dV \\
\bar{I}_{yy}^{(n)} &= \int\limits_{V_n} \rho\left[\left(\bar{x}_{P/A}^{(n)}\right)^2 + \left(\bar{z}_{P/A}^{(n)}\right)^2\right] dV \\
\bar{I}_{zz}^{(n)} &= \int\limits_{V_n} \rho\left[\left(\bar{x}_{P/A}^{(n)}\right)^2 + \left(\bar{y}_{P/A}^{(n)}\right)^2\right] dV \\
\bar{I}_{xy}^{(n)} = \bar{I}_{yx}^{(n)} &= -\int\limits_{V_n} \rho \bar{x}_{P/A}^{(n)} \bar{y}_{P/A}^{(n)} dV \\
\bar{I}_{xz}^{(n)} = \bar{I}_{zx}^{(n)} &= -\int\limits_{V_n} \rho \bar{x}_{P/A}^{(n)} \bar{z}_{P/A}^{(n)} dV \\
\bar{I}_{yz}^{(n)} = \bar{I}_{zy}^{(n)} &= -\int\limits_{V_n} \rho \bar{y}_{P/A}^{(n)} \bar{z}_{P/A}^{(n)} dV
\end{aligned}
\tag{8.1.9}
$$

which are all constants during the motion. When point A is again at the center of mass we have

$$
\bar{\mathbf{L}}_C^{(n)} = \bar{\mathbf{I}}_C^{(n)} \, {}_I\bar{\boldsymbol{\omega}}_B^{(n)} \tag{8.1.10}
$$

In Appendix B we discuss the parallel axis theorem for transforming the moment of inertia matrix with respect to a set of axes at the center of mass point to a set of parallel axes through a separate point. The parallel axis theorem is particularly easy to obtain in terms of the matrix-vector forms seen here. We let $\mathbf{x}_{P/A}^{(n)} = \mathbf{x}_{C/A}^{(n)} + \mathbf{x}_{P/C}^{(n)}$. We have

$$
\begin{aligned}
\mathbf{I}_A^{(n)} &= \int\limits_{V_n} \rho \mathbf{X}_{P/A}^{(n)} \left(\mathbf{X}_{P/A}^{(n)}\right)^T dV = \int\limits_{V_n} \rho\left[\mathbf{X}_{C/A}^{(n)} + \mathbf{X}_{P/C}^{(n)}\right]\left[\mathbf{X}_{C/A}^{(n)} + \mathbf{X}_{P/C}^{(n)}\right]^T dV \\
&= m^{(n)} \mathbf{X}_{C/A}^{(n)} \left(\mathbf{X}_{C/A}^{(n)}\right)^T + \mathbf{X}_{C/A}^{(n)} \int\limits_{V_n} \rho \left(\mathbf{X}_{P/C}^{(n)}\right)^T dV + \left(\int\limits_{V_n} \rho \mathbf{X}_{P/C}^{(n)} dV\right)\left(\mathbf{X}_{C/A}^{(n)}\right)^T \\
&\quad + \int\limits_{V_n} \rho \mathbf{X}_{P/C}^{(n)} \left(\mathbf{X}_{P/C}^{(n)}\right)^T dV
\end{aligned}
\tag{8.1.11}
$$

But the two middle terms vanish since the coordinates in those integrals are all measured from the center of mass and we recognize the last term as just the moment of inertia matrix with respect to the center of mass, so we have finally

$$\mathbf{I}_A^{(n)} = \mathbf{I}_C^n + m^{(n)}\mathbf{X}_{C/A}^{(n)}\left(\mathbf{X}_{C/A}^{(n)}\right)^T \tag{8.1.12}$$

which is just the parallel axis theorem of Appendix B in matrix form. The "shifting" matrix in Eq. (8.1.12) is just

$$\mathbf{X}_{C/A}^{(n)}\left(\mathbf{X}_{C/A}^{(n)}\right)^T = \begin{bmatrix} \left(y_{C/A}^{(n)}\right)^2 + \left(z_{C/A}^{(n)}\right)^2 & -x_{C/A}^{(n)}y_{C/A}^{(n)} & -x_{C/A}^{(n)}z_{C/A}^{(n)} \\ -x_{C/A}^{(n)}y_{C/A}^{(n)} & \left(x_{C/A}^{(n)}\right)^2 + \left(z_{C/A}^{(n)}\right)^2 & -y_{C/A}^{(n)}z_{C/A}^{(n)} \\ -x_{C/A}^{(n)}z_{C/A}^{(n)} & -y_{C/A}^{(n)}z_{C/A}^{(n)} & \left(x_{C/A}^{(n)}\right)^2 + \left(y_{C/A}^{(n)}\right)^2 \end{bmatrix} \tag{8.1.13}$$

which has the same components as described in Appendix B.

8.1.2 Angular Momentum About a General Point

If we want to compute the angular momentum about a general moving point B that may or may not be moving with the body (Fig. 8.2) then

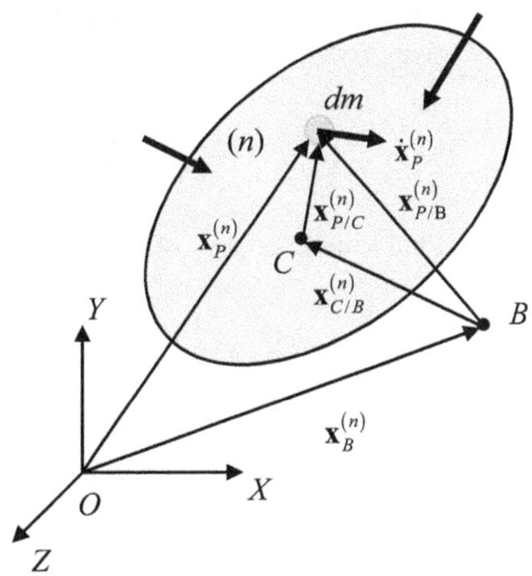

Fig. 8.2 Geometry of a body and an arbitrary moving reference point B

$$\mathbf{L}_B^{(n)} = \int_{V_n} \rho \mathbf{x}_{P/B}^{(n)} \times \dot{\mathbf{x}}_P^{(n)} dV \tag{8.1.14}$$

and we can let $\mathbf{x}_{P/B}^{(n)} = \mathbf{x}_{C/B}^{(n)} + \mathbf{x}_{P/C}^{(n)}$ (see Fig. 8.2) which gives

$$\mathbf{L}_B^{(n)} = \mathbf{x}_{C/B}^{(n)} \times \int_{V_n} \rho \dot{\mathbf{x}}_P^{(n)} dV + \int_{V_n} \rho \mathbf{x}_{P/C}^{(n)} \times \dot{\mathbf{x}}_P^{(n)} dV \tag{8.1.15}$$

The first integral is $m^{(n)}\dot{\mathbf{x}}_C^{(n)}$, a linear momentum term, and the second integral is just the angular momentum about the center of mass point, C, so

$$\mathbf{L}_B^{(n)} = m^{(n)}\mathbf{x}_{C/B}^{(n)} \times \dot{\mathbf{x}}_C^{(n)} + \mathbf{L}_C^{(n)} \tag{8.1.16}$$

Note that Eq. (8.1.16) is a general relationship since point B can be any moving point and the body need not be a rigid body. It is the generalization, for a continuous body, of the angular momentum expression for a system of particles, Eq. (3.2.21), in Chap. 3.

8.2 Kinetic Energy

When using Lagrange's equation to obtain the equations of motion we need to obtain general expressions for the kinetic energy of a rigid body as the kinetic energy is one of the key parameters that generate the equations of motion in generalized coordinates. In this section we derive the kinetic energy expressions.

From our discussion of particle motion we know that if an element of mass $dm = \rho dV$ of a continuous rigid body is treated as a particle it has a kinetic energy, $dT^{(n)}$, given by

$$dT^{(n)} = \frac{1}{2} dm \mathbf{v}_P^{(n)} \cdot \mathbf{v}_P^{(n)}$$

so the total kinetic energy of the body is

$$T^{(n)} = \frac{1}{2} \int_{V_n} \rho \mathbf{v}_P^{(n)} \cdot \mathbf{v}_P^{(n)} dV = \frac{1}{2} \int_{V_n} \rho \left(\mathbf{v}_P^{(n)}\right)^T \mathbf{v}_P^{(n)} dV \tag{8.2.1}$$

Using the relative velocity expression of Eq. (4.8.16a) we have

$$T^{(n)} = \frac{1}{2}\int_{V_n} \rho\Big(\left(\mathbf{v}_A^{(n)}\right)^T \mathbf{v}_A^{(n)} - \left(\mathbf{v}_A^{(n)}\right)^T \mathbf{X}_{P/A}^{(n)} {}_I\boldsymbol{\omega}_B^{(n)} - \left({}_I\boldsymbol{\omega}_B^{(n)}\right)^T \left(\mathbf{X}_{P/A}^{(n)}\right)^T \mathbf{v}_A^{(n)} \\ + \left({}_I\boldsymbol{\omega}_B^{(n)}\right)^T \left(\mathbf{X}_{P/A}^{(n)}\right)^T \mathbf{X}_{P/A}^{(n)} {}_I\boldsymbol{\omega}_B^{(n)}\Big) dV \tag{8.2.2}$$

which reduces to

$$T^{(n)} = \frac{1}{2} m^{(n)} \left(\mathbf{v}_A^{(n)}\right)^T \mathbf{v}_A^{(n)} - \frac{1}{2}\left(\mathbf{v}_A^{(n)}\right)^T \left(\int_{V_n} \rho \mathbf{X}_{P/A}^{(n)} dV\right) {}_I\boldsymbol{\omega}_B^{(n)} \\ - \frac{1}{2}\left({}_I\boldsymbol{\omega}_B^{(n)}\right)^T \left(\int_{V_n} \rho \left(\mathbf{X}_{P/A}^{(n)}\right)^T dV\right) \mathbf{v}_A^{(n)} + \frac{1}{2}\left({}_I\boldsymbol{\omega}_B^{(n)}\right)^T \left(\int_{V_n} \rho \left(\mathbf{X}_{P/A}^{(n)}\right)^T \mathbf{X}_{P/A}^{(n)} dV\right) {}_I\boldsymbol{\omega}_B^{(n)} \tag{8.2.3}$$

where we can recognize

$$\mathbf{X}_{C/A}^{(n)} = \frac{1}{m^{(n)}} \int_{V_n} \rho \mathbf{X}_{P/A}^{(n)} dV, \quad \mathbf{I}_A^{(n)} = \int_{V_n} \rho \left(\mathbf{X}_{P/A}^{(n)}\right)^T \mathbf{X}_{P/A}^{(n)} dV \tag{8.2.4}$$

where C is the center of mass and

$$\mathbf{X}_{C/A}^{(n)} = \begin{bmatrix} 0 & -Z_{C/A}^{(n)} & Y_{C/A}^{(n)} \\ Z_{C/A}^{(n)} & 0 & -X_{C/A}^{(n)} \\ -Y_{C/A}^{(n)} & X_{C/A}^{(n)} & 0 \end{bmatrix} \tag{8.2.5}$$

so that

$$T^{(n)} = \frac{1}{2} m^{(n)} \left(\mathbf{v}_A^{(n)}\right)^T \mathbf{v}_A^{(n)} - \frac{1}{2} m^{(n)} \left(\mathbf{v}_A^{(n)}\right)^T \mathbf{X}_{C/A}^{(n)} {}_I\boldsymbol{\omega}_B^{(n)} - \frac{1}{2} m^{(n)} \left({}_I\boldsymbol{\omega}_B^{(n)}\right)^T \left(\mathbf{X}_{C/A}^{(n)}\right)^T \mathbf{v}_A^{(n)} \\ + \frac{1}{2}\left({}_I\boldsymbol{\omega}_B^{(n)}\right)^T \mathbf{I}_A^{(n)} {}_I\boldsymbol{\omega}_B^{(n)} \tag{8.2.6}$$

If we combine the velocity of point A and the angular velocity into a single-column vector $\dot{\mathbf{q}}^{(n)} = \left[\mathbf{v}_A^{(n)}, \ {}_I\boldsymbol{\omega}_B^{(n)}\right]^T$ then we can write the total kinetic energy as

$$T^{(n)} = \frac{1}{2}\left(\dot{\mathbf{q}}^{(n)}\right)^T \mathbb{M}^{(n)} \dot{\mathbf{q}}^{(n)} \tag{8.2.7}$$

in terms of a "mass" matrix, $\mathbb{M}^{(n)}$, where

$$\mathbb{M}^{(n)} = \begin{bmatrix} m^{(n)}\mathbf{1} & -m^{(n)}\mathbf{X}_{C/A}^{(n)} \\ -m^{(n)}\left(\mathbf{X}_{C/A}^{(n)}\right)^T & \mathbf{I}_A^{(n)} \end{bmatrix} \tag{8.2.8}$$

where $\mathbf{1}$ is the 3×3 identity matrix. However, note that $\dot{\mathbf{q}}^{(n)}$ is not really a generalized velocity because the angular velocity vector is not the derivative of a corresponding rotational vector but instead is related to derivatives of the rotation matrix or derivatives of angular parameters such as the Euler angles. Thus, when we use Euler angles or other angular variables such as Euler parameters the associated mass matrix will be different from Eq. (8.2.8).

As with the angular momentum, if we take point A to be the center of mass point, C, the kinetic energy expression simplifies. We have

$$T^{(n)} = \frac{1}{2}m^{(n)}\left(\mathbf{v}_C^{(n)}\right)^T\mathbf{v}_C^{(n)} + \frac{1}{2}\left({}_I\boldsymbol{\omega}_B^{(n)}\right)^T\mathbf{I}_C^{(n)}\,{}_I\boldsymbol{\omega}_B^{(n)} \tag{8.2.9}$$

so the kinetic energy is the kinetic energy due to the translation of the body of total mass m with the velocity of the center of mass plus the rotational kinetic energy about the center of mass. The mass matrix then reduces to

$$\mathbb{M}^{(n)} = \begin{bmatrix} m^{(n)}\mathbf{1} & \mathbf{0} \\ \mathbf{0} & \mathbf{I}_C^{(n)} \end{bmatrix} \tag{8.2.10}$$

8.3 Impulse-Momentum and Work-Energy Principles for a Rigid Body

As we will see in the next section the Newton-Euler equations for a continuous rigid body can be written as

$$\begin{aligned} \dot{\mathbf{p}}_C^{(n)} &= \mathbf{F}^{(n)} \\ \dot{\mathbf{L}}_C^{(n)} &= \mathbf{M}_C^{(n)} \end{aligned} \tag{8.3.1}$$

where $\mathbf{p}_C^{(n)} = m^{(n)}\dot{\mathbf{x}}_C^{(n)}$ is a linear momentum associated with the motion of the center of mass and $\mathbf{L}_C^{(n)}$ is the angular momentum with respect to the center of mass. These are identical to the equations of motion found for a system of rigidly constrained particles so that the impulse-momentum principles are also identical, giving

$$\begin{aligned}\Delta \mathbf{p}_C^{(n)} &= \mathbf{p}_C^{(n)}(t_2) - \mathbf{p}_C^{(n)}(t_1) = \int_{t_1}^{t_2} \mathbf{F}^{(n)}(t)dt \\ \Delta \mathbf{L}_C^{(n)} &= \mathbf{L}_C^{(n)}(t_2) - \mathbf{L}_C^{(n)}(t_1) = \int_{t_1}^{t_2} \mathbf{M}_C^{(n)}(t)dt\end{aligned} \tag{8.3.2}$$

These momentum principles are useful for dealing with the collisions of rigid bodies of the same type as discussed in Chap. 3 for systems of particles and in analyzing systems where certain momenta are conserved. For rigid bodies, conservation of generalized momenta, as discussed in Chap. 2 for particles, also is useful for analyzing rigid body motions. Recall that Lagrange's equations for a problem where all the forces were derivable from a potential are given by

$$\frac{d}{dt}\left(\frac{\partial L}{\partial \dot{q}_k}\right) - \frac{\partial L}{\partial q_k} = 0 \tag{8.3.3}$$

so that if a specific generalized coordinate, q_K, is not present in a problem then we have

$$\frac{d}{dt}\left(\frac{\partial L}{\partial \dot{q}_K}\right) = 0 \tag{8.3.4}$$

and the generalized momenta, p_K, associated with that coordinate, defined as

$$p_K = \frac{\partial L}{\partial \dot{q}_K} = \frac{\partial T}{\partial \dot{q}_K} \tag{8.3.5}$$

are conserved. We can replace the Lagrangian, $L = T - V$, with the kinetic energy, T, in Eq. (8.3.5) since the potential energy is typically not dependent on the generalized velocities. When conservation of momenta exists in rigid body problems this information can be used to analyze and understand the nature of the motion.

Work-energy principles for the three-dimensional motion of a rigid body are also closely related to those principles already discussed for systems of particles and for rigid bodies in plane motion. For example, consider a rigid body acted upon by a set of N forces, as shown in Fig. 8.3. The differential work, $dW^{(n)}$, done by these forces is

$$dW^{(n)} = \sum_{j=1}^{N} \mathbf{F}_j^{(n)} \cdot d\mathbf{x}_j^{(n)} \tag{8.3.6}$$

or, in terms of the rate of work (power),

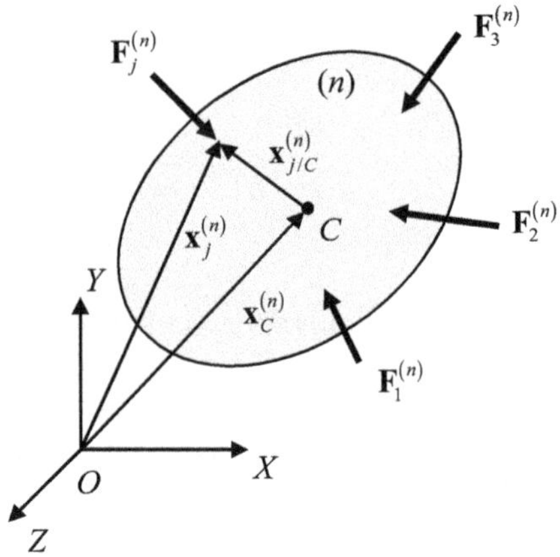

Fig. 8.3 A rigid body acted upon by a set of forces

$$\frac{dW^{(n)}}{dt} = \sum_{j=1}^{N} \mathbf{F}_j^{(n)} \cdot \mathbf{v}_j^{(n)} \tag{8.3.7}$$

But we can relate the velocities at where the forces act to the velocity of the center of mass through the relative velocity equation for the rigid body, i.e.,

$$\mathbf{v}_j^{(n)} = \mathbf{v}_C^{(n)} + \boldsymbol{\omega}^{(n)} \times \mathbf{x}_{j/C}^{(n)} \tag{8.3.8}$$

Placing that velocity relationship into Eq. (8.3.7) then gives

$$\begin{aligned} \frac{dW^{(n)}}{dt} &= \sum_{j=1}^{N} \mathbf{F}_j^{(n)} \cdot \mathbf{v}_C^{(n)} + \sum_{j=1}^{N} \mathbf{F}_j^{(n)} \cdot \left(\boldsymbol{\omega}^{(n)} \times \mathbf{x}_{j/C}^{(n)}\right) \\ &= \mathbf{F}^{(n)} \cdot \mathbf{v}_C^{(n)} + \sum_{j=1}^{N} \boldsymbol{\omega}^{(n)} \cdot \left(\mathbf{x}_{j/C}^{(n)} \times \mathbf{F}_j^{(n)}\right) \\ &= \mathbf{F}^{(n)} \cdot \mathbf{v}_C^{(n)} + \mathbf{M}_C^{(n)} \cdot \boldsymbol{\omega}^{(n)} \end{aligned} \tag{8.3.9}$$

where use was made of the vector identity $\mathbf{a} \cdot (\mathbf{b} \times \mathbf{c}) = \mathbf{b} \cdot (\mathbf{c} \times \mathbf{a})$ and we define the net external moment with respect to the center of mass, acting on the rigid body, as

$$\mathbf{M}_C^{(n)} = \sum_{j=1}^{N} \mathbf{x}_{j/C}^{(n)} \times \mathbf{F}_j^{(n)} \tag{8.3.10}$$

When the forces acting on the rigid body are derivable from a potential, then

$$\frac{dW^{(n)}}{dt} = -\sum_{j=1}^{N} \nabla V_j^{(n)} \cdot \frac{d\mathbf{x}_j^{(n)}}{dt} = -\sum_{j=1}^{N} \frac{dV_j^{(n)}}{dt} = -\frac{dV^{(n)}}{dt} \tag{8.3.11}$$

where $V^{(n)}$ is the total potential energy of all the forces. The kinetic energy expression can also be written as (see Eq. (8.2.9))

$$T = \frac{1}{2} m^{(n)} \mathbf{v}_C^{(n)} \cdot \mathbf{v}_C^{(n)} + \frac{1}{2} \boldsymbol{\omega}^{(n)} \cdot \mathbf{I}_C^{(n)} \boldsymbol{\omega}^{(n)} \tag{8.3.12}$$

Differentiating the translational kinetic energy term on time is easy. For the rotational part we can write the term in principal body-fixed coordinates, where the moment of inertia matrix is a constant diagonal matrix and we find

$$\begin{aligned}
&\frac{1}{2}\frac{d}{dt}\left(\bar{I}_{C1}^{(n)}\left(\bar{\omega}_1^{(n)}\right)^2 + \bar{I}_{C2}^{(n)}\left(\bar{\omega}_2^{(n)}\right)^2 + \bar{I}_{C3}^{(n)}\left(\bar{\omega}_3^{(n)}\right)^2\right) \\
&= \bar{I}_{C1}^{(n)}\bar{\omega}_1^{(n)}\dot{\bar{\omega}}_1^{(n)} + \bar{I}_{C2}^{(n)}\bar{\omega}_2^{(n)}\dot{\bar{\omega}}_2^{(n)} + \bar{I}_{C2}^{(n)}\bar{\omega}_3^{(n)}\dot{\bar{\omega}}_3^{(n)} = \bar{\boldsymbol{\omega}}^{(n)} \cdot \dot{\bar{\mathbf{L}}}_C^{(n)} \\
&= \boldsymbol{\omega}^{(n)} \cdot \dot{\mathbf{L}}_C^{(n)}
\end{aligned} \tag{8.3.13}$$

a result which is valid in any coordinate system since it is a scalar. Thus,

$$\begin{aligned}
\frac{dT^{(n)}}{dt} &= m^{(n)} \mathbf{v}_C^{(n)} \cdot \dot{\mathbf{v}}_C^{(n)} + \boldsymbol{\omega}^{(n)} \cdot \dot{\mathbf{L}}_C^{(n)} \\
&= \mathbf{v}_C^{(n)} \cdot \mathbf{F}^{(n)} + \boldsymbol{\omega}^{(n)} \cdot \mathbf{M}_C^{(n)} \\
&= \frac{dW^{(n)}}{dt}
\end{aligned} \tag{8.3.14}$$

showing that the time rate of change of the work done (power) is equal to the time rate of change of the kinetic energy of the rigid body. When the forces are derivable from a potential it then follows that

$$\frac{dE^{(n)}}{dt} = \frac{d\left(T^{(n)} + V^{(n)}\right)}{dt} = 0 \tag{8.3.15}$$

i.e., the total energy, $E^{(n)} = T^{(n)} + V^{(n)}$, is conserved.

8.4 Newton-Euler Equations of Motion

8.4.1 Governing Equations: General Case

We will first derive the governing equations for the case where the moment equation is taken with respect to a general point B. Point B can be moving and it is not necessarily attached to the body (see Fig. 8.2). Following the discussion in Chap. 5, for a continuous body we have

$$\begin{aligned}&\int_{V_n} \rho \ddot{\mathbf{x}}_P^{(n)} dV = \mathbf{F}^{(n)}\\&\int_{V_n} \rho \left[\mathbf{x}_{P/\mathrm{B}}^{(n)} \times \left(\ddot{\mathbf{x}}_P^{(n)}\right)\right] dV = \mathbf{M}_B^{(n)}\end{aligned} \tag{8.4.1}$$

where $\ddot{\mathbf{x}}_P^{(n)}$ is the acceleration (in an inertial system) of an arbitrary point P in the body, $\mathbf{x}_{P/\mathrm{B}}^{(n)}$ is the position of P relative to the reference point B, and $\mathbf{F}^{(n)}$ and $\mathbf{M}_B^{(n)}$ are the net external force and moment about point B acting on the body. For the force equation we will relate the motion of point P to the motion of the center of mass, C, through $\mathbf{x}_P^{(n)} = \mathbf{x}_C^{(n)} + \mathbf{x}_{P/C}^{(n)}$, giving

$$\ddot{\mathbf{x}}_C^{(n)} \int_{V_n} \rho dV + \int_{V_n} \rho \ddot{\mathbf{x}}_{P/C}^{(n)} dV = \mathbf{F}^{(n)} \tag{8.4.2}$$

But the second integral on the left side of Eq. (8.2.2) vanishes since the integral of the relative position vector $\mathbf{x}_{P/C}^{(n)}$ or any of its derivatives vanishes (see the corresponding case for the system of particles treated in Chap. 3, Eq. (3.2.18)) so we have

$$m^{(n)} \ddot{\mathbf{x}}_C^{(n)} = \mathbf{F}^{(n)} \tag{8.4.3}$$

which is just the equation of motion of the center of mass. Now, consider the moment equation in Eq. (8.4.1) and let $\mathbf{x}_{P/B}^{(n)} = \mathbf{x}_{C/B}^{(n)} + \mathbf{x}_{P/C}^{(n)}$. Then

$$\mathbf{M}_B^{(n)} = \mathbf{x}_{C/B}^{(n)} \times \int_{V_n} \rho \ddot{\mathbf{x}}_P^{(n)} dV + \int_{V_n} \rho \mathbf{x}_{P/C}^{(n)} \times \ddot{\mathbf{x}}_P^{(n)} dV \tag{8.4.4}$$

The first integral in Eq. (8.4.4) is just the mass times the acceleration of the center of mass and the second integral can be expressed as the sum of two integrals, i.e.,

$$\mathbf{M}_B^{(n)} = m^{(n)} \mathbf{x}_{C/B}^{(n)} \times \ddot{\mathbf{x}}_C^{(n)} + \frac{d}{dt}\left(\int_{V_n} \rho \mathbf{x}_{P/C}^{(n)} \times \dot{\mathbf{x}}_P^{(n)} dV\right) - \int_{V_n} \rho \dot{\mathbf{x}}_{P/C}^{(n)} \times \dot{\mathbf{x}}_P^{(n)} dV \tag{8.4.5}$$

The first integral in Eq. (8.4.5) can be recognized as the derivative of the angular momentum of the body with respect to the center of mass. Letting

$\dot{\mathbf{x}}_{P}^{(n)} = \dot{\mathbf{x}}_{C}^{(n)} + \dot{\mathbf{x}}_{P/C}^{(n)}$ in the second integral it is just the sum of two integrals that are both zero so

$$\mathbf{M}_B^{(n)} = m^{(n)} \mathbf{x}_{C/B}^{(n)} \times \ddot{\mathbf{x}}_C^{(n)} + \dot{\mathbf{L}}_C^{(n)} \tag{8.4.6}$$

If we differentiate Eq. (8.1.16), which expresses the angular momentum with respect to the center of mass to the angular momentum about a general point, we have

$$\dot{\mathbf{L}}_B^{(n)} = \dot{\mathbf{L}}_C^{(n)} + m^{(n)} \mathbf{x}_{C/B}^{(n)} \times \ddot{\mathbf{x}}_C^{(n)} + m^{(n)} \dot{\mathbf{x}}_{C/B}^{(n)} \times \dot{\mathbf{x}}_C^{(n)} \tag{8.4.7}$$

so that using Eq. (8.4.7) in Eq. (8.4.6) we find

$$\begin{aligned} \mathbf{M}_B^{(n)} &= \dot{\mathbf{L}}_B^{(n)} - m^{(n)} \dot{\mathbf{x}}_{C/B}^{(n)} \times \dot{\mathbf{x}}_C^{(n)} \\ &= \dot{\mathbf{L}}_B^{(n)} + m^{(n)} \dot{\mathbf{x}}_B^{(n)} \times \dot{\mathbf{x}}_C^{(n)} \end{aligned} \tag{8.4.8}$$

which is the generalization of the corresponding moment equation found in Chap. 5 for planar problems, Eq. (5.5.18b). Together, Eqs. (8.4.3) and (8.4.8) are the governing equations for the most general case of a continuous body, even one that is not rigid, and for a general reference point *B* used in the moment equation. One can also use Eq. (8.4.6) for the moment equation if one wants to relate the moments about point *B* to changes of the angular momentum about the center of mass.

8.4.2 Governing Equations for a Rigid Body: Use of a Body-Fixed Point

Now, consider the equations of motion for a continuous rigid body labeled (n) as shown in Fig. 8.4 and where the moments are taken with respect to a point *A* that moves with the rigid body. In this case Eq. (8.4.1) becomes

$$\begin{aligned} &\int_{V_n} \rho \ddot{\mathbf{x}}_P^{(n)} dV = \mathbf{F}^{(n)} \\ &\int_{V_n} \rho \left[\mathbf{x}_{P/A}^{(n)} \times \left(\ddot{\mathbf{x}}_P^{(n)} \right) \right] dV = \mathbf{M}_A^{(n)} \end{aligned} \tag{8.4.9}$$

where $\ddot{\mathbf{x}}_P^{(n)}$ is the acceleration (in an inertial system) of an arbitrary point *P* in the rigid body, $\mathbf{x}_{P/A}^{(n)}$ is the position of *P* relative to a reference point *A* that moves with the rigid body, and $\mathbf{F}^{(n)}$ and $\mathbf{M}_A^{(n)}$ are the net external force and moment about point *A* acting

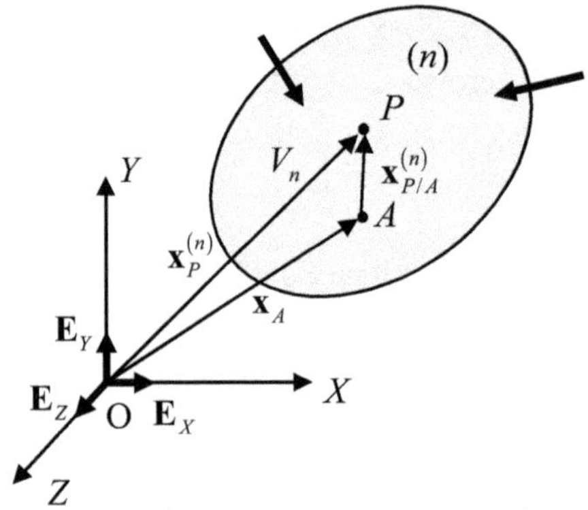

Fig. 8.4 The geometry of a rigid body. Point P is an arbitrary point in the body and point A is a reference point moving with the body

on the rigid body. Let's consider the force equation first. Placing Eq. (4.9.8) (the acceleration relationship between points in a rigid body) into that equation we find

$$\ddot{\mathbf{x}}_A^{(n)} \int\limits_{V_n} \rho dV - \left(\int\limits_{V_n} \rho \mathbf{X}_{P/A}^{(n)} dV \right) {}_I\boldsymbol{\alpha}_B^{(n)} + {}_I\boldsymbol{\Omega}_B^{(n)}\, {}_I\boldsymbol{\Omega}_B^{(n)} \int\limits_{V_n} \rho \mathbf{x}_{P/A}^{(n)} dV = \mathbf{F}^{(n)} \tag{8.4.10}$$

in terms of inertial coordinates which can be written more compactly as

$$m^{(n)}\mathbf{a}_A^{(n)} - m^{(n)}\mathbf{X}_{C/A}^{(n)}\, {}_I\boldsymbol{\alpha}_B^{(n)} + m^{(n)}\, {}_I\boldsymbol{\Omega}_B^{(n)}\, {}_I\boldsymbol{\Omega}_B^{(n)}\mathbf{x}_{C/A}^{(n)} = \mathbf{F}^{(n)} \tag{8.4.11}$$

where $\mathbf{a}_A^{(n)} = \ddot{\mathbf{x}}_A^{(n)}$ is the acceleration of A as measured in the inertial system. You may have seen this equation written in a purely vector form. Replacing the skew symmetric matrices with cross products we have

$$m^{(n)}\mathbf{a}_A^{(n)} + m^{(n)}\, {}_I\boldsymbol{\alpha}_B^{(n)} \times \mathbf{x}_{C/A}^{(n)} + m^{(n)}\, {}_I\boldsymbol{\omega}_B^{(n)} \times \left({}_I\boldsymbol{\omega}_B^{(n)} \times \mathbf{x}_{C/A}^{(n)} \right) = \mathbf{F}^{(n)} \tag{8.4.12}$$

For point A at the center of mass point C all of these forms reduce, as before, to simply

$$m^{(n)}\mathbf{a}_C^{(n)} = \mathbf{F}^{(n)} \tag{8.4.13}$$

Now, consider the moment equation. From the relative acceleration equation, Eq. (4.9.8), and the moment equation in Eq. (8.4.9) we have

$$\left(\int_{V_n}\rho\mathbf{x}_{P/A}^{(n)}dV\right)\times\ddot{\mathbf{x}}_A^{(n)}-\int_{V_n}\rho\mathbf{x}_{P/A}^{(n)}\times\mathbf{X}_{P/A}^{(n)}{}_I\boldsymbol{\alpha}_B^{(n)}dV+\int_{V_n}\rho\mathbf{x}_{P/A}^{(n)}$$
$$\times\,{}_I\boldsymbol{\Omega}_B^{(n)}{}_I\boldsymbol{\Omega}_B^{(n)}\mathbf{x}_{P/A}^{(n)}dV=\mathbf{M}_A^{(n)} \tag{8.4.14}$$

which can be written as

$$m^{(n)}\mathbf{x}_{C/A}^{(n)}\times\ddot{\mathbf{x}}_A^{(n)}+\left(\int_{V_n}\rho\left(\mathbf{X}_{P/A}^{(n)}\right)^T\mathbf{X}_{P/A}^{(n)}dV\right){}_I\boldsymbol{\alpha}_B^{(n)}+\int_{V_n}\rho\mathbf{x}_{P/A}^{(n)}$$
$$\times\,{}_I\boldsymbol{\Omega}_B^{(n)}{}_I\boldsymbol{\Omega}_B^{(n)}\mathbf{x}_{P/A}^{(n)}dV=\mathbf{M}_A^{(n)} \tag{8.4.15}$$

Consider the integrand of third term on the left side of Eq. (8.4.15). It can be written in purely vector terms as

$$J=\rho\mathbf{x}_{P/A}^{(n)}\times\left({}_I\boldsymbol{\omega}_B^{(n)}\times\left({}_I\boldsymbol{\omega}_B^{(n)}\times\mathbf{x}_{P/A}^{(n)}\right)\right)$$

But we have the general vector relationship $\mathbf{a}\times(\mathbf{b}\times\mathbf{c})=(\mathbf{a}\cdot\mathbf{c})\mathbf{b}-(\mathbf{a}\cdot\mathbf{b})\mathbf{c}$ so

$$\begin{aligned}J&=\rho\left[\left\{\mathbf{x}_{P/A}^{(n)}\cdot\left({}_I\boldsymbol{\omega}_B^{(n)}\times\mathbf{x}_{P/A}^{(n)}\right)\right\}{}_I\boldsymbol{\omega}_B^{(n)}-\left\{\mathbf{x}_{P/A}^{(n)}\cdot{}_I\boldsymbol{\omega}_B^{(n)}\right\}{}_I\boldsymbol{\omega}_B^{(n)}\times\mathbf{x}_{P/A}^{(n)}\right]\\&=-\rho\left[\left\{\mathbf{x}_{P/A}^{(n)}\cdot{}_I\boldsymbol{\omega}_B^{(n)}\right\}{}_I\boldsymbol{\omega}_B^{(n)}\times\mathbf{x}_{P/A}^{(n)}\right]\end{aligned}$$

since $\mathbf{x}_{P/A}^{(n)}\cdot\left({}_I\boldsymbol{\omega}_B^{(n)}\times\mathbf{x}_{P/A}^{(n)}\right)=0$. Now, consider the triple cross product, K, given below, evaluated in the same fashion:

$$\begin{aligned}K&=\rho_I\boldsymbol{\omega}_B^{(n)}\times\left(\mathbf{x}_{P/A}^{(n)}\times\left(\mathbf{x}_{P/A}^{(n)}\times{}_I\boldsymbol{\omega}_B^{(n)}\right)\right)\\&=\rho\left[\left\{\mathbf{x}_{P/A}^{(n)}\cdot\left(\mathbf{x}_{P/A}^{(n)}\times{}_I\boldsymbol{\omega}_B^{(n)}\right)\right\}x_{P/A}^{(n)}-\left\{{}_I\boldsymbol{\omega}_B^{(n)}\cdot\mathbf{x}_{P/A}^{(n)}\right\}\mathbf{x}_{P/A}^{(n)}\times{}_I\boldsymbol{\omega}_B^{(n)}\right]\\&=\rho\left[\left\{{}_I\boldsymbol{\omega}_B^{(n)}\cdot\mathbf{x}_{P/A}^{(n)}\right\}{}_I\boldsymbol{\omega}_B^{(n)}\times\mathbf{x}_{P/A}^{(n)}\right]\end{aligned}$$

Thus we see $K=\ -J$ so that we can rewrite the integrand as

$$\begin{aligned}J&=-\rho_I\boldsymbol{\omega}_B^{(n)}\times\left(\mathbf{x}_{P/A}^{(n)}\times\left(\mathbf{x}_{P/A}^{(n)}\times{}_I\boldsymbol{\omega}_B^{(n)}\right)\right)\\&=\rho\,{}_I\boldsymbol{\omega}_B^{(n)}\times\left(\mathbf{X}_{P/A}^{(n)}\right)^T\left(\mathbf{X}_{P/A}^{(n)}\right)\boldsymbol{\omega}_B^{(n)}\end{aligned}$$

and Eq. (8.4.15) becomes

$$m^{(n)}\mathbf{x}_{C/A}^{(n)} \times \ddot{\mathbf{x}}_A^{(n)} + \left(\int_{V_n} \rho \left(\mathbf{X}_{P/A}^{(n)} \right)^T \mathbf{X}_{P/A}^{(n)} dV \right) {}_I\boldsymbol{\alpha}_B^{(n)}$$
$$+ {}_I\boldsymbol{\omega}_B^{(n)} \times \left(\int_{V_n} \rho \left(\mathbf{X}_{P/A}^{(n)} \right)^T \mathbf{X}_{P/A}^{(n)} dV \right) {}_I\boldsymbol{\omega}_B^{(n)} = \mathbf{M}_A^{(n)} \tag{8.4.16}$$

where we can recognize the integrals as moment of inertia matrices, $\mathbf{I}_A^{(n)}$, giving the moment equation as

$$m^{(n)}\mathbf{x}_{C/A}^{(n)} \times \mathbf{a}_A^{(n)} + \mathbf{I}_A^{(n)}{}_I\boldsymbol{\alpha}_B^{(n)} + {}_I\boldsymbol{\omega}_B^{(n)} \times \left(\mathbf{I}_A^{(n)}{}_I\boldsymbol{\omega}_B^{(n)} \right) = \mathbf{M}_A^{(n)} \tag{8.4.17a}$$

or, in matrix-vector form,

$$m^{(n)}\mathbf{X}_{C/A}^{(n)}\mathbf{a}_A^{(n)} + \mathbf{I}_A^{(n)}{}_I\boldsymbol{\alpha}_B^{(n)} + {}_I\boldsymbol{\Omega}_B^{(n)}\mathbf{I}_A^{(n)}{}_I\boldsymbol{\omega}_B^{(n)} = \mathbf{M}_A^{(n)} \tag{8.4.17b}$$

Equations (8.4.17a) and (8.4.17b) are moment equations about any point A that moves with the rigid body. However, the moment equation is seldom used in this form in terms of inertial coordinates because the moment of inertia matrix is then time dependent. Instead, if the moment equation is written in body-fixed coordinates we have the equivalent moment equations

$$m^{(n)}\bar{\mathbf{x}}_{C/A}^{(n)} \times \bar{\mathbf{a}}_A^{(n)} + \bar{\mathbf{I}}_A^{(n)}{}_I\bar{\boldsymbol{\alpha}}_B^{(n)} + {}_I\bar{\boldsymbol{\omega}}_B^{(n)} \times \left(\bar{\mathbf{I}}_A^{(n)}{}_I\bar{\boldsymbol{\omega}}_B^{(n)} \right) = \bar{\mathbf{M}}_A^{(n)} \tag{8.4.18a}$$

and in terms of skew symmetric matrices

$$m^{(n)}\bar{\mathbf{X}}_{C/A}^{(n)}\bar{\mathbf{a}}_A^{(n)} + \bar{\mathbf{I}}_A^{(n)}{}_I\bar{\boldsymbol{\alpha}}_B^{(n)} + {}_I\bar{\boldsymbol{\Omega}}_B^{(n)}\bar{\mathbf{I}}_A^{(n)}{}_I\bar{\boldsymbol{\omega}}_B^{(n)} = \bar{\mathbf{M}}_A^{(n)} \tag{8.4.18b}$$

Using moments about a general point moving with the rigid body makes the moment equation particularly complex. In many applications choosing A to be the center of mass point C is convenient, where Eq. (8.4.18b) becomes

$$\bar{\mathbf{I}}_C^{(n)}{}_I\bar{\boldsymbol{\alpha}}_B^{(n)} + {}_I\bar{\boldsymbol{\Omega}}_B^{(n)}\bar{\mathbf{I}}_C^{(n)}{}_I\bar{\boldsymbol{\omega}}_B^{(n)} = \bar{\mathbf{M}}_C^{(n)} \tag{8.4.19}$$

which can be solved in conjunction with the force equation, Eq. (8.4.13).In other cases choosing point A to be a fixed point O about which the rigid body is rotating is a good choice, leading to the moment equation

$$\bar{\mathbf{I}}_O^{(n)}{}_I\bar{\boldsymbol{\alpha}}_B^{(n)} + {}_I\bar{\boldsymbol{\Omega}}_B^{(n)}\bar{\mathbf{I}}_O^{(n)}{}_I\bar{\boldsymbol{\omega}}_B^{(n)} = \bar{\mathbf{M}}_O^{(n)} \tag{8.4.20}$$

which can again be solved in conjunction with Eq. (8.4.13). Thus, when A is either the center of mass or a fixed point about which the body rotates, the moment equation is

$$\bar{\mathbf{I}}_P^{(n)}\,{}_I\bar{\boldsymbol{\alpha}}_B^{(n)} + {}_I\bar{\boldsymbol{\Omega}}_B^{(n)}\bar{\mathbf{I}}_P^{(n)}\,{}_I\bar{\boldsymbol{\omega}}_B^{(n)} = \bar{\mathbf{M}}_P^{(n)} \quad (P = C, O) \tag{8.4.21}$$

Because of the presence of the inertia matrix, the terms in Eq. (8.4.21) are still complicated. As a consequence, many applications choose the body-fixed axes to be principal axes of the rigid body where the moment of inertia matrix is diagonal. If we let the moments of inertia about a set of principal body-fixed axes $(\bar{x}_1, \bar{x}_2, \bar{x}_3)$ be $\left(\bar{I}_{P;1}^{(n)}, \bar{I}_{P;2}^{(n)}, \bar{I}_{P;3}^{(n)}\right)$ and let the corresponding body-fixed angular velocity and angular acceleration components be $\left(\bar{\omega}_1^{(n)}, \bar{\omega}_2^{(n)}, \bar{\omega}_3^{(n)}\right)$ and $\left(\bar{\alpha}_1^{(n)}, \bar{\alpha}_2^{(n)}, \bar{\alpha}_3^{(n)}\right)$, respectively, then the moment equations are

$$\begin{aligned}
\bar{I}_{P;1}^{(n)}\bar{\alpha}_1^{(n)} - \left(\bar{I}_{P;2}^{(n)} - \bar{I}_{P;3}^{(n)}\right)\bar{\omega}_2^{(n)}\bar{\omega}_3^{(n)} &= \bar{M}_{P;1}^{(n)} \\
\bar{I}_{P;2}^{(n)}\bar{\alpha}_2^{(n)} - \left(\bar{I}_{P;3}^{(n)} - \bar{I}_{P;1}^{(n)}\right)\bar{\omega}_1^{(n)}\bar{\omega}_3^{(n)} &= \bar{M}_{P;2}^{(n)} \quad (P = C, O) \\
\bar{I}_{P;3}^{(n)}\bar{\alpha}_3^{(n)} - \left(\bar{I}_{P;1}^{(n)} - \bar{I}_{P;2}^{(n)}\right)\bar{\omega}_1^{(n)}\bar{\omega}_2^{(n)} &= \bar{M}_{P;3}^{(n)}
\end{aligned} \tag{8.4.22}$$

Equations (8.4.22) are the form of the moment equation commonly called *Euler's equations*. Together with Eq. (8.4.13) we have the *Newton-Euler equations* for a rigid body. Although Euler's equations cannot be directly integrated to obtain the rotations of the rigid body they can be part of the integration process, as we will see shortly, and they can be used to deduce important properties of the motion. They can be expressed in terms of generalized coordinates such as Euler angles or any other set of parameters that does allow them to be integrated to determine the motion of the rigid body. Before illustrating the use of Euler's equations we should highlight the meaning of the angular acceleration terms in Eqs. (8.4.21) and (8.4.22) and how they are calculated. If we have the angular velocity of the body, ${}_I\boldsymbol{\omega}_B^{(n)}$, measured in inertial frame components, then we can differentiate that angular velocity to obtain the angular acceleration, ${}_I\boldsymbol{\alpha}_B^{(n)}$, and also measure it in an inertial frame. Following the discussion in Chap. 4 we can make explicit both the frame in which the time derivative is taken and the frame in which the resulting vector is computed by writing

$$\left[{}_I\boldsymbol{\alpha}_B^{(n)}\right]_I = \left[\frac{{}^I d\,{}_I\boldsymbol{\omega}_B^{(n)}}{dt}\right]_I \tag{8.4.23}$$

which we have generally written in the equivalent but less explicit form

$${}_I\boldsymbol{\alpha}_B^{(n)} = {}_I\dot{\boldsymbol{\omega}}_B^{(n)} \tag{8.4.24}$$

But we can also transform this angular acceleration to body-fixed coordinates with the rotation matrix via

$$_I\bar{\boldsymbol{\alpha}}_B^{(n)} = \left(\mathbf{R}^{(n)}\right)^T {_I\boldsymbol{\alpha}_B^{(n)}} \tag{8.4.25}$$

and this is the angular acceleration term appearing in Eqs. (8.4.21) and (8.4.22). However, this is not the most convenient way to compute this acceleration term. We would like instead to do the computation directly in body-fixed coordinates. This is possible since recall from Chap. 4 that we can relate the time derivative taken in one frame to the derivative taken in another frame with the transport theorem. In this case, we can write that theorem in the inertial frame as

$$\begin{aligned}\left[\frac{^I d\,_I\boldsymbol{\omega}_B^{(n)}}{dt}\right]_I &= \left[\frac{^B d\,_I\boldsymbol{\omega}_B^{(n)}}{dt}\right]_I + \left[{_I\boldsymbol{\omega}_B^{(n)}} \times {_I\boldsymbol{\omega}_B^{(n)}}\right]_I \\ &= \left[\frac{^B d\,_I\boldsymbol{\omega}_B^{(n)}}{dt}\right]_I\end{aligned} \tag{8.4.26}$$

since the cross product term vanishes. But the derivative of the angular velocity components in the body-fixed frame is just what we have called $_I\dot{\bar{\boldsymbol{\omega}}}_B^{(n)}$ and transforming those derivatives to the inertial frame is done with the rotation matrix as $\mathbf{R}^{(n)}\,_I\dot{\bar{\boldsymbol{\omega}}}_B^{(n)}$ so Eq. (8.4.26) can be written as

$$_I\boldsymbol{\alpha}_B^{(n)} \equiv \left[\frac{^I d\,_I\boldsymbol{\omega}_B^{(n)}}{dt}\right]_I = \mathbf{R}^{(n)}\,_I\dot{\bar{\boldsymbol{\omega}}}_B^{(n)} \tag{8.4.27}$$

and from Eq. (8.4.25) we have finally

$$_I\bar{\boldsymbol{\alpha}}_B^{(n)} = {_I\dot{\bar{\boldsymbol{\omega}}}_B^{(n)}} \tag{8.4.28}$$

so that the angular acceleration components appearing in Euler's equations can be calculated by simply taking time derivatives of the corresponding angular velocity vector components expressed in body-fixed coordinates.

8.5 Solutions of Euler's Equations for Rotational Motion

Simulation 8.1: Fixed Gyro-Pendulum (Spherical Pendulum)

As a first relatively simple application of Euler's equations, consider a thin cylindrical bar of negligible mass and length L attached to a thin circular disk of mass m and radius R, as shown in Fig. 8.5a, where an applied moment acting in the Z-direction turns a rotating support (whose mass is also neglected) and where the

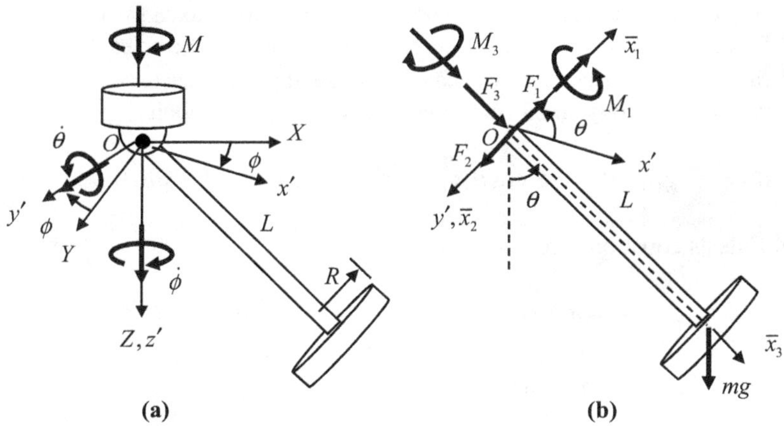

Fig. 8.5 (**a**) A thin disk attached to a bar of negligible mass and that swings about a pinned support which is driven by a given moment, M, and (**b**) the free body diagram of the assembly showing the constraint forces and moments in a set of body-fixed axes

support is pinned to the end of the bar at O. If the disk is allowed to rotate about the bar axis then this assembly will act as a gyro-pendulum. In this example, however, we will assume that the disk is rigidly fixed to the bar so that there are no gyroscopic effects present. Thus, the fixed gyro-pendulum is essentially a spherical pendulum. A free body diagram of the bar is shown in Fig. 8.5b. There are constraint forces along all three body-fixed axes $(\bar{x}_1, \bar{x}_2, \bar{x}_3)$ but only constraint moments acting along the $(\bar{x}_1, \bar{x}_3)$ axes since the $\bar{x}_2$-axis is always aligned with the axis of the smooth pin at O, about which the bar is free to rotate. It is convenient to take moments about point O because then the constraint forces are eliminated. The $(\bar{x}_1, \bar{x}_2, \bar{x}_3)$ axes are principal axes for the bar/disk. From Appendix B and the use of the parallel axis theorem, the moments of inertia are

$$\bar{I}_{O;3} = \frac{mR^2}{2}, \quad \bar{I}_{O;1} = \bar{I}_{O;2} = m\left(L^2 + \frac{R^2}{4}\right)$$

Since the pinned support prevents rotating about the $\bar{x}_3$-axis (which is the reason for the existence of the constraint moment, M_3) there are only two angular degrees of freedom (ϕ, θ) for the system. These are like the 3-2 components of a set of 3-2-3 Euler angles but where the final 3 rotation is absent. The angular velocity of the bar in body-fixed coordinates is (dropping the (n) superscript since there is only one body)

$$\begin{aligned} {}_I\boldsymbol{\omega}_B &= \dot{\phi}\,\mathbf{E}_Z + \dot{\theta}\,\mathbf{e}_2 \\ &= \dot{\phi}\,(\cos\theta\mathbf{e}_3 - \sin\theta\mathbf{e}_1) + \dot{\theta}\,\mathbf{e}_2 \end{aligned} \tag{8.5.1}$$

where $(\mathbf{E}_X, \mathbf{E}_Y, \mathbf{E}_Z)$ are unit vectors along the fixed inertial axes (X, Y, Z) shown in Fig. 8.5a and $(\mathbf{e}_1, \mathbf{e}_2, \mathbf{e}_3)$ are unit vectors along the body-fixed axes $(\bar{x}_1, \bar{x}_2, \bar{x}_3)$ of Fig. 8.5b. As just discussed, we can calculate the angular acceleration as the derivative of these body-fixed components of the angular velocity, giving

$$ {}_I\bar{\boldsymbol{\alpha}}_B = (-\ddot{\phi}\sin\theta - \dot{\phi}\dot{\theta}\cos\theta)\mathbf{e}_1 + \ddot{\theta}\mathbf{e}_2 + (\ddot{\phi}\cos\theta - \dot{\phi}\dot{\theta}\sin\theta)\mathbf{e}_3 \tag{8.5.2} $$

so that Euler's equations are

$$ \begin{aligned} &\bar{I}_{O;1}(-\ddot{\phi}\sin\theta - \dot{\phi}\dot{\theta}\cos\theta) - (\bar{I}_{O;2} - \bar{I}_{O;3})(\dot{\theta}\dot{\phi}\cos\theta) = \bar{M}_1 \\ &\bar{I}_{O;2}\ddot{\theta} - (\bar{I}_{O;3} - \bar{I}_{O;1})(-\dot{\phi}^2\sin\theta\cos\theta) = -mgL\sin\theta \\ &\bar{I}_{O;3}(\ddot{\phi}\cos\theta - \dot{\phi}\dot{\theta}\sin\theta) = \bar{M}_3 \end{aligned} \tag{8.5.3} $$

which reduce to

$$ \begin{aligned} &-m\left(L^2 + \frac{R^2}{4}\right)\ddot{\phi}\sin\theta - 2mL^2(\dot{\theta}\dot{\phi}\cos\theta) = \bar{M}_1 \\ &m\left(L^2 + \frac{R^2}{4}\right)\ddot{\theta} + m\left(\frac{R^2}{4} - L^2\right)\dot{\phi}^2\sin\theta\cos\theta = -mgL\sin\theta \\ &m\frac{R^2}{2}(\ddot{\phi}\cos\theta - \dot{\phi}\dot{\theta}\sin\theta) = \bar{M}_3 \end{aligned} \tag{8.5.4} $$

Since the mass of the rotating support is also negligible the support is in dynamic "equilibrium" so the sum of the moments must vanish (see Fig. 8.6). Setting the sum of the moments along the Z-axis to be zero gives

$$ M(t) - \bar{M}_3\cos\theta + \bar{M}_1\sin\theta = 0 \tag{8.5.5} $$

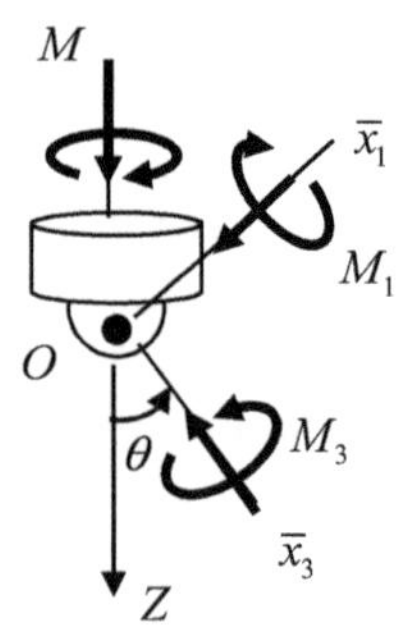

Fig. 8.6 The support of the assembly of Fig. 8.5 showing the driving and constraint moments. Note that these constraint moments are opposite to those shown in Fig. 8.5 since these are the moments that the assembly exerts on the support while the moments in Fig. 8.5 are the equal and opposite moments that the support exerts on the assembly

Multiplying the third equation in Eq. (8.5.4) by cosθ and the first equation by $-\sin\theta$ and adding the resulting two equations then give the two equations of motion:

$$\begin{aligned} &m\left[\left(L^2+\frac{R^2}{4}\right)\sin^2\theta+\frac{R^2}{2}\cos^2\theta\right]\ddot{\phi}+2m\left(L^2-\frac{R^2}{4}\right)\left(\dot{\theta}\dot{\phi}\sin\theta\cos\theta\right)=M(t)\\ &m\left(L^2+\frac{R^2}{4}\right)\ddot{\theta}-m\left(L^2-\frac{R^2}{4}\right)\dot{\phi}^2\sin\theta\cos\theta=-mgL\sin\theta \end{aligned} \tag{8.5.6}$$

which we will write in the equivalent form

$$\begin{aligned} &\left[\left(1+\frac{R^2}{4L^2}\right)\sin^2\theta+\frac{R^2}{2L^2}\cos^2\theta\right]\ddot{\phi}+2\left(1-\frac{R^2}{4L^2}\right)\left(\dot{\theta}\dot{\phi}\sin\theta\cos\theta\right)=\frac{M(t)}{mL^2}\\ &\left(1+\frac{R^2}{4L^2}\right)\ddot{\theta}-\left(1-\frac{R^2}{4L^2}\right)\dot{\phi}^2\sin\theta\cos\theta=-\frac{g}{L}\sin\theta \end{aligned} \tag{8.5.7}$$

A MATLAB® function fixed_gyro_pend was written that implements these equations of motion and uses the MATLAB® solver ode45 to calculate the motion. The calling sequence for fixed_gyro_pend is

```
[tout, xout] = fixed_gyro_pend(tspan, x0, R ,L, m, g, rtol, atol);
```

where tspan is the time interval over which the calculations take place, x0 is a 1×4 vector of initial conditions for $[\theta, \phi, \dot{\theta}, \dot{\phi}]$, R is the radius of the disk, L is the length of the bar, m is the mass, and g is the acceleration of gravity. In all the examples the driving moment, *M(t)*, is set equal to zero in the equations of motion but it can be easily added back in, if necessary. The variables rtol and atol are the relative and absolute error tolerances used by ode45. The output, tout, of the function is a column vector of times at which the solution was obtained and xout is a matrix which contains the four variables $(\theta, \phi, \dot{\theta}, \dot{\phi})$ in its columns at the evaluation times. A script fixed_gyro_pend_sim was also written which sets up default parameters for the parameters of the system given by R = 0.1 m, L = 0.5 m, m = 1 kg, g = 9.81 m/s^2, tspan = [0 11] s, and x0 = [45*pi/180, 0, 0, 3], where the angles are in rad and the angular velocities are in rad/s. The default error tolerances are set as rtol = 1e-6 and atol = 1e-6. A marker parameter is also present which allows the path of the free end of the bar-disk assembly to be plotted. The default value is marker = 'on.' The script calls the fixed_gyro_pend function that returns the solution for $(\theta, \phi, \dot{\theta}, \dot{\phi})$. Using the definitions of the angles (θ, ϕ), as shown in Fig. 8.7, the location of the endpoint, *P*, of the bar-disk assembly in inertial (X_P, Y_P, Z_P) coordinates was calculated as

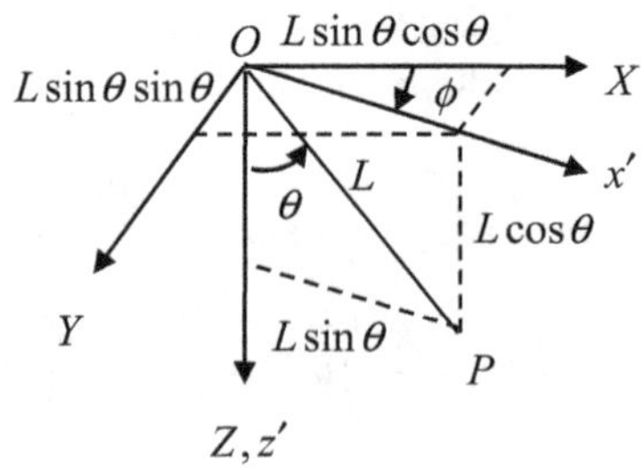

Fig. 8.7 Geometry for determining the location of point P, the end of the bar-disk assembly, in terms of the (X, Y, Z) coordinates

$$\begin{aligned} X_P &= L\sin\theta\cos\phi \\ Y_P &= L\sin\theta\sin\phi \\ Z_P &= L\cos\theta \end{aligned} \tag{8.5.8}$$

Alternatively, we could use the rotation matrix between the body-fixed axes and the inertial axes to transform the body-fixed coordinates of P, $(\bar{x}_1, \bar{x}_2, \bar{x}_3) = (0, 0, L)$, to (X_P, Y_P, Z_P). Since the resulting motion can be rather complex, a 3-D animation of the motion of the system (shown as simply the motion of the line from O to P) is generated by the script. Figure 8.8a shows a plot of the motion when the default values are used. The bar-disk assembly performs a combination of precessional ϕ motions as well as nutational θ motions, resulting in a path that takes the shape of an undulating bowl-like figure as seen in Fig. 8.8a. If the workspace is cleared and the initial values are changed to x0 = [45*pi/180, 0, 0, 6], the higher precessional velocity causes the assembly to perform a mostly conical motion (Fig. 8.8b) (see Problem P8.1 which discusses conditions for pure conical motion). Finally, if the workspace is cleared and the initial values are given as x0 = [45*pi/180, 0, 0, 0], the assembly acts as a two-dimensional pendulum with its motion in the X–Z plane as shown in Fig. 8.8c.

If the disk in the assembly is allowed to rotate about its central axis (Fig. 8.9), then it acts as a true gyro-pendulum. In this case a set of body-fixed axes $(\bar{x}_1, \bar{x}_2, \bar{x}_3)$ attached to the disk also have an angular velocity component, $\dot{\psi}$, as well as a rotation ψ, taken about the third coordinate of the previous body-fixed axes of the fixed gyro-pendulum assembly. Those previous body-fixed axes are now a set of intermediate axes which are called (x_1^a, x_2^a, x_3^a) as shown in Fig. 8.9. In this case the angular velocity components as computed in the $(\bar{x}_1, \bar{x}_2, \bar{x}_3)$ body-fixed axes of the gyro-pendulum are

$$\begin{aligned} \overline{\omega}_1 &= -\dot{\phi}\sin\theta\cos\psi + \dot{\theta}\sin\psi \\ \overline{\omega}_2 &= \dot{\phi}\sin\theta\sin\psi + \dot{\theta}\cos\psi \\ \overline{\omega}_3 &= \dot{\phi}\cos\theta + \dot{\psi} \end{aligned} \tag{8.5.9}$$

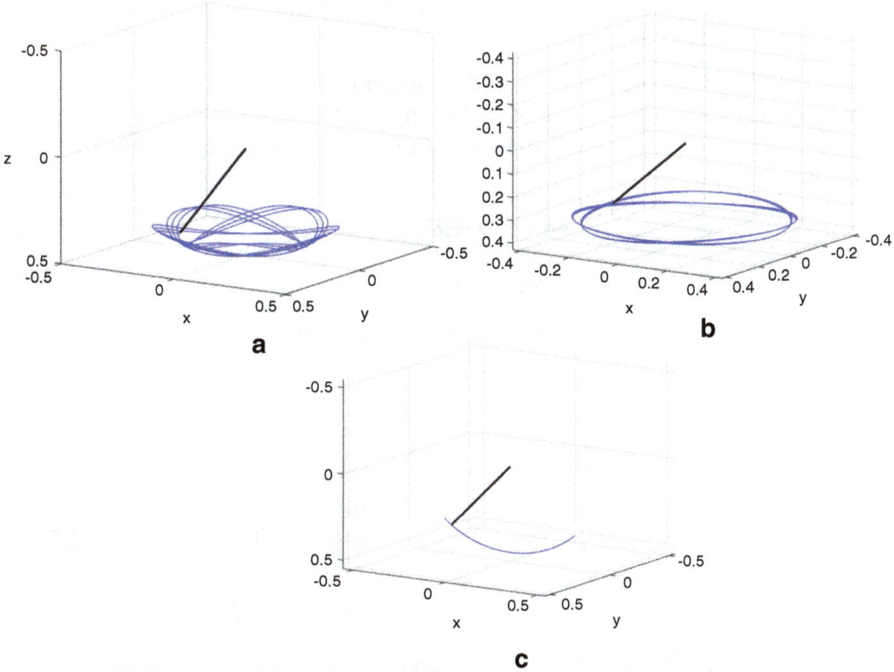

Fig. 8.8 Motions of the bar-disk assembly (shown as a moving line from the origin O to the endpoint P) under different starting conditions with no driving moment. (**a**) When the bar-disk is started at $\theta = 45°, \phi = 0$ with an angular velocity $\dot{\phi} = 3$ rad/s, $\dot{\theta} = 0$, tspan = [0, 11] s (the default conditions in the script fixed_gyro_pend_sim). (**b**) When the bar-disk is started at $\theta = 45°$, $\phi = 0$, with an angular velocity $\dot{\phi} = 6$ rad/s, $\dot{\theta} = 0$, for the time interval tspan = [0 11] s. (**c**) When the bar-disk is started at $\theta = 45°$, $\phi = 0$ with an angular velocity $\dot{\theta} = \dot{\phi} = 0$ rad/s for a time interval tspan = [0 11] s

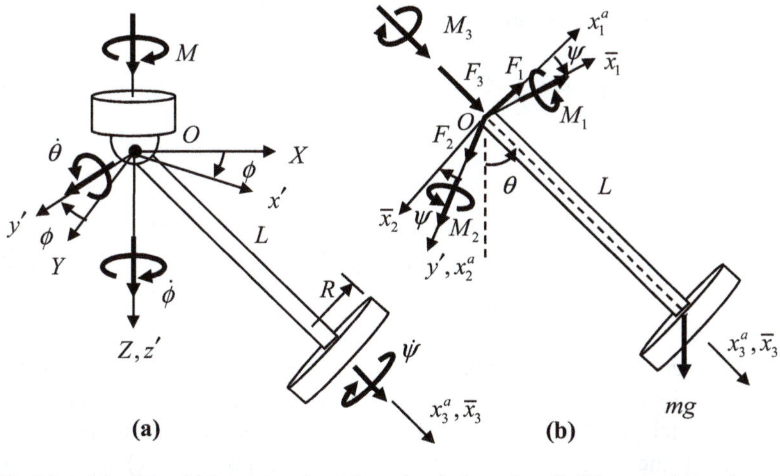

Fig. 8.9 (**a**) A thin disk which rotates about the axis of a bar of negligible mass that swings about a pinned support, and where the support is driven by a given moment, M, and (**b**) the free body diagram of this gyro-pendulum showing the constraint forces and moments in a set of body-fixed axes

Note that Eq. (8.5.9) reduces to Eq. (8.5.1) when $\dot{\psi} = \psi = 0$. The (ϕ, θ, ψ) angles correspond to using a 3-2-3 set of Euler angles. We can alternatively write the derivatives of these Euler angles in terms of the angular velocities to find

$$\begin{aligned}
\dot{\phi} &= \frac{1}{\sin\theta}\left(-\overline{\omega}_1 \cos\psi + \overline{\omega}_2 \sin\psi\right) \\
\dot{\theta} &= \overline{\omega}_1 \sin\psi + \overline{\omega}_2 \cos\psi \\
\dot{\psi} &= \frac{1}{\sin\theta}\left(\overline{\omega}_1 \cos\theta\cos\psi - \overline{\omega}_2 \cos\theta \sin\psi\right) + \overline{\omega}_3
\end{aligned} \tag{8.5.10}$$

or $\dot{\boldsymbol{\theta}} = \bar{\mathbf{H}}^{-1}\overline{\boldsymbol{\omega}}$ where (see Eq. (4.11.6))

$$\bar{\mathbf{H}}^{-1} = \begin{bmatrix} -\dfrac{\cos\psi}{\sin\theta} & \dfrac{\sin\psi}{\sin\theta} & 0 \\ \sin\psi & \cos\psi & 0 \\ \dfrac{\cos\theta\cos\psi}{\sin\theta} & \dfrac{-\cos\theta\sin\psi}{\sin\theta} & 1 \end{bmatrix} \tag{8.5.11}$$

Equation (8.5.11) shows explicitly the singularities that are always present at certain angles when using Euler angles, Here, we see that for the 3-2-3 set of angles the derivative expressions are unbounded when $\theta = 0$ or $\theta = \pi$ due to the presence of the sine term in the denominators of these expressions.

Simulation 8.2: Gyro-Pendulum and Euler Angles

There are a number of possible ways to set up the equations of motion for the gyro-pendulum. One way is to directly write Eq. (8.4.21), which is written in body-fixed coordinates, in terms of the Euler angles, giving

$$\bar{\mathbf{I}}_O^{(n)}\bar{\mathbf{H}}^{(n)}\ddot{\boldsymbol{\theta}}^{(n)} + \bar{\mathbf{I}}_O^{(n)}\dot{\bar{\mathbf{H}}}^{(n)}\dot{\boldsymbol{\theta}}^{(n)} + \bar{\mathbf{H}}^{(n)}\dot{\boldsymbol{\theta}}^{(n)} \times \bar{\mathbf{I}}_O^{(n)}\bar{\mathbf{H}}^{(n)}\dot{\boldsymbol{\theta}}^{(n)} = \bar{\mathbf{M}}_O^{(n)} \tag{8.5.12}$$

where

$$\dot{\bar{\mathbf{H}}} = \begin{bmatrix} \left(-\cos\theta\cos\psi\dot{\theta} + \sin\theta\sin\psi\dot{\psi}\right) & \cos\psi\dot{\psi} & 0 \\ \left(\cos\theta\sin\psi\dot{\theta} + \sin\theta\cos\psi\dot{\psi}\right) & -\sin\psi\dot{\psi} & 0 \\ -\sin\theta\dot{\theta} & 0 & 0 \end{bmatrix} \tag{8.5.13}$$

This leads to very complex-looking equations and is not usually an approach found in traditional dynamics texts which concentrate on expressing the equations of motion in explicit, analytical forms. However, for a numerical solution one only needs to set up the necessary matrices and vectors in Eq. (8.5.12) so this can be a viable method. A second approach is to write Euler's equation in terms of the $\left(x_1^a, x_2^a, x_3^a\right)$ axes (called the a-frame) used in the fixed gyro-pendulum problem instead of body-fixed coordinates. This choice is attractive since the moments of inertia of the disk are still constants in this coordinate system and we are generally interested in following the motion of the bar in the assembly as opposed to a point as it rotates on the disk. Use of the $\left(x_1^a, x_2^a, x_3^a\right)$ coordinates also makes the expressions for the equations of motion much less complex than when using body-fixed coordinates. A third choice is to write the moment equation in terms of derivatives in the a-frame directly, leading to what is called the modified Euler equations. We will consider the last two of these choices in the problems at the end of this chapter. A fourth choice, that we will implement, is to leave Euler's equations in terms of the angular velocity components in the body-fixed coordinates and combine those equations with Eq. (8.5.10) for the derivatives of the Euler angles in terms of the angular velocities. Thus, the six parameters to be solved for in this case are $(\overline{\omega}_1, \overline{\omega}_2, \overline{\omega}_3, \phi, \theta, \psi)$. Note that both Euler's equations and Eq. (8.5.10) involve only first-order derivatives so it is easy to place these equations in the form needed for a numerical solution. The MATLAB® function gyro_pend and the script gyro_pend_sim implement the gyro-pendulum problem in this manner for the case where $M(t) = 0$. Now, there are six variables to calculate but after the solution is obtained, one can still follow the point P at the end of the bar-disk assembly by using Eq. (8.5.8) since the motion of that point is not directly dependent on the angle, ψ. To make the problem formulation similar to the fixed gyro-pendulum case we will give the initial conditions in terms of the six Euler angle variables and their derivatives, $(\phi, \theta, \psi, \dot{\phi}, \dot{\theta}, \dot{\psi})$, and transform them, using Eq. (8.5.9), into initial conditions for $(\overline{\omega}_1, \overline{\omega}_2, \overline{\omega}_3, \phi, \theta, \psi)$. The first-order differential equations of motion here are

$$
\begin{aligned}
\dot{\overline{\omega}}_1 - \left(\bar{I}_{O;2} - \bar{I}_{O;3}\right)\overline{\omega}_2\overline{\omega}_3/\bar{I}_{O;1} &= -M(t)\sin\theta\cos\psi/\bar{I}_{O;1} - mgL\sin\theta\sin\psi/\bar{I}_{O;1} \\
\dot{\overline{\omega}}_2 - \left(\bar{I}_{O;3} - \bar{I}_{O;1}\right)\overline{\omega}_1\overline{\omega}_3/\bar{I}_{O;2} &= M(t)\sin\theta\sin\psi/\bar{I}_{O;2} - mgL\sin\theta\cos\psi/\bar{I}_{O;2} \\
\dot{\overline{\omega}}_3 - \left(\bar{I}_{O;1} - \bar{I}_{O;2}\right)\overline{\omega}_1\overline{\omega}_2/\bar{I}_{O;3} &= M(t)\cos\theta/\bar{I}_{O;3} \\
\dot{\phi} &= \frac{1}{\sin\theta}\left(-\overline{\omega}_1\cos\psi + \overline{\omega}_2\sin\psi\right) \\
\dot{\theta} &= \overline{\omega}_1\sin\psi + \overline{\omega}_2\cos\psi \\
\dot{\psi} &= \frac{1}{\sin\theta}\left(\overline{\omega}_1\cos\theta\cos\psi - \overline{\omega}_2\cos\theta\sin\psi\right) + \overline{\omega}_3
\end{aligned}
\tag{8.5.14}
$$

where we have resolved the driving moment into its body-fixed components by first resolving the moment into its components along the $\left(x_1^a, x_2^a, x_3^a\right)$ coordinates, as seen in Fig. 8.6, and then finding the components in the body-fixed axes (Fig. 8.10). From the dynamic moment equilibrium of the support in the $\left(x_1^a, x_3^a\right)$ directions (Fig. 8.10a) we have

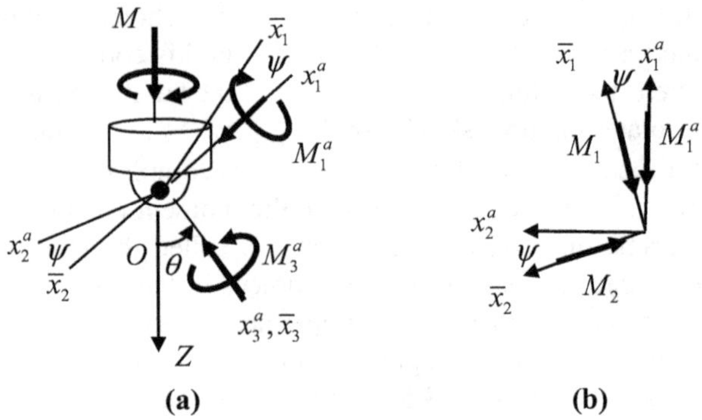

Fig. 8.10 (**a**) The support of the assembly of Fig. 8.8 showing the driving moment, which acts along the Z-axis, and the constraint moment components along the (x_1^a, x_2^a, x_3^a) axes. (**b**) The transformation of the constraint moments into components along the body-fixed $(\bar{x}_1, \bar{x}_2, \bar{x}_3)$ axes

$$\begin{aligned} M_3^a &= M\cos\theta \\ M_1^a &= -M\sin\theta \end{aligned} \tag{8.5.15}$$

and in the body-fixed coordinates $(\bar{x}_1, \bar{x}_2, \bar{x}_3)$ as seen in Fig. 8.10b

$$\begin{aligned} \bar{M}_1 &= M_1^a\cos\psi = -M(t)\sin\theta\cos\psi \\ \bar{M}_2 &= -M_1^a\sin\psi = M(t)\sin\theta\sin\psi \\ \bar{M}_3 &= M_3^a = M(t)\cos\theta \end{aligned} \tag{8.5.16}$$

which are the components seen in Eq. (8.5.14). The moment contribution from the weight was also calculated by noting that the moment is just $-mgL\sin\theta$ along the x_2^a -axis (see Eq. (8.5.6)) so that if we simply resolve that moment into its $(\bar{x}_1, \bar{x}_2)$ components we obtain the terms seen in Eq. (8.5.14).

The script gyro_pend_sim has defaults for the parameters of the system given by R = 0.1 m, L = 0.5 m, m = 1 kg, tspan = linspace(0,10,1000); s, and x0 = [0, 90*pi/180, 0, 0, 0, 200], where the angles are in rad and the angular velocities are in rad/s. Thus, the gyro-pendulum is released from rest with the bar in a horizontal position. The default error tolerances are set as rtol =1e-6 and atol = 1e-6. The script solves the system of equations and then animates the 3-D motion by showing the motion of the bar in the assembly. If the script is executed with the defaults the animation shows a precessional motion combined with a nutation that forms a series of cusps as seen in Fig. 8.11. The specification of the time interval as 1000 points over a time interval of 10 s allows the animation to proceed at a reasonable rate. However, if we had used tspan = [0,10] the animation would be very slow since the MATLAB® ode45 solver calculates approximately 40,000 positions of the gyro-pendulum, taking very small time steps, so that displaying all those small position changes is slow. Note that, as we have mentioned before, even by specifying tspan = linspace(0, 10, 1000) the solver still calculates the same number of points during the solution process. You can examine other examples of the use of gyro_pend_sim but we will stop further discussion here and instead implement the same problem with Euler parameters.

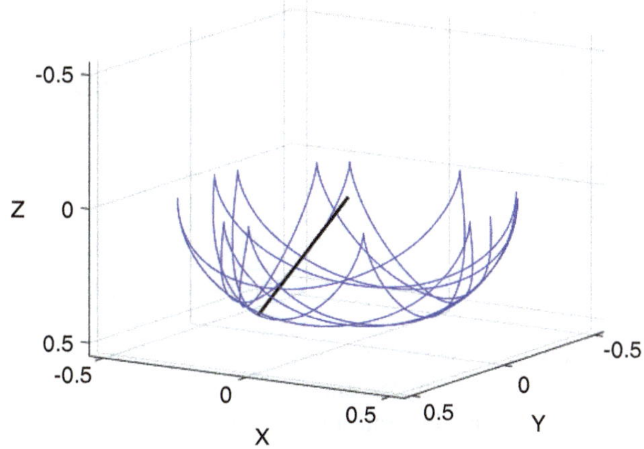

Fig. 8.11 Motion of the gyro-pendulum (solved using Euler angles) for R = 0.1 m, L = 0.5 m, m = 1 kg, tspan = linspace (0,10,1000); s, and x0 = [0, 90*pi/180, 0, 0, 0, 200]. The angles all are in rad and the angular velocities are all in rad/s. The default error tolerances are set as rtol =1e-6 and atol = 1e-6

Although we did not use Lagrange's equations to obtain the equations of motion for the gyro-pendulum, examining the Lagrangian does give us significant information. Using the symmetry of the gyro-pendulum we have

$$L = \frac{1}{2}\bar{I}_{O;1}\left(\dot{\phi}^2 \sin^2\theta + \dot{\theta}^2\right) + \frac{1}{2}\bar{I}_{O;3}\left(\dot{\phi}\cos\theta + \dot{\psi}\right)^2 + mgL\cos\theta \tag{8.5.17}$$

which shows that the Lagrangian does not depend explicitly on both ϕ and ψ; that is, these are ignorable coordinates. Thus the generalized momenta associated with these coordinates are constants, so we have

$$\begin{aligned} p_\phi &= \frac{\partial L}{\partial \dot{\phi}} = \bar{I}_{O;1}\dot{\phi}\sin^2\theta + \bar{I}_{O;3}\left(\dot{\psi} + \dot{\phi}\cos\theta\right)\cos\theta = \text{constant} \\ p_\psi &= \frac{\partial L}{\partial \dot{\psi}} = \bar{I}_{O;3}\left(\dot{\phi}\cos\theta + \dot{\psi}\right) = \text{constant} \end{aligned} \tag{8.5.18}$$

From Eq. (8.5.9) we see that $p_\psi = \bar{I}_{O;3}\bar{\omega}_3 = \bar{L}_3$, so the angular momentum component about the body-fixed axis $\bar{x}_3$ is a constant. It can also be shown that $p_\phi = L_Z$, where L_Z is the angular momentum about the inertial Z-axis. We can write both $\dot{\phi}$ and $\dot{\psi}$ in terms of the constants $\bar{L}_3, L_Z$ as

$$\begin{aligned} \dot{\phi} &= \frac{L_Z - \bar{L}_3\cos\theta}{\bar{I}_{O;1}\sin^2\theta} \\ \dot{\psi} &= \frac{\bar{L}_3}{\bar{I}_{O;3}} - \frac{\left(L_Z - \bar{L}_3\cos\theta\right)\cos\theta}{\bar{I}_{O;1}\sin^2\theta} \end{aligned} \tag{8.5.19}$$

It follows that the total energy, E, which is conserved in this problem, is given by

$$E = \frac{1}{2}\bar{I}_{O;1}\left(\dot{\phi}^2 \sin^2\theta + \dot{\theta}^2\right) + \frac{1}{2}\bar{I}_{O;3}\left(\dot{\phi}\cos\theta + \dot{\psi}\right)^2 - mgL\cos\theta \tag{8.5.20}$$

which can also be written as

$$E = \frac{1}{2}\bar{I}_{O;1}\dot{\theta}^2 + U(\theta) \tag{8.5.21}$$

where $U(\theta)$ acts as the effective potential:

$$U(\theta) = \frac{\left(L_Z - \bar{L}_3 \cos\theta\right)^2}{2\bar{I}_{O;1}\sin^2\theta} + \frac{\bar{L}_3^2}{2\bar{I}_{O;3}} - mgL\cos\theta \tag{8.5.22}$$

Many texts use these expressions for the generalized momenta and energy to infer the behavior of the gyro-pendulum without actually solving the equations of motion. Because we can illustrate the behavior of this system directly with the numerical solutions (as we will do shortly) we will not go into those discussions but simply refer you to a reference such as [1] for more details. Problem P8.3 at the end of this chapter, however, does consider verifying that the numerical solution satisfies the conservation of these momenta and the total energy.

Simulation 8.3: Gyro-Pendulum and Euler Parameters
In Chap. 4 we showed that the singularities present in the Euler angles could be avoided by using Euler parameters instead to describe the angular motion. For the gyro-pendulum, one formulation is to simply replace the six coordinates $(\bar{\omega}_1, \bar{\omega}_2, \bar{\omega}_3, \phi, \theta, \psi)$ by the seven coordinates $(\bar{\omega}_1, \bar{\omega}_2, \bar{\omega}_3, \theta_0, \theta_1, \theta_2, \theta_3)$, where $(\theta_0, \theta_1, \theta_2, \theta_3)$ are the Euler parameters described in Chap. 4. Thus, we can write a set of seven first-order differential equations instead as

$$\begin{aligned}
&\dot{\bar{\omega}}_1 - \left(\bar{I}_{O;2} - \bar{I}_{O;3}\right)\bar{\omega}_2\bar{\omega}_3/\bar{I}_{O;1} = 2M(t)(\theta_1\theta_3 - \theta_0\theta_2)/\bar{I}_{O;1} - 2mgL(\theta_2\theta_3 + \theta_0\theta_1)/\bar{I}_{O;1}\\
&\dot{\bar{\omega}}_2 - \left(\bar{I}_{O;3} - \bar{I}_{O;1}\right)\bar{\omega}_1\bar{\omega}_3/\bar{I}_{O;2} = 2M(t)(\theta_2\theta_3 + \theta_0\theta_1)/\bar{I}_{O;2} + 2mgL(\theta_1\theta_3 - \theta_0\theta_2)/\bar{I}_{O;2}\\
&\dot{\bar{\omega}}_3 - \left(\bar{I}_{O;1} - \bar{I}_{O;2}\right)\bar{\omega}_1\bar{\omega}_2/\bar{I}_{O;3} = M(t)\left(1 - 2\theta_1^2 - 2\theta_2^2\right)/\bar{I}_{O;3}\\
&\dot{\theta}_0 = \frac{1}{2}(-\bar{\omega}_1\theta_1 - \bar{\omega}_2\theta_2 - \bar{\omega}_3\theta_3)\\
&\dot{\theta}_1 = \frac{1}{2}(\bar{\omega}_1\theta_0 - \bar{\omega}_2\theta_3 + \bar{\omega}_3\theta_2)\\
&\dot{\theta}_2 = \frac{1}{2}(\bar{\omega}_1\theta_3 + \bar{\omega}_2\theta_0 - \bar{\omega}_3\theta_1)\\
&\dot{\theta}_3 = \frac{1}{2}(-\bar{\omega}_1\theta_2 + \bar{\omega}_2\theta_1 + \bar{\omega}_3\theta_0)
\end{aligned} \tag{8.5.23}$$

where we have used Eq. (4.10.22) and have written the moment terms in terms of the Euler parameters by comparing matrix elements in the rotation matrix, given in terms of the Euler parameters, Eq. (4.10.11), with the same elements written in terms of the Euler angle, Eq. (4.11.5). We will still give the initial conditions in terms of the Euler angles and their derivatives, $(\phi, \theta, \psi, \dot{\phi}, \dot{\theta}, \dot{\psi})$, and then generate the initial conditions for our seven parameters by using Eq. (8.5.9) and Table 4.1, i.e.,

$$\begin{aligned}
\bar{\omega}_1 &= -\dot{\phi} \sin\theta \cos\psi + \dot{\theta} \sin\psi \\
\bar{\omega}_2 &= \dot{\phi} \sin\theta \sin\psi + \dot{\theta} \cos\psi \\
\bar{\omega}_3 &= \dot{\phi} \cos\theta + \dot{\psi} \\
\theta_0 &= \cos\left(\frac{\phi + \psi}{2}\right) \cos\left(\frac{\theta}{2}\right) \\
\theta_1 &= \sin\left(\frac{\psi - \phi}{2}\right) \sin\left(\frac{\theta}{2}\right) \\
\theta_2 &= \cos\left(\frac{\psi - \phi}{2}\right) \sin\left(\frac{\theta}{2}\right) \\
\theta_3 &= \sin\left(\frac{\phi + \psi}{2}\right) \cos\left(\frac{\theta}{2}\right)
\end{aligned} \tag{8.5.24}$$

In solving the seven first-order differential equations of Eq. (8.5.23) the Euler parameters are related to each other by the constraint

$$\theta_0^2 + \theta_1^2 + \theta_2^2 + \theta_3^2 = 1 \tag{8.5.25}$$

But this constraint does not produce a generalized constraint force so that it does not change the equations of motion. We will prove this fact and discuss this issue more later. However, the constraint still must be satisfied so that we have to either incorporate it into the solution process or monitor the constraint during the numerical solution to ensure that it is not being violated. The MATLAB® function gyro_pend_ep and the script gyro_pend_ep_sim implement the solution of the gyro-pendulum problem using Eqs. (8.5.23) and (8.5.24) with $M(t) = 0$. Again the point P is followed, whose inertial coordinates can be calculated using the rotation matrix, written in terms of the Euler parameters, (Eq. (4.10.11)), to transform the body-fixed coordinates of P, $(\bar{x}_1, \bar{x}_2, \bar{x}_3) = (0, 0, L)$ to (X_P, Y_P, Z_P), giving

$$\begin{aligned}
X_P &= -2L(\theta_1\theta_3 - \theta_0\theta_2) \\
Y_P &= 2L(\theta_2\theta_3 + \theta_0\theta_1) \\
Z_P &= L\left(1 - 2\theta_1^2 - 2\theta_2^2\right)
\end{aligned} \tag{8.5.26}$$

The MATLAB® function gyro_pend_ep_sim and the script gyro_pend_ep_sim implement this Euler parameter model of the equations of motion, following very similar steps to those used in the Euler angle model. The default parameters in the script are identical to those used in the Euler angle case, namely R = 0.1 m, L = 0.5 m, m = 1 kg, tspan = linspace(0, 10,1000), and x0 = [0, 90*pi/180, 0, 0, 0, 200], where again the angles are in rad and the angular velocities are in rad/s. The default error tolerances are set as rtol = 1e-6 and atol = 1e-6. In this case the initial Euler angles and their derivatives in x0 are transformed into a set of initial conditions for $(\varpi_1, \varpi_2, \varpi_3, \theta_0, \theta_1, \theta_2, \theta_3)$ that are used by the ode45 solver. Running the gyro_pend_ep_sim script results in the animation results shown in Fig. 8.12a, which is identical to the simulation using Euler angles. If the initial conditions are changed to x0 = [0, 90*pi/180, 0, −0.9, 0, 200], with all other parameters unchanged, then the animation produces a smoother "nodding" behavior during the precession, as shown in Fig. 8.12b. If instead we set x0 = [0, 90*pi/180, 0, 0.9, 0, 200], then the motion is as shown in Fig. 8.12c where the axis of the system exhibits loops during the precession. In problem P8.6 you are asked to examine the satisfaction of the constraint equation and its derivative during these motions.

As discussed previously, many texts give detailed analytical discussions of the behaviors seen in Fig. 8.12 from an analysis of the conservation of momenta and

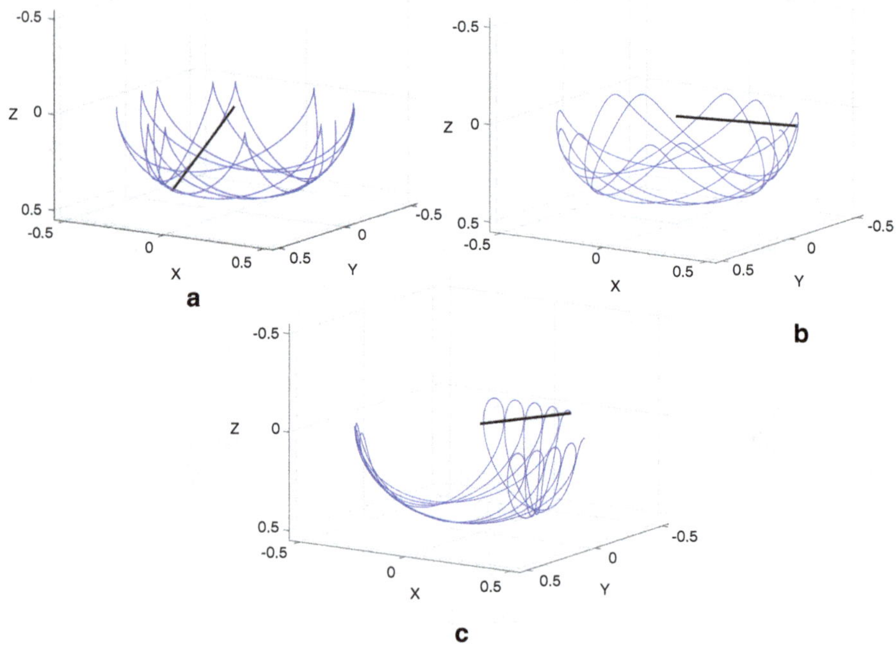

Fig. 8.12 (**a**) Motion of the gyro-pendulum (solved using Euler parameters) for the default parameters R = 0.1 m, L = 0.5 m, m = 1 kg, tspan = linspace (0,10,1000);s, and rtol = 1e-6, atol = 1e-6, and x0 = [0, 90*pi/180, 0, 0, 0, 200], and (**b**) the solution with the initial conditions changed to x0 = [0, 90*pi/180, 0, −0.9, 0, 200], and finally (**c**) the solution with the initial conditions changed to x0 = [0, 90*pi/180, 0, 0.9, 0, 200]

energy. By solving the problem numerically and animating that solution we can directly see the fascinating motions of the gyro-pendulum. Such motions are also characteristic of other spinning bodies such as tops and gyros.

8.6 Rotational Motion and the Euler Parameter Constraint

Let us return to the constraint condition, Eq. (8.5.25), and the solution of the equations of motion. This constraint comes from describing the three degrees of freedom of the angular motion in terms of four parameters so it is a result of that over-determinacy, not a result of any physical constraint. Thus, we might expect that this constraint will be different from the physical constraints we have previously discussed in many examples. This is indeed the case since if we try to incorporate the constraint into the rotational equations of motion we can show that it does not produce a generalized "force." To see this, consider writing Euler's equations, Eq. (8.4.21), in terms of Euler parameters. Using Eq. (4.10.25) and the identities of Eqs. (4.10.27a) and (4.10.27b) we have

$$\bar{\mathbf{I}}_P^{(n)}\bar{\mathbf{H}}^{(n)}\ddot{\boldsymbol{\theta}}^{(n)} + \frac{1}{2}\bar{\mathbf{H}}^{(n)}\left(\dot{\bar{\mathbf{H}}}^{(n)}\right)^T\bar{\mathbf{I}}_P^{(n)}\bar{\mathbf{H}}^{(n)}\dot{\boldsymbol{\theta}}^{(n)} = \bar{\mathbf{M}}_P^{(n)} \tag{8.6.1}$$

We cannot incorporate the constraint on the four Euler parameters directly into these equations since there are only three equations for the four unknown Euler parameters. However, if we pre-multiply Eq. (8.6.1) by $\left(\bar{\mathbf{H}}^{(n)}\right)^T$ we generate a set of four equations that are subject to these constraints:

$$\begin{aligned}&\left(\bar{\mathbf{H}}^{(n)}\right)^T\bar{\mathbf{I}}_P^{(n)}\bar{\mathbf{H}}^{(n)}\ddot{\boldsymbol{\theta}}^{(n)} + \frac{1}{2}\left(\bar{\mathbf{H}}^{(n)}\right)^T\bar{\mathbf{H}}^{(n)}\left(\dot{\bar{\mathbf{H}}}^{(n)}\right)^T\bar{\mathbf{I}}_P^{(n)}\bar{\mathbf{H}}^{(n)}\dot{\boldsymbol{\theta}}^{(n)}\\ &\quad = \left(\bar{\mathbf{H}}^{(n)}\right)^T\bar{\mathbf{M}}_P^{(n)}\end{aligned} \tag{8.6.2}$$

Since the constraint can be written as $\boldsymbol{\theta}^T\boldsymbol{\theta} = 1$ we can use a Lagrange multiplier, λ, and incorporate it into Eq. (8.6.2) as

$$\begin{aligned}&\left(\bar{\mathbf{H}}^{(n)}\right)^T\bar{\mathbf{I}}_P^{(n)}\bar{\mathbf{H}}^{(n)}\ddot{\boldsymbol{\theta}}^{(n)} + \frac{1}{2}\left(\bar{\mathbf{H}}^{(n)}\right)^T\bar{\mathbf{H}}^{(n)}\left(\dot{\bar{\mathbf{H}}}^{(n)}\right)^T\bar{\mathbf{I}}_P^{(n)}\bar{\mathbf{H}}^{(n)}\dot{\boldsymbol{\theta}}^{(n)} + \boldsymbol{\theta}^{(n)}\lambda\\ &\quad = \left(\bar{\mathbf{H}}^{(n)}\right)^T\bar{\mathbf{M}}_P^{(n)}\end{aligned} \tag{8.6.3}$$

However, if we pre-multiply Eq. (8.6.3) by $(\boldsymbol{\theta}^{(n)})^T$ and use the fact that $\bar{\mathbf{H}}^{(n)}\boldsymbol{\theta}^{(n)} = 0$ we find

$$\left(\boldsymbol{\theta}^{(n)}\right)^T \boldsymbol{\theta}^{(n)} \lambda = 0 \tag{8.6.4}$$

so that $\lambda = 0$ and there is no generalized "force" associated with this constraint. Thus, we can write the moment equation in terms of the Euler parameters in its original form, Eq. (8.6.2), or pre-multiply Eq. (8.6.2) by $\bar{\mathbf{H}}$ and use the relationship between the Euler parameters and the angular velocity and acceleration to recover the original Euler's equations, Eq. (8.4.21). In either case the constraint does not appear directly in the governing equations of motion. There are other ways to write the equations of motion in forms where the Lagrange multiplier, λ, is not zero. However, those forms all result in the same equations for $\ddot{\boldsymbol{\theta}}$. See [2] for a very clear discussion of the different forms of the rotational equations of motion.

Note that Eq. (8.6.2) can be written in the form

$$\mathbb{I}_P^{(n)} \ddot{\boldsymbol{\theta}}^{(n)} + \frac{1}{2}\left(\bar{\mathbf{H}}^{(n)}\right)^T \bar{\mathbf{H}}^{(n)} \left(\dot{\bar{\mathbf{H}}}^{(n)}\right)^T \bar{\mathbf{I}}_P^{(n)} \bar{\mathbf{H}}^{(n)} \dot{\boldsymbol{\theta}}^{(n)} = \left(\bar{\mathbf{H}}^{(n)}\right)^T \bar{\mathbf{M}}_P^{(n)} \tag{8.6.5}$$

where the generalized "inertia" matrix, $\mathbb{I}_P^{(n)}$, is

$$\mathbb{I}_P^{(n)} = \left(\bar{\mathbf{H}}^{(n)}\right)^T \bar{\mathbf{I}}_P^{(n)} \bar{\mathbf{H}}^{(n)} \tag{8.6.6}$$

If we multiply Eq. (8.6.5) by the matrix

$$\mathbb{J}_P^{(n)} = \left(\bar{\mathbf{H}}^{(n)}\right)^T \left(\bar{\mathbf{I}}_P^{(n)}\right)^{-1} \bar{\mathbf{H}}^{(n)} \tag{8.6.7}$$

then since

$$\mathbb{J}_P^{(n)} \mathbb{I}_P^{(n)} = 16\left(\mathbf{I}_4 - \boldsymbol{\theta}^{(n)}\left(\boldsymbol{\theta}^{(n)}\right)^T\right) \tag{8.6.8}$$

we have

$$\begin{aligned} 16\ddot{\boldsymbol{\theta}}^{(n)} - 16\boldsymbol{\theta}^{(n)}\left(\boldsymbol{\theta}^{(n)}\right)^T \ddot{\boldsymbol{\theta}}^{(n)} &= -2(\bar{\mathbf{H}}^{(n)})^T \left(\bar{\mathbf{I}}_P^{(n)}\right)^{-1} \bar{\mathbf{H}}^{(n)} \left(\dot{\bar{\mathbf{H}}}^{(n)}\right)^T \bar{\mathbf{I}}_P^{(n)} \bar{\mathbf{H}}^{(n)} \dot{\boldsymbol{\theta}}^{(n)} \\ &\quad + 4(\bar{\mathbf{H}}^{(n)})^T \left(\bar{\mathbf{I}}_P^{(n)}\right)^{-1} \bar{\mathbf{M}}_P^{(n)} \end{aligned} \tag{8.6.9}$$

where we have also used $\bar{\mathbf{H}}^{(n)}\left(\bar{\mathbf{H}}^{(n)}\right)^T = 4\mathbf{I}$. Differentiating the constraint equation twice on time gives

$$\left(\boldsymbol{\theta}^{(n)}\right)^T\ddot{\boldsymbol{\theta}}^{(n)} + \left(\dot{\boldsymbol{\theta}}^{(n)}\right)^T\dot{\boldsymbol{\theta}}^{(n)} = 0 \tag{8.6.10}$$

so that the equations of motion become finally

$$\begin{aligned}\ddot{\boldsymbol{\theta}}^{(n)} = &-\frac{1}{8}\left(\bar{\mathbf{H}}^{(n)}\right)^T\left(\bar{\mathbf{I}}_P^{(n)}\right)^{-1}\bar{\mathbf{H}}^{(n)}\left(\dot{\bar{\mathbf{H}}}^{(n)}\right)^T\bar{\mathbf{I}}_P^{(n)}\bar{\mathbf{H}}^{(n)}\dot{\boldsymbol{\theta}}^{(n)} \\ &+\frac{1}{4}\left(\bar{\mathbf{H}}^{(n)}\right)^T\left(\bar{\mathbf{I}}_P^{(n)}\right)^{-1}\bar{\mathbf{M}}_P^{(n)} - \boldsymbol{\theta}^{(n)}\left(\dot{\boldsymbol{\theta}}^{(n)}\right)^T\dot{\boldsymbol{\theta}}^{(n)}\end{aligned} \tag{8.6.11}$$

which is in a form that can easily be turned into a system of first-order differential equations for the Euler parameters. During the numerical solution we still need to ensure that the constraint equation $\boldsymbol{\theta}^T\boldsymbol{\theta} = 1$ is satisfied. As with ordinary physical-based constraints we could try to include satisfaction of this constraint (and its derivative) as part of the solution process, but further discussion of this topic is left to other more advanced texts.

8.7 Solving Systems of Rigid Bodies

In solving for the motion of a rigid body, one must choose a set of generalized coordinates. If we know the motion of a given point in the body and the behavior of a set of angles that define the rotational motion, then we can find the motion of any other point in the rigid body so consider using the Cartesian components of the motion of a point in the body, $\mathbf{x}_A = (X_A, Y_A, Z_A)$, and a set of either Euler angles or Euler parameters, $\boldsymbol{\theta}$, a choice which is similar to the one used in planar problems. Thus, our generalized coordinates for body (n) will be $\mathbf{q}^{(n)} = \left(\mathbf{x}_A^{(n)}, \boldsymbol{\theta}^{(n)}\right)$. We want to write the equations of motion in those generalized coordinates in the form

$$\mathbf{M}^{(n)}\ddot{\mathbf{q}}^{(n)} = \mathbf{Q}^{(n);in} + \mathbf{Q}^{(n);e} \tag{8.7.1}$$

where $\mathbf{M}^{(n)}$ is a mass matrix and $\mathbf{Q}^{(n);in}$ and $\mathbf{Q}^{(n);e}$ are the generalized inertial and applied external forces, respectively. To obtain these terms we can go back to the Newton-Euler equations given in Eqs. (8.4.11) and (8.4.18b):

$$\begin{aligned}m^{(n)}\ddot{\mathbf{x}}_A^{(n)} - m^{(n)}\mathbf{X}_{C/A}^{(n)}{}_I\boldsymbol{\alpha}_B^{(n)} + m^{(n)}{}_I\boldsymbol{\Omega}_B^{(n)}{}_I\boldsymbol{\Omega}_B^{(n)}\mathbf{x}_{C/A}^{(n)} &= \mathbf{F}^{(n)} \\ m^{(n)}\bar{\mathbf{X}}_{C/A}^{(n)}\ddot{\bar{\mathbf{x}}}_A^{(n)} + \bar{\mathbf{I}}_A^{(n)}{}_I\bar{\boldsymbol{\alpha}}_B^{(n)} + {}_I\bar{\boldsymbol{\Omega}}_B^{(n)}\bar{\mathbf{I}}_A^{(n)}{}_I\bar{\boldsymbol{\omega}}_B^{(n)} &= \bar{\mathbf{M}}_A^{(n)}\end{aligned} \tag{8.7.2}$$

which are in a mixed form involving both inertial and body-fixed axes. To make them consistently use the acceleration of point A, $\ddot{\mathbf{x}}_A$, in inertial coordinates and to ensure that all terms that involve the relative positions of two points in the body such as $\mathbf{x}_{C/A}^{(n)}, \mathbf{X}_{C/A}^{(n)}, \mathbf{I}_A^{(n)}$ are written in body-fixed axes so they remain constants during the motion, we will use the relationship between inertial and body-fixed coordinates given by

$$\begin{aligned}
&\mathbf{x}_{C/A}^{(n)} = \mathbf{R}^{(n)}\bar{\mathbf{x}}_{C/A}^{(n)}, \quad {}_I\boldsymbol{\alpha}_B^{(n)} = \mathbf{R}^{(n)}{}_I\bar{\boldsymbol{\alpha}}_B^{(n)} \\
&\mathbf{X}_{C/A}^{(n)} = \mathbf{R}^{(n)}\bar{\mathbf{X}}_{C/A}^{(n)}\left(\mathbf{R}^{(n)}\right)^T, \quad {}_I\boldsymbol{\Omega}_B = \mathbf{R}^{(n)}{}_I\bar{\boldsymbol{\Omega}}_B\left(\mathbf{R}^{(n)}\right)^T \\
&\ddot{\bar{\mathbf{x}}}_A^{(n)} = \left(\mathbf{R}^{(n)}\right)^T\ddot{\mathbf{x}}_A^{(n)}
\end{aligned} \tag{8.7.3}$$

to rewrite these equations as

$$\begin{aligned}
&m^{(n)}\ddot{\mathbf{x}}_A^{(n)} - m^{(n)}\mathbf{R}^{(n)}\bar{\mathbf{X}}_{C/A}^{(n)}{}_I\bar{\boldsymbol{\alpha}}_B^{(n)} + m^{(n)}\mathbf{R}^{(n)}{}_I\bar{\boldsymbol{\Omega}}_B^{(n)}{}_I\bar{\boldsymbol{\Omega}}_B^{(n)}\bar{\mathbf{x}}_{C/A}^{(n)} = \mathbf{F}^{(n)} \\
&m^{(n)}\bar{\mathbf{X}}_{C/A}^{(n)}\left(\mathbf{R}^{(n)}\right)^T\ddot{\mathbf{x}}_A^{(n)} + \bar{\mathbf{I}}_A^{(n)}{}_I\bar{\boldsymbol{\alpha}}_B^{(n)} + {}_I\bar{\boldsymbol{\Omega}}_B^{(n)}\bar{\mathbf{I}}_A^{(n)}{}_I\bar{\boldsymbol{\omega}}_B^{(n)} = \bar{\mathbf{M}}_A^{(n)}
\end{aligned} \tag{8.7.4}$$

In terms of either Euler angles or Euler parameters we have ${}_I\bar{\boldsymbol{\omega}}_B^{(n)} = \bar{\mathbf{H}}^{(n)}\dot{\boldsymbol{\theta}}^{(n)}$ and ${}_I\bar{\boldsymbol{\alpha}}_B^{(n)} = \bar{\mathbf{H}}^{(n)}\ddot{\boldsymbol{\theta}}^{(n)} + \dot{\bar{\mathbf{H}}}^{(n)}\dot{\boldsymbol{\theta}}^{(n)}$ (but note that for Euler parameters $\dot{\bar{\mathbf{H}}}^{(n)}\dot{\boldsymbol{\theta}}^{(n)} = 0$ (see Eq. (4.10.27b))) so placing these relations into Eq. (8.7.4) gives

$$\begin{aligned}
&m^{(n)}\ddot{\mathbf{x}}_A^{(n)} - m^{(n)}\mathbf{R}^{(n)}\bar{\mathbf{X}}_{C/A}^{(n)}\left(\bar{\mathbf{H}}^{(n)}\ddot{\boldsymbol{\theta}}^{(n)} + \dot{\bar{\mathbf{H}}}^{(n)}\dot{\boldsymbol{\theta}}^{(n)}\right) + m^{(n)}\mathbf{R}^{(n)}{}_I\bar{\boldsymbol{\Omega}}_B^{(n)}{}_I\bar{\boldsymbol{\Omega}}_B^{(n)}\bar{\mathbf{x}}_{C/A}^{(n)} = \mathbf{F}^{(n)} \\
&m^{(n)}\bar{\mathbf{X}}_{C/A}^{(n)}\left(\mathbf{R}^{(n)}\right)^T\ddot{\mathbf{x}}_A^{(n)} + \bar{\mathbf{I}}_A^{(n)}\left(\bar{\mathbf{H}}^{(n)}\ddot{\boldsymbol{\theta}}^{(n)} + \dot{\bar{\mathbf{H}}}^{(n)}\dot{\boldsymbol{\theta}}^{(n)}\right) + {}_I\bar{\boldsymbol{\Omega}}_B^{(n)}\bar{\mathbf{I}}_A^{(n)}{}_I\bar{\boldsymbol{\omega}}_B^{(n)} = \bar{\mathbf{M}}_A^{(n)}
\end{aligned} \tag{8.7.5}$$

Equation (8.7.5) is almost in the final form we need. For Euler parameters we need four equations for these parameters, which we can obtain by multiplying the moment equation by $\left(\bar{\mathbf{H}}^{(n)}\right)^T$. This will also make the mass matrix symmetrical for both the Euler angle and Euler parameter choices, as we will see shortly. Thus, performing this multiplication and rearranging terms we find finally

$$\begin{aligned}
&m^{(n)}\ddot{\mathbf{x}}_A^{(n)} - m^{(n)}\mathbf{R}^{(n)}\bar{\mathbf{X}}_{C/A}^{(n)}\bar{\mathbf{H}}^{(n)}\ddot{\boldsymbol{\theta}}^{(n)} \\
&\quad = \mathbf{F}^{(n)} - m^{(n)}\mathbf{R}^{(n)}{}_I\bar{\boldsymbol{\Omega}}_B^{(n)}{}_I\bar{\boldsymbol{\Omega}}_B^{(n)}\bar{\mathbf{x}}_{C/A}^{(n)} + m^{(n)}\mathbf{R}^{(n)}\bar{\mathbf{X}}_{C/A}^{(n)}\dot{\bar{\mathbf{H}}}^{(n)}\dot{\boldsymbol{\theta}}^{(n)} \\
&m^{(n)}\left(\bar{\mathbf{H}}^{n}\right)^T\bar{\mathbf{X}}_{C/A}^{(n)}\left(\mathbf{R}^{(n)}\right)^T\ddot{\mathbf{x}}_A^{(n)} + \left(\bar{\mathbf{H}}^{(n)}\right)^T\bar{\mathbf{I}}_A^{(n)}\bar{\mathbf{H}}^{(n)}\ddot{\boldsymbol{\theta}}^{(n)} \\
&\quad = \left(\bar{\mathbf{H}}^{(n)}\right)^T\bar{\mathbf{M}}_A^{(n)} - \left(\bar{\mathbf{H}}^{(n)}\right)^T\bar{\mathbf{I}}_A^{(n)}\dot{\bar{\mathbf{H}}}^{(n)}\dot{\boldsymbol{\theta}}^{(n)} - \left(\bar{\mathbf{H}}^{(n)}\right)^T{}_I\bar{\boldsymbol{\Omega}}_B^{(n)}\bar{\mathbf{I}}_A^{(n)}{}_I\bar{\boldsymbol{\omega}}_B^{(n)}
\end{aligned} \tag{8.7.6}$$

which we can write in the form of Eq. (8.7.1) as

$$\begin{bmatrix} \mathbf{M}_{xx}^{(n)} & \mathbf{M}_{x\theta}^{(n)} \\ \mathbf{M}_{\theta x}^{(n)} & \mathbf{M}_{\theta\theta}^{(n)} \end{bmatrix}\begin{Bmatrix} \ddot{\mathbf{x}}_A \\ \ddot{\boldsymbol{\theta}} \end{Bmatrix} = \begin{Bmatrix} \mathbf{Q}_x^{(n);in} \\ \mathbf{Q}_\theta^{(n);in} \end{Bmatrix} + \begin{Bmatrix} \mathbf{Q}_x^{(n);e} \\ \mathbf{Q}_\theta^{(n;e)} \end{Bmatrix} \tag{8.7.7}$$

where

$$\begin{aligned}
\mathbf{M}_{xx}^{(n)} &= m^{(n)}\mathbf{I}, \quad \mathbf{M}_{x\theta}^{(n)} = \left(\mathbf{M}_{\theta x}^{(n)}\right)^T = -m^{(n)}\mathbf{R}^{(n)}\bar{\mathbf{X}}_{C/A}^{(n)}\bar{\mathbf{H}}^{(n)} \\
\mathbf{M}_{\theta\theta}^{(n)} &= \left(\bar{\mathbf{H}}^{(n)}\right)^T\bar{\mathbf{I}}_A^{(n)}\bar{\mathbf{H}}^{(n)} \\
\mathbf{Q}_x^{(n);e} &= \mathbf{F}^{(n)}, \; \mathbf{Q}_\theta^{(n);e} = \left(\bar{\mathbf{H}}^{(n)}\right)^T\bar{\mathbf{M}}_A^{(n)} \\
\mathbf{Q}_x^{(n);in} &= -m^{(n)}\mathbf{R}^{(n)}{}_I\bar{\boldsymbol{\Omega}}_B^{(n)}{}_I\bar{\boldsymbol{\Omega}}_B^{(n)}\bar{\mathbf{x}}_{C/A}^{(n)} + m^{(n)}\mathbf{R}^{(n)}\bar{\mathbf{X}}_{C/A}^{(n)}\dot{\bar{\mathbf{H}}}^{(n)}\dot{\boldsymbol{\theta}}^{(n)} \\
\mathbf{Q}_\theta^{(n);in} &= -\left(\bar{\mathbf{H}}^{(n)}\right)^T\bar{\mathbf{I}}_A^{(n)}\dot{\bar{\mathbf{H}}}^{(n)}\dot{\boldsymbol{\theta}}^{(n)} - \left(\bar{\mathbf{H}}^{(n)}\right)^T{}_I\bar{\boldsymbol{\Omega}}_B^{(n)}\bar{\mathbf{I}}_A^{(n)}{}_I\bar{\boldsymbol{\omega}}_B^{(n)}
\end{aligned} \tag{8.7.8}$$

and $\mathbf{I}$ is the 3×3 identity matrix. These terms are rather complex since there is coupling between the motion of A and the rotations in both the mass matrix and the generalized inertial force. However, if we choose point A to be the mass center, C, we arrive at a much simpler result where

$$\begin{aligned}
\mathbf{M}_{xx}^{(n)} &= m^{(n)}\mathbf{I}, \quad \mathbf{M}_{x\theta}^{(n)} = \left(\mathbf{M}_{\theta x}^{(n)}\right)^T = \mathbf{0} \\
\mathbf{M}_{\theta\theta}^{(n)} &= \left(\bar{\mathbf{H}}^{(n)}\right)^T\bar{\mathbf{I}}_C^{(n)}\bar{\mathbf{H}}^{(n)} \\
\mathbf{Q}_x^{(n);e} &= \mathbf{F}^{(n)}, \quad \mathbf{Q}_\theta^{(n);e} = \left(\bar{\mathbf{H}}^{(n)}\right)^T\bar{\mathbf{M}}_C^{(n)} \\
\mathbf{Q}_x^{(n);in} &= 0 \\
\mathbf{Q}_\theta^{(n);in} &= -\left(\bar{\mathbf{H}}^{(n)}\right)^T\bar{\mathbf{I}}_C^{(n)}\dot{\bar{\mathbf{H}}}^{(n)}\dot{\boldsymbol{\theta}}^{(n)} - \left(\bar{\mathbf{H}}^{(n)}\right)^T{}_I\bar{\boldsymbol{\Omega}}_B^{(n)}\bar{\mathbf{I}}_C^{(n)}{}_I\bar{\boldsymbol{\omega}}_B^{(n)}
\end{aligned} \tag{8.7.9}$$

Equation (8.7.7) is for a single unconstrained rigid body. For a system of N_b interconnected rigid bodies we can collect these terms into a system mass matrix and corresponding generalized inertial and external forces and include the constraint forces, $\mathbf{Q}^{(n);\,c}$, acting on each rigid body as well in the form

$$\begin{bmatrix} \mathbf{M}^{(1)} & 0 & \cdots & 0 \\ 0 & \mathbf{M}^{(2)} & & \vdots \\ \vdots & & \ddots & 0 \\ 0 & \cdots & 0 & \mathbf{M}^{(N_b)} \end{bmatrix} \begin{Bmatrix} \ddot{\mathbf{q}}^{(1)} \\ \ddot{\mathbf{q}}^{(2)} \\ \vdots \\ \ddot{\mathbf{q}}^{(N_b)} \end{Bmatrix} = \begin{Bmatrix} \mathbf{Q}^{(1);in} \\ \mathbf{Q}^{(2);in} \\ \vdots \\ \mathbf{Q}^{(N_b);in} \end{Bmatrix} + \begin{Bmatrix} \mathbf{Q}^{(1);e} \\ \mathbf{Q}^{(2);e} \\ \vdots \\ \mathbf{Q}^{(N_b);e} \end{Bmatrix} + \begin{Bmatrix} \mathbf{Q}^{(1);c} \\ \mathbf{Q}^{(2);c} \\ \vdots \\ \mathbf{Q}^{(N_b);c} \end{Bmatrix} \tag{8.7.10}$$

If the generalized constraint forces are ideal constraints and we use the Lagrange prescription for these constraints then we see that the equations of motion and constraints of a complex system of rigid bodies in three-dimensional motion can be placed into the same form we have seen before for particles, systems of particles, and rigid bodies in plane motion, namely

$$\begin{aligned} \mathbf{M}\ddot{\mathbf{q}} &= \mathbf{Q}^{in} + \mathbf{Q}^{e} + \mathbf{A}^{T}\boldsymbol{\lambda} \\ \mathbf{A}\ddot{\mathbf{q}} &= \mathbf{b} \end{aligned} \tag{8.7.11}$$

and these equations can be solved for the equations of motion and generalized constraint forces by the augmented approach. Details of solving such complex systems are beyond the scope of this book. You can find more details in texts on multibody dynamics (see, in particular, references [4–7]).

We should mention some details in computing the generalized forces in Eq. (8.7.9). The force $\mathbf{F}^{(n)}$ is the total external force acting on the rigid body. It comes from whatever individual forces, $\mathbf{F}_j^{(n)}$, are acting, as shown in Fig. 8.13. Thus, we have for the generalized force

$$\mathbf{Q}_x^{(n);e} = \sum_{j=1}^{n_f} \mathbf{F}_j^{(n)} \tag{8.7.12}$$

where n_f is the total number of forces. Note that all of these forces are computed in an inertial frame. The generalized force, $\mathbf{Q}_\theta^{(n);e}$, however, is in terms of the moment about the center of mass C, $\bar{\mathbf{M}}_C^{(n)}$, calculated in body-fixed axes. The moment about C in terms of inertial coordinates, $\mathbf{M}_C^{(n)}$, comes from the moments about C due to the external forces plus the moments due to any couples, $\mathbf{C}_j^{(n)}$, acting on the body, i.e.,

$$\mathbf{M}_C^{(n)} = \sum_{j=1}^{n_f} \mathbf{x}_{j/C}^{(n)} \times \mathbf{F}_j^{(n)} + \sum_{j=1}^{n_c} \mathbf{C}_j^{(n)} \tag{8.7.13}$$

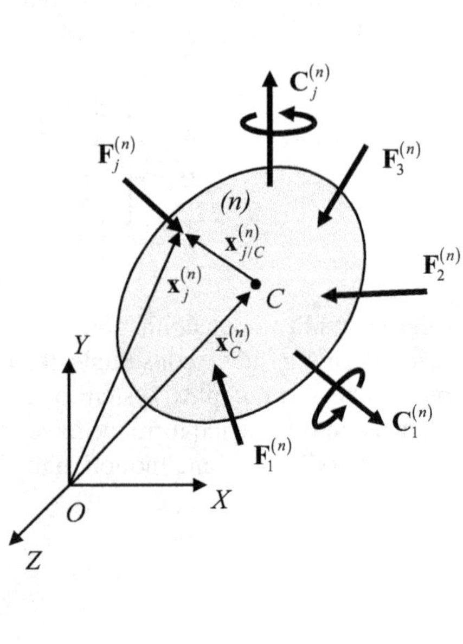

Fig. 8.13 External forces and couples acting on a rigid body

(see Fig. 8.13) where n_c is the total number of couples. In body-fixed coordinates, therefore, we have

$$\bar{\mathbf{M}}_C^{(n)} = \sum_{j=1}^{n_f} \bar{\mathbf{x}}_{j/C}^{(n)} \times \bar{\mathbf{F}}_j^{(n)} + \sum_{j=1}^{n_c} \bar{\mathbf{C}}_j^{(n)} \tag{8.7.14}$$

which is the moment appearing in Eq. (8.7.9).

8.7.1 Lagrange's Equations

The use of the motion of a body-fixed point as measured in Cartesian components and angular variables such as Euler angles or Euler parameters to describe the rotational motion is a good choice for generalized coordinates, $\mathbf{q}$, to describe even complex systems, as shown in the last section. However, one can also express the kinetic energy, T, in terms of a set of more general $\mathbf{u}$ coordinates where $\mathbf{q} = \mathbf{q}(\mathbf{u}, t)$ and use that expression to generate the equations of motion. We examined this case in Chap. 5 for planar motion problems. For 3-D spatial dynamics problems the details follow closely that of the planar case so we will not give the details here (see [8] for a complete discussion). For a single body under p ideal constraints one finds

$$\frac{d}{dt}\left(\frac{\partial T}{\partial \dot{u}_k}\right) - \frac{\partial T}{\partial u_k} = \tilde{Q}_k^e + \sum_{m=1}^{p} A_{mk}\lambda_m \tag{8.7.15}$$

with

$$\tilde{Q}_k^e = \mathbf{F}^e \cdot \frac{\partial \mathbf{v}_C}{\partial \dot{u}_k} + \mathbf{M}_C^e \cdot \frac{\partial_I \boldsymbol{\omega}_B}{\partial \dot{u}_k} \tag{8.7.16}$$

Note that the kinetic energy terms on the left side of Eq. (8.7.15) produce both the mass matrix term in the equations of motion as well as the generalized inertia forces. For the important special case where all the external forces (and moments) are derivable from a potential energy function, $V(\mathbf{q}, t) = \tilde{V}(\mathbf{u}, t)$, one finds

$$\frac{d}{dt}\left(\frac{\partial T}{\partial \dot{u}_k}\right) - \frac{\partial T}{\partial u_k} = -\frac{\partial \tilde{V}}{\partial u_k} + \sum_{m=1}^{p} A_{mk}\lambda_m \tag{8.7.17}$$

Dynamics texts often use Eq. (8.7.17) since one can go directly from the scalar kinetic energy and potential energy expressions to obtain explicit expressions for the terms in Eq. (8.7.17) by performing all the appropriate derivatives, thus bypassing the vectors and matrices present in the approach described earlier in this section. For constrained bodies, however, the unknown Lagrange multipliers are also present so they must be obtained or eliminated before we have the equations of motion described as a system of ordinary differential equations. This is often difficult to do by hand in spatial dynamics problems, but the augmented approach gives us the required equations directly once Eq. (8.7.17) is written in the form of Eq. (8.7.11).

8.8 The Rolling Disk

One of the classical problems that demonstrate three-dimensional motion with gyroscopic effects is where a thin disk of radius R rolls without slipping on a horizontal plane, as shown in Fig. 8.14. This problem also illustrates a case where there are non-holonomic constraints. To solve this problem we will use a set of 3-1-3 Euler angles, (ϕ, θ, ψ), where ϕ defines the precessional motion of the disk as it rolls around the plane, θ defines the nutational "nodding" of the disk, and ψ defines the angle of spin of the disk about a central axis normal to the disk. However, rather than using a set of body-fixed axes $(\bar{x}_1, \bar{x}_2, \bar{x}_3)$ we will consider a set of intermediate (x_1^a, x_2^a, x_3^a) axes (also called the a-frame) similar to what we used in the fixed gyro-pendulum problem, where here the (x_1^a, x_2^a, x_3^a) axes result from the 3-1 rotations; that is, the third (ψ) rotation is absent. Thus, we can obtain results in these intermediate axes by using expressions for a set of 3-1-3 Euler angles and setting $\psi = 0$. In terms of these intermediate coordinates from Eqs. (4.11.2) and (4.11.4) we find the angular velocity of the disk as

$$ {}_I\tilde{\boldsymbol{\omega}}_B \equiv \left[{}_I\boldsymbol{\omega}_B\right]_a = \dot{\theta}\,\mathbf{e}_1^a + \sin\theta\dot{\phi}\,\mathbf{e}_2^a + \left(\dot{\phi}\,\cos\theta + \dot{\psi}\right)\mathbf{e}_3^a \tag{8.8.1} $$

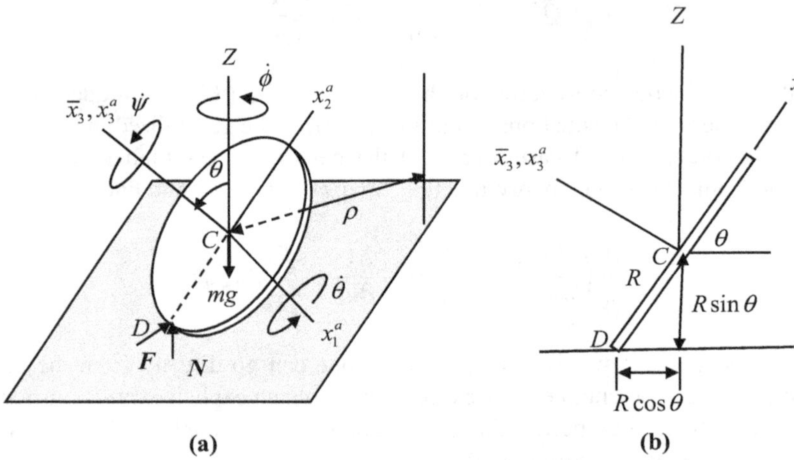

Fig. 8.14 (**a**) A disk rolling without slipping on a horizontal plane. (**b**) A side view

where the tilde and bracket notation $[]_a$ indicate the evaluation of this angular velocity in the a-frame. Note that the angular velocity of the intermediate coordinate a-frame, ${}_I\boldsymbol{\omega}_a$, can similarly be written as

$$ {}_I\tilde{\boldsymbol{\omega}}_a \equiv \left[{}_I\boldsymbol{\omega}_a\right]_a = \dot{\theta}\,\mathbf{e}_1^a + \sin\theta\dot{\phi}\,\mathbf{e}_2^a + \dot{\phi}\,\cos\theta\mathbf{e}_3^a \tag{8.8.2} $$

In solving this problem, consider $(X_C, Y_C, Z_C, \phi, \theta, \psi)$ as generalized coordinates, where (X_C, Y_C, Z_C) are the coordinates of the center of mass, C, as measured in a set of fixed (X, Y, Z) axes and (ϕ, θ, ψ) are our 3-1-3 Euler angles. Since we have assumed the disk rolls without slipping these generalized coordinates are not independent. The point of contact, D, of the disk with the plane must have zero velocity for no slipping so we must have

$$ \mathbf{v}_D = \mathbf{v}_C + {}_I\boldsymbol{\omega}_B \times \mathbf{x}_{D/C} = 0 \tag{8.8.3a} $$

which we can write more explicitly using inertial coordinates to evaluate the velocities and the a-frame coordinates for the cross product terms as

$$ \left[\mathbf{v}_D\right]_I = \left[\mathbf{v}_C\right]_I + \left[{}_I\tilde{\boldsymbol{\omega}}_B \times \tilde{\mathbf{x}}_{D/C}\right]_I = 0 \tag{8.8.3b} $$

where $[]_I$ means we must transform the cross product in the a-frame to inertial coordinates. Thus, we find

$$ \dot{X}_C\mathbf{E}_X + \dot{Y}_C\mathbf{E}_Y + \dot{Z}_C\mathbf{E}_Z + \left[\left(\dot{\theta}\,\mathbf{e}_1^a + \sin\theta\dot{\phi}\,\mathbf{e}_2^a + (\dot{\phi}\,\cos\theta + \dot{\psi}\,)\mathbf{e}_3^a\right) \times \left(-R\mathbf{e}_2^a\right)\right]_I = 0 \tag{8.8.4} $$

or

$$ \dot{X}_C\mathbf{E}_X + \dot{Y}_C\mathbf{E}_Y + \dot{Z}_C\mathbf{E}_Z + \left[-R\dot{\theta}\,\mathbf{e}_3^a + R(\dot{\phi}\,\cos\theta + \dot{\psi}\,)\mathbf{e}_1^a\right]_I = 0 \tag{8.8.5} $$

But from Eq. (4.11.3), using $\mathbf{R}^T$ with $\theta_3 = \psi = 0$ to relate $\mathbf{e}_n^a$ to $\mathbf{E}_n$ we find

$$ \begin{aligned} \mathbf{e}_1^a &= \cos\phi\mathbf{E}_X + \sin\phi\mathbf{E}_Y \\ \mathbf{e}_2^a &= -\cos\theta\sin\phi\mathbf{E}_X + \cos\theta\cos\phi\mathbf{E}_Y + \sin\theta\mathbf{E}_Z \\ \mathbf{e}_3^a &= \sin\theta\sin\phi\mathbf{E}_X - \sin\theta\cos\phi\mathbf{E}_Y + \cos\theta\mathbf{E}_Z \end{aligned} \tag{8.8.6} $$

which gives the constraint equations in inertial coordinates as

$$ \begin{aligned} &(\dot{X}_C - R\dot{\theta}\,\sin\theta\sin\phi + R(\dot{\phi}\,\cos\theta + \dot{\psi}\,)\cos\phi)\mathbf{E}_X \\ &+(\dot{Y}_C + R\dot{\theta}\,\sin\theta\cos\phi + R(\dot{\phi}\,\cos\theta + \dot{\psi}\,)\sin\phi)\mathbf{E}_Y \\ &+(\dot{Z}_C - R\dot{\theta}\,\cos\theta)\mathbf{E}_Z = 0 \end{aligned} \tag{8.8.7} $$

The first two of these constraints are non-holonomic:

$$\begin{aligned} \dot{X}_C - R\dot{\theta}\,\sin\theta\sin\phi + R\big(\dot{\phi}\,\cos\theta + \dot{\psi}\,\big)\cos\phi = 0 \\ \dot{Y}_C + R\dot{\theta}\,\sin\theta\cos\phi + R\big(\dot{\phi}\,\cos\theta + \dot{\psi}\,\big)\sin\phi = 0 \end{aligned} \tag{8.8.8}$$

while the third equation is holonomic since

$$\dot{Z}_C - R\dot{\theta}\,\cos\theta = 0 \tag{8.8.9}$$

can be integrated to give

$$Z_C = R\sin\theta \tag{8.8.10}$$

(see Fig. 8.14b) so that we can use this constraint to eliminate Z_C (i.e., embed this constraint) and describe the problem in terms of the five generalized coordinates $(q_1, q_2, q_3, q_4, q_5) = (X_C, Y_C, \phi, \theta, \psi)$. If we use Lagrange's equations and Lagrange multipliers to represent the ideal constraint forces, as discussed in the last section, we have

$$\begin{aligned} \frac{d}{dt}\left(\frac{\partial T}{\partial \dot{q}_k}\right) - \frac{\partial T}{\partial q_k} = -\frac{\partial V}{\partial q_k} + \sum_{p=1}^{2} A_{pk}\lambda_p \quad (k = 1, \ldots, 5) \\ \sum_{k=1}^{5} A_{jk}\ddot{q}_k = b_j \quad (j = 1, 2) \end{aligned} \tag{8.8.11}$$

Rather than solving this system of equations explicitly with an augmented approach (see Prob P8.7 where you are asked to obtain the augmented solution), we will follow the steps given by Baruh [3] which will allow us to analyze an important special case where the disk rolls in a circle with no nutational motion. First, write the equations of motion in terms of the Lagrange multiplier, which is easy to do since the **A** matrix appearing in the constraint equations involving the accelerations is the same matrix appearing in our velocity constraints in Eq. (8.8.8), which we can write as

$$\sum_{k=1}^{5} A_{jk}\dot{q}_k = 0 \quad (j = 1, 2) \tag{8.8.12}$$

where

$$[A] = \begin{bmatrix} 1 & 0 & R\cos\theta\cos\phi & -R\sin\theta\sin\psi & R\cos\phi \\ 0 & 1 & R\cos\theta\sin\phi & R\sin\theta\cos\phi & R\sin\psi \end{bmatrix} \tag{8.8.13}$$

The kinetic and potential energy is given as (see Eq. (8.8.1) for the angular velocity components)

$$\begin{aligned} T &= \frac{1}{2}m\left(\dot{X}_C^2 + \dot{Y}_C^2 + \dot{Z}_C^2\right) + \frac{1}{2}I_3\left(\dot{\phi}\cos\theta + \dot{\psi}\right)^2 + \frac{1}{2}I_1\dot{\theta}^2 + \frac{1}{2}I_2\sin^2\theta\dot{\phi}^2 \\ &= \frac{1}{2}m\left(\dot{X}_C^2 + \dot{Y}_C^2 + R^2\dot{\theta}^2\cos^2\theta\right) + \frac{1}{2}I_3\left(\dot{\phi}\cos\theta + \dot{\psi}\right)^2 + \frac{1}{2}I_1\dot{\theta}^2 + \frac{1}{2}I_2\sin^2\theta\dot{\phi}^2 \\ V &= mgR\sin\theta \end{aligned} \tag{8.8.14}$$

For a thin homogeneous disk of mass m and radius R we have

$$I_3 = \frac{1}{2}mR^2, \quad I_2 = I_1 = \frac{1}{4}mR^2 \tag{8.8.15}$$

Using Eq. (8.8.14) in the Lagrange equations of Eq. (8.8.11) then gives

$$\begin{aligned} &m\ddot{X}_C = \lambda_1 \\ &m\ddot{Y}_C = \lambda_2 \\ &\left(I_1\sin^2\theta + I_3\cos^2\theta\right)\ddot{\phi} + I_3\cos\theta\ddot{\psi} + (2I_1 - 2I_3)\sin\theta\cos\theta\dot{\theta}\dot{\phi} - I_3\sin\theta\dot{\theta}\dot{\psi} \\ &\quad = R\cos\theta\cos\phi\lambda_1 + R\cos\theta\sin\phi\lambda_2 \\ &\left(I_1 + mR^2\cos^2\theta\right)\ddot{\theta} - mR^2\sin\theta\cos\theta\dot{\theta}^2 - (I_1 + I_3)\sin\theta\cos\theta\dot{\phi}^2 - I_3\sin\theta\dot{\phi}\dot{\psi} \\ &\quad = -mgR\cos\theta - R\sin\theta\sin\phi\lambda_1 + R\sin\theta\cos\phi\lambda_2 \\ &I_3\cos\theta\ddot{\phi} + I_3\ddot{\psi} - I_3\sin\theta\dot{\theta}\dot{\phi} = R\cos\phi\lambda_1 + R\sin\phi\lambda_2 \end{aligned} \tag{8.8.16}$$

which show us that the Lagrange multipliers (λ_1, λ_2) are just the constraint forces acting in the (X, Y) directions, respectively. If we differentiate the constraint equations, Eq. (8.8.8), on time and use them in the first two equations of motion, we can write the Lagrange multipliers in terms of the (ϕ, θ, ψ) coordinates and their derivatives and then place those expressions into the last three equations in Eq. (8.8.16). When done by hand, this process is essentially the same as using the explicit expressions for the constraint forces in the augmented approach. The details of this lengthy process we will not describe here but instead simply give the end result as the three equations obtained by Baruh [3]:

$$\begin{aligned} &\left(I_3 + mR^2\right)\left(\ddot{\psi} + \cos\theta\ddot{\phi} - \sin\theta\dot{\theta}\dot{\phi}\right) - mR^2\sin\theta\dot{\theta}\dot{\phi} = 0 \\ &\left(I_1 + mR^2\right)\ddot{\theta} + \left(I_3 + mR^2\right)\sin\theta\dot{\phi}\left(\cos\theta\dot{\phi} + \dot{\psi}\right) - I_1\sin\theta\cos\theta\dot{\phi}^2 = -mgR\cos\theta \\ &I_1\sin\theta\ddot{\phi} + 2I_1\cos\theta\dot{\theta}\dot{\phi} - I_3\dot{\theta}\left(\cos\theta\dot{\phi} + \dot{\psi}\right) = 0 \end{aligned} \tag{8.8.17}$$

These are still rather complex equations but we can use them to consider the following special case. Let the nutation angle, θ, be a constant. Then Eq. (8.8.17) reduces to

$$\begin{aligned}&(I_3 + mR^2)(\ddot{\psi} + \cos\theta\ddot{\phi}) = 0\\&+(I_3 + mR^2)\sin\theta\dot{\phi}\left(\cos\theta\dot{\phi} + \dot{\psi}\right) - I_1\sin\theta\cos\theta\dot{\phi}^2 = -mgR\cos\theta\\&I_1\sin\theta\ddot{\phi} = 0\end{aligned} \tag{8.8.18}$$

The third equation in Eq. (8.8.18) shows that the precessional rate, $\dot{\phi}$, is a constant. Placing this result back into the first equation in Eq. (8.8.18) gives that the spin rate, $\dot{\psi}$, is also a constant. A motion that satisfies these conditions is when the center of the disk, C, follows a circular path with the precessional velocity being the rotational rate of the radial line that defines that circular path. To see this, try a solution for the motion of C that satisfies

$$\begin{aligned}X_C &= \rho\sin\phi\\Y_C &= -\rho\cos\phi\end{aligned} \tag{8.8.19}$$

where ρ is the constant radius of the path (see Fig. 8.13a). Placing this relationship into Eq. (8.8.8), we see that the constraints are satisfied if

$$\dot{\psi} = -\dot{\phi}\left(\frac{\rho}{R} + \cos\theta\right) \tag{8.8.20}$$

Placing Eq. (8.8.20) into the second equation of Eq. (8.8.18) then gives

$$\dot{\phi}^2 = \frac{mgR^2\cot\theta}{(I_3 + mR^2)\rho + I_1R\cos\theta} \tag{8.8.21}$$

which can be expressed in a more compact form if we let $I_3 = m\kappa^2, I_1 = m\kappa^2/2$, where κ is a radius of gyration for the disk. We find

$$\dot{\phi}^2 = \frac{2gR^2\cot\theta}{2(\kappa^2 + R^2)\rho + \kappa^2R\cos\theta} \tag{8.8.22}$$

For a homogeneous disk $\kappa = R/\sqrt{2}$ and Eq. (8.8.22) reduces to

$$\dot{\phi}^2 = \frac{4g\cot\theta}{6\rho + R\cos\theta} \tag{8.8.23}$$

To obtain a numerical solution for the generalized coordinates $(X_C, Y_C, \phi, \theta, \psi)$ for the case where the disk does follow a circular path with no nutational motion, we must first specify the parameters $(\rho, R, \theta = \theta_0)$, which fixes the radius of the path, the size of the disk, and the constant nutation angle. We also need to know the initial conditions for $(X_C, Y_C, \phi, \theta, \psi)$ and their velocities $\left(\dot{X}_C, \dot{Y}_C, \dot{\phi}, \dot{\theta}, \dot{\psi}\right)$ which will lead to such circular motion. Obviously for a constant nutation angle we have $\theta(0) = \theta_0$ and $\dot{\theta}(0) = 0$. From Eqs. (8.8.23) and (8.8.20) we find

$$\begin{aligned} \dot{\phi}(0) &= \sqrt{\frac{4g\cot\theta_0}{6\rho + R\cos\theta_0}} \\ \dot{\psi}(0) &= -\dot{\phi}(0)\left(\frac{\rho}{R} + \cos\theta_0\right) \end{aligned} \tag{8.8.24}$$

If we specify the initial precessional and spin angles as $\phi(0) = \phi_0$ and $\psi(0) = \psi_0$ then from Eq. (8.8.19) we must have

$$\begin{aligned} X_C(0) &= \rho\sin\phi_0 \\ Y_C(0) &= -\rho\cos\phi_0 \end{aligned} \tag{8.8.25}$$

Finally, from the constraint equations, Eq. (8.8.8), we must have

$$\begin{aligned} \dot{X}_C(0) &= -R\left(\dot{\phi}(0)\cos\theta_0 + \dot{\psi}(0)\right)\cos\phi_0 \\ \dot{Y}_C(0) &= -R\left(\dot{\phi}(0)\cos\theta_0 + \dot{\psi}(0)\right)\sin\phi_0 \end{aligned} \tag{8.8.26}$$

Rolling in a circular path with no nutational angle changes is a special case that can be used to check the numerical solution of the augmented system of equations for the five generalized coordinates. In Prob. P8.7 you are asked to do just such a check for this problem.

The constraint forces acting on the rolling-disk problem are frictional forces so that in order to have rolling without slipping those forces must not exceed their maximum allowable values. It is useful to examine those frictional forces in an (X', Y', Z') system which is obtained after the first Euler angle rotation (Fig. 8.15). There are two components of the frictional force in this coordinate system. If the constraint maintained by force F_1 is lost the spin of the disk is decoupled from its forward motion, a condition called spin slip. If, however, the constraint maintained by F_2 is lost instead the disk will slide on the surface, resulting in what is called sliding slip. These conditions of course do not occur separately in general, where for no slip of either type we must have

$$\sqrt{F_1^2 + F_2^2} \geq \mu N \tag{8.8.27}$$

where N is the normal force and μ is the coefficient of friction. It may seem that by embedding the Z-constraint in our analysis we will have difficulty in examining this no-slip condition since the normal force never appears in the analysis. However, the normal force can easily be found since the missing force equation in the Z-direction is just

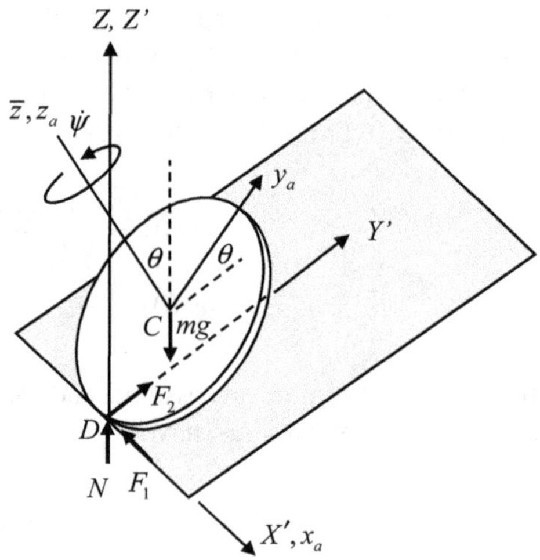

Fig. 8.15 Geometry for examining the friction forces in the rolling-disk problem

$$N - mg = m\ddot{Z} = m\frac{d}{dt}\left(R\cos\theta\dot{\theta}\right) \tag{8.8.28}$$

To obtain the frictional forces we will again follow a series of steps outlined by Baruh [3]. We note that the relationship between the acceleration of the center of mass and these forces in the (X', Y', Z') system is just

$$\begin{aligned} ma_{CX'} &= F_{X'} = -F_1 \\ ma_{CY'} &= F_{Y'} = F_2 \\ ma_{CZ'} &= N - mg \end{aligned} \tag{8.8.29}$$

where the last equation in Eq. (8.8.29) is just Eq. (8.8.28). The relationship between the unit vectors in the (X', Y', Z') system and the a-frame is just

$$\begin{aligned} \mathbf{e}_1^a \equiv \mathbf{e}_{x_a} &= \mathbf{e}_{X'} \\ \mathbf{e}_2^a \equiv \mathbf{e}_{y_a} &= \cos\theta\mathbf{e}_{Y'} + \sin\theta\mathbf{e}_{Z'} \\ \mathbf{e}_3^a \equiv \mathbf{e}_{z_a} &= -\sin\theta\mathbf{e}_{Y'} + \cos\theta\mathbf{e}_{Z'} \end{aligned} \tag{8.8.30}$$

since these axes only differ by the θ-rotation (see Fig. 8.14). Thus, if we write the acceleration of the center of mass in terms of the a-frame components we can use Eq. (8.8.30) to place these accelerations into their (X', Y', Z') components and so obtain expressions for the forces in Eq. (8.8.29). First, recall we already know the angular velocity of the disk and the angular velocity of the a-frame from Eqs. (8.8.1) and (8.8.2), which we rewrite here:

$$
\begin{aligned}
{}_I\tilde{\boldsymbol{\omega}}_B &= \dot{\theta}\,\mathbf{e}_1^a + \sin\theta\dot{\phi}\,\mathbf{e}_2^a + (\dot{\phi}\,\cos\theta + \dot{\psi})\mathbf{e}_3^a \\
{}_I\tilde{\boldsymbol{\omega}}_a &= \dot{\theta}\,\mathbf{e}_1^a + \sin\theta\dot{\phi}\,\mathbf{e}_2^a + \dot{\phi}\,\cos\theta\mathbf{e}_3^a
\end{aligned}
\tag{8.8.31}
$$

In obtaining the acceleration of the center of mass we will have to perform some time derivatives which will become rather complex algebraically, so let us first rewrite Eq. (8.8.31) in the simpler forms:

$$
\begin{aligned}
{}_I\tilde{\boldsymbol{\omega}}_B &= \tilde{\omega}_1\mathbf{e}_1^a + \tilde{\omega}_2\mathbf{e}_2^a + \tilde{\omega}_3\mathbf{e}_3^a \\
{}_I\tilde{\boldsymbol{\omega}}_a &= \tilde{\omega}_1\mathbf{e}_1^a + \tilde{\omega}_2\mathbf{e}_2^a + \frac{\tilde{\omega}_2}{\tan\theta}\mathbf{e}_3^a
\end{aligned}
\tag{8.8.32}
$$

Because of the no-slip condition the velocity of the center of mass in a-frame coordinates, $\tilde{\mathbf{v}}_C$, is therefore

$$
\begin{aligned}
\tilde{\mathbf{v}}_C &= {}_I\tilde{\boldsymbol{\omega}}_B \times R\mathbf{e}_2^a \\
&= R\omega_1\mathbf{e}_3^a - R\omega_3\mathbf{e}_1^a
\end{aligned}
\tag{8.8.33}
$$

We can obtain the acceleration in a-frame coordinates by using the transport theorem of Chap. 4 as

$$
\begin{aligned}
\tilde{\mathbf{a}}_C &= \frac{d^a\tilde{\mathbf{v}}_C}{dt} + {}_I\tilde{\boldsymbol{\omega}}_a \times \tilde{\mathbf{v}}_C \\
&= \left(R\dot{\tilde{\omega}}_1\mathbf{e}_3^a - R\dot{\tilde{\omega}}_3\mathbf{e}_1^a\right) + \left(\tilde{\omega}_1\mathbf{e}_1^a + \tilde{\omega}_2\mathbf{e}_2^a + \frac{\tilde{\omega}_2}{\tan\theta}\mathbf{e}_3^a\right) \times \left(R\tilde{\omega}_1\mathbf{e}_3^a - R\tilde{\omega}_3\mathbf{e}_1^a\right) \\
&= R\left(\tilde{\omega}_1\tilde{\omega}_2 - \dot{\tilde{\omega}}_3\right)\mathbf{e}_1^a - R\left(\tilde{\omega}_1^2 + \frac{\tilde{\omega}_2\tilde{\omega}_3}{\tan\theta}\right)\mathbf{e}_2^a + R\left(\dot{\tilde{\omega}}_1 + \tilde{\omega}_2\tilde{\omega}_3\right)\mathbf{e}_3^a
\end{aligned}
\tag{8.8.34}
$$

Replacing $(\tilde{\omega}_1, \tilde{\omega}_2, \tilde{\omega}_3)$ by their Euler angle expressions and carrying out the necessary derivatives, we find

$$
\begin{aligned}
\tilde{\mathbf{a}}_C = {} & R\left(-\ddot{\phi}\cos\theta + 2\dot{\phi}\dot{\theta}\,\sin\theta - \ddot{\psi}\right)\mathbf{e}_1^a - R\left(\dot{\phi}\,\cos\theta\left(\dot{\phi}\,\cos\theta + \dot{\psi}\right) - \dot{\theta}^2\right)\mathbf{e}_2^a \\
& + R\left(\ddot{\theta} + \dot{\phi}\,\sin\theta\left(\dot{\phi}\,\cos\theta + \dot{\psi}\right)\right)\mathbf{e}_3^a
\end{aligned}
\tag{8.8.35}
$$

This gives us the acceleration in terms of a-frame components so if we use Eq. (8.8.30) we can obtain the corresponding (X', Y', Z') components and therefore explicit expressions for the force equations in the (X', Y', Z') system. The final result is

$$
\begin{aligned}
F_1 &= mR\frac{d}{dt}(\dot{\phi}\cos\theta + \dot{\psi}) - mR\dot{\phi}\dot{\theta}\sin\theta \\
F_2 &= -mR\dot{\phi}\,(\dot{\phi}\cos\theta + \dot{\psi}) - mR\frac{d}{dt}(\dot{\theta}\sin\theta) \\
N - mg &= mR\frac{d}{dt}(\dot{\theta}\cos\theta)
\end{aligned}
\tag{8.8.36}
$$

For the special case of rolling without nutation in a circular path, since θ is a constant and $\dot{\phi}\cos\theta + \dot{\psi} = -\dot{\phi}\,\rho/R$, which is also a constant, we find simply

$$
\begin{aligned}
F_1 &= 0 \\
F_2 &= m\rho\dot{\phi}^2 = \frac{mv^2}{\rho} \\
N &= mg
\end{aligned}
\tag{8.8.37}
$$

where $v = \rho\dot{\phi}$ is the constant speed of the center of mass along its circular path. Using these results Prob. P8.8 explores the no-slip condition for this special case.

8.9 Problems

P8.1. For the fixed gyro-pendulum shown in Fig. 8.5, show that if the assembly is released at an angle $\theta < \pi/2$ with $\dot{\theta} = 0$ and $\phi = 0$ then the assembly will execute a purely conical motion (no nutation) with a constant precessional angular velocity, $\dot{\phi}$, if initially we have

$$
\dot{\phi} = \sqrt{\frac{g}{L\cos\theta(1 - R^2/4L^2)}}
\tag{P8.1}
$$

Demonstrate that conical motion by running the MATLAB® script fixed_gyro_pend_sim with $\theta = 45^\circ$ and this initial $\dot{\phi}$ while retaining all of the other default parameters.

P8.2. For the fixed gyro-pendulum shown in Fig. 8.5, determine the driving moment that must be applied in order to produce a constant precessional angular velocity $\dot{\phi} = \omega_0$. This is now a one degree of freedom problem. From the equation of motion for θ determine the possible equilibrium positions of the assembly and find the conditions for them being stable or unstable. Write a MATLAB® function and MATLAB® script that solves for the θ motion numerically and use that software to examine that motion about an unstable equilibrium position.

P8.3. Verify that the total energy, E, and angular momentum, p_ψ, are conserved for the motion of the gyro-pendulum of Simulation 8.2.

P8.4. In this problem we will solve the gyro-pendulum problem of Simulation 8.2 by writing Euler's equations in terms of the intermediate $\left(x_1^a, x_2^a, x_3^a\right)$ coordinates, as used in the fixed gyro-pendulum problem, called the a-frame. To explain in detail what we mean we will give a rather long introduction to the problem so bear with me. In terms of a-frame coordinates the Euler's equations become (dropping the (n) superscript as we only have a single body here)

$$\tilde{\mathbf{I}}_{OI}\tilde{\boldsymbol{\alpha}}_B^{(n)} + {}_I\tilde{\boldsymbol{\Omega}}_B\tilde{\mathbf{I}}_{OI}\tilde{\boldsymbol{\omega}}_B = \tilde{\mathbf{M}}_O \tag{P8.2}$$

where the tilde over all these terms means that they are measured in the a-frame just as the overbar in Eq. (8.4.20) indicated that these terms were in the body-fixed frame. Note that a-frame components are identical with body-fixed frame components when $\psi = 0$. In terms of 3-2-3 Euler angles the angular velocity of the body and the angular velocity of the a-frame are given by (see the fixed-gyro-pendulum problem for similar terms where $\dot{\psi} = 0$)

$$\begin{aligned} {}_I\tilde{\boldsymbol{\omega}}_B &= -\sin\theta\dot{\phi}\,\mathbf{e}_1^a + \dot{\theta}\,\mathbf{e}_2^a + \left(\dot{\phi}\,\cos\theta + \dot{\psi}\,\right)\mathbf{e}_3^a \\ {}_I\tilde{\boldsymbol{\omega}}_a &= -\sin\theta\dot{\phi}\,\mathbf{e}_1^a + \dot{\theta}\,\mathbf{e}_2^a + \dot{\phi}\,\cos\theta\mathbf{e}_3^a \end{aligned} \tag{P8.3}$$

Note that

$$\begin{aligned} {}_I\tilde{\boldsymbol{\omega}}_B &= {}_I\tilde{\boldsymbol{\omega}}_a + \dot{\psi}\,\mathbf{e}_3^a \\ &= {}_I\tilde{\boldsymbol{\omega}}_a + \tilde{\boldsymbol{\omega}}_s \end{aligned} \tag{P8.4}$$

where $\tilde{\boldsymbol{\omega}}_s = \dot{\psi}\,\mathbf{e}_3^a$ is the spin of the disk as seen in the a-frame.
The components of angular velocity of the body in Eq. (P8.3) are those that also appear in the matrix ${}_I\tilde{\boldsymbol{\Omega}}_B$. The moment of inertia matrix in the a-frame is identical to that seen in body-fixed coordinates and both are principal axes so

$$\tilde{\mathbf{I}}_O = \begin{bmatrix} \tilde{I}_{O;1} & 0 & 0 \\ 0 & \tilde{I}_{O;1} & 0 \\ 0 & 0 & \tilde{I}_{O;3} \end{bmatrix} = \begin{bmatrix} I_{O;1} & 0 & 0 \\ 0 & I_{O;1} & 0 \\ 0 & 0 & I_{O;3} \end{bmatrix} \tag{P8.5}$$

The $I_{O;\,3}$ term is the same in both frames since the x_3^a-axis is the same as the body-fixed $\bar{x}_3$. The moments of inertia $(I_{O;1}, I_{O;2} = I_{O;1})$ along the $\left(x_1^a, x_2^a\right)$-axes, respectively, are the same as those about the body-fixed $\left(\bar{x}_1, \bar{x}_2\right)$-axes since even though these two sets of axis differ by the spin angle, ψ, the rotational symmetry of the body means that these moments of inertia in the two frames are independent of this angle and hence equal.

The external moment in Eq. (P8.2) can be obtained directly by writing the components directly in the a-frame, or one can simply set $\psi = 0$ for the moment terms seen in Eq. (8.5.14), giving

$$\begin{aligned}
\tilde{M}_{O;1} &= -M(t)\sin\theta \\
\tilde{M}_{O;2} &= -mgL\sin\theta \\
\tilde{M}_{O;3} &= M(t)\cos\theta
\end{aligned} \tag{P8.6}$$

The only remaining term we need to find is the angular acceleration term, ${}_I\boldsymbol{\alpha}_B$. As done with the corresponding body-fixed term (see the discussion in Sect. 8.4) we can find this term with the transport theorem. We will repeat the use of that theorem when using the a-frame components. Note that in Sect. 8.4 we used a very explicit notation that we will repeat here. First, the angular acceleration of the body in inertial coordinates was written as

$${}_I\boldsymbol{\alpha}_B \equiv \left[{}_I\boldsymbol{\alpha}_B\right]_I = \left[\frac{d^I\,{}_I\boldsymbol{\omega}_B}{dt}\right]_I \tag{P8.7}$$

to make it clear that the derivative of the angular velocity of the body was being taken in inertial coordinates and the result of that differentiation was also being expressed in inertial coordinates.

We transformed this angular acceleration to/from a body-fixed frame through the rotation matrix, $\mathbf{R}$, where

$${}_I\bar{\boldsymbol{\alpha}}_B = \mathbf{R}^T\,{}_I\boldsymbol{\alpha}_B, \qquad {}_I\boldsymbol{\alpha}_B = \mathbf{R}\;{}_I\bar{\boldsymbol{\alpha}}_B \tag{P8.8}$$

and the overbar indicated that this acceleration was in body-fixed coordinates. The $\mathbf{R}$ matrix is here represented by the 3-2-3 set of Euler angles, i.e., $\mathbf{R} = \mathbf{R}(\phi, \theta, \psi)$. We can transform the angular acceleration between a set of inertial axes to/from the a-frame in exactly the same fashion, namely

$${}_I\tilde{\boldsymbol{\alpha}}_B = \mathbf{R}_a^T\,{}_I\boldsymbol{\alpha}_B, \quad {}_I\boldsymbol{\alpha}_B = \mathbf{R}_a\;{}_I\bar{\boldsymbol{\alpha}}_B \tag{P8.9}$$

where the rotation matrix, $\mathbf{R}_a$, only involves the Euler angles associated with the 3-2 rotations, i.e., $\mathbf{R}_a = \mathbf{R}(\phi, \theta, \psi = 0)$. Now consider the transport theorem for the angular acceleration, using time derivatives ${}^I d/dt$ in an inertial frame and ${}^a d/dt$ in an a-frame and where the final results of all terms are expressed in the inertial frame:

$$
{}_I\boldsymbol{\alpha}_B \equiv \left[\frac{{}^I d_I\boldsymbol{\omega}_B}{dt}\right]_I = \left[\frac{{}^a d_I\boldsymbol{\omega}_B}{dt}\right]_I + \left[{}_I\boldsymbol{\omega}_a \times {}_I\boldsymbol{\omega}_B\right]_I \tag{P8.10}
$$

Thus, using Eq. (P8.9)

$$
\begin{aligned}
{}_I\tilde{\boldsymbol{\alpha}}_B &= \mathbf{R}_a^T\,{}_I\boldsymbol{\alpha}_B = \mathbf{R}_a^T\left[\frac{{}^I d_I\boldsymbol{\omega}_B}{dt}\right]_I = \mathbf{R}_a^T\left[\frac{{}^a d_I\boldsymbol{\omega}_B}{dt}\right]_I + \mathbf{R}_a^T\left[{}_I\boldsymbol{\omega}_a \times {}_I\boldsymbol{\omega}_B\right]_I \\
&= \left[\frac{{}^a d_I\boldsymbol{\omega}_B}{dt}\right]_a + \left[{}_I\boldsymbol{\omega}_a \times {}_I\boldsymbol{\omega}_B\right]_a \\
&= {}_I\dot{\tilde{\boldsymbol{\omega}}}_B + {}_I\tilde{\boldsymbol{\omega}}_a \times {}_I\tilde{\boldsymbol{\omega}}_B = {}_I\dot{\tilde{\boldsymbol{\omega}}}_B + {}_I\tilde{\boldsymbol{\omega}}_a \times \tilde{\boldsymbol{\omega}}_s
\end{aligned} \tag{P8.11}
$$

Here ${}_I\dot{\tilde{\boldsymbol{\omega}}}_B$ is the time derivative of the angular velocity, ${}_I\tilde{\boldsymbol{\omega}}_B$,written in a-frame coordinates, holding the unit vectors along the axes of that a-frame, $(\mathbf{e}_1^a, \mathbf{e}_2^a, \mathbf{e}_3^a)$, fixed and where the result of that differentiation is left in the a-frame. Thus, we see that ${}_I\tilde{\boldsymbol{\alpha}}_B \neq {}_I\dot{\tilde{\boldsymbol{\omega}}}_B$ whereas in body-fixed coordinates we had ${}_I\bar{\boldsymbol{\alpha}}_B = {}_I\dot{\bar{\boldsymbol{\omega}}}_B$.

(a) Using the results just discussed obtain the equations of motion of the gyro-pendulum for the 3-2-3 Euler angles (ϕ, θ, ψ) using Euler's equations (Eq. (P8.2)) written in a-frame coordinates.
(b) Solve the equations of motion obtained in (a) numerically for the same parameters considered in Simulation 8.2 and show that your results agree with those in that simulation.

P8.5. In this problem we will solve the gyro-pendulum problem of Simulation 8.2 by writing the moment equation directly in a-frame components, leading to what is called the modified Euler's equation. In this case the moment equation written in terms of the angular momentum in inertial coordinates is, using the very explicit notation described in Chap. 4 (see the use of this notation in Prob. P8.4 also),

$$
\left[\frac{{}^I d\mathbf{L}_O}{dt}\right]_I = \left[\mathbf{M}_O\right]_I \tag{P8.12}
$$

If we use the transport theorem to relate the inertial time derivatives to derivatives in the a-frame we have

$$
\left[\frac{{}^I d\mathbf{L}_O}{dt}\right]_I = \left[\frac{{}^a d\mathbf{L}_O}{dt}\right]_I + \left[{}_I\boldsymbol{\omega}_a \times \mathbf{L}_O\right]_I = \left[\mathbf{M}_O\right]_I \tag{P8.13}
$$

If we multiply this equation by the rotation matrix $\mathbf{R}_a^T$ (see Prob. P8.4) to transform to the a-frame coordinates we have

$$\left[\frac{{}^{I}d\mathbf{L}_O}{dt}\right]_a = \left[\frac{{}^{a}d\mathbf{L}_O}{dt}\right]_a + \left[{}_I\boldsymbol{\omega}_a \times \mathbf{L}_O\right]_a = \left[\mathbf{M}_O\right]_a \tag{P8.14}$$

or, using the notation of Prob. P8.4 where a tilde indicates evaluation in the a-frame coordinates,

$$\dot{\tilde{\mathbf{L}}}_O + {}_I\tilde{\boldsymbol{\omega}}_a \times \tilde{\mathbf{L}}_O = \tilde{\mathbf{M}}_O \tag{P8.15}$$

and $\dot{\tilde{\mathbf{L}}}_O$ denotes taking the derivative of the angular momentum, written in a-frame coordinates, holding the unit vectors along the axes of that a-frame, $\left(\mathbf{e}_1^a, \mathbf{e}_2^a, \mathbf{e}_3^a\right)$, fixed and expressing the result of that differentiation in the a-frame.

(a) Obtain the equations of motion of the gyro-pendulum for the 3-2-3 Euler angles (ϕ, θ, ψ) using Eq. (P8.15). If you have also solved Prob. P8.4, show that the two sets of equations agree.
(b) Solve the equations of motion obtained in (a) numerically for the same parameters considered in Simulation 8.2 and show that your results agree with those in that simulation.
(c) Write the angular momentum in the a-frame as

$$\tilde{\mathbf{L}}_O = \tilde{\mathbf{I}}_O \, {}_I\tilde{\boldsymbol{\omega}}_B \tag{P8.16}$$

and assume that the body is rotationally symmetric in the a-frame so that the moment of inertia is a diagonal matrix with diagonal terms $\left(\tilde{I}_{O;1}, \tilde{I}_{O;2} = \tilde{I}_{O;1}, \tilde{I}_{O;3}\right)$. Also write (see Eq. (P8.4))

$$\begin{aligned} {}_I\tilde{\boldsymbol{\omega}}_B &= \tilde{\omega}_1\mathbf{e}_1^a + \tilde{\omega}_2\mathbf{e}_2^a + \tilde{\omega}_3\mathbf{e}_3^a \\ {}_I\tilde{\boldsymbol{\omega}}_a &= \tilde{\omega}_{a1}\mathbf{e}_1^a + \tilde{\omega}_{a2}\mathbf{e}_2^a + \tilde{\omega}_{a3}\mathbf{e}_3^a \\ &= \tilde{\omega}_1\mathbf{e}_1^a + \tilde{\omega}_2\mathbf{e}_2^a + \tilde{\omega}_{a3}\mathbf{e}_3^a \end{aligned} \tag{P8.17}$$

Show that the moment equation then becomes

$$\begin{aligned} &\tilde{I}_{O;1}\dot{\tilde{\omega}}_1 + \tilde{\omega}_2\left(\tilde{I}_{O;3}\tilde{\omega}_3 - \tilde{I}_{O;1}\tilde{\omega}_{a3}\right) = \tilde{M}_{O;1} \\ &\tilde{I}_{O;1}\dot{\tilde{\omega}}_2 - \tilde{\omega}_1\left(\tilde{I}_{O;3}\tilde{\omega}_3 - \tilde{I}_{O;1}\tilde{\omega}_{a3}\right) = \tilde{M}_{O;2} \\ &\tilde{I}_{O;3}\dot{\tilde{\omega}}_3 = \tilde{M}_{O;3} \end{aligned} \tag{P8.18}$$

which is known as the *modified Euler's equations*. Use these equations directly to obtain the equations of motion for the gyro-pendulum and show that they agree with those obtained in part (a). Note that since the moment equation written in principal coordinates about the center of mass is of the same form as the moment equation used here about a fixed point, we can also use these modified equations when we replace the fixed point O by the center of mass point, C.

P8.6. Monitor the satisfaction of the constraint equation and its derivative in the solution obtained in Simulation 8.3.

P8.7.

(a) Solve the rolling-disk problem of Sect. 8.8 numerically for the generalized coordinates $(X_C, Y_C, \phi, \theta, \psi)$ using the augmented method. As an initial check that your solution is valid, consider the special case discussed in Sect. 8.8 where the disk rolls around in a circular path with no changes of the nutational angle and constant precessional and spin angular velocities. Let R = 0.25 m, ρ = 5R, $\theta(0) = \pi/3$ rad, and $\dot{\theta}(0) = 0$ (note that the mass drops out of the equations). Also take $\phi(0) = \psi(0) = 0$ and find the remaining initial conditions as discussed in Sect. 8.8. Solve for and plot the nutational angle, $\theta(t)$, and the spin angle, $\psi(t)$, to show that they are constant and linearly varying in time, respectively, over a time interval of 0–10 s. Also plot the X- versus Y-coordinates of the center of the disk to show that the motion is indeed that of a circle.

(b) Repeat the solution of part (a) by changing the initial nutational angle to $\theta(0) = \pi/4$ and by setting the initial values for the generalized coordinates at time $t = 0$ to be $(X_C, Y_C, \phi, \theta, \psi) = (0, \; -5R, \; 0, \; \pi/4, \; 0)$, which differs from part (a) only in the initial nutational angle. For the initial velocities take $\left(\dot{\phi}(0), \dot{\psi}(0)\right)$ to be the same values as found in part (a) but for the initial nutational angular velocity take $\dot{\theta}(0) = \dot{\psi}(0)/2$. Finally, use the constraint equations, Eq. (8.8.8), to calculate consistent values for $\left(\dot{X}_C(0), \dot{Y}_C(0)\right)$. Plot the same parameters as in part (a). What can you deduce about the motion?

(c) The rolling-disk problem is a conservative problem so examine the total energy during the motion obtained in parts (a) and (b) to ensure that the total energy is conserved during the motion.

P8.8. Consider the special case examined in Sect. 8.8 where a disk rolls in a circular path with a constant nutational angle. If the disk is a thin homogeneous disk of radius 10 cm and it is given an initial speed $v = \rho\dot{\phi} = 6$ m/s at a nutation angle $\theta_0 = 20°$, determine the radius of the circular path, ρ (note that you will have to solve a quadratic equation), and find the minimum coefficient of friction needed to prevent slipping. Is this a realistic value? (b) Repeat the analysis of part (a) where the initial nutation angle is $\theta_0 = 60°$.

P8.9. The single-axis gyroscope of Fig. P8.1 is a device that can be used to measure an angular velocity component of a vehicle. The gyroscope consists of a single gimbal that can rotate about one axis and a spinning disk. A linear torsional spring and torsional dashpot are attached to the axis of the gimbal to resist the

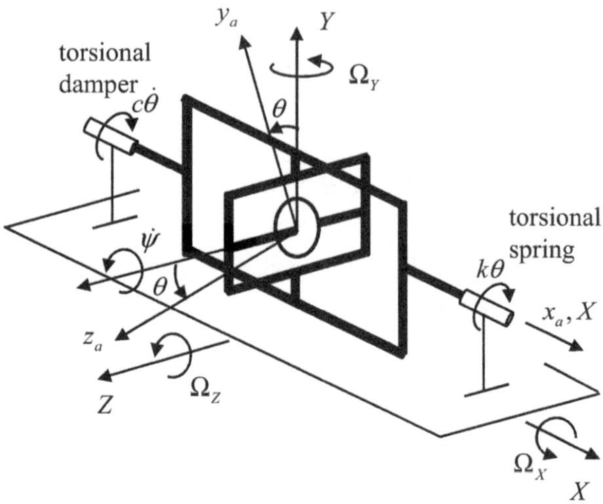

Fig. P8.1 A single-axis gyroscope

angular motion of the gimbal. Let (X, Y, Z) be a reference frame attached to the vehicle which has angular velocity components $(\Omega_X, \Omega_Y, \Omega_Z)$. The angular motion of the gimbal relative to the vehicle is defined by the angle θ and the spin of the disk is $\dot{\psi}$.

(a) Express the total angular velocity of the rotor in the a-frame, (x_a, y_a, z_a), and write the kinetic energy due to rotation only (neglect any translational contributions). Model the spring and dashpot through the potential energy and the Rayleigh dissipation function (see Chap. 7) as

$$V = \frac{1}{2}k\theta^2, \quad D = \frac{1}{2}c\dot{\theta}^2 \tag{P8.19}$$

Use Lagrange's equations to obtain the equations of motion for θ and ϕ. Show that the momentum, p_ψ, is a constant and obtain the equation of motion for θ in terms of this constant momentum and other variables.

(b) Assuming that θ is small and the spin of the disk is much larger than the angular velocity components of the vehicle, linearize the equation of motion. For the case when the angular velocity components of the vehicle are slowly varying so they can be considered to be constants, show that the steady-state solution of the linearized equation of motion is just

$$\theta = \frac{p_\psi}{k}\Omega_Y \tag{P8.20}$$

so that deflection of the gimbal is proportional to the angular velocity, Ω_Y. By placing three such gyroscopes on the *XY*, *XZ*, and *YZ* planes we thus

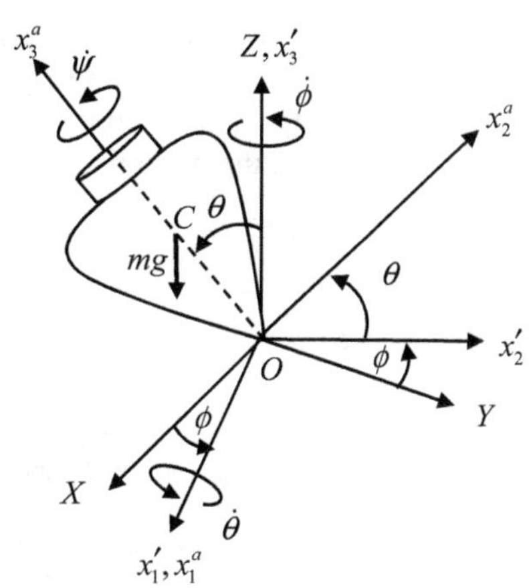

Fig. P8.2 Geometry of a spinning top and a set of 3-1-3 Euler angle rotations

have a way to measure all three angular velocity components of the vehicle.

P8.10. Consider the spinning top shown in Fig. P8.2. As with the gyro-pendulum problem discussed in Sect. 8.5 there are a number of ways to obtain the equations of rotational motion for this problem. In this problem (and in problem P8.11) we will explore several ways. We will use a set of 3-1-3 Euler angles (ϕ, θ, ψ) for this problem (note that although the symbols for these angles are the same as used for the gyro-pendulum problems, they are in fact a different set). Let the mass of the top be $m = 2$ kg and the mass moment of inertia along the top spin axis of symmetry be $I_3 = 0.25\ kg - m^2$. The mass moment of inertia perpendicular to the axis of symmetry is $I_1 = 0.5\ kg - m^2$ (as measured at O) and the distance, d, from the origin O to the center of mass C is $d = 0.2$ m.

(a) Obtain the equations of motion of the top by writing Euler's equations in terms of the intermediate $\left(x_1^a, x_2^a, x_3^a\right)$ coordinates, called the a-frame (see Fig. P8.2). This follows the same steps outlined in Prob. P8.4 for the gyro-pendulum.
(b) Using the equations obtained in (a) numerically solve for the motion of the top. Choose different initial conditions as done in Simulation 8.3 to illustrate cases where different types of motion for the top are seen (see Fig. 8.12 for the corresponding gyro-pendulum motions).

P8.11. Consider again the spinning top of Prob. P8.10. Use the top properties in Prob. P8.10. and the same 3-1-3 Euler angles to define the initial conditions.

(a) Obtain the equation of motion of the top by using Euler parameters and Eq. (8.6.11). Leave your results in terms of $\bar{\mathbf{H}}, \bar{\mathbf{I}}_\mathbf{O}$ and carry out all the matrix multiplications numerically.
(b) Using the equations obtained in (a) numerically solve for the motion of the top. Choose different initial conditions as done in Simulation 8.3 to

illustrate cases where different types of motion for the top are seen (see Fig. 8.12 for the corresponding gyro-pendulum motions).

(c) Verify that the constraint equation for the Euler parameters is satisfied during the motion.

References

1. D.T. Greenwood, *Principles of Dynamics* (Prentice-Hall, Upper Saddle River, 1965)
2. K. Sherif, K. Nachbagauer, W. Steiner, On the rotational equations of motion in rigid body dynamics when using Euler parameters. Nonlinear Dynamics **81**, 343–352 (2015)
3. H. Baruh, *Analytical Dynamics* (McGraw-Hill, New York, 1999)
4. A.A. Shabana, *Computational Dynamics*, 3rd edn. (John Wiley and Sons, Chichester, 2010)
5. O.A. Bachau, *Flexible Multibody Dynamics* (Springer, New York, 2011)
6. F. Amirouche, *Fundamentals of Multibody Dynamics* (Birkhauser, Boston, 2006)
7. A.A. Shabana, *Dynamics of Multibody Systems*, 4th edn. (Cambridge University Press, NewYork, 2013)
8. O.M. O'Reilly, *Intermediate Dynamics for Engineers* (Cambridge University Press, New York, 2008)

Chapter 9
Dynamics of Deformable Bodies

Dynamics texts typically follow a sequence where the motion of a particle is considered first, then the motion of systems of particles, and finally the motion of continuous, rigid bodies. We have followed that same traditional path in this book. Treatments of the motion of continuous deformable bodies, however, are often left to more specialized texts at a higher undergraduate and graduate level. In this chapter we give an introduction to the dynamics of deformable bodies that can serve as a bridge to more detailed studies.

We use the one-dimensional motion of a continuous, elastic rod as a simple example of a continuous system. The motion of the rod will be shown to be governed by the wave equation, a partial differential equation that is one of the most fundamental equations found in physics and engineering. To solve partial differential equations typically requires numerical methods so we discuss the solution of the wave equation using two important numerical approaches—the method of finite differences and the finite element method. The finite element method in particular has become an indispensable tool for engineers in a wide range of applications including the statics of structures, flow of heat, propagation of sound, and electrical and magnetic behavior of materials, to name a few, so that we hope that this brief look at finite elements will encourage you to study this numerical method in more depth elsewhere.

9.1 Longitudinal Wave Motion

As an example of the dynamics of a deformable body, we will consider the case of the longitudinal (axial) motion of a rod, where the mass is distributed in the rod uniformly and continuously with a mass density, ρ, having the dimensions of mass/length^3. To model the motion of this rod we will start with a model of the rod as a large number of particles, each of mass, Δm, connected to each other by linear springs, each having a spring constant, K_s, where K_s has the dimensions of force/unit

L. W. Schmerr, *Engineering Dynamics 2.0*, Solid Mechanics and Its Applications 254,
https://doi.org/10.1007/978-3-319-98470-4_9

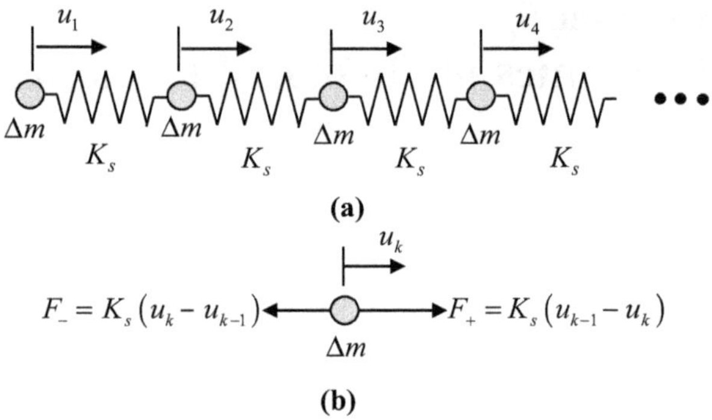

Fig. 9.1 (**a**) A system of masses separated by linear springs, and (**b**) the free body diagram of the kth mass in the system

length (see Fig. 9.1a). The masses are all assumed to have displacements, u, only along the x-axis. The generalized coordinates for this system of particles are just the individual displacements, u_n, ($n = 1, 2, ...$) of each mass.

If we examine the free body diagram of the kth mass, as shown in Fig. 9.1b, we see that Newton's second law for this mass is just

$$\Delta m\, \ddot{u}_k = K_s(u_{k+1} - u_k) - K_s(u_k - u_{k-1}) \tag{9.1.1}$$

which we will write in the equivalent form

$$\Delta m\, \ddot{u}_k = K_s \Delta x \left[\left(\frac{u_{k+1} - u_k}{\Delta x} \right) - \left(\frac{u_k - u_{k-1}}{\Delta x} \right) \right] \tag{9.1.2}$$

where Δx is the separation between any two particles in their equilibrium positions. If the number of particles is large and the separation, Δx, is very small we expect that this collection of particles will approximate a continuous rod where the displacement will also be a continuous function of both the x-position and time, i.e., $u = u(x, t)$. When Δx is small we have

$$\begin{aligned} \frac{u_{k+1} - u_k}{\Delta x} &\cong \left. \frac{\partial u(x,t)}{\partial x} \right|_{x = k\Delta x + \Delta x/2} \\ \frac{u_k - u_{k-1}}{\Delta x} &\cong \left. \frac{\partial u(x,t)}{\partial x} \right|_{x = k\Delta x - \Delta x/2} \end{aligned} \tag{9.1.3}$$

so that we can write Eq. (9.1.2) as

$$\Delta m \ddot{u}_k \cong K_s \Delta x^2 \left[\frac{ \left. \frac{\partial u(x,t)}{\partial x} \right|_{x = k\Delta x + \Delta x/2} - \left. \frac{\partial u(x,t)}{\partial x} \right|_{x = k\Delta x - \Delta x/2} }{\Delta x} \right] \tag{9.1.4}$$

But the quantity in brackets in Eq. (9.1.4) we recognize as approximately the second partial derivative of the displacement with respect to x so that we find

$$\ddot{u}_k \cong \frac{K_s \Delta x^2}{\Delta m} \frac{\partial^2 u(x,t)}{\partial x^2}\bigg|_{x=k\Delta x} \tag{9.1.5}$$

If the cross-sectional area of the rod is A, then we can define a mass density, ρ, for the rod as $\rho = \Delta m/A\Delta x$. We will also define the elastic stiffness of a small section of the rod of length Δx, E, as $E = K_s \Delta x/A$, where E has the dimensions of force/unit area and is called *Young's modulus*. In terms of those parameters we have

$$\ddot{u}_k \cong \frac{E}{\rho} \frac{\partial^2 u(x,t)}{\partial x^2}\bigg|_{x=k\Delta x} \tag{9.1.6}$$

Now consider the limit as $\Delta m, \Delta x \to 0$ but with E and ρ both remaining finite constants. Since we also have $\ddot{u}_k \to \partial^2 u(x,t)/\partial t^2$ we obtain a partial differential equation

$$\frac{1}{c^2} \frac{\partial^2 u}{\partial t^2} = \frac{\partial^2 u}{\partial x^2} \tag{9.1.7}$$

which is called the 1-D wave equation, with $c = \sqrt{E/\rho}$, where we will see that c is the wave speed of disturbances. You might have noticed that if E is a finite constant but also $E = K_s \Delta x/A$ then as $\Delta x \to 0$ the spring constant, K_s, must go to infinity while $K_s \Delta x$ remains constant. This makes sense since if you take a linear spring with a given unstretched length D and stiffness K_s and cut it into two pieces, each piece will have a stiffness, K, which is twice as stiff as the original spring ($K = 2K_s$) and have a new length $d = D/2$ but the product Kd will be the same as the original spring; that is, $Kd = K_s D$ will remain constant. What about the internal force acting in the rod? If we look at a small segment of the rod of length Δx the elongation of this segment will also be small so let that elongation be Δu. Then the force acting on the small element will be $F = K_s \Delta u = EA\Delta u/\Delta x$ and as the segment gets smaller, in the limit as $\Delta x \to 0$, we have $F = EA\partial u/\partial x$. The quantity $\varepsilon_x = \partial u/\partial x$ is the elongation per unit length in the continuous rod, a dimensionless quantity called the axial strain, ε_x. For more general deformations of a deformable body there will be multiple strains. Strains are important parameters as they measure the local deformation in a continuous body. If we divide the total force, F, by the cross-sectional area of the rod we have $\sigma_x = F/A = E\partial u/\partial x$, where σ_x is the axial stress (force/unit area) in the rod and we have the stress-strain relation $\sigma_x = E\varepsilon_x$, which is called Hooke's law. For more complex deformations there will be multiple stresses within a deformable body and multiple strains. For a linear elastic body the stresses and strains are related through a generalized Hooke's law. Strength of materials and other mechanics courses cover those more general descriptions for deformable bodies. Figure 9.2 shows an example

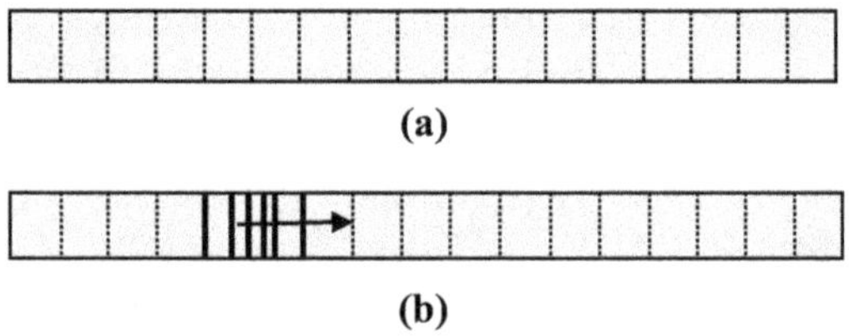

Fig. 9.2 (**a**) A stationary rod, whose length is shown divided into multiple segments, and (**b**) the same rod when it is excited so that a longitudinal wave is traveling in the rod causing segments to change length as the wave propagates

where an initially stationary rod that is divided into multiple segments (shown in Fig. 9.2a) is excited, causing a wave to propagate that changes the spacing of the segments of the rod locally as the wave moves along the rod (Fig. 9.2b). In reality a wave propagating along the rod will also cause the cross-sectional area to change but we have not shown those changes (which can be expressed in terms of the x-displacement of the rod) in Fig. 9.2.

As mentioned in the introduction to this chapter, the wave equation appears in a wide range of problems and governs, for example, the propagation of electromagnetic, seismic, and acoustic disturbances. For the problem that we are considering here, the wave equation governs the propagation of longitudinal elastic waves in the rod. The quantity c is the wave speed of those disturbances. While the dynamics of rigid bodies leads to ordinary differential equations, the motion of deformable bodies requires the solution of partial differential equations like the wave equation. A general solution for the 1-D wave equation is in the form

$$u = f(t - x/c) + g(t + x/c) \tag{9.1.8}$$

where f and g are arbitrary functions, where $f(t - x/c)$ represents a wave of constant form propagating in the positive x-direction with wave speed, c, and $g(t+x/c)$ represents a wave of constant form propagating in the negative x-direction with that same wave speed. You can easily verify that these are exact solutions by plugging them into the wave equation. We will see how to obtain these types of solutions for different wave motion problems numerically. Unlike the dynamics of rigid bodies, which often leads to nonlinear equations, the wave equation is linear. Thus, while solving partial differential equations is substantially harder than solving ordinary differential equations, having linear partial differential equations like the wave equation means that there are a wide variety of analytical tools that can be used to help solve those equations. However, for deformable bodies of complex shapes and varying material properties numerical solutions are required. Two commonly used numerical methods are the methods of finite differences and the method of finite elements. In the next subsection we examine the method of finite differences.

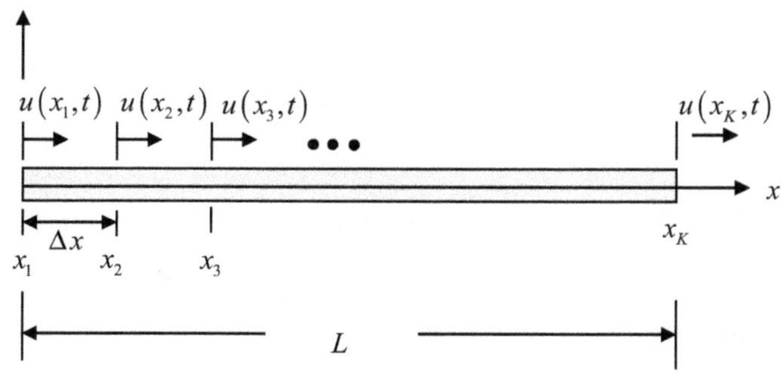

Fig. 9.3 Modeling the motion of a rod at discrete locations in the finite differences method

9.1.1 The Method of Finite Differences

The method of finite differences uses a discrete model to solve for the motion of a rod at specific locations along the rod, $x_k = (k - 1)\Delta x \quad (k = 1, 2, \ldots, K)$, where $k = 1$ corresponds to the left end of the rod at $x = 0$ and $k = K$ is at the right-hand end $x = L$, which is divided into *K-1* small segments of length Δx (Fig. 9.3). If you go back to a calculus book you will see that the second-order derivatives of a function can be approximated by a central difference formula that relates the second derivative to values of the function at discrete locations. For a function of one variable, *f*(*x*), for example, we have

$$\frac{d^2f(x)}{dx^2} \cong \frac{f(x+\Delta x) - 2f(x) + f(x-\Delta x)}{\Delta x^2} \tag{9.1.9}$$

This approximation can be used on both the time and spatial derivatives for the displacement of the rod to give

$$\begin{aligned} \frac{\partial^2 u(x_k, t)}{\partial t^2} &\cong \frac{u(x_k, t+\Delta t) - 2u(x_k, t) + u(x_k, t-\Delta t)}{\Delta t^2} \\ \frac{\partial^2 u(x_k, t)}{\partial x^2} &\cong \frac{u(x_k+\Delta x, t) - 2u(x_k, t) + u(x_k-\Delta x, t)}{\Delta x^2} \end{aligned} \tag{9.1.10}$$

which, when placed back into the wave equation, gives

$$\begin{aligned} &\frac{u(x_k, t+\Delta t) - 2u(x_k, t) + u(x_k, t-\Delta t)}{\Delta t^2} \\ &= c^2 \frac{u(x_k+\Delta x, t) - 2u(x_k, t) + u(x_k-\Delta x, t)}{\Delta x^2} \end{aligned} \tag{9.1.11}$$

If we call time t the current time, we see in Eq. (9.1.11) that there is a term at the future time, $t+\Delta t$, and a term at the past time, $t-\Delta t$, as well as terms at the current time. We can solve for the future time term as a function of the current and past time terms as

$$u(x_k, t+\Delta t) = r^2[u(x_k+\Delta x, t) + u(x_k-\Delta x, t)] + 2(1-r^2)u(x_k, t) - u(x_k, t-\Delta t) \tag{9.1.12}$$

or, in equivalent form,

$$u(x_k, t+\Delta t) = r^2[u(x_{k+1}, t) + u(x_{k-1}, t)] + 2(1-r^2)u(x_k, t) - u(x_k, t-\Delta t) \tag{9.1.13}$$

where $r = c\Delta t/\Delta x$. Now let's use Eq. (9.1.13) to solve a specific problem for the rod, where we drive the left end of the rod with a known displacement and hold the right end fixed (zero displacement). In this case we have $u(x_1, t) = h(t)$ and $u(x_K, t) = 0$. The objective is to determine the displacements as a function of time at all of the other locations, x_k, for $k = 2,3,\ldots,K-1$. Equation (9.1.13) is an updating formula that allows us to calculate those unknown displacements. To describe how this updating works it is instructive to set up the problem in a form that can be coded in MATLAB®. In the updating formula of Eq. (9.1.13) we see that there are values of the displacements at past, current, and future times so let's place those values in three arrays called past, current, and future. The past array will hold the values at time $t-\Delta t$, the current array will hold the values at time t, and the array future will hold the values at time $t+\Delta t$. The updating formula, Eq. (9.1.13), then can be written for elements $k = 2, 3,\ldots, K$-1 in these arrays as

$$\text{future}(k) = r^2[\text{current}(k+1) + \text{current}(k-1)] + 2(1-r^2)\text{current}(k) - \text{past}(k) \tag{9.1.14}$$

To begin with the rod is stationary (i.e., we have zero initial conditions) so we will initialize all the elements of the past, current, and future arrays with zeros. In MATLAB® this gives

```
past = zeros(1, K);
current = zeros(1, K);
future = zeros(1, K);
```

We now need to calculate the elements of the future array. This is accomplished with the updating formula, starting at time $t = 0$ and extending to some final time, t_{max}, specified by the user. At each time we need to apply the updating for all the elements k = 2, 3,. . ., K-1. One could use loops to perform both the space and time updates but in MATLAB® we can use array operations, which are faster, to handle all the spatial components of the arrays and only use an explicit loop to step through all the times. Thus, our updating formulae (and subsequent actions) are:

```
for t = 0:dt:tmax
% perform updating over all interior points

future(2:K-1) = r^2*(current(3:K) - current(1:K-2))+...
2*(1-r^2)*current(2:K-1) - past(2:K-1);

% specify driving and boundary conditions (assuming h(t) is known)
future(1) = h(t);
future(K) = 0;

% initialize arrays in preparation for updating at next time
past = current;
current = future;
end
```

We see that at any time, *t*, the values of all of the interior points in the future array, going from point 2 to point K-1, are obtained from the updating formula. Then the end values at point 1 and point K are set equal to their known values. We have performed the interior point updating first before applying the values at the boundaries since in some cases, as we will see, the boundary conditions may depend on the values at the interior points. After the all updating occurs at any given time *t* then the values of the current array are assigned to those of the past array and the values of the future array are assigned to those of the current array. This initializes those arrays with their proper values for the next time, *t* + d*t*, before performing the next updating to calculate the next future array values. Obviously, this entire process continues for *t* + 2d*t*, *t* + 3d*t*, etc. over all the specified times.

Simulation 9.1: Longitudinal Elastic Wave Propagating in a Rod

The MATLAB® script wave_1D_sim executes the steps just outlined for a rod being driven at one end ($x = 0$) with a specified displacement and being held fixed at the other end ($x = L$). We will consider, for example, a 100 mm long rod made of steel. The driving function h(t) is specified as the function sin(t).*(t < pi), where *t* is in μs, which produces a single half cycle of a sine curve. Default values for the input parameters are specified in the script unless the user sets those parameters in the command window before executing the script. The defaults are $dx = 0.1$ mm, $dt = 0.01$ μs, $c = 5$ mm/μs, $L = 100$ mm, and $t_{max} = 60$ μs, where dx is the spacing between points where the displacements are being calculated, dt is the spacing between evaluation times, c is the wave speed, L is the length of the rod, and t_{max} is the time at which the simulation ends. We have used mm for the lengths and μs for the times since the wave speed of elastic waves in a steel rod is very large ($c = \sqrt{E/\rho} \cong 5000$ m/s) so that it makes sense to use both small length and timescales. The t_{max} parameter was chosen so that the wave will travel the length of the rod three times during the calculations. If dx and L are chosen such that the number of segments K-1 $= L/dx$ is not an integer (it is an integer for the default values) then L/dx will be rounded to the nearest integer. This will change the value of dx slightly so this new value of dx being used will be displayed in the command window. The default r parameter, which is called the *Courant number*, in this case is $r = cdt/dx = (5)(0.01)/(0.1) = 0.5$. This is a very important parameter since it can be shown that the updating procedure we are using is only stable if $r \leq 1$. Thus, when choosing dx and dt values that are different from the default values it is important that the Courant number satisfies this stability criterion. The script will check this condition and generate an error if $r > 1$. The script wave_1D_sim sets up the default parameters; initializes the past, current, and future arrays; and then implements the iterative updating over the time interval $[0, t_{max}]$. Since a large number of iterations occur, the script plots the current displacements in the rod only after every *Nframe* iterations (i.e., the number of time "frames" calculated), where the default value for *Nframe* is *Nframe* = 10. This will generate an animation of the motion. A small pause between plots is specified by the parameter *tplot*. The default value is *tplot* = 0.01 s but this can be changed (as well as *Nframe*), if necessary, to adjust the speed of the animation.

Executing the script will show a waveform in the form of a positive half-cycle sine wave propagating to the right. When the wave reaches the end $x = L$ it will be reflected, producing a waveform of the same shape but with the opposite sign traveling to the left. The wave will then be reflected off the end $x = 0$ and travel to the right again as a positive waveform until it reaches the end $x = L$ again, at which time the simulation stops. With these default parameters the finite difference solution will be very good but you will notice a small "ringing" in the tail of the traveling waveform at later times which is due to numerical errors (see Fig. 9.4a). There is also a slight rounding of the leading edge response of the waveform which is not as evident. Remarkably, we can eliminate all the numerical errors completely and generate an exact numerical solution to the 1-D wave equation by choosing the parameters such that $r = 1$. Thus, if we change dx to dx = 0.05 in the command

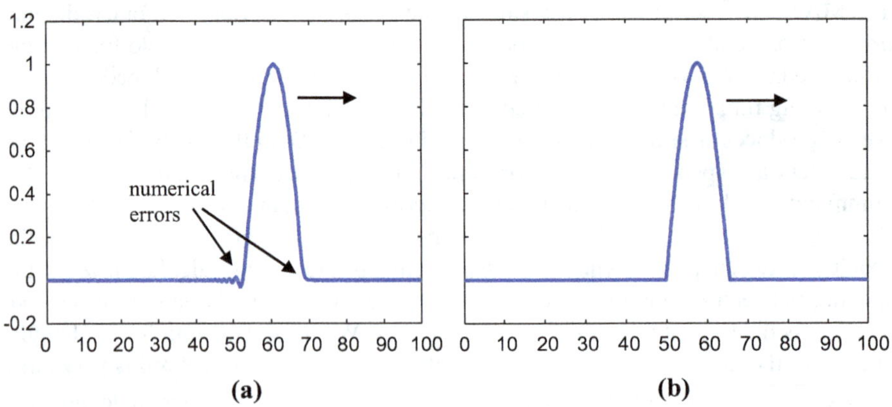

Fig. 9.4 (**a**) Finite difference solution of the 1-D wave equation, $r = 0.5$, showing small numerical errors, and (**b**) the finite difference solution for $r = 1.0$, which satisfies the "magic time step" value where the numerical solution is exact. Both waves are shown at approximately the same time after the wave has been reflected from both the right and left ends

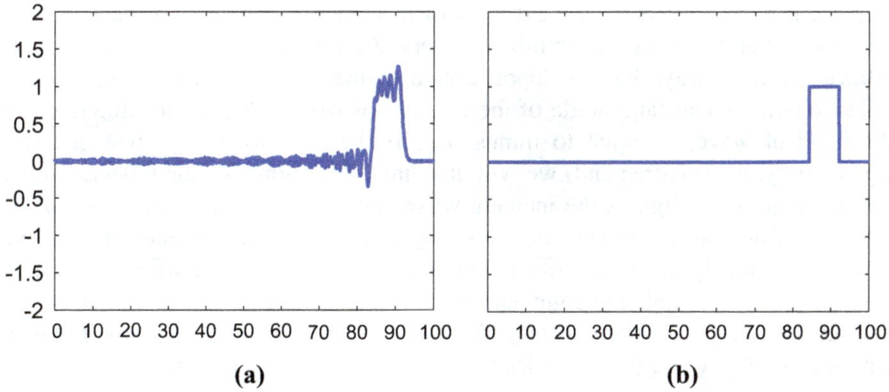

Fig. 9.5 (**a**) Finite difference solution of the 1-D wave equation, $r = 0.5$, showing large distortions of a traveling "box" function due to numerical errors, and (**b**) the finite difference solution for $r = 1.0$, which produces an exact solution. Both waves are shown at the final time, t_{max}

window and leave all the other parameters the same, we will have $r = 1$ and when executing the script under this condition even the small numerical errors previously seen will vanish. This happens because for $r = 1$ we have $\Delta x = c\Delta t$ so that each time calculation is done at exactly the time it takes a wave traveling with the wave speed c to reach the next evaluation point. Having numerical solutions generate exact results is rare and so this choice is called the *magic time step*. As Fig. 9.4b shows, at the magic time step the waveform is indeed a propagating undistorted half-cycle sine wave. A more dramatic example of the effects of the magic time step can be seen if we choose a driving function that has discontinuities. For example, if we set future(1) = (t < pi) – (t < pi/2), this will generate a "box" function. If we modify the wave_1D_sim script with this change and use the default parameters ($r = 0.5$) we will see a highly distorted propagating box waveform, shown at the final time t_{max} in Fig. 9.5a. However, if we change the dx value to $dx = 0.05$ to generate the magic time step again before executing the script, we will indeed see a perfect propagating box function as shown in Fig. 9.5b at the final time t_{max}. We could, of course, have instead refined the grid and time steps for the $r = 0.5$ case to try to reduce these numerical errors but for functions with discontinuities in value or slope we cannot expect to eliminate them entirely if the magic time step is not used. Unfortunately, for wave propagation in 2-D and 3-D it is more difficult to set up similar exact conditions for the finite difference grid of points and times since waves can travel in many different directions and thus their distance of propagation in a single time step will not always be matched to the distance between grid points. One can, however, greatly reduce numerical errors such as those shown in Fig. 9.5a even when r is not equal to one by choosing driving functions that are very smooth (i.e., have continuous values and derivatives). Gaussian functions, for example, have this property and so are a popular choice.

Our simulations show that specifying a zero displacement condition at a boundary causes the waveform to reflect from that boundary. When executing wave_1D_sim with its default parameters, reflections off both ends of the rod satisfy zero displacement conditions since the driving displacement at the left end is also zero at the times when the waves reach this boundary. At a fixed boundary we see that the reflected wave always has an opposite sign to the incident wave so we say the reflection coefficient (amplitude of the reflected wave divided by the amplitude of the incident wave) is equal to minus one. If the rod has a force-free end (or, equivalently, a stress-free end) we will also have reflections but the reflected wave will have the same sign as the incident wave and the reflection coefficient will be plus one. Since at a free end the stress, σ_x (force/unit area), is zero and, as we showed previously, $\sigma_x = E\partial u/\partial x$ it follows that at a free end $\partial u/\partial x = 0$, which we can satisfy at, say, the right end of the rod, $x = L$, by approximating the derivative there (at $x = x_K$) by $[u(x_K, t) - u(x_{K-1}, t)]/\Delta x = 0$ at any time t, which would give $u(x_K, t) = u(x_{K-1}, t)$ leading to the boundary condition

$$\text{future}(K) = \text{future}(K-1) \tag{9.1.15}$$

If we change the script wave_1D_sim to use Eq. (9.1.15) and run the changed script you will see that the wave reflects but does not change sign at $x = L$ but it still reflects with a change of sign at $x = 0$ since the displacement is zero there again at the time of reflection. Fig. 9.6a shows the half-cycle sine curve wave after it has reflected from $x = L$, and Fig. 9.6b shows the wave after having reflected at the $x = 0$ boundary. Note that if we have a problem where the left-hand end of the rod is stress free we can similarly approximate the boundary condition there as $u(x_2, t) = u(x_1, t)$ and use this condition to obtain

$$\text{future}(1) = \text{future}(2) \tag{9.1.16}$$

Under most boundary conditions the boundaries reflect waves. A finite difference grid will always have boundaries since the mesh of evaluation points will, like the

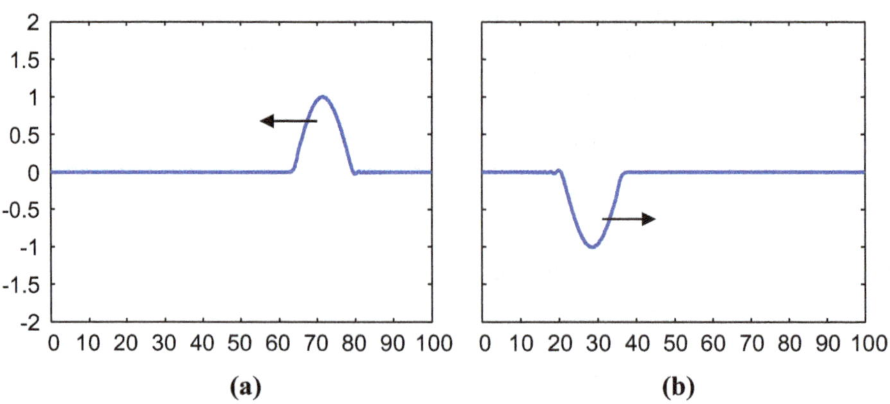

Fig. 9.6 (**a**) The wave reflected from a stress-free end at $x = 100$ mm, and (**b**) the wave reflected from $x = 0$, which acts as a fixed end since the driving displacement is zero during the reflection at this end

rod, always be finite. This is not a problem if we want to include the boundary reflections in our simulation, but it complicates matters if we want to look at wave interactions within a region that do not include reflections from the boundaries of the mesh. We can, of course, in principle simply use a very large mesh so that any reflected waves do not arrive at the points of interest until very long times, but this is often impractical because of limited computational resources. An alternative is to try to define a boundary condition that does not reflect any wave incident on it, i.e., to generate an *absorbing boundary* condition. For the 1-D wave equation we can actually design a boundary condition, called the *Mur boundary condition* that is a perfect absorber. We will not go into the details of how the Mur boundary condition is obtained (see Prob. P9.1) but simply state it here. For the right boundary, $x_K = L$, for example, we would have

$$u(x_K, t + \Delta t) = u(x_{K-1}, t) + \frac{r-1}{r+1}[u(x_{K-1}, t + \Delta t) - u(x_K, t)] \tag{9.1.17}$$

which would translate to the updating expression in terms of our arrays as

$$\text{future}(K) = \text{current}(K-1) + \frac{r-1}{r+1}[\text{future}(K-1) - \text{current}(K)] \tag{9.1.18}$$

At the other boundary, $x_1 = 0$, the absorbing boundary condition would be

$$\text{future}(1) = \text{current}(2) + \frac{r-1}{r+1}[\text{future}(2) - \text{current}(1)] \tag{9.1.19}$$

Notice that at the magic time step $r = 1$ so these absorbing boundary conditions become

$$\begin{aligned} \text{future}(K) &= \text{current}(K-1) \\ \text{future}(1) &= \text{current}(2) \end{aligned} \tag{9.1.20}$$

which makes sense since in that case the wave travels from an interior point immediately adjacent to the boundary to the boundary point in exactly one time step and, if no reflection occurs at the boundaries, the boundary values must be the same as the values in the rightward-propagating or leftward-propagating wave at those immediately adjacent interior points and one time step earlier.

You can easily demonstrate an absorbing boundary condition with wave_1D_sim. Replace the fixed boundary condition at $x = L$ with the absorbing boundary condition of Eq. (9.1.18) and save the changed script. Change the value of t_{max} to $t_{max} = 25$ in the command window since we only want to examine the wave as it traverses the rod once. Finally, execute the script. You should see that the wave simply crosses the rod and disappears without any reflections taking place. Figure 9.7, for example, shows the wave at $t = 21$ μs showing that the boundary indeed acts as if it were transparent to the incident wave.

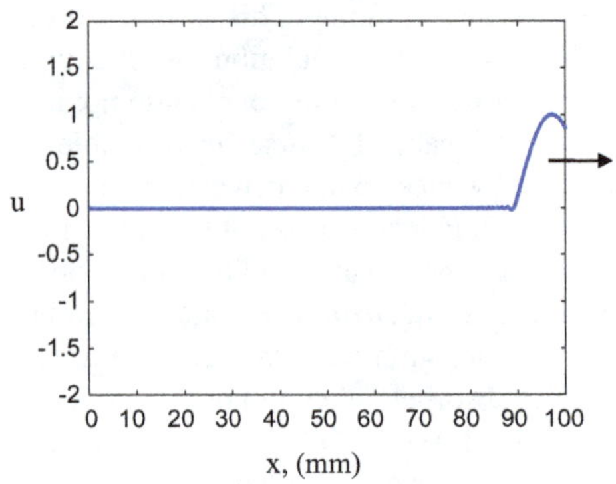

Fig. 9.7 Wave motion in an elastic rod with an absorbing boundary condition at $x = 100$ mm, showing a half-cycle sine wave passing through the right end without reflection

In obtaining the solution to the wave equation numerically we replaced both spatial and time derivatives with discrete differences. If we just had replaced the spatial terms by differences, instead of Eq. (9.1.11), we would have

$$\frac{d^2u(x_k,t)}{dt^2} = \frac{c^2}{\Delta x^2}[u(x_{k+1},t) - 2u(x_k,t) + u(x_{k-1},t)] \tag{9.1.21}$$

which, for $k = 2$ to $k = K$-1, is just a set of ordinary differential equations for the unknown interior displacements. We can write this system as

$$\frac{d^2}{dt^2}\begin{Bmatrix} u_2 \\ u_3 \\ \vdots \\ u_{K-1} \end{Bmatrix} = \frac{c^2}{\Delta x^2}\begin{bmatrix} -2 & 1 & 0 & \dots & \dots & 0 \\ 1 & -2 & 1 & 0 & \dots & 0 \\ \vdots & & & & & \vdots \\ 0 & \dots & 0 & 1 & -2 & 1 \\ 0 & \dots & & 0 & 1 & -2 \end{bmatrix}\begin{Bmatrix} u_2 \\ u_3 \\ \vdots \\ u_{K-1} \end{Bmatrix} + \frac{c^2}{\Delta x^2}\begin{Bmatrix} u_1(t) \\ 0 \\ \vdots \\ 0 \\ u_K(t) \end{Bmatrix} \tag{9.1.22}$$

where $u_k(t) \equiv u(x_k, t)$. If we define the vector $\mathbf{U} = [u_2, u_3, \ldots, u_{K-1}]^T$ and the vector $\mathbf{b} = (c^2/\Delta x)[u_1, 0, \ldots, 0, u_K]^T$ and let $\mathbf{B}$ be $(c^2/\Delta x)$ times the 2×2 matrix appearing in Eq. (9.1.22), then this system of second-order equations can be expressed simply as

$$\ddot{\mathbf{U}} = \mathbf{BU} + \mathbf{b} \tag{9.1.23}$$

Assuming that $u_1(t)$ and $u_K(t)$ are known (and both are not zero so we have a driving term) this is a linear system of second-order ordinary differential equations that can easily be placed into a system of first-order ordinary differential equations and solved, under appropriate initial conditions, by a solver such as ode45, as we have done before. However, because the system is a linear system and because the matrix $\mathbf{B}$ is a sparse tridiagonal matrix (i.e., it contains nonzero elements only along its main diagonal and the two immediately adjacent diagonals), there are other options for solvers that might be more efficient. Equation (9.1.21) is called a semi-discrete form of the wave equation in contrast to the fully discrete form used in the method of finite differences. Solving Eq. (9.1.22) involves solving a very large system of equations so that it could be a challenge. We can view the difference formula used for the time variable in the finite differences method as a specific choice of an ordinary differential equation solver that uses an explicit time stepping method to efficiently obtain the solution to Eq. (9.1.23).

Modeling the rod in the semi-discrete form is essentially going back to the particle model that we started out with for the rod since Eq. (9.1.2) for the system of particles can be written as

$$\Delta m \ddot{u}_k = K_s[u_{k+1} - 2u_k + u_{k-1}] \tag{9.1.24}$$

which is just Eq. (9.1.21) expressed in terms of different constants. The finite difference formulation is thus also just a fully discrete form of the particle model. However, the finite difference method can be applied to many other partial differential equations and more complex cases so the method has broad applicability to other problems. Problem P9.2, for example, shows how to use finite differences to solve for the 1-D waves in an inhomogeneous bar where the material properties vary along its length.

9.1.2 The Finite Element Method

Since the 1960s, when computers first became capable of solving significant engineering problems, the method of finite elements has grown to become one of the most widely used numerical approaches in a range of engineering applications, including dynamics. Like finite differences, finite elements are applicable to the solution of the partial differential equations that describe the motion of continuous bodies. To see the differences (and similarities) with the finite difference method, we will use the one-dimensional wave equation for a rod again as a specific example. To solve the wave equation in the rod the finite element method, like the finite difference method, will solve for the displacements in the rod at a number of discrete positions, $u_k(t) = u(x_k, t)$ (see Fig. 9.3). In finite elements these displacements are called the *nodal displacements*. Unlike finite differences these discrete positions need not be equally spaced in the finite element method, but to compare finite elements with finite differences we will assume that $x_{k+1} - x_k = \Delta x$, where Δx is a constant. To approximate the continuous function $u(x, t)$ in terms of the nodal displacements we will use a set of approximating functions, $\phi_k(x)$, that are nonzero only over small finite portions (i.e., elements) of the rod but which collectively can be used to describe the motion of the entire rod, i.e., we take

$$u(x,t) = \sum_{k=1}^{K} u_j(t)\,\phi_k(x) \tag{9.1.25}$$

The $\phi_k(x)$ are called *shape functions*. Often these shape functions are expressed in terms of polynomials. For example, one can take combinations of linear functions where

$$\phi_k(x) = \begin{cases} \dfrac{\varepsilon}{\Delta x} + 1 & -\Delta x \le \varepsilon \le 0 \\ 1 - \dfrac{\varepsilon}{\Delta x} & 0 \le \varepsilon \le \Delta x \\ 0 & \text{otherwise} \end{cases} \tag{9.1.26}$$

where $\varepsilon = x - x_k$ is the distance measured from the node point $x = x_k$. These functions represent "hat" functions (Fig. 9.8a) that we can use to cover the entire region of the rod (Fig. 9.8b). Note that these shape functions satisfy $\phi_k(x_k) = 1$, $\phi_k(x_j) = 0$ $(j \neq k)$, so that if we multiply the shape functions by the nodal displacements, as done in Eq. (9.1.25), we obtain a piecewise linear approximation to the continuous displacement, $u(x, t)$ (Fig. 9.9) where $u(x_k, t) = u_k(t)$. We will use these functions to determine the governing equations for the motion for all the interior points of the rod, u_k $(k = 2, \ldots, K - 1)$, again assuming that the rod is driven at the left end ($u_1 = h(t)$) and fixed at the right end ($u_K = 0$). First, multiply the wave equation by the interior shape functions ϕ_j $(j = 2, \ldots, K - 1)$ and integrate over the length of the rod to obtain

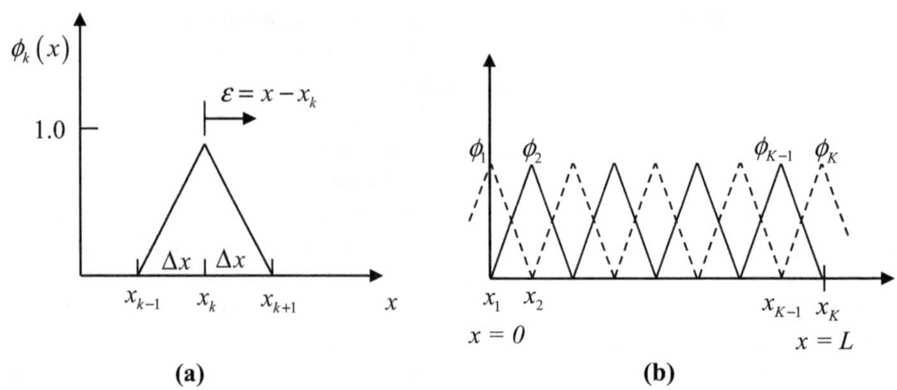

Fig. 9.8 (**a**) A localized linear shape function, and (**b**) a covering of the entire rod length with these shape functions (the functions are shown alternately as solid and dashed lines to better distinguish the individual functions, which are overlapping)

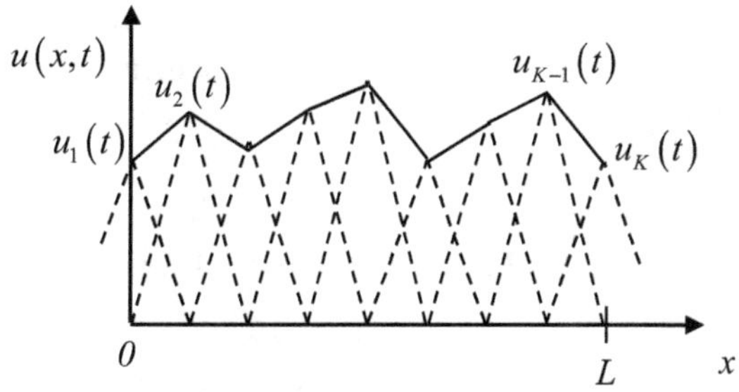

Fig. 9.9 A piecewise linear approximation of the continuous displacement in the rod obtained by multiplying the shape functions by the nodal displacements

$$\int_0^L \left(\frac{\partial^2 u}{\partial t^2} - c^2 \frac{\partial^2 u}{\partial x^2} \right) \phi_j(x) dx = 0 \quad (j = 2, \ldots, K-1) \tag{9.1.27}$$

In this form we cannot use linear shape functions to approximate the displacement since all the second derivatives would then vanish, but if we integrate Eq. (9.1.27) by parts and use the fact that $\phi_j(0) = \phi_j(L) = 0$ for all $j = (2, 3, \ldots, K\text{-}1)$ we find what is called the weak form of Eq. (9.1.27), namely

$$\int_0^L \left(\frac{\partial^2 u}{\partial t^2} \phi_j + c^2 \frac{\partial u}{\partial x} \frac{d\phi_j}{dx} \right) dx = 0 \qquad (j = 2, \ldots, K-1) \tag{9.1.28}$$

If we now place our approximation for the displacement, Eq. (9.1.25), into Eq. (9.1.28) we obtain

$$\sum_{k=1}^{K} \frac{d^2 u_k(t)}{dt^2} \int_0^L \phi_k(x) \phi_j(x) dx + c^2 \sum_{k=1}^{K} u_k(t) \int_0^L \frac{d\phi_k}{dx} \frac{d\phi_j}{dx} dx$$
$$= 0 \quad (j = 2, \ldots, K-1) \tag{9.1.29}$$

which we can write in matrix form as

$$[M]\{\ddot{u}\} + [K]\{u\} = 0 \tag{9.1.30}$$

where $\{u\} = [u_1, u_2, \ldots, u_K]^T$, $\{\ddot{u}\} = [\ddot{u}_1, \ddot{u}_2, \ldots, \ddot{u}_K]^T$ and $[M]$ and $[K]$ are $(K\text{-}2) \times (K)$ matrices where

$$\begin{aligned} M_{jk} &= \int_0^L \phi_j(x) \phi_k(x) dx \quad (j = 2, \ldots, K-1), (k = 1, \ldots, K) \\ K_{jk} &= c^2 \int_0^L \frac{d\phi_j}{dx} \frac{d\phi_k}{dx} dx \quad (j = 2, \ldots, K-1), (k = 1, \ldots, K) \end{aligned} \tag{9.1.31}$$

Because the shape functions overlap only with their immediately adjacent functions, in general the only nonzero elements of these matrices are $M_{j(j-1)}$,M_{jj},$M_{j(j+1)}$ and $K_{j(j-1)}$,K_{jj},$K_{j(j+1)}$. For example, consider the evaluation of

$$
\begin{aligned}
M_{jj} &= \int_0^L \phi_j^2(x)dx = \int_{-\Delta x}^{0} \left(1+\frac{\varepsilon}{\Delta x}\right)^2 d\varepsilon + \int_0^{\Delta x} \left(1-\frac{\varepsilon}{\Delta x}\right)^2 d\varepsilon \\
&= \Delta x \int_0^1 v^2 dv - \Delta x \int_1^0 w^2 dw = \frac{2}{3}\Delta x
\end{aligned}
$$

where we have let $v = 1+\varepsilon/\Delta x$ and $w = 1 - \varepsilon/\Delta x$. A similar evaluation of the other two nonzero terms will give $M_{j(j-1)} = M_{j(j-1)} = \Delta x/6$. Similarly for K_{jj} we have

$$
\begin{aligned}
K_{jj} &= c^2 \int_0^L \left(\frac{d\phi_j}{dx}\right)^2 dx = \frac{c^2}{\Delta x^2} \int_{-\Delta x}^{0} d\varepsilon + \frac{c^2}{\Delta x} \int_0^{\Delta x} d\varepsilon \\
&= \frac{2c^2}{\Delta x}
\end{aligned}
$$

and for the other two terms we would find $K_{j(j-1)} = K_{j(j+1)} = -c^2/\Delta x$. If we look at these matrices, therefore, we would have

$$
[M]\{\ddot{u}\} = \Delta x \begin{bmatrix} 1/6 & 2/3 & 1/6 & 0 & \dots & & & 0 \\ 0 & 1/6 & 2/3 & 1/6 & 0 & \dots & & 0 \\ \vdots & & & & & & & \vdots \\ 0 & & & & 1/6 & 2/3 & 1/6 & 0 \\ 0 & 0 & \dots & & & 1/6 & 2/3 & 1/6 \end{bmatrix} \begin{Bmatrix} \ddot{u}_1 \\ \ddot{u}_2 \\ \vdots \\ \ddot{u}_K \end{Bmatrix} \tag{9.1.32}
$$

and

$$
[K]\{u\} = \frac{c^2}{\Delta x} \begin{bmatrix} -1 & 2 & -1 & 0 & \dots & & & 0 \\ 0 & -1 & 2 & -1 & 0 & \dots & & 0 \\ \vdots & & & & & & & \vdots \\ 0 & & & & -1 & 2 & -1 & 0 \\ 0 & 0 & \dots & & & -1 & 2 & -1 \end{bmatrix} \begin{Bmatrix} u_1 \\ u_2 \\ \vdots \\ u_K \end{Bmatrix} \tag{9.1.33}
$$

Note that these are still non-square matrices. The matrix $[M]$ is called the "mass" matrix since it multiplies the nodal acceleration terms. This matrix is also called a *consistent mass matrix* since it was obtained by weighting the acceleration terms at the nodes consistent with the assumed shape functions. In the particle model of the rod, however, each actual mass was only associated with an individual nodal acceleration so the mass matrix was diagonal. In using finite elements to solve dynamics problems it is common practice to replace the consistent mass matrix by a lumped mass matrix consisting only of diagonal matrix terms. This is often done for computational efficiency, but it can also be viewed as using an equivalent particle model of the continuous rod. To obtain the lumped mass matrix here we merely have to sum the values (1/6, 2/3, 1/6) in each row and replace the middle value (2/3) by that sum, which here is 1/6 + 2/3 + 1/6 = 1. Thus, in terms of a lumped mass matrix, $[\tilde{M}]$, we have

$$\begin{aligned}[\tilde{M}]\{\ddot{u}\} &= \Delta x\begin{bmatrix} 0 & 1 & 0 & 0 & \dots & & & 0 \\ 0 & 0 & 1 & 0 & 0 & \dots & & 0 \\ \vdots & & & & & & & \vdots \\ 0 & & & & 0 & 1 & 0 & 0 \\ 0 & 0 & \dots & & & 0 & 1 & 0 \end{bmatrix}\begin{Bmatrix} \ddot{u}_1 \\ \ddot{u}_2 \\ \vdots \\ \ddot{u}_K \end{Bmatrix} \\ &= \Delta x\begin{bmatrix} 1 & 0 & 0 & \dots & 0 \\ 0 & 1 & & & 0 \\ \vdots & & 1 & & \vdots \\ 0 & & & 1 & 0 \\ 0 & 0 & \dots & 0 & 1 \end{bmatrix}\begin{Bmatrix} \ddot{u}_2 \\ \ddot{u}_3 \\ \vdots \\ \ddot{u}_{K-1} \end{Bmatrix} = \Delta x\begin{Bmatrix} \ddot{u}_2 \\ \ddot{u}_3 \\ \vdots \\ \ddot{u}_{K-1} \end{Bmatrix}\end{aligned} \tag{9.1.34}$$

where the first and Kth columns of the lumped mass matrix are all zeros so that we can eliminate them and obtain a square (K-2) × (K-2) identity matrix involving only the interior accelerations $(\ddot{u}_2, \ddot{u}_3, \dots, \ddot{u}_{K-1})$. The [$K$] matrix is called the "stiffness" matrix since it multiplies the displacements at the nodes just as a discrete set of spring constants would do in a spring-particle model. This non-square matrix term can also be reduced to a square (K-2) × (K-2) matrix involving only the interior displacements $(u_2, u_3, \dots, u_{K-1})$ by removing the first and Kth columns, which involve only the known end displacements, and placing them into a separate known vector so that we have

$$[K]\{u\} = \frac{c^2}{\Delta x}\begin{bmatrix} 2 & -1 & 0 & & 0 \\ -1 & 2 & -1 & & \vdots \\ \vdots & & & & \\ 0 & & -1 & 2 & -1 \\ 0 & 0 & \dots & -1 & 2 \end{bmatrix}\begin{Bmatrix} u_2 \\ u_3 \\ \vdots \\ u_{K-1} \end{Bmatrix} - \frac{c^2}{\Delta x}\begin{Bmatrix} u_1 \\ 0 \\ \vdots \\ 0 \\ u_K \end{Bmatrix} \tag{9.1.35}$$

If we place Eq. (9.1.34) and Eq. (9.1.35) into Eq. (9.1.30) and divide by Δx we find

$$\begin{Bmatrix} \ddot{u}_2 \\ \ddot{u}_3 \\ \vdots \\ \ddot{u}_{K-1} \end{Bmatrix} = \frac{c^2}{\Delta x^2} \begin{bmatrix} -2 & 1 & 0 & \dots & \dots & 0 \\ 1 & -2 & 1 & 0 & \dots & 0 \\ \vdots & & & & & \vdots \\ 0 & \dots & 0 & 1 & -2 & 1 \\ 0 & \dots & & 0 & 1 & -2 \end{bmatrix} \begin{Bmatrix} u_2 \\ u_3 \\ \vdots \\ u_{K-1} \end{Bmatrix} + \frac{c^2}{\Delta x^2} \begin{Bmatrix} u_1(t) \\ 0 \\ \vdots \\ 0 \\ u_K(t) \end{Bmatrix} \tag{9.1.36}$$

which is identical to the semi-discrete form of Eq. (9.1.22) that we obtained previously with the method of finite differences. A matrix form of Eq. (9.1.36) is

$$[\tilde{M}]\{\ddot{u}\} + [\tilde{K}]\{u\} = \{\tilde{f}\} \tag{9.1.37}$$

where $[\tilde{M}]$ and $[\tilde{K}]$ are square mass and stiffness matrices ($[\tilde{M}]$ here is just the identity matrix) and $\{\tilde{f}\}$ is a known "forcing" vector. This same form (but with different matrix terms) is also obtained with the finite element method for much more complex problems involving the small 3-D time-dependent deformations of elastic bodies. Note that the 1-D displacement, $u(t)$, of a particle of mass, m, connected to a spring whose spring constant is k and being driven by a known force $f(t)$ (see Fig. 9.10) is governed by the equation of motion

$$m\ddot{u} + ku = f \tag{9.1.38}$$

so that the finite element model is a matrix-based model that is a generalization of this very simple dynamical system, which is one we have seen previously in a variety of contexts.

Obviously, if we generated a fully discrete form of Eq. (9.1.36) by approximating the second-order time derivatives by the same difference expressions used in the finite difference method this finite element model will be identical to the finite difference expression used previously. The coincidence of the finite element and finite difference models only occurred because we were using linear shape functions defined on a uniform mesh and a lumped "mass" matrix. Finite elements, however, can easily use higher order terms in the shape functions and nonuniform meshes of

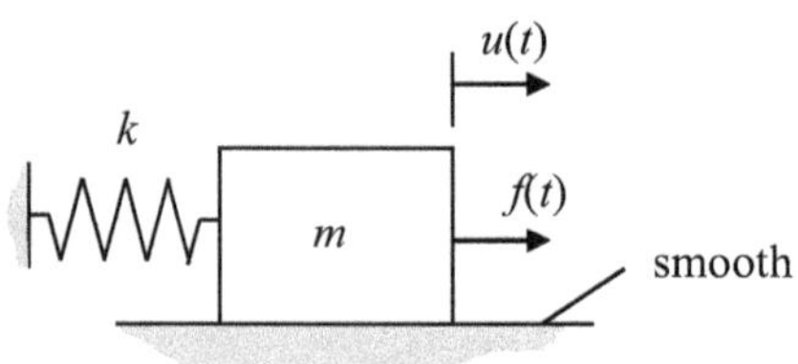

Fig. 9.10 A spring-mass system being driven by a known force, $f(t)$, where the displacement, $u(t)$, is measured from the unstretched length of the spring

nodal points so that generally it is more flexible than finite differences in representing complex geometries and behaviors.

To obtain this finite element model for the rod we used what is called a weak form of a *Galerkin approach* where we satisfied the wave equation in a weighted integral sense (see Eq. (9.1.27)) using the shape functions as the weighting functions and then represented the unknown displacement in terms of those same shape functions (Eq. (9.1.25)). There are also other ways to derive finite element models, including using virtual work principles, which are very popular in the engineering literature. Regardless of the approach used, however, a finite element method is distinguished by its use of localized shape functions to approximate the solution of a system. The method of finite differences, in contrast, is distinguished by its use of discrete approximations to derivatives (i.e., operators) appearing in the system model.

9.2 Problems

P9.1. The Mur perfectly absorbing boundary condition can be obtained by noting that the second-order wave equation can be factored into two first-order operators as

$$\frac{\partial^2 u}{\partial x^2} - \frac{1}{c^2}\frac{\partial^2 u}{\partial t^2} = \left(\frac{\partial}{\partial x} - \frac{1}{c}\frac{\partial}{\partial t}\right)\left(\frac{\partial}{\partial x} + \frac{1}{c}\frac{\partial}{\partial t}\right)u = 0 \tag{P9.1}$$

A wave traveling in the +x-direction, $u_+(x - ct)$, satisfies the first-order equation

$$\frac{\partial u_+}{\partial x} + \frac{1}{c}\frac{\partial u_+}{\partial t} = 0 \tag{P9.2}$$

while a wave traveling in the –x direction, $u_-(x+ct)$, satisfies

$$\frac{\partial u_-}{\partial x} - \frac{1}{c}\frac{\partial u_-}{\partial t} = 0 \tag{P9.3}$$

Now consider a wave traveling in the +x-direction. If we discretize the governing first-order equation, Eq. (P9.2), we must be careful to evaluate the derivatives at the same points. Thus, consider the following averaging of discrete values in approximating Eq. (P9.2):

$$\begin{aligned}&\frac{1}{2}\left(\frac{u(x_K,t_m)-u(x_{K-1},t_m)}{\Delta x}+\frac{u(x_K,t_{m+1})-u(x_{K-1},t_{m+1})}{\Delta x}\right)\\&=-\frac{1}{2c}\left(\frac{u(x_K,t_{m+1})-u(x_K,t_m)}{\Delta t}+\frac{u(x_{K-1},t_{m+1})-u(x_{K-1},t_m)}{\Delta t}\right)\end{aligned}\tag{P9.4}$$

These averages represent satisfying Eq. (P9.2) in the following fashion:

$$\begin{aligned}&\frac{1}{2}(u(x_K-\Delta x/2,t_m)+u(x_K-\Delta x/2,t_{m+1}))\\&=-\frac{1}{2c}(u(x_K,t_m+\Delta t/2)+u(x_{K-1},t_m+\Delta t/2))\end{aligned}\tag{P9.5}$$

which are both approximations for Eq. (P9.2) at the same point and time, namely

$$\frac{\partial u}{\partial x}(x_K-\Delta x/2,t_m+\Delta t/2)=-\frac{1}{c}\frac{\partial u}{\partial t}(x_K-\Delta x/2,t_m+\Delta t/2)\tag{P9.6}$$

(a) Show that Eq. (P9.4) yields the Mur boundary condition for a 1-D wave traveling in the +x direction at the boundary point $x = x_K$.
(b) Derive the corresponding Mur boundary condition for a wave traveling in the –x direction at the boundary $x = 0$.

P9.2. The rod examples considered so far were for a homogeneous rod, i.e., one whose properties were constant. Numerical methods such as finite differences, however, can also handle wave propagation in inhomogeneous media. Consider, for example, a rod whose Young's modulus, E, and density, ρ, are both functions of position, x. Recall that the internal force, F, acting in the rod is given in terms of the displacement, u, by $F = EA\partial u/\partial x$ where A is the cross-sectional area of the rod and E is Young's modulus. Thus, if we isolate a small element of length dx at a point x in the rod, as shown in Fig. P9.1, the net force acting on this element is

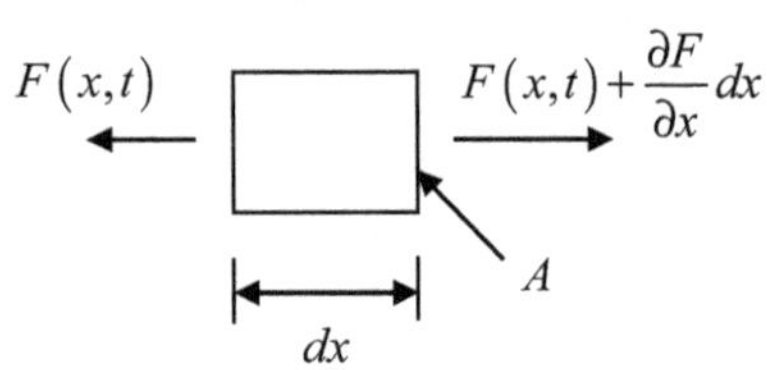

Fig. P9.1 A small section of a rod of cross-sectional area, A, showing the internal forces

$$\begin{aligned} F(x+dx,t) - F(x,t) &\cong F(x,t) + \frac{\partial F}{\partial x}dx - F(\mathrm{x},t) \\ &= \frac{\partial F}{\partial x}dx \end{aligned} \tag{P9.7}$$

and this net force must by Newton's second law be equal to the mass, dm, of this small element times its acceleration so that we have

$$\frac{\partial F}{\partial x}dx = dm\frac{\partial^2 u}{\partial t^2} = \rho A dx \frac{\partial^2 u}{\partial t^2} \tag{P9.8}$$

Placing the relationship between the force and displacement in Eq. (P9.8) and using the fact that the area A is a constant, we find

$$\frac{\partial}{\partial x}\left(E\frac{\partial u}{\partial x}\right) = \rho\frac{\partial^2 u}{\partial t^2} \tag{P9.9}$$

which is the equation for the motion of the rod when $E = E(x)$, $\rho = \rho(x)$. When the rod is homogeneous this is just the 1-D wave equation.
We can replace Eq. (P9.9) by its approximation in terms of finite differences at equally spaced discrete locations, x_k, and times, t_n. To keep the expressions manageable, let us write

$$\begin{aligned} \rho(x_k) &= \rho_k \\ E(x_k) &= E_k \\ u(x_k, t_n) &= u_k^n \end{aligned} \tag{P9.10}$$

We can approximate the right side of Eq. (P9.9) as

$$\rho\frac{\partial^2 u}{\partial t^2} = \frac{\rho_k}{\Delta t^2}\left(u_k^{n+1} - 2u_k^n + u_k^{n-1}\right) \tag{P9.11}$$

which is very similar to the homogeneous case. The left-hand side of Eq. (P9.9) is a bit more complex since we are mixing derivatives of both u and E. Note that if E was a constant we could approximate this left side as

$$\begin{aligned} &\frac{E\left(u_{k+1}^n - u_k^n\right) - E\left(u_k^n - u_{k-1}^n\right)}{\Delta x^2} \\ &= \frac{E[u(x_{k+1}, t_n) - u(x_k, t_n)] - E[u(x_k, t_n) - u(x_{k-1}, t_n)]}{\Delta x^2} \end{aligned} \tag{P9.12}$$

which is the same as the homogeneous case. However, E is not a constant so we need to bring the changes of this quantity into account. We see that Eq. (P9.12) involves evaluating the displacement as differences in the two intervals (x_k, x_{k+1}) and (x_k, x_{k-1}) so let us specify the E values multiplying these differences as the average value between the endpoints of these intervals, i.e., compute the left side of Eq. (P9.9) as

$$\frac{E_k^+\left[u_{k+1}^n - u_k^n\right] - E_k^-\left[u_k^n - u_{k-1}^n\right]}{\Delta x^2} \tag{P9.13}$$

where

$$\begin{aligned} E_k^+ &\equiv \frac{E(x_{k+1}) + E(x_k)}{2} = \frac{E_{k+1} + E_k}{2} \\ E_k^- &\equiv \frac{E(x_k) + E(x_{k-1})}{2} = \frac{E_k + E_{k-1}}{2} \end{aligned} \tag{P9.14}$$

With these approximations we obtain the updating (in time) equation

$$u_k^{n+1} = r^2\left[\left(\frac{\hat{E}_k^+}{\hat{\rho}_k}\right)\left(u_{k+1}^n - u_k^n\right) - \left(\frac{\hat{E}_k^-}{\hat{\rho}_k}\right)\left(u_k^n - u_{k-1}^n\right)\right] + 2u_k^n - u_k^{n-1} \tag{P9.15}$$

where

$$\begin{aligned} r^2 &= \frac{\Delta t^2}{\Delta x^2} C_{ref}^2 \\ \hat{E}_k^+ &= \frac{E_k^+}{\rho_{ref} C_{ref}^2}, \quad \hat{E}_k^- = \frac{E_k^-}{\rho_{ref} C_{ref}^2} \\ \hat{\rho}_k &= \frac{\rho_k}{\rho_{ref}} \end{aligned} \tag{P9.16}$$

In Eq. (P9.16) ρ_{ref} is a reference density in the rod and $C_{ref} = \sqrt{\max(E_k/\rho_k)}$ is a reference wave speed that is the largest wave speed in the rod. Equation (P9.15) is the generalization for the inhomogeneous rod of the updating equation for the homogeneous case, Eq. (9.1.13), which can be written in our current notation as

$$u_k^{n+1} = r^2\left[u_{k+1}^n - 2u_k^n + u_{k-1}^n\right] + 2u_k^n - u_k^{n-1} \tag{P9.17}$$

Equation (P9.15) reduces to Eq. (P9.17) when E and ρ are constants.

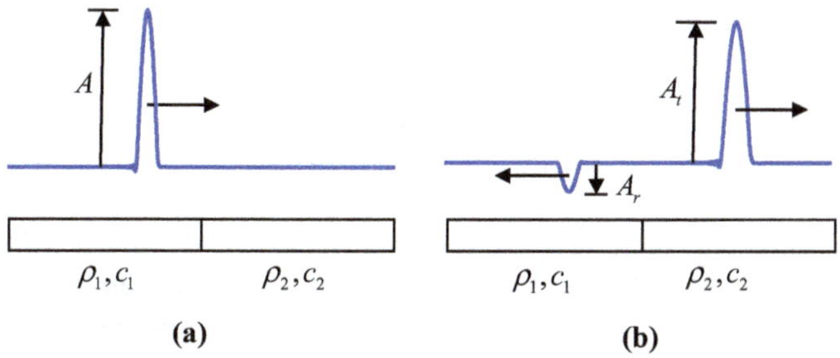

Fig. P9.2 (**a**) An incident wave traveling in a rod whose density and wave speed in the first half are (ρ_1, c_1) and (ρ_2, c_2) in the second half. (**b**) The reflected and transmitted waves generated by the interaction of the incident wave with the interface at the center of the rod

(a) Modify the MATLAB® script wave_1D_sim so that it generates a finite difference solution to an inhomogeneous rod. Keep the same default values for dx, dt, L, Nframe, and tplot but change the default $t_{\max}$ to 40 μs. Define the Courant number as $r = C_{ref}\Delta t/\Delta x$. Note that when performing the updating in Eq. (P9.15) we must multiply the displacements with normalized density and Young's modulus values. These multiplications must be done element by element using the MATLAB® .* if the updating is written in the form used in script wave_1D_sim.

To apply your modified script consider a rod with a uniform normalized density $\hat{\rho} = \rho/\rho_{ref} = 1.0$ for the entire rod but where $\hat{E} = E/\rho C_{ref}^2 = 0.5$ from $x = 0$ to $x = L/2$ and $\hat{E} = E/\rho C_{ref}^2 = 1.0$ from $x = L/2$ to L, and where $C_{ref} = 5.0$ mm/μs, which is the same wave speed in the homogeneous case. This corresponds to a rod with a sudden change of properties at $x = L/2$ (Fig. P9.2) where the wave speed in the first half of the rod is $\sqrt{12.5}$ mm/μs and the wave speed in the second half is 5 mm/μs. This discontinuity in wave speed will cause a reflected wave to appear and modify the amplitude of a wave that is transmitted through the interface at $x = L/2$, as the animation should reveal (see Fig. P9.2b). The $t_{\max} = 40$ μs value was chosen so that the wave does not reach the end of the rod but you can make that value larger if you want to see subsequent end reflections.

(b) When a 1-D incident wave of displacement amplitude A is reflected and transmitted at an interface between a homogeneous material with density and wave speed (ρ_1, c_1) and a second homogeneous material with density and wave speed (ρ_2, c_2) the amplitude ratio, A_r/A, of the reflected wave displacement to the incident wave displacement and the amplitude ratio, A_t/A, of the transmitted wave displacement to the incident wave displacement can be shown to be given by (see problem P9.3)

$$\frac{A_r}{A} = \frac{\rho_1 c_1 - \rho_2 c_2}{\rho_2 c_2 + \rho_1 c_1}$$
$$\frac{A_t}{A} = \frac{2\rho_1 c_1}{\rho_2 c_2 + \rho_1 c_1} \tag{P9.18}$$

These ratios are called reflection and transmission coefficients. [Note that the displacement amplitudes in these ratios are all based on displacements in the +*x*-direction. Many wave propagation books define the reflection coefficient in terms of ratio of displacements of the waves as measured in their directions of propagation, which will give a reflection coefficient that is the negative of the one seen in Eq. (P9.18) since the reflected wave is traveling in the –*x*-direction.] Run your modified script and show that the amplitudes seen in the animation are in agreement with these ratios (see Fig. P9.2).

P9.3. A 1-D wave traveling in the +*x* direction has the form $f(t - x/c)$ while a wave traveling in the –*x* direction has the form $g(t+x/c)$ where f and g are arbitrary functions. Show that the Fourier transforms of these waves (see Appendix E) have the forms $F(\omega) \exp(ikx)$ and $G(\omega) \exp(-ikx)$, respectively, in terms of the circular frequency, ω, where the wave number $k = \omega/c$ and $F(\omega)$ and $G(\omega)$ are the Fourier transforms of $f(t)$ and $g(t)$. The inverse Fourier transforms in terms of ω then have integrands of the forms $[F(\omega)/2\pi] \exp(-ikx+i\omega t)$ and $[G(\omega)/2\pi] \exp(ikx+i\omega t)$. These integrands represent traveling plane harmonic waves. Likewise, show that in taking the inverse Fourier transforms of these harmonic waves we recover the original traveling waves.

P9.4. Consider an incident plane harmonic wave $u_{inc} = A \exp(-ik_1x+i\omega t)$ traveling in a material whose density and wave speed are (ρ_1, c_1), respectively, where $k_1 = \omega/c_1$ is the wave number for the material whose wave speed is c_1. If this wave strikes a plane interface at $x = 0$ with another material whose density and wave speed are (ρ_2, c_2), respectively (see Fig. P9.3), a reflected wave $u_r = A_r \exp(ik_1x+i\omega t)$ traveling in the first medium and a transmitted wave $u_t = A_t \exp(-ik_2x+i\omega t)$ traveling in the second medium are generated, where $k_2 = \omega/c_2$ in terms of the wave speed, c_2, of the second medium. At the interface $x = 0$ the total displacement, u, is continuous, as is the internal force/unit area (stress), σ_x, where $\sigma_x = \rho c^2 \partial u/\partial x$. Show that the reflection and

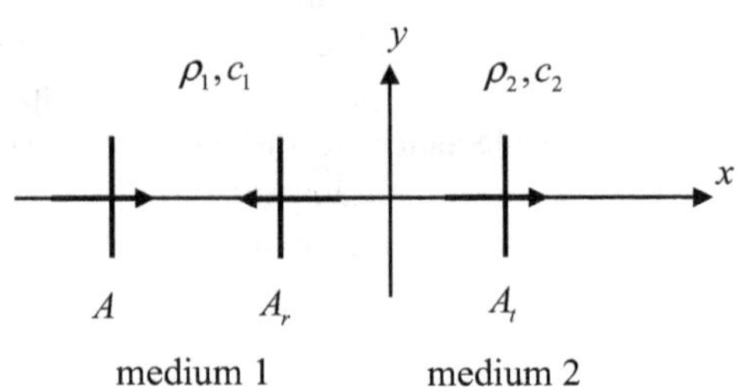

Fig. P9.3 A plane harmonic wave of amplitude *A* incident on a plane interface between two media at $x = 0$, generating both transmitted and reflected waves with amplitudes as shown

transmission coefficients A_r/A and A_t/A are given by Eq. (P9.18). The quantity ρc appearing in these coefficients is called the mechanical or acoustic impedance of a material.

P9.5. Explore solving the system of ordinary differential equations, Eq. (9.1.22), numerically. This equation can be placed in the form given in Eq. (9.1.37) so you can find many discussions on the web or in the literature on such numerical solutions.

Appendix A: Matrices

Introductory courses in dynamics have been traditionally taught with the use of scalars, such as lengths and masses, and vector quantities, such as forces, moments, velocities, and accelerations. In modern treatments of these subjects, however, it is extremely useful to also consider matrices. In fact, quantities that are normally considered as just a collection of scalars, such as mass moments of inertia, are really matrix quantities. Modern software packages such as MATLAB® can easily operate with matrices, even those with many elements, in a very efficient manner so that engineers can use these software tools to manipulate matrices with ease. In this chapter, we give a brief introduction to matrices, using MATLAB®, to show specific examples. More detailed treatments can be found in many references such as [1, 2].

Basic Matrix Algebra

A matrix is represented in MATLAB® as a square or rectangular 2-D array of numbers. For example, a matrix, A, can be constructed in MATLAB® by entering an array of values within brackets ([ ]) via the command

```
A = [1 3; 3 5; 2 4; 6 7]
A =     1  3
        3  5
        2  4
        6  7
```

where the semicolons are used to indicate the start of a new row in the matrix. The dimensions of a matrix are N × M, where N is the number of rows and M is the number of columns in the matrix. In the above example, matrix A has the dimensions (4 × 2). In many books, matrices are written either as symbols in brackets such as [A] or as bold symbols, such as **A**, or in some cases as simply an upper case symbol

L. W. Schmerr, *Engineering Dynamics 2.0*, Solid Mechanics and Its Applications 254,
https://doi.org/10.1007/978-3-319-98470-4

such as A when it is clear that the quantity being described is a matrix. All of these notations are used in this book.

We can add or subtract matrices by simply adding or subtracting each element in the arrays located at the same position (row, column). For example, consider the two matrices, A and B, defined as

```
A = [ 1 2; 3 4]
A =     1   2
        3   4
B = [3 3; 5 5]
B =   3     3
      5     5
```

Then the sum of these two matrices, C = A + B, is just the sum of their elements, i.e.,

```
C = A+B
C =
    4  5
    8  9
```

Note that to add or subtract two matrices they must have the same dimensions.

To multiply (or divide) matrices by a scalar we simply multiply (or divide) each element by the value of that scalar. For example, to obtain ten times the above matrix C we would simply write

```
D = 10*C
D =    40  50
       80  90
```

We can also define multiplication for matrices in the following manner. Let A_{mn} be the element at the mth row and nth column position in an array, A, of dimensions (M $\times$ N). Similarly for a matrix, B, of dimensions (N $\times$ K) let B_{nk} be the element at the nth row and kth column position. Then the matrix product of A and B, C = AB, is defined for each element of C as

$$C_{mk} = \sum_{n=1}^{N} A_{mn} B_{nk} \quad (m = 1, 2, \ldots, M) \ (k = 1, 2, \ldots, K) \tag{A.1.1}$$

and C is an (M $\times$ K) matrix. Many books write Eq. (A.1.1) for two matrices A and B and their product matrix, C, as simply

$$\mathrm{C} = \mathrm{AB} \ \text{ or } \ \mathbf{C} = \mathbf{AB} \text{ or } [C] = [A][B]$$

Note that for this multiplication to be meaningful we must have the number of columns of A be equal to the number of rows of B but the other dimensions of A and

B can be different. Symbolically, for example, we could multiply a 2 × 3 A with a 3 × 2 matrix B to obtain a 2 × 2 matrix C:

$$\begin{bmatrix} C_{11} & C_{12} \\ C_{21} & C_{22} \end{bmatrix} = \begin{bmatrix} A_{11} & A_{12} & A_{13} \\ A_{21} & A_{22} & A_{23} \end{bmatrix} \begin{bmatrix} B_{11} & B_{12} \\ B_{21} & B_{22} \\ B_{31} & B_{32} \end{bmatrix} \tag{A.1.2}$$

and from Eq. (A.1.1) we see that

$$\begin{aligned} C_{11} &= A_{11}B_{11} + A_{12}B_{21} + A_{13}B_{31} \\ C_{12} &= A_{11}B_{12} + A_{12}B_{22} + A_{13}B_{32} \\ C_{21} &= A_{21}B_{11} + A_{22}B_{21} + A_{23}B_{31} \\ C_{22} &= A_{21}B_{12} + A_{22}B_{22} + A_{23}B_{32} \end{aligned} \tag{A.1.3}$$

Multiplying matrices typically involves the multiplication of quite a few individual elements but packages such as MATLAB® can do all of this algebra for us. For example, consider the two 2 × 2 matrices:

```
A = [ 2 3; 1 5 ];
B = [ 2 2; 6 3 ];
```

where the semicolons at the end of each line in MATLAB® prevents the echoing of the entered result, as seen in our earlier examples. In MATLAB® matrix multiplication is again indicated by the " * " symbol so to multiply the above matrices together we write

```
C = A*B
C =  22  13
     32  17
```

which can be seen consistent with our multiplication rule, Eq. (A.1.1), since by that rule C is a 2 × 2 matrix with elements:

$$\begin{aligned} C_{11} &= A_{11}B_{11} + A_{12}B_{21} = (2)(2) + (3)(6) = 22 \\ C_{12} &= A_{11}B_{12} + A_{12}B_{22} = (2)(2) + (3)(3) = 13 \\ C_{21} &= A_{21}B_{11} + A_{22}B_{21} = (1)(2) + (5)(6) = 32 \\ C_{22} &= A_{21}B_{12} + A_{22}B_{22} = (1)(2) + (5)(3) = 17 \end{aligned}$$

It is important to realize, however, that even when the two matrices, A and B, are of the same size, as in the previous example, ordinary matrix multiplication, A*B, is *not* the multiplication of each element of A by the corresponding element of B, but still follows the rule of Eq. (A.1.1). For equal-sized matrices, MATLAB® does also define such an element-by-element multiplication by using the special multiplication symbol ".* ". Thus, for the two 2 × 2 matrices, A and B, just considered, this element-by-element multiplication yields

```
D = A.*B
D =   4  6
      6  15
```

which is obviously different from our previously obtained matrix product A*B. This type of multiplication does not appear in ordinary matrix algebra but is very useful in tasks such as forming up and plotting the values of functions at multiple points. For example, if we have the function $y = x\,e^x$ and a series of x-values which we place in a 1 × 4 matrix:

```
x = [ 1 2 3 4 ];
```

we can find and plot the values of this function at those x-values by simply writing

```
y = x.*exp(x);
plot(x, y)
```

which generates a vector y containing the products of each component x_n of $x = [x_1, x_2, \ldots, x_n]$, and the corresponding values of $[\exp(x_1), \exp(x_2), \ldots, \exp(x_n)]$, and then plots the y versus x values.

Unlike multiplication of scalars, where the result is independent of the order of the multiplication, order does make a difference in matrix multiplication. Thus, B*A is not the same as A*B. For the two 2 × 2 matrices we last considered, for example, we find

```
B*A

Ans = 6   16
      15  33
```

which indeed is different from the C = A*B we found earlier.

Other algebraic operations like the division of two matrices or the raising of matrices to powers other than powers that are positive integers are generally not defined for matrices (except in special cases) as they are for scalars. One special case that is important in dynamics is when the matrix is a *positive definite* matrix. In that case we can define quantities like the square roots or inverse square roots of a positive definite matrix, M, which we write as $M^{1/2}$ and $M^{-1/2}$, respectively. For example, if we have a diagonal matrix composed of positive elements such as

$$\mathrm{M} = \begin{bmatrix} m_1 & 0 & 0 \\ 0 & m_2 & 0 \\ 0 & 0 & m_3 \end{bmatrix}$$

then the square root and inverse square root of this positive definite matrix are just

$$\mathrm{M}^{1/2} = \begin{bmatrix} \sqrt{m_1} & 0 & 0 \\ 0 & \sqrt{m_2} & 0 \\ 0 & 0 & \sqrt{m_3} \end{bmatrix}, \quad \mathrm{M}^{-1/2} = \begin{bmatrix} 1/\sqrt{m_1} & 0 & 0 \\ 0 & 1/\sqrt{m_2} & 0 \\ 0 & 0 & 1/\sqrt{m_3} \end{bmatrix}$$

These matrices have some of the same properties we expect for their scalar counterparts, namely under matrix multiplication rules we have $\mathrm{M}^{1/2}\mathrm{M}^{1/2} = \mathrm{M}$ and $\mathrm{M}^{-1/2}\mathrm{M}^{1/2} = \mathrm{I}$, where I is the identity matrix with ones on its diagonal and zeros elsewhere. These diagonal matrices are only special cases of a positive definite matrix. For a more extensive discussion of positive definite matrices, see [3].

One matrix operation that is used often is that of transpose. By definition, the transpose of an M × N matrix A is the N × M matrix, A^T, where the rows of matrix A are the columns of A^T, i.e., if we have the (2 × 4) matrix A:

```
A = [ 1 2 3 4; 5 6 7 8]
A =    1  2   3  4
       5  6   7  8
```

then the transpose of A, A^T, which is denoted in MATLAB® as A', is given as a (4 × 2) matrix:

```
A'
Ans =  1  5
       2  6
       3  7
       4  8
```

Thus, we see if the (M × N) components of a matrix D are D_{mn} $(m = 1, 2, \ldots, M)$, $(n = 1, 2, \ldots, N)$, then the components of the transpose of D, D^T, are the (N × M) values, D_{nm} $(n = 1, 2, \ldots, N)$ $(m = 1, 2, \ldots, M)$, or more explicitly in matrix notation for a 2 × 2 matrix, D, for example, we have

$$\mathrm{D} = \begin{bmatrix} D_{11} & D_{12} \\ D_{21} & D_{22} \end{bmatrix}, \quad \mathrm{D}^T = \begin{bmatrix} D_{11} & D_{21} \\ D_{12} & D_{22} \end{bmatrix}$$

Notice that if we have a square matrix D that is symmetrical, i.e., $D_{mn} = D_{nm}\,(m \neq n)$, then $\mathrm{D} = \mathrm{D}^\mathrm{T}$, which we can see easily in this example if $D_{12} = D_{21}$.

One property of the transpose that we often find is useful to use is the relation

$$(\mathrm{AB})^T = \mathrm{B}^T\mathrm{A}^T \tag{A.1.4}$$

i.e., the transpose of the product of two matrices is the product of the transpose of those individual matrices, but taken in opposite order. We will not prove that result here but you can experiment with some matrices in MATLAB® to see that this relationship is true. In MATLAB®, the right-hand side of Eq. (A.1.4) would be written as B'*A' and the left-hand side would be written as (A*B)'.

One important note about calculating the transpose with MATLAB® is that we have been assuming that the elements of the matrix whose transpose we seek are all real when we use A' to calculate the transpose. But if elements of A are complex then computing A' also takes the complex conjugate of the elements as well as interchanging of the rows and columns of A. Thus, if we want to ensure that we only do the interchange of rows and columns, leaving the original elements unchanged, in MATLAB® we must write the transpose as A.', i.e., we must include a "dot" as well.

Vectors as Matrices

Vectors can also be considered to be one-dimensional matrices. If the vector has N components then it can be treated either as a matrix of (1 × N) dimensions (called a row vector) or (N × 1) dimensions (called a column vector). Obviously, row and column vectors are just two different matrix representations of the same vector. For example, a vector could be written either as a row matrix, $[v_1\ v_2\ \ldots\ v_N]$, or as a column matrix, i.e., $[v_1 \quad v_2 \quad \ldots \quad v_N]^T$, since by the properties of the transpose

$$[v_1 \quad v_2 \quad \ldots \quad v_N]^T = \begin{bmatrix} v_1 \\ v_2 \\ \ldots \\ v_N \end{bmatrix}$$

Generally, it is useful to distinguish if a vector is being represented as a column or row matrix without stating it explicitly or giving its elements. This will be accomplished in this book by writing a column vector in "squiggly" brackets as $\{v\}$ and a row vector as $\{v\}^T$. Following the example just given a row vector quantity can be written explicitly as [v1, v2,. . .,vn]. Then $[\mathrm{v1, v2,\ldots,vn}]^\mathrm{T}$ will indicate, respectively, a corresponding single-column matrix. Also, a notation where the bracketed symbols are replaced by lower case bold symbols such as $\mathbf{v}$ and $\mathbf{v}^\mathrm{T}$ will be used, or in some cases just a lower case symbol such as v and v^T will be used where it will be implicitly assumed that the vectors **v** or v are column vectors. Occasionally, we may also write a vector as a combination of other vectors such as $\mathbf{v} = [\mathbf{u}^T\ \mathbf{w}^T]^T$. Here **v** is being represented as a column vector through the use of the bracket and transpose symbols, $[]^T$, and $\mathbf{u}$ and $\mathbf{w}$ are column vectors whose transposes, $\mathbf{u}^T$ and $\mathbf{w}^T$, are row vectors. These two row vectors have been combined into the single-row vector $[\mathbf{u}^T \ \ \mathbf{w}^T]$ whose transpose produces the column vector, $\mathbf{v}$. The dimension of $\mathbf{v}$ is the sum of the dimensions of $\mathbf{u}$ and $\mathbf{w}$. We can write the same result in a much simpler fashion as

$$\mathbf{v} = \begin{Bmatrix} \mathbf{u} \\ \mathbf{w} \end{Bmatrix}$$

where a column vector, $\mathbf{v}$, is the concatenation of the column vectors $\mathbf{u}$ and $\mathbf{w}$ so again the dimension of $\mathbf{v}$ is the sum of the dimensions of $\mathbf{u}$ and $\mathbf{w}$. However, writing the relationship as $\mathbf{v} = [\mathbf{u}^T \ \ \mathbf{w}^T]^T$ allows it to fit easier within a paragraph of text.

Vectors can also be easily generated in MATLAB®. For example, to make a 1×4 (row) vector, v1, we have

```
v1 = [ 1 2 3 5]
v1 = 1   2   3   5
```

In contrast, to generate a column vector, we can do it directly as

```
v2 = [1;2;3;5]
v2 =     1
         2
         3
         5
```

or, equivalently, using the transpose:

```
v2 = [1 2 3 5] '
v2 =     1
         2
         3
         5
```

We can multiply two vectors, a and b, to form a product, a*b, with matrix multiplication as long as we adhere to the matrix multiplication rule that the number of columns of a equals the number of rows of b in the product ab. For example

```
v1 = [ 1 2 3 4];
v2 = [1;2;3;4];
d = v1*v2
d = 30
```

where we have multiplied a (1×4) matrix (row vector) by a (4×1) matrix (column vector) to obtain a (1×1) matrix (scalar). Here the product is the same as the dot product:

$$\mathbf{u} \cdot \mathbf{v} = [u_x \quad u_y \quad u_z][v_x \quad v_y \quad v_z]^T \\ = u_x v_x + u_y v_y + u_z v_z, \tag{A.2.1}$$

In MATLAB®, the dot product can also be implemented with the built-in function dot, i.e.,

```
v3 = [ 1 2 3 4];
v4 = [1 2 3 4];
d = dot ( v3, v4)
d = 30
```

Now, consider what happens when we reverse the order and instead multiply the previous column vector, v2, by the row vector v1 to obtain the product v2*v1:

```
v1 = [ 1 2 3 4];
v2 = [1;2;3;4];
p = v2*v1
p = 1   2   3   4
    2   4   6   8
    3   6   9  12
    4   8  12  16
```

This multiplication is possible since we are multiplying the (4 × 1) matrix v2 with the (1 × 4) matrix v1, which according to our matrix multiplication rules yields a (4 × 4) matrix p. The elements, p_{mn}, of the product matrix p, where

$$\mathrm{p} = \begin{bmatrix} p_{11} & p_{12} & p_{13} & p_{14} \\ p_{21} & p_{22} & p_{23} & p_{24} \\ p_{31} & p_{32} & p_{33} & p_{34} \\ p_{41} & p_{42} & p_{43} & p_{44} \end{bmatrix}$$

are given in terms of the products of the elements of v1 and v2, as

$$p_{mn} = v2_m v1_n \quad (m = 1, 2, 3, 4) \ (n = 1, 2, 3, 4) \tag{A.2.2}$$

Determinants and Cofactors

The determinant of a square matrix A, often written as |A|, is a scalar determined from products of various elements of A [3]. For example, for a 2 × 2 matrix we have

$$|\mathrm{A}| = \begin{vmatrix} A_{11} & A_{12} \\ A_{21} & A_{22} \end{vmatrix} = A_{11}A_{22} - A_{12}A_{21} \tag{A.3.1}$$

In MATLAB®, you can evaluate the determinant, D, of a square matrix A with the det function, i.e.,

```
D = det(A);
```

We can also evaluate determinants of subsets of various elements of a square matrix A, as long as those elements also form up a square array. For example, the minor, M_{ij}, of a matrix A is defined as the determinant obtained from the matrix obtained by deleting the ith row and the jth column of A. The cofactor matrix of the matrix A, C_{ij}, is also defined as

$$C_{ij} = (-1)^{i+j} M_{ij} \tag{A.3.2}$$

The determinant of a NxN matrix, A, can then be expanded in terms of the elements of its ith row and cofactors of the same row as

$$|\mathrm{A}| = \sum_{j=1}^{N} A_{ij} C_{ij} \tag{A.3.3}$$

As a specific example, consider a 3 × 3 matrix and expand its determinant about the elements of its first row ($i = 1$). Then we have

$$|\mathrm{A}| = \begin{vmatrix} A_{11} & A_{12} & A_{13} \\ A_{21} & A_{22} & A_{23} \\ A_{31} & A_{32} & A_{33} \end{vmatrix} = A_{11}(-1)^{1+1} M_{11} + A_{12}(-1)^{1+2} M_{12} + A_{13}(-1)^{1+3} M_{13}$$

$$= A_{11}(-1)^{1+1} \begin{vmatrix} A_{22} & A_{23} \\ A_{32} & A_{33} \end{vmatrix} + A_{12}(-1)^{1+2} \begin{vmatrix} A_{21} & A_{23} \\ A_{31} & A_{33} \end{vmatrix} + A_{13}(-1)^{1+3} \begin{vmatrix} A_{21} & A_{22} \\ A_{31} & A_{32} \end{vmatrix}$$

$$= A_{11}(A_{22}A_{33} - A_{23}A_{32}) - A_{12}(A_{21}A_{33} - A_{23}A_{31}) + A_{13}(A_{21}A_{32} - A_{22}A_{31})$$

where we have also expanded the 2 × 2 sub-determinants present. In Eq. (A.3.3) the same i appears in both the A_{ij} and C_{ij} because we are using the same rows in both matrices. However, if the cofactors are taken from any other row than the ith row in the expansion then the result is zero. We can summarize both of these results as

$$\sum_{j=1}^{N} A_{ij} C_{kj} = \begin{cases} |\mathrm{A}| & i = k \\ 0 & i \neq k \end{cases} \tag{A.3.4}$$

Similarly, if one expands the determinant of an NxN matrix A in terms of the elements of the jth column of A and the kth column of C, one finds

$$\sum_{i=1}^{N} A_{ij}C_{ik} = \begin{cases} |\mathrm{A}| & j = k \\ 0 & j \neq k \end{cases} \tag{A.3.5}$$

We can simply write both Eq. (A.3.4) and Eq. (A.3.5) in matrix forms, as

$$[\mathrm{A}][\mathrm{C}]^T = [\mathrm{C}]^T[\mathrm{A}] = |\mathrm{A}|[\mathrm{I}] \tag{A.3.6}$$

where [I] is a N × N identity matrix whose diagonal elements are all equal to one and whose off-diagonal elements are zero. You can generate a N × N identity matrix, I, in MATLAB® via the eye function:

```
I = eye(N);
```

Inverses and Solutions of Linear Equations

The above discussion of determinants, minors, and cofactors may seem rather abstract to you but in fact those quantities have great importance in practice. This is because many problems in mechanics lead to a set of simultaneous linear algebraic equations for a set of unknowns which can be written in matrix-vector form as

$$[A]\{x\} = \{b\} \tag{A.4.1}$$

If there are N equations for N unknowns then [A] is a square N × N matrix of known coefficients, $\{b\}$ is a N × 1 column vector of known constants, and $\{x\}$ is a N × 1 column vector of unknowns. If we premultiply Eq. (A.4.1) by the transpose of the cofactor of A, $[\mathrm{C}]^T$, which is also called the adjugate of A, then from Eq. (A.3.6) we have

$$[\mathrm{C}]^T[\mathrm{A}]\{x\} = |\mathrm{A}|[\mathrm{I}]\{x\} = |\mathrm{A}|\{x\} = [\mathrm{C}]^T\{b\} \tag{A.4.2}$$

which then can be directly solved for $\{x\}$ as

$$\{x\} = \frac{[\mathrm{C}]^T}{|\mathrm{A}|}\{b\} \tag{A.4.3}$$

The adjugate of A divided by its determinant is called the inverse of the matrix A, $[\mathrm{A}]^{-1}$. From Eq. (A.3.6) we see that

$$[\mathrm{A}]^{-1}[\mathrm{A}] = [\mathrm{A}][\mathrm{A}]^{-1} = [\mathrm{I}] \tag{A.4.4}$$

and that the solution of the N × N simultaneous equations of Eq. (A.4.1) is just

$$\{x\} = [\mathrm{A}]^{-1}\{b\} \tag{A.4.5}$$

Note that since

$$[\mathrm{A}]^{-1} = \frac{[\mathrm{C}]^T}{|\mathrm{A}|}$$

for the inverse to exist the determinant of A cannot be zero. If the determinant of A is zero then A is said to be singular and the solution of Eq. (A.4.5) is not valid. This solution fails because the condition $|\mathrm{A}| = 0$ essentially means that all N equations are not independent of each other.

In MATLAB® if we let B be the inverse of a square matrix A, this inverse can be obtained with the inv function:

```
B = inv(A) ;
```

If we have a $\mathrm{N} \times \mathrm{N}$ matrix A and a $\mathrm{N} \times 1$ column vector b, we can then solve the simultaneous equations of Eq. (A.4.1) in the form

```
x = inv(A)*b ;
```

However, this is not the most efficient way to typically solve such linear equations and MATLAB® can also obtain the solution with the use of the "backslash" operator (\ ) where the solution x is obtained via

```
x = A\b ;
```

which is generally the preferred method of solution. We will not cover the details of how these two methods differ here.

The inverse shares the same property as the transpose when we take the inverse of the product of two non-singular square matrices, namely (see Eq. (A.1.4))

$$(\mathrm{AB})^{-1} = \mathrm{B}^{-1}\mathrm{A}^{-1} \tag{A.4.6}$$

In writing a system of simultaneous linear equations we often write it in the form of Eq. (A.4.1) where $\{x\}$ and $\{b\}$ are both column vectors, but we can also write the same system of equations with corresponding row vectors by taking the transpose of both sides of Eq. (A.4.1) and then using Eq. (A.1.4) to obtain

$$\{x\}^T[\mathrm{A}]^T = \{b\}^T \tag{A.4.7}$$

but note that in this form we must use the transpose of the matrix of coefficients [A] instead of [A] itself. To solve Eq. (A.4.7) we can post-multiply both sides by $(\mathrm{A}^{-1})^T$ and, after the use of Eqs. (A.1.4) and (A.4.4), obtain the solution for the row vector $\{x\}^T$ as

$$\{x\}^T[\mathrm{A}]^T\left[\mathrm{A}^{-1}\right]^T = \{x\}^T\left(\mathrm{A}^{-1}\mathrm{A}\right)^T = \{x\}^T[\mathrm{I}]^T = \{x\}^T = \{b\}^T\left[\mathrm{A}^{-1}\right]^T$$

which we can also obtain by simply taking the transpose of both sides of Eq. (A.4.5) and using Eq. (A.1.4). In MATLAB®, we can solve this system of equations in this fashion using the transpose of the inverse, but again MATLAB® has more efficient methods which are preferred. If $\mathrm{y} = \{x\}^T$ is a row vector of unknowns, $\mathrm{d} = \{b\}^T$ is a row vector of coefficients, and the matrix $\mathrm{C} = \mathrm{A}^T$, then the solution of the system of equations $\mathrm{yC} = \mathrm{d}$ in MATLAB® is solved with the "forward slash" operator (/) in the form

```
y = d/C ;
```

We sometimes encounter problems where we need to solve a set of linear equations where there are more equations than there are unknowns. If there are M equations and N unknowns and $\mathrm{M} > \mathrm{N}$, the equations $[\mathrm{A}]\{x\} = \{b\}$ are said to be *overdetermined.* A common example of an overdetermined system is when we try to fit a model's behavior to a large number of experimental measurements. Note that in this case [A] is a non-square matrix with dimensions $\mathrm{M} \times \mathrm{N}$. There is no a unique solution to the set of overdetermined equations unless we also specify some additional conditions on the problem. One common condition is to ask to find a solution where $|[\mathrm{A}]\{x\} - \{b\}|$ is a minimum. In this case we can find a unique solution given by

$$\{x\} = \left([\mathrm{A}]^T[\mathrm{A}]\right)^{-1}[\mathrm{A}]^T\{b\} \tag{A.4.8}$$

as long as the inverse is non-singular so it exists. The matrix $[\mathrm{A}]^T[\mathrm{A}]$ is a square $\mathrm{N} \times \mathrm{N}$ matrix so it can have an ordinary inverse. A singular inverse typically indicates that some of the rows of [A] are not independent. In that case the solution given by Eq. (A.4.8) fails.

A similar situation occurs when there are fewer equations than unknowns, i.e., $\mathrm{M} < \mathrm{N}$. In that case the equations $[\mathrm{A}]\{x\} = \{b\}$ are *underdetermined.* This type of situation occurs frequently in dynamics where the equations are constraint equations for the unknown coordinates that define the motion of a system. There must be fewer constraints than there are unknowns since otherwise the system would be unable to move. Again, there is not a unique solution to an underdetermined system unless we place some additional conditions on the solution. If we seek a *minimum norm* solution, for example, then we ask to find the solution where $|\{x\}|$ is a minimum. There is a unique minimum norm solution given by

$$\{x\} = [\mathrm{A}]^T\left([\mathrm{A}][\mathrm{A}]^T\right)^{-1}\{b\} \tag{A.4.9}$$

as long as the inverse is non-singular so that it exists. In this case the matrix $[A][A]^T$ is a square MxM matrix so that it can have an ordinary inverse. A singular inverse again typically indicates that some of the rows of [A] are not independent; that is, the constraints on the motion of the system are not independent. Remarkably, a unique minimum norm solution also exists when the constraints are dependent and the ordinary inverse in Eq. (A.4.9) does not exist. In that case, a unique solution can still be obtained to the undetermined system given by

$$\{x\} = [\mathrm{A}]^T \left([\mathrm{A}][\mathrm{A}]^T\right)^{\dagger} \{b\} \tag{A.4.10}$$

where $()^{\dagger}$ denotes the *Moore-Penrose pseudoinverse* of the square $[\mathrm{A}][\mathrm{A}]^T$ matrix. Just as $[\mathrm{A}]^{-1}$ can be implemented for a matrix A in MATLAB® with the function inv(A), the pseudoinverse can also be implemented with the MATLAB® function pinv(A). Going into the details of inverses and pseudoinverses is beyond the scope of this book, but if you are interested in digging deeper into this subject, see [4] for example. The minimum norm solution of an underdetermined system of linear equations is important for dynamics since as shown in this book Gauss's principle of least constraint (or a generalization of his principle for nonideal constraint forces) is equivalent to the laws of dynamics of rigid bodies and can be expressed as just such a minimum norm problem.

References

1. H. Sneider, G.P. Barker, *Matrices and Linear Algebra*, 2nd edn. (Dover Publications, New York, 1973)
2. S.J. Chapman, *MATLAB® Programming for Engineers*, 4th edn. (Thomson, Toronto, 2008)
3. R. Horn, C. Johnson, *Matrix Analysis* (Cambridge University Press, Cambridge, 1985)
4. C. Lawson, R. Hanson, *Solving Least Squares Problems* (Prentice-Hall, Englewood Cliffs, 1974)

Appendix B: Mass Moments and Products of Inertia

The governing equations for the rotational motion of rigid bodies inherently involve the mass moments and products of inertia of those bodies. We examined these types of quantities for rigidly connected systems of particles in Chap. 3, for planar motion problems of continuous rigid bodies in Chap. 5, and for 3-D motion of continuous rigid bodies in Chap. 8. In this Appendix we want to provide an overview of the moments and products of inertia involved in general 3-D motions of continuous rigid bodies. Moments of inertia of some common shapes are given in Fig. B.5.

Definitions

The mass moments and products of inertia can be defined with respect to any coordinates but normally the coordinates used will move and rotate with the rigid body so that these inertia values of the body do not change during the motion. For example, Fig. B.1 shows a rigid body (labeled as (n)) and an element of mass $dm = \rho dV$ at point P in that body located with respect to coordinates $(x^{(n)}, y^{(n)}, z^{(n)})$ that rotate with the body and whose origin is a point A that translates with the body, and where ρ is the density of the body (mass/vol) and dV is a differential volume element. The mass moments and products of inertia can be defined as simply the following integrals:

L. W. Schmerr, *Engineering Dynamics 2.0*, Solid Mechanics and Its Applications 254,
https://doi.org/10.1007/978-3-319-98470-4

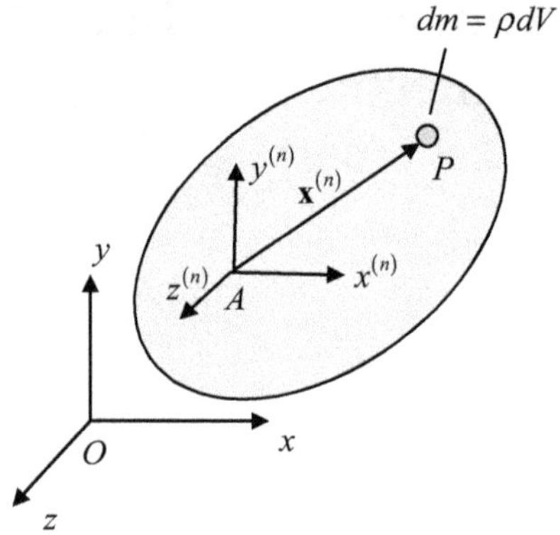

Fig. B.1 The geometry of a rigid body and a coordinate system with origin at A moving with the rigid body. The location of a mass element at point P is located relative to A

$$
\begin{aligned}
I_{xx}^{(n)} &= \int_{V_n} \rho\left[\left(y^{(n)}\right)^2 + \left(z^{(n)}\right)^2\right] dV \\
I_{yy}^{(n)} &= \int_{V_n} \rho\left[\left(x^{(n)}\right)^2 + \left(z^{(n)}\right)^2\right] dV \\
I_{zz}^{(n)} &= \int_{V_n} \rho\left[\left(x^{(n)}\right)^2 + \left(y^{(n)}\right)^2\right] dV \\
I_{xy}^{(n)} &= I_{yx}^{(n)} = -\int_{V_n} \rho x^{(n)} y^{(n)} dV \\
I_{xz}^{(n)} &= I_{zx}^{(n)} = -\int_{V_n} \rho x^{(n)} z^{(n)} dV \\
I_{yz}^{(n)} &= I_{zy}^{(n)} = -\int_{V_n} \rho y^{(n)} z^{(n)} dV
\end{aligned}
\tag{B.1.1}
$$

where $\left(I_{xx}^{(n)}, I_{yy}^{(n)}, I_{zz}^{(n)}\right)$ are called moments of inertia and $\left(I_{xy}^{(n)}, I_{xz}^{(n)}, I_{yz}^{(n)}\right)$ are called products of inertia. We can place all these quantities in an inertia matrix, $\left[I_A^{(n)}\right]$, given by

$$
\left[I_A^{(n)}\right] = \begin{bmatrix} I_{xx}^{(n)} & I_{xy}^{(n)} & I_{xz}^{(n)} \\ I_{yx}^{(n)} & I_{yy}^{(n)} & I_{yz}^{(n)} \\ I_{zx}^{(n)} & I_{zy}^{(n)} & I_{zz}^{(n)} \end{bmatrix} \tag{B.1.2}
$$

where since $I_{yx}^{(n)} = I_{xy}^{(n)}, I_{zx}^{(n)} = I_{xz}^{(n)}$, and $I_{zy}^{(n)} = I_{yz}^{(n)}$ the inertia matrix is symmetric. One can also show that this inertia matrix, like the mass matrix, is positive definite. This property, as we will see, has some important consequences for the behavior of the

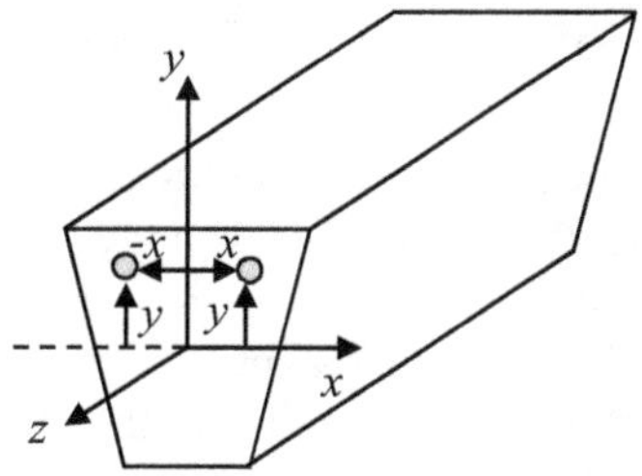

Fig. B.2 A body which has the y–z plane as a plane of symmetry

matrix. All the moments and products of inertia are in nonzero in general. However, if any two coordinate axes (a, b) form a plane of symmetry for the body, then the products of inertia associated with a coordinate axis (c) normal to that plane of symmetry will vanish, i.e., $I_{ac}^{(n)} = I_{bc}^{(n)} = 0$. For example, consider the body shown in Fig. B.2 which has the y–z plane as a plane of symmetry ($a = y, b = z$). Then ($c = x$) and we have $I_{xy}^{(n)} = I_{xz}^{(n)} = 0$. We can see in Fig. B.2 why the $I_{xy}^{(n)}$ in particular should vanish since the body is always composed of matching elements where one element is located at (x, y) and an identical element is located at $(-x, y)$ so their contributions to the product of inertia are $xy\,dm$ and $-xy\,dm$, which cancel.

The inertia matrix can be expressed in matrix-vector form in several ways. We can write

$$\left[I_A^{(n)}\right] = \int_{V_n} \left(\left\{x^{(n)}\right\}^T \left\{x^{(n)}\right\}[1] - \left\{x^{(n)}\right\}\left\{x^{(n)}\right\}^T\right)\rho dV \tag{B.1.3}$$

where $\{x^{(n)}\} = [x^{(n)}, y^{(n)}, z^{(n)}]^T$, and [1] is the identity matrix,which elsewhere in this book we have called [I]. In this Appendix we will use [1] to avoid confusing this identity matrix with an inertia matrix. Note that $\{x^{(n)}\}^T\{x^{(n)}\} = (x^{(n)})^2 + (y^n)^2 + (z^{(n)})^2$. Another, more compact form is

$$\left[I_A^{(n)}\right] = \int_{V_n} \left[X^{(n)}\right]^T \left[X^{(n)}\right]\rho dV = \int_{V_n} \left[X^{(n)}\right]\left[X^{(n)}\right]^T \rho dV \tag{B.1.4}$$

where $[X^{(n)}]$ is the skew matrix formed from the elements of $\{x^{(n)}\}$, i.e.,

$$\left[X^{(n)}\right] = \begin{bmatrix} 0 & -z^{(n)} & y^{(n)} \\ z^{(n)} & 0 & -x^{(n)} \\ -y^{(n)} & x^{(n)} & 0 \end{bmatrix} \tag{B.1.5}$$

Parallel Axis Theorem

The moments and products of inertia found in tables are often calculated with respect to the centroid of a body, which is also the center of mass of the body if it is homogeneous. See the table at the end of this Appendix for some examples. However, in a dynamics problem we may need to calculate the moment of inertia about another point. In the physical pendulum problem, for example, it is useful to calculate the moment of inertia about the axis through the pendulum's support rather than its center of mass since the moment equation about the support gives the equation of motion directly. If we know the values of inertia matrix with respect to a set of axes at the center of mass then we can calculate the values with respect to any other set of axes that are parallel to the center of mass axes through the *parallel axis theorem*. Consider, for example, the body shown in Fig. B.3 where we assume that we know the moments and products of inertia about the center of mass axes at point C but we want to calculate the values instead about a set of parallel axes through point A. From the geometry, we can locate an arbitrary element at point P with respect to C or A through the vector relationship

$$\{x_{P/C}\} = \{x_{P/A}\} + \{d\} \tag{B.2.1}$$

where $\{d\} = [d_x, d_y, d_z]^T$ has components that are location of point A as measured in the coordinate system at C. Then by the definition of the inertia matrix about point A we have

$$\begin{aligned}
[I_A] &= \int_{V_n} \left(\{x_{P/A}\}^T\{x_{P/A}\}[1] - \{x_{P/A}\}\{x_{P/A}\}^T\right)\rho dV \\
&= \int_{V_n} \left(\left(\{x_{P/C}\}^T\{x_{P/C}\} + \{d\}^T\{d\} - 2\{x_{P/C}\}^T\{d\}\right)[1]\right. \\
&\left. -\{x_{P/C}\}\{x_{P/C}\}^T - \{d\}\{d\}^T - 2\{x_{P/C}\}\{d\}\right)\rho dV
\end{aligned} \tag{B.2.2}$$

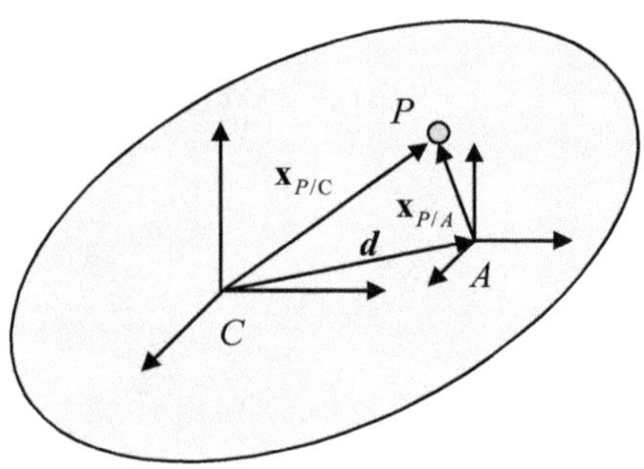

Fig. B.3 An arbitrary point P in a rigid body as seen in a coordinate system at the center of mass, C, or in a parallel coordinate system at point A

where we have dropped the (n) label on all the terms to make the expressions less complex. We can collect the first and fourth terms of the integrand and recognize it as the inertia matrix about point C and note that integration of the third and sixth terms involving $\{x_{P/C}\}$ vanishes because the origin is the center of mass, leaving

$$[I_A] = [I_C] + \int_{V_n} \left(\{d\}^T\{d\}[1] - \{d\}\{d\}^T\right)\rho dV \tag{B.2.3}$$

but all the matrix elements are constants in the integration of Eq. (B.2.3), giving finally

$$[I_A] = [I_C] + m\left(\{d\}^T\{d\}[1] - \{d\}\{d\}^T\right) \tag{B.2.4}$$

in terms of the mass, m, of the body. This equation can be expanded out more explicitly as

$$[I_A] = [I_C] + m\begin{bmatrix} d_y^2 + d_z^2 & -d_x d_y & -d_x d_z \\ -d_x d_y & d_x^2 + d_z^2 & -d_y d_z \\ -d_x d_z & -d_y d_z & d_x^2 + d_y^2 \end{bmatrix} \tag{B.2.5}$$

Equation (B.2.5) is the *parallel axis theorem* that allows us to shift the inertia matrix values from those calculated about the center of mass C to those calculated through a parallel set of axes at A. It is particularly useful when a given body can be decomposed into simpler parts. Such bodies are called composite bodies. See Chap. 5 where a composite body problem was considered for a simple planar problem.

Rotation of Axes

The moments and products of inertia can change with a rotation of axes at a given point. This is shown schematically in Fig. B.4. If we let the position vector from a point A to a general point P in the body in (x, y, z) coordinates be $\{x\}$ and let $\{x'\}$ represent the coordinates (x', y', z') of the same point P with respect to A in a rotated coordinate system, then we can relate $\{x\}$ and $\{x'\}$ through a rotation matrix, $[Q]$, where

$$\{x'\} = [Q]\{x\} \tag{B.3.1}$$

[Note: The rotation matrix Q here is not the rotation matrix considered in the text that relates the body-fixed coordinates to inertial coordinates. The matrix Q is simply a change from one coordinate system at A to another.] The inertia matrix in the rotated coordinates, $[I'_A]$, is

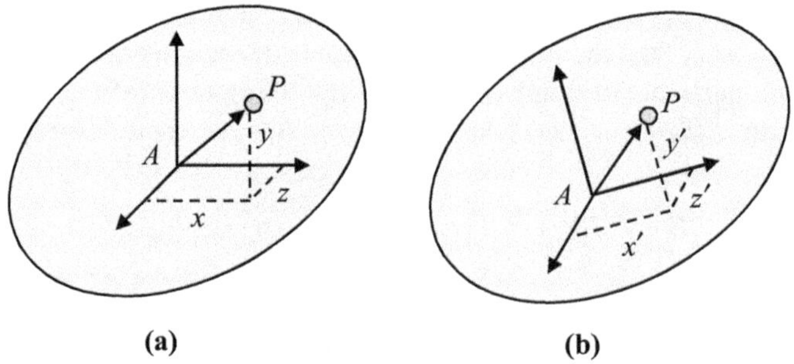

Fig. B.4 (**a**) A given coordinate system for computing the moments and products of inertia with respect to point A, and (**b**) a coordinate system at A that has a different orientation

$$\begin{aligned}[I'_A] &= \int_{V_n} \left(\{x'\}^T\{x'\}[1] - \{x'\}\{x'\}^T\right)\rho dV \\ &= \int_{V_n} \left(\{x\}^T\{x\}[1] - [Q]\{x\}\{x\}^T[Q]\right)\rho dV \\ &= \int_{V_n} \left([Q]\{x\}^T\{x\}[1][Q]^T - [Q]\{x\}\{x\}^T[Q]\right)\rho dV\end{aligned} \tag{B.3.2}$$

where we have used the fact that the rotation matrix is orthogonal, $[Q]^T[Q] = [Q][Q]^T = [I]$, and rewritten the first term since $\{x\}^T\{x\}[1] = |x|^2[1] = |x|^2[Q][1][Q]^T = [Q]\{x\}^T\{x\}[1][Q]^T$. But we can take the rotation matrices outside the integration so we find

$$\begin{aligned}[I'_A] &= [Q]\left(\int_{V_n} \left(\{x\}^T\{x\}[1] - \{x\}\{x\}^T\right)\rho dV\right)[Q]^T \\ &= [Q][I_A][Q]^T\end{aligned} \tag{B.3.3}$$

which is the rotational transformation relationship. Note that point A can be any point.

We can invert this relationship and also write

$$[I_A] = [Q]^T\left[I'_A\right][Q] \tag{B.3.4}$$

Principal Moments of Inertia

The inertia matrix is a symmetrical, positive definite matrix so that mathematically it has real, positive eigenvalues and its eigenvectors are real. We have seen this behavior before when considering vibration problems in Chap. 7. There the eigenvalues were associated with natural frequencies and the eigenvectors represented natural mode shapes of the vibration. In this case what does an eigenvalue problem mean physically? If a body has three axes of symmetry, such as an ellipsoid, for example, then the inertia matrix is diagonal and all the products of inertia vanish. Suppose we rotate axes at a given point and try to find a set of axes, like the ellipsoid, where the inertia matrix $\left[I'_A\right]$ is diagonal. Can we do that? The answer is yes. If $\left[I'_A\right]$ is diagonal let its diagonal components be $(\lambda_1, \lambda_2, \lambda_3)$. Now, consider Eq. (B.3.4) and post-multiply both sides of that relationship by $[Q]^T$. Then we find

$$[I_A][Q]^T = [Q]^T \begin{bmatrix} \lambda_1 & 0 & 0 \\ 0 & \lambda_2 & 0 \\ 0 & 0 & \lambda_3 \end{bmatrix} \tag{B.4.1}$$

If we let the vector $\{u\}$ be one of the columns of the matrix $[Q]^T$ we see that Eq. (B.4.1) is equivalent to solving the eigenvalue problem:

$$[I_A]\{u\} = \lambda\{u\} \tag{B.4.2}$$

where λ is one of the diagonal inertia matrix values. As we have seen before the eigenvalue problem has a nontrivial solution if

$$\det([I_A] - \lambda[1]) = 0 \tag{B.4.3}$$

which is a third-order polynomial in λ that has three real, positive roots corresponding to the diagonal moment of inertia, where the three values λ_k $(k = 1, 2, 3)$ are called the *principal moments* of inertia. The three corresponding eigenvectors $\{u_k\}$ $(k = 1, 2, 3)$ are real and represent the three directions along which the principal moments act, called the *principal directions*. We can follow the same steps outlined in Chap. 7 for vibration problems and show that the eigenvectors satisfy orthogonality relations, which here are

$$\{u_i\}^T\{u_j\} = 0 \quad \{u_i\}^T[I_A]\{u_j\} = 0 \quad (i \neq j) \tag{B.4.4}$$

and the eigenvectors can be normalized such that $\{u_k\}^T\{u_k\} = 1$ $(k = 1, 2, 3)$. It then follows that we also have $\{u_i\}^T[I]\{u_i\} = \lambda_i$ $(i = 1, 2, 3)$. Also as done in the vibration problem we can place the eigenvectors as columns in an eigenvector matrix, $[U]$, where because of the normalization of the eigenvectors we have $[U]^T[U] = [1]$, i.e., the eigenvector matrix is orthogonal. In terms of this eigenvector matrix we see that the eigenvalue problem can be written in matrix form as

$$[I_A][U] = [U][\lambda] \tag{B.4.5}$$

where $[\lambda]$ is a diagonal matrix containing the eigenvalues. Comparing Eq. (B.4.5) with Eq. (B.4.1) it appears that we simply have $[U] = [Q]^T$. This, however, may not always be the case. If $\{u_i\}$ is an eigenvector then so is $-\{u_i\}$ so there is some ambiguity in those vectors and, hence, in the matrix $[U]$. If we choose the signs such that ($\{u_1\}$, $\{u_2\}$, $\{u_3\}$) form a right-handed system, then det$[U] = +1$ and the $[U]$ matrix is a proper orthogonal transformation matrix, as is $[Q]^T$. With that choice for $[U]$ we can set $[U] = [Q]^T$.

Some Moments of Inertia (Fig. B.5)

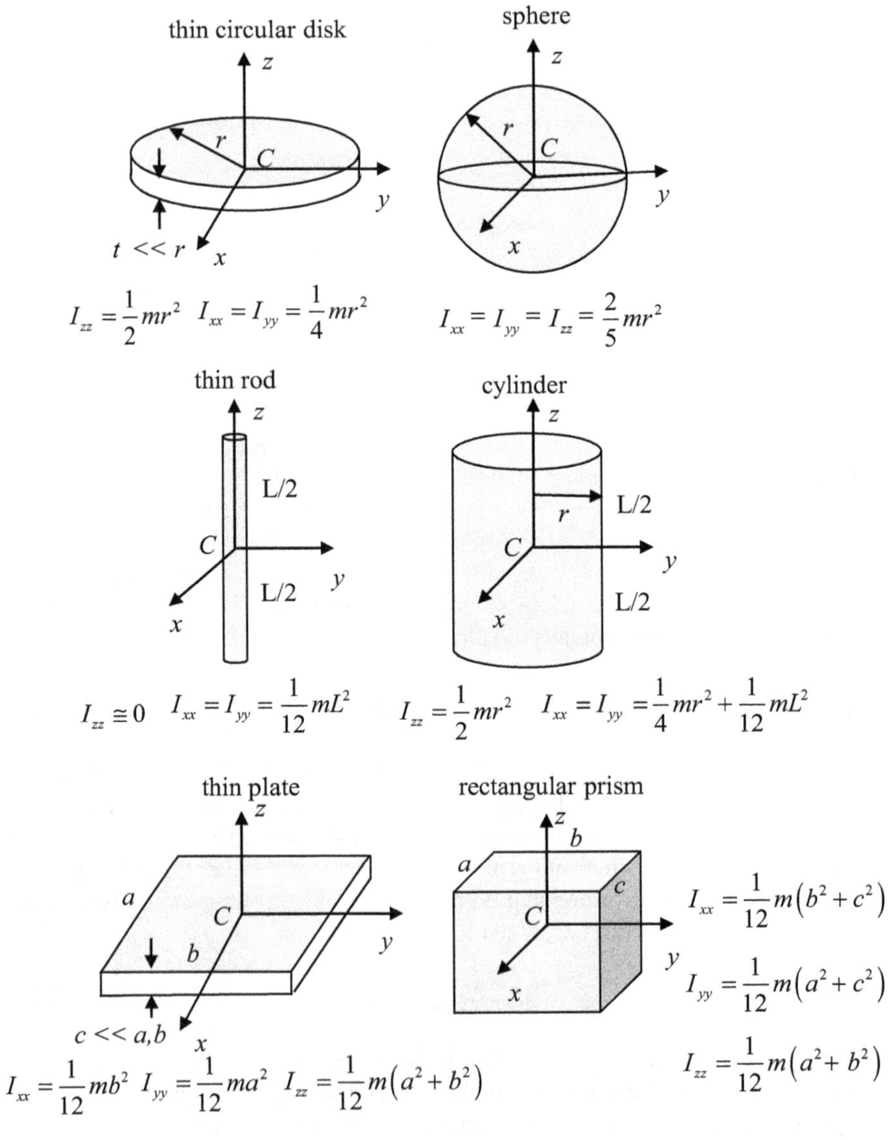

Fig. B.5 The moments of inertia with respect to the centroids (center of mass) of some simple homogeneous bodies

Appendix C: Numerical Methods

Numerical Solutions of Ordinary Differential Equations

Solving the equations of motion of dynamical systems usually requires the use of numerical methods. Normally, as we have seen, before we attempt to solve the equations of motion we write them as a system of first-order ordinary differential equations of the form

$$\begin{aligned} \dot{\mathbf{x}} &= \mathbf{f}(\mathbf{x}, t) \\ \mathbf{x}(t_0) &= \mathbf{x}_0 \end{aligned} \tag{C.1.1}$$

Equation (C.1.1) is called an initial value problem, i.e., where we start with the known values of the initial solution, $\mathbf{x}_0$, at time t_0, and we want to determine $\mathbf{x}$ as a function of the later times, t. There are numerous numerical methods for obtaining such a solution, methods that you can study in detail in books and courses on solving ordinary differential equations. Here, our objectives are much more modest and we only want to give you a basic understanding of how a particular class of solvers work that are used in the MATLAB® functions ode45 and ode4. Thus, we examine in this Appendix only methods called single-step explicit methods. Specifically, we describe single-step explicit *Runge-Kutta methods*, which are some of the most popular solvers used for ordinary differential equations, and the ones that are used exclusively in this book. The idea behind single-step explicit solvers is simple. If we know the solution at a given time, t_n, we numerically integrate the ordinary differential equations of Eq. (C.1.1) to obtain the solution at time t_{n+1} and then repeat this same process from time t_{n+1} to t_{n+2}, t_{n+2} to t_{n+3}, etc. over the entire time interval we want to have the solution. To formally integrate on time the equations of motion in Eq. (C.1.1) from t_n to t_{n+1}, for example, we have

L. W. Schmerr, *Engineering Dynamics 2.0*, Solid Mechanics and Its Applications 254,
https://doi.org/10.1007/978-3-319-98470-4

$$\mathbf{x}(t_{n+1}) = \mathbf{x}(t_n) + \int_{t_n}^{t_{n+1}} \mathbf{f}(\mathbf{x}(t), t)dt \tag{C.1.2}$$

We will examine this integration, assuming that we use a fixed step size $\Delta t = t_{n+1} - t_n$ for all the time intervals. As you might expect, a given numerical solver makes a particular choice of how the integration in Eq. (C.1.2) is performed. A very simple choice would be to assume that the function $\mathbf{f}$ is slowly varying over the time interval and simply take the integral to be the value of the function at the starting time, t_n, times the length of the interval, i.e.,

$$\int_{t_n}^{t_{n+1}} \mathbf{f}(x(t), t)dt \approx \Delta t f(\mathbf{x}(t_n), t_n) \tag{C.1.3}$$

giving

$$\mathbf{x}(t_{n+1}) = \mathbf{x}(t_n) + \Delta t \mathbf{f}(\mathbf{x}(t_n), t_n) \tag{C.1.4}$$

A more accurate estimate might take the integral to be the average of its two values at the endpoints of the interval time times the length of the interval, i.e.,

$$\int_{t_n}^{t_{n+1}} \mathbf{f}(\mathbf{x}(t), t)dt \approx \frac{\Delta t}{2}[\mathbf{f}(\mathbf{x}(t_n), t_n) + \mathbf{f}(\mathbf{x}(t_{n+1}), t_{n+1})] \tag{C.1.5}$$

which you might recall from calculus is called the trapezoidal rule. If we use the trapezoidal rule in Eq. (C.1.2) we obtain

$$\mathbf{x}(t_{n+1}) = \mathbf{x}(t_n) + \frac{\Delta t}{2}[\mathbf{f}(\mathbf{x}(t_n), t_n) + \mathbf{f}(\mathbf{x}(t_{n+1}), t_{n+1})] \tag{C.1.6}$$

Note that since we are trying to use Eq. (C.1.6) to obtain the value of the solution, $\mathbf{x}(t_{n+1})$, at time t_{n+1}, that solution is only given implicitly by Eq. (C.1.6) since $\mathbf{x}(t_{n+1})$ is also present within the function $\mathbf{f}$ on the right side of that equation. However, if we use Eq. (C.1.4) to evaluate $\mathbf{x}(t_{n+1})$ on the right side of Eq. (C.1.6) we have

$$\mathbf{x}(t_{n+1}) = \mathbf{x}(t_n) + \frac{\Delta t}{2}[\mathbf{f}(\mathbf{x}(t_n), t_n) + \mathbf{f}(\mathbf{x}(t_n) + \Delta t \mathbf{f}(\mathbf{x}(t_n), t_n), t_{n+1})] \tag{C.1.7}$$

which is now an explicit expression for $\mathbf{x}(t_{n+1})$. If we let $\mathbf{x}(t_n) = \mathbf{x}_n$ and $\mathbf{x}(t_{n+1}) = \mathbf{x}_{n+1}$, we can write Eq. (C.1.7) compactly as

$$\mathbf{x}_{n+1} = \mathbf{x}_n + \frac{\Delta t}{2}[\mathbf{f}(\mathbf{x}_n, t_n) + \mathbf{f}(\mathbf{x}_n + \Delta t\, \mathbf{f}(\mathbf{x}_n, t_n), t_{n+1})] \tag{C.1.8}$$

or even more compactly as

$$\mathbf{x}_{n+1} = \mathbf{x}_n + \frac{1}{2}\mathbf{k}_1 + \frac{1}{2}\mathbf{k}_2 \tag{C.1.9}$$

where

$$\begin{aligned} \mathbf{k}_1 &= \Delta t \mathbf{f}(\mathbf{x}_n, t_n) \\ \mathbf{k}_2 &= \Delta t \mathbf{f}(\mathbf{x}_n + \mathbf{k}_1, t_n + \Delta t) \end{aligned} \tag{C.1.10}$$

Equations (C.1.9) and (C.1.10) are called the classical second-order Runge-Kutta integration method. It uses a fixed step size, Δt, for each interval, and we can apply it to any first-order system of ordinary differential equations by simply supplying the appropriate function $\mathbf{f}(\mathbf{x}(t), t)$ and evaluating that function at the argument values given in Eq. (C.1.10). This method is normally not developed through the steps shown above but instead through a Taylor series expansion of the function $\mathbf{f}$ in the time interval, but we will not give the details of such a derivation here. Since the expansion extends to two derivatives this is called a second-order method where the local truncation error used in approximating the integral is $O(\Delta t^3)$. It is also called a two-stage method because it requires, for each time interval, evaluations of the two functions $\mathbf{k}_1$ and $\mathbf{k}_2$. Much more accurate and popular than the second-order method, however, is the classical fourth-order Runge-Kutta method, which is given by

$$\mathbf{x}_{n+1} = \mathbf{x}_n + \frac{1}{6}\mathbf{k}_1 + \frac{1}{3}\mathbf{k}_2 + \frac{1}{3}\mathbf{k}_3 + \frac{1}{6}\mathbf{k}_4 \tag{C.1.11}$$

where

$$\begin{aligned} \mathbf{k}_1 &= \Delta t \mathbf{f}(\mathbf{x}_n, t_h) \\ \mathbf{k}_2 &= \Delta t \mathbf{f}\left(\mathbf{x}_n + \frac{1}{2}\mathbf{k}_1, t_n + \frac{\Delta t}{2}\right) \\ \mathbf{k}_3 &= \Delta t \mathbf{f}\left(\mathbf{x}_n + \frac{1}{2}\mathbf{k}_2, t_n + \frac{\Delta t}{2}\right) \\ \mathbf{k}_4 &= \Delta t \mathbf{f}(\mathbf{x}_n + \mathbf{k}_3, t_n + \Delta t) \end{aligned} \tag{C.1.12}$$

It is a four-stage method and the local truncation error for this method is $O(\Delta t^5)$. We will also not derive this fourth-order method here, but as you can see it follows a very similar structure to the second-order method. This fourth-order Runge-Kutta method is a versatile method that, like the second-order method, can be easily applied to a system of ordinary differential equations by simply supplying the appropriate function $\mathbf{f}$. The MATLAB® function ode4 used in this book is based on this classical fourth-order, fixed step size, explicit Runge-Kutta method.

These second- and fourth-order Runge-Kutta methods are only two examples of a whole class of different Runge-Kutta approximations which can be described in terms of coefficients c_i, b_i, a_{ij} $(i,j = 1,\ldots,s)$ and where

$$\begin{aligned} \mathbf{g}_i &= \mathbf{x}_n + \sum_{j=1}^{s} a_{ij}\mathbf{k}_j \\ \mathbf{k}_i &= \Delta t\mathbf{f}(\mathbf{g}_i, t_n + c_i\Delta t) \\ \mathbf{x}_{n+1} &= \mathbf{x}_n + \sum_{j=1}^{s} b_i\mathbf{k}_i \end{aligned} \tag{C.1.13}$$

Here s is the number of stages. For example, for the classical second-order Runge-Kutta method we have $s = 2$ and $[c_1, c_2] = [0, 1]$, $[b_1, b_2] = [1/2, 1/2]$, and $a_{21} = 1$ with $a_{11} = a_{12} = a_{22} = 0$. For the classical fourth-order Runge-Kutta method $s = 4$ and $[c_1, c_2, c_3, c_4] = [0, 1/2, \ 1/2, \ 1]$, $[b_1, b_2, b_3, b_4] = [1/6, 1/3, 1/3, 1/6]$, and $a_{21} = a_{32} = 1/2, a_{43} = 1$ with all the other $a_{ij} = 0$.

Fixed-step methods have a major disadvantage from the standpoint of numerical efficiency in that they must use the small step size needed to produce accurate results in regions where the function **f** is rapidly changing even in other regions where the step size could potentially be larger. Also, to estimate the numerical errors present it is necessary, with fixed-step methods, to calculate the solution twice, using different step sizes, which is inefficient. For these reasons, modern solvers often instead use an adaptive step size to control the step sizes based on a specified error tolerance in the solution. The MATLAB® function ode45 is an example of this type of solver. The ode45 function implements a Dormand-Prince scheme which uses two Runge-Kutta methods of the type shown in Eq. (C.1.13). It involves the seven functions

$$\begin{aligned} \mathbf{k}_1 &= \Delta t\mathbf{f}(\mathbf{x}_n, t_n) \\ \mathbf{k}_2 &= \Delta t\mathbf{f}\left(\mathbf{x}_n + \frac{1}{5}\mathbf{k}_1, t_n + \frac{\Delta t}{5}\right) \\ \mathbf{k}_3 &= \Delta t\mathbf{f}\left(\mathbf{x}_n + \frac{3}{40}\mathbf{k}_1 + \frac{9}{40}\mathbf{k}_2, t_n + \frac{3}{10}\Delta t\right) \\ \mathbf{k}_4 &= \Delta t\mathbf{f}\left(\mathbf{x}_n + \frac{44}{45}\mathbf{k}_1 - \frac{56}{15}\mathbf{k}_2 + \frac{32}{9}\mathbf{k}_3, t_n + \frac{4}{5}\Delta t\right) \\ \mathbf{k}_5 &= \Delta t\mathbf{f}\left(\mathbf{x}_n + \frac{19372}{6561}\mathbf{k}_1 - \frac{25360}{2187}\mathbf{k}_2 + \frac{64448}{6561}\mathbf{k}_3 - \frac{212}{729}\mathbf{k}_4, t_n + \frac{8}{9}\Delta t\right) \\ \mathbf{k}_6 &= \Delta t\mathbf{f}\left(\mathbf{x}_n + \frac{9017}{3168}\mathbf{k}_1 - \frac{355}{33}\mathbf{k}_2 - \frac{46732}{5247}\mathbf{k}_3 + \frac{49}{176}\mathbf{k}_4 - \frac{5103}{18656}\mathbf{k}_5, t_n + \Delta t\right) \\ \mathbf{k}_7 &= \Delta t\mathbf{f}\left(\mathbf{x}_n + \frac{35}{384}\mathbf{k}_1 + \frac{500}{1113}\mathbf{k}_3 + \frac{125}{192}\mathbf{k}_4 - \frac{2187}{6784}\mathbf{k}_5 + \frac{11}{84}\mathbf{k}_6, t_n + \Delta t\right) \end{aligned} \tag{C.1.14}$$

where the next step value, $\mathbf{x}_{n+1}$, is calculated from a Runge-Kutta method of order four given by

$$\mathbf{x}_{n+1} = \mathbf{x}_n + \frac{35}{384}\mathbf{k}_1 + \frac{500}{1113}\mathbf{k}_3 + \frac{125}{192}\mathbf{k}_4 - \frac{2187}{6784}\mathbf{k}_5 + \frac{11}{84}\mathbf{k}_6 \tag{C.1.15}$$

Note that although $\mathbf{k}_2$ does not appear explicitly in Eq. (C.1.15) it is needed to calculate $\mathbf{k}_3$ and so forth. From these same $\mathbf{k}$ functions we can determine a different next step value, $\tilde{\mathbf{x}}_{n+1}$, by a Runge-Kutta method of order 5 as

$$\tilde{\mathbf{x}}_{n+1} = \mathbf{x}_n + \frac{5179}{57600}\mathbf{k}_1 + \frac{7571}{16695}\mathbf{k}_3 + \frac{393}{640}\mathbf{k}_4 - \frac{92097}{339200}\mathbf{k}_5 + \frac{187}{2100}\mathbf{k}_6 + \frac{1}{40}\mathbf{k}_7 \tag{C.1.16}$$

The magnitude of the difference of the two values, $|\tilde{\mathbf{x}}_{n+1} - \mathbf{x}_{n+1}|$, is then

$$|\tilde{\mathbf{x}}_{n+1} - \mathbf{x}_{n+1}| = \left|\frac{71}{57600}\mathbf{k}_1 - \frac{71}{16695}\mathbf{k}_3 + \frac{71}{1920}\mathbf{k}_4 - \frac{17253}{339200}\mathbf{k}_5 + \frac{22}{525}\mathbf{k}_6 - \frac{1}{40}\mathbf{k}_7\right| \tag{C.1.17}$$

This difference is considered to be the error, e_n, in $\mathbf{x}_{n+1}$. In the MATLAB® ode45 function we have seen that there are two error tolerances, an absolute tolerance, atol, and a relative tolerance, rtol, where rtol defines an error based on the size of the solution. From these two tolerances we can define a maximum allowable error, $e_{\max}$, as

$$e_{\max} = \max\{\text{rtol}|x_n|, \text{atol}\} \tag{C.1.18}$$

If $e_n = |\tilde{\mathbf{x}}_{n+1} - \mathbf{x}_n| \le e_{\max} = \max\{\text{rtol}|x_n|, \text{atol}\}$ we keep the current step size, Δt, but if not, we adjust the step size to satisfy this error criterion. In this fashion we can have an adaptive step size based on the user-supplied error tolerances. We will not go into any further detail as to the explicit workings of the ode45 function, but this should give you at least a general understanding of how this function solves a system of ordinary differential equations and the fundamental role that rtol and atol play in that solution.

Numerical Solutions of Nonlinear Algebraic Equations

To solve for the motion in kinematically driven dynamical systems one has to find the solution for the values of a set of coordinates, x_k $(k = 1, 2, \ldots, n)$, that satisfy the equations $f_i(x_k) = 0$ $(i = 1, 2, \ldots, n)$. In Chap. 4 we showed how the Newton-Raphson method provides a numerical solution to these nonlinear algebra equations.

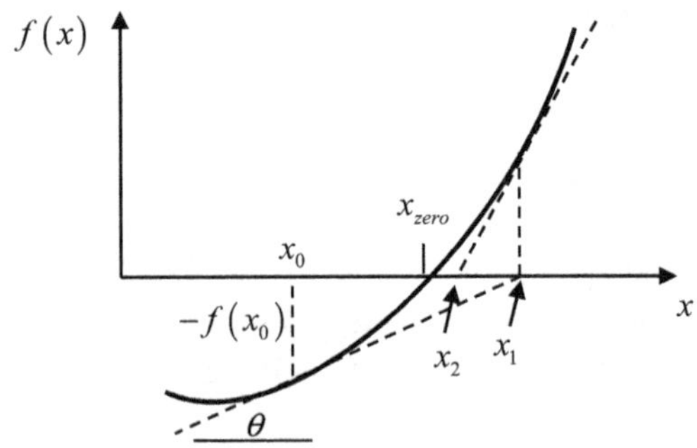

Fig. C.1 Geometry for determining the zero of a function with Newton's method

The basis of the Newton-Raphson method can be understood by considering the simpler scalar problem of finding an x that is the solution to the equation $f(x) = 0$. One numerical solution to this problem is called *Newton's method.* Newton's method can be understood graphically by examining the function $f(x)$ in a neighborhood of the value, x , where $f(x_{\text{zero}}) = 0$ as shown in Fig. C.1. The method requires that we have three things: (1) an initial "good" approximation to the solution, which we will call x_0; (2) the value of the function at this initial approximation and all subsequent approximations, $f(x_k) \quad (k = 0, 1, 2, \ldots)$; and (3) the value of the derivative of the function f at this initial approximation and all subsequent approximations, $f'(x_k) \quad (k = 0, 1, 2, \ldots)$. With the initial approximation values we can evaluate the function at a new value, x_1, which we will take as the next approximation for x_{zero} so in a first-order Taylor series expansion near x_0 we have

$$f(x_1) = f(x_0) + f'(x_0)(x_1 - x_0) = 0 \tag{C.2.1}$$

which can be solved for x_1 as

$$x_1 = x_0 - \frac{f(x_0)}{f'(x_0)} \tag{C.2.2}$$

Graphically, this corresponds to equating the slope of the curve at x_0 to the tangent of the angle that the straight-line tangent to the curve at x_0 makes with the x-axis (see Fig. C.1), i.e.,

$$f'(x_0) = \tan\theta = \frac{-f(x_0)}{(x_1 - x_0)} \tag{C.2.3}$$

where x_1 is the point on the straight line that intersects the x-axis. This same process can be continued at x_1 to obtain a new approximation, x_2, as shown in Fig. C.1, and so on. The general method, therefore, is an iterative scheme where

$$x_{k+1} = x_k - \frac{f(x_k)}{f'(x_k)} \quad (k = 0, 1, 2, \ldots) \tag{C.2.4}$$

and the method can be stopped whenever the magnitude of the changes to the approximation falls below some small error tolerance value, tol, i.e.,

$$\left|\frac{f(x_k)}{f'(x_k)}\right| \leq \text{tol} \tag{C.2.5}$$

An implementation of this method is contained in the MATLAB® function newton_m. The calling sequence for the function is

```
[xzero, iter] = newton_m(funh, dfunh, x0, tol);
```

where funh and dfunh are function handles for the function and its derivative, x0 is the initial guess for the solution, and tol is an error tolerance which stops the iterations when the magnitude of the difference between two approximations is less than this specified value. As a simple example, consider $f(x) = x^3 - 6x^2 - 2x + 12$ whose derivative is $f'(x) = 3x^2 - 12x - 2$. This function has zeros at $x = \pm\sqrt{2}, x = 6$ as shown in Fig. C.2. The following MATLAB® commands generate handles to these functions in MATLAB® and give an initial guess of x0 = 8, with a tolerance of 1e-6. Executing newton_m produces the solution at approximately $x = 6$ in five iterations:

```
fun1 = @ (x) x.^3 -6*x.^2 -2*x +12;
dfun1 = @ (x) 3*x.^2 -12*x -2;
 [xzero, iter] =newton_m( fun1, dfun1, 8, 1e-6)
xzero = 6.0000
iter = 5
```

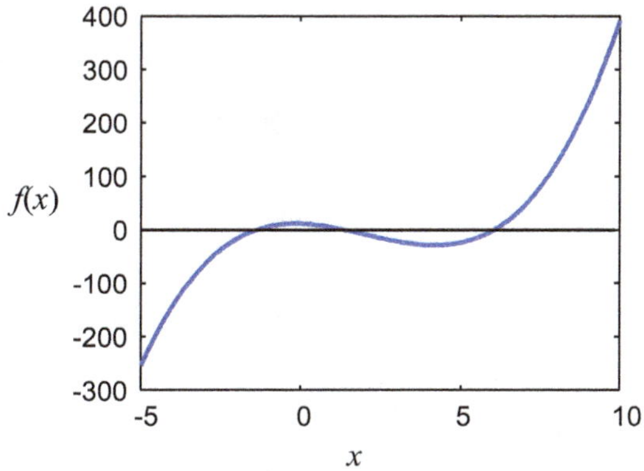

Fig. C.2 A graph of the function $x^3 - 6x^2 - 2x + 12$ that has zeros at $x = 6$, $x = \pm\sqrt{2}$

However, if the initial guess is changed to x0 = 4, Newton's method finds approximately the solution $x = -\sqrt{2}$ in eight iterations. If the initial guess is x0 = 2 the method finds approximately the root $x = \sqrt{2}$ in three iterations. Thus, Newton's method can find a solution accurately and rapidly if the initial guess is close enough to the "expected" solution but if there are competing solutions then a poor guess may result in convergence to an unexpected solution. A poorly behaved function can also result in very slow convergence or no convergence.

The Newton-Raphson method outlined in Chap. 4 follows this same solution procedure for the solution of the system of equations $\mathbf{f}(\mathbf{x}) = 0$, where there are n functions and n coordinates $\mathbf{x} = (x_1, x_2, \ldots, x_n)$. Satisfying that equation at an approximation $\mathbf{x}_{k+1}$ through the use of a first-order Taylor series expansion at $\mathbf{x}_k$ gives

$$\mathbf{f}(\mathbf{x}_{k+1}) = \mathbf{f}(\mathbf{x}_k) + \nabla\mathbf{f}(\mathbf{x}_{k+1} - \mathbf{x}_k) = 0 \tag{C.2.6}$$

in terms of $\mathbf{J} = \nabla\mathbf{f}$, which is the Jacobian matrix of derivatives of the functions. The components of $\mathbf{J}$ are $J_{ij} = \partial f_i(\mathbf{x}_k)/\partial x_j \quad (i, j = 1, 2, \ldots, n)$. Eq. (C.2.6) is the basis for the iterative improvement relation

$$\mathbf{x}_{k+1} = \mathbf{x}_k - \mathbf{J}^{-1}\mathbf{f}(\mathbf{x}_k) \tag{C.2.7}$$

The function newton_raph discussed in Chap. 4 implements this iterative solution process. The close similarity of the Newton-Raphson method to Newton's method for the scalar case means that the Newton-Raphson method shares many of the same features (and failings) of Newton's method.

Appendix D: Vibrations of One Degree of Freedom Systems

Chapter 7 discusses the small vibrations of multiple degree of freedom systems. This Appendix will lay the foundation for many of the results obtained in Chap. 7 by examining the behavior of a single degree of freedom vibrating system.

General Solutions

In Chap. 7 we showed that the small-amplitude vibration of many one degree of freedom systems can be characterized as equivalent to the mass-spring-dashpot system seen in Fig. D.1a. The governing equation for this system from the free body diagram of Fig. D.1b is

$$m\ddot{x}(t) + c\dot{x}(t) + kx(t) = F(t) \tag{D.1.1}$$

subject to initial conditions on how the motion starts at $t = 0$ given as $x(0) = x_0$, $\dot{x}(0) = \dot{x}_0$. Here m is a mass, c is a dashpot (damping) constant, k is a spring constant, and F is a driving force. A system that satisfies Eq. (D.1.1) is an example of what is called a *linear, time-shift invariant* (LTI) system. LTI systems can be viewed in general as a system where some operator $L[\cdot]$ acts on an input to the system, $i(t)$, and produces an output, $o(t)$, i.e.,

$$o(t) = L[i(t)] \tag{D.1.2}$$

This is shown schematically in Fig. D.2. The vibrating system characterized by Eq. (D.1.1) can be viewed as an LTI system where the input is the driving force, $i(t) = F(t)$, and the output is the displacement, $o(t) = x(t)$. The operator, L, is essentially the operation we must perform to solve the differential equation of Eq. (D.1.1). In LTI system the operator L is a linear operator which means that for any two inputs $(i_1(t), i_2(t))$ and their corresponding outputs $(o_1(t), o_2(t))$ we must have

L. W. Schmerr, *Engineering Dynamics 2.0*, Solid Mechanics and Its Applications 254,
https://doi.org/10.1007/978-3-319-98470-4

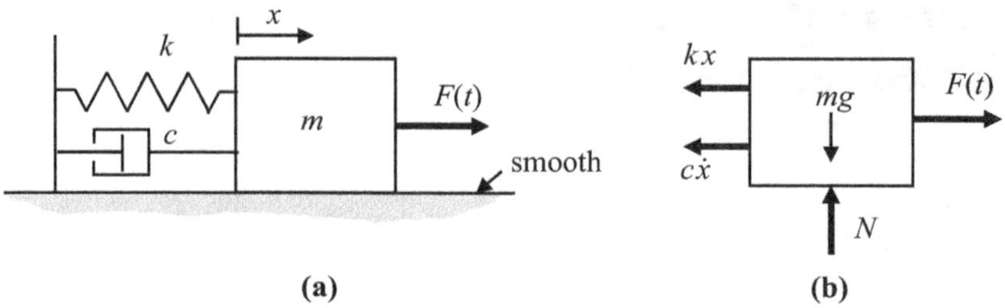

Fig. D.1 (**a**) A one degree of freedom vibrating system and (**b**) its free body diagram

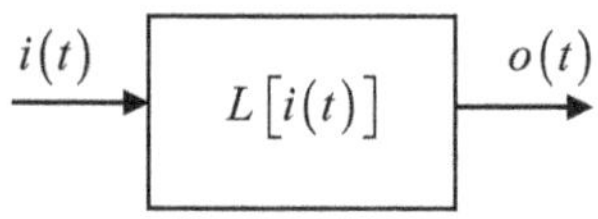

Fig. D.2 A linear time-shift invariant system that takes an input function, $i(t)$, and performs a linear time-shift invariant operation $L[\cdot]$ on that input to generate an output $o(t) = L[i(t)]$

$$\begin{aligned} o(t) &= L[a_1 i_1(t) + a_2 i_2(t)] \\ &= a_1 L[i_1(t)] + a_2 L[i_2(t)] \end{aligned} \tag{D.1.3}$$

where (a_1, a_2) are arbitrary constants. Thus, in linear systems if we multiply the input by a given amount, the output is multiplied by that same amount and if we superimpose two inputs, the output is also the superposition of the outputs generated by the individual inputs. If we examine Eq. (D.1.1) we see that the vibrating system it describes does have these properties. The time-shift invariance property of LTI systems means that if we advance or delay the input by a certain amount then the output will also be advanced or delayed by the same amount, a property we can express through the relationship

$$o(t - t_0) = L[i(t - t_0)] \tag{D.1.4}$$

where t_0 is some constant time. Our vibrating system of Eq. (D.1.1) also satisfies Eq. (D.1.4) since we see if the force is changed to $F(t - t_0)$ then the displacement $x(t - t_0)$ satisfies the differential equation. We have described an LTI system in these general terms since it is precisely these properties that LTI systems possess that allows us to obtain many of the analytical results we will obtain in this Appendix for our vibrating system of Eq. (D.1.1).

Equation (D.1.1) is a second-order linear differential equation with constant coefficients. To obtain a solution of this equation we can write the solution as the sum of two terms called the homogeneous and particular solutions as $x(t) = x_h(t)+x_p(t)$. The homogeneous solution, $x_h(t)$, satisfies Eq. (D.1.1) with $F(t) = 0$, i.e.,

$$m\ddot{x}_h(t) + c\dot{x}_h(t) + kx_h(t) = 0 \tag{D.1.5}$$

The particular solution, $x_p(t)$, is a solution of Eq. (D.1.1) that generates the force term so

$$m\ddot{x}_p(t) + c\dot{x}_p(t) + kx_p(t) = F(t) \tag{D.1.6}$$

In general, neither the homogeneous nor the particular solution will satisfy the initial conditions but their sum must so that

$$\begin{aligned} x_h(0) + x_p(0) &= x_0 \\ \dot{x}_h(0) + \dot{x}_p(0) &= \dot{x}_0 \end{aligned} \tag{D.1.7}$$

Homogeneous Solutions

First, let's examine the homogeneous solution. We will find it convenient first to write Eq. (D.1.1) in different terms. If we divide that equation by the mass m we can rewrite it as

$$\ddot{x}(t) + 2\varsigma\omega_n\dot{x}(t) + \omega_n^2 x(t) = 0 \tag{D.1.8}$$

where $\omega_n = \sqrt{k/m}$ and $\varsigma = c/2\sqrt{km}$ where ω_n is a characteristic frequency (in rad/s) and ς is a normalized damping factor. To solve Eq. (D.1.8) we will assume a solution in the form

$$x(t) = C\exp(\lambda t) \tag{D.1.9}$$

Placing this solution into Eq. (D.1.8) gives

$$\left(\lambda^2 + 2\varsigma\omega_n\lambda + \omega_n^2\right)C\exp(\lambda t) = 0 \tag{D.1.10}$$

Equation (D.1.10) can be satisfied nontrivially for all times t only if

$$\lambda^2 + 2\varsigma\omega_n\lambda + \omega_n^2 = 0 \tag{D.1.11}$$

which is a quadratic equation which has two roots that we can write as

$$\lambda_{1,2} = -\varsigma\omega_n \pm i\omega_n\sqrt{1-\varsigma^2} \tag{D.1.12}$$

The damping factor, as we will see, plays a crucial role in determining the nature of the solution. There are four different cases:

1. When $\varsigma = 0$ (no damping) the two roots are purely imaginary and the solution (D.1.9) can be written as

$$x(t) = C_1 \exp(i\omega_n t) + C_2 \exp(-i\omega_n t) \tag{D.1.13a}$$

or, in purely real terms,

$$x(t) = A\cos(\omega_n t - \phi) \tag{D.1.13b}$$

In either case there are two unknown constants (C_1, C_2) or (A, ϕ). This *undamped case* is just a constant-amplitude, oscillatory motion with an undamped natural frequency, ω_n (Fig. D.3a).

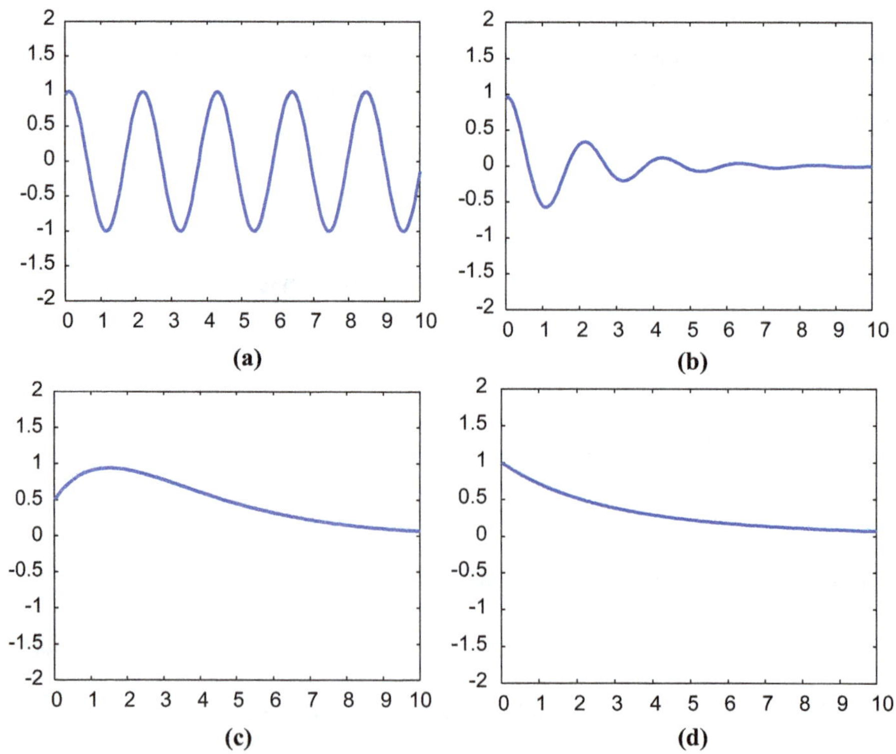

Fig. D.3 Homogeneous solution for a 1-D system for (**a**) undamped case, (**b**) underdamped case, (**c**) critically damped case, and (**d**) overdamped case

2. When $0 < \varsigma < 1$ the two roots are complex with negative real parts. The solution then can be written as

$$x(t) = \exp(-\varsigma\omega_n t)[C_1 \exp(i\omega_d t) + C_2 \exp(-i\omega_d t)] \tag{D.1.14a}$$

or in real terms as

$$x(t) = A\exp(-\varsigma\omega_n t) \cos(\omega_d t - \phi) \tag{D.1.14b}$$

where $\omega_d = \omega_n\sqrt{1-\varsigma^2}$ is called the damped natural frequency. The motion in this case has exponentially decaying oscillations at the damped natural frequency and is called the *underdamped case* (Fig. D.3b).
3. When $\varsigma = 1$ the two roots coalesce into a single negative real root $\lambda_1 = \lambda_2 = -\omega_n$. In this case the form of the solution given by Eq. (D.1.9) is only one part of the total solution. The complete solution in this case can be shown to be

$$x(t) = (C_1 + C_2 t)\exp(-\omega_n t) \tag{D.1.14c}$$

which may initially increase but which eventually decays in time without oscillation. The motion in this case is said to be *critically damped* (Fig. D.3c).
4. When $\varsigma > 1$ there are two real negative roots:

$$\begin{aligned}\lambda_{1,2} &= -\varsigma\omega_n \pm \omega_n = -\varsigma\omega_n\left(1 \pm \sqrt{1-1/\varsigma^2}\right)\\ &= -\varsigma\omega_n\mu_{1,2}\end{aligned}$$

where $\mu_{1,2} = 1 \pm \sqrt{1-1/\varsigma^2}$ are both real, positive values and the solution is

$$x(t) = C_1\exp(-\varsigma\omega_n\mu_1 t) + C_2\exp(-\varsigma\omega_n\mu_2 t) \tag{D.1.14d}$$

which is a purely non-oscillatory exponential decaying response. This case is called *overdamped* (Fig. D.3d).

There is also the possibility that $\varsigma < 0$ but this would correspond to a negative damping coefficient and produce an unstable growing rather than a damped solution so it is not a case we will consider.

Free Vibration Solutions

The homogeneous solutions developed in the last section are only a part of the total solution for a forced vibration problem. However if the driving force is absent the system can vibrate as a free vibration problem. In this case the homogeneous solution is all that we need and we can find the unknown constants in that solution from the initial conditions. We will not examine all four cases here but will only consider the underdamped case since that case is usually the one of most physical interest where the damping is relatively small. In that case the solution in terms of the initial conditions can be shown to be [1]

$$x(t) = \exp(-\varsigma\omega_n t)\left[x_0 \cos(\omega_d t) + \frac{(\dot{x}_0 + \varsigma\omega_n x_0)}{\omega_d}\sin(\omega_d t)\right] \tag{D.1.15}$$

The free vibration undamped case can be obtained directly by letting $\varsigma = 0$ which gives

$$x(t) = x_0 \cos(\omega_n t) + \frac{\dot{x}_0}{\omega_n}\sin(\omega_n t) \tag{D.1.16}$$

Impulse Response and a Particular Solution as a Convolution Integral

To obtain an expression for the particular part of the general solution of Eq. (D.1.1) we first will introduce the concept of the unit impulse response. Consider a system with zero initial conditions and an applied force that has a constant value, F_0, only over a very small time interval, ε, near $t = 0$ (Fig. D.4). Let us integrate Eq. (D.1.1) from $t = 0$ to $t = \varepsilon$. We find

$$m\dot{x}(\varepsilon) - m\dot{x}(0) + \int_0^{\varepsilon} c\dot{x}\,dt + \int_0^{\varepsilon} kx dt = F_0\varepsilon \tag{D.1.17}$$

But the initial conditions are assumed to be zero so $\dot{x}(0)$ vanishes. Now, consider letting $\varepsilon \to 0$ but simultaneously let the force $F_0 \to \infty$ such that the product $I = F_0\varepsilon$

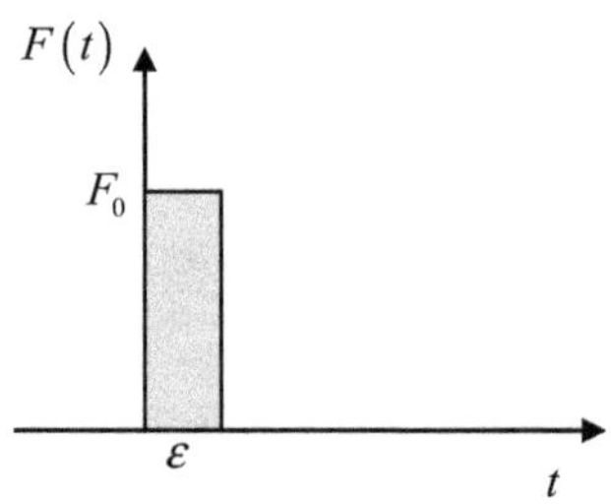

Fig. D.4 A large force acting over a small time interval as a model of an impulsive force remains a constant. Such a force is called an impulsive force and I is the magnitude of the impulse. As $\varepsilon \to 0$ the integrals in Eq. (D.1.17) go to zero so that in the limit we are left with

$$\begin{aligned} x(0^+) &= 0 \\ m\dot{x}(0^+) &= I \end{aligned} \quad \text{(D.1.18)}$$

where 0^+ indicates that we are evaluating the response immediately after the impulsive force is applied at $t = 0$. Thus we see that the response of a system to an impulsive force at $t = 0$ can also be considered to be a free vibration problem where $F = 0$ for $t > 0$ but where the initial conditions are given by Eq. (D.1.18). If we let $I = 1$ then we say that we have a *unit impulse response* of our system. Solving for the unit impulse response then is equivalent to solving the free vibration problem:

$$\begin{aligned} m\ddot{x}(t) + c\dot{x}(t) + kx(t) = 0 \\ x(0^+) = 0, \quad m\dot{x}(0^+) = 1 \end{aligned} \quad \text{(D.1.19)}$$

For an underdamped system, therefore, we have the unit impulse response of the system, $g(t)$, from Eq. (D.1.15) as simply

$$g(t) = \begin{cases} \dfrac{1}{m\omega_d} \exp(-\varsigma\omega_n t) \sin(\omega_d t) & t \geq 0 \\ 0 \quad t < 0 \end{cases} \quad \text{(D.1.20)}$$

and for the undamped system

$$g(t) = \begin{cases} \dfrac{1}{m\omega_n} \sin(\omega_d t) & t \geq 0 \\ 0 \quad t < 0 \end{cases} \quad \text{(D.1.21)}$$

where we have included the values for $t < 0$ to emphasize that the unit impulse response does not begin until the impulse is applied. What makes the unit impulse response so useful is that it can be used as a building block to form up a particular solution that satisfies

$$m\ddot{x}_p(t) + c\dot{x}_p(t) + kx_p(t) = F(t) \quad \text{(D.1.22)}$$

To see this, Fig. D.5 shows a general force as a function of time and a small portion of that force at a time τ. The shaded area in Fig. D.5 thus acts like an impulse of strength $F(\tau)\Delta\tau$. If we let $x_p(t, \tau)$ be the response of the system to the force in that shaded area we have

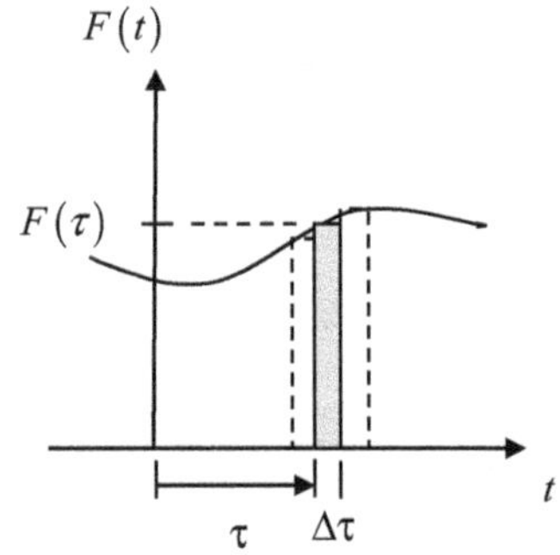

Fig. D.5 Approximation of the area under the force curve, $F(t)$, as a series of impulses

$$x_p(t,\tau) = F(\tau)\Delta\tau g(t-\tau) \tag{D.1.23}$$

Notice that in obtaining Eq. (D.1.23) we have implicitly used the fact that our system is an LTI system since the output is scaled by the same amount that the unit impulse input is scaled (the factor $F\Delta\tau$) and the output is shifted in time by the same amount as the shift in the input (the impulse acts at a delayed time $t = \tau$) by writing the output in terms of the shifted impulse response $g(t - \tau)$. Using the properties of LTI systems again the total response to the force can be represented as the superposition of the responses to impulses such as those shown as the dashed areas in Fig. D.5, so we have

$$x_p(t) = \sum_{\tau} x_p(t,\tau) = \sum_{\tau} F(\tau)\Delta\tau g(t-\tau) \tag{D.1.24}$$

or in the limit as $\Delta\tau \to 0$ we obtain the integral over all times τ as

$$x_p(t) = \int_{-\infty}^{\infty} F(\tau) g(t-\tau) d\tau \tag{D.1.25}$$

The integral in Eq. (D.1.25) is called a *convolution integral.* Often we see this integral expressed in a form which accounts for the properties of the functions F and g. For example the impulse response function satisfies $g(t - \tau) = 0$ for $\tau > t$ so we can replace the upper limit of $\tau = +\infty$ by $\tau = t$. Also, in most cases the driving force satisfies $F(\tau) = 0$ for $\tau < 0$ (i.e., the driving force vanishes before the start of the motion) so we can replace the lower limit of $\tau = -\infty$ by $\tau = 0$. These two conditions then allow us to write the convolution integral as

$$x_p(t) = \int_0^t F(\tau)g(t-\tau)d\tau \tag{D.1.26a}$$

By a change of variables it can also be expressed in the equivalent form

$$x_p(t) = \int_0^t F(t-\tau)g(\tau)d\tau \tag{D.1.26b}$$

Equations (D.1.26a) and (D.1.26b) are the forms of the convolution integral one sees most often in texts but Eq. (D.1.25) is the most general form. Although we derived Eq. (D.1.25) with our vibrating system of Eq. (D.1.1) in mind, it is also valid for any LTI system if written in the form

$$o(t) = \int_{-\infty}^{+\infty} i(\tau)g(t-\tau)d\tau \tag{D.1.27}$$

where $g(t)$ is the unit impulse response of the system, i.e., the response to an input of magnitude $i_0 = 1/\varepsilon$ over a small time ε near $t = 0$, in the limit as $\varepsilon \to 0$. Thus, LTI systems have the remarkable property that we can find a solution for any input to those systems by knowing only the system impulse response, $g(t)$.

A convolution integral solution, Eqs. (D.1.26a) and (D.1.26b), is a particular solution to Eq. (D.1.22) that also satisfies zero initial conditions, i.e., $x_p(0) = \dot{x}_p(0) = 0$. The first initial condition is obvious if we let $t = 0$ in Eqs. (D.1.26a) and (D.1.26b) since the integral then vanishes. Proving that $\dot{x}_p(0) = 0$ is a bit trickier since both the upper limit of the integral and the integrand are functions of time. A theorem from calculus, called Leibniz's rule, shows how to handle such cases. Namely, for any function $f(t, \tau)$ we have

$$\frac{d}{dt}\int_0^t f(t,\tau)d\tau = f(t,t) + \int_0^t \frac{\partial f(t,\tau)}{\partial t}d\tau \tag{D.1.28}$$

so letting $f(t, \tau) = F(\tau)g(t - \tau)$ we have

$$\dot{x}_p(t) = F(t)g(0) + \int_0^t F(\tau)g'(t-\tau)d\tau \tag{D.1.29}$$

where $g'(t-\tau) = dg(t-\tau)/d(t-\tau)$. Examining Eq. (D.1.29), from the properties of the impulse response we know $g(0) = 0$ and the integral goes to zero as t goes to zero, so we also have $\dot{x}_p(0) = 0$.

Since this particular solution in the form of a convolution integral satisfies zero initial conditions when we form up the total solution $x(t) = x_h(t) + x_p(t)$ the homogeneous solution itself must then satisfy whatever initial conditions are present; that is, it must be a free vibration solution as already found, and we can write the solution for a forced, underdamped system, therefore, as (see Eq. (D.1.15))

$$\begin{aligned} x(t) = {} & \exp(-\varsigma\omega_n t)\left[x_0 \cos(\omega_d t) + \frac{(\dot{x}_0 + \varsigma\omega_n x_0)}{\omega_d}\sin(\omega_d t)\right] \\ & + \frac{1}{m\omega_d}\int_0^t F(t-\tau)\exp(-\varsigma\omega_n \tau)\sin(\omega_d \tau)d\tau \end{aligned} \tag{D.1.30a}$$

or, equivalently,

$$\begin{aligned} x(t) = {} & \exp(-\varsigma\omega_n t)\left[x_0 \cos(\omega_d t) + \frac{(\dot{x}_0 + \varsigma\omega_n x_0)}{\omega_d}\sin(\omega_d t)\right] \\ & + \frac{1}{m\omega_d}\int_0^t F(\tau)\exp(-\varsigma\omega_n (t-\tau))\sin(\omega_d (t-\tau))d\tau \end{aligned} \tag{D.1.30b}$$

and for the undamped case (see Eq. (D.1.16))

$$x(t) = x_0 \cos(\omega_n t) + \frac{\dot{x}_0}{\omega_n}\sin(\omega_n t) + \frac{1}{m\omega_n}\int_0^t F(t-\tau)\sin(\omega_n \tau)d\tau \tag{D.1.31a}$$

or, equivalently,

$$x(t) = x_0 \cos(\omega_n t) + \frac{\dot{x}_0}{\omega_n}\sin(\omega_n t) + \frac{1}{m\omega_n}\int_0^t F(\tau)\sin(\omega_n (t-\tau))d\tau \tag{D.1.31b}$$

The Steady-State Response of One Degree of Freedom Systems

In many cases the driving force is a periodic force of a very long duration. Mass imbalances in motors are an example of such forces. As long as there is some damping in the system the homogeneous solution will decay with time, leaving only the particular solution to the periodic driving force. The long time response, therefore, is only due to the particular solution in this case, which is called the *steady-state response*. We can often obtain the steady-state response directly from the equations of motion. For example, consider a driving force given by $F(t) = F_0 \exp(i\omega t)$, where F_0 is a (complex-valued) amplitude and ω is the driving frequency. The equation of motion is then

$$m\ddot{x}(t) + c\dot{x}(t) + kx(t) = F_0 \exp(i\omega t) \tag{D.1.32}$$

and we can find a steady-state particular solution, $x_{ss}(t)$, by simply letting $x_{ss}(t) = X \exp(i\omega t)$, where X is the steady-state amplitude response. If we place this solution into the equation of motion we find

$$X = \frac{F_0}{k + i\omega c - m\omega^2} = \frac{F_0}{k} \frac{1}{1 - \omega^2/\omega_n^2 + 2i\varsigma\omega/\omega_n} \tag{D.1.33}$$

which we can also write as

$$X = |X| \exp(-i\psi) \tag{D.1.34}$$

where the magnitude and phase are

$$|X| = \frac{F_0}{k} \frac{1}{\sqrt{\left(1 - \omega^2/\omega_n^2\right)^2 + \left(2\varsigma\omega/\omega_n\right)^2}}, \quad \psi = \tan^{-1}\left(\frac{2\varsigma\omega/\omega_n}{1 - \omega^2/\omega_n^2}\right) \tag{D.1.35}$$

These functions are plotted in Fig. D.6. The magnitude shown in Fig. D.6a has a peak near the natural resonant frequency $\omega_n = \sqrt{k/m}$ of the system. Higher damping factors reduce the peak amplitude. In the undamped case (not shown) the amplitude goes to infinity at the natural frequency. The phase angle plots for different damping factors are shown in Fig. D.6b.[Note: In calculating the phase

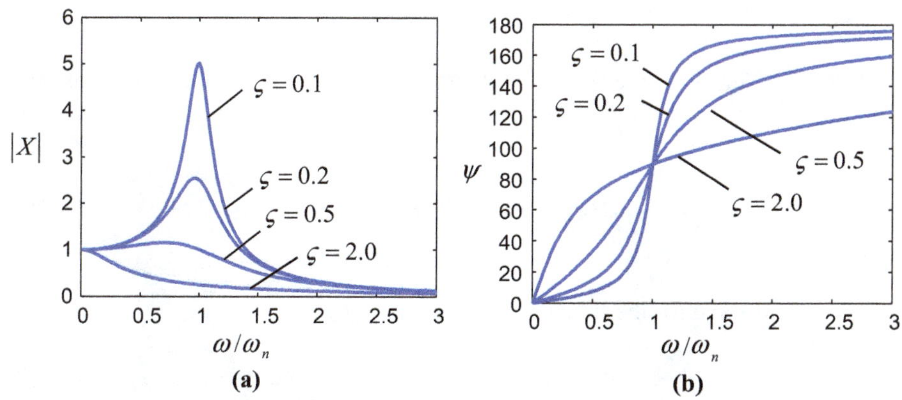

Fig. D.6 (**a**) The magnitude of the steady-state response as a function of the damping factor, and (**b**) the corresponding behavior of the phase of the steady-state response

use the MATLAB® function atan2 since it produces angles over a range of $(0, \pi)$ whereas the function atan will produce angles over $(-\pi 2, +\pi/2)$ and so atan produces artificial jumps in the phase when the phase exceeds those limits.]

A zero-degree phase corresponds to positive and negative forces and displacements occurring in synchronization while a 180° phase corresponds to the force and the displacement having opposite signs (and hence directions) while in synchronization. For the undamped case (not shown) the phase is zero below the natural frequency and 180° above the natural frequency so the force and displacement are completely in phase when $\omega/\omega_n < 1$ and completely out of phase when $\omega/\omega_n > 1$ in undamped systems. It is easy to demonstrate this behavior by holding in your hand a long elastic band with a weight on its end. If you move your hand up and down slowly the mass will move up and down in synchronization with your hand while if you move your hand up and down quickly the mass will move up when your hand moves down and it will move down when your hand moves up.

Writing the steady-state solution as

$$x_{ss}(t) = G(\omega)F_0 \exp(i\omega t) \tag{D.1.36}$$

the function G is called the *transfer function* of the system, where

$$G(\omega) = \frac{1}{k + i\omega c - m\omega^2} \tag{D.1.37}$$

This transfer function is closely related to the unit impulse response of the system. To see this relationship consider a force $F(t) = F_0 \exp(i\omega t)$ that has been in existence from $t = -\infty$. Since $F(t)$ is not zero for $t < 0$ we must write the convolution integral response to this function as (see Eq. (D.1.25))

$$x(t) = \int_{-\infty}^{t} F(t-\tau)g(\tau)d\tau = F_0 \exp(i\omega t) \int_{-\infty}^{t} \exp(-i\omega\tau)g(\tau)d\tau \tag{D.1.38}$$

As $t \to \infty$ this response should be the steady-state response since any transients will have died out at large times giving

$$x_{ss}(t) = G(\omega)F_0 \exp(i\omega t) = F_0 \exp(i\omega t) \int_{-\infty}^{+\infty} \exp(-i\omega\tau)g(\tau)d\tau \tag{D.1.39}$$

which we can solve for the transfer function as

$$G(\omega) = \int_{-\infty}^{+\infty} g(\tau)\exp(-i\omega\tau)d\tau \tag{D.1.40}$$

The integral that appears in Eq. (D.1.40) is called the *Fourier transform* of $g(\tau)$. One of the properties of Fourier transforms is that we also have an *inverse Fourier transform* relationship given by

$$g(\tau) = \frac{1}{2\pi}\int_{-\infty}^{+\infty} G(\omega)\exp(i\omega\tau)d\omega \tag{D.1.41}$$

so that we can obtain the transfer function from the impulse response and the impulse response from the transfer function through these Fourier transform relations. This same relationship is true for any well-behaved function, $f(t)$, and its Fourier transform, $F(\omega)$, i.e.,

$$\begin{aligned} F(\omega) &= \int_{-\infty}^{+\infty} f(t)\exp(-i\omega t)dt \\ f(t) &= \frac{1}{2\pi}\int_{-\infty}^{+\infty} F(\omega)\exp(i\omega t)d\omega \end{aligned} \tag{D.1.42}$$

The Fourier transform of $f(t)$, $F(\omega)$, is also called the *frequency spectrum* of the function or the *Fourier domain* characterization of the function in contrast to $f(t)$ itself which describes the behavior of the function in the *time domain*. An equivalent form that is often used in numerical calculations writes the frequency in terms of f cyles/s rather than ω rad/s. Since there are 2π radians per cycle we have $\omega = 2\pi f$ and the forward and inverse Fourier transforms in terms of f become

$$\begin{aligned} F(f) &= \int_{-\infty}^{+\infty} f(t)\exp(-2\pi i f t)dt \\ f(t) &= \int_{-\infty}^{+\infty} F(f)\exp(2\pi i f t)df \end{aligned} \tag{D.1.43}$$

Fourier transforms play an extremely important role in analyzing linear dynamical systems and they are used in numerous areas of science and engineering. One property of Fourier transforms that is very useful for LTI systems is embodied in the following theorem:

If the Fourier transform of $f_1(t)$ is $F_1(f)$ and the Fourier transform of $f_2(t)$ is $F_2(f)$ then the Fourier transform of the convolution of f_1 and f_2 given by

$$u(t) = \int_{-\infty}^{+\infty} f_1(\tau) f_2(t-\tau) d\tau$$

is the product of $F_1(f)$ and $F_2(f)$ so $U(f) = F_1(f)F_2(f)$, where $U(f)$ is the Fourier transform of $u(t)$.

As a consequence of this theorem, one can perform the convolution of two functions by multiplying their Fourier transforms together and then performing an inverse Fourier transform on the result of that multiplication. Because Fourier transforms are in general complex functions, their multiplication must be done as a complex multiplication. What makes this approach attractive is that there are well-developed algorithms, called *fast Fourier transform* (FFT) algorithms, that can do the Fourier and inverse Fourier transforms numerically with high speed and accuracy. As discussed in Appendix E we have written two MATLAB® functions Ftrans and IFtrans that implement the FFT algorithms for generating the Fourier transform and its inverse given in Eq. (D.1.43).

Fourier transforms are important tools for modeling and analyzing dynamical systems. While we cannot give a comprehensive discussion of those transforms at the level of this book the details found in Appendix E should be sufficient to give you a basic understanding of them.

Combining Homogeneous and Particular Solutions

Writing the particular solution as a convolution integral gives us a general expression that can be used for any force, $F(t)$. However, we saw in the last section that for simple driving forces such as harmonic forces we can obtain a particular solution such as a steady-state solution directly. Unlike the convolution integral expression, such particular solutions do not satisfy any particular initial conditions and contain no arbitrary constants so that if we want to obtain a complete solution that also satisfies some given initial conditions we must combine the particular solution with a homogeneous solution written in terms of two arbitrary constants and determinethose constants so as to satisfy the initial conditions. For example, take the case of the periodic driving force of the last section. If we take the real part of that driving force we have $F(t) = F_0 \cos(\omega t)$ and the steady-state particular solution is

$$x_p(t) = |X| \cos(\omega t - \psi) \tag{D.1.44}$$

We can combine this solution with the homogeneous solution of Eq. (D.1.14b) for the underdamped case to find the total solution as

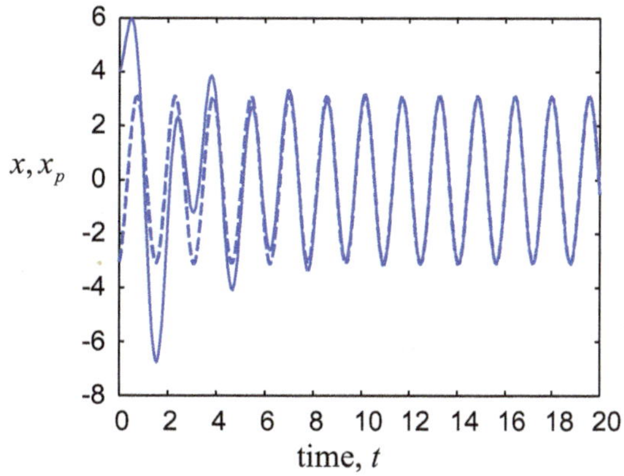

Fig. D.7 The response of a 1-D underdamped system driven by a sinusoidal driving force. Solid line—total response, $x(t)$. Dashed line—steady-state response, $x_p(t)$

$$x(t) = A\exp(-\varsigma\omega_n t)\cos(\omega_d t - \phi) + |X|\cos(\omega t - \psi) \tag{D.1.45}$$

so that at $t = 0$ we have

$$\begin{aligned} x(0) &= x_0 = A\cos\phi + |X|\cos\psi \\ \dot{x}(0) &= \dot{x}_0 = -\varsigma\omega_n A\cos\phi + \omega_d A\sin\phi + \omega|X|\sin\psi \end{aligned} \tag{D.1.46}$$

which can be solved for (A, ϕ) as

$$\phi = \tan^{-1}(F/D), \quad A = \sqrt{F^2 + D^2} \tag{D.1.47}$$

where

$$\begin{aligned} F &= \dot{x}_0/\omega_d + \varsigma(\omega_n/\omega_d)(x_0 - |X|\cos\psi) - (\omega/\omega_d)|X|\sin\psi \\ D &= x_0 - |X|\cos\psi \end{aligned} \tag{D.1.48}$$

The total solution of Eq. (D.1.45) is plotted in Fig. D.7 along with the steady-state solution of Eq. (D.1.44) for $k = 4$ N/m, $m = 1$ kg, $\omega = 4$ rad/s, $\varsigma = 0.2$, $F_0/k = 1$, and $x_0 = 4$ m and $\dot{x}_0 = 0$. The total solution is the solid line in Fig. D.7 and the steady-state solution is the dashed line. In this case it can be seen that the transient effects of the homogeneous solution are rapidly attenuated and the steady-state solution is all that remains after approximately $t = 10$ s.

Reference

1. H. Baruh, *Applied Dynamics* (CRC press, Boca Raton, 2015)

Appendix E: Fourier Transforms

Fourier transforms are frequently used in analyzing dynamical systems and they are also tools that are used in a wide variety of other areas of science and engineering. In this Appendix we outline the important properties of Fourier transforms and describe how they are calculated numerically using a fast Fourier transform (FFT) algorithm.

Fourier Transforms and Discrete Fourier Transforms

In Chap. 7 and in Appendix D Fourier transforms were introduced for vibrating systems. The forward and inverse Fourier transforms were expressed as integrals over a frequency variable, *f*, measured in cycles/s or Hertz (Hz) as

$$\begin{aligned} V(f) &= \int_{-\infty}^{+\infty} v(t)\exp(-2\pi i f t)dt \\ v(t) &= \int_{-\infty}^{+\infty} V(f)\exp(2\pi i f t)df \end{aligned} \tag{E.1.1}$$

where $v(t)$ is a function of the time, *t*, and $V(f)$ is its Fourier transform as a function of the frequency, *f*. While there are other forms that are used to define these transforms (see Sect. E.3) we will use the forms of Eq. (E.1.1) exclusively. Notice that because of the $i = \sqrt{-1}$ appearing in the exponential function even if $v(t)$ is a real function of time the Fourier transform, $V(f)$, will usually be a complex-valued function. However, when we integrate those Fourier transform complex values over frequency in the inverse Fourier transform integral we will recover the original real $v(t)$ function. For some simple functions we can perform these transforms analytically but in most cases they will have to be done numerically. Those numerical evaluations are normally done by evaluating (sampling) the continuous time and

L. W. Schmerr, *Engineering Dynamics 2.0*, Solid Mechanics and Its Applications 254,
https://doi.org/10.1007/978-3-319-98470-4

frequency functions ($v(t)$, $V(f)$) at equally spaced discrete times, t_n, and equally spaced frequencies, f_n, ($n = 1,2,\ldots,N$) and rewriting the Fourier transforms as discrete forward and inverse transforms of those sampled values given by

$$\begin{aligned} V_p(f_n) &= \Delta t \sum_{k=1}^{N} v_p(t_k)\exp(-2\pi i(k-1)(n-1)/N) \quad (n = 1,2,\ldots,N) \\ v_p(t_k) &= \frac{1}{N\Delta t}\sum_{n=1}^{N} V_p(f_n)\exp(2\pi i(k-1)(n-1)/N) \quad (k = 1,2,\ldots,N) \end{aligned} \tag{E.1.2}$$

where N is the number of samples of these functions and $\Delta t = t_{k+1} - t_k$ is the time interval between sampling times. Equation (E.1.2) can be derived directly from the Fourier transforms of Eq. (E.1.1) without approximation and the expressions in Eq. (E.1.2) are called the forward and inverse *discrete Fourier transforms*. You may wonder how discrete Fourier transforms can arise directly from Fourier transforms since Eq. (E.1.1) involves infinite integrals but Eq. (E.1.2) involves finite sums. The answer is in the subscript p that appears on $v_p(t_k)$ and $V_p(f_n)$. This p stands for "periodic" indicating that the functions appearing in the discrete transforms are repeated periodically an infinite number of times. This is illustrated for the time function in Fig. E.1 which shows a given function in Fig. E.1a and its periodically repeated version in Fig. E.1b. The period of the repeating function is $T = N\Delta t$. We will not prove the equivalence of Eqs. (E.1.1) and (E.1.2) here but you can find it in texts that cover Fourier transforms in depth. Figure E.1b also shows where the N sampled values of the time function lie within a given period. The first sample is at $t = 0$ and the last (Nth) sample is at $t = T - \Delta t$. If we look at a sample at $t = T$, which is not part of the sum in Eq. (E.1.2), it would just be the same value as the one at $t = 0$. A similar situation holds in the frequency domain as shown in Fig. E.2. The magnitude of a Fourier transform is plotted in Fig. E.2a since the Fourier transform itself is usually complex valued. The same periodically repeated function is shown in Fig. E.2b. Figure E.2b also shows where the sampled values of the frequency

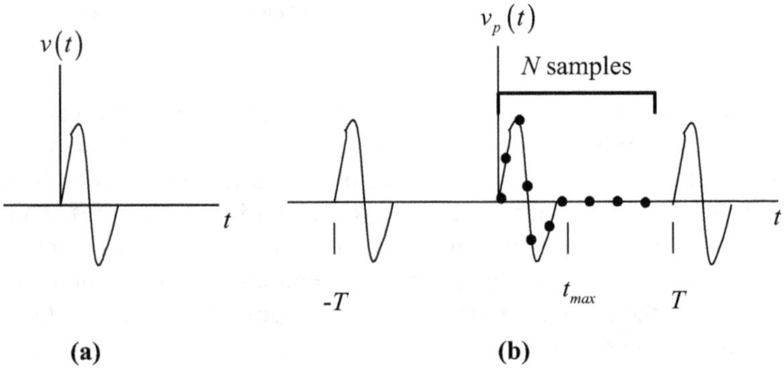

Fig. E.1 (**a**) A transient function. (**b**) The same function periodically repeated and the sampled values used in a discrete Fourier transform

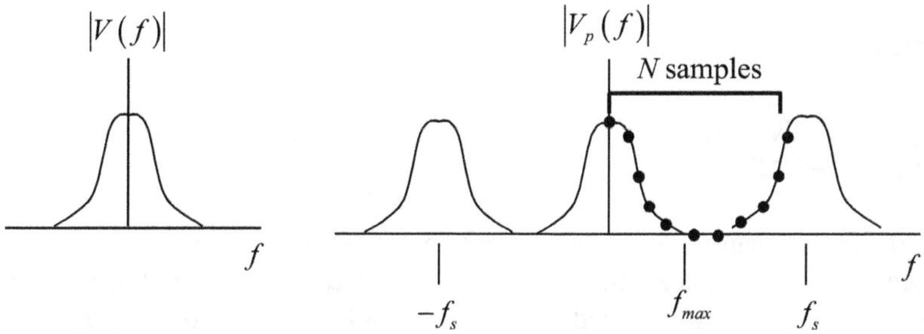

Fig. E.2 (**a**) The magnitude of the Fourier transform. (**b**) The same function periodically repeated and the sampled values used in the inverse discrete Fourier transform

function lie within a given period. The first sample is at $f = 0$ and the last (Nth) sample is at $f = f_s - \Delta f$ where $f_s = 1/\Delta t$ is the sampling frequency and where $\Delta f = f_s/N = 1/T$ is the frequency interval between samples. If we look at a sample at $f = f_s$, which is not part of the sum in Eq. (E.1.2), it would just be the same value as the one at $f = 0$. One difference between the frequency functions and the time functions is that the time function is zero for $t < 0$ (assuming that we have transient functions that start at some given time that we can always take as $t = 0$) but the frequency functions have values at both positive and negative frequencies. In the sampled frequency values we see that the values at negative frequencies actually are in the upper half of the sampled values from $n = N/2$ to $n = N$. In Fig. E.1 we have also indicated a time, $t_{\max}$, that denotes the time at which the time signal has become sufficiently small so that it is essentially zero for $t > t_{\max}$. Similarly, in Fig. E.2 there is a frequency, $f_{\max}$, that indicates the highest frequency at which the Fourier transform magnitude has any significant value. Thus, if $T \geq t_{\max}$ and $f_s \geq 2f_s$ the sampled values will be just those of the original time and frequency functions and we can drop the p subscript in Eq. (E.1.2). This last requirement is a famous criterion called the *Nyquist criterion*, which states:

The sampling frequency, $f_s = 1/\Delta t$, must be at least twice the maximum significant frequency, $f_{\max}$, contained in the waveform being sampled.

If the Nyquist criterion is not satisfied, then the periodically repeated function may not be representative of the Fourier transform we seek and fictitious frequency values (called aliases) can appear. Thus, the time function must be sampled at sufficiently small time intervals to prevent such behavior, which is also called *aliasing*. Although the Nyquist criterion only requires one to sample at a factor of two times $f_{\max}$ in practice one often uses higher factors.

If we assume that we have satisfied the Nyquist criterion and that the sampled values are representative of the original nonperiodic functions, then we can drop the p subscript and simply write Eq. (E.1.2) as

$$\begin{aligned} V(f_n) &= \Delta t \sum_{k=1}^{N} v(t_k) \exp(-2\pi i (k-1)(n-1)/N) \quad (n = 1, 2, \ldots, N) \\ v(t_k) &= \frac{1}{N\Delta t} \sum_{n=1}^{N} V(f_n) \exp(2\pi i (k-1)(n-1)/N) \quad (k = 1, 2, \ldots, N) \end{aligned} \tag{E.1.3}$$

To obtain the sampled values present in Eq. (E.1.3) we need a function that generates N evenly spaced time samples spaced $\Delta t = T/N$ apart from $t = 0$ to $t = T - \Delta t$ or N evenly spaced frequency samples spaced $\Delta f = f_s/N$ apart from $f = 0$ to $f = f_s - \Delta f$. Unfortunately, the built-in MATLAB® function linspace does not give us those samples. If we let t = linspace(0, T, N) or let f = linspace(0, f_s, N) we will generate N samples but they will include the repeated values at $t = T$ or $f = f_s$ and we would improperly have $\Delta t = T/(N - 1)$ and $\Delta f = f_s/(N - 1)$. Thus, we have written a MATLAB® sampling function s_space that does do the proper sampling. It is a very simple function that we list here:

```
function y = s_space(xstart, xend, num)
% S_SPACE(XSTART,XEND, NUM) generates num evenly spaced sampled
% values from xstart to (xend - dx), where dx is the sample
% spacing. This is useful in FFT analysis where we generate
% sampled periodic functions. Example: generate 1000 sampled
% frequencies from 0 to 100Hz via f=s_space(0,100,1000);
% In this case the last value of f will be 99.9 Hz and the
% sampling interval will be 100/1000 =0.1 Hz.

ye = linspace(xstart, xend, num+1);
y = ye(1:num);
```

You should use the function s_space to generate any of the sampled time or frequency values needed when doing Fourier transform analysis.

Fast Fourier Transforms and Numerical Fourier Analysis

We could evaluate Fourier transforms or inverse Fourier transforms directly from Eq. (E.1.3) but there are faster methods than simply implementing the sum of all the products of terms appearing in Eq. (E.1.3). These methods are called fast Fourier transform (FFT) methods. Basically FFTs take advantage of the periodic nature of the exponentials in Eq. (E.1.3) to reduce significantly the number of multiplications needed. We will not describe how that works here but again you can find the details in books specifically covering Fourier transforms. MATLAB® has two built-in functions called fft and ifft that use such an FFT algorithm. Unfortunately, those built-in functions compute the following terms:

$$\text{MATLAB}^{\circledR}\text{ fft: } \sum_{k=1}^{N} v(t_k)\exp[-2\pi i(k-1)(n-1)/N]$$

$$\text{MATLAB}^{\circledR}\text{ ifft: } \frac{1}{N}\sum_{n=1}^{N} V(f_n)\exp[2\pi i(k-1)(n-1)/N]$$

so that they do not include the Δt parameter in their definition. Thus, we have written two new MATLAB® functions, Ftrans and IFtrans, that simply use fft and ifft, respectively, and include the appropriate Δt factors appearing in Eq. (E.1.3). The calling sequence for these functions is

```
yf = Ftrans(y, dt);  % calculate the Fourier transform of the y samples
y = IFtrans(yf, dt); % calculate the inverse Fourier transform of the yf
samples
```

where dt $= \Delta t$.

In performing Fourier transforms numerically there are a few issues that need to be considered. First, in the definition of the inverse Fourier transform the integration is done over both positive and negative frequencies. If the time domain function, $v(t)$, is real, however, its Fourier transform, $V(f)$, satisfies $V(-f) = V^*(f)$, where $()^*$ denotes the complex conjugate, where recall for a complex number $a+ib$, $(a+ib)^* = a-ib$. Thus, the integration over the negative frequencies can be found from the integration over the positive values. The integration over the negative frequencies, however, is present in order to produce a real $v(t)$ function from the inverse Fourier transform. What happens if we calculate the inverse Fourier transform only over positive frequencies? Then the resulting function will be complex and we can in fact show that we obtain

$$\frac{v(t)}{2} - \frac{i}{2}H[v(t)] = \int_{0}^{+\infty} V(f)\exp(2\pi ift)df \qquad \text{(E.2.1)}$$

where $H[v(t)]$ is the *Hilbert transform* of $v(t)$ given by

$$H[v(t)] = \frac{1}{\pi}\int_{-\infty}^{+\infty} \frac{v(\tau)d\tau}{(\tau - t)} \qquad \text{(E.2.2)}$$

Thus, we can find the time domain function $v(t)$ from only the positive frequency components of $V(f)$ if we take twice the real part of the integration over those positive components, i.e.,

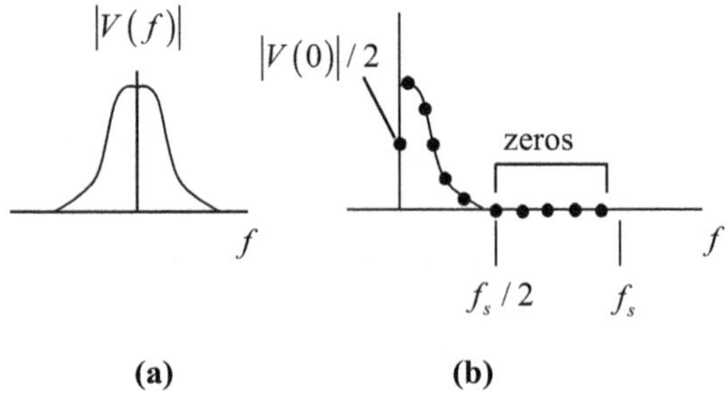

Fig. E.3 (**a**) The magnitude of a spectrum of a real function, *v*(*t*), and (**b**) the samples needed for computing *v*(*t*) with the inverse Fourier transform using only the positive frequency components of the spectrum

$$v(t) = 2\text{Re}\left\{\int_0^\infty V(f)\exp(2\pi ift)df\right\} \tag{E.2.3}$$

where Re indicates "real part of." Figure E.3 illustrates the process. When computing the inverse Fourier transform with an FFT algorithm the positive sampled frequency components of the spectrum are placed in the first half of the samples going from zero to $f_s/2$ and the components from $f_s/2$ to f_s (which is where the sampled negative frequency components would ordinarily be) are filled with zeros. There is one more step, however, before these positive frequency components and zeros can be inverted. When we throw out the negative frequencies there is the issue of what to do with the zero frequency component since it is neither a positive nor a negative frequency. From Fig. E.3b we see that if the negative frequencies are removed then the remaining frequency spectrum has a jump at zero frequency. If you have studied Fourier series you will know that if you represent a periodic function that has a jump from, say, V_1 to V_2 at some location then the Fourier series will converge to the mean $(V_1+V_2)/2$ at that location. The same is true for the discrete Fourier transforms. Thus, the first sampled value of $V(f)$, which is $V(0)$, must be divided by 2 before the spectrum can be inverted. This is shown in Fig. E.3b.

As an example of the steps needed in the calculation of an inverse Fourier transform using only positive frequency components, consider the steady-state transfer function for a 1-D system obtained in Appendix D (see Eq. (D.1.37)). In terms of the frequency, *f*, that transfer function can be written as

$$G(f) = \frac{1}{k + 2\pi icf - 4\pi^2 mf^2} \tag{E.2.4}$$

Now, let the mass $m = 1$ kg, the spring constant $k = 2$ N/m, and the damping constant $c = 0.03$ kg/s. The plot of the magnitude and phase of this function for frequencies from $f = 0$ Hz to $f = 4$ Hz (a Hertz (Hz) is one cycle/s) are shown in Fig. E.4. We can see that the function is very small above frequencies of approximately 1.0 Hz and has a strong peak near the undamped natural frequency of 0.225 Hz. The phase goes continuously from 0 to $-180°$. These behaviors are

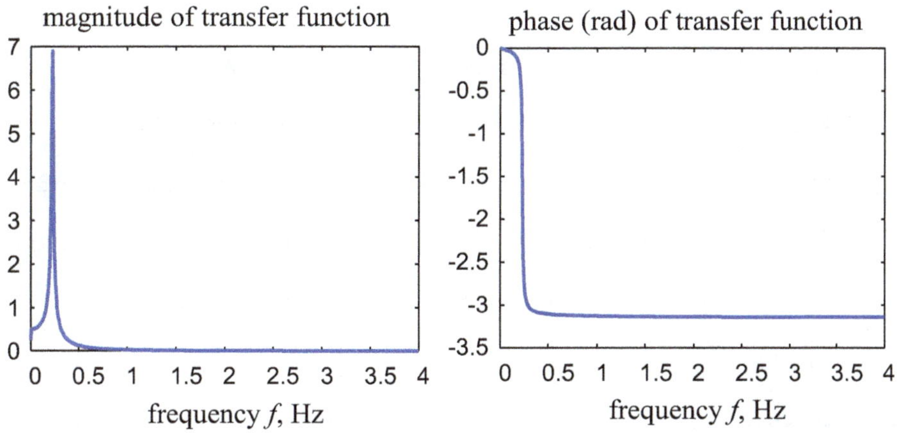

(a) **(b)**

Fig. E.4 (**a**) The magnitude of the transfer function at positive frequencies for a system with $m = 1$ kg, $k = 2$ N/m, and $c = 0.1$ kg/s and (**b**) the phase (rad) of the transfer function

consistent with those seen in Appendix D in Fig. D.6, but notice that the phase angle ψ in Fig. D.6 is the negative of the angle shown in Fig. E.4. These phases, as mentioned in Appendix D, must be calculated with the MATLAB® atan2 function.

Simulation E.1: Numerical Evaluation of the Impulse Response

As seen in Appendix D, we can calculate the impulse response function, $g(t)$, for this system from the steady-state transfer function by using the inverse fast Fourier transform function. Here, we will perform that transform numerically with IFtrans. First, we sample the transfer function over a frequency range $(0, f_s)$, where f_s must be at least twice the highest frequency present in the transfer function. Since the function is small for frequencies greater than about 1.0 Hz, we conservatively take $f_s = 4.0$ Hz, and use the function s_space to generate a set of $N = 1024$ equally spaced frequencies over the interval $(0, f_s)$. We keep the first 512 of the samples of the transfer function and add another 512 zeros to complete the sampled spectrum. Note that the total time span, T, over which the impulse response function will be calculated (called the time window) is $T = N\Delta t = N/f_s = \Delta f$, where $\Delta f = f_s/N$ is the sampling spacing between frequencies. To generate the impulse response function we use the function IFtrans(G, dt), where G is the sampled transfer function and dt is the time interval between time samples. First, however, we divide the zero frequency component of the transfer function by two and then compute twice the real part of the inverse Fourier transform of the modified G. The complete set of steps needed in MATLAB® to numerically evaluate the impulse response function and compare it with the exact impulse response given in Appendix D by Eq. (D.1.20) are

```
m = 1; k = 2; c= 0.1;              % give system parameters

fs = 4.0;                          % choose sampling frequency

f = s_space(0, fs, 1024);          % create N = 1024 frequencies over (0, fs)

G = 1./(k +2*pi*1i*f*c - m*(2*pi*f).^2);   % evaluate transfer function

G = [G(1:512), zeros(1,512)];         % place zeros in upper part of spectrum

dt = 1/fs                             % calculate sampling time interval

T = 1024*dt;                          % calculate time window

t = s_space(0, T, 1024);              % create 1024 time values over (0, T)

G(1) =G(1)/2;                      % take one half of the zero frequency
                                   % component

g = 2*real(IFtrans(G,dt));         % take 2x real part of inverse Fourier
                                   % transform

plot(t(t<25), g(t<25))             % plot the impulse response function

wn = sqrt(k/m);                    % undamped natural frequency

d = c/(2*sqrt(k*m));               % damping factor

wd = wn*sqrt(1-d^2);               % damped natural frequency

ge = (1/(m*wd))*exp(-wn*d*t).*sin(wd*t);     % exact impulse response

hold on

plot(t(t< 25), ge(t<25), '--');              % plot exact response
                                             % comparison

hold off
```

These steps are placed in the MATLAB® script impulse_response. We have listed all of them here so that you can see explicitly how the evaluation proceeds. The plot generated by the script is shown in Fig. E.5 where we can see that the numerical solution (solid line) is nearly coincident with the exact solution (dashed line) on this scale. Note that the plotted time interval is only about one-tenth of the total time window, T. This abbreviated plot was generated by using a logical statement to act as a "mask" to determine the values we want plotted. Thus a MATLAB expression such as $t(t < 25)$ selects only those indices in the vector t that satisfy the logical expression, yielding a plot up to $t = 25$ s only.

It is instructive to take the numerically generated impulse function and use Ftrans to recalculate the transfer function. This can be easily done after running the impulse-response script with the commands

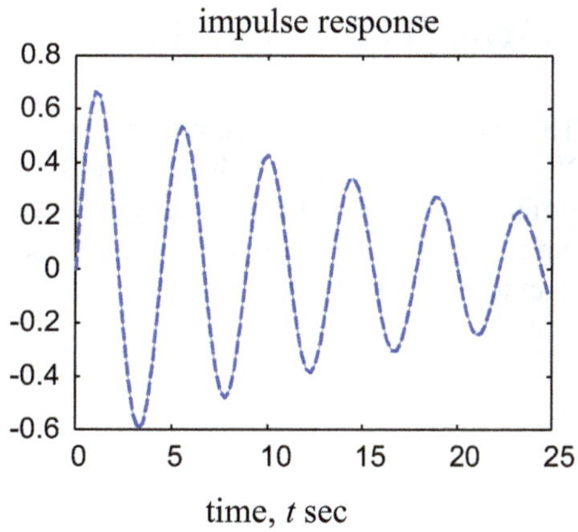

Fig. E.5 The impulse response calculated numerically (solid line) and the exact impulse response (dashed line). The two results are nearly indistinguishable

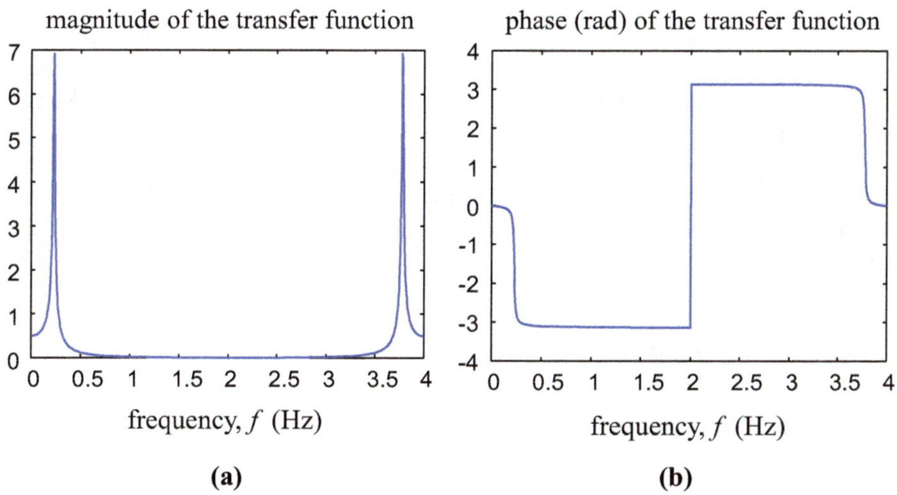

Fig. E.6 (**a**) The magnitude of the transfer function, and (**b**) its phase

```
Ga = Ftrans(g, dt);
plot(f, abs(Ga))
figure(2)
plot(f, atan2(imag(Ga), real(Ga)));
```

where Ga is the numerically generated transfer function. The results are shown in Fig. E.6. In that figure we now see both the positive and negative frequency components of the transfer function in the lower and upper half, respectively, of the samples plotted.

Different Forms of the Fourier Transform

This Appendix should give you the basic tools you will need to do Fourier analysis. You should be aware of the fact, however, that other books and authors may use different definitions for Fourier transforms and discrete Fourier transforms. We can define a general pair of Fourier and inverse Fourier transforms in the forms (using the circular frequency, ω)

$$\begin{aligned} V(\omega) &= n_1 \int_{-\infty}^{+\infty} v(t)\exp(\pm i\omega t)dt \\ v(t) &= n_2 \int_{-\infty}^{+\infty} V(\omega)\exp(\mp i\omega t)d\omega \end{aligned} \tag{E.3.1}$$

as long as the coefficients(n_1, n_2) satisfy $n_1 n_2 = 1/2\pi$. Many math books, for example, like to write these transforms in symmetrical forms where $n_1 = n_2 = 1/\sqrt{2\pi}$. Our choice of transforms used $n_1 = 1$ and $n_2 = 1/2\pi$. As can be seen in Eq. (E.3.1) there also may be different choices made for the signs in the complex exponentials. For our choice of the transforms we used a minus sign in the forward transform and a plus sign in the inverse transform. Similar differences may also appear in the discrete Fourier transforms and hence in FFT analyses. In general we can write those transforms as

$$\begin{aligned} V_n &= n_1 \sum_{k=1}^{N} v_k \exp(\pm 2\pi i(k-1)(n-1)/N) \quad (n = 1, 2, \ldots, N) \\ v_k &= n_2 \sum_{n=1}^{N} V_n \exp(\mp 2\pi i(k-1)(n-1)/N) \quad (k = 1, 2, \ldots, N) \end{aligned} \tag{E.3.2}$$

as long as $n_1 n_2 = 1/N$, where N is the total number of samples. We used $n_1 = \Delta t$ and $n_2 = 1/N\Delta t$ to ensure that the discrete transforms were consistent with the forward and inverse Fourier transforms we used. Recall that MATLAB® used $n_1 = 1$ and $n_2 = 1/N$ which is why we had to modify the MATLAB® definitions of fft and ifft to get a consistent set of discrete FFT transform functions. We chose a minus sign in the forward transform and a plus sign in the inverse transform, which are the same choices made in the MATLAB® fft and ifft functions. We should mention that some authors use values and sums which range over $(0, N-1)$ rather than $(1, N)$ seen in Eq. (E.3.2) but this is just a simple renumbering of the samples. MATLAB® vectors and arrays cannot have zero indices so we need to use the range $(1, N)$ in MATLAB®. Finally, some disciplines, such as electrical engineering, for example, prefer to define $j = \sqrt{-1}$ instead of the $i = \sqrt{-1}$ we have used throughout.

Appendix F: MATLAB® Functions and Scripts

There are many MATLAB® functions and scripts that are defined and used in this book. In this section we have organized listings of these programs according to the chapter and simulation example in which they appear. Most of the simulations and animations are performed with scripts whose names end in either _sim or _anim, respectively. The functions that define the first-order system of ordinary differential equations used by ode45 to perform the numerical solution usually have names ending in _eqs. In some cases these functions are nested inside other functions, as described in the text. The MATLAB® m-files for the functions and scripts developed in this book are available at https://doi.org/10.1007/978-3-319-98470-4_2 on the web. Note that there are differences between the line widths in the MATLAB editor and this text, causing some lines in the listings to be wrapped around. I have corrected the comments where this happens but have left the actual code unchanged so that it is identical in syntax to what is in the m-files.

Chapter 2

Simulation 2.1: A Projectile with Drag

1. projectile_w_drag_eqs

```
function [xdot] = projectile_w_drag_eqs(t, x, c, g)
% [xdot] = projectile_w_drag_eqs(t, x, c, g) implements the equations of
% motion for a particle where there is a drag force proportional to the
% square of the particle velocity. The input parameters are the time, t, a
% vector of system parameters [x(1), x(2), x(3), x(4)], and the
% normalized
% drag coefficient c (of dimensions 1/ length) and the acceleration of
```


L. W. Schmerr, *Engineering Dynamics 2.0*, Solid Mechanics and Its Applications 254,
https://doi.org/10.1007/978-3-319-98470-4

```
% gravity, g. The function returns a column vector [ xdot(1), xdot(2),
% xdot(3), xdot(4)]' containing the first derivatives on time of the
% system
% parameters.

% set up a column vector place holder for the derivatives
xdot = zeros(4,1);
% rename the system variables to make the equations easier to read
x1 = x(1); % x
x2 = x(2); % dx/dt
x3 = x(3); % y
x4 = x(4); % dy/dt
% define the system of equations
xdot(1) = x2;
xdot(2) = -c*sqrt(x2.^2 + x4.^2).*x2;
xdot(3) = x4;
xdot(4) = -c*sqrt(x2.^2 + x4.^2).*x4 -g;
end
```

2. projectile_w_drag_sim

```
% projectile_w_drag_sim script
% solve an example projectile problem with drag numnerically and
% compare solution to the case where the drag force is absent.

clear
close all
% input parameters
tspan = [0, 7];     % time interval for solution (sec)
x0 = [0 30 0 30];   % initial values of [x,xdot,y,ydot] (m and m/sec)
c = 0.006;          % normalized drag coefficient (1/m)
g = 9.81;           % acceleration of gravity (m/sec^2)

%use function projectile_w_drag_eqs to numerically solve equations
%  of motion
[tout, xout] = ode45(@projectile_w_drag_eqs, tspan, x0, [], c, g);

% rename output variables and plot path of projectile
x = xout(:,1);
xdot = xout(:, 2);
y = xout(:,3);
ydot = xout(:, 4);
plot( x, y)

% consider case with zero drag and also plot on the same graph
hold on
c2 = 0.0;
[tout, xout] = ode45(@projectile_w_drag_eqs, tspan, x0, [], c2, g);
plot(xout(:,1), xout(:,3), '--')
xlabel('x, meters');ylabel('y, meters')
hold off
```

Simulation 2.2: Bead on a Straight Wire

3. bead_on_s_wire_eqs

```
function [xdot] = bead_on_s_wire_eqs(t,x, m ,g, ang)
% [xdot] = bead_on_s_wire_eqs(t , x, m ,g, ang) generates the first order
% differential equations for a bead sliding on a smooth, inclined wire. The
% input variables are the time, t, in sec) the system parameters, [x(1),
% x(2),x(3), x(4)], the mass m (in kg), the acceleration of gravity, g, (in
% m/sec^2), and the angle of the wire with respect to the x-axis, ang, (in
% rad). The output is a column vector xdot = [ xdot(1), xdot(2), xdot(3),
% xdot(4)]', of derivatives of the system parameters.

% generate column vector for output vector
xdot = zeros(4,1);

% rename variables to make them easier to read
xc = x(1);    % x coordinate
vx = x(2);    % velocity in x-direction
yc = x(3);    % y coordinate
vy = x(4);    % velocity in y-direction

% specify applied force as a function of time, t
F = (20-10*t).*(t <= 2);

% form up equations of motion
xdot(1) = vx;
xdot(2) = cos(ang).*(F - m*g*sin(ang))/m;
xdot(3) = vy;
xdot(4) = sin(ang).*(F- m*g*sin(ang))/m;
end
```

4. bead_on_s_wire_sim

```
% script bead_on_s_wire_sim
% This script solves the problem of a bead sliding on a smooth,
% straight inclined wire and subject to an applied force, F(t), acting
% along the wire.

close all

% default parameters unless otherwise specified
if ~exist('m','var'), m = 1.0; end  % mass (kg)
if ~exist('g','var'), g = 9.81; end  % acceleration of gravity (m/sec^2)
if ~exist('ang','var'), ang = 30*pi/180; end % angle of the wire (rad)
if ~exist('tspan','var'), tspan = [0, 10]; end % time interval
if ~exist('x0','var'), x0 = [ 0, 0, 0, 0]; end %initial conditions
```

```
% solve the equations of motion
[tout,xout] = ode45(@bead_on_s_wire_eqs, tspan, x0, [], m,g,ang);

% rename the outputs
t = tout;
x = xout(:,1);
vx = xout(:,2);
y = xout(:,3);
vy = xout(:,4);

% calculate the bead position along the wire as a function of time
xp = cos(ang)*x +sin(ang)*y;

% evaluate the constraint and its first derivative
C = y*cos(ang)-x*sin(ang) ;
Cd = vy*cos(ang)-vx*sin(ang);

% plot position versus time along the wire, the value of the constraint
% and
% its first derivative
figure (1)
plot(t, xp)
xlabel('t, sec'); ylabel('x''');
figure (2)
plot(t, C)
xlabel('t, sec');ylabel('C');
axis([ 0 10 -1e-13 1e-13])
figure (3)
plot(t, Cd)
axis([ 0 10 -1e-13 1e-13])
xlabel('t, sec'); ylabel('dC/dt');
```

Simulation 2.3: Bead on a Parabolic Wire

5. bead_on_p_wire_eqs

```
function [xdot] = bead_on_p_wire_eqs(t, x, a, g)
% [xdot] = bead_on_p_wire_eqs(t, x, a, g) implements the equations of
% motion for a particle sliding on the smooth parabolic wire given by
% y = ax^2. The input parameters are the time, t, a vector of system
% parameters, [x(1), x(2)], the constant a (of dimensions 1/ length), and
% the acceleration of gravity, g. The function returns a column vector
% [ xdot(1), xdot(2)]' containing the first derivatives on time of the
% system parameters.
```

```
% set up a column vector place holder for the derivatives
xdot = zeros(2,1);

% rename the system variables to make them easier to read
x1 = x(1); % x
x2 = x(2); % dx/dt

% define the system of equations
xdot(1) = x2;
xdot(2) = -2*a*x1*(g+2*a*x2.^2)./(1+4*a^2*x1.^2);
```

6. bead_on_p_wire_sim

```
% script bead_on_p_wire_sim
% This script solves the equations of motion for a bead sliding on a
% smooth, parabolic wire whose shape is y = a*x^2 as a one degree of
% freedom system

close all

% default input parameters unless otherwise specified
if ~exist('a','var'), a = 0.1; end          % parabola constant
if ~exist('g','var'), g = 9.81; end         % acceleration of gravity
if ~exist('tspan','var'), tspan = [0, 10]; end % time interval
if ~exist('x0','var'), x0 = [0, 10]; end     % initial conditions
if ~exist( 'rtol','var'), rtol = 1e-3;end    % relative error tolerance
if ~exist('atol','var'), atol = 1e-6; end    % absolute error tolerance

% place error tolerances in options
options = odeset('RelTol', rtol, 'AbsTol', atol);

% solve the system of equations numerically
[tout, xout] = ode45(@bead_on_p_wire_eqs, tspan, x0, options, a, g);

% rename the output variables
t = tout;
x = xout(:, 1);
vx = xout(:,2);

% plot x and normal force divided by the weight (Nw = N/mg) versus time
plot(t, x)
xlabel('t, sec'); ylabel('x');
Nw = (1 + 2*a*vx.^2/g)./sqrt(1+4*a^2*x.^2);
figure(2)
plot(t, Nw)
axis([ 0 10 0 3.5])
xlabel('t, sec'); ylabel('N/mg');
```

7. sliding_bead_energy

```
% script sliding_bead_energy
% This script generates phase plots for a bead sliding on the parabolic
% shaped wire given by y = a*x^2, where a = 0.1.

clear
close all
% define input parameters
a = 0.1;
g = 9.81;

% define grid of x-coordinate values, xx, and x-velocity values, xxd
x0 = linspace(-8,8,500);
x0d = linspace (-10, 10, 500);
[xx, xxd] = meshgrid(x0, x0d);

% calculate E/m, the total energy,E, divided by the mass,m,
% and plot constant E/m contours
Em = 0.5*(1+4*a^2*xx.^2).*xxd.^2 +g*a*xx.^2;
contour(xx, xxd, Em, 10, 'k')
xlabel('x');ylabel('dx/dt');
```

Simulation 2.4: Bead on a Wire with Damping

8. bead_on_p_wire_eqs2

```
function [xdot] = bead_on_p_wire_eqs2(t, x, a, cm, g)
% [xdot] = bead_on_p_wire_eqs2(t, x, a, g) implements the equations of
% motion for a particle sliding on the smooth parabolic wire given by
% y = ax^2 with damping present. The input parameters are the time, t, a
% vector of system parameters, [x(1), x(2)], the constant a
% (of dimensions 1/ length), the damping coefficient cm = c/m, and
% the acceleration of gravity, g. The function returns a column vector
% [xdot(1), xdot(2)]'containing the first derivatives on time of the system
% parameters.

% set up a column vector place holder for the derivatives
xdot = zeros(2,1);

% rename the system variables to make them easier to read
x1 = x(1);  % x
x2 = x(2);  % dx/dt

% define the system of equations
xdot(1) = x2;
xdot(2) = -2*a*x1*(g+2*a*x2.^2)./(1+4*a^2*x1.^2)-cm *x2;
```

9. bead_on_p_wire_sim2

```
% script bead_on_p_wire_sim2
% This script solves the equations of motion for a bead sliding on a
% smooth, parabolic wire whose shape is y = a*x^2 as a one degree of
% freedom system, where damping is present.

close all

% default input parameters unless otherwise specified
if ~exist('a','var'), a = 0.1; end          % parabola constant
if ~exist('g','var'), g = 9.81; end          % acceleration of gravity
if ~exist('tspan','var'), tspan = [0, 10]; end % time interval
if ~exist('x0','var'), x0 = [0, 10]; end      % initial conditions
if ~exist( 'rtol','var'), rtol = 1e-3;end    % relative error tolerance
if ~exist('atol','var'), atol = 1e-6; end    % absolute error tolerance
if ~exist('cm','var'), cm = 0.5; end         % damping term

% place error tolerances into options
options = odeset('RelTol', rtol, 'AbsTol', atol);

% solve system of equations numerically
[tout, xout] = ode45(@bead_on_p_wire_eqs2, tspan, x0, options, a, cm, g);

% rename output variables
t = tout;
x = xout(:, 1);
vx = xout(:,2);

% plot x and normal force divided by the weight (Nw = N/mg) versus time
plot(t, x)
xlabel('t, sec');ylabel('x');
Nw = (1 + 2*a*vx.^2/g)./sqrt(1+4*a^2*x.^2);
figure(2)
plot(t, Nw)
axis([ 0 10 0 3.5])
xlabel('t, sec');ylabel('N/mg');
```

Simulation 2.5: A Roller Coaster

10. roller_coaster_1D (and nested function roller_cart)

```
function [tout, xout] = roller_coaster_1D(tspan,x0,g, A, L,rtol,atol)
% [tout, xout] = roller_coaster_1D(tspan, x0, A,L) solves for the motion of
% a roller coaster cart on a sinusoidal track whose height, h, is given by
% h = A*(1-cos(pi*x/L))/2 (the MATLAB solver is ode45). tspan is the time
% interval over which we want to obtain the solution and x0 is the vector
% of initial conditions on the x-coordinate, x(0) and x-velocity, vx(0)
```

```
% so that x0 = [ x(0) vx(0)] . rtol and atol are input error tolerances.
% The outputs of the function are the times, tout, over which the
% solution is calculated, and a matrix, xout, which holds the values of x
% and dx/dt evaluated at those times in its columns.
% It is assumed that x is in meters (m) and dx/dt is in m/sec. g is the
% acceleration of gravity.

% place error tolerances into options

 options = odeset('RelTol', rtol, 'AbsTol', atol);

% solve equations of motion. The inputs g, A, L are shared with the
% nested function roller_cart

[tout,xout] = ode45(@roller_cart, tspan, x0,options);

% the nested function to generate the equations of motion

function xdot = roller_cart( t, x)

% generate column vector for outputs
xdot = zeros(2,1);

%rename inputs to make them easier to read
x1 = x(1);
x2 = x(2);

% define various parameters in equations of motion
hp = A*pi*sin(pi*x1/L)/(2*L); % calculate h'(x) = dh/dx
D = 1 + hp.^2;
hpp = A*(pi^2)*cos(pi*x1/L)/(2*L^2); % calculate h''(x)
U = g + hpp.*(x2.^2);

% form equations of motion
xdot(1) = x2;
xdot(2) = -(hp./D).*U;
end

end
```

11. roller_coaster_sim

```
% script roller_coaster_sim
% This script solves for the motion of a roller coaster cart on a
% sinusoidal track whose height, h, is given by h= A*(1-cos(pi*x/L))/2.
% The script plots the cart path and the normalized normal force.

close all

% default parameters unless otherwise specified
```

```
if ~exist('g','var'), g = 9.81; end      % the acceleration of gravity
if ~exist('L','var'), L = 5; end           % length parameter for track
if ~exist('A', 'var'), A = 4;end          % amplitude parameter for track
if ~exist('tspan', 'var'), tspan = [0, 10]; end % time span
if ~exist('x0','var'), x0 = [0, 8.5]; end      % initial conditions
if ~exist('rtol','var'), rtol = 1e-3; end     % relative error tolerance
if ~exist('atol','var'), atol = 1e-6; end     % absolute error tolerance

% solve the equations of motion numerically
 [tout, xout] = roller_coaster_1D(tspan,x0,g, A, L, rtol, atol);

 % rename output variables
 x = xout(:, 1);
 vx = xout(:,2);
 y = A*(1-cos(pi*x/L))/2;
 vy = pi*A*sin(pi*x/L).*vx/(2*L);

 % plot path of the particle
 figure(1)
 plot(x, y)
 xlabel('x'); ylabel('y');
 % Evaluate normalized normal force and plot it versus time
 hp = A*pi*sin(pi*x/L)/(2*L);
 hpp = A*(pi^2)*cos(pi*x/L)/(2*L^2);
 Nmg = (1./(sqrt(1+hp.^2))).*(g + hpp.*vx.^2)/g; % N divided by mg
 figure(2)
 plot(tout, Nmg)
 xlabel('t, sec'); ylabel('N/mg');
```

12. roller_energy

```
% script roller_energy
% This script uses the output values of roller_coaster_sim to plot the
% normalized energy Em/Em(1). Em is the value of 2E/m, where E is the total
% energy and m is the mass so that Em(t)/Em(0) = E(t)/E(0)is the
% normalized
% total energy. Note the energy Em at t = 0 is the MATLAB value Em(1)

% The values of A, L, g, x, vx, tout  must be present in the workspace from
% the execution of roller_coaster_sim. Close any open windows

close all

% calculate E/m, plot normalized values
Em = (1 +(pi*A*sin(pi*x/L)/(2*L)).^2).*vx.^2 +g*A*(1-cos(pi*x/L));
plot(tout, Em./Em(1))
axis( [0 max(tout) 0.50 1.50])
xlabel('t, sec'); ylabel('E(t)/E(0)');
```

Simulation 2.6: Spring Pendulum

13.spring_pendulum (and nested function spring_pend_eqs)

```
function [tout, xout] = spring_pendulum(tspan, x0, K, rtol, atol)
% [tout, xout] = spring_pendulum(tspan, x0, K, rtol, atol) solves for the
% motion of a spring-pendulum using the MATLAB solver ode45. tspan is the
% time interval over which we want the solution and x0 is the vector of
% initial conditions for the normalized spring length, its velocity, the
% angular position, and the angular velocity. The constant K defines the
% one free parameter of the system. rtol and atol are the error tolerances
% used in ode45. The outputs of the function are the times, tout, at which
% the solution is calculated, and a matrix, xout, which holds the values of
% the normalized length of the pendulum, its time derivative,the angular
% position, and its time derivatives at those times (in the columns of
% xout).

% place the error tolerances in options

options = odeset('RelTol', rtol, 'AbsTol', atol);

% solve for the motion of the spring-pendulum using ode45

[tout, xout] = ode45(@spring_pend_eqs, tspan, x0, options);

% the nested function to generate the equations of motion

  function xdot = spring_pend_eqs(t, x)
    % create column vector for outputs
    xdot = zeros(4,1);

    % rename input parameters
    x1 = x(1); % rn
    x2 = x(2); % rndot
    x3 = x(3); % theta
    x4 = x(4); % thetadot

    % generate equations of motion
    xdot(1) = x2;
    xdot(2) = (1+x1).*x4.^2 - K*(1-cos(x3))-x1;
    xdot(3) = x4;
    xdot(4) = -2*x2.*x4./(1+x1) -K*sin(x3)./(1+x1);
  end

end
```

14. spring_pendulum_sim

```
% script spring_pendulum_sim
% This script solves the equations of motion for a spring-pendulum and
% plots the normalized radial length and angle of the pendulum versus
% time.
% Also plotted is a normalized energy versus time for this conservative
% system.

close all

% default parameters unless otherwise specified

if ~exist('K','var'), K = 0.1; end                        % K = mg/kLe (see text)
if ~exist('tspan', 'var'), tspan = [0, 50]; end     % time interval
if ~exist('x0', 'var'), x0 = [0.1, 0, 0.05, 0];end % initial conditions
if ~exist('rtol', 'var'), rtol = 1e-5; end       % relative error tolerance

if ~exist('atol', 'var'), atol = 1e-10; end     % absolute error tolerance

% solve the equations of motion using ode45
[tout, xout] = spring_pendulum(tspan, x0, K, rtol, atol);

% rename angle and normalized radius outputs and plot them versus time
ang = xout(:, 3); % angle (rad)
rn = xout(:, 1);  % normalized radial length
figure(1)
plot(tout, ang);
xlabel('time, \tau'); ylabel('\theta');
figure(2)
plot(tout, rn)
xlabel('time, \tau'); ylabel('r_n');

% calculate the normalized energy,R, (see text) and plot versus time
% Note that R ranges from -1 to infinity. The parameters R and mu determine
% if chaos is present or not.

mu = 1/K;             % parameter used to describe R in the text
rnd = xout(:,2);  % derivative of the normalized radial length
angd = xout(:,4); % derivative of the angle

R = (1./(2.*mu-1)).*(mu^2*(rnd.^2 + ((1+rn).^2).*angd.^2) ...
    -2.*mu.*(1+rn).*cos(ang) + mu.^2.*(rn+1./mu).^2);
figure(3)
plot(tout , R)
xlabel('time, \tau'); ylabel('R');
```

15. spring_pendulum_anim3

```
% script spring_pendulum_anim3
% This script solves for the motion of the spring-pendulum and then
% generates an animation of the motion.
```

```
close all

% default parameters unless otherwise specified
if ~exist('K','var'), K = 0.25; end            % parameter K = mg/kLe
if ~exist('tspan', 'var'), tspan = 0:0.1: 400; end  % time interval
if ~exist('x0', 'var'), x0 = [0.1, 0, 0.025, 0];end % initial conditions
if ~exist('rtol', 'var'), rtol = 1e-5; end %relative error tolerance

if ~exist('atol','var'), atol = 1e-10; end % absolute error tolerance

if ~exist('marker', 'var'), marker = 'off'; end    % marker to show path
if ~exist('sldwn','var'),sldwn = 1; end % slow down factor for animation
if ~exist('Le','var'), Le = 1; end  % equilibrium length

% solve equations of motion with ode45
[tout, xout] = spring_pendulum(tspan, x0, K, rtol, atol);

% rename outputs
t = tout;
th = xout(:, 3); % the angle, theta, in rad
rn = xout(:, 1); % the normalized radius

% change from normalized radius to actual radius of the spring

r = rn*Le + Le;
% define arrays with x- and y-coordinates for the endpoints of the spring
x_array(:,1) = zeros(size(r));
x_array(:,2) = r.*sin(th);
y_array(:,1) = zeros(size(r));
y_array(:,2)= -r.*cos(th);          % positive y is up

% find largest and smallest x- and y-values of the points being evaluated

xbig = max(max(x_array));
ybig = max(max(y_array));
xsmall = min(min(x_array));
ysmall = min(min(y_array));

% x- and y-locations of the center of the region where the points are
% being plotted and the lengths of that region in the x- and y-directions
xcenter = (xbig + xsmall)/2;
ycenter = (ybig + ysmall)/2;
lx = abs(xbig - xsmall);
ly = abs(ybig - ysmall);
%define limits of a somewhat larger region for the plot
xmin = xcenter - 1.5*lx/2;
xmax = xcenter + 1.5*lx/2;
ymin = ycenter - 1.5*ly/2;
ymax = ycenter +1.5*ly/2;
```

```
% define handle of the figure window and clear it
h = figure(1);
clf(h);

% define parameters for drawing the pendulum support and mass
a = 0:pi/100:2*pi;
s = [ -1 -1 1 1 -1; -0.25 0.25 0.25 -0.25 -0.25];
maxr = max(r);
xscircle = maxr/50*sin(a);
yscircle = maxr/50*cos(a);
xssquare = maxr/50*s(1,:);
yssquare = maxr/50*s(2,:);

fill(xssquare,yssquare, 'k ');

hold on

% define path, mass , and spring objects at t = 0
if strcmp(marker, 'on')
  path = plot(x_array(1,2), y_array(1 ,2), 'k.', 'MarkerSize', 2);
end
mass = fill(xscircle + x_array(1,2), yscircle + y_array(1,2),'k');
spring = plot([x_array(1,1), x_array(1,2)], [y_array(1,1),y_array
(1,2)],...
  'r', 'LineWidth', 2);

% set up plot region properties
axis equal
grid off
axis([xmin xmax ymin ymax]);
xlabel('x-axis')
ylabel('y-axis')

% draw mass, spring and path marker at their successive locations in time
for i = 1:length(t)
 set(mass,'XData',xscircle +x_array(i,2), 'YData', yscircle +y_array
(i,2));
 set(spring,'XData', [x_array(i,1), x_array(i,2)], 'YData', ...
   [y_array(i,1), y_array(i,2)]);
 if strcmp(marker, 'on')
  set(path, 'XData', x_array(1:i, 2), 'YData' , y_array(1:i, 2));
 end

  if i < length(t)
    pause(0.01*sldwn);

  end
end
hold off

% define the normalized energy, R, and plot versus time
mu = 1/K;
```

```
rnd = xout(:,2);
thd = xout(:,4);
R = (1./(2.*mu-1)).*(mu^2*(rnd.^2 + ((1+rn).^2).*thd.^2) ...
    -2.*mu.*(1+rn).*cos(th) + mu.^2.*(rn+1./mu).^2);
h2 = figure(2);
clf(h2);
plot(t, R)
xlabel('time, \tau'); ylabel('R');
```

Simulation 2.7: Rotating Hoop

16. rotating_hoop (and nested function rotating_hoop_eqs)

```
function [tout, xout] = rotating_hoop(tspan, x0, omega, R, g, rtol,
atol)
% [tout, xout] = rotating_hoop(tspan, x0, omega, R, rtol, atol) solves the
% equations of motion for a bead sliding on a circular, rotating hoop. The
% inputs are tspan, the time interval for the calculations, x0, the initial
% conditions, omega, the angular velocity of the hoop, R, the radius of the
% hoop, and rtol, atol, the relative and absolute error tolerances used by
% ode45 to solve the equations. The output tout, is a column vector which
% holds the times atwhich the solution is evaluated and xout is a matrix
% which contains the values of the angle and its derivative at those times
% in its columns.

% place error tolerances into options
options = odeset('RelTol', rtol, 'AbsTol', atol);

% solve equations of motion
[tout, xout] = ode45(@rotating_hoop_eqs, tspan, x0, options);

% function that defines the equations of motion
  function xdot = rotating_hoop_eqs(t,x)

    % generate column vector for outputs
    xdot = zeros(2,1);

    % rename inputs to make them more readable
    x1 = x(1); % the angle, theta
    x2 = x(2); % the angular velocity of the bead, thetadot

    % the system of equations
    xdot(1) = x2;
    xdot(2) = sin(x1).*(omega^2.*cos(x1)-g/R);
  end
end
```

17. rotating_hoop_sim

```
% rotating_hoop_sim script
% This script solves the equations of motion for a bead sliding on a
% smooth, circular hoop that is also rotating about its axis in a vertical
% plane. The script plots the angle, theta, of the bead on the hoop versus
% time and a phase plot for the angle and its derivative.

close all

% default inputs unless otherwise specified
if ~exist('R','var'), R = 0.5; end          %radius of the hoop
if ~exist('omega','var'), omega = 2; end    %angular velocity of the hoop
if ~exist('rtol','var'), rtol = 1e-6; end  % relative error tolerance

if ~exist('atol','var'), atol = 1e-6; end   % absolute error tolerance

if ~exist('tspan','var'), tspan = [0, 10]; end %time interval
if ~exist('x0','var'), x0 = [ 0, 0.1]; end       %initial conditions
if ~exist('g','var'), g = 9.81; end %acceleration of gravity
% solve equations of motion

[tout, xout] = rotating_hoop (tspan, x0, omega, R, g, rtol, atol);

% rename outputs and plot
ang = xout(:,1)*180/pi;  % angle of the bead, theta
angd = xout(:,2)*180/pi; % angular velocity, thetadot, around the hoop
figure(1)
plot(tout, ang)
xlabel('t, sec'); ylabel(' \theta, deg');
figure (2)
plot(ang, angd)
xlabel('\theta, deg'); ylabel('d\theta/dt, deg/sec');
```

Simulation 2.8: Forced Vibration of a Mass Attached to a Spring and Dashpot

18. moving_wedge (and nested function wedge_func)

```
function [xout, tout] = moving_wedge(tspan, x0, km, cm, Am, t0,rtol,
atol)
% [xout, tout] =moving_wedge(tspan, x0, km , cm, Am, t0, rtol, atol)solves
% the problem of a mass connected to a spring and dashpot on a wedge which
% has a specified motion. This problem is  equivalent to mass attached to a
% spring and dashpot where the mass is moving on a horizontal surface under
% the action of a known applied force, a common problem found in many
% dynamics texts. This problem can be solved analytically.
```

```
% place the error tolerances into options
options = odeset('RelTol', rtol, 'AbsTol', atol);

% solve the equations of motion
[tout, xout] = ode45(@wedge_func ,tspan, x0,options);

% the nested function that defines the system of equations
  function xdot = wedge_func(t,x)

  % make a column vector to hold the outputs
  xdot = zeros(2,1);

  % rename the system variables to make them easier to read
  x1 = x(1); % the displacement of the mass
  x2 = x(2); % the velocity of the mass

  % define the system of equations
  fm = Am.*(t>= 0 & t <= t0);  % the applied force divided by the mass
  xdot(1) = x2;
  xdot(2) = -cm.*x2 -km.*x1 + fm;
  end

end
```

19. moving_wedge_sim

```
% moving_wedge_sim script
% This script solves the problem of a mass attached to a spring and dashpot
% that slides on a moving wedge which has a specified motion. This problem
% is equivalent to a mass connected to a spring and dashpot, moving on a
% smooth, horizontal surface under the action of a known applied force, a
% problem that can be solved analytically but which is solved here
% numerically with ode45. The script plots the displacenent of the mass
% versus time as well as a phase plot of the displacement versus velocity.

close all

% default inputs unless otherwise specified
if ~exist('km','var'), km = 10; end    % normalized spring constant k/m
if ~exist('cm','var'), cm = 0.05; end % normalized damping constant, c/m
if ~exist('rtol','var'), rtol = 1e-6; end % relative error tolerance

if ~exist('atol','var'), atol = 1e-6; end  % absolute error tolerance


if ~exist('tspan','var'), tspan = [0, 10]; end % time interval
if ~exist('x0','var'), x0 = [ 0, 0.1]; end        % initial conditions
if ~exist('Am','var'), Am = 10; end      % normalized amplitude of applied
force
if ~exist('t0','var'), t0 = 3.0; end  % time duration of applied force
```

```
% solve equations of motion
[ xout, tout] = moving_wedge(tspan, x0, km, cm, Am, t0,rtol,atol);

% rename outputs and plot displacement versus time as well as a phase plot
d = xout(:,1);      % displacement of the mass
ddot = xout(:,2); % velocity of the mass
figure(1)
plot(tout, d)
xlabel('time, t'); ylabel('displacement');
figure (2)
plot(d, ddot)
xlabel('displacement'); ylabel('velocity');
```

Simulation 2.9: A Roller Coaster with Coulomb Friction

20. roller_coaster_f2 (and nested functions roller_cartf and stick)

```
function [tout, xout, te, xe, ie]= roller_coaster_f2(tspan,x0, A,...
  L,mu,rtol,atol, vtol)
% [tout, xout, te, xe, ie] =
% roller_coaster_f2(tspan, x0, A,L, mu, rtol, atol,vtol) solves for the
% motion of a cart on a rough sinusoidal track whose height, h, is given by
% h= A*(1-cos(pi*x/L))/2 (the MATLAB solver is ode45). tspan is the time
% interval over which we want to obtain the solution and x0 is the vector
% of initial conditions. mu is the coefficient of friction. rtol and atol
% are the specified error tolerances. vtol is a tolerance on the magnitude
% of the velocity when the event of "sticking" has occurred. The outputs of
% the function are the times, tout, over which the solution is calculated,
% and xout, the values of x and dx/dt at those times. The value of te
% is the time within tspan at which sticking occurs and the integration
% stops, and xe is the vector xe = [ x, dx/dt ] at time te. The output ie
% is an indicator of which event (if there are more than one events) that
% the numerical solver detected. In this case there is only one event.
% It is assumed that x is in meters (m) and dx/dt is in m/sec.

% place error tolerances into options

 options = odeset('Events', @stick,'RelTol', rtol, 'AbsTol', atol);

% acceleration of gravity in m/s^2
g = 9.81;

% solve using ode45
[tout,xout, te, xe, ie] = ode45(@roller_cartf, tspan, x0,options);
```

```
% the nested function to generate the equations of motion
% -----------------------------------------------------------------
function xdot = roller_cartf( t, x)

% create column vector for outputs
xdot = zeros(2,1);

% rename variables to make them easier to read
x1 = x(1);
x2 = x(2);

%define terms in equations
hp = A*pi*sin(pi*x1/L)/(2*L); % calculate h'(x) = dh/dx
D = 1 + hp.^2;
hpp = A*(pi^2)*cos(pi*x1/L)/(2*L^2); % calculate h''(x)
U = g + hpp.*(x2.^2);

% generate system of equations
xdot(1) = x2;
xdot(2) = -(hp./D).*U- mu.*(1./D).*norm(U).*sign(x2);
end

% the nested function that defines the "sticking" event
% -----------------------------------------------------------------
  function [value, isterminal, direction]=stick( t,x)

    % calculate wave speed magnitude
    hp = A*pi*sin(pi*x(1)/L)/(2*L);
    vmag = sqrt((x(2).^2).*(1 +hp.^2));

    % value is equal to one until the velocity and slope are in a
    % "sticking" region

    value = 1 -real(vmag < vtol && abs(sin(pi*x(1)/L)) <2*mu*L/(pi*A));

    % terminate integration when value changes from one to zero
    isterminal = 1;
    direction = -1;
  end
end
```

21. roller_coaster_f2_sim

```
% script roller_coaster_f2_sim
% this script solves the problem of the motion of a cart on a rough
% sinusoidal track. It plots the path of the cart, the normal force
% (normalized by the weight mg of the cart) versus time, and gives
% a phase plot of the motion.

 close all

 % default parameters unless otherwise specified
```

```
if ~exist('g','var'), g = 9.81; end % the acceleration of gravity
if ~exist('L','var'), L = 5; end
if ~exist('A', 'var'), A = 1;end
if ~exist('tspan', 'var'), tspan = [0, 30]; end
if ~exist('x0','var'), x0 = [0, 8]; end
if ~exist('rtol','var'), rtol = 1e-5; end
if ~exist('atol','var'), atol = 1e-6; end
if ~exist('mu','var'),mu = 0.1; end
if ~exist('vtol','var'), vtol = 1e-3; end

% solve the equations of motion numerically
 [tout, xout, te, xe, ie] = roller_coaster_f2(tspan,x0, A, L,mu, ...
   rtol, atol, vtol);

 % rename output variables
 x = xout(:, 1);
 vx = xout(:,2);
 y = A*(1-cos(pi*x/L))/2;
 vy = pi*A*sin(pi*x/L).*vx/(2*L);

 % plot path of the particle
 figure(1)
 plot(x, y)
 xlabel('x'); ylabel('y');
 axis equal

 % evaluate normal force and plot it versus time
 hp = A*pi*sin(pi*x/L)/(2*L);
 hpp = A*(pi^2)*cos(pi*x/L)/(2*L^2);
 Nmg = (1./(sqrt(1+hp.^2))).*(g + hpp.*vx.^2)/g; % N divided by mg
 figure(2)
 plot(tout, Nmg)
 xlabel('t, sec'); ylabel('N/mg');

 % phase plot of the motion (x and dx/dt)
 figure(3)
 plot(x, vx)
 xlabel('x'); ylabel('dx/dt');
```

Simulation 2.10: Motion of a Pendulum with Constraint Satisfaction

22. MATLAB® ode4

```
function Y = ode4(odefun,tspan,y0,varargin)
%ODE4 Solve differential equations with a non-adaptive method of order 4.
%  Y = ODE4(ODEFUN,TSPAN,Y0) with TSPAN = [T1, T2, T3, ... TN] integrates
%  the system of differential equations y' = f(t,y) by stepping from T0 to
%  T1 to TN. Function ODEFUN(T,Y) must return f(t,y) in a column vector.
```

```
% The vector Y0 is the initial conditions at T0. Each row in the solution
% array Y corresponds to a time specified in TSPAN.
%
% Y = ODE4(ODEFUN,TSPAN,Y0,P1,P2...) passes the additional parameters
% P1,P2... to the derivative function as ODEFUN(T,Y,P1,P2...).
%
% This is a non-adaptive solver. The step sequence is determined by TSPAN
% but the derivative function ODEFUN is evaluated multiple times per step.
% The solver implements the classical Runge-Kutta method of order 4.
%
% Example
%     tspan = 0:0.1:20;
%     y = ode4(@vdp1,tspan,[2 0]);
%     plot(tspan,y(:,1));
%   solves the system y' = vdp1(t,y) with a constant step size of 0.1,
%   and plots the first component of the solution.
%

if ~isnumeric(tspan)
 error('TSPAN should be a vector of integration steps.');
end

if ~isnumeric(y0)
 error('Y0 should be a vector of initial conditions.');
end

h = diff(tspan);
if any(sign(h(1))*h <= 0)
 error('Entries of TSPAN are not in order.')
end

try
 f0 = feval(odefun,tspan(1),y0,varargin{:});
catch
 msg = ['Unable to evaluate the ODEFUN at t0,y0. '];
 error(msg);
end

y0 = y0(:);  % make a column vector.
if ~isequal(size(y0),size(f0))
 error('Inconsistent sizes of Y0 and f(t0,y0).');
end

neq = length(y0);
N = length(tspan);
Y = zeros(neq,N);
F = zeros(neq,4);

Y(:,1) = y0;
for i = 2:N
 ti = tspan(i-1);
 hi = h(i-1);
 yi = Y(:,i-1);
```

```
  F(:,1) = feval(odefun,ti,yi,varargin{:});
  F(:,2) = feval(odefun,ti+0.5*hi,yi+0.5*hi*F(:,1),varargin{:});
  F(:,3) = feval(odefun,ti+0.5*hi,yi+0.5*hi*F(:,2),varargin{:});
  F(:,4) = feval(odefun,tspan(i),yi+hi*F(:,3),varargin{:});
  Y(:,i) = yi + (hi/6)*(F(:,1) + 2*F(:,2) + 2*F(:,3) + F(:,4));
end
Y = Y.';
```

23. pendulum_cs_eqs

```
function [xdot] = pendulum_cs_eqs(t, x, g,L, dt, e)
%[xdot] = pendulum_cs_eqs(t, x, g, l, dt, e) generates the equations of
% motion for a simple pendulum using an augmented approach where the
% satisfaction of the constraint function and its derivative are included
% in the formulation. The inputs are the time t, the vector x = [ x, y,
%dx/dt, dy/dt],the acceleration of gravity, g, the length of the pendulum,
%L, the fixed time interval used in the numerical solution, and a constant
% ,e, which is used to either include constraint satisfaction (e = 1) or
% leave it out(e = 0).

% generate column vector for outputs
xdot = zeros(4,1);

% rename variables so they are easier to read
x1 = x(1); % coordinate x
x2 = x(2); % coordinate y
x3 = x(3); % dx/dt
x4 = x(4); % dy/dt

% define expressions in equations of motion
f1 = -2*x1.*x3 -2*x2.*x4 - (x1.^2 + x2.^2 -L^2)/dt;
f2 = -2*(x3.^2 + x4.^2) -2*g*x2 -2*e*(x1.*x3 + x2.*x4)/dt;
c1 = x1./(2*L^2);
c2 = x2./(2*L^2);

% form up equations of motion
xdot(1) = x3 + e*c1.*f1;
xdot(2) = x4 + e*c2.*f1;
xdot(3) = c1.*f2;
xdot(4) = g + c2.*f2;

end
```

24. pendulum_cs_sim

```
% script pendulum_cs_sim
% This script solves for the motion of a simple pendulum using a fixed time
% step solver, ode4. It allows one to satisfy the constraint function
% and its derivative explicitly during the solution. The value of the
% constraint function and its derivative as well as the total energy
% divided by the mass are plotted versus time.
```

```
close all

% default parameters unless otherwise specified
if ~exist('g','var'), g = 9.81; end    % the acceleration of gravity
if ~exist('L','var'), L = 1; end       % length of the pendulum
if ~exist('tspan', 'var'), tspan = 0:0.001: 900; end % time interval
if ~exist('x0','var'), x0 = [L, 0,0,0]; end        % initial conditions
if ~exist('e','var'), e = 0.0; end % constraint satisfaction (e = 0 no
                        % satisfaction, e = 1 satisfaction)

% fixed time interval for calculations
dt = tspan(2)-tspan(1);

% obtain the numerical solution with ode4
[xout] = ode4(@pendulum_cs_eqs, tspan, x0, g , L, dt,e);

% rename outputs
x = xout(:,1);  % x-coordinate
y = xout(:,2);  % y-coordinate
vx = xout(:,3); % vx = dx/dt
vy = xout(:,4); % vy = dy/dt

% compute the constraint function and its derivative
C = x.^2 + y.^2 - L^2;     %constraint function
Cd = 2*x.*vx + 2*y.*vy; % derivative of the constraint function

% plot the constraint function, its derivative, and the total energy, E,
% divided by the mass, m
figure(1)
plot(tspan', C)
xlabel('t, sec'); ylabel('f');
figure (2)
plot(tspan' , Cd)
xlabel('t, sec'); ylabel('df/dt');
figure(3)
plot(tspan', (vx.^2 +vy.^2)/2-g*y)  % E/m
xlabel('t, sec'); ylabel('E/m');
```

Chapter 3

Simulation 3.1: Multiple-Particle Rigid Body Pendulum

25. MP_pendulum (and nested function MP_eqs)

```
function [tout, xout] = MP_pendulum( x0, tspan, c1, c2, lr, rtol,atol)
% [ tout, xout] = MP_pendulum( x0, tspan, c1, c2, lr, rtol, atol) solves
% the equations of motion for a multi-particle pendulum, where the input
% parameters are x0, the initial conditions, tspan, the time interval for
```

```
% the calculations, c1 and c2, which are normalized angular damping
% coefficients, lr ,which is a ratio of lengths, and rtol, atol are the
% relative and absolute error tolerances. The solution for the motion is
% obtained with ode45. The outputs are tout, the times of the calculations,
% and xout, which is a matrix containing the first order system variables
% at those times in columns.

% place error tolerances into options
options = odeset('RelTol', rtol, 'AbsTol', atol);

% solve the equations of motion
[tout, xout] = ode45(@MP_eqs, tspan, x0, options);

% nested function that defines the first order system of equations
function [xdot] = MP_eqs(t,x)

% create column vector to hold the outputs
xdot = zeros(4,1);

% rename variables to make them easier to read.
x1 = x(1); %theta
x2 = x(2); %thetadot
x3 = x(3); %phi
x4 = x(4); %phidot

% form equations of motion as a first order system
xdot(1) = x2;
xdot(2) = -4*lr^2*c1*(x2+x4);
xdot(3) = x4;
xdot(4) = -c2*x4 - c1*(x2 + x4) - sin(x3);
end
end
```

26. MP_pendulum_sim

```
% script MP_pendulum_sim
% This script solves for the motion of a multi-particle pendulum.
% It plots the theta and phi angles of the pendulum verus time.
% Note that the angular inputs in the initial conditions are in
% units of degrees and degrees/sec.

close all

% default parameters unless otherwise specified
if ~exist('x0', 'var'), x0 = [30, 10, 20, 0]; end % initial conditions
% for [theta, thetadot, phi, phidot] in degrees and degrees/sec
if ~exist('tspan', 'var'), tspan = [0, 50];end % time interval
if ~exist('c1', 'var'),c1 = 0.2; end      % angular damping coefficient 1
if ~exist('c2', 'var'), c2 = 0.2;end      % angular damping coefficient 2
if ~exist('lr', 'var'), lr = 2; end       % length ratio
if ~exist('rtol', 'var'), rtol = 1e-6;end % relative error tolerance
```

```
if ~exist('atol', 'var'), atol = 1e-6; end  % absolute error tolerance

% change initial conditions to radians and radians/sec
x0 = (pi/180)*x0;

% solve for motion of the multi-particle pendulum
[tout, xout] = MP_pendulum( x0, tspan, c1, c2, lr, rtol,atol);

% rename output variables and change units back into degrees
t = tout;
theta = (180/pi)*xout(:,1);
phi = (180/pi)*xout(:,3);

% plot theta and phi versus time
figure(1)
plot(t, theta);
xlabel('time, \tau'); ylabel('\theta');
figure(2)
plot(t, phi)
xlabel('time, \tau'); ylabel('\phi');
```

Simulation 3.2: The Double Pendulum

27. double_pend (and nested function dp_eqs)

```
function [tout, xout] = double_pend( x0, tspan, m1, m2, l1,l2,g, rtol,
atol)
% [tout, xout] = double_pend( x0, tspan, m1, m2, l1, l2, g, rtol, atol)
% solves for the motion of a double pendulum. The inputs are the initial
% conditions x0 = [ theta, thetadot, phi, phidot], measured in rad and
% rad/sec, tspan, the time interval for the calculations, m1 and m2, the
% two masses in the pendulum, l1, l2, the lengths of the two massless rods
% connecting the masses, g, the acceleration of gravity, and rtol, atol,
% the relative and absolute error tolerances used in the solution. The
% outputs are tout, the times at which the solution is obtained, and xout,
% a matrix that contains the first order system variables at those times,
% given in columns.

% place error tolerances into options
options = odeset('RelTol', rtol, 'AbsTol', atol);

% solve the system of equations
[tout, xout] = ode45(@dp_eqs, tspan, x0, options);

% nested function that defines the first order system
function [xdot] = dp_eqs(t,x)
```

```
% create column vector to hold outputs
xdot = zeros(4,1);

% rename variables to make them easier to read
x1 = x(1); % theta
x2 = x(2); % thetadot
x3 = x(3); % phi
x4 = x(4); % phidot

% define terms used in the equations
M11 = (m1+m2).*l1.^2;
M12 = m2.*l1.*l2*cos(x1 -x3);
M21 = M12;
M22 = m2.*l2.^2;
t1 = -m2.*l1.*l2.*sin(x1-x3).*x4.^2 - (m1+m2).*g.*l1.*sin(x1);
t2 = m2.*l1.*l2.*sin(x1-x3).*x2.^2-m2.*g.*l2.*sin(x3);
D = M11.*M22-M12.^2;

% generate system of first order equations
xdot(1) = x2;
xdot(2) = (M22.*t1-M12.*t2)./D;
xdot(3) = x4;
xdot(4) = (M11.*t2-M21.*t1)./D;
end
end
```

28. double_pendulum_sim

```
% script double_pendulum_sim
% This script solves for the motion of a double pendulum. It plots the
% angles of the two masses versus time, and calculates and plots the
% normalized total energy for this conservative system. Finally, the x-y
% coordinate motion of the second mass is plotted.

close all

% default input parameters unless otherwise specified
if ~exist('m1','var'), m1 = 0.1; end % mass 1
if ~exist('m2','var'), m2 = 0.1; end % mass 2
if ~exist('l1','var'), l1 = 0.1; end % massless rod length 1
if ~exist('l2','var'), l2 = 0.1; end % massless rod length 2
if ~exist('g','var'), g = 9.81; end  % acceleration of gravity
if ~exist('rtol','var'), rtol = 1e-10; end % relative error tolerance

if ~exist('atol','var'), atol = 1e-10; end % absolute error tolerance

if ~exist('tspan','var'), tspan = [0, 4]; end % time interval
if ~exist('x0','var'), x0 = [ 50*pi/180, 0,25*pi/180,0]; end % initial
% conditions for [theta1, theta1dot, theta2, theta2dot] in rad and rad/sec

% solve equations of motion
[tout, xout] = double_pend(x0,tspan, m1, m2, l1,l2,g, rtol, atol);
```

```
% rename outputs
t = tout;
theta1 = xout(:,1);
theta1dot = xout(:,2);
theta2 = xout(:,3);
theta2dot = xout(:,4);

% plot angles versus time
figure(1)
plot(t, theta1)
xlabel('time, t'); ylabel('\theta_1');
figure (2)
plot(t, theta2)
xlabel('time, t'); ylabel('\theta_2');

% compute and plot the magnitude of the normalized total energy E(t)/E(0)
E = (m1/2)*l1^2*theta1dot.^2 +(m2/2)*(l1^2*theta1dot.^2 + ...
    l2^2*theta2dot.^2 + 2*l1*l2*cos(theta1 -theta2).*theta1dot.* ...
    theta2dot) -(m1+m2)*g*l1*cos(theta1)- m2*g*l2*cos(theta2);
figure(3)
plot(t,abs(E/E(1)))
axis([tspan(1) tspan(end) 0 1.5])
xlabel('time, t'); ylabel('E(t)/E(0)');

% compute the x2- and y2- coordinates of mass 2 and plot (x2, y2)
x2 = l1*sin(theta1) +l2*sin(theta2);
y2 = -l1*cos(theta1)- l2*cos(theta2); %y2 here is positive upwards
figure(4)
plot(x2,y2)
xlabel('x_2'); ylabel('y_2');
```

29. double_pendulum_anim2

```
% script double_pendulum_anim2
% This script solves the equations of motion for the double pendulum,
% using the same formulation as in the script double_pendulum_sim and then
% generates an animation of the motion.

close all

% default input parameters unless otherwise specified
if ~exist('m1','var'), m1 = 0.1; end % mass 1
if ~exist('m2','var'), m2 = 0.1; end % mass 2
if ~exist('l1','var'), l1 = 0.1; end % length of massless rod 1
if ~exist('l2','var'), l2 = 0.1; end % length of massless rod 2
if ~exist('g','var'), g = 9.81; end  % acceleration of gravity
if ~exist('rtol','var'), rtol = 1e-10; end % relative error tolerance

if ~exist('atol','var'), atol = 1e-10; end % absolute error tolerance

if ~exist('tspan','var'), tspan = [0, 4]; end % time interval
if ~exist('x0','var'), x0 = [ 50*pi/180, 0,25*pi/180,0]; end % initial
```

```
% conditions [theta1, theta1dot, theta2, theta2dot] measured in rad and
% rad/sec
if ~exist('marker', 'var'), marker ='on';end % marker to show path of m2
if~exist('sldwn','var'), sldwn=1; end% "slowdown" factor for animation

% solve the equations of motion
[tout, xout] = double_pend(x0,tspan, m1, m2, l1,l2,g, rtol, atol);

% rename outputs
t = tout;
theta1 = xout(:,1);
theta1dot = xout(:,2);
theta2 = xout(:,3);
theta2dot = xout(:,4);

% define arrays holding the endpoints of the two lines between the masses
% note: y1, y2 here are positive upwards

x_array(:,1) = zeros(size(theta1));
x_array(:,2) = l1*sin(theta1);
x_array(:,3) = l1*sin(theta1) +l2*sin(theta2);
y_array(:,1) = zeros(size(theta1));
y_array(:,2) = -l1*cos(theta1);
y_array(:,3) = -l1*cos(theta1) - l2*cos(theta2);

% define the limits of the plotting region for the animation
% find largest and smallest x- and y-values of the points being evaluated

xbig = max(max(x_array));
ybig = max(max(y_array));
xsmall = min(min(x_array));
ysmall = min(min(y_array));

% find x- and y-locations of the center of the region where the points are
% being plotted and the lengths of that region in the x- and y-directions
xcenter = (xbig + xsmall)/2;
ycenter = (ybig + ysmall)/2;
lx = abs(xbig - xsmall);
ly = abs(ybig - ysmall);
%define limits of a somewhat larger region for the plot
xmin = xcenter - 1.5*lx/2;
xmax = xcenter + 1.5*lx/2;
ymin = ycenter - 1.5*ly/2;
ymax = ycenter +1.5*ly/2;

% define handle of the figure window and clear it
h=figure(1);
clf(h);

% define parameters for drawing the support and the masses
a = 0:pi/100:2*pi;
s = [-1 -1 1 1 -1; -0.25 0.25 0.25 -0.25 -0.25];
```

```
maxr = l1+l2;
xscircle = maxr/50*sin(a);
yscircle = maxr/50*cos(a);
xssquare = maxr/50*s(1,:);
yssquare = maxr/50*s(2,:);

fill(xssquare,yssquare,'k');
hold on

% define path, massless rods connecting the particles, and the masses as
% objects at t = 0.
if strcmp(marker, 'on')
  path = plot(x_array(1,3), y_array(1,3), 'k.', 'MarkerSize',2);
end

r1 = plot([x_array(1,1),x_array(1,2)],[y_array(1,1),y_array(1,2)],
...
  'r', 'LineWidth', 2);
r2 = plot([x_array(1,2),x_array(1,3)], [y_array(1,2), y_array(1,3)],
...
  'r', 'LineWidth', 2);
mass1 = fill(xscircle +x_array(1,2), yscircle + y_array(1,2), 'k');
mass2 = fill(xscircle + x_array(1,3), yscircle + y_array(1,3), 'k');
axis equal
grid off
axis([ xmin xmax ymin ymax]);
xlabel('x-axis')
ylabel( 'y-axis')

% draw rods, masses, and path marker at their successive locations in time
for i = 1:length(t)

  set(r1,'XData',[x_array(i,1), x_array(i,2)], 'YData', [y_array
  (i,1) ...
    , y_array(i,2)]);
  set(r2,'XData',[x_array(i,2), x_array(i,3)], 'YData', [y_array
  (i,2) ...
    , y_array(i,3)]);
  set(mass1, 'XData', xscircle +x_array(i,2), 'YData', yscircle + ...
    y_array(i,2));
  set(mass2, 'XData', xscircle +x_array(i,3), 'YData', yscircle + ...
    y_array(i,3));
  if strcmp(marker, 'on')
    set(path,'XData', x_array(1:i,3), 'YData', y_array(1:i,3));
  end
  if i <length(t)
    pause(0.01*sldwn);
  end
end
hold off
```

Simulation 3.3: Linear Oscillations of the Double Pendulum

30. linear_dp

```
% script linear_dp
% This script solves analytically for the small amplitude motion of the
% double pendulum using a superposition of the two normal modes of the
% system. The script plots the angles of the two masses in the double
% pendulum and also provides all of the inputs needed to run the script
% double_pendulum_sim with the same parameters.

clear
close all

% input parameters
m1 = 0.1;  % mass 1
m2 = 0.1;  % mass 2
l1 = 0.2;  % massless rod length 1
l2 = 0.2;   % massless rod length 2
A1 = 0.1;   % amplitude of first normal mode
A2 = 0.07;  % amplitude of second normal mode
d1 = 0;     % starting phase of first normal mode
d2 = pi/2;  % starting phase of second normal mode
g = 9.81;   % acceleration of gravity
w0 = sqrt(g/l1);             % simple pendulum natural frequency
w1 = sqrt(2-sqrt(2))*w0;  % first natural frequency of double pendulum
w2 = sqrt(2+sqrt(2))*w0;  % second natural frequency of double pendulum
t = linspace(0, 10, 300); % times for calculation
tspan = t; % place times into tspan for use in double_pendulum_sim

% define initial conditions for theta1, theta1dot, theta2, theta2dot
theta10 = A1*cos(d1)/sqrt(3) + A2*cos(d2)/sqrt(3);
theta20 = (sqrt(2)/sqrt(3))*(A1*cos(d1) -A2*cos(d2));
th1d0 = -w1*A1*sin(d1)/sqrt(3) - w2*A2*sin(d2)/sqrt(3);
th2d0 = (sqrt(2)/sqrt(3))*(-w1*A1*sin(d1)+ w2*A2*sin(d2));

% place initial conditions into vector x0 for use in double_pendulum_sim
x0 = [theta10 , th1d0, theta20, th2d0];

% solution of linear equations of motion
theta1 = A1*cos(w1*t+d1)/sqrt(3) +A2*cos(w2*t +d2)/sqrt(3);
theta2 = (sqrt(2)/sqrt(3))*(A1*cos(w1*t+d1) - A2*cos(w2*t+d2));

% plot the angles theta1 and theta2 versus time
figure (1)
plot(t, theta1)
xlabel('time, t'); ylabel('\theta_1');
```

```
figure (2)
plot(t, theta2)
xlabel('time, t'); ylabel('\theta_2');
```

Simulation 3.4: The Double Pendulum with a Mass Matrix

31. double_pend_m (and nested functions MM and dp_m_eqs)

```
function [tout, xout] = double_pend_m( x0, tspan, m1, m2,...
  l1,l2,g, rtol,atol)
% [tout, xout] = double_pend_m( x0, tspan, m1, m2, l1, l2, g, rtol, atol)
% solves the equations of motion of a double pendulum where a mass matrix
% is included in the formulation. The inputs are x0, the initial
% conditions, tspan, the time interval for the calculations, m1 and m2, the
% two masses, l1 and l2, the lengths of the two massless rods connecting
% the particles, g, the acceleration of gravity, and rtol, atol, the error
% tolerances used in the calculations. The outputs are tout, a column
% vector of the times at which the motion was evaluated, and xout, a matrix
% of the first order system variables [ theta1, theta1dot, theta2,
% theta2dot], calculated at those times,in the columns of the matrix.

% place error tolerances into options
options = odeset('RelTol', rtol, 'AbsTol', atol, 'Mass', @MM);

% solve equations of motion
[tout, xout] = ode45(@dp_m_eqs, tspan, x0, options);

% nested function that defines the mass matrix
  function  M = MM(t,x)

   % generate rows of mass matrix
   r1 = [1 0 0 0];
   r2 = [ 0 (m1+m2)*l1 0 m2*l2*cos(x(1)-x(3))];
   r3 = [ 0 0 1 0];
   r4 = [0 m1*l1*cos(x(1)-x(3)) 0 m2*l2];

   % form up mass matrix
  M = [r1;r2;r3;r4];
  end

% nested function that generates the equations of motion
function [f] = dp_m_eqs(t,x)

% create column vector to hold outputs
f = zeros(4,1);
```

```
% rename variables to make them easier to read
x1 = x(1); %theta1
x2 = x(2); %theta1dot
x3 = x(3); %theta2
x4 = x(4); %theta2dot

% right side of the first order system of equations with mass matrix
f(1) = x2;
f(2) = -(m2*l2*x4^2)*sin(x1-x3)-(m1+m2)*g*sin(x1);
f(3) = x4;
f(4) =(m2*l1*x2^2)*sin(x1-x3) -m2*g*sin(x3);
end

end
```

32. double_pendulum_m_sim

```
% script double_pendulum_m_sim
% This script solves the double pendulum problem using a formulation that
% includes a mass matrix in the solution. Like the script
% double_pendulum_sim this script plots the angles of the two masses
% versus
% time, evaluates and plots the normalized total energy versus time, and
% plots the (x2, y2) coordinates of the mass m2 during the motion.

close all

% default parameters unless otherwise specified
if ~exist('m1','var'), m1 = 0.1; end        % mass 1
if ~exist('m2','var'), m2 = 0.1; end        % mass 2
if ~exist('l1','var'), l1 = 0.1; end        % length of massless rod 1
if ~exist('l2','var'), l2 = 0.1; end        % length of massless rod 2
if ~exist('g','var'), g = 9.81; end         % acceleration of gravity
if ~exist('rtol','var'), rtol = 1e-10; end   % relative error tolerance

if ~exist('atol','var'), atol = 1e-10; end   % absolute error tolerance

if ~exist('tspan','var'), tspan = [0, 4]; end % time interval
if ~exist('x0','var'), x0 = [ 50*pi/180, 0,25*pi/180,0]; end % initial
% conditions for [ theta1, theta1dot, theta2, theta2dot]

% solve equations of motion
[tout, xout] = double_pend_m(x0,tspan, m1, m2, l1,l2,g, rtol, atol);

% rename output variables
t = tout;
theta1 = xout(:,1);
theta1dot = xout(:,2);
theta2 = xout(:,3);
theta2dot = xout(:,4);
```

```
% plot the two angles in the double pendulum
figure(1)
plot(t, theta1)
xlabel('time, t'); ylabel('\theta_1');

figure (2)
plot(t, theta2)
xlabel('time, t'); ylabel('\theta_2');

% calculate the total energy E(t) and plot E(t)/E(0) versus time
E = (m1/2)*l1^2*theta1dot.^2 + (m2/2)*(l1^2*theta1dot.^2 + ...
  l2^2*theta2dot.^2 + 2*l1*l2*cos(theta1 -theta2).*theta1dot.* ...
  theta2dot) - (m1+m2)*g*l1*cos(theta1) - m2*g*l2*cos(theta2);
figure(3)
plot(t,abs(E/E(1)))
axis([ tspan(1)  tspan(end)  0 1.5])
xlabel('time, t'); ylabel('E(t)/E(0)');

% calculate the (x2, y2) coordinates of mass m2 and plot
x2 = l1*sin(theta1) +l2*sin(theta2);
y2 = -l1*cos(theta1)- l2*cos(theta2); %y2 here is up
figure(4)
plot(x2,y2)
xlabel('x_2'); ylabel('y_2');
```

Simulation 3.5: Animation Function for the Double Pendulum

33. animate_lines

```
function animate_lines(x_array, y_array, sldwn, marker)
% animate_lines( x_array, y_array, sldwn, marker) generates an
% animation
% of the lines drawn between points contained in the x_array(:, m) and
% y_array(:, m) where m = 1:npts and nlines = npts -1, where npts is the
% number of points and nlines is the number of lines. The variable sldwn
% injects a pause of 0.01*sldwn seconds between draws in the animation.
% marker is a string ('on' or 'off') to indicate if the path of the end
% point of the last line is to be plotted as a series of dots.

% find number of line end points, npts, and the number of values in those
% end points, nvals, to be plotted
[nr,nc] = size(x_array);
npts = nc;
nvals = nr;

% number of lines
nlines = npts -1;
```

```
% find largest and smallest x- and y-values of the points being plotted and
% define a region slightly larger for the animation.
xbig = max(max(x_array));
ybig = max(max(y_array));
xsmall = min(min(x_array));
ysmall = min(min(y_array));

xcenter = (xbig + xsmall)/2;
ycenter = (ybig + ysmall)/2;
lx = abs(xbig - xsmall);
ly = abs(ybig - ysmall);

xmin = xcenter - 1.5*lx/2;
xmax = xcenter + 1.5*lx/2;
ymin = ycenter - 1.5*ly/2;
ymax = ycenter +1.5*ly/2;

% define handles for the lines and marker
hh = figure(1);
clf(hh);
hold on

for nn = 1:nlines

  h(nn) = plot([x_array(1,nn), x_array(1, nn+1)], [y_array(1,nn), ...
    y_array(1,nn+1)], 'r', 'LineWidth', 2);
end

if strcmp(marker, 'on')
  path=plot(x_array(1, npts), y_array(1, npts), 'k.','MarkerSize', ...
    2);
end

% set up plotting region and plotting parameters
axis equal
grid off
axis([ xmin xmax ymin ymax]);
xlabel('x-axis')
ylabel('y-axis')

% go through the animation
for ii = 1:nvals

 for nn = 1:nlines
  set(h(nn),'XData', [x_array(ii, nn), x_array(ii,nn+1)], 'YData', ...
    [y_array(ii, nn) , y_array( ii, nn+1)]);
 end

 if strcmp(marker,'on')
  set(path, 'XData', x_array(1:ii, npts), 'YData', y_array(1:ii, npts));
 end
```

```
   if ii <= nvals
     pause(.01*sldwn);
   end

end

hold off

end
```

34. double_pendulum_m_anim

```
% script double_pendulum_m_anim
% This script solves the double pendulum problem using a formulation that
% includes a mass matrix in the solution. Like the script
% double_pendulum_anim this script animates the motion of the masses but
% in this case the animation is done by calling a 2-D animation function,
% animate_lines.

close all

% default parameters unless otherwise specified
if ~exist('m1','var'), m1 = 0.1; end        % mass 1
if ~exist('m2','var'), m2 = 0.1; end        % mass 2
if ~exist('l1','var'), l1 = 0.1; end        % length of massless rod 1
if ~exist('l2','var'), l2 = 0.1; end        % length of massless rod 2
if ~exist('g','var'), g = 9.81; end            % acceleration of gravity
if ~exist('rtol','var'), rtol = 1e-10; end  % relative error tolerance

if ~exist('atol','var'), atol = 1e-10; end   % absolute error tolerance
if ~exist('tspan','var'), tspan = linspace(0, 4, 1000); end % time interval
if ~exist('x0','var'), x0 = [ 50*pi/180, 0,25*pi/180,0]; end % initial
% conditions for [ theta1, theta1dot, theta2, theta2dot]
if ~exist('sldwn','var'), sldwn = 1; end    % slow down animation factor
if ~exist('marker','var'), marker = 'off'; end % plotting marker

% solve equations of motion
[tout, xout] = double_pend_m(x0,tspan, m1, m2, l1,l2,g, rtol, atol);

% rename output variables
t = tout;
theta1 = xout(:,1);
theta1dot = xout(:,2);
theta2 = xout(:,3);
theta2dot = xout(:,4);

% calculate the (x1, y1) coordinates of mass m1 and the (x2, y2)
% coordinates of mass m2 , store them in arrays and animate
x1 = l1*sin(theta1);
y1 = -l1*cos(theta1); % positive y1 here is measured upwards
x2 = l1*sin(theta1) +l2*sin(theta2);
```

```
y2 = -l1*cos(theta1) - l2*cos(theta2); %positive y2 also measured upwards
% put in arrays and animate
xb = zeros(size(x1)); % fixed base point x-coordinates
yb = zeros(size(y1)); % fixed base point y-coordinates
x_array(:,1) = xb;
x_array(:,2) = x1;
x_array(:,3) = x2;
y_array(:,1) = yb;
y_array(:,2) = y1;
y_array(:,3) = y2;
animate_lines(x_array, y_array, sldwn, marker)
```

Simulation 3.6: The Rolling Axle

35. rolling_axle (and the nested function axle2)

```
function [tout, xout] = rolling_axle( x0, tspan, rtol, atol);
% [tout, xout] = rolling_axle(x0, tspan, rtol, atol) solves the problem of
% a rolling axle where there are three constraints, one of which is
% holonomic and the other two are non-holonomic. The constraints are not
% independent so the solution is obtained by use of the pseudo-inverse. The
% inputs are x0, the initial conditions, tspan, the time interval for the
% calculations, and rtol, atol the error tolerances used in the solution.
% The outputs are tout, the times at which the motion is calculated, and
% xout, a matrix of system variables [ x1, y1, x2, y2, vx1, vy1, vx2, vy2]
% evaluated at those times, in the columns of the matrix.

% place error tolerances into options
options = odeset('RelTol', rtol, 'AbsTol', atol);

% solve equations of motion
[tout, xout] = ode45(@axle2, tspan, x0, options);

% nested function that defines the equations of motion
function xdot = axle2(t, x)

% create column vectors to hold the outputs and the accelerations, which
% are parts of the first order system of equations
xdot = zeros(8,1);
accel = zeros(4,1);

% generate first order system of equations involving the eight variables
% [ x1, y1, x2, y2, vx1, vy1, vx2, vy2]

% the four velocities
xdot(1) = x(5);  % vx1
```

```
xdot(2) = x(6);  % vy1
xdot(3) = x(7);  % vx2
xdot(4) = x(8);  % vy2

% use pseudo-inverse to solve for the accelerations
A = [ -(x(3)-x(1)) -(x(4) -x(2)) (x(3) - x(1)) (x(4) - x(2)); ...
   (x(3) - x(1)) (x(4) - x(2))  0  0;...
   0  0  (x(3)- x(1)) (x(4) - x(2))];
b = [ -(x(7) - x(5)).^2-(x(8) - x(6)).^2; ...
   -(x(7) - x(5)).*x(5)-(x(8) - x(6)).*x(6); ...
   -(x(7) - x(5)).*x(7)-(x(8) - x(6)).*x(8)];
 accel = A.'*pinv(A*A.')*b;

 % the four accelerations
 xdot(5) = accel(1); % ax1
 xdot(6) = accel(2); % ay1
 xdot(7) = accel(3); % ax2
 xdot(8) = accel(4); % ay2
end
end
```

36. rolling_axle_sim

```
% script rolling_axle_sim
% This script solves for the motion of a rolling axle where the mass of the
% system is concentrated at the ends of the axle. This problem is an
% example of a combination of holonomic and non-holonomic constraints
% where
% the constraints are not independent. The solution is obtained with the
% use of the pseudo-inverse of a matrix. The path of the center of mass of
% the axle is plotted. There is an analytical solution to this problem that
% can be used to verify the numerical solution.

close all

% default parameters unless otherwise specified
if ~exist('L','var'), L = 0.5; end  % length of the axle
if ~exist('v0','var'), v0 = 2; end % initial y-velocity of center of mass
if ~exist('w','var'), w = 3; end   % initial angular velocity of the axle
if ~exist('rtol','var'), rtol = 1e-5; end    % relative error tolerance

if ~exist('atol','var'), atol = 1e-6; end    % absolute error tolerance
if ~exist('tspan','var'), tspan = [0, 2]; end % time interval
if ~exist('x0','var'), ...
 x0 = [ -L/2, 0, L/2, 0 , 0, (v0-w*L/2),0 , (v0 +w*L/2) ]; end % the
% initial conditions for [x1, y1, x2, y2, vx1, vy1, vx2, vy2]. Note: these
% must be consistent with the specification of L, v0 and w.

% solve the equations of motion
[tout, xout] = rolling_axle(x0,tspan, rtol, atol);
```

```
% rename the output variables
t = tout;
x1 = xout(:,1);
x2  = xout(:,3);
y1 = xout(:,2);
y2 = xout(:,4);

% calculate coordinates of the center of mass and plot its path
xc = (x1 + x2)/2;
yc = (y1 +y2)/2 ;
plot (xc, yc)
axis equal
xlabel('x_c'); ylabel('y_c');
```

Chapter 4

Simulation 4.1: The Slider-Crank

37. slider_crank

```
% slider_crank script
% This script solves for the angular positions, velocities, and the
% accelerations of the slider-crank mechanism by writing general
% analytical expressions for the positions and then differentiating those
% expressions two times analytically to obtain linear equations for the
% velocities and accelerations. The angular motion of the crank is a
% specified function of time.

clear
close all

% define the problem parameters
l1 = 5;          % length of crankshaft, in.
l2 = 13;         % length of connecting rod, in.
tmax = 1.0;      % maximum time for calculations
numpts = 500;    % number of times between [0, tmax]

% set up calculation times and specify angular position, angular velocity,
% , and angular acceleration of the crankshaft as a function of time
t = linspace(0, tmax, numpts);
[nr, nc] = size(t);
ang1 = -6*t;
dang1 = -6*ones(1,nc);
ddang1 = zeros(1,nc);
```

```
% position analysis
ang2 = asin(l1*sin(ang1)/l2);       % angle of connecting rod
xp = l1*cos(ang1)+ sqrt(l2.^2 -(l1.^2)*sin(ang1).^2); % position of
% piston

% velocity analysis
% solving the linear system [A]{vels} = {bv} at the times t
% {vels} = [ vp ; dang2] where vp is the velocity of the piston and
% dang2 is the angular velocity of the connecting rod
for nn = 1: numpts
A = [1 l2*sin(ang2(nn)) ; 0  l2*cos(ang2(nn))];
bv = [-l1*dang1(nn)*sin(ang1(nn)); l1*dang1(nn)*cos(ang1(nn))];
vels = A\bv;
vp(nn) = vels(1);
dang2(nn) = vels(2);
end

% acceleration analysis
% solving the linear system [A]{acels} = {ba} at the times t
% {acels} = [ ap ; ddang2] where ap is the acceleration of the piston and
% ddang2 is the angular acceleration of the connecting rod. The column
% vector {ba} =[ b1 ; b2]

% calculate b1, b2 separately for all times t
b1 = -(l2.*dang2.^2).*cos(ang2) -l1.*ddang1.*sin(ang1) ...
     -(l1.*dang1.^2).*cos(ang1);
b2 = (l2.*dang2.^2).*sin(ang2) +l1.*ddang1.*cos(ang1) ...
     -(l1.*dang1.^2).*sin(ang1);

for mm = 1:numpts
 A = [1 l2*sin(ang2(mm)) ; 0  l2*cos(ang2(mm))];
 ba = [b1(mm); b2(mm)];
 acels = A\ba;
 ap(mm) = acels(1);
 ddang2(mm) = acels(2);

end

% plot crankshaft angle (deg) versus time
figure(1)
plot(t,180*ang1/pi)
xlabel('time, t'); ylabel('\theta_1');

% plot velocity of piston versus time (in/sec)
figure(2)
plot(t, vp)
xlabel('time, t'); ylabel('v_p');
% plot angular velocity (rad/sec) of connecting rod versus time
figure (3)
plot(t, dang2)
xlabel('time, t'); ylabel('d\theta_2/dt');
% plot acceleration of piston versus time (in/sec^2)
```

```
figure(4)
plot(t, ap)
xlabel('time, t'); ylabel('a_p');
%plot angular acceleration of the connecting rod (rad/sec^2) versus time
figure(5)
plot(t, ddang2)
xlabel('time, t'); ylabel('d^2\theta/dt^2');
```

Simulation 4.2: Numerical Solution of the Positional Constraints of a Kinematically Driven Rigid Body System

38. slider_crank_f

```
function [f] = slider_crank_f(t, x,d1, d2, ang10)
% [f] = slider_crank_f(t,x,d1, d2, ang10) computes the driving and
% kinematic constraints for the slider-crank problem at the time, t, and
% for x = [ ang1, ang2, xp], where ang1 is the angle of the crankshaft,
% ang2 is the angle of the connecting rod, and xp is the displacement
% of the piston. The imputs d1, d2, are the lengths of the crankshaft and
% connecting rod, respectively, and ang10 is the starting angle of the
% crankshaft. All angles are in rad. f is returned as a 3x1 column vector.

% the three positional equations for the slider-crank
f1 = x(1) + 6*t - ang10;
f2 = d1*cos(x(1))+d2*cos(x(2)) -x(3);
f3 = d1*sin(x(1)) -d2*sin(x(2));

% place into an output column vector
f = [f1;f2;f2];
```

39. slider_crank_J

```
function [J] = slider_crank_J(t,x,d1,d2,ang10)
% [J] = slider_crank_J(t,x,d1,d2,ang10) computes the Jacobian, J, for the
% slider-crank problem, at time,t, and for x-values x = [ang1, ang2, xp],
% where ang1 is the angle of the crankshaft, ang2 is the angle of the
% connecting rod, and xp is the displacement of the slider (piston). The
% distances d1, d2, are the lengths of the crankshaft and connecting
% rods, respectively, and ang10 is the starting angle for the crankshaft.
% (which is not used in this function). All angles are in rad. The output
% Jacobian is a 3x3 matrix.

% values of the derivatives to place in Jacobian
J1 = [ 1, 0 ,0];
J2 = [ -d1*sin(x(1)),-d2*sin(x(2)), -1];
J3 = [d1*cos(x(1)), -d2*cos(x(2)), 0];
```

```
% form up the Jacobian matrix
J = [J1;J2;J3];
```

40. newton_raph

```
function [xout, iter] = newton_raph( fh, Jh,t, x0 ,ftol, varargin)
% [xout, iter] = newton_raph(fh, Jh, t , x0, ftol , varargin) implements
% the Newton-Raphson method to solve approximately a system of n
% non-linear equations f(t, x, q1, q2 ...) = 0 for the n values of the
% vector,x. The input argument, fh, is a function handle for a function
% f(t,x,q1,q2,...)which returns the function values [f1; f2; ... ;fn]
% at time t and for x = [x1;x2; ...;xn].The input argument Jh is a function
% handle to a function J(t, x, q1, q2, ...)which returns the Jacobian
% matrix of derivatives d(f1, f2, ..., fn)/d(x1, x2, ..., xn)at time t and
% the x-values. The input argument,x0,is a row vector for an initial guess
% [x01, x02, ...,x0n] of the solution.  ftol is a tolerance value for the
% magnitude of the functions that is used to determine when to stop the
% iterative solution. varargin is a set of additional input argument
% variables(q1, q2, ...)that may be needed to define the functions f and J.
% The full set of these variables, varargin{:}, must be arguments of both
% functions f and J. The newton_raph function returns the approximate
% value
% of x ,xout,that satisfies the system of equations. xout is a column
% vector. iter is the number of iterations needed to reach a solution.

% internal variables
jtol = 1e-10;  % tolerance value to detect singular Jacobian
maxval = 10000.;% maximum value of the functions used to detect divergence
Niter = 100;   % maximum number of iterations

NN = Niter;   % initialize iteration count
xx = x0(:);   % initialize initial guess as a column vector

% perform iterative Newton-Raphson solution

while (NN>0)
  JJ = feval(Jh, t, xx, varargin{:});% evaluate Jacobian at current guess
  if abs(det(JJ)) < jtol    % check to ensure Jacobian is non-singular
     error('newtonrm-Jacobian is singular');
     abort;
  end

  xn = xx - JJ\feval(fh,t,xx,varargin{:}); % update guess

  if  norm(feval(fh,t, xn, varargin{:})) < ftol % if converged

     xout = xn;                   % then return current
     iter = Niter - NN;           % guess and number of iterations
     return;
  end;
```

```
  if abs(feval(fh, t,xx,varargin{:})) > maxval   % check for divergence
    iter = 100-NN;
    disp(['iterations = ', num2str(iter)]);
    error('solution diverges');
    abort;
  end

  NN = NN -1;              % update iteration count and start
  xx = xn;                 % new iteration with updated guess
end
disp([num2str(Niter),'iterations exceeded']) % stop, indicate number of
error('no convergence' );                    % iterations which exceeds
abort;                                       % Niter
end
```

41. slider_crank_sol2

```
function [xout, numit] = slider_crank_sol2(t, x0, d1, d2,ang10, ftol)
% [xout, numit] = slider_crank_sol2(t, x0, d1, d2, ang10, ftol) determines
% approximate values for the variables describing the motion of a slider-
% crank problem  at the specified times, t. The input x0 = [anga10,
% ang2a0, xpa0 ] are guesses for the initial values of ang1, the angle of
% the crankshaft, ang2, the angle of the connecting rod, and xp, the
% displacement of the slider (piston). The inputs d1, d2, are the lengths
% of the crankshaft and connecting rod, respectively, and ang10 is the
% starting angle for the crankshaft driving constraint. The initial guess
% for anga10 need not be equal to the exact input ang10. The input ftol
% is an error tolerance factor used to stop the iterative solution. The
% stopping criterion is |f| < ftol, where |f| is the magnitude of the
% constraint function vector. The output matrix xout = [ang1a, ang2a, xpa],
% is a matrix of approximate values for (ang1, ang2, xp) obtained with the
% Newton-Raphson method,where these values are placed into the 3 columns
% of the xout matrix. The output numit is a vector containing the
% number of iterations required at the specified times.

% determine number of time calculations
numtimes = length(t);

% use the Newton-Raphson method to solve the nonlinear constraint
% functions
% for approximate values ang1a (angle of the crankshaft), ang2a, (angle of
% the connecting rod), and xpa (displacement of the slider (piston) at the
% times, t. Also, compute the number of iterations, numit, for each time
% value.

for nn = 1:numtimes
    [xn, iter] = newton_raph(@slider_crank_f, ...
     @slider_crank_J,t(nn),x0, ftol,d1,d2,ang10);
   ang1a(nn) = xn(1);
   ang2a(nn) = xn(2);
   xpa(nn) = xn(3);
   numit(nn) = iter;
   x0 = xn;
end
```

```
%place the angles and displacements as column vectors in the matrix,xout.
xout = [ang1a', ang2a', xpa'];

end
```

42. slider_crank_newton_sim2

```
% slider_crank_newton_sim2
%This script solves the positional equations for the slider-crank problem
%numerically using the Newton-Raphson method. It generates plots for the
% the connecting rod and piston motions, both from the numerical solution
% and from evaluating anaytical expressions.

close all

% default parameters unless otherwise specified
if ~exist('d1','var'), d1 = 5; end              % length of the crank
if ~exist('d2','var'), d2 = 13; end             % length of connecting rod
if ~exist('ftol','var'), ftol = 1e-10; end % tolerance used in newton-
% raphson
if ~exist('tspan','var'),tspan = linspace(0,1,500);end % time interval
if ~exist('x0','var'), x0 = [0, 0, 18];end % initial conditions for ang1,
% the angle of the crank, ang 2, the angle of the connecting rod, and xp,
% the x-coordinate of the piston
if ~exist('ang10','var'), ang10 = 0 ; end % initial guess for ang1

%evaluate exact positional equations (used for comparison with numerical
% solution)
ang1 = -6*tspan + ang10;
ang2 = asin(d1*sin(ang1)/d2);
xp = d1*cos(ang1)+sqrt(d2^2 - (d1^2).*sin(ang1).^2);

% solve positional equations with Newton-Raphson
[xout, numit] = slider_crank_sol2(tspan, x0, d1, d2,ang10, ftol);

% rename outputs
ang1a = xout(:,1);
ang2a = xout(:,2);
xpa = xout(:,3);

% plot exact and approximate results for ang2 and xp
figure (1)
plot(tspan, ang2)
xlabel('time, sec'); ylabel('exact \theta_2');
figure (2)
plot(tspan,ang2a)
xlabel('time, sec'); ylabel('approx \theta_2');
figure(3)
plot(tspan, xp)
xlabel('time, sec'); ylabel('exact X_p');
```

```
figure (4)
plot(tspan, xpa)
xlabel('time, sec'); ylabel('approx X_p');
```

Simulation 4.3: Complete Numerical Solution of a Kinematically Driven Rigid Body System

43. slider_crank_sim3

```
% slider_crank_sim3 is a script to determine the motion variables of a
% kinematically driven slider-crank mechanism, where the angle, ang1, of
% the crankshaft (in rad) is specified as ang1 =ang10 -6*t, and t is the
% time in seconds. The script determines approximate values for ang1, (the
% angle of the crankshaft, ang2 (the angle of the connecting rod), and xp
% (piston displacement) versus time using the Newton-Raphson method. All
% angles are in rad. The script also determines the time derivatives of
% these three variables as a function of time (velocities) as well as
% their second time derivatives (accelerations). The script plots the
% two unknown variables (ang2, xp) and their first and second time
% derivatives versus time.

close all

% default parameters unless otherwise specified

if ~exist('d1','var'), d1 = 5; end        % length of the crank
if ~exist('d2','var'), d2 = 13; end       % length of the connecting rod
if ~exist('ftol','var'), ftol = 1e-10; end % stopping tolerance for the
%Newton-Raphson method
if ~exist('tspan','var'),tspan = linspace(0,1,500);end % calculation
% times
if ~exist('x0','var'), x0 = [0, 0, 18];end % initial conditions for
% (ang1, ang 2, xp)
if ~exist('ang10','var'), ang10 = 0 ; end % starting angle guess

% solve positional equations with the Newton-Raphson method
[xout, numit] = slider_crank_sol2(tspan, x0, d1, d2,ang10, ftol);

% these are the numerical solutions for the generalized coordinates
ang1 = xout(:,1);  % angular position of the crank
ang2 = xout(:,2);  % angular position of the connecting rod
xp = xout(:,3);    % position of the piston

% number of time calculations for determining the velocities and
% accelerations
num = length(tspan);
```

```
for nn = 1: num

% set up linear equations for the velocities and solve
ft = [-6;0;0];
[J] = slider_crank_J(tspan(nn), [ang1(nn), ang2(nn)],d1,d2,ang10);
vels = J\ft;
ang1d(nn) = vels(1); % angular velocity of the crank
ang2d(nn) = vels(2); % angular velocity of connecting rod
xpd(nn) = vels(3);   % velocity of the piston

% set up linear equations for the accelerations and solve
bv1 = 0;
bv2 = d1*cos(ang1(nn))*ang1d(nn)^2 +d2*cos(ang2(nn))*ang2d(nn)^2;
bv3 = d1*sin(ang1(nn))*ang1d(nn)^2 - d2*sin(ang2(nn))*ang2d(nn)^2;
bv = [bv1 ; bv2 ; bv3];

accels = J\bv;

ang1dd(nn) = accels(1); % angular acceleration of the crank
ang2dd(nn) = accels(2); % angular acceleration of the connecting rod
xpdd(nn) = accels(3);   % acceleration of the piston
end

% plot the kinematic variables of the connecting rod and piston
figure (1)
plot(tspan, ang2)
xlabel('time, sec'); ylabel('\theta_2');
figure(2)
plot(tspan, ang2d)
xlabel('time, sec'); ylabel('d\theta_2/dt');
figure (3)
plot(tspan, xp)
xlabel('time, t'); ylabel('X_p');
figure(4)
plot(tspan, xpd)
xlabel('time, sec'); ylabel('V_p');
figure(5)
plot(tspan, ang2dd)
xlabel('time, sec'); ylabel('d^2\theta_2/dt^2');
figure(6)
plot(tspan, xpdd)
xlabel('time, sec'); ylabel('A_p');
```

Simulation 4.4: Foucault Pendulum

44. foucalt (and nested function f_pendulum)

```
function [tout, xout]= foucault(tspan,x0, L,lat, rtol,atol)
% [tout, xout] = foucault(tspan, x0, L, lat, rtol, atol) obtains a
% numerical solution for the motion of the Foucault pendulum using the
% linearized equations of motion. The inputs are tspan and x0, the time
% interval for the calculations and the initial conditions, respectively,
% L, the length of the pendulum, lat, the latitude in degrees, and rtol,
% atol, the relative and absolute error tolerances. The function outputs
% are tout, a column vector containing the times of the calculations and
% xout, a matrix of the variables (x, xdot, y, ydot) for the system,
% calculated at the times in tout and located in the columns of the matrix.

% place error tolerances into options
 options = odeset('RelTol', rtol, 'AbsTol', atol);

% angular velocity of the earth (rad/sec)
omega = 7.272e-5;

% acceleration of gravity in m/s^2
g = 9.81;

% solve equations of motion using ode45
[tout,xout] = ode45(@f_pendulum, tspan, x0,options);

% nested function to generate the equations of motion
function xdot = f_pendulum( t, x)

% create column vector to hold outputs
xdot = zeros(4,1);

% rename inputs to make them easier to read
x1 = x(1); % x
x2 = x(2); % xdot
x3 = x(3); % y
x4 = x(4); % ydot

% form up equations of motion
xdot(1) = x2;
xdot(2) = 2*x4*omega*sind(lat) -g*x1/L;
xdot(3) = x4;
xdot(4) = -2*x2*omega*sind(lat)-g*x3/L;
end

end
```

45. foucault_sim

```
% script foucault_sim
% This script solves for the motion of the Foucault pendulum numerically
% over a long time interval (10,000 sec) to demonstrate the rotation of the
% plane of motion of the pendulum over time. The script also plots the
% normalized total energy over this time interval.

close all

% default parameters unless otherwise specified
if ~exist('L','var'), L = 10; end                  % length of the pendulum (m)
if ~exist('lat','var'), lat = 45; end              % latitude (degrees)
if ~exist('rtol','var'), rtol = 1e-10; end         % relative error tolerance

if ~exist('atol','var'), atol = 1e-10; end         % absolute error tolerance

if ~exist('tspan','var'), tspan = [0, 10000]; end % time interval
if ~exist('x0','var'), x0 = [0.01*L, 0, 0, 0];end % initial conditions for
% (x, xdot, y, ydot)
if ~exist('g', 'var'), g = 9.81; end % acceleration of gravity

% solve equations of motion
[tout, xout] = foucault(tspan,x0, L, lat, rtol, atol);

% rename output variables
t = tout;
x = xout(:,1);   % x-position
vx = xout(:,2);  % vx = xdot = x-velocity
y = xout(:,3);   % y-position
vy = xout(:,4);  % vy = ydot = y-velocity

% obtain limits on (x,y) values
ym = max(y);
xm = max(x);

% plot x versus y motion of the pendulum
figure(1)
plot(x, y)
axis([ -1.1*xm 1.1*xm -1.1*ym 1.1*ym]);
xlabel('x'); ylabel('y');

% calculate normalized energy and plot versus time
E = vx.^2/2 +vy.^2/2 + (g/L)*(x.^2/2 + y.^2/2);
figure(2)
plot(t,E/E(1))
axis([ tspan(1)  tspan(2)  .95 1.05])
xlabel('t, sec'); ylabel('E(t)/E(0)');
```

Simulation 4.5: Matrix Kinematics of a Slider-Crank

46. slider_crank_f_matrix

```
function [f] = slider_crank_f_matrix(t, x, d1, d2, ang10)
% [f] = slider_crank_f_matrix(t, x, d1, d2, ang10) generates a column
% vector of constraint function values, f, for the slider-crank problem.
% The inputs for this function are the time,t, the vector of generalized
% coordinates ( X1, Y1, ang1, X2, Y2, ang 2, X3, Y3, ang3) where (X1, Y1,
% ang1) are the coordinates of the center of mass of the crank and its
% angle (measured positive counterclockwise from the positive inertial
% X-axis), and (X2, Y2, ang2) and (X3, Y3, ang3) are the corresponding
% values for the connecting rod and piston, respectively. d1 and d2 are the
% lengths of the crank and connecting rod, respectively, and ang10 is the
% initial angle of the crank at t = 0.

% constraint functions
f1 = x(1) -d1*cos(x(3))/2;
f2 = x(2) - d1*sin(x(3))/2;
f3 = x(4) - d2*cos(x(6))/2 - x(1) - d1*cos(x(3))/2;
f4 = x(5) - d2*sin(x(6))/2 - x(2) -d1*sin(x(3))/2;
f5 = x(7) - x(4) - d2*cos(x(6))/2;
f6 = x(8) - x(5) - d2*sin(x(6))/2;
f7 = x(8);
f8 = x(9);
f9 = x(3) +6*t -ang10;

% place constraint function values in a column vector
f = [f1; f2; f3; f4; f5; f6; f7; f8; f9];
```

47. slider_crank_J_matrix

```
function [J] = slider_crank_J_matrix(t, x, d1, d2, ang10)
% [J] = slider_crank_J_matrix(t, x, d1, d2, ang10) generates the
% Jacobian matrix, J, for the slider-crank problem as part of a
% matrix-based kinematics solution.
% The inputs for this function are the time,t, the vector of generalized
% coordinates ( X1, Y1, ang1, X2, Y2, ang 2, X3, Y3, ang3) where (X1, Y1,
% ang1) are the coordinates of the center of mass of the crank and its
% angle (measured positive counterclockwise from the positive inertial
% X-axis), and (X2, Y2, ang2) and (X3, Y3, ang3) are the corresponding
% values for the connecting rod and piston, respectively. d1 and d2 are the
% lengths of the crank and connecting rod, respectively, and ang10 is the
% initial angle of the crank at t = 0 (which is not used in this function).

J1 = [ 1 0 d1*sin(x(3))/2 0 0 0 0 0 0];
J2 = [ 0 1 -d1*cos(x(3))/2 0 0 0 0 0 0];
J3 = [ -1 0 d1*sin(x(3))/2 1 0 d2*sin(x(6))/2 0 0 0 ];
J4 = [ 0 -1 -d1*cos(x(3))/2 0 1 -d2*cos(x(6))/2 0 0 0];
```

```
J5 = [ 0  0  0  -1  0  d2*sin(x(6))/2  1  0  0];
J6 = [ 0  0  0  0  -1  -d2*cos(x(6))/2  0  1  0];
J7 = [ 0  0  0  0  0  0  0  1  0];
J8 = [ 0  0  0  0  0  0  0  0  1];
J9 = [ 0  0  1  0  0  0  0  0  0];

J = [J1; J2; J3; J4; J5; J6; J7; J8; J9];
```

48. slider_crank_matrix_sol

```
function [xout, numit] = slider_crank_matrix_sol(t, x0, d1, d2,ang10, ftol)
% [xout, numit] = slider_crank_sol2(t, x0, d1, d2, ang10, ftol) determines
% approximate values for the variables describing the motion of a slider-
% crank problem at the specified times, t. The input x0 = [anga10,
% ang2a0, xpa0 ] are guesses for the initial values of ang1, the angle of
% the crankshaft, ang2, the angle of the connecting rod, and xp, the
% displacement of the slider (piston). The inputs d1, d2, are the lengths
% of the crankshaft and connecting rod, respectively, and ang10 is the
% starting angle for the crankshaft driving constraint. The initial guess
% for anga10 need not be equal to the exact input ang10. The input ftol
% is an error tolerance factor used to stop the iterative solution. The
% stopping criterion is |f| < ftol, where |f| is the magnitude of the
% constraint function vector. The output matrix xout = [ang1a, ang2a, xpa],
% is a matrix of approximate values for (ang1, ang2, xp) obtained with the
% Newton-Raphson method, where these values are placed into the 3 columns
% of the xout matrix. The output numit is a vector containing the
% number of iterations required at the specified times.

% determine number of time calculations
numtimes = length(t);

% use the Newton-Raphson method to solve the nonlinear constraint
% functions
% for approximate values ang1a (angle of the crankshaft), ang2a, (angle of
% the connecting rod), and xpa (displacement of the slider (piston) at the
% times, t. Also, compute the number of iterations, numit, for each time
% value.

for nn = 1:numtimes
   [xn, iter] = newton_raph(@slider_crank_f_matrix, ...
    @slider_crank_J_matrix,t(nn),x0, ftol,d1,d2,ang10);
   x1(nn) = xn(1);
   y1(nn) = xn(2);
   ang1(nn) = xn(3);
   x2(nn) = xn(4);
   y2(nn) = xn(5);
   ang2(nn) = xn(6);
   x3(nn) = xn(7);
   y3(nn) = xn(8);
   ang3(nn) = xn(9);
```

```
   numit(nn) = iter;
   x0 = xn;
end

% place the angles and displacements as column vectors in the matrix, xout.
xout = [x1', y1', ang1', x2',y2',ang2',x3',y3',ang3'];

end
```

49. slider_crank_matrix_sim

```
% slider_crank_matrix_sim
% This script sets up matrix-vector equations for a kinematically driven
% crank-slider for the positions, velocities, and accelerations and solves
% them numerically, starting with the use of the Newton-Raphson method to
% solve the non-linear positional constraints and then solving the
% linear equations for the velocities and accelerations in succession.

close all

% default values unless otherwise specified
if ~exist('d1','var'), d1 = 5; end  % length of crank
if ~exist('d2','var'), d2 = 13; end % length of connecting rod
if ~exist('ftol','var'), ftol = 1e-10; end % function tolerance for use in
% Newton-Raphson method
if ~exist('tspan','var'),tspan = linspace(0,1,500);end % evaluation
% times
% initial conditions for (X1, Y1, ang1, X2, Y2, ang2, X3, Y3, ang3)
if ~exist('x0','var'), x0 = [d1/2, 0, 0, d1+d2/2, 0,0,d1+d2, 0, 0];end
if ~exist('ang10','var'), ang10 = 0 ; end % initial angle of crank

% solve constraint equations numerically with Newton-Raphson method
[xout, numit] = slider_crank_matrix_sol(tspan, x0, d1, d2,ang10,
ftol);

% rename outputs of positional analysis
x1 = xout(:,1);
y1 = xout(:,2);
ang1 = xout(:,3);
x2 = xout(:,4);
y2 = xout(:,5);
ang2 = xout(:,6);
x3 = xout(:,7);
y3 = xout(:,8);
ang3 = xout(:,9);

% determine number of evaluation times
num = length(tspan);

% set up and solve the velocity and acceleration equations
for nn = 1: num
```

```
% set up linear equations for the velocities and solve
xin = [ x1(nn), y1(nn), ang1(nn), x2(nn), y2(nn), ang2(nn), x3(nn),...
    y3(nn), ang3(nn)];
ft = [0;0;0;0;0;0;0;0;-6]; % this is minus ft
[J] = slider_crank_J_matrix(tspan(nn),xin,d1,d2,ang10); % Jacobian
% matrix
% solve linear system
vels = J\ft;

% place outputs into velocity vectors
x1d(nn) = vels(1);
y1d(nn) = vels(2);
ang1d(nn) = vels(3);
x2d(nn) = vels(4);
y2d(nn) = vels(5);
ang2d(nn) = vels(6);
x3d(nn) = vels(7);
y3d(nn) = vels(8);
ang3d(nn) = vels(9);

% set up linear equations for the accelerations and solve
bv1 = -(ang1d(nn)^2)*d1*cos(ang1(nn))/2;
bv2= -(ang1d(nn)^2)*d1*sin(ang1(nn))/2;
bv3 = -(ang2d(nn)^2)*d2*cos(ang2(nn))/2 -(ang1d(nn)^2)*d1*cos(ang1
(nn))/2;
bv4 = -(ang2d(nn)^2)*d2*sin(ang2(nn))/2 -(ang1d(nn)^2)*d1*sin(ang1
(nn))/2;
bv5 = -(ang2d(nn)^2)*d2*cos(ang2(nn))/2;
bv6 = -(ang2d(nn)^2)*d2*sin(ang2(nn))/2;
bv7 = 0;
bv8 = 0;
bv9 = 0;
bv = [bv1 ; bv2 ; bv3; bv4; bv5; bv6; bv7; bv8; bv9];

% solve linear system
accels = J\bv;

% place outputs into acceleration vectors
x1dd(nn) = accels(1);
y1dd(nn) = accels(2);
ang1dd(nn) = accels(3);
x2dd(nn) = accels(4);
y2dd(nn) = accels(5);
ang2dd(nn) = accels(6);
x3dd(nn) = accels(7);
y3dd(nn) = accels(8);
ang3dd(nn) = accels(9);
end

% plot the kinematic variables of the connecting rod and piston
figure (1)
plot(tspan, ang2)
```

```
xlabel('time, t'); ylabel('\theta_2');
figure(2)
plot(tspan, ang2d)
xlabel('time, t'); ylabel('d\theta_2/dt');
figure (3)
plot(tspan, ang2dd)
xlabel('time, t'); ylabel('d^2\theta_2/dt^2');
figure(4)
plot(tspan, x3)
xlabel('time, t'); ylabel('X');
figure(5)
plot(tspan, x3d)
xlabel('time, t'); ylabel('V');
figure(6)
plot(tspan, x3dd)
xlabel('time, t'); ylabel('A') ;
```

Chapter 5

Simulation 5.1: Rolling Without Slipping

50. rolling_disk (and nested function disk_eqs)

```
function [tout, xout] = rolling_disk(tspan, x0, rtol, atol, g, R,r)
% [ tout, xout] = rolling_disk(tspan, x0, rtol, atol, g, R , r) generates
% the equations of motion for a disk rolling without slipping on a
% circular track. The inputs are tspan and x0, the time interval for the
% calculations, and the initial conditions, respectively, rtol and atol,
% the relative and absolute error tolerances used in the numerical
% solution, and g, the acceleration of gravity, R, the radius of the
% circular track, and r, the radius of the disk. The outputs are tout, a
% column vector of the times at which the motion was calculated, and xout,
% which is a matrix that contains the first order system variable and
% its derivative in its columns.

% place error tolerances into options
options = odeset('RelTol', rtol, 'AbsTol', atol);

% solve the system of equations
[tout, xout] = ode45(@disk_eqs, tspan, x0, options);

% nested function for equations of motion of the rolling disk
function xdot = disk_eqs(t,x)

x1 = x(1); % the angle phi
x2 = x(2); % the angular velocity phidot
```

```
% create column vector to hold the outputs
xdot = zeros(2,1);

% generate system of equations
xdot(1) = x2;
xdot(2) = -2*g*sin(x1)/(3*(R-r));
end
end
```

51. rolling_disk_sim

```
% script rolling_disk_sim
% This  script solves for the motion of a circular disk of radius r that
% rolls without slipping on a circular track of radius R. The script plots
% the angular motion versus time and plots both the friction force and
% the maximum allowable frictional force to ensure that slipping has not
% occurred.

close all

% default parameters unless otherwise specified
if ~exist('m','var'), m = 1.0; end     % the mass
if ~exist('g','var'), g = 9.81; end   % the acceleration of gravity
if ~exist('R','var'), R = 2.0; end     % radius of the circular track
if ~exist('r','var'), r = 0.1; end     % radius of the disk
if ~exist('mu','var'), mu = 0.2; end % coefficient of friction
if ~exist('rtol','var'), rtol = 1e-5; end      % relative error tolerance

if ~exist('atol','var'), atol = 1e-10; end     % asolute error tolerance

if ~exist('tspan','var'), tspan = [0, 10]; end % time interval
if ~exist('x0','var'), x0 = [ 0, 0.9]; end      % initial conditions for
% phi, phidot, where phi is the angle that a line from the center of the
% track to the center of mass  of the disk makes with respect to a vertical
% axis.

% solve equations of motion
[tout, xout] = rolling_disk(tspan, x0,rtol,atol,g, R, r);

% rename outputs
t = tout;
phi = xout(:,1);
phidot = xout(:,2);

% compute friction force and normal force
F = m*g*sin(phi)/3;
N = m*g*(cos(phi) +(R-r)*phidot.^2/g);

% plot angle phi versus time and plot the magnitude of the friction force,
% |F|, together with mu times the normal force, N, to examine if slipping
% has occurred.
```

```
figure(1)
plot(t, phi)
xlabel('time, t'); ylabel('\phi');
figure (2)
plot(t, abs(F))
hold on
plot(t, mu*N, '--')
xlabel('time, t'); ylabel(' |F| and \muN');
hold off
```

Chapter 7

Simulation 7.1: Vibration of Masses on a Stretched String

52. vibrating_string_anim

```
% vibrating_string_anim script
% This script solves for the free, vertical vibrating of a series of masses
% that are held together by an elastic string and where the ends of the
% string are held fixed. The natural frequencies (eigenvalues) and
% eigenfunctions of the system are calculated and then used to express the
% solution to the problem as a superposition of modal coordinates. The
% system of masses is given a set of initial displacements and velocities
% and the solution for the subsequent free vibration motion is animated.

close all

% default input parameters unless otherwise specified
if ~exist('m','var'), m = 2.0; end      % all masses have same value, m
if ~exist('ke','var'), ke = 10; end     % effective stiffness of the string
if ~exist('num', 'var'), num = 5; end % number of masses (coordinates)
if ~exist('q0','var'), q0 = [1, 1.6,2,1.6,1]; end % initial displacements
if ~exist('v0','var'), v0 = [0, 0, 0, 0, 0]; end  % initial velocities
if ~exist('tmax','var'), tmax = 20; end  % maximum time for calculations
if ~exist('dt','var'), dt = .1; end       %time interval

% check to ensure initial condition vectors are compatible with number of
% masses
lq0 = length(q0);
lv0 = length(v0);
if (lq0 ~= num || lv0 ~= num)
  error('initial conditions not consistent with number of coordinates')
end

% form mass matrix
mv = ones(1,num)*m;
M = diag(mv);
```

```
% form tridiagonal stiffness matrix
vd = ones(1,num)*(2*ke);
vu = ones(1,num-1)*(-ke);
vl = vu;
K = diag(vd) + diag(vu,1) + diag(vl,-1);

% find eigenvectors and eigenvalues
[V,D] = eig(K,M);

% determine natural frequencies from eigenvalues (column vector)
w = sqrt(diag(D));

% add very small constant to frequencies in case one is zero
w = w + eps;

% calculate natural mode initial conditions
% note transpose on q0 and v0 to make them column vectors
z0 = V.'*M*q0';
zd0 = V.'*M*v0';

% initialize generalized coordinate vector
q = zeros(num,1);

% set up plot of the string for animation
h = figure(1);
clf(h);
x = 0:num+1; %set up x-coordinates of masses, including fixed end points
qp = zeros(1,7); % create empty vector to hold generalized coordinates for
% plotting the string
ph = plot(x, qp, 'k');  % set up and hold plot in figure window
axis([ 0 num+1 -3 3]);

a = 0:pi/100:2*pi;
xcircle = 0.05*cos(a);
ycircle = 0.05*sin(a);

hold on
% generate small point masses
mass1 = fill(xcircle +1,ycircle,'b');
mass2 = fill(xcircle +2,ycircle,'b');
mass3 = fill(xcircle +3,ycircle,'b');
mass4 = fill(xcircle +4,ycircle,'b');
mass5 = fill(xcircle +5,ycircle,'b');

% calculate motion and animate
for t = 0:dt:tmax
  z = z0.*cos(w*t)+zd0.*sin(w*t)./w; % modal coordinates (column vector)
    q = V*z;             % generalized coordinates (column vector)

   qp = [0 q.' 0];   % include string endpoints for plotting
   set(ph,'YData',qp); % update y-data in the plot to show string
```

```
  % update drawing of masses
  set(mass1,'YData', ycircle +q(1));
  set(mass2,'YData', ycircle +q(2));
  set(mass3,'YData', ycircle +q(3));
  set(mass4,'YData', ycircle +q(4));
  set(mass5,'YData', ycircle +q(5));

 pause(.07);
end
hold off
```

Simulation 7.2: Free Vibrations of an Underdamped Two Degree of Freedom System

53. free_vibs_2D

```
% script free_vibs_2D
% This script solves for the free vibrations of a damped, two degree of
% freedom system where two masses are connected to each other and to two
% fixed supports by springs and dashpots. It sets up the problem as a
% system of first order differential equations, solves for the eigenvalues
% and eigenvectors of the system, which are then used to find the constants
% in the homogeneous solution that satisfy the initial conditions. The
% script plots the displacement versus time of the first mass.
% NOTE ****** This solution assumes that the system is an underdamped
% system without any rigid body modes *******

close all

% default inputs unless otherwise specified
if ~exist('m1','var'), m1 = 2.0; end  % mass 1
if ~exist('m2','var'), m2 = 4.0; end  % mass 2
if ~exist('k1','var'), k1 = 2; end    % spring constant 1
if ~exist('k2','var'), k2 = 3; end    % spring constant 2
if ~exist('k3','var'), k3 = 2; end    % spring constant 3
if ~exist('c1','var'), c1 = 0.25; end % damping coefficient 1
if ~exist('c2','var'), c2 = 0.5; end  % damping coefficient 2
if ~exist('c3','var'), c3 = 0.25; end % damping coefficient 3
if ~exist('x0','var'), x0 = [1, 0, 0, 0]; end % initial conditions for
% [x1, x2, x1dot, x2dot]
if ~exist('tmax','var'), tmax = 50; end  % maximum time of calculations

% set up times and time step
t = linspace(0, tmax, 500);
dt = t(2) -t(1);
```

```
% form mass, stiffness, and damping matrices
M = [ m1 0; 0 m2];
K = [(k1+k2) -k2 ; -k2 (k2+k3)];
C = [(c1+c2) -c2; -c2 (c2+c3)];

% set up coefficient matrices for the first order system A*xdot = B
I = [ 1 0; 0 1];
Z = [0 0; 0 0];
A = [ I Z ; Z M];
B = [ Z I ; -K -C];

% solve for the eigenvectors and eigenvalue of the first order system
[V,D] = eig(B,A);

% solve the system [V]{N} = {x0} for the coefficients {N}
N = V\x0';

% initialize nn (serves as an indice value for the output vectors)
nn = 1;
for tt = 0:dt:tmax

  % form diagonal matrix with eigenvalues
  vd = [ exp(D(1,1)*tt) exp(D(2,2)*tt) exp(D(3,3)*tt) exp(D(4,4)*tt)];
  G = diag(vd);

  % free vibration solution
  x = V*G*N;

  % place displacements and velocities into output vectors
  x1(nn) = x(1);
  x2(nn) = x(2);
  vx1(nn) = x(3);
  vx2(nn) = x(4);
  nn = nn+1;

end
% plot displacement of mass 1. Take real part to eliminate any small
% imaginary components
plot(t, real(x1))
xlabel('time, t'); ylabel('x_1');
```

Simulation 7.3: Approximate Free Vibrations of Underdamped Systems

54. free_vibs_2D_approx

```
% script free_vibs_2D_approx
% This script solves for the free vibrations of a damped, two degree of
```

```
% freedom system where two masses are connected to each other and to two
% fixed supports by springs and dashpots. It solves the problem
% by assuming the solution can be expanded in terms of the undamped natural
% modes of the system and then forces the damping matrix in the modal
% coordinates to be diagonal by ignoring the off-diagonal terms. This
% decouples the modal equations but is only an approximation. The
% script plots the displacement versus time of the first mass.
% NOTE ****** This solution assumes that the system is an underdamped
% system without any repeated natural frequencies *******

close all

% default inputs unless otherwise specified
if ~exist('m1','var'), m1 = 2.0; end      % mass 1
if ~exist('m2','var'), m2 = 4.0; end      % mass 2
if ~exist('k1','var'), k1 = 2; end        % spring constant 1
if ~exist('k2','var'), k2 = 3; end        % spring constant 2
if ~exist('k3','var'), k3 = 2; end        % spring constant 3
if ~exist('c1','var'), c1 = 0.25; end     % damping coefficient 1
if ~exist('c2','var'), c2 = 0.5; end      % damping coefficient 2
if ~exist('c3','var'), c3 = 0.25; end     % damping coefficient 3
if ~exist('x0','var'), x0 = [1, 0,]; end % initial conditions for x1, x2
if ~exist('v0','var'), v0 = [0, 0,]; end % initial conditions for vx1, vx2
if ~exist('tmax','var'), tmax = 50; end  % maximum time for calculations

% set up times and time step
t=linspace(0, tmax, 500);
dt = t(2) -t(1);

% form mass, stiffness, and damping matrices
M = [ m1 0; 0 m2];
K = [(k1+k2) -k2 ; -k2 (k2+k3)];
C = [(c1+c2) -c2; -c2 (c2+c3)];

% solve eigenvalue problem for an undamped system ([K] - w^2[M]){u} =0
[V,D] = eig(K,M);

% calculate undamped natural frequencies
w1 = sqrt(D(1,1));
w2 = sqrt(D(2,2));

% calculate damping matrix for natural modes
ct = V.'*C*V;

% define damping factors in terms of diagonal components of ct, ignoring
% the off-diagonal terms
d1 = ct(1,1)/(2*w1);
d2 = ct(2,2)/(2*w2);

% calculate the damped natural frequencies
w1d = w1*sqrt(1-d1^2);
w2d = w2*sqrt(1-d2^2);
```

```
% initialize nn (serves as an indice value for the output vectors)
nn = 1;
for tt = 0:dt:tmax
  % place underdamped modal solutions into matrices
  vd1 = [ exp(-d1*w1*tt)*(cos(w1d*tt)+d1*(w1/w1d)*sin(w1d*tt)), ...
    exp(-d2*w2*tt)*(cos(w2d*tt) +d2*(w2/w2d)*sin(w2d*tt))];
  G1 = diag(vd1);

  vd2 = [ exp(-d1*w1*tt)*sin(w1d*tt)/w1d , exp(-d2*w2*tt)*sin
(w2d*tt)/w2d];
  G2 = diag(vd2);
  % form up free vibration solution
  x = V*G1*V.'*M*x0.' + V*G2*V.'*M*v0.';
  x1(nn) = x(1); % displacement of mass 1
  x2(nn) = x(2); % displacement of mass 2

  nn = nn+1;

end
% plot motion of mass 1
plot(t, x1, '--')
xlabel('time, t'); ylabel('x_1');
```

Chapter 8

Simulation 8.1: Fixed Gyro-Pendulum (Spherical Pendulum)

55. fixed_gyro_pend

```
function [tout, xout] = fixed_gyro_pend(tspan, x0, R , L, m, g,rtol,atol)
% [tout, xout] = fixed_gyro_pend(tspan, x0, R, L ,m, g, rtol, atol) solves
% the equations of motion for a fixed gyro-pendulum (spherical pendulum).
% The inputs are tspan, the time interval for the calculations, x0, the
% initial conditions for (theta, phi, thetadot, phidot),R, the radius of
% the disk, L, the length of the weightless arm, m, the mass of the disk,
% g, the acceleration of gravity,and rtol, atol, the relative and absolute
% error tolerances. The outputs are tout, a column vector of the evaluation
% times, and xout, a matrix that contains the system variables evaluated
% at those times and placed into columns of that matrix.

% place error tolerances into options
options = odeset('RelTol', rtol, 'AbsTol', atol);

% define terms appearing in equations of motion
r1 = 1+R^2/(4*L^2);
r2 = 1-R^2/(4*L^2);
r3 = R^2/(2*L^2);
```

```
% solve Euler equations for angular motion
[tout, xout] = ode45(@f_gyro_p_eqs, tspan, x0, options);

% nested function for the fixed gyro-pendulum
function xdot = f_gyro_p_eqs(t,x)

% create column vector to hold outputs
xdot = zeros(4,1);

% rename variables
theta = x(1);
phi = x(2);
thdot = x(3);
phdot = x(4);

M = 0;  %M(t) = 0, i.e. no driving moment in this simulation (see text)

%another term appearing in equations of motion
m1 = r1*(sin(theta)).^2 + r3*(cos(theta)).^2;

% define the system of equations
xdot(1) = thdot;
xdot(2) = phdot;
xdot(3) = (-g*sin(theta)/L + r2*(phdot).^2.*sin(theta).*cos(theta))/
r1;
xdot(4) = (M/(m*L^2) -2*r2*thdot.*phdot.*sin(theta).*cos(theta))/m1;
end
end
```

56. fixed_gyro_pend_sim

```
% script fixed_gyro_pend_sim
% This script simulates the motion of a bar-disk assembly that is pinned
% at one end to a support that is free to rotate about a vertical axis.
% The assembly acts like a spherical pendulum since the disk is rigidly
% fixed to the bar. The output of the script is an animation of the motion
% of the end of the bar-disk assembly.

close all

% default input parameters unless otherwise specified
if ~exist('R','var'), R = 0.1; end % radius of the disk
if ~exist('L','var'), L = 0.5; end % length of the weightless bar
if ~exist('m', 'var'), mass = 1; end  %mass of the disk
if ~exist('g', 'var'), g = 9.81; end % acceleration of gravity
if ~exist('tspan','var'), tspan = [0, 11]; end % time interval
if ~exist('x0','var'), x0 = [45*pi/180, 0, 0, 3]; end % initial conditions
% for [theta, phi, thd, phd]
if ~exist( 'rtol','var'), rtol = 1e-6;end % relative error tolerance

if ~exist('atol','var'), atol = 1e-6; end % absolute error tolerance
```

```
if ~exist('marker','var'), marker = 'on';end % marker to show the path of
% motion of the end of the bar in the animation

% solve equations of motion
[tout, xout] = fixed_gyro_pend(tspan, x0, R ,L, mass, g, rtol, atol);

% rename the evaulation times
t = tout;
num = length(t);

% rename outputs as angles and their derivatives
theta = xout(:,1);
phi = xout(:,2);
thd = xout(:,3);
phd = xout(:,4);

% calculate location of the end of the bar-disk assembly
xp = L*sin(theta).*cos(phi);
yp = L*sin(theta).*sin(phi);
zp = L*cos(theta);
% make these row vectors
xp = xp.';
yp = yp.';
zp = zp.';

% ---------------- animation ----------------------------%

% set up limits of animation region

% find max and min values of (x,y,z)
mxx = max(xp); mxy = max(yp); mxz = max(zp);
mnx = min(xp); mny = min(yp); mnz = min(zp);
%determine largest absolute values of (x,y,z)
maxx = max(abs([mxx,mnx]));
maxy = max(abs([mxy, mny]));
maxz = max(abs([mxz, mnz]));
% take 1.1 times largest of all three axes as limits of the plot axes
% which are set later below
r = 1.1*max([maxx, maxy, maxz]);

% clear any existing figure
clf

% generate default marker object with starting values
r1 = line(xp(1), yp(1),zp(1), 'Marker', '.', 'MarkerSize', 0.5, ...
    'Color', 'b');

% set up 3-D plotting region with X-Y-Z axes oriented and viewed as shown
% in the text and label the axes
view(3)
grid on
```

```
set(gca,'ZDir', 'reverse')
set(gca,'YDir', 'reverse')
view(33,15)
xlabel('x')
ylabel('y')
zlabel('z')
axis([-r, r, -r, r, -r, r]);

% generate default bar object with starting values
r2 = line([0,xp(1)],[0,yp(1)],[0, zp(1)], 'LineWidth', 2, 'Color',
'k');

% perform animation over times previously calculated
for i = 1:length(t)
  % update plot of bar assembly
  set(r2,'XData',[0, xp(i)], 'YData', [0, yp(i)], 'ZData', [0, zp(i)]);
  % plot marker that follows the free end of the bar assembly
  if strcmp(marker, 'on')
     set(r1,'XData', xp(1:i), 'YData', yp(1:i), 'ZData', zp(1:i));
  end

 pause(0.01);
end
```

Simulation 8.2: Gyro-Pendulum and Euler Angles

57. gyro_pend

```
function [tout, xout] = gyro_pend(tspan, x0, R , L, m,g,rtol,atol)
% [tout, xout] = gyro_pend( tspan, x0, R , L, m, g, rtol, atol) solves the
% equations of motion for the gyro-pendulum. The inputs are tspan, the
% times at which the motion is evaluated, x0, a vector of initial
% conditions for (w1, w2, w3, ph, th, psi), R, the radius of the disk, L,
% the length of the weightless arm, m, the mass, g, the acceleration of
% gravity, and rtol, atol, the relative and absolute error tolerances.
% The outputs are tout, a column vector of times at which the motion is
% evaluated and xout, a matrix containing the system variables
% evaluated at those times, and located in the columns of the matrix.

% place error tolerances into options
options = odeset('RelTol', rtol, 'AbsTol', atol);

% define moments of inertia
I1 = m*(L^2 + R^2/4);
I2 = I1;
I3 = m*R^2/2;
```

```
% solve system of equations
[tout, xout] = ode45(@gyro_p_eqs, tspan, x0, options);

% nested function to define equations of motion
function xdot = gyro_p_eqs(t,x)

% create column vector for outputs
xdot = zeros(6,1);

% rename inputs
w1 = x(1); % omega1 angular velocity
w2 = x(2); % omega2 angular velocity
w3 = x(3); % omega3 angular velocity
ph = x(4); % angle phi
th = x(5); % angle theta
psi =x(6); % angle psi

% M(t) = 0, i.e. no driving moment (see text)

% generate system of equations
xdot(1) = (I2 - I3).*w2.*w3/I1 -m*g*L.*sin(th).*sin(psi)./I1;
xdot(2) = (I3 - I1).*w1.*w3./I2 -m*g*L.*sin(th).*cos(psi)./I2;
xdot(3) = (I1 - I2).*w1.*w2./I3;
xdot(4) =(-w1.*cos(psi) + w2.*sin(psi))./sin(th);
xdot(5) = w1.*sin(psi) + w2.*cos(psi);
xdot(6) = (w1.*cos(th).*cos(psi) - w2.*cos(th).*sin(psi))./sin(th) + w3;
end
end
```

58. gyro_pend_sim

```
% script gyro_pend_sim
% This script simulates the motion of a bar-disk assembly that is pinned
% at one end to a support that is free to rotate about a vertical axis.
% The disk is free to spin so the assembly acts as a gyro-pendulum. The
% script solves for the motion of the system using Euler's equation,
% written in terms of the angular velocities, and the time derivatives of
% Euler angles (the 3-2-3 set) written in terms of the angles and the
% angular velocities. The output of the script is an animation of the
% motion of the end of the arm of the gyro-pendulum.

close all

% default input parameters unless otherwise specified

if ~exist('R','var'), R = 0.1; end   % radius of the disk
if ~exist('L','var'), L = 0.5; end   % length of the weightless arm
if ~exist('m','var'), m = 1; end     % mass
if ~exist('g','var'), g = 9.81; end % acceleration of gravity
if ~exist('tspan','var'), tspan = linspace(0, 10, 1000); end %times
% initial values for [phi, theta, psi, phidot, thetadot, psidot]
```

```
if ~exist('x0','var'), x0 = [0, 90*pi/180, 0, 0, 0 ,200]; end
if ~exist( 'rtol','var'), rtol = 1e-6;end % relative error tolerance
if ~exist('atol','var'), atol = 1e-6; end % absolute error tolerance
if ~exist('marker','var'), marker = 'on';end % marker for animation

% change to new coordinates x0 = [ w1, w2, w3, phi, theta, psi]

w10 = -x0(4)*sin(x0(2))*cos(x0(3)) + x0(5)*sin(x0(3));
w20 = x0(4)*sin(x0(2))*sin(x0(3)) +x0(5)*cos(x0(3));
w30 = x0(4)*cos(x0(2)) + x0(6);

% change initial conditions to match new coordinates
xs = [ w10 , w20, w30, x0(1), x0(2), x0(3)];

% solve equations of motion
[tout, xout] = gyro_pend(tspan, xs, R ,L, m, g, rtol, atol);

% rename times
t = tout;
num = length(t); % number of calculation times

% rename outputs as angles and their derivatives
w1 = xout(:,1);
w2 = xout(:,2);
w3 = xout(:,3);
phi = xout(:,4);
theta = xout(:,5);
psi = xout(:,6);

% calculate location of the end of the bar-disk assembly
xp = L*sin(theta).*cos(phi);
yp = L*sin(theta).*sin(phi);
zp = L*cos(theta);
% make these row vectors
xp = xp.';
yp = yp.';
zp = zp.';

%---------------- animation ----------------------------%

% set up limits of animation region

% find max and min values of (x,y,z)
mxx = max(xp); mxy = max(yp); mxz = max(zp);
mnx = min(xp); mny = min(yp); mnz = min(zp);
%determine largest absolute values of (x,y,z)
maxx = max(abs([mxx,mnx]));
maxy = max(abs([mxy, mny]));
maxz = max(abs([mxz, mnz]));
% take 1.1 times largest of all three axes as limits of the plot axes
```

```
% which are set later below
r = 1.1*max([maxx, maxy, maxz]);

% clear any existing figure
clf

% generate default marker object with starting values
r1 = line(xp(1), yp(1),zp(1), 'Marker', '.', 'MarkerSize', 0.5, ...
    'Color', 'b');

% set up 3-D plotting region with X-Y-Z axes oriented and viewed as shown
% in the text and label the axes
view(3)
grid on
set(gca,'ZDir', 'reverse')
set(gca,'YDir', 'reverse')
view(33,15)
xlabel('X')
ylabel('Y')
zlabel('Z')
axis([-r, r, -r, r, -r, r]);

% generate default bar object with starting values
r2 = line([0,xp(1)], [0,yp(1)], [0, zp(1)], 'LineWidth', 2, 'Color', 'k');

% perform animation over times previously calculated
for i = 1:length(t)
  % update plot of bar assembly
  set(r2,'XData', [0, xp(i)], 'YData', [0, yp(i)], 'ZData', [0, zp(i)]);
  % plot marker that follows the free end of the bar assembly
  if strcmp(marker, 'on')
     set(r1,'XData', xp(1:i), 'YData', yp(1:i), 'ZData', zp(1:i));
  end

 pause(0.01);
end
```

Simulation 8.3: Gyro-Pendulum and Euler Parameters

59. gyro_pend_ep

```
function [tout, xout] = gyro_pend_ep(tspan, x0, R , L, m, g, rtol, atol)
% [tout, xout] = gyro_pend_ep( tspan, x0, R , L, m, g, rtol, atol) solves
% the equations of motion for the gyro-pendulum using Euler's equations,
% written in terms of the angular velocities, in conjunction with the
% derivatives of the Euler parameters, written in terms of those
```

```
% parameters
% and the angular velocities. The inputs are tspan,the times at which the
% motion is calculated, x0, the initial conditions for the system
% variables
% (w1, w2, w3, e0, e1, e2, e3), R, the radius of the disk, L, the length of
% the weightless arm, m, the mass, g, the acceleration of gravity, and
% rtol, atol, the relative and absolute error tolerances. The outputs are
% tout, a column vector of the times at which the solution is calculated,
% xout, a matrix which contains the system variables evaluated at those
% times and located in the columns of the matrix.

% place error tolerances into options
options = odeset('RelTol', rtol, 'AbsTol', atol);

% define moments of inertia
I1 = m*(L^2 + R^2/4);
I2 = I1;
I3 = m*R^2/2;

% solve equations of motion
[tout, xout] = ode45(@gyro_p_ep_eqs, tspan, x0, options);

% nested function for the gyro-pendulum in terms of Euler parameters
function xdot = gyro_p_ep_eqs(t,x)

% create column vector for the outpus
xdot = zeros(7,1);

% rename inputs
% angular velocities
w1 = x(1); % omega1
w2 = x(2); % omega2
w3 = x(3); % omega3
% Euler parameters
e0 = x(4); % theta0
e1 = x(5); % theta1
e2 = x(6); % theta2
e3 = x(7); % theta3

% M(t) = 0, i.e. no driving moment in this simulation (see text)

% form up system of equations
xdot(1) = (I2 - I3).*w2.*w3/I1 -m*g*L.*(2*(e2*e3 +e0*e1))./I1;
xdot(2) = (I3 - I1).*w1.*w3./I2 -m*g*L.*(-2*(e1*e3 -e0*e2))./I2;
xdot(3) = (I1 - I2).*w1.*w2./I3;
xdot(4) = 0.5*(-w1*e1 -w2*e2 -w3*e3);
xdot(5) = 0.5*(w1*e0 -w2*e3 +w3*e2);
xdot(6) = 0.5*(w1*e3+w2*e0-w3*e1);
xdot(7) = 0.5*(-w1*e2+w2*e1 +w3*e0);
end
end
```

60. gyro_pend_ep_sim

```
% script gyro_pend_ep_sim
% This script simulates the motion of a bar-disk assembly that is pinned
% at one end to a support that is free to rotate about a vertical axis.
% The disk is free to spin so the assembly acts as a gyro-pendulum. The
% script solves for the motion of the system using Euler's equation,
% written in terms of the angular velocities, and the time derivatives of
% the Euler parameters, written in terms  of those parameters and the
% angular velocities. The output of the script is an animation of the
% motion of the end of the arm of the gyro-pendulum.

close all

% default input parameters unless otherwise specified
if ~exist('R','var'), R = 0.1; end  % radius of the disk
if ~exist('L','var'), L = 0.5; end  % length of the massless arm
if ~exist('m','var'), m = 1; end      % the mass
if ~exist('g', 'var'), g = 9.81; end % acceleration of gravity
if ~exist('tspan','var'), tspan = linspace(0,10,1000); end % times
% initial conditions for the 3-2-3 Euler angles and their derivatives
% [phi, theta, psi, phidot, thetadot, psidot]
if ~exist('x0','var'), x0 = [0, 90*pi/180, 0, 0, 0 ,200]; end
if ~exist( 'rtol','var'), rtol = 1e-3;end     % relative error tolerance
if ~exist('atol','var'), atol = 1e-6; end     % absolute error tolerance
if ~exist('marker','var'), marker = 'on';end  % marker for animation

% change to new coordinates [ w1, w2, w3, e0, e1, e2,e3]
% initial angular velocities
w10 = -x0(4)*sin(x0(2))*cos(x0(3)) + x0(5)*sin(x0(3));
w20 = x0(4)*sin(x0(2))*sin(x0(3)) +x0(5)*cos(x0(3));
w30 = x0(4)*cos(x0(2)) + x0(6);
% initial Euler parameters
e0 = cos((x0(1)+x0(3))/2)*cos(x0(2)/2);
e1 = sin((x0(3)-x0(1))/2)*sin(x0(2)/2);
e2 = cos((x0(3)-x0(1))/2)*sin(x0(2)/2);
e3 = sin((x0(1)+x0(3))/2)*cos(x0(2)/2);

% initial conditions for new coordinates
xs = [ w10 , w20, w30, e0, e1, e2, e3];

% solve equations of motion
[tout, xout] = gyro_pend_ep(tspan, xs, R ,L, m, g, rtol, atol);

% rename times
t = tout;
% number of time calculations
num = length(t);

% rename outputs as anglular velocities and their derivatives
% angular velocities
w1 = xout(:,1); % omega1
```

```
w2 = xout(:,2); % omega2
w3 = xout(:,3); % omega3
% Euler parameters
e0 = xout(:,4); % theta0
e1 = xout(:,5); % theta1
e2 = xout(:,6); % theta2
e3 = xout(:,7); % theta3

% calculate location of the end of the bar-disk assembly
xp = 2*L*(e1.*e3+e0.*e2);
yp = 2*L*(e2.*e3 -e0.*e1);
zp = L*(1-2*e1.^2 -2*e2.^2);
% make these row vectors
xp = xp.';
yp = yp.';
zp = zp.';

% --------------- animation ---------------------------%

% set up limits of animation region

% find max and min values of (x,y,z)
mxx = max(xp); mxy = max(yp); mxz = max(zp);
mnx = min(xp); mny = min(yp); mnz = min(zp);
% determine largest absolute values of (x,y,z)
maxx = max(abs([mxx,mnx]));
maxy = max(abs([mxy, mny]));
maxz = max(abs([mxz, mnz]));
% take 1.1 times largest of all three axes as limits of the plot axes
% which are set later below
r = 1.1*max([maxx, maxy, maxz]);

% clear any existing figure
clf

% generate default marker object with starting values
r1 = line(xp(1), yp(1),zp(1), 'Marker', '.', 'MarkerSize', 0.5, ...
    'Color', 'b');

% set up 3-D plotting region with X-Y-Z axes oriented and viewed as shown
% in the text and label the axes
view(3)
grid on
set(gca,'ZDir', 'reverse')
set(gca,'YDir', 'reverse')
view(33,15)
xlabel('X')
ylabel('Y')
zlabel('Z')
axis([-r, r, -r, r, -r, r]);
```

```
% generate default bar object with starting values
r2 = line([0,xp(1)],[0,yp(1)],[0, zp(1)], 'LineWidth', 2, 'Color',
'k');

% perform animation over times previously calculated
for nn = 1:length(t)
  % update plot of bar assembly
  set(r2,'XData',[0, xp(nn)], 'YData', [0, yp(nn)], 'ZData', ...
    [ 0, zp(nn)]);
  % plot marker that follows the free end of the bar assembly
  if strcmp(marker, 'on')
    set(r1,'XData', xp(1:nn), 'YData', yp(1:nn), 'ZData', zp(1:nn));
  end

 pause(0.01);
end
```

Chapter 9

Simulation 9.1: Longitudinal Elastic Wave Propagating in a Rod

61. wave_1D_sim

```
% wave_1D_sim script
% This script generates a finite difference solution to the 1-D wave
% equation that simulates the longitudinal waves propagating in an
% elastic
% rod. The rod is driven by specifying the displacement as a function of
% time at one end and holding the other end fixed. An animation of the
% traveling waveform is generated. close all

% default input parameters unless otherwise specified
if ~exist('dx','var'), dx = 0.1; end        % element length in mm
if ~exist('dt','var'), dt = 0.01; end       % time step in micro-sec
if ~exist('c','var'), c = 5; end            % wave speed in mm/micro-sec
if ~exist('L','var'), L = 100; end          % rod length in mm
if ~exist('tmax','var'), tmax = 60; end     % max time in micro-sec
if ~exist('Nframe', 'var'), Nframe = 10; end  % number of frames to skip
if ~exist('tplot', 'var'), tplot = 0.01; end  % time delay in animation

% calculate number of locations in rod where displacements are to be
% found
% and the r parameter
K = round(L/dx) +1; % number of points so that x(1) = 0, x(K) = L.
new_dx = L/(K-1);
if round(L/dx) ~= L/dx
  display(['To make K an integer, setting dx =', num2str(new_dx)]);
end
```

```
r = c*dt/new_dx; % Courant number
if r >1
  error('Unstable. The r value is too large');
  abort;
end

% initialize arrays with zero displacements
past = zeros(1,K);
current = zeros(1,K);
future = zeros(1,K);

% perform updating to calculate future array of displacements at all
% specified times

for t = 0:dt:tmax

  % perform updating over all interior points
  future(2:K-1) = r^2*(current(1:K-2) +current(3:K))+ ...
   2*(1-r^2)*current(2:K-1)-past(2:K-1);

  % specify driving and boundary conditions
  future(1) = sin(t).*(t <= pi);
  future(K) = 0 ;

 % initialize arrays in preparation for updating at next time
   past = current;
  current = future;

%plot the current displacements in the rod after every Nframe iterations
% in the loop and pause tplot seconds to allow the plot to be displayed
if mod(t/dt, Nframe) == 0
  plot([0:new_dx:L], current, 'b', 'LineWidth',1.5)
  axis([0 L -2 2])
  pause(tplot)
end
end
```

Appendix C

62. newton_m

```
function [xzero, iter] = newton_m(funh, dfunh, x0, tol)

% internal variables
dtol = 1e-10; % tolerance on derivative
```

```
maxval=10000; % maximum size of function (used to detect divergence);
Niter = 100;  % maximum number of iterations

% initialize iteration count and initial approximation
NN = Niter;
xx = x0;

% perform iterative solution
while NN > 0

% check to see if derivative is too small
df = feval(dfunh,xx);
if abs(df) < dtol
  error('derivative is zero');
end

% if not, update approximation
xn = xx - feval(funh, xx)/df;

% check to see if change of approximation is less than tolerance. If so,
% return current value as the solution, number of iterations
if abs(xn - xx) < tol
  xzero = xn;
  iter = Niter -NN;
  return;
end

% if not, check to make sure solution is not diverging
 if abs(feval(funh,xx)) > maxval
   iter = 100-NN;
   disp(['iterations = ', num2str(iter)]);
   error('solution diverges');
 end

 % if no divergence, update iteration count and approximation
 NN = NN-1;
 xx = xn;
end

% if NN goes to zero then number of iterations exceeded
disp([num2str(Niter), 'iterations exceeded']);
error('no convergence');
abort;
end
```

Appendix E

Simulation E.1: Numerical Evaluation of the Impulse Response

63. s_space

```
function y = s_space(xstart, xend, num)
% S_SPACE(XSTART,XEND, NUM) generates num evenly spaced sampled
% values from xstart to (xend - dx), where dx is the sample
% spacing. This is useful in FFT analysis where we generate
% sampled periodic functions. Example: generate 1000
% sampled frequencies from 0 to 100MHz via f = s_space(0,100,1000);
% In this case the last value of f will be 99.9 MHz and the
% sampling interval will be 100/1000 = 0.1 MHz.

ye = linspace(xstart, xend, num+1);
y = ye(1:num);
```

64. Ftrans

```
function [y] = Ftrans(x,dt)
% Ftrans(x,dt) computes forward FFT of x with sampling time interval dt
% Ftrans approximates the Fourier transform where the integrand of the
% transform is x*exp(-2*pi*i*f*t).

y = dt*fft(x);
end
```

65. IFtrans

```
function [y] =IFtrans(x, dt)
% IFtrans(x,dt)computes the inverse FFT of x,for a sampling time
% interval
% dt. IFtrans assumes the integrand of the inverse transform is given by
% x*exp(2*pi*i*f*t)
% The first half of the sampled values of x are the spectral components for
% positive frequencies ranging from 0 to the Nyquist frequency 1/(2*dt)
% The second half of the sampled values are the spectral components for
% the corresponding negative frequencies. If these negative frequency
% values are set equal to zero then to recover the inverse FFT of x we must
% replace x(1) by x(1)/2 and then compute 2*real(IFourierT(x,dt)).

y = (1/dt)*ifft(x);
end
```

66. impulse_response

```
% script impulse_response

close all

m = 1; k = 2; c = 0.1; % give system parameters
fs = 4.0;  % choose sampling frequency
f = s_space(0, fs, 1024);  % create N = 1024 frequencies over (0, fs)
G = 1./(k +2*pi*1i*f*c - m*(2*pi*f).^2);  % evaluate transfer function
G = [ G(1:512), zeros(1,512) ]; %place zeros in upper part of spectrum
dt = 1/fs; % calculate sampling time interval
T = 1024*dt;  % calculate time window
t = s_space(0, T, 1024);  % create 1024 time values over (0, T)
G(1) = G(1)/2; % take one half of the zero frequency component value
g = 2*real(IFtrans(G,dt)); %take 2x real part of inverse Fourier transform
plot(t(t<25), g(t<25)) % plot the impulse response function
xlabel('time, t sec'); ylabel('g(t)');
wn = sqrt(k/m);  % undamped natural frequency
d = c/(2*sqrt(k*m));  % damping factor
wd = wn*sqrt(1-d^2);     % damped natural frequency
ge = (1/(m*wd))*exp(-wn*d*t).*sin(wd*t);% exact impulse response
% function
hold on
plot(t(t< 25), ge(t<25), '--');   %plot exact response comparison
hold off
```

Index

A

Absolute error tolerances (atol), 78, 130, 161, 210, 515, 635, 637, 639, 641, 642, 644–646, 653–656, 661, 664, 666, 675, 676, 681, 682, 689, 691, 693, 695, 696

Addition theorem (angular velocities), 245, 246

Adjugate (matrix), 584

Allowed differential displacements, 124, 126

Analytical mechanics, vi, 105, 107

Angular acceleration matrix, 286, 318

Angular accelerations, 240, 241, 246, 248, 251–253, 255, 257, 258, 269, 270, 285, 286, 289, 298, 318, 324, 325, 329, 331, 333, 511, 512, 514, 542, 667, 669, 674

Angular impulse, 182–185

Angular momentum
- particle, 2, 19, 92, 93, 120, 174, 181, 182, 188–190, 497
- rigid body in planar motion, 190, 376, 379, 495
- rigid body in three dimensional motion, 495
- system of particles, 180, 182, 188–190, 194, 495, 497, 500
- system of rigidly constrained particles, 182, 502

Angular velocity, 9–12, 14, 15, 87, 110, 111, 116, 117, 120, 164, 165, 185, 188, 194, 195, 223, 235, 241–248, 250–253, 256, 257, 269–272, 274, 276, 284, 285, 289, 298, 301, 308–317, 322–326, 329, 331–333, 369, 374, 402, 407, 443, 454, 501, 502, 511–513, 516, 517, 519, 526, 532, 535, 538, 540–543, 545, 546, 640, 644, 645, 666–668, 674, 675, 681, 692

Angular velocity matrix, 283, 284, 309–311, 314, 322

Augmented approach, 73, 86, 102–105, 121, 122, 149, 152–161, 164, 167, 168, 223, 232, 233, 235, 339, 355–362, 382, 386, 387, 389–392, 530, 532, 534, 535, 651

Autonomous systems, 404

B

Bead on a wire
- parabolic wire, 47–52, 99, 166, 167, 449, 634–636
- solved by augmented approach, 167
- solved by explicit embedding, 99
- solved with Lagrange's equations, 121
- straight wire, 37, 40–46, 633–634
- with viscous damping, 52

Bifurcations, 13, 412, 415–436

Buckling load, 415, 417, 446

Buckling mode, 426

C

Center of mass, 176–185, 188–190, 194, 195, 197, 199, 222, 223, 225–227, 230, 231, 234, 248, 291, 293, 298, 300, 302, 330, 333, 335, 341, 343, 345–352, 354, 357, 364–367, 370, 377–382, 387, 389–392, 497–502, 504, 506–508, 510, 530, 533, 538–540, 545, 547, 592, 593, 596, 666, 667, 677, 682

Centrifugal effects, 110, 111

Centripetal acceleration, 7, 8, 14, 225, 241, 272, 454

L. W. Schmerr, *Engineering Dynamics 2.0*, Solid Mechanics and Its Applications 254,
https://doi.org/10.1007/978-3-319-98470-4

Chaotic behavior, 83, 85, 210, 212
Characteristic equation, 405–412, 456, 469, 478
Circulatory matrix, 455
Cofactors (matrix), 582–584
Conservation of angular momentum, 92, 182
Conservation of energy, 30, 162, 163, 193, 387–390
Conservation of linear momentum, 120, 178
Conservative systems, 29, 30, 80, 162, 210, 397–402, 407, 412, 414, 444, 445, 655
Constrained motion, vii, 9, 13, 37–43, 45–61, 73, 75, 85–90, 94, 105, 133, 135, 229, 261
Constraint drift, 42, 43, 73, 74, 161, 167, 382
Constraint forces, v, 13, 37, 41, 43, 46, 57, 62–66, 68–75, 86–88, 91, 93–95, 102–107, 120–123, 125–128, 131–144, 146, 149–151, 153, 154, 156, 157, 164, 173, 174, 176, 177, 186, 200, 201, 222–224, 226, 228, 229, 232, 235, 339, 352–369, 374, 381, 384–393, 513, 517, 523, 529, 530, 534, 535, 537
Constraint satisfaction, 153–161, 649–652
Convolution integrals, 475, 486–488, 493, 610, 612–614, 616, 618
Coriolis acceleration, 7, 8, 241, 272, 274
Coulomb friction force, 66
Courant number, 556, 571, 699
Critical load, *see* Buckling load
Cylindrical coordinates, 5–7, 11, 26, 167, 241

D

D'Alembert-Lagrange principle, 125, 127, 128, 200, 412, 413
Damping factors, 481, 483, 607, 608, 615, 628, 687, 702
Degrees of freedom, 41, 93–95, 106, 114, 146, 151, 174–176, 182, 194, 205, 222, 223, 227, 235, 268, 340, 345, 354, 397, 399, 401, 412, 424, 428, 450, 456, 513, 525
Dependent constraints, 227–232
Dependent coordinates, 41, 42, 149, 151, 268, 354
Determinants (matrix), 11, 39, 46, 214, 215, 259, 263, 399, 402, 415, 425, 426, 430, 431, 456, 469, 478, 582–585
Differential algebraic equation (DAE), 38, 169
Direction cosine matrix, 304, 308
Double pendulum, 81, 204–221, 389, 400–402, 408, 410, 451, 457, 654, 659–665
Drag forces, 22, 23, 30, 33–36, 52, 141, 631, 632
Dynamic potential energy, 111, 395, 404

E

Eigenvalue problems, 217, 405, 427, 428, 430, 456–458, 462, 479, 480, 595, 687
Embedding of constraints, 94–102, 146–153
Equations of motion, 3, 21, 176, 248, 352, 404, 449, 495
Equivalent force systems, 356, 365
Euler angles, vii, 237, 312, 314–317, 319, 321, 324, 325, 327, 502, 511, 513, 518–524, 527, 528, 531, 532, 537, 539, 541–544, 547, 691, 692, 696
Euler parameters, vii, 237, 314, 319–325, 502, 520, 522–528, 531, 547, 694, 695, 697
Euler parameters constraint, 525–527
Euler's equations, 511–525, 541, 543, 544, 547, 692, 696
Expansion theorem, 462

F

Fast Fourier transform (FFT), 618, 624, 630, 701
Finite differences method, 553
Finite element method, vii, 561–567
Foucault's pendulum, 3, 277
Fourier transforms, 14, 488, 572, 617, 618, 621–628, 630, 701
Frames, 81, 221, 237, 244–248, 270, 281, 284, 286, 288–291, 310, 311, 314, 316, 322, 325, 327–329, 336, 338, 392, 511, 512, 519, 530, 533, 538, 539, 541, 542, 544, 546, 547, 556
Frobenius conditions, 93

G

Galerkin approach, 567
Gauss's law of least constraint, 133–136, 228
Gaussian virtual acceleration, 132, 137
Generalized coordinates, 75, 94–108, 110, 111, 114, 119–121, 125, 136, 140, 146–149, 152, 153, 194, 200–223, 227–229, 231–235, 253, 260, 262, 263, 268, 291, 292, 294, 295, 301, 343, 355, 357, 364, 382, 386, 389, 395, 397, 404, 412, 414, 452, 462–464, 474, 476, 482, 487, 500, 503, 511, 527, 531, 533, 534, 536, 537, 545, 550, 673, 684

Generalized momenta, 119, 120, 503, 521, 522
Gibbs-Gauss variational equation, 132
Gyroscopic matrix, 454, 484

H
Hilbert transform, 625
Holonomic constraints, 92–94, 101, 102, 105, 106, 122, 123, 131, 132, 146, 147, 149, 152, 154, 168, 173, 186, 200, 201, 205, 221, 223, 227, 235, 268, 354, 363
Homogeneous solutions, 475, 476, 478–480, 487, 488, 607–609, 614, 618, 619, 685
Hooke's Law, 551
Hypocycloid, 275, 276

I
Ideal constraint forces, 63, 66, 70–74, 93, 95, 102, 103, 106, 120, 122, 123, 125–127, 131, 135–140, 154, 164, 173, 187, 201, 223, 228, 374, 386, 390, 534
Ignorable coordinates, 119, 120, 165, 521
Impulse response, 484–488, 612–614, 616, 617, 627–629, 701, 702, 610–ENF
Independent coordinates, 41, 42, 73, 93, 94, 99, 101–103, 106, 109, 121, 146, 149, 151–153, 183, 268, 354
Inertial coordinate system, 3, 4, 237, 238, 241, 246
Inertial force, 75, 91, 98, 104, 122, 140, 148, 149, 229, 269, 343, 344, 346, 362, 386, 529
Initial conditions, 21–23, 35, 37, 40–42, 47, 50, 53, 59–61, 73, 77–79, 81, 83, 84, 86, 93, 118, 130, 156, 162, 163, 165–167, 194, 198, 210, 212–214, 216, 223, 225, 230–235, 265, 273, 275, 277, 325, 331, 360, 368, 395, 396, 401, 415, 444–446, 457, 467, 470, 475, 476, 478–485, 487, 491–494, 515, 519, 523, 524, 537, 545, 547, 554, 560, 605, 607, 610, 611, 613, 614, 618, 633, 635, 637, 639–642, 645–647, 650, 652, 653, 659, 665, 666, 672, 673, 675, 676, 679, 681–685, 687–689, 693, 695, 696
Initial value problem, 22, 486, 597
Internal forces, 171–179, 186–188, 191, 201, 371, 413, 551, 568, 569, 572
Internal potential energy, 192, 413, 464
Inverse (matrix), 70, 71, 133, 134, 149, 153, 208, 228, 261, 429, 486, 488, 578, 579, 584–586, 666

J
Jacobian matrix, 123, 147, 261–263, 268, 294, 604, 670, 677, 680
Jacobi integral, 115, 165, 277, 402
Jourdain variational equation, 132
Jourdain virtual velocity, 131, 132

K
Kepler's laws, 88, 89, 91
Kinematically driven systems, 13, 253–258, 260, 268, 271, 291–296, 298, 299
Kinetic energy
  particle, 13, 30, 31, 106, 107, 109, 168, 186–193, 203, 204, 369, 371, 500
  rigid body in plane motion, 369–371
  rigid body in three dimensional motion, 13
  system of particles, 187–193, 204, 369
Kronecker delta, 304, 461, 463, 485

L
Lagrange-Dirichlet stability theorem, 397–402
Lagrange multiplier, 63, 70, 149, 150, 152, 155, 164, 223–225, 525, 526, 531, 534, 535
Lagrange prescription, 63, 386, 529
Lagrange's equations, 13, 105–117, 119–122, 125, 126, 164, 168, 207, 223, 339, 382–386, 398, 402, 414, 451–453, 455, 465, 500, 503, 521, 531, 532, 534, 546
Limit load, 13, 412, 432–440
Linear impulse, 178, 182, 183
Linear momentum, 2, 18, 19, 120, 178, 182, 183, 500, 502
Linear time-shift invariant (LTI) systems, 487, 605, 606, 612, 613, 617
Loop closure equation, 253
Lyapunov stability, 396

M
Magic time step, 556, 557, 559
MATLAB® ode4 solver, 156, 597, 599, 600, 649–651
MATLAB® ode45 solver, vi, 14, 35, 38, 43, 47, 48, 57, 78, 116, 142, 156, 162, 210, 277, 515, 520, 597, 637, 640, 647
Modal coordinates, 463, 464, 468, 471, 472, 474, 475, 477, 481, 491, 492, 683, 684
Modified Euler's equations, 543, 544
Modified potential energy, *see* Dynamic potential

Moment equation
particle, 2, 18–20, 176, 179, 194, 339, 375
rigid body in plane motion, 339
rigid body in three dimensional motion, 194
system of particles, 20, 194, 339
Moment of inertia, 189, 196, 197, 345, 348–352, 355, 377, 380, 381, 496–499, 505, 510, 511, 541, 544, 547, 592, 595
Moore-Penrose pseudoinverse, *see* Pseudoinverse
Multiparticle pendulum, 198–200, 235, 400, 652–654
Mur boundary condition, 559, 568

N

Natural frequency, 198, 214, 215, 232, 274, 410, 468, 482, 483, 492, 608, 609, 615, 616, 626, 628, 659, 702
Natural modes, *see* Normal modes
Natural systems, 109, 115, 397–402, 407, 412, 454–456, 474, 491–493, 687
Nested function, 57–59, 78, 116, 130, 142, 143, 198, 210, 219, 231, 368, 637, 638, 640, 644–648, 652–655, 660, 665, 675, 681, 682, 689, 692, 695
Newton-Euler equations, 2, 3, 12, 73, 107, 108, 176, 182, 194–196, 200, 237, 340, 343, 345, 352–355, 420, 495, 502, 505–512, 527
Newton-Raphson method, 13, 14, 151, 237, 260, 262–265, 268, 295, 297, 329, 331, 333, 334, 601, 604, 670–673, 678, 679
Newton's law of gravitation, 28, 88, 91
Newton's method, 602, 604
Newton's three laws
first law, 18
second law, 2, 17–21, 23, 24, 31–33, 37, 38, 41, 45, 46, 55, 56, 62, 67, 70, 74, 75, 86, 91, 94, 99, 103, 104, 107, 108, 111, 121, 123, 125, 132, 136, 137, 156, 157, 164, 176, 229, 269, 273, 376, 450, 458, 550, 569
third law, 171, 173, 174, 177, 179, 186, 191, 340
Non-holonomic constraints, 13, 92, 93, 102, 103, 122, 131–133, 151, 152, 168, 222–224, 235, 532, 666
Nonideal constraint forces, 66, 136–144, 146, 587
Non-independent constraints, 93
Nonlinear parametric resonance, 80, 82
Nonnatural systems, 109, 402, 403
Normal modes, 215–218, 410, 456, 659
Nutation angle, 536, 537, 545
Nyquist criterion, 623

O

Ode4 solver, *see* MATLAB® ode4 solver
Ode45 solver, *see* MATLAB® ode45 solver
Ordinary differential equations (ODEs), vi, 12, 22, 33, 35, 38, 78, 161, 218, 368, 479, 532, 552, 560, 561, 573, 597–601
Orthonormality relations, 461–463, 475, 481, 489–491, 493

P

Parallel axis theorem, 350, 351, 354, 367, 498, 499, 513, 592, 593
Partitioning of generalized coordinates, 152, 153
Pendulum, 3, 64, 198, 237, 352, 400
Phase plots, 49–54, 87, 88, 117–119, 130, 144, 145, 165, 166, 406, 444–446, 645–649
Physical pendulum
augmented approach, 157, 355, 651
Newton-Euler approach, 352, 354
Polar coordinates, 5–7, 15, 27, 74–94, 98, 103–105, 108, 110, 153
Positive definite matrix, 134, 138, 399, 415, 430, 457, 578, 579, 594
Potential energy, 24–29, 32, 50, 80, 107, 111–114, 121, 165, 172, 188, 191–193, 223, 226, 374, 395, 397–402, 404, 407, 412–415, 417–419, 421–423, 428–430, 432–434, 436–439, 441, 444, 445, 451, 452, 455, 464–466, 503, 505, 531, 535, 546
Powers, vi, 31, 193, 374, 503, 505, 578
Precession angle, 312
Principal coordinates, *see* Modal coordinates
Principal directions, 595
Principal moments of inertia, 594, 595
Principle of virtual work, 412–414
Projectile problem
with drag force, 31, 33–36, 632
without drag force, 162
Propagating wave, 559
Pseudoinverse (matrix), 229, 587

R

Rayleigh dissipation function, 454, 455, 546
Relative error tolerance (rtol), 48, 49, 59, 635, 637, 639, 641, 642, 645, 646, 653, 655, 656, 661, 664, 666, 676, 682, 689, 693, 696
Relative velocity and acceleration, 237–243, 247, 248, 250, 258, 260, 269, 279, 286–289, 299, 329, 331
Rodriguez formula, 321
Roller-coaster
  with Coulomb friction, 142–144, 146, 647–649
  frictionless, 140
Rolling without slipping, 222, 363–369, 375, 376, 380, 532, 537, 681–683
Rotating hoop, 114–119, 165, 277, 402, 403, 443, 450, 454, 644, 645
Rotation matrix, 280–283, 289, 291, 292, 303, 304, 306–310, 312–314, 319, 321, 322, 325, 328, 329, 338, 502, 511, 512, 516, 523, 542, 543, 593, 594
Runge-Kutta methods, 156, 597, 599–601, 650

S

Shape functions, 562–567
Singular configurations, 258–260, 262, 268, 319
Skew matrix, 11, 14, 591
Slider-crank, 250, 251, 253–263, 267, 269–271, 291–293, 295, 297–299, 329–331, 333, 335, 667–669, 671–674, 677–681
Snap-through instability, 412, 438, 439, 441–443
Spherical coordinates, 7–9, 11, 15, 26, 27, 75, 165, 241, 242
Spin angles, 312, 537, 541, 545
Spring Pendulum, 75–85, 164, 210, 451, 640
Standing wave, 474
Steady-state solutions, 476, 484, 488, 546, 616, 618, 619
Stiffness matrix, 427, 428, 451, 453, 454, 457, 459, 462, 467, 565, 684
Stokes' theorem, 32
Strain energy, *see* Internal potential energy
System coordinates, 41, 42, 73, 94, 102, 106, 146–148, 158, 456
System of first-order equations, 33, 166, 224, 478, 481, 655

T

Terminal velocity, 36
Total energy, 29, 30, 50, 54, 55, 61, 80, 81, 83, 87, 114, 115, 159, 160, 162, 163, 166–168, 188, 199, 210, 277, 374–376, 397, 419, 444, 445, 505, 521, 522, 540, 545, 636, 639, 652, 655, 656, 661, 662, 676
Transfer matrix, 484, 488
Transport theorem, 243–269, 289–292, 329, 512, 539, 542, 543
Transpose (matrix), 9, 125, 135, 280, 283, 308, 311, 337, 338, 358, 579–581, 584–586, 684

U

Unconstrained motion, 21–36, 42, 75–85, 94, 106, 135, 194
Unit impulse response, *see* Impulse response
Universal law of gravitation, *see* Newton's law of gravitation

V

Virtual displacements, vi, 13, 122, 124–128, 131, 132, 201, 413
Virtual work, 13, 32, 122–130, 201, 372, 412–414, 567
Viscous damping force, 52, 128, 455

W

Wave equation, vii, 472, 473, 551–553, 556, 557, 559–562, 567, 568
Wavelength, 473
Work-energy relation
  elastically connected system of particles, 191–193
  particle, 31, 186, 193
  rigid body in plane motion, 371–375
  system of particles, 188–190
Work of a force, 32

Y

Young's modulus, 441, 472, 551, 568, 571

www.ingramcontent.com/pod-product-compliance
Lightning Source LLC
Chambersburg PA
CBHW052318080925
32316CB00003B/40
9783319984698